# Theoretical Inorganic Chemistry

# Reinhold Chemistry Textbook Series

CONSULTING EDITORS

HARRY H. SISLER          CALVIN A. VANDERWERF
*University of Florida*     *Hope College*
*Gainesville, Florida*      *Holland, Michigan*

Bonner and Castro — *Essentials of Modern Organic Chemistry*
Day and Selbin — *Theoretical Inorganic Chemistry, Second Edition*
Drago — *Physical Methods in Inorganic Chemistry*
Fairley and Kilgour — *Essentials of Biological Chemistry, Second Edition*
Fieser and Fieser — *Advanced Organic Chemistry*
Fieser and Fieser — *Topics in Organic Chemistry*
Heftmann — *Chromatography, Second Edition*
Heftmann and Mosettig — *Biochemistry of Steroids*
Klingenberg and Reed — *Introduction to Quantitative Chemistry*
Lingane — *Analytical Chemistry of Selected Metallic Elements*
Meyer — *Food Chemistry*
Mortimer — *Chemistry: A Conceptual Approach*
Neckers — *Mechanistic Organic Photochemistry*
Reid — *Principles of Chemical Thermodynamics*
Sanderson — *Chemical Periodicity*
Sanderson — *Inorganic Chemistry*
Smith — *Chemical Thermodynamics: A Problems Approach*
Smith and Cristol — *Organic Chemistry*

## Selected Topics in Modern Chemistry

SERIES EDITORS

HARRY H. SISLER *and* CALVIN A. VANDERWERF

Brey — *Physical Methods for Determining Molecular Geometry*
Cheldelin and Newburgh — *The Chemistry of Some Life Processes*
Eyring and Eyring — *Modern Chemical Kinetics*
Hildebrand — *An Introduction to Molecular Kinetic Theory*
Kieffer — *The Mole Concept in Chemistry*
Moeller — *The Chemistry of the Lanthanides*
Morris — *Principles of Chemical Equilibrium*
Murmann — *Inorganic Complex Compounds*
O'Driscoll — *The Nature and Chemistry of High Polymers*
Overman — *Basic Concepts of Nuclear Chemistry*
Rochow — *Organometallic Chemistry*
Ryschkewitsch — *Chemical Bonding and the Geometry of Molecules*
Sisler — *Electronic Structure, Properties, and the Periodic Law*
Sisler — *Chemistry in Non-Aqueous Solvents*
Sonnessa — *Introduction to Molecular Spectroscopy*
Strong and Stratton — *Chemical Energy*
VanderWerf — *Acids, Bases, and the Chemistry of the Covalent Bond*
Vold and Vold — *Colloid Chemistry*

# Consulting Editors' Statement

The enthusiastic reception of the first edition of Professors Day and Selbin's text by teachers and research workers in the field of inorganic chemistry provides abundant evidence of the soundness of the author's treatment of the theory of inorganic chemistry. In this new edition, the authors have refined their material, deleted certain topics (e.g., nuclear theory) and have added much new information and data, particularly in the area of the study of the crystalline state and of coordination chemistry. This new edition will, without question, make an even greater contribution than the first to the training of graduate and advanced undergraduate students in the important area of theoretical inorganic chemistry. We are proud to present this new edition of an important member of the Reinhold Chemistry Textbook Series.

HARRY H. SISLER
CALVIN A. VANDERWERF

**M. CLYDE DAY, JR.**
*Professor of Chemistry*
*Louisiana State University*
*Baton Rouge, Louisiana*

**JOEL SELBIN**
*Professor of Chemistry*
*Louisiana State University*
*Baton Rouge, Louisiana*

# Theoretical Inorganic Chemistry

## SECOND EDITION

**VAN NOSTRAND REINHOLD COMPANY**
NEW YORK   CINCINNATI   TORONTO   LONDON   MELBOURNE

# Preface to the Second Edition

The rapid growth of chemistry, in which we see the research literature currently doubling in volume every eight to nine years, has its good and its bad aspects. One of the latter is the need by authors to revise their books, and we have found ourselves faced with this problem. Although there are several topics covered in "Theoretical Inorganic Chemistry" that are quite basic and not actually subject to change, this, for example, is certainly not true for the topics in the chapters on coordination chemistry and nonaqueous solvents. Somewhat surprisingly perhaps, there have also been some significant advances in the field of periodic properties and, to a lesser extent, there has been a new dimension added to the discussion of acids and bases because of the emphasis on the hard and soft acid-base concept. These chapters have been extensively revised or added to with the greatest change occurring in the treatment of coordination chemistry. Here we have expanded the material to the extent that it is now covered in two chapters each roughly as long as the original one. Except for Chapter 1, all of the remaining chapters have been altered to some extent to improve clarity. There have also been some new topics added and new data have been incorporated where this seemed desirable, notably in the chapter on stereochemistry.

Because of comments we have received over the years since "Theoretical Inorganic Chemistry" was first published in 1962, the chapter on the theory of the nucleus has been dropped. Whereas it may be reasonably argued that a chapter on the nucleus is not actually appropriate in an inorganic text, it may equally be argued that a discussion of the solid state does belong in such a book. Consequently a chapter on this topic has been added. Secondly, we have received many requests since the first edition was published to add problems at the ends of the chapters. With some misgivings we have acceded to this request, but it should be pointed out that these problems vary in difficulty from the almost trivial to those that require considerable outside study. In fact, some of these problems may have no direct answer. Questions of this type are posed to help the student to recognize the limitations of the field by expanding his exposure to the frontiers of our knowledge. Finally, an appendix has been added in which several topics are discussed at a level beyond that considered necessary in the body of the text; the most important example of this is a comparison between the Schoenflies and Hermann-Mauguin point symmetry notation.

*Baton Rouge, Lousiana*                                         M. Clyde Day, Jr.
                                                               Joel Selbin

# Preface to the First Edition

There was a time when any attempt to comprehend natural chemical processes in terms of theoretical concepts was generally relegated to the physical chemist. We trust that this attitude is no longer prevalent. With the recent renaissance in inorganic chemistry, we find an ever-increasing emphasis on the theoretical aspects of the field. This is as it should be. Since a knowledge of theory permits the scientist to take the maximum advantage of his experimental data, we can see that the chemist would be derelict if he did not utilize every available tool in his research efforts. Obviously, such an attitude is not original with us. Most of the inorganic textbooks that have been written during the past decade have attempted to emphasize both the theoretical and the descriptive aspects of the subject. However, it is our personal feeling that it is not possible to adequately treat the theoretical side with only a portion of an inorganic chemistry textbook.

For some years, a course has been offered at Louisiana State University under the title of "Theoretical Inorganic Chemistry." It is generally intended as an advanced undergraduate and first-year graduate level course. As a guide for the course, the theoretical portions of several of the available textbooks have been used, but none of these seemed adequate. It was for this reason, along with a feeling that our experiences were not unique, that we were prompted to write this book. There will certainly be those who will question its classification as a textbook of inorganic chemistry, but on the other hand, we feel that there will be a large number of teachers and students who will welcome it. Even among those who generally favor this approach, there will be disagreement concerning the emphasis placed on various topics as well as the omission of others. In support of our choice of topics, we can only say that we have treated those which seemed to us to be most relevant to the modern inorganic chemist.

It might be noted that the level of mathematics we have used is somewhat higher than is customary in an inorganic textbook. However, nowhere have we used the mathematics for its own sake. Throughout the book we have attempted to develop the underlying principles of inorganic chemistry. So often these principles have no meaning except through the language of mathematics. In spite of the mathematical formalism, we feel that the complexity of the treatment is well within the abilities of a senior chemistry major. At the other extreme, we have given a relatively low-level treatment in Chapters 3 and 4. Although much of this material is covered in earlier courses, we feel that it is so fundamental to an understanding of inorganic chemistry that it is worthy of review.

It may seem somewhat inconsistent to have large sections of the book devoted to historical material and at the same time to place such heavy emphasis on advanced theoretical treatments. There are two reasons for this balance. To begin with, we feel that a well-rounded chemist should be aware of the historical as well as of the modern aspects of his field. Secondly, there are many concepts in modern chemistry that cannot be well appreciated unless an understanding is first had of the work that led to them. This is particularly true of the concepts underlying quantum mechanics. It is for this reason that the first chapter was written. This chapter is not intended to give the student a detailed discussion of the quantum theory. Rather, it is intended to present the ideas that inevitably led to quantum mechanics.

In writing a book of this nature, there are always many individuals who make contributions, both large and small. It would be very difficult to give credit without overlooking some one or more of these persons. As is undoubtedly the case with all authors, we sincerely appreciate their assistance and encouragement. We would like especially to single out Professors Paul Delahay, Sean McGlynn, and Robert Nauman, as well as Drs. Loys Nunez and Mohd. Quereshi for reading various chapters and offering valuable comments and suggestions, and Professor Harry Sisler who read the entire manuscript. We would also like to express our sincere appreciation to Messrs. Adnon Shiblak and Samir Shurbaji for the drafting work, and to Mrs. Camille Delaquis for the preparation of the manuscript. Finally, we must recognize the patience and understanding of our wives.

*Baton Rouge, Louisiana*                                          M. Clyde Day, Jr.
*February, 1962*                                                      Joel Selbin

# Contents

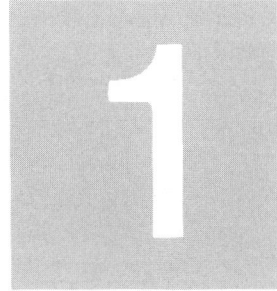

# Origin of the
# Quantum Theory

One of the most interesting and also one of the most important problems in the early development of chemistry and physics was the nature of radiant energy. Throughout the eighteenth century, the vast majority of physicists accepted the idea that visible light consists of small particles that are emitted from the source like bullets. Such a corpuscular theory had been proposed by Sir Isaac Newton in a communication to the Royal Society in 1675, and the result was almost universal acceptance of his views throughout the scientific world. Yet there was some dissension. Even before the work of Newton, Huygens had proposed an undulatory theory of light which was supported about that time by Hooke. They proposed that light possesses a vibrational character analogous to that of a water wave. As it turned out, some of the strongest support that is now known for the wave theory was used at that time to discredit it, and the corpuscular theory of Newton reigned until the nineteenth century.

There were few significant changes in the ideas of the nature of light until Thomas Young published his first attack on the corpuscular theory in 1800. At that time, he showed the superiority of the wave theory in explaining reflection and refraction. Then, in 1801, he discovered the phenomenon of interference and utilized it to explain the existence of Newton rings which Newton had earlier explained in terms of the corpuscular theory. Actually, the idea of interference was not completely original with Young, for Newton himself had used it in his theory of the tides. For the case of light, Young found that if a source of monochromatic light is focused on a double slit in a diaphragm such as that shown in Figure 1–1, a series of lines can be observed on a screen behind the slits. The positions of these lines can readily be explained in terms of the ideas of wave motion by means of interference and reinforcement. As the light rays pass through the two openings, the waves spread outward. When a crest of one wave coincides with a crest of another wave a reinforcement results, giving a bright line on the screen. However, when a crest of one wave coincides with a hollow of another wave, the result is complete de-

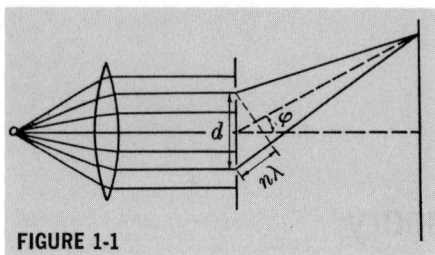

**FIGURE 1-1**

Diffraction of a Monochromatic Beam of Light at Two Narrow Slits.

struction and, therefore, a dark line appears on the screen. From the geometry of the system, it is relatively easy to calculate where the lines will occur. If the difference in the paths from the two slits is an integral multiple of the wavelength, the waves will be in phase and there will be a reinforcement. But when the path lengths differ by half a wavelength, the waves will be exactly out of phase and there will be complete destructive interference. Thus, we find that a bright line will be observed when $n \lambda = d \sin \varphi$, and a dark line will occur when $(n + \frac{1}{2}) \lambda = d \sin \varphi$, where $n$ is an integer such as $0, \pm 1, \pm 2, \ldots, \pm n$.

It would seem that the discovery of the phenomenon of interference should have been sufficient in itself to turn the tide in favor of the wave theory. However, in spite of his continued attacks on the corpuscular theory of radiation, Thomas Young failed to make any significant headway, and his reward was mostly ridicule by his fellow scientists. Then, in 1815, Fresnel rediscovered the phenomenon of interference and proceeded to put the wave theory on a firm mathematical basis. His work proved to be too much for the proponents of the corpuscular theory of radiation, and Huygens' wave theory was finally accepted more than a century after its author had died.

Throughout the remainder of the nineteenth century numerous other experiments were carried out on the nature of light, and they all gave further support to the wave theory. Thus, at the beginning of the twentieth century there was little doubt in the world of science that light is of a wave character, just as one hundred years earlier there had been little doubt that it is corpuscular in character.

At the close of the nineteenth century there was a predominant sense of completion among the physicists. The classical fields of physics such as mechanics and electrodynamics appeared capable of describing all observable phenomena, and there appeared to be no new worlds to conquer. Then, suddenly out of this attitude of complacency came a succession of experimental discoveries of tremendous importance. Between 1895 and 1898, Roentgen discovered X rays, Becquerel discovered radioactivity, and Thompson discovered the electron. In three years physics had become a science that no one could have dreamed of a few short years before.

Of equal importance, it became apparent shortly thereafter that our views of electromagnetic radiation were inadequate. Starting with *black body* radiation, a preponderance of experimental observations were accumulated that could not be explained in terms of the wave theory. This led to the development of a new quantum theory which has now permeated virtually all phases of physics and chemistry, not only from a structural and mechanistic standpoint, but from a philosophical point of view as well.

## BLACK BODY RADIATION

When a body is heated it emits *thermal* radiation, the nature of which depends on the temperature of the emitting body. Thermal radiation is a form of electromagnetic radiation just as is visible light. However, it usually consists of wavelengths that are longer, and therefore of lower energy, than visible light. It has been noted that the energy of the radiation from a heated body is spread over a continuous spectrum that is dependent upon the temperature of the body. At lower temperatures the spectrum consists mostly of low-energy radiation in the infrared region. However, as the temperature is raised it shifts toward the higher-energy region. This is evident from the fact that a body begins to radiate in the visible region as it becomes hotter. At first it becomes red, and as the temperature is raised, it approaches white, such as may be observed in an incandescent light.

In order to study such radiation, it was found that a particularly desirable system is one known as a black body. When radiation falls on a surface, some of the radiation is reflected and some is absorbed. The *absorptivity* of a surface is defined as the fraction of the light incident on the surface that is absorbed, and a black body is a surface that has an absorptivity of unity. That is, it absorbs all of the radiation that is incident upon it. In addition, it has been shown (Kirchhoff's law) that the ratio of *emissive power*, $E$, to the absorptivity, $A$, is a constant for a given temperature,

$$\frac{E_0}{A_0} = \frac{E_1}{A_1}$$

Now, since the absorptivity of a black body has been defined as unity ($A_0 = 1$), we see that the total emissive power of any surface must be given by

$$E = AE_0$$

where $E_0$ is the total emissive power of a black body. Since $A$ is necessarily less than unity for any surface other than a black body, it is obvious that no surface can emit more strongly than a black body. Therefore, it is evident that a black body is both the most efficient absorber and also the most efficient emitter of radiant energy.

Ordinarily the apparatus used for the study of black body radiation consists of a well-insulated cavity with a small opening in one of the walls. For instance, a long tube heated by means of an electric current flowing through a wire wrapped around the tube is often used. The radiation is observed as it passes out through a small hole in one of the walls. If this furnace is maintained at constant temperature, the enclosure is known as an *isothermal enclosure*. Through thermodynamic arguments, it has been shown that the radiation field inside such an enclosure has some very special characteristics. In 1859, Kirchhoff was able to show that if the walls and contents of the cavity are kept at a common temperature at equilibrium, the stream of radiation in any one direction must be the same as that in any other direction, it must be the same at any point in the enclosure, and it makes no difference of what material the walls are composed. If this were not true, the second law of thermodynamics would be violated. Now, if we consider an isothermal enclosure that also approximates a black body radiator, it obviously will have some very interesting properties.

Before the turn of the century, a considerable amount of work was being done on the problem of black body radiation. As early as 1879, Stefan had given an empirical relation for the rate of emission of radiant energy per unit area of a surface. This is expressed by the equation

(1-1)          $E = e\sigma T^4$

where $E$ is the rate of emission of radiant energy per unit area or the total emissive power, $T$ is the absolute temperature, $\sigma$ is a constant known as the Stefan-Boltzmann constant, and $e$ is the *emissivity* of the surface. The emissivity is defined as $E/E_0$. It is therefore seen that for a black body the emissivity has a value of unity.

A problem that was of considerable interest at the time was the distribution of energy in the spectrum as a function of wavelength and temperature. Here, the monochromatic emissive power, $E_\lambda$, is of interest. This is the energy emitted between wavelengths $\lambda$ and $\lambda + d\lambda$. Before any experimental determination of the energy distribution was made for a black body, theoretical attempts were made to calculate the shapes of the energy spectra as a function of the wavelength. In an attempt to find an expression for the monochromatic emissive power, Wien utilized the classical methods of thermodynamics to obtain the equation

(1-2)          $E_\lambda = \dfrac{a}{\lambda^5} f(\lambda T)$

where $a$ is a constant and $f(\lambda T)$ is some function of the wavelength and the absolute temperature. In order to determine the function, $f(\lambda T)$, it was necessary to consider the mechanism by which the radiation is emitted. Since Kirchhoff had shown that the nature of the walls, and therefore the nature of the radiator, is not important in an isothermal enclosure, any reasonable model could be chosen. Wien chose oscillators of molecular size and applied the laws of classical electromagnetic theory. As a result, he obtained the equation

$$(1\text{-}3) \qquad E_\lambda = \frac{a}{\lambda^5}\, e^{-b/\lambda T}$$

where $a$ and $b$ are constants. It was later shown that the distribution curves calculated from the Wien equation fit the experimental curves in the high-energy region (short wavelengths) extremely well, but the Wien equation failed completely to give a correct spectral distribution over all wavelengths.

Another theoretical attempt to determine a distribution law was made in 1900 by Rayleigh in which he applied the classical principles of equipartition of energy. The result was the equation known as the Rayleigh-Jeans equation,

$$(1\text{-}4) \qquad E_\lambda = \frac{2\pi k T}{c\lambda^4}$$

This equation was found to give fair agreement with the observed spectral distribution in the low-energy region (long wavelengths), but it failed even to approach the observed data in the high-energy region of the spectrum.

It was in 1899 that Lummer and Pringsheim made the experimental determination of the energy distribution from a black body at various values of the temperature. Their results are shown in Figure 1–2. As can be seen in Figure 1–3, the Wien equation gives excellent agreement with experiment in the region of short wavelengths and the Rayleigh-Jeans equation appears to be asymptotically correct at very long wavelengths. However, neither of the equations is consistent with the experimental curves over the complete spectral range.

In an attempt to fit the total experimental spectrum for black body radiation, an empirical formula was at first sought that would fit the data from $\lambda \rightarrow 0$ to $\lambda \rightarrow \infty$. An example of one such formula is

$$E = cT^{5-\mu}\, \lambda^{-\mu}\, e^{-b/(\lambda T)^\nu}$$

from which, for $\mu = 5$ and $\nu = 1$, the Wien equation is obtained; and for $\mu = 4$ and $b = 0$, the Rayleigh-Jeans equation is obtained. Although a usable equation may be found, such an approach gives little intellectual satisfaction. Nevertheless, it was out of an attempt to determine such an empirical relation that Max Planck arrived at what was possibly the most revolutionary hypothesis of this era.[1]*

For the same reason that Wien was able to choose any type of reasonable

---

* The superscript numbers refer to bibliographical references at the end of the chapter.

energy radiator that he wished, Planck also was able to make such a choice. It had to be a system capable of emitting and absorbing radiation; and among the simplest types for the purpose of calculation is a set of

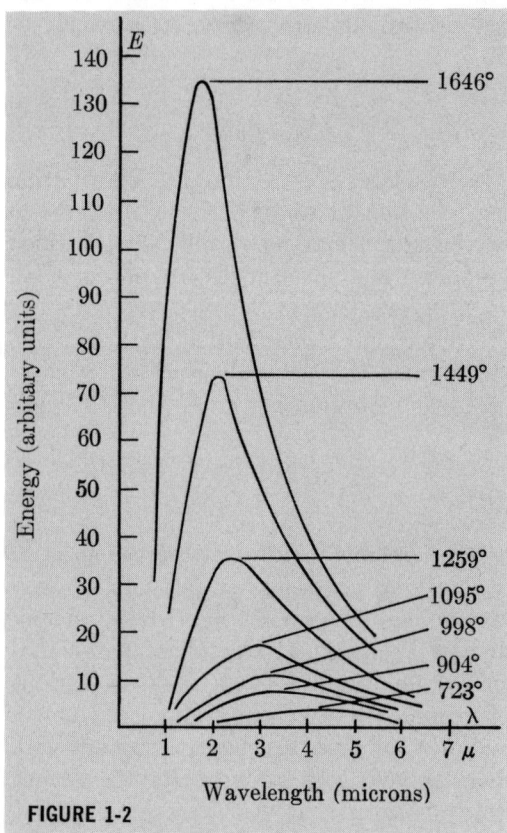

**FIGURE 1-2**

Energy Spectrum of a Black Body at Various Temperatures.

simple harmonic oscillators. Now, according to classical ideas, an oscillator must take up energy continuously and emit energy continuously. However, in order to find a formula that would fit the experimentally determined spectrum of a black body radiator, Planck found it necessary to postulate that such an oscillator cannot take up energy continuously as demanded by classical theory, but rather it must take up energy in discrete amounts. These amounts are integral multiples of a fundamental energy unit $\epsilon_0$, that is, $0, \epsilon_0, 2\epsilon_0, 3\epsilon_0, \ldots, n\epsilon_0$.

Using this idea, Planck was able to derive the equation

$$(1\text{--}5) \qquad E = \frac{2\pi c}{\lambda^4} \frac{\epsilon_0}{e^{\epsilon_0/kT} - 1}$$

for the monochromatic emissive power of a black body. Here $c$ is the velocity of light and $k$ is the Boltzmann constant. Since Eq. (1-2) is of thermodynamic origin, and therefore basically correct, it is necessary for

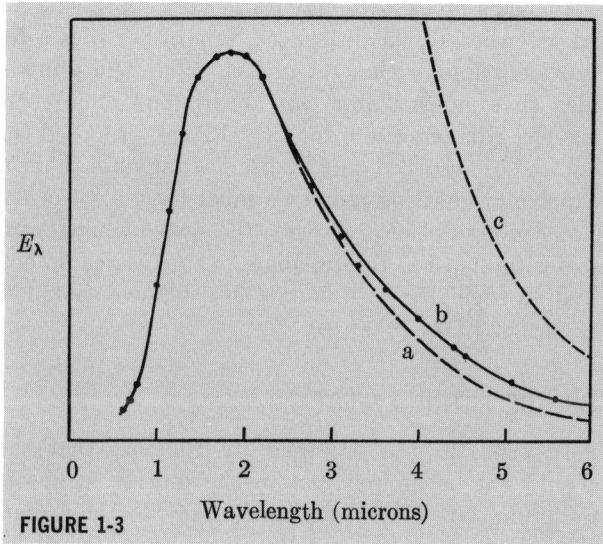

**FIGURE 1-3**    Wavelength (microns)

A Comparison with Experiment of the Three Radiation Laws: (a) Wien; (b) Planck; (c) Rayleigh-Jeans. The dots represent the experimental points. (From F. K. Richtmeyer and E. H. Kennard, "Introduction to Modern Physics," 5th ed., copyright 1955, McGraw-Hill Book Co., Inc., New York.)

the distribution law of Planck to contain the temperature in the combination $\lambda T$, or since $\lambda = c/\nu$, the combinations $T/\nu$ or $\nu/T$ are also acceptable. Consequently, it can be seen from Eq. (1-5) that the quantum of energy $\epsilon_0$ must be proportional to $1/\lambda$, or what amounts to the same thing, proportional to $\nu$. Therefore, we find that $\epsilon_0 = h\nu$, where $h$ is a new constant known as Planck's constant, the presently accepted value being $6.62 \times 10^{-27}$ erg-sec. By making the substitution for $\epsilon_0$, Planck's distribution law may now be expressed as

$$(1-6) \qquad E_\lambda = \frac{2\pi hc^2}{\lambda^5} \frac{1}{e^{ch/\lambda kT} - 1}$$

Whereas the energy distribution laws for black body radiation deduced from classical concepts had consistently failed to explain the experimental data, the quantum hypothesis of Planck succeeded. In Figure 1-3 is seen a comparison of the distribution curves of Wien, Rayleigh-Jeans, and Planck with the actual experimental data. The solid line represents the theoretical curve as determined by Planck, and the experimental curve is represented by the dots. Here it is seen that the theoretical curve determined by Planck coincides exactly with the experimental curve, while the classical curves as determined by Wien and Rayleigh fail at either one end of the spectrum or the other.

Planck's break with classical theory represents a real break. It involves no extension of classical ideas, but rather it is a radical change from the prevalent line of thought of that time. Quite in contrast to the classical idea that an oscillator can absorb and emit energy continuously from wavelengths of zero to infinity, Planck proposed that the energy must be emitted or absorbed only in discrete amounts. This implies that any system capable of emitting radiation must have a set of energy states, and emission can take place only when the system changes from one of these energy states to another. Intermediate energy states do not occur. Thus, we may find an oscillator with energy $2h\nu$, but we should never find one with an energy of $1.5\ h\nu$.

## PHOTOELECTRIC EFFECT

In 1905, it was proposed by Einstein that the quantum properties should not be limited just to the process of absorption and emission of radiation, but they should also apply to the radiation itself. This means that electromagnetic radiation consists of particles, which we now call *photons*, that have an energy $h\nu$ and shoot through space with the velocity of light. Such radical changes in thought as proposed first by Planck and then by Einstein could not have been accepted without considerable experimental evidence to support them. This experimental evidence was available, and the success of the quantum theory was too great to be denied.

A satisfactory explanation by Einstein of the photoelectric effect was among the first triumphs of the quantum theory.[2] It had been found by Hertz, as early as 1887, that if ultraviolet light is focused onto a metal surface, the surface becomes positively charged. This, of course, means that in some manner negative charge is being removed. Then, shortly after the discovery of the electron, it was shown that this charge is being carried away by electrons.

There are two important features of the electrons in the photoelectric effect that can be experimentally observed. These are the energy and the number of electrons emitted from the metal surface. When these were observed under controlled conditions, some very serious problems arose with regard to their interpretation. According to classical electromagnetic theory, the energy of the emitted electron should increase with an increase in intensity of the light used. Also, it would be expected that if light were permitted to shine upon the surface for a sufficient length of time, electrons would be emitted regardless of the frequency of the incident light. However, quite the contrary is observed. An increase in intensity fails to increase the energy at all, but rather it increases the number of emitted electrons. In addition it is observed that if the frequency of the incident light is not above a certain value no electrons are emitted regardless of how long the light is allowed to shine on the surface.

In Figure 1-4, it is seen that for a given intensity of light $I_1$ or $I_2$, a

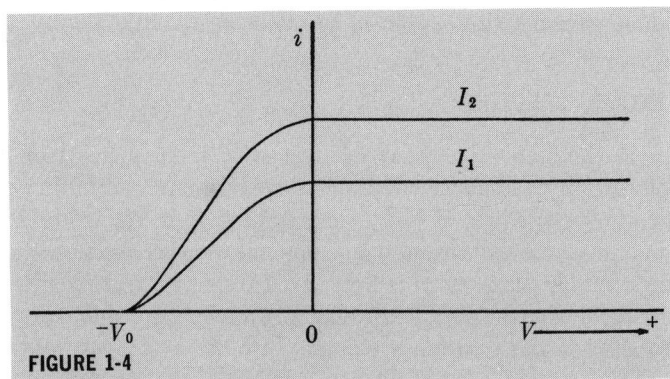

**FIGURE 1-4**

The Effect of Potential on the Photoelectric Current Produced by a Monochromatic Light Beam.

certain current, $i$, is obtained if a positive potential is placed on the electron collecting plate. However, as the potential is decreased to zero and then becomes negative, a point is finally reached where the current drops to zero. It is found that for light of a given wavelength, no matter what the intensity, the current will drop to zero at some point $V_0$. Thus, we see that the number of photoelectrons emitted in unit time by the surface is directly proportional to the intensity of the light, but the stopping potential, $V_0$, is independent of the intensity. These observations are in complete disagreement with the predictions of classical theory.

For the photoelectric effect, as for black body radiation, the classical theory fails to explain the experimental observations. On the other hand, Einstein was able to apply the quantum ideas of Planck with great success. According to the quantum theory, when a photon is incident on the surface of the metal it transfers its total energy to an electron in the metal surface. The electron then escapes from the metal surface with a kinetic energy equal to the energy of the incident photon minus the energy, $w$, necessary to escape from the surface. Since the energy of the photon is $h\nu$, the expression for the energy of the photoelectron becomes

(1-7)          $KE = \frac{1}{2} mv^2 = h\nu - w$

In terms of the quantum theory the curves in Figure 1-4 now become quite reasonable. If the energy of the incident photon is greater than $w$, a photoelectron will be emitted, and the negative potential necessary to stop the photoelectron will be $V_0$. When the intensity of the light is increased, such as $I_1$ to $I_2$, a larger current, $i$, will be observed because the surface is being bombarded by a greater number of photons, which yields a greater number of photoelectrons. If, on the other hand, the energy of the incident photon is less than $w$, no photoelectrons will be emitted. This is true regardless of the intensity or how long a period of time the light shines on the surface merely because an electron does not receive sufficient energy from a single photon to break away.

## ATOMIC SPECTRA

At the same time that interest was being focused on the problems of black body radiation, a similar development was taking place in the field of atomic spectra. It had been found that if, for instance, an electric discharge is passed through an element in the gaseous state, light will be emitted. Analysis of this light by a prism or grating spectrometer gives a series of sharp lines of a definite wavelength which prove to be characteristic of the particular element. In the case of a light element such as hydrogen, this line spectrum turns out to be fairly simple, as can be seen in Figure 1–5. However, for the heavier elements it is more likely to be extremely complex.

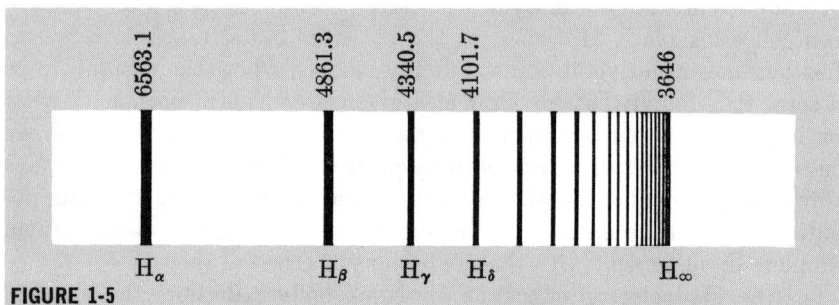

FIGURE 1-5

Spectral Lines of Atomic Hydrogen in the Visible Region.

During these first years a considerable amount of spectral data was accumulated, and as is commonly the practice, one of the first interests was to obtain an empirical relation to predict the sequence of lines. It was evident that the lines are not haphazard, but rather that they follow some sort of order. As early as 1883, Liveing and Dewar had realized that several possible series exist in the spectra of the alkali and alkaline earth metals. Although they were not able to discover an empirical equation to predict this order, they did recognize various repetitions of groups of lines and relations between groups that appear to be either *sharp* or *diffuse*. Then, in 1885, Balmer discovered the equation

$$(1\text{–}8) \qquad \lambda = \lambda_0 \frac{m^2}{m^2 - 4}$$

which relates the nine lines of the hydrogen spectrum that were known at that time. In this case $\lambda_0$ is a constant with a value of 3646 Å, and $m$ is a variable integer which can take on the values of 3, 4, 5, . . . , $\infty$. The agreement between the observed values of the lines in the hydrogen spectrum and their values calculated by the Balmer formula turns out to be extremely good.

Ordinarily, these equations are expressed in terms of wave number, $\bar{\nu}$, instead of wavelength. The frequency of a wave is given by $\nu = c/\lambda$ and represents the number of vibrations per second, and the wave number is $\bar{\nu} = 1/\lambda$ which is the number of vibrations per centimeter. In terms of wave number, the Balmer equation assumes the more familiar form

(1–9) $\qquad \bar{\nu} = R \left( \dfrac{1}{2^2} - \dfrac{1}{m^2} \right)$

Or more generally, this can be expressed as

(1–10) $\qquad \bar{\nu} = R \left( \dfrac{1}{a^2} - \dfrac{1}{m^2} \right)$

where $R$ is a constant known as the *Rydberg constant*, $a$ is a constant that depends on the particular spectral series, and $m$ is still a variable integer as it was in Eq. (1–8). The Rydberg constant has been found to be a constant for a given element and very nearly constant for all elements. The difference in its value is due to the atomic weight of the element, and it has been found to have a value of 109,677.58 cm$^{-1}$ for hydrogen.

At the time the Balmer series was discovered, the only portion of the electromagnetic spectrum that was known was the visible region. We now know that this represents an almost infinitesimally small portion of the total spectrum, ranging in wavelength from about 4000 to 8000 Å, as can be seen in Figure 1–6. Thus, after the discovery of the Balmer series, it

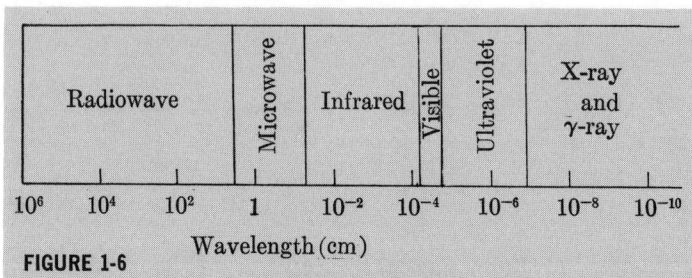

**FIGURE 1-6**

A Simple Division of the Electromagnetic Spectrum.

is not too surprising that other series of the same general type were subsequently discovered in the hydrogen spectrum. The *Lyman series* was found in the ultraviolet region and the *Paschen, Brackett,* and *Pfund series* were found in the infrared. The form of the equation describing each series is the same as that of the Balmer equation, the only difference being the value of the parameter, $a$, and the minimum value of the parameter, $m$, in Eq. (1–10). For $a = 1$, we have the Lyman series, and the Paschen, Brackett, and Pfund series arise from $a = 3$, 4, and 5, respectively.

Shortly after the discovery of the Balmer formula, Rydberg sought to find an equation of more general character. By 1890, he was able to show that a large number of the observed series can be represented by the formula

$$(1\text{–}11) \qquad \bar{\nu}_n = \bar{\nu}_\infty - \frac{R}{(n + b)^2}$$

where $b$ and $\bar{\nu}_\infty$ are constants depending on the particular series, and $n$ is a parameter that can take on successive integral values. This means that the wave number of each line can be represented by the difference of two terms, one of which is constant. In this case $\bar{\nu}_\infty$ is the constant term. Following up the work of Liveing and Dewar, Rydberg was able to classify a large number of series in the spectra of the more complex elements such as the alkali metals. He found it possible to distinguish between certain of these series that have lines that are quite sharp and others that have lines that are more diffuse than ordinary, and he named them accordingly. In addition, he noted another type of series in which the lines tend to be brighter than in other cases, and he called this the *principal series*. All of these series were found to be related by a formula of the type

$$(1\text{–}12) \qquad \bar{\nu} = \frac{R}{(m + a)^2} - \frac{R}{(n + b)^2}$$

Four of these series are of particular interest, mainly due to the continued use of the symbols $S$, $P$, $D$, and $F$ in subsequent work. These are:

Principal series: $\quad \bar{\nu} = \dfrac{R}{(1 + S)^2} - \dfrac{R}{(n + P)^2}, \, n = 2,3,4, \ldots,$

Sharp series: $\quad \bar{\nu} = \dfrac{R}{(2 + P)^2} - \dfrac{R}{(n + S)^2}, \, n = 2,3,4, \ldots,$

Diffuse series: $\quad \bar{\nu} = \dfrac{R}{(2 + P)^2} - \dfrac{R}{(n + D)^2}, \, n = 3,4,5, \ldots,$

Fundamental series: $\bar{\nu} = \dfrac{R}{(3 + D)^2} - \dfrac{R}{(n + F)^2}, \, n = 4,5,6, \ldots,$

Here $S$, $P$, $D$, and $F$ are constants that are characteristics of the particular series, and $R$ has its usual significance. An example of the sharp, principal, and diffuse series is seen in Figure 1–7, where a diagram of a portion of the spectrum of the sodium atom is shown.

## ATOMIC MODELS

Although the early developments in atomic spectra were significant, they were nevertheless empirical. For the most part, they were restricted to classifying and correlating observed data by means of empirical relations, and there was no concept at all of the mechanism by which these spectral lines arose. It might have been a reasonable assumption that they come from atoms, but how an atom is able to emit such lines could hardly

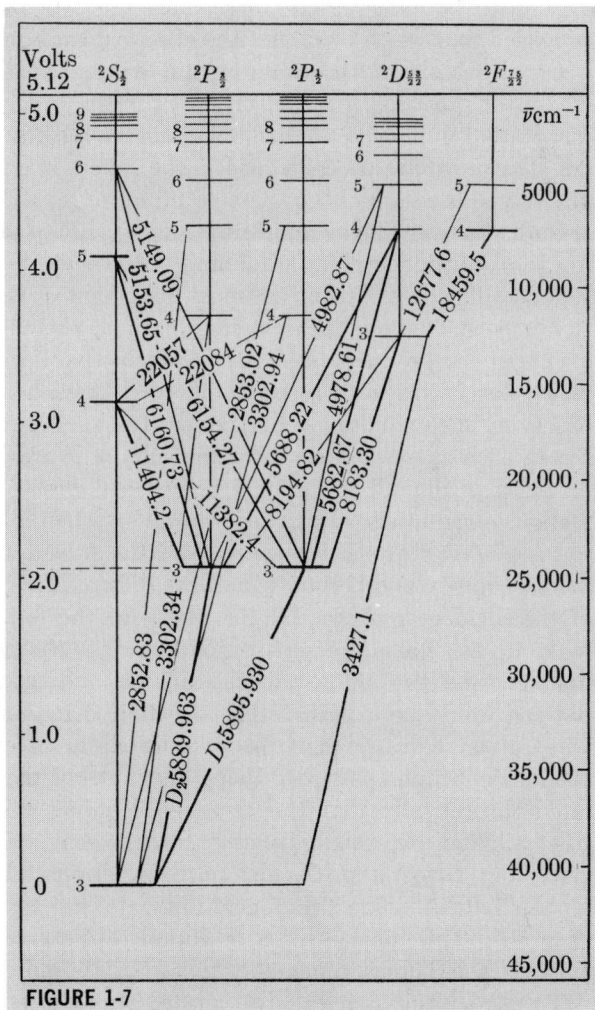

**FIGURE 1-7**

Energy Level Diagram of Sodium. Wavelengths of the various transitions are given in angstroms along the lines.

be a point of speculation since there was no satisfactory concept of the structure of an atom.

This situation, of course, did not last long. With the discovery of the emission of both positive and negative particles from an atom through radioactive decay along with other associations of the electron with the atom, it was realized that an atom must be composed in some manner of these newly found particles. The questions then naturally arise as to how many of each of these particles exist in a given atom and how they are arranged. The answers to these questions will depend on which of the proposed models best satisfies the observed experimental data.

On the basis of the experimental data available at that time, J. J. Thomson proposed a model of the atom in which the positive charge is distributed uniformly throughout a sphere with a diameter of about $10^{-8}$ cm. The electrons are embedded in the sphere in equilibrium positions and allowed to oscillate about these equilibrium positions when disturbed. As crude as this model may appear to us today, it had some merit in accounting for the occurrence of spectral lines. However, it also met with several serious difficulties. One of the most important of these was the interpretation it permitted for the scattering of alpha particles.

One of the products observed in radioactive decay is the *alpha particle*, which has been shown to be a doubly charged helium ion. One means of observing such particles is by the scintillations they cause on a fluorescent screen such as one coated with zinc sulfide. If a collimated beam of alpha particles is allowed to strike a fluorescent screen, an image of the cross section of the beam is observed. However, when a thin film such as a gold foil is placed between the source and the screen, the pattern is found to increase in size and become somewhat diffuse. This is due to the scattering of the incident particles by the atoms of the foil. Since the atoms that make up the foil are composed of electrical charges distributed in some manner, and the alpha particles are also charged, some change in the pattern would be expected. The question that immediately arises is how will a given distribution of charge in the atom affect the scatter pattern of the incident alpha particles. Using his model of the atom, Thomson calculated theoretically that the average deflection of alpha particles should be small, and the probability of large-angle scattering should be essentially zero.[3] Yet, Geiger and Marsden noted experimentally that about 1 in 8000 alpha particles incident on a gold foil is deflected through an angle greater than 90°.[4] This, of course, is in complete disagreement with the predictions of the Thomson model, which predicts only small-angle scattering.

To resolve this problem, Rutherford proposed a new model of an atom[5] in which the positive charge is concentrated in a small volume at the center of the atom. The electrons are then assumed to move around this center of positive charge in various orbits as the planets in the solar system. This model is an improvement over the Thomson model, since it gives a distribution of positive and negative charge in the atom that is in agreement with the observed scattering of alpha particles. Nevertheless, it also met with some serious problems. To begin with, the electrons could not be considered to be stationary because the unlike charges of the electron and the nucleus would cause them to come together. On the other hand, if the electrons were considered to be moving around the nucleus, another problem arose. When an electric charge is accelerated it emits or absorbs radiation. If the electrons are pictured as moving around the nucleus, they are subject to centripetal accelerations. According to the principles of electromagnetic theory, the electrons, therefore, must radiate energy. The only place for this continuous supply of energy to come from is the atom itself, and eventually the electron should spiral into the nucleus and, in essence, run down. Since we have no evidence to indicate that atoms run down, we are forced to the conclusion that the Rutherford atom is not the final answer.

The problems existing in atomic structure at that time were not solely problems concerning the distribution of the electrons and the nucleus in the atom. Even with a given distribution, it had yet to be determined how an atom is able to give discrete spectral lines, if they come from atoms at all. Neither Thomson nor Rutherford had been able to satisfactorily solve this problem. An important contribution was made by Conway in 1907, when he made probably the first attempt to explain the phenomenon in terms of quantum ideas. Without the aid of an atomic model, Conway concluded that an atom produces spectral lines one at a time and that a complete spectrum results from an extremely large number of atoms, each of which has to be in an excited state involving one electron.

## THE BOHR ATOM

As is the case in any phase of physics or chemistry, numerous theoretical models have also been proposed for the atom, and certainly there will be more. Each one is usually found to be superior to the previous ones in some manner. However, there have been few models in any field of physics or chemistry that have attracted such universal recognition as the one proposed in 1913 by Niels Bohr for the hydrogen-like atom.[6] Using the structural ideas of the Rutherford atom, Bohr was successful in quantitatively applying the concepts of quantum theory to explain the origin of line spectra as well as the stability of the atom.

It has been seen that a major problem with the Rutherford atom is that of continual radiation of energy as the electron moves about the nucleus. Bohr was able to overcome this problem by applying the quantum concept of *discrete* energy states. He maintained that the electron in an atom is restricted to move in a particular stable orbit, and as long as it remains in this orbit it will not radiate energy. Then, using the quantum principle that an oscillator will not emit energy except as a result of a jump from one of these energy states to another, Bohr postulated that when the electron jumps from a stable energy state of energy $W_1$ to a state of lower energy, $W_2$, a quantum of radiation is emitted with an energy equal to the energy difference of the two states. Mathematically, this is given by

$$(1\text{--}13) \qquad h\nu = W_1 - W_2$$

Once it is decided that the electron remains in some stable orbit around the nucleus, the question of the size and shape of this orbit arises. In the final form of his theory, Bohr assumed the orbits to be circular with a size such as to satisfy the quantum condition that the angular momentum, $p$, of the electron is an integral multiple of the quantity $h/2\pi$. Thus, we obtain

$$(1\text{--}14) \qquad p = \frac{nh}{2\pi} = mvr$$

where $m$ and $v$ are the mass and the velocity of the electron, $r$ is the radius of the orbit, $h$ is Planck's constant, and $n$ is a positive integer known as a quantum number. This leads to the picture of an atom as shown in Figure 1–8. For different values of the quantum number we find different orbits available for the electron. The lowest orbit, for which $n = 1$, is the most stable orbit for a hydrogen-like atom; and an electron in this orbit is said to be in its *ground state*. The emission of radiation can then be seen to

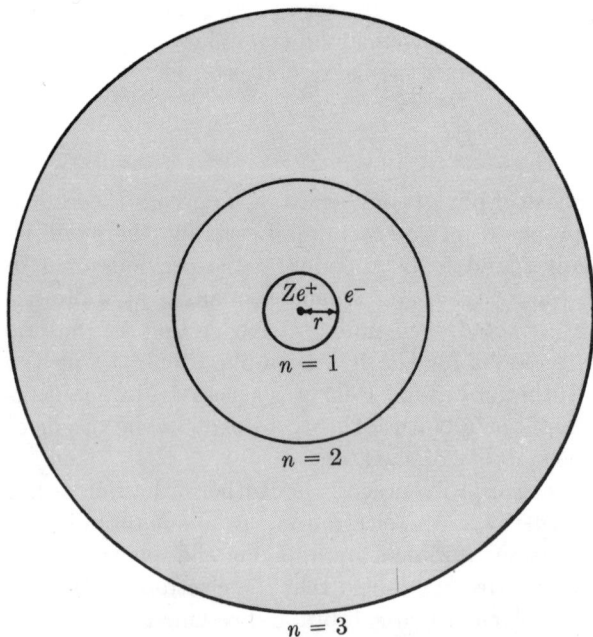

**FIGURE 1-8**
A Simple Representation of the Bohr Atom Showing the First Three Energy Levels.

result when the electron is raised by some means to an excited state and then drops back to one of the available lower energy states.

The acceptance of the Bohr model resulted primarily from its success in quantitatively explaining the line spectra of hydrogen-like atoms. This success was necessary, for there were several facets of the Bohr model that made it difficult to accept. It is interesting to note that there is no mechanism provided for the electron to radiate energy in the Bohr model. When Bohr threw out the idea of continuous radiation by an accelerated charge, he threw out the only known means for a charged particle to radiate energy. According to the Bohr model, the radiation results from a change in energy state of the electron, but how this takes place is not answered. In addition, Bohr arbitrarily utilized both classical and quantum ideas as he found necessary in order to obtain his final result. Therefore, we see that the quite remarkable agreement between the theoretical calculations of the Bohr model and the experimentally determined data was the justification of the Bohr approach.

For a quantitative treatment of a one-electron system, the force of attraction between the electron and the nucleus is considered to arise from the electrostatic attraction between the positive charge of the nucleus and the negative charge of the electron. Thus,

$$(1\text{--}15) \qquad F = \frac{Ze^2}{r^2}$$

where $Z$ is the atomic number of the element and $r$ is the distance between the electron and the nucleus. This electrostatic attractive force must equal the centripetal force resulting from the motion of the electron about the nucleus. This leads to the relations

$$(1\text{--}16) \qquad F = ma = \frac{mv^2}{r} = \frac{Ze^2}{r^2}$$

If this expression is solved for the radius, $r$, we obtain

$$(1\text{--}17) \qquad r = \frac{Ze^2}{mv^2}$$

But from Eq. (1–14) it is seen that

$$(1\text{--}18) \qquad v = \frac{nh}{2\pi mr}$$

If the value of $v$ from Eq. (1–18) is now substituted into Eq. (1–17), it is found that the radius of the electron orbit is given by

$$(1\text{--}19) \qquad r = \frac{n^2h^2}{4\pi^2 mZe^2}$$

For the hydrogen atom, $Z = 1$, and if we consider the electron to be in its ground state ($n = 1$), the radius of the atom can readily be calculated to be $r = 0.529 \times 10^{-8}$ cm or 0.529 Å, which is of the correct order of magnitude as compared with $r$ determined from other sources. Here, we see an early success of the Bohr atom.

The energy of the electron in the atom, however, is the problem of primary importance. The total energy of the electron is made up of its kinetic energy and its potential energy. If the zero of potential energy is defined as the energy of the electron when it is at rest at an infinite distance from the nucleus, its potential energy with respect to the nucleus at any distance $r$ is found to be

$$(1\text{--}20) \qquad V = \int_{\infty}^{r} F \cdot dr = \int_{\infty}^{r} \frac{Ze^2}{r^2} \, dr = -\frac{Ze^2}{r}$$

Using Eq. (1–16), it is seen that the kinetic energy is given by

$$(1\text{--}21) \qquad T = \tfrac{1}{2} mv^2 = \frac{Ze^2}{2r}$$

Since the total energy of the electron is the sum of the kinetic and potential energy, we find that

$$(1\text{--}22) \qquad W = T + V = \frac{Ze^2}{2r} - \frac{Ze^2}{r} = -\frac{Ze^2}{2r}$$

Now, by substituting the value found for $r$ in Eq. (1–19) into Eq. (1–22), the energy of the electron in the $n$th quantum state is

$$(1\text{--}23) \qquad W_n = -\frac{2\pi^2 m e^4 Z^2}{n^2 h^2}$$

It is important to note that the quantum number $n$ is, in a sense, a measure of the energy of the electron. When the electron is in the first shell $(n = 1)$, it has a maximum stability, but as $n$ increases, the electron energy with respect to the nucleus decreases until it approaches zero energy as $n \to \infty$.

It was shown earlier that the energy of the radiation emitted by the atom is equal to the difference in energy of two given quantum states. Thus, for a transition between two quantum states of energy $W_{n_1}$ and $W_{n_2}$ the frequency of the emitted radiation is

$$(1\text{--}24) \qquad \nu = \frac{W_{n_1} - W_{n_2}}{h}$$

Now, if the expression for $W_n$ is substituted into Eq. (1–24), and the frequency is converted to wave number by the relation $\bar{\nu} = \nu/c$, the wave number is given by

$$(1\text{--}25) \qquad \bar{\nu} = \frac{2\pi^2 m e^4}{ch^3} Z^2 \left( \frac{1}{n_2^2} - \frac{1}{n_1^2} \right)$$

If the parameter $n_2$ is given the value of 2, it is seen that Eq. (1–25) is of exactly the same form as the Balmer equation, Eq. (1–9). This then offers a real challenge to the Bohr theory. The constant term in Eq. (1–25), $2\pi^2me^4/ch^3$, must give a reasonable value for the Rydberg constant in order to be in agreement with experiment. If the Bohr theory had failed in this test, it would have been necessary to start looking for a new atomic model. As it turned out, the agreement was found to be very good, and if further refinements are made, the calculated value of $R$ can be brought into even better agreement.

The most significant improvement can be made by considering the finite masses of the nucleus and the electron. Thus far, it has been assumed that the mass of the nucleus is infinitely great with regard to that of the electron. With this assumption it was possible to neglect the motion of the nucleus and to consider it to exist at the exact center of the atom. However, for a nucleus of finite mass, it is necessary to consider the motion both of the nucleus and of the electron around a common center of mass, as seen in Figure 1–9. This leads to the following expression for $R$:

$$(1\text{–}26) \qquad R = \frac{2\pi^2me^4}{ch^3} Z^2 \frac{1}{1 + \dfrac{m}{M}}$$

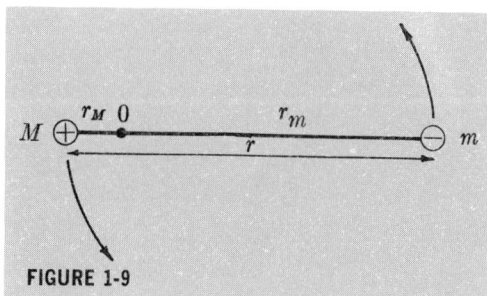

**FIGURE 1-9**

Motion Around the Center of Mass of a Hydrogen Atom.

Using the *reduced mass* and the best known values of $c$, $h$, $m$, and $e$, it is found that $R = 109{,}681$ cm$^{-1}$ which is in excellent agreement with the experimentally determined value of $109{,}677.58$ cm$^{-1}$.

Now that it has been shown that the equation for the wave number developed by Bohr is the same as that found by Balmer, it is possible to understand the origin of the spectral series. For the Balmer series, the constant, $a$, in Eq. (1–10) is found to equal 2. From Figure 1–10, it is seen that the value of $a = 2$ arises from the fact that the electron transitions are to the second shell. In a similar manner, an analogous relation exists between $a = 1$ and the Lyman series, and $a = 3, 4$, and 5 for the Paschen, Brackett, and Pfund series respectively. At the time the Bohr atom was developed, only the Balmer and the Paschen series were known. The completion of the five series, all of which could be predicted by the Bohr theory, came with the discovery of the Pfund series in 1924.

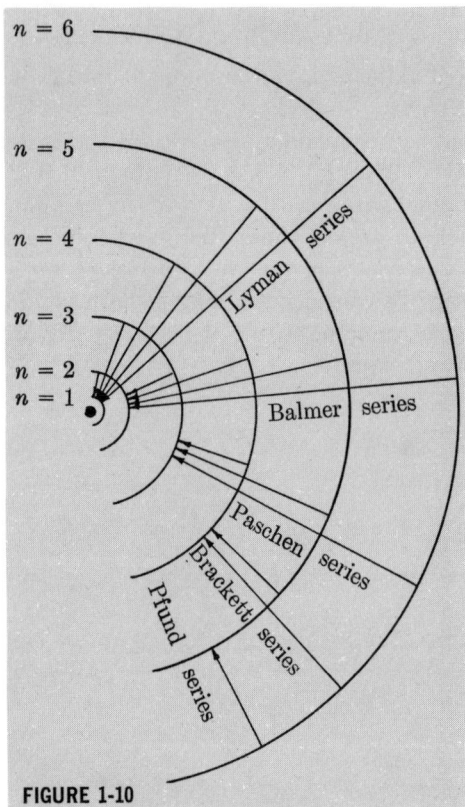

**FIGURE 1-10**

Transitions Leading to the Five Spectral Series of the Hydrogen Atom.

## EXTENSIONS OF THE BOHR THEORY

It is certainly true that the Bohr theory met with many successes, such as the quantitative prediction of the energies of the line spectra of the hydrogen-like atoms. Yet it also met with some difficulties. One of the first of these was the problem of the *fine structure* in the line spectrum of the hydrogen-like atom. The Bohr theory had explained the existence of the various lines in the hydrogen spectrum, but it predicted that only a series of single lines exist. At that time, this is exactly what had been observed. However, as better instruments and techniques were developed, it was realized that what had been thought to be single lines were actually a collection of several lines very close together. This implies that there are several energy levels close together rather than a single level for each quantum number $n$. This would then require the existence of new quantum numbers, and there is no way to obtain them directly from the Bohr model.

This problem was solved to some extent by Sommerfeld when he considered in detail the effect of *elliptical* orbits for the electron. Bohr had admitted the possibility of elliptical orbits in his original work but had carried it no further. For the case of circular orbits for the electron, the only coordinate that varies is the angle of revolution, $\varphi$. However, for an elliptical orbit, it can be seen in Figure 1–11 that both the angle, $\varphi$, and

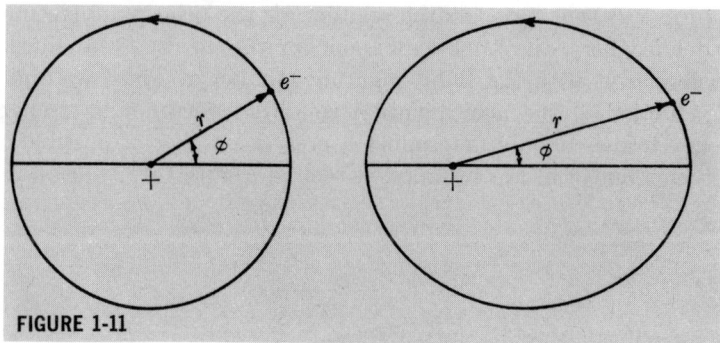

**FIGURE 1-11**

Effect of an Elliptical Orbit on the Variables $r$ and $\varphi$.

the radius vector, $r$, can vary. With two degrees of freedom the possibility of two quantum conditions arises. In order to quantize both degrees of freedom, Sommerfeld generalized the quantum condition of Bohr that $p = h/2\pi$, to the condition

$$(1\text{--}27) \qquad \oint p_i \, dq_i = n_i h$$

Here $p_i$ is the momentum for the given coordinate, $q_i$. In terms of the two variables, $\varphi$ and $r$, the following two integrals result:

$$\oint p_\varphi \, d\varphi = n_\varphi h$$

$$\oint p_r \, dr = n_r h$$

Since the angular momentum of an isolated system is a constant, the integral for the angular momentum gives the same result as obtained by Bohr,

$$(1\text{--}28) \qquad p_\varphi = n_\varphi \frac{h}{2\pi}$$

The quantum number $n_\varphi$ is known as the *azimuthal* or *angular momentum* quantum number. The solution to the radial integral is not nearly so simple as that for the angular momentum. The solution contains a relation between the azimuthal quantum number, the radial quantum number, $n_r$, and the eccentricity of the ellipse.

In carrying out his treatment, Sommerfeld was then able to show that the energy of the electron depends on the principal quantum number which can be defined as

$$n = n_\varphi + n_r$$

It turns out that by using the relation for the principal quantum number as defined here, the same expression for the energy is obtained as for a circular orbit with the Bohr quantum number $n$. Thus, we find that the introduction of the new quantum condition does not in itself give new energy terms. It only determines a greater number of possible orbits for a given value of $n$. For instance, as seen in Figure 1–12, when $n = 3$, there

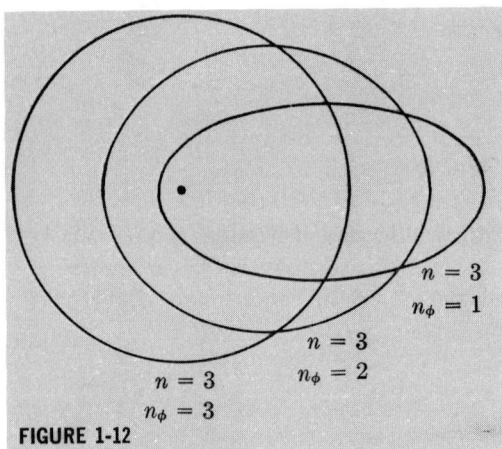

**FIGURE 1-12**

Possible Electron Orbits Arising from the Quantum Number $n = 3$.

is the possibility of a circular orbit where $n_\varphi = 3$ and $n_r = 0$, or two different ellipses for $n_\varphi = 2$ and $n_r = 1$, and $n_\varphi = 1$ and $n_r = 2$.

When one quantum number is sufficient to determine the energy states of a system with two or more degrees of freedom, the system is said to be *degenerate*. In order to explain the fine structure in the hydrogen-like spectrum it was necessary to remove this degeneracy. This means that at least two quantum numbers will have to make a contribution to the energy of the system. Sommerfeld found that the degeneracy in his atomic model can be removed by considering the *relativistic* change in the mass of the electron during its motion around the nucleus. As the electron revolves around the nucleus, its velocity changes continuously, depending on its proximity to the nucleus. From the special theory of relativity it is known that the mass of a particle increases as its velocity increases. If this effect is taken into consideration, a small difference in energy is found to exist between a circular orbit and an elliptical orbit. This energy difference is a function of the azimuthal quantum number $n_\varphi$, and can be related to the physical picture of energy levels in the Bohr atom by considering each major energy level to be composed of several sublevels lying very close together. With this picture, a rather reasonable agreement with the observed fine structure in the hydrogen spectrum was obtained.

In the presence of a magnetic field, it is found that the spectral lines are split even more. This effect, known as the *Zeeman effect*, is illustrated in Figure 1–13, with the Zeeman pattern for the sodium principal doublet.

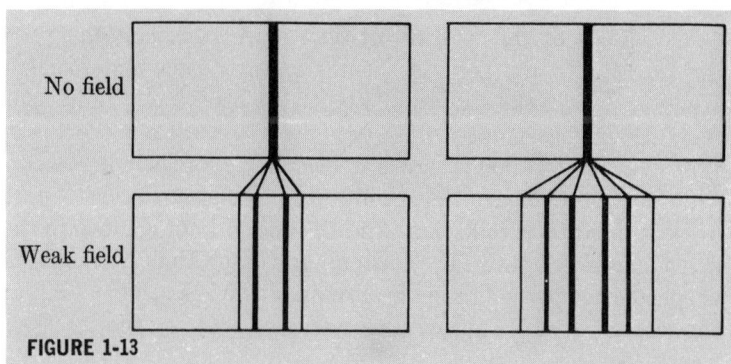

**FIGURE 1-13**

Zeeman Pattern for the Sodium Principal Doublet in the Presence of a Weak Magnetic Field.

An explanation of this phenomenon requires the introduction of a third quantum number, $m$, the *magnetic* quantum number. An electron in space requires three coordinates to describe its position. This results in three degrees of freedom and should require three quantum numbers to describe its energy. Without a spatial reference, the arrangement of the orbital plane of the electron is completely arbitrary, and this third degree of freedom is degenerate. However, in the presence of an external field, the orbital plane of the electron will *precess* about the direction of the field and thereby remove the degeneracy. This third quantum condition is similar to that of the angular momentum, giving

$$(1\text{–}29) \qquad p_z = m\frac{h}{2\pi}$$

Thus, we now see the necessity of using three quantum numbers to describe the energy of the electron. Each new quantum number had to be introduced to meet the demands of experiment. Yet even with three quantum numbers a complete explanation of atomic spectra is not possible. For instance, the effect of a weak magnetic field gives the so-called *anom-*

*alous* Zeeman effect, and this can not be understood in terms of the Bohr-Sommerfeld model. In addition to this shortcoming, there are numerous other points where the Bohr atom and its modifications fail. Among the more important of these is the inability to apply the system to more complicated atoms. Application of the theory to the spectrum of an atom as simple as helium met with complete failure, and all attempts to understand the basis of the periodic system in terms of the Bohr model were unsuccessful. This indicates that the treatment is valid only for a one-electron system. Such a limitation is unreasonable, and we therefore see the need for something better.

## STATUS OF THE QUANTUM THEORY

We have seen the concepts of the nature of radiant energy change from one extreme to another and then apparently back again since the time of Newton. Before the studies of Planck on black body radiation, all experimental evidence pointed unquestionably toward a wave theory of radiation. Yet, since 1900, there has been an overwhelming accumulation of experimental evidence that points just as surely toward a particle theory of electromagnetic radiation. And the end did not come with the particular cases considered thus far. Einstein, and later Debye, treated the problem of specific heats of solids in terms of quantum ideas, and Compton explained the scattering of X rays with electrons by treating the collisions as if they were between billiard balls.

With such a preponderance of evidence in favor of the quantum theory, one might be prone to feel that the cycle is complete and we are back again to the basic ideas of Newton. But this is not really true, for we cannot neglect the fact that electromagnetic radiation has been shown to have wave character just as surely as it has been shown to have particle character. This leaves us faced with a dilemma: Is a photon a wave or is it a particle? The problem is one that is not easily reconciled, and the answer does not lie in a simple chemical or physical approach. For here we see a new side to natural science. It must now take on a definite philosophical character.

Possibly the root of the problem lies in the nature of the macroscopic world in which we live. Here we observe only two types of motion, one of which is of a wave nature and the other of a particle nature. If a baseball is thrown, it appears to be a particle and its motion can be described by Newton's laws of motion. On the other hand, if a pebble is dropped into a pool of water we see a motion that must be described by a wave equation. Nowhere in our experience do we see a motion that is in between these two extremes. However, this does not mean that it does not exist. Nevertheless, it is extremely difficult to conceive of something that is foreign to our experience, and it is impossible to conceive of a type of motion that actually contradicts our experience. Yet this is what we are faced with, and out of this dilemma must come our new approach to the problems of chemistry and physics.

## References

1. M. Planck, *Ann. d. Physik*, **4**, 553 (1901).
2. A. Einstein, *Ann. d. Physik*, **17**, 132 (1905).
3. J. J. Thomson, *Proc. Phil. Soc. (Cambridge)*, **15**, 465 (1910).
4. H. Geiger and E. Marsden, *Proc. Roy. Soc. (London)*, **82**A, 495 (1909).
5. E. Rutherford, *Phil. Mag.*, **21**, 669 (1911).
6. N. Bohr, *Phil. Mag.*, **26**, 1 (1913).

## Suggested Supplementary Reading

M. R. Wehr, and J. A. Richards, Jr., "Introductory Atomic Physics," Addison-Wesley Publishing Company, Inc., Reading, Mass., 1962.

F. K. Richtmeyer, and E. H. Kennard, "Introduction to Modern Physics," 5th ed., McGraw-Hill, Inc., New York, 1955.

Sir Edmund Whittaker, "A History of the Theories of Aether and Electricity," Thomas Nelson & Sons, New York, 1951.

H. Semat, "Introduction to Atomic Physics," 3rd ed., Longmans, Green and Co., Ltd., London, 1949.

G. Herzberg, "Atomic Spectra and Atomic Structure," Dover Publications, Inc., New York, 1946.

M. Born, "Atomic Physics," Hafner Publishing Co., Inc., New York, N.Y., 1946.

I. Kaplan, "Nuclear Physics," Addison-Wesley Publishing Company, Inc., Reading, Mass., 1956.

## Problems

1. Calculate the (a) radius, (b) velocity, and (c) energy of an electron in the fourth Bohr orbit of the hydrogen atom.
2. Determine the wavelength of a photon resulting from the transition $n = 4 \rightarrow n = 1$ in the hydrogen atom.
3. Use the Bohr model to determine the energy of the ground state of $Li^{2+}$.
4. Calculate the ionization potential of the hydrogen atom using the Balmer equation.
5. It was stated that because of its thermodynamic origin, the temperature must occur in Planck's equation in the form $(\lambda T)$. What is so unique about thermodynamics to require this?
6. One might propose that the problem of wave-particle duality for a photon can be solved by simply imagining the photon to move through space along a wave path. How does this contradict our conventional ideas of wave motion?
7. What is the effect of wave-particle duality on present-day acceptance of the three "Laws of thought" in Aristotelian logic?

# Wave Mechanics

It was seen in the last chapter that electromagnetic radiation exhibits a duality in character. Under certain experimental conditions it is found to behave as a wave, and at other times it takes on the unmistakable nature of a particle. A behavior such as this is in complete contradiction to all physical experience. We have always observed a particle to be constrained within certain finite boundaries, whereas a wave tends to dissipate itself throughout space. Any attempt to construct an ordinary physical picture of such a system cannot help but fail, and we are forced to admit the existence of a situation which defies any understanding in terms of our classical ideas of nature.

## MATTER WAVES

To add to the dilemma, Louis de Broglie proposed in 1924,[1] that this duality should apply not only to radiant energy but to matter itself, thereby leading to matter waves. The duality of radiant energy was a serious problem, but not nearly so difficult as the acceptance of matter waves. Our experience with radiant energy is only indirect, but we deal directly with matter and feel much more familiar with its properties. For instance, a rock is a particle, and we feel quite confident that it will remain as such. Yet, if it has wave characteristics, it should show the features of wave motion. This means that it must be dissipating itself throughout space if our common ideas of wave motion are to be retained. Obviously, such a classical approach to the problem is inadequate.

Of course, the mere proposal of the existence of wave character for a material particle would in itself be insignificant. Some demonstration of this existence must be found before so radical a view could be acceptable. We have seen that a link does exist between the wave and particle character of a photon in the expression for its energy,

(2-1) $E = h\nu$

The frequency is certainly a variable that is associated with wave motion, but the energy of a system can be expressed in terms of particle concepts such as mass and velocity. In terms of relativity theory, the energy of a particle of mass, $m$, is given by

(2-2) $$E = mc^2$$

where $c$ is the velocity of light. By equating these two expressions, we obtain

$$h\nu = mc^2$$

or

$$\frac{h\nu}{c} = mc = p$$

where, in this case, $p$ is the momentum of a photon. In terms of wavelength, this becomes for a photon,

(2-3) $$\lambda = \frac{h}{mc} = \frac{h}{p}$$

The development of Eq. (2–3) is legitimate. However, the extension to a particle cannot be justified in terms of well-founded relationships. The difficulty arises from the use of Eq. (2–2). Here the velocity of light occurs regardless of the nature of the particle. Yet, if we are dealing with the movement of a conventional particle, its velocity will not be $c$, but rather some lesser value, $v$. Nevertheless, de Broglie proposed that wave nature exists for a particle of mass $m$ and velocity $v$ given by the analogous relation

(2-4) $$\lambda = \frac{h}{mv} = \frac{h}{p}$$

Due to its origin, the wavelength of a *particle* wave is often referred to as the *de Broglie* wavelength. Thus, for any particle of mass $m$ and a known velocity, the de Broglie wavelength of the particle can be calculated. If we consider the case of an electron with an energy of about $1.6 \times 10^{-10}$ erg, which is a rather low energy, the de Broglie wavelength is found to be of the order of 1.2 Å. This distance is of the same order of magnitude as the spacings in a crystal lattice.

Taking advantage of the similarity in magnitude of crystal spacings and the de Broglie wavelength of an electron in this energy range, Davisson and Germer[2] were able to show that the electron actually does possess a wave character. Using the crystal spacings in a nickel crystal as a diffraction grating, they were able to obtain diffraction patterns that can readily be understood in terms of wave motion of the electron. Although some question might arise as to the true particle character of an electron, wave properties have also been observed for such unquestionably material particles as the neutron and the helium atom.

The question of whether these waves are true waves in terms of our mechanical analogs such as water waves and vibrating strings is difficult to answer. It is possible that the only similarity is in their mathematical behavior. Throughout the nineteenth century the tendency was to reduce a physical phenomenon to a mechanical picture in terms of observable physical experience. However, the scientific advances of the twentieth century seem to make this no longer possible, and it may be that a mathematical understanding is the best for which we can hope. We may rationalize the difficulty to some extent by recalling the development of our ideas of atomic structure. Starting with the Thomson atom and proceeding on through the Bohr atom we see a consistent trend toward a superior model. In each of these models we may have felt that the approach to *true objective reality* had been attained. However, such an attitude is belief, not science. There is no scientific proof that further advancement will not be made, and from a scientific point of view it is necessary to keep an open mind to any such possible advancement. In fact, if an atomic model were proposed that happened to be identical with some sort of *true* atom, we could not be aware of this good fortune. The superiority of each of our models lies in its consistency with a greater number of experimental observations than the previous model. If this model defies description in terms of a reasonable physical picture, then we must be satisfied with a mathematical picture.

## THE UNCERTAINTY PRINCIPLE

In the wave properties of the electron we find the first of the two underlying precepts of wave mechanics. The second of these is the *Heisenberg uncertainty principle*, which finds its expression in the statistical nature of scientific observation. We have seen that before the advent of wave mechanics, it was customary to construct a model of atomic-sized systems in terms of familiar everyday concepts. With the wave-particle dilemma, the impossibility of constructing such a *deterministic* model was first realized. This might lead one to question the validity of even considering the wave character of a particle. However, we might, at the same time, question if it is really justifiable to feel that a strict *particle* treatment will allow the construction of such a model. It is certainly possible that in the domain of the atom the situation might be quite different from what it is in the macroscopic world.

If we are to maintain that a certain *thing* is a particle, we should be able to measure the particle properties of this *thing*, such as its momentum and position. This would not be a difficult problem if the *thing* were a baseball, but it might be worth considering in some detail the nature of such measurements on an electron. To carry out these measurements, the position of the electron could be determined by use of a microscope such as that shown schematically in Figure 2–1. It has been found that a limit of accuracy in the position measured by a microscope exists which is dependent upon the resolving power of the instrument. The $x$-component of this error is given by the expression $\Delta x \sim \lambda/\sin \alpha$. Since we are interested in determining the exact position of the electron, it is obvious that the error in position $\Delta x$ must be made as small as possible. From the expression for $\Delta x$ it is readily seen that this can be accomplished merely by using an illuminating light of very short wavelength, $\lambda$. For this purpose a gamma-ray source might be used. However, this offers a new problem. When a high-energy photon such as a gamma ray collides with an electron, there will be a resultant Compton recoil. The gamma ray will be scattered by the electron, and its resultant momentum will be of the order of magnitude $p = h/\lambda$. But immediately we realize that there is a degree of uncertainty in this momentum. In order for the electron to be observed, it will be necessary for the scattered gamma ray to enter the microscope. On reference to Figure 2–1, we see that there is a considerable $x$-component in which the photon may enter the microscope and still be observed. This leads to an uncertainty in the $x$-component of momentum of the electron given by

$$\Delta p_x \sim \frac{h}{\lambda} \sin \alpha$$

If we now take the product of the error in position and the error in momentum, we obtain the approximate relation

(2–5) $$\Delta x \Delta p \sim \left(\frac{\lambda}{\sin \alpha}\right)\left(\frac{h}{\lambda} \sin \alpha\right) \sim h$$

This is the well-known Heisenberg uncertainty principle[3] which was first enunciated by Heisenberg in 1927.

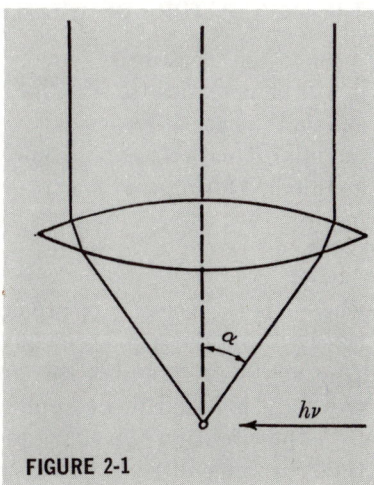

**FIGURE 2-1**

Diagrammatic Operation of a Gamma Ray Microscope.

On the basis of the uncertainty principle, we see that it is not possible to say that an electron is here or there with a known velocity; but rather, we are restricted to speak only in terms of probability. If the position of the electron is known at a given instant, we can only speak of a probable value of its momentum; and if the momentum of the electron is known, we cannot hope to know simultaneously its exact position. Thus, we find that even without the difficulties of the wave-particle dualism, a deterministic model of an atom in the classical sense is actually in contradiction to a fundamental scientific principle.

It is important that this point be appreciated in approaching wave mechanics. The concepts that we will use are not those of our everyday experience, for these familiar concepts have been seen to be inconsistent with our observations of nature in the realm of the atom. It is very possible that the wave-particle dilemma is completely illusionary. The difficulty may arise from the fact that in all of our previous experience only two types of motion have been observed, and it is only natural to attempt to explain the motion of an atom or electron in terms of this experience. The only thing that can really be said is that the electron behavior can be described by an equation that is of the same general form as that found for wave motion. Nevertheless, regardless of what philosophical conclusion one might come to with regard to the characteristics of an atom, we must admit that it is no longer possible to construct a deterministic model in the classical sense, and whatever type of model we use, it must be consistent with our observations of nature. This means that we must recognize (1) the wave-like behavior of the system, and (2) the probability character of our observations.

## THE WAVE NATURE OF THE ELECTRON

Since the electron is found to behave in the same manner as a wave, it will be necessary to describe its motion by a wave equation. Ordinarily, the mathematical treatment of wave motion involves a second-order differential equation. For instance, the transmission of a disturbance along a stretched string can be expressed by the equation

$$(2\text{--}6) \qquad \frac{\partial^2 \Phi}{\partial x^2} = \frac{1}{c^2} \frac{\partial^2 \Phi}{\partial t^2}$$

where $c$ is the velocity of propagation of the wave. The wave function, $\Phi$, is seen to be the displacement of the string as a function of the variable, $x$, at any time, $t$, and is therefore an amplitude function. An equation of this form is found to be applicable to virtually all forms of wave motion from the vibration of a string to the transmission of electromagnetic radiation. In three-dimensional cartesian space, the wave equation becomes

$$(2\text{--}7) \qquad \frac{\partial^2 \Phi}{\partial x^2} + \frac{\partial^2 \Phi}{\partial y^2} + \frac{\partial^2 \Phi}{\partial z^2} = \frac{1}{c^2} \frac{\partial^2 \Phi}{\partial t^2}$$

or more simply

$$\nabla^2 \Phi = \frac{1}{c^2} \frac{\partial^2 \Phi}{\partial t^2}$$

$\nabla^2$ is the *Laplacian operator*, which in cartesian coordinates is given by

$$\nabla^2 = \left( \frac{\partial^2}{\partial x^2} + \frac{\partial^2}{\partial y^2} + \frac{\partial^2}{\partial z^2} \right)$$

A typical example of such a wave function, $\Phi$, is the familiar sine function

(2–8)        $$\Phi = A \sin \frac{2\pi}{\lambda} (x - ct)$$

This, of course, could just as readily be a cosine function or any other function that is still a solution to the differential equation of wave motion. This is a type of wave motion with which we are familiar. However, its extension to matter waves is not necessarily straightforward. We have never encountered the problem of matter waves before, and we can only guess at the type of equation that will successfully describe their properties. The validity of our ultimate choice can be known only by the results we obtain.

## INTERPRETATION OF THE WAVE FUNCTION

Although we have not, as yet, decided the exact manner in which the wave character of the electron will be expressed, we are certain that it must be with a wave equation. This necessitates a wave function to describe something about the electron. For the familiar forms of wave motion, it has been possible to give a physical interpretation to the wave function that is both reasonable and useful. However, the question of what significance a wave function of a particle might have is not so readily answered. The success of wave mechanics was well demonstrated by Erwin Schroedinger before an acceptable interpretation of the wave function was given. This might indicate that the wave function has only a mathematical significance and no interpretation in a physical sense is really necessary. This would appear to be particularly true in the light of the conceptual difficulties of the wave-particle duality. Such a viewpoint should have appeal to those who feel that any attempt to give a physical picture to all natural processes is a hindrance to scientific advancement. However, there is also much to be said for a description of natural processes in terms of concepts that maintain some link with our physical world.

Max Born utilized the probability concept of the uncertainty principle to give us our presently accepted idea of the wave function.[4] According to Born, the wave function of a particle is not an amplitude function in the common sense used for ordinary waves, but rather, it is a measure of the probability of a mechanical event. When the wave amplitude is large, the probability of the event is large, and when the amplitude is small, the probability of the event is correspondingly small. To some extent, we have lost sight of the physical world in this interpretation, for this wave motion is not the wave motion with which we are familiar. Yet such a concept is consistent with reasonable interpretations of certain quantum considerations of electromagnetic wave motion.

If a beam of light is incident upon a cross section perpendicular to its path, the intensity of the beam of light can be thought of as the number of photons that pass through a unit area, $dxdy$, of this cross section per second. Since the velocity of the photon is a constant, in a given time it will have traveled a distance, $dz$, and therefore will have defined a volume element, $dxdydz$. The intensity, as defined by the number of photons passing through the unit cross section, is proportional to the photon density, the number of photons in the volume element. Now, according to wave theory, the intensity, $I$, of a light beam has been shown to be proportional to the square of the amplitude of the electric vector, $E$,

$$I = \frac{c}{4\pi} E^2$$

This indicates that a link exists between the density of particles and the square of the wave function.

More correctly, the density can be considered to be a probability density. An appreciation of the probability character of such a system can readily be had by considering the diffraction of a light beam by a narrow slit. If a photographic plate is placed behind the slit, a diffraction pattern is observed on the plate after exposure in which there are alternate regions of dark and light corresponding to high and low intensities, respectively. Where the intensity of the incident photons is great, the film will be dark due to exposure, and where the intensity is small, there will be a light region. If we now consider a beam of very low intensity, it is seen that we cannot be certain exactly where the photons will strike the film. Where the film had been observed to be the darkest, we could say that the probability will be the greatest that they should strike in that region. However, each region is not sharply defined, and this leads to the possibility of an infinite number of positions at which a photon might strike the film. Thus, our knowledge of the position of the photon can be expressed only in terms of probability, and we are led to the conclusion that the wave function squared expresses the probability of finding a photon in a given element of volume $dxdydz$.

There is certainly an analogy between the diffraction of a light beam and the diffraction of a beam of electrons. It might, then, be expected that a quantum interpretation that seems quite reasonable for a photon should also hold for an electron. This leads to the postulate that the square of the wave function of an electron is proportional to the probability of finding the electron in a given volume element, $dxdydz$. Such an interpretation is just a postulate and may or may not be legitimate. Thus far it appears to be consistent with experimental observation. One of the most significant indications of its validity lies in the treatment of directional bonding in molecules. The positions at which the electron density of the bonding electrons is calculated to be the greatest are where the bonded atoms are found to be located. For instance, in the molecule $H_2S$, the hydrogen atoms lie at an angle of about 92° with respect to each other, and according to simple theoretical calculations the electron density is a maximum at an angle of 90°.

The symbol $\psi$ is usually used to denote the wave function of an electron, and very often $\psi$ contains the imaginary quantity, $i$, the square root of $-1$. Since the probability that an electron is in a given volume element must be a real quantity, the product $\psi\psi^*$ is used rather than $\psi^2$, where $\psi^*$ is the complex conjugate of $\psi$. The product $\psi\psi^*$ will always be real, whereas $\psi^2$ can possibly be imaginary. As an example, $\psi$ can be considered the complex quantity $a \pm ib$. The complex conjugate of $\psi$ can then be obtained by changing $i$ to $-i$, giving $a \mp ib$. The product $\psi\psi^*$ will then be $a^2 + b^2$, which is a real quantity. If $\psi$ turns out to be a real quantity initially, then $\psi$ and its complex conjugate are the same.

## NORMALIZED AND ORTHOGONAL WAVE FUNCTIONS

The square of the wave function is said to be proportional rather than equal to the probability that the electron is in a given volume element, $dxdydz$. This arises from the fact that if $\psi$ is a solution to the wave equation, multiplication by any constant such as $A$ will give a wave function $A\psi$, which is also a solution to the wave equation. This means that it is not possible to say in general that $\int\psi\psi^*dxdydz$ is equal to the probability, but only that it is proportional to the probability that the electron is in the given volume element. However, since multiplication by a constant is possible, it is usually convenient to multiply the wave function by a constant that will make the square of the resultant wave function equal to the probability.

The probability of a certainty is defined as unity. Thus, if it is a known fact that the electron is in a given volume element, $dxdydz$, then we can say that the probability that it is in this volume element is unity. This leads to the relation

$$(2\text{–}9) \qquad \int \psi\psi^*dxdydz = 1$$

If a wave function satisfies this relation, it is said to be *normalized*, and if the electron is in the volume element, $dxdydz$, then $\int\psi\psi^*dxdydz$ will be equal to the probability that the electron is in this volume element. Very often $\psi$ is not a normalized wave function. However, since it is possible to multiply $\psi$ by a constant, $A$, to give a new wave function, $A\psi$, which is also a solution to the wave equation, the problem becomes one of choosing a value for $A$ which will make the new wave function a normalized function. In order for the new wave function, $A\psi$, to be a normalized function, it must meet the requirement

$$\int A\psi\, A\psi^*\, dxdydz = 1$$

Since $A$ is a constant, it can be removed from under the integral sign giving

$$A^2 \int \psi\psi^*\, dxdydz = 1$$

or

$$(2\text{--}10) \qquad A^{-2} = \int \psi\psi^*\, dxdydz$$

$A$ is known as a *normalizing constant* and can be determined from Eq. (2–10), and the new wave function, $A\psi$, will be a normalized wave function.

If we represent two different acceptable wave functions of a given system by $\psi_i$ and $\psi_j$, the wave functions will be normalized if they meet the requirement that

$$\int \psi_i\,\psi_i^*\, dxdydz = 1 \qquad \text{and} \qquad \int \psi_j\,\psi_j^*\, dxdydz = 1$$

If, on the other hand, it is found that they behave such that

$$\int \psi_i\,\psi_j^*\, dxdydz = 0 \qquad \text{or} \qquad \int \psi_i^*\,\psi_j\, dxdydz = 0$$

they are said to be mutually *orthogonal*.

## THE WAVE EQUATION

In the same sense that Newton's equations of motion have no derivation, the equation of motion of an electron should have no derivation. Both are consistent mathematical descriptions of certain processes of nature. However, in the case of the electron we find that the final form of the equation is sufficiently complex to make it rather difficult to see directly. This difficulty probably arises because it is really a combination of observations. In our final equation, we find that it is necessary to incorporate two basic observations, (1) the wave character of the electron, and (2) the probability character of our measurements. These observations lead us to use a wave equation and attempt to introduce particle character through the de Broglie relation.

To introduce the wave character into our equation, the general partial differential equation of wave motion

(2–7)
$$\frac{\partial^2\Phi}{\partial x^2} + \frac{\partial^2\Phi}{\partial y^2} + \frac{\partial^2\Phi}{\partial z^2} = \frac{1}{c^2}\frac{\partial^2\Phi}{\partial t^2}$$

or more simply

$$\nabla^2\,\Phi = \frac{1}{c^2}\frac{\partial^2\Phi}{\partial t^2}$$

is used. $\Phi$ was seen to be the amplitude function of the wave and $c$ is the velocity of light if the wave motion is that of an electromagnetic wave. Ordinarily the symbol $\Psi$ is used for the wave function of an electron, and if the velocity of a particle, $v$, is substituted for $c$, the analogous equation for the wave motion of a particle of velocity $v$ is obtained.

The wave function, $\Psi$, which has been used here is a function of both the space coordinates and the time. For the most part, it is found that an equation giving *standing waves* will be of more interest to us. This requires an equation that does not contain time as a variable. The *time-dependent* wave equation finds its application in the field of *radiation*, whereas the problems concerning the energy of the electron system utilize the *time-independent* equation.

In order to obtain the wave equation in a form that does not depend on time, the assumption is made that the function, $\Psi_{(xyzt)}$, can be replaced by a product of functions such as $\Psi_{(xyzt)} = \psi_{(xyz)}g_{(t)}$, where $\psi_{(xyz)}$ is a function of the space coordinates only and $g_{(t)}$ is a function only of the time. The assumption that the variables are separable by means of such a substitution is a standard approach to the solution of a partial differential equation and will be used on numerous occasions. In order to successfully separate the time dependence from the wave equation, several possible wave functions may be chosen for $g_{(t)}$, such as $\exp(2\pi i v t)$ or $\sin 2\pi v t$. In this particular case, it is not too difficult to find a function that will separate the time coordinate from the space coordinates. If the substitution $\Psi_{(xyzt)} = \psi_{(xyz)}\exp(2\pi i v t)$ is made, the equation

$$\nabla^2\,\Psi = \frac{1}{v^2}\frac{\partial^2\Psi}{\partial t^2}$$

becomes

$$\nabla^2\,\psi_{(xyz)}\,e^{2\pi i v t} = \frac{1}{v^2}\frac{\partial^2}{\partial t^2}\,\psi_{(xyz)}\,e^{2\pi i v t}$$

Since the operator, $\nabla^2$, contains only the space coordinates and not the time, the time function on the left side of the equation can be considered a constant. On the right side of the equation, $\psi_{(xyz)}$ can be considered a constant with regard to the operator, $\partial^2/\partial t^2$. Rearranging the equation, we obtain

$$e^{2\pi i v t}\,\nabla^2\,\psi_{(xyz)} = \frac{1}{v^2}\,\psi_{(xyz)}\,\frac{\partial^2}{\partial t^2}\left(e^{2\pi i v t}\right)$$

which on two successive differentiations with respect to time gives

$$e^{2\pi i \nu t} \nabla^2 \psi_{(xyz)} = \frac{1}{v^2} \psi_{(xyz)} \frac{\partial}{\partial t} \left( 2\pi i \nu e^{2\pi i \nu t} \right)$$

$$e^{2\pi i \nu t} \nabla^2 \psi_{(xyz)} = \frac{1}{v^2} \psi_{(xyz)} \left( -4\pi^2 \nu^2 \right) e^{2\pi i \nu t}$$

$$(2\text{--}11) \qquad \nabla^2 \psi_{(xyz)} = -\frac{4\pi^2 \nu^2}{v^2} \psi_{(xyz)}$$

It is seen that the variable, $t$, cancels, and we have succeeded in separating out the time dependence thereby leaving a wave equation that depends only on the space coordinates of the system.

Now that the wave portion of the equation is in the proper form, the particle character must be introduced. If the particle analog of the expression $c = \lambda \nu$ is taken, we obtain $v = \lambda \nu = h\nu/p$. Making this substitution into Eq. (2–11) gives

$$(2\text{--}12) \qquad \nabla^2 \psi_{(xyz)} = -\frac{4\pi^2 p^2}{h^2} \psi_{(xyz)}$$

The momentum, $p$, can be related to the kinetic energy, $T$, as follows:

$$T = \tfrac{1}{2} m v^2 = \frac{(mv)^2}{2m} = \frac{p^2}{2m}$$

Substituting for $p^2$, we obtain

$$\nabla^2 \psi_{(xyz)} = -\frac{4\pi^2 (2Tm)}{h^2} \psi_{(xyz)}$$

Since the kinetic energy is equal to the difference of the total energy, $E$, and the potential energy, $V$, that is, $T = E - V$, it is now possible to express the wave equation in the form

$$\nabla^2 \psi_{(xyz)} = -\frac{8\pi^2 m}{h^2} (E - V) \psi_{(xyz)}$$

or

$$(2\text{--}13) \qquad \nabla^2 \psi_{(xyz)} + \frac{8\pi^2 m}{h^2} (E - V) \psi_{(xyz)} = 0$$

This is the well-known Schroedinger time-independent wave equation[5] which we have obtained by taking the general differential equation for wave motion, separating out the space-dependent part, and using the de Broglie relation to introduce particle character.

## THE PRINCIPLE OF SUPERPOSITION

If we return to an equation of the form of Eq. (2–11),

$$\nabla^2 \psi_{(xyz)} = -\frac{4\pi^2 \nu^2}{v^2} \psi_{(xyz)}$$

it is readily seen that in one dimension this equation will become

$$\frac{d^2\psi_{(x)}}{dx^2} = -\frac{4\pi^2\nu^2}{v^2}\,\psi_{(x)}$$

This may be expressed more simply by setting

$$\alpha^2 = \frac{4\pi^2\nu^2}{v^2}$$

thereby giving

(2–14) $$\frac{d^2\psi_{(x)}}{dx^2} + \alpha^2\,\psi_{(x)} = 0$$

One solution of this differential equation is

(2–15) $$\psi_{(x)} = A \sin \alpha\, x + B \cos \alpha\, x$$

This can easily be verified by carrying out the successive differentiations of the function, thus:

$$\psi_{(x)} = A \sin \alpha\, x + B \cos \alpha\, x$$

$$\frac{d\psi_{(x)}}{dx} = \alpha\,(A \cos \alpha\, x - B \sin \alpha\, x)$$

$$\frac{d^2\psi_{(x)}}{dx^2} = -\alpha^2\,(A \sin \alpha\, x + B \cos \alpha\, x) = -\alpha^2\psi_{(x)}$$

or, on rearranging

$$\frac{d^2\psi_{(x)}}{dx^2} + \alpha^2\,\psi_{(x)} = 0$$

It is interesting to note that

$$\psi_{(x)} = A \sin \alpha\, x$$

as well as

$$\psi_{(x)} = B \cos \alpha\, x$$

and

$$\psi_{(x)} = C\, e^{i\alpha x}$$

are also solutions to Eq. (2–14). *According to the principle of superposition, any linear combination of solutions is also a solution.* For instance,

$$\psi_{(x)} = A \sin \alpha\, x \quad\text{and}\quad \psi_{(x)} = C\, e^{i\alpha x}$$

are both solutions to the differential equation

$$\frac{d^2\psi_{(x)}}{dx^2} + \alpha^2\,\psi_{(x)} = 0$$

Therefore,

$$\psi_{(x)} = A \sin \alpha\, x \pm C e^{i\alpha x}$$

is also a solution to the differential equation.

This is a general solution to the differential equation as evidenced by the fact that the coefficients $A$, $B$, and $C$ can be any constants and, for Eq. (2–14), which we are actually considering, $\alpha$ must also be evaluated. Thus, there is essentially an infinite number of solutions to this differential equation. However, we will see that a particular solution can be obtained by considering the boundary conditions of a given problem. In this manner specific values of the coefficients and parameters can be determined.

## THE PARTICLE IN A ONE-DIMENSIONAL BOX

One of the simplest applications of wave mechanics is found in the treatment of a particle confined to move within a box. A rectangular box with dimensions $abc$ lying along the $x$, $y$, and $z$ axes, respectively, is chosen, and the particle is restricted so as to move only inside the box. That is, it has no existence outside the box. Such a restriction may be met by allowing the potential energy to go to infinity at the sides of the box. This results in a *reflection* of the particle as it comes in contact with a side of the box rather than a possible *penetration*. Anywhere inside of the box the particle experiences a zero potential energy.

For the sake of simplicity, a one-dimensional box will first be considered. In the three-dimensional box, the wave function is represented by $\psi_{(xyz)}$ and in a one-dimensional box by $\psi_{(x)}$. Since the particle is to be some sort of realistic particle such as an electron, our wave function must be a function that does things a real particle will do. Such a function is known as a *well-behaved* function or one of class $Q$. In general, this requires that it be everywhere *continuous*, *finite*, and *single-valued*.

To solve a problem in wave mechanics, it is necessary to solve the wave equation

$$\nabla^2 \psi_{(xyz)} + \frac{8\pi^2 m}{h^2} (E - V) \psi_{(xyz)} = 0$$

for the particular problem at hand. For the case of the one-dimensional system, the wave equation reduces to

$$\frac{d^2\psi_{(x)}}{dx^2} + \frac{8\pi^2 m}{h^2} (E - V) \psi_{(x)} = 0$$

It is assumed that while the particle remains in the box, it has zero potential energy. Thus, the wave equation will reduce further to

$$(2\text{–}16) \qquad \frac{d^2\psi_{(x)}}{dx^2} + \frac{8\pi^2 m}{h^2} E \psi_{(x)} = 0$$

This can be simplified to

$$\frac{d^2\psi_{(x)}}{dx^2} + \alpha^2 \psi_{(x)} = 0$$

by letting $\alpha^2 = 8\pi^2 mE/h^2$. We now have an equation that is identical to Eq. (2–14), and the solution was shown to be

$$\psi_{(x)} = A \sin \alpha x + B \cos \alpha x$$

This, then, is a solution to the wave equation for the particle in our one-dimensional box.

As such, the general solution to the differential equation gives very little information. However, we know certain restrictions that apply to this particular system. These are known as *boundary conditions*. For instance, since the particle must not exist outside the box, it is necessary for the wave function, $\psi_{(x)}$, to go to zero at the walls of the box. This means

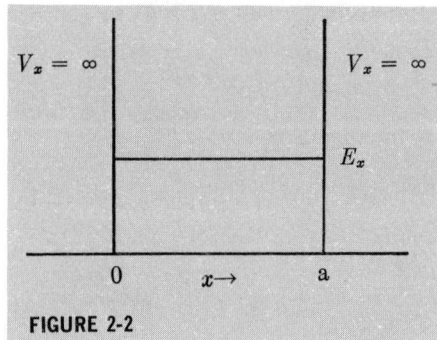

$V_x = \infty$     $V_x = \infty$

$E_x$

0     $x \rightarrow$     a

**FIGURE 2-2**

Particle of Energy $E_x$ in a One-dimensional Box.

that for the one-dimensional box shown in Figure 2–2, $\psi_{(x)} = 0$ at the point $x = 0$. Thus, we find that at the point $x = 0$,

$$0 = A \sin \alpha 0 + B \cos \alpha 0$$

$$0 = A\,(0) + B\,(1)$$

In order for the equality to hold, it is obvious that the constant, $B$, must equal zero. As a result of this boundary condition, the wave function reduces to

$$\psi_{(x)} = A \sin \alpha x$$

At the other wall, it is seen that the wave function must again go to zero. Therefore at the point $x = a$, it is again necessary that $\psi_{(x)} = 0$. This condition offers two possible solutions. For the point $x = a$, the wave function becomes

$$0 = A \sin \alpha a$$

The right side of the equation may be forced to equal zero by letting $A$ equal zero. This would maintain the identity, but it would accomplish nothing towards a useful solution. Such a solution is a *trivial* solution.

However, there is another way in which the identity may be maintained. The sine of an angle is zero at any integral multiple of $\pi$. Thus, if $\alpha = n\pi/a$, where $n$ is an integer, the identity can still be satisfied. As a result of applying these boundary conditions, the wave equation for the particle now becomes

$$(2\text{-}17) \qquad \psi_{(x)} = A \sin \frac{n\pi}{a} x$$

The only term yet to be determined is the coefficient, $A$. This can be determined by normalizing the wave function. Since it is known that the particle must be in the box, the probability that it is in the box is unity. Knowing that this probability is represented by the square of the wave function, we can say that

$$\int_0^a \psi\psi^* \, dx = 1$$

which leads to

$$\int_0^a A^2 \sin^2 \alpha \, x \, dx = 1$$

or

$$\frac{1}{A^2} = \int_0^a \sin^2 \alpha \, x \, dx$$

If this expression is solved for $A$ and the results substituted into the wave equation, the complete normalized wave function for the particle in a one-dimensional box is found to be

$$(2\text{-}18) \qquad \psi_{(x)} = \sqrt{\frac{2}{a}} \sin \frac{n\pi}{a} x$$

It is apparent that the wave function does not have to be determined in order to find the energy of the particle. This can be done once the value of $\alpha$ is known. The parameter can be removed from the expression $\alpha^2 = 8\pi^2 mE/h^2$ by noting also that $\alpha = n\pi/a$. By equating the two values of $\alpha$, we obtain

$$\frac{8\pi^2 mE}{h^2} = \frac{n^2\pi^2}{a^2}$$

which can be solved for the energy, giving

$$(2\text{-}19) \qquad E = \frac{n^2 h^2}{8ma^2}$$

There are two significant features to be noted with regard to the energy of the particle. To begin with, it is seen that the energy is quantized. Since the parameter, $n$, can have only integral values, the energy takes on the same type of discontinuous character that quantum theory has demanded since its rebirth with Max Planck in 1900. One of the beauties of wave mechanics is that this discontinuity results from a limited number of basic postulates rather than an *ad hoc* proposal as was necessary in the Bohr model of the atom. Secondly, a relation between the size of the box and the energy of the particle is seen to exist. The smaller the box becomes, the greater is the energy of the particle.

## THE PARTICLE IN A THREE-DIMENSIONAL BOX

For the particle in a three-dimensional box, the wave function will be a function of all three space coordinates. The wave equation for such a particle moving in a region of zero potential energy is

$$(2\text{-}20) \qquad \nabla^2 \psi_{(xyz)} + \frac{8\pi^2 m}{h^2} E \, \psi_{(xyz)} = 0$$

This is a partial differential equation containing three variables, and the standard approach to the solution of such an equation is about the same as that used to separate the time and space parts of the time-dependent wave equation. The first step is to assume that the variables are separable into three individual equations, each containing only one variable. Each of these equations will then be a total differential equation since it will contain only one variable, and it is often possible to find some sort of solution to these final equations.

It is not always possible to find an expression that will allow the variables to be separated, but the assumption that the total wave function can be represented as a product of wave functions is the usual place to start. For the particle in a three-dimensional box, it is then assumed that

$$\psi_{(xyz)} = X_{(x)} Y_{(y)} Z_{(z)}$$

where $X_{(x)}$ represents a wave function that depends only on the variable, $x$; $Y_{(y)}$ represents a wave function that depends only on the variable, $y$; and so on. If this expression is now substituted for $\psi_{(xyz)}$ in the wave equation, we obtain

$$\left( \frac{\partial^2}{\partial x^2} + \frac{\partial^2}{\partial y^2} + \frac{\partial^2}{\partial z^2} \right) X_{(x)} Y_{(y)} Z_{(z)} + \frac{8\pi^2 m}{h^2} E \, X_{(x)} Y_{(y)} Z_{(z)} = 0$$

Since the operator $\partial^2/\partial x^2$ has no effect on $Y_{(y)}$ and $Z_{(z)}$, and the operator $\partial^2/\partial y^2$ has no effect on $X_{(x)}$ and $Z_{(z)}$, etc., the wave equation may be rearranged to give

$$Y_{(y)} Z_{(z)} \frac{\partial^2 X_{(x)}}{\partial x^2} + X_{(x)} Z_{(z)} \frac{\partial^2 Y_{(y)}}{\partial y^2} + X_{(x)} Y_{(y)} \frac{\partial^2 Z_{(z)}}{\partial z^2}$$

$$(2\text{-}21) \qquad + \frac{8\pi^2 mE}{h^2} X_{(x)} Y_{(y)} Z_{(z)} = 0$$

If we now divide Eq. (2–21) by $X_{(x)}Y_{(y)}Z_{(z)}$, it is found that

$$(2\text{--}22) \qquad \frac{1}{X_{(x)}}\frac{\partial^2 X_{(x)}}{\partial x^2} + \frac{1}{Y_{(y)}}\frac{\partial^2 Y_{(y)}}{\partial y^2} + \frac{1}{Z_{(z)}}\frac{\partial^2 Z_{(z)}}{\partial z^2} = -\frac{8\pi^2 m}{h^2}E$$

It is to be noted that each term on the left side of Eq. (2–22) is a function of only one variable, and the sum of these terms is the constant $-8\pi^2 mE/h^2$. If we now keep the variables $y$ and $z$ constant and allow $x$ to vary, it is seen that the sum of the three terms is still the same constant. Such a situation can exist only if the term $\dfrac{1}{X_{(x)}}\dfrac{\partial^2 X_{(x)}}{\partial x^2}$ is independent of $x$, and is therefore itself a constant. The same argument will apply equally well to the $y$ and the $z$ terms. Thus, each variable is seen to be independent of the other variables, and we have succeeded in our separation. This is not a particularly unusual situation. For instance, the simple differential $\dfrac{d}{dx}(Ax)$ is a constant and is, therefore, independent of the variable $x$. If the wave function $X_{(x)}$ is substituted into the expression $\dfrac{1}{X_{(x)}}\dfrac{\partial^2 X_{(x)}}{\partial x^2}$ and the indicated differentiation carried out, this term can also be seen to be independent of the variable $x$.

Now, if the constants are represented by $-\alpha_x^2$ for the $x$ term, $-\alpha_y^2$ for the $y$ term, and $-\alpha_z^2$ for the $z$ term, the following three total differential equations are obtained:

$$(2\text{--}23a) \qquad \frac{1}{X_{(x)}}\frac{d^2 X_{(x)}}{dx^2} = -\alpha_x^2$$

$$(2\text{--}23b) \qquad \frac{1}{Y_{(y)}}\frac{d^2 Y_{(y)}}{dy^2} = -\alpha_y^2$$

$$(2\text{--}23c) \qquad \frac{1}{Z_{(z)}}\frac{d^2 Z_{(z)}}{dz^2} = -\alpha_z^2$$

From the development of our equations, it is seen that

$$(2\text{--}24) \qquad \alpha_x^2 + \alpha_y^2 + \alpha_z^2 = \frac{8\pi^2 m}{h^2}E$$

Thus, each degree of freedom can make its own contribution, such that

$$(2\text{--}25a) \qquad \alpha_x^2 = \frac{8\pi^2 m}{h^2}E_x$$

$$(2\text{--}25b) \qquad \alpha_y^2 = \frac{8\pi^2 m}{h^2}E_y$$

$$(2\text{--}25c) \qquad \alpha_z^2 = \frac{8\pi^2 m}{h^2}E_z$$

Now that the variables have been separated, it is necessary to solve each of the equations. In this particular problem, all three of the resulting equations are of the same form. Thus, the solution of one is sufficient to demonstrate the method. If the equation in $x$ is used as an example, it is seen that on rearrangement it is of the exact form as the wave equation we have just solved for the one-dimensional box,

$$\frac{d^2 X_{(x)}}{dx^2} + \alpha_x^2 X_{(x)} = 0$$

The normalized solution then is

$$X_{(x)} = \sqrt{\frac{2}{a}} \sin \frac{n_x \pi}{a} x$$

and an analogous solution would be obtained for the $y$ and $z$ equations. Since $\psi_{(xyz)} = X_{(x)} Y_{(y)} Z_{(z)}$, the total wave function is given by

$$(2\text{-}26) \qquad \psi_{(xyz)} = \sqrt{\frac{8}{abc}} \sin \frac{n_x \pi}{a} x \sin \frac{n_y \pi}{b} y \sin \frac{n_z \pi}{c} z$$

It is significant to note that there is a quantum number for each degree of freedom. This same idea was emphasized in the Sommerfeld quantization of the hydrogen atom, but here the quantization is a natural consequence of the mathematics.

Now that it is seen that $\alpha_x = n_x \pi / a$, $\alpha_y = n_y \pi / b$, and $\alpha_z = n_z \pi / c$, the total energy for the particle in the three-dimensional box can be expressed as

$$(2\text{-}27) \qquad E = E_x + E_y + E_z = \frac{h^2}{8m} \left( \frac{n_x^2}{a^2} + \frac{n_y^2}{b^2} + \frac{n_z^2}{c^2} \right)$$

Here, again, it is seen that the energy of the particle is quantized. This might lead one to wonder at the success of the classical approach to the mechanics of atoms and molecules as found in the kinetic theory of gases. Actually, no conflict exists between the two approaches. If quantum numbers and containers of reasonable size are chosen, it is found that the separation of energy levels is so small that the energy distribution will essentially be *continuous*.

## DEGENERACY

For a complete description of the energy states of a particle in a three-dimensional box, we see that it is necessary to consider three quantum numbers. This, of course, is what one should expect. The idea of quantum numbers in atomic spectra, for instance, came from an attempt to understand the positions of the spectral lines, that is, the energies they represent. The observation of new lines necessarily led to a new quantum number which could be associated with the corresponding new energy levels. Thus, we are prone to conclude that each quantum number represents a contribution to the energy of the system. However, it is frequently found that for various reasons a particular set of the quantum numbers may not be unique in defining the energy of the particle. If, as an example, the particle in a three-dimensional box is considered, we can again say that the energy is given by

$$(2\text{--}27) \qquad E = E_x + E_y + E_z = \frac{h^2}{8m} \left( \frac{n_x^2}{a^2} + \frac{n_y^2}{b^2} + \frac{n_z^2}{c^2} \right)$$

But if we now choose a box that is cubical in shape such that $a = b = c$, the energy can be expressed by

$$E = \frac{h^2}{8ma^2} \left( n_x^2 + n_y^2 + n_z^2 \right)$$

For the lowest quantum state, (111), in which $n_x$, $n_y$, and $n_z$, respectively, equal unity, it is seen that $E = 3h^2/8ma^2$. There is only one set of quantum numbers that gives this energy state, and this level is said to be *non-degenerate*. If we now consider the second energy state as shown in Figure 2–3, it is seen that there are three sets (112), (121), and (211) of the quantum numbers $n_x$, $n_y$, and $n_z$ that will give the same energy level, $E = 3h^2/4ma^2$. Such a level is said to be *degenerate*, and in this particular case, the level is *triply* degenerate. For a cubical box, it can be concluded from Figure 2–3 that virtually all of the energy levels are degenerate to some degree.

The mathematics of systems with degenerate energy states and the

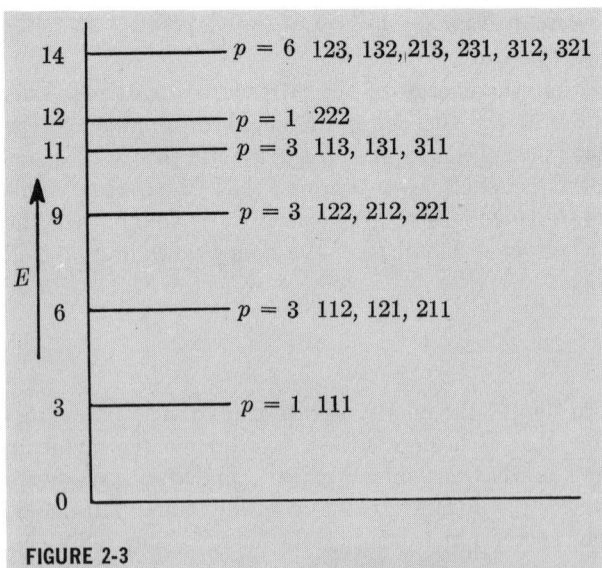

| 14 | ———— | $p = 6$  123, 132, 213, 231, 312, 321 |
| 12 | ———— | $p = 1$  222 |
| 11 | ———— | $p = 3$  113, 131, 311 |
| 9 | ———— | $p = 3$  122, 212, 221 |
| $E$ | | |
| 6 | ———— | $p = 3$  112, 121, 211 |
| 3 | ———— | $p = 1$  111 |
| 0 | | |

**FIGURE 2-3**

Degeneracy of the Energy Levels in a Cubical Box. The energy scale is in units of $h^2/8ma^2$.

means of removing the degeneracy are often problems of considerable importance. For the particle in a three-dimensional box, the degeneracy can be removed by using a box of different dimensions. If no integral relationship exists between the sides of the box, $a$, $b$, and $c$, the energy levels will all be nondegenerate. Thus, it is a relatively simple problem to obtain nondegenerate energy states for the particle in a box. However, in the problems of atoms and molecules this is not always so.

## THE HYDROGEN ATOM

One would like to think that wave mechanics can, in principle, offer a solution to all the theoretical problems of chemistry and physics. And this may be the case. However, from a practical standpoint this is not so. No matter how far quantum mechanics might go toward this end, a practical barrier always arises. It is usually possible to set up the differential equation for a particular problem, but it is then found that the resulting differential equation can rarely be solved without resorting to approximation methods. In fact, there are very few quantum mechanical problems that can be solved without some form of approximation, and the hydrogen-like atom is one of these. This fact alone is sufficient to emphasize the importance of the hydrogen atom problem. Yet, in addition to this, there are also numerous principles and concepts that carry over to any future wave mechanical treatment.

The problem, of course, is to solve the Schroedinger equation as set up for the hydrogen atom. Thus far, we have expressed the Schroedinger equation in the form

$$\nabla^2 \psi_{(xyz)} + \frac{8\pi^2 m}{h^2} (E - V) \psi_{(xyz)} = 0$$

which has proved satisfactory for the motion of a single particle of mass $m$. In the hydrogen atom there are two particles, the electron and the nucleus. For such a system, it will be found convenient to put the wave equation in the form

$$(2\text{--}28) \qquad \frac{1}{m} \nabla^2 \psi_{(xyz)} + \frac{8\pi^2}{h^2} (E - V) \psi_{(xyz)} = 0$$

Now, when we consider the motion of the two particles in the hydrogen atom, the wave equation becomes

$$(2\text{--}29) \qquad \frac{1}{m_1} \nabla_1^2 \psi_T + \frac{1}{m_2} \nabla_2^2 \psi_T + \frac{8\pi^2}{h^2} (E - V) \psi_T = 0$$

where $m_1$ is the mass of the electron and $m_2$ is the mass of the nucleus.

In order to evaluate the potential energy term, it is necessary to consider the coulombic attraction between the electron and the nucleus. In this case, the potential energy is defined as the work necessary to take the electron to infinity from its equilibrium distance, $r$, with respect to the nucleus. Since work is represented by the product of force times distance, the potential energy may be determined as follows:

$$V = \int F \cdot dr = \int_r^\infty \frac{q_1 q_2}{r^2} \, dr = -\frac{q_1 q_2}{r} \Big]_r^\infty = \frac{q_1 q_2}{r}$$

and considering the sign of the electron charge

(2–30) $$V = -\frac{e^2}{r}$$

Here $+e$ is the nuclear charge and $-e$ is the charge on the electron. If the potential energy term is now introduced, the equation for the hydrogen atom becomes

(2–31) $$\frac{1}{m_1} \nabla_1^2 \, \psi_T + \frac{1}{m_2} \nabla_2^2 \, \psi_T + \frac{8\pi^2}{h^2} \left( E + \frac{e^2}{r} \right) \psi_T = 0$$

## Transformation of Coordinates

A new problem arises when we note that the total energy, $E$, in the wave equation is made up of two parts: (1) the translational motion of the atom as a whole, and (2) the energy of the electron with respect to the proton. It is this latter portion of the energy in which we are interested. This leads us again to the problem of separation of variables. In order to obtain the desired equation, it will be necessary to separate out and discard the translational portion of the total wave equation. To carry out this particular separation, it is necessary to introduce a new set of variables $x$, $y$, and $z$, which are cartesian coordinates of the *center of mass* of the hydrogen atom, and the variables $r$, $\theta$, and $\varphi$, which are polar coordinates of the electron with respect to the nucleus.

A coordinate of the center of mass of a system is, in general, given by

(2–32) $$q = \frac{\sum\limits_i m_i q_i}{\sum\limits_i m_i}$$

where $m_i$ is the mass of the $i^{th}$ particle and $q_i$ is the $q$ coordinate of the $i^{th}$ particle. For the hydrogen atom, the cartesian coordinates of the center of mass will be given by

(2–33a) $$x = \frac{m_1 x_1 + m_2 x_2}{m_1 + m_2}$$

(2–33b) $$y = \frac{m_1 y_1 + m_2 y_2}{m_1 + m_2}$$

(2–33c) $$z = \frac{m_1 z_1 + m_2 z_2}{m_1 + m_2}$$

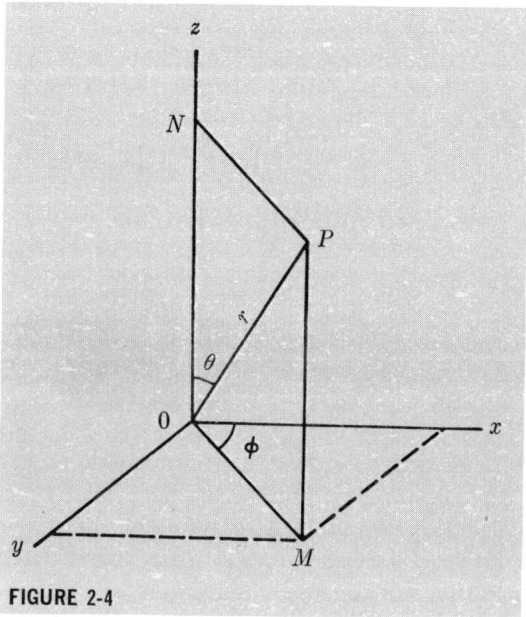

**FIGURE 2-4**

Transformation Diagram for Spherical Coordinates.

and the transformation to spherical coordinates can be seen from Figure 2–4 to be

(2–34a) $\qquad r \sin \theta \cos \varphi = x_2 - x_1$

(2–34b) $\qquad r \sin \theta \sin \varphi = y_2 - y_1$

(2–34c) $\qquad r \cos \theta = z_2 - z_1$

By using these transformation equations, it is a straightforward procedure to obtain the wave equation in terms of the cartesian coordinates of the center of mass of the system and the polar coordinates $r$, $\theta$, and $\varphi$. The $x$, $y$, $z$ coordinates of the center of mass of the atom obviously relate to the translational motion of the atom as a whole, and the $r$, $\theta$, $\varphi$ coordinates are seen to relate the coordinates of the electron $(x_1, y_1, z_1)$ to the coordinates of the nucleus $(x_2, y_2, z_2)$.

As an example of the procedure, consider the $z$ coordinate. Solving Eq. (2–33c) for $z_2$, it is seen that

$$z_2 = \left(\frac{m_1 + m_2}{m_2}\right) z - \frac{m_1}{m_2} z_1$$

If the value of $z_2$ is now substituted into Eq. (2–34c), it is found that

$$r \cos \theta = \left(\frac{m_1 + m_2}{m_2}\right) z - \frac{m_1}{m_2} z_1 - z_1$$

or

$$z_1 = z - \left(\frac{m_2}{m_1 + m_2}\right) r \cos \theta$$

If we now multiply through by the term $m_1/m_1$, we obtain the expression

(2–35)     $$z_1 = z - \frac{\mu}{m_1} r \cos \theta$$

where $\mu$ is the reduced mass of the system and is given by

(2–36)     $$\mu = \frac{m_1 m_2}{m_1 + m_2}$$

Using this procedure, a transformation equation can be found for each of the coordinates, and when the proper substitutions are made, the wave equation is obtained in terms of the new variables $x$, $y$, $z$, $r$, $\theta$, and $\varphi$. Although the procedure is rather tedious, it is nevertheless straightforward. In terms of the new variables, it is found to be

$$\frac{1}{m_1 + m_2} \left(\frac{\partial^2 \psi_T}{\partial x^2} + \frac{\partial^2 \psi_T}{\partial y^2} + \frac{\partial^2 \psi_T}{\partial z^2}\right) + \frac{m_1 + m_2}{m_1 m_2} \left\{\frac{1}{r^2} \frac{\partial}{\partial r}\left(r^2 \frac{\partial \psi_T}{\partial r}\right) + \frac{1}{r^2 \sin^2 \theta} \frac{\partial^2 \psi_T}{\partial \varphi^2}\right.$$

(2–37)     $$\left. + \frac{1}{r^2 \sin \theta} \frac{\partial}{\partial \theta}\left(\sin \theta \frac{\partial \psi_T}{\partial \theta}\right)\right\} + \frac{8\pi^2}{h^2} [E - V_{(r)}] \psi_T = 0$$

The wave function, $\psi_T$, is a function of the variables $x$, $y$, $z$, $r$, $\theta$, and $\varphi$, and the energy, $E$, contains the translational energy of the atom as well as the energy of the electron with respect to the proton.

The purpose of this transformation to new coordinates is to make a separation of variables possible. In principle, the separation is carried out in the same manner as it was in the particle in a box problem. However, in this particular case, the algebra is somewhat more complex. In the usual manner, the total wave function, $\psi_{(xyzr\theta\varphi)}$, is assumed to be expressible as the product of two wave functions such that

$$\psi_{(xyzr\theta\varphi)} = F_{(xyz)} \psi_{(r\theta\varphi)}$$

When this expression is substituted into Eq. (2–37), the following two equations are obtained:

(2–38)     $$\frac{\partial^2 F_{(xyz)}}{\partial x^2} + \frac{\partial^2 F_{(xyz)}}{\partial y^2} + \frac{\partial^2 F_{(xyz)}}{\partial z^2} + \frac{8\pi^2(m_1 + m_2)}{h^2} E_{\text{trans}} F_{(xyz)} = 0$$

$$\frac{1}{r^2} \frac{\partial}{\partial r}\left(r^2 \frac{\partial \psi_{(r\theta\varphi)}}{\partial r}\right) + \frac{1}{r^2 \sin^2 \theta} \frac{\partial^2 \psi_{(r\theta\varphi)}}{\partial \varphi^2}$$

(2–39)     $$+ \frac{1}{r^2 \sin \theta} \frac{\partial}{\partial \theta}\left(\sin \theta \frac{\partial \psi_{(r\theta\varphi)}}{\partial \theta}\right) + \frac{8\pi^2 \mu}{h^2} [E - V_{(r)}] \psi_{(r\theta\varphi)} = 0$$

The first of these equations contains only the variables $x$, $y$, and $z$, with no potential energy term. This is identical to the wave equation for a free particle, and therefore represents the translational energy of the atom as a whole. The second equation, which relates the electron to the proton, is the equation of particular interest to us.

## Separation of Variables

Since it is the second part of the total wave equation that is of interest, the translational part will be discarded. We now have the desired wave equation for the electron with respect to the nucleus. This equation is a second-order partial differential equation, and the standard methods already used must again be employed to obtain its solution. This will require that the variables be separated such that three independent equations are obtained, each containing only one of the three variables.

In order to separate the variables, it is necessary to assume that the wave function $\psi_{(r\theta\varphi)}$ may be represented by the product of three wave functions, each containing one and only one of the three variables $r$, $\theta$, and $\varphi$. If we let

$$\psi_{(r\theta\varphi)} = R_{(r)}\,\Theta_{(\theta)}\,\Phi_{(\varphi)}$$

and make this substitution into Eq. (2–39), we obtain

$$\frac{1}{r^2}\frac{\partial}{\partial r}\left(r^2\frac{\partial}{\partial r}R_{(r)}\,\Theta_{(\theta)}\,\Phi_{(\varphi)}\right) + \frac{1}{r^2\sin^2\theta}\frac{\partial^2}{\partial\varphi^2}R_{(r)}\,\Theta_{(\theta)}\,\Phi_{(\varphi)}$$

$$+ \frac{1}{r^2\sin\theta}\frac{\partial}{\partial\theta}\left(\sin\theta\frac{\partial}{\partial\theta}R_{(r)}\,\Theta_{(\theta)}\,\Phi_{(\varphi)}\right) + \frac{8\pi^2\mu}{h^2}[E - V_{(r)}]\,R_{(r)}\,\Theta_{(\theta)}\,\Phi_{(\varphi)} = 0$$

which, on dividing by $R_{(r)}\,\Theta_{(\theta)}\,\Phi_{(\varphi)}$ gives

$$\frac{1}{r^2R_{(r)}}\frac{\partial}{\partial r}\left(r^2\frac{\partial R_{(r)}}{\partial r}\right) + \frac{1}{\Phi_{(\varphi)}\,r^2\sin^2\theta}\frac{\partial^2\Phi_{(\varphi)}}{\partial\varphi^2}$$

$$+ \frac{1}{\Theta_{(\theta)}\,r^2\sin\theta}\frac{\partial}{\partial\theta}\left(\sin\theta\frac{\partial\Theta_{(\theta)}}{\partial\theta}\right) + \frac{8\pi^2\mu}{h^2}[E - V_{(r)}] = 0$$

If we multiply by $r^2\sin^2\theta$, we obtain

$$\frac{\sin^2\theta}{R_{(r)}}\frac{\partial}{\partial r}\left(r^2\frac{\partial R_{(r)}}{\partial r}\right) + \frac{1}{\Phi_{(\varphi)}}\frac{\partial^2\Phi_{(\varphi)}}{\partial\varphi^2} + \frac{\sin\theta}{\Theta_{(\theta)}}\frac{\partial}{\partial\theta}\left(\sin\theta\frac{\partial\Theta_{(\theta)}}{\partial\theta}\right)$$

$$+ \frac{8\pi^2\mu r^2\sin^2\theta}{h^2}[E - V_{(r)}] = 0$$

or

$$\frac{\sin^2\theta}{R_{(r)}}\frac{\partial}{\partial r}\left(r^2\frac{\partial R_{(r)}}{\partial r}\right) + \frac{\sin\theta}{\Theta_{(\theta)}}\frac{\partial}{\partial\theta}\left(\sin\theta\frac{\partial\Theta_{(\theta)}}{\partial\theta}\right)$$

$$(2\text{–}40) \qquad + \frac{8\pi^2\mu r^2\sin^2\theta}{h^2}[E - V_{(r)}] = -\frac{1}{\Phi_{(\varphi)}}\frac{\partial^2\Phi_{(\varphi)}}{\partial\varphi^2}$$

This leads to a situation that is analogous to that which arose for the particle in a three-dimensional box. The left side of Eq. (2–40) contains only the variables $r$ and $\theta$, whereas the right side of the equation contains only the variable $\varphi$. No matter what values $r$, $\theta$, or $\varphi$ might independently take, the sum of the terms on the left must always equal the term on the right. This can be true only if each side of the equation is equal to the same constant. If we let this constant be $m^2$, it is seen that the variable $\varphi$ can immediately be separated from Eq. (2–40), giving

$$(2\text{–}41) \qquad \frac{1}{\Phi_{(\varphi)}} \frac{d^2\Phi_{(\varphi)}}{d\varphi^2} = -m^2$$

Thus, we find that the first of the three variables has been successfully separated from the original wave equation. The problem now is to carry out the separation of the remaining two variables $r$ and $\theta$. By equating the portion of the equation containing $R_{(r)}$ and $\Theta_{(\theta)}$ to the constant $m^2$, it is seen that

$$\frac{\sin^2\theta}{R_{(r)}} \frac{\partial}{\partial r}\left(r^2 \frac{\partial R_{(r)}}{\partial r}\right) + \frac{\sin\theta}{\Theta_{(\theta)}} \frac{\partial}{\partial\theta}\left(\sin\theta \frac{\partial\Theta_{(\theta)}}{\partial\theta}\right)$$

$$+ \frac{8\pi^2 \mu r^2 \sin^2\theta}{h^2}[E - V_{(r)}] = m^2$$

On division by $\sin^2\theta$, this becomes

$$\frac{1}{R_{(r)}} \frac{\partial}{\partial r}\left(r^2 \frac{\partial R_{(r)}}{\partial r}\right) + \frac{1}{\Theta_{(\theta)}\sin\theta} \frac{\partial}{\partial\theta}\left(\sin\theta \frac{\partial\Theta_{(\theta)}}{\partial\theta}\right) - \frac{m^2}{\sin^2\theta}$$

$$+ \frac{8\pi^2 \mu r^2}{h^2}[E - V_{(r)}] = 0$$

or, on rearranging

$$\frac{1}{R_{(r)}} \frac{\partial}{\partial r}\left(r^2 \frac{\partial R_{(r)}}{\partial r}\right) + \frac{8\pi^2 \mu r^2}{h^2}[E - V_{(r)}] = \frac{m^2}{\sin^2\theta} - \frac{1}{\Theta_{(\theta)}\sin\theta} \frac{\partial}{\partial\theta}\left(\sin\theta \frac{\partial\Theta_{(\theta)}}{\partial\theta}\right)$$

Again, since each side of the equation contains only one variable, they both must be equal to the same constant. If the right side of the equation is set equal to the constant $\beta$, this gives on multiplication by $\Theta_{(\theta)}$,

$$(2\text{–}42) \qquad \frac{m^2 \Theta_{(\theta)}}{\sin^2\theta} - \frac{1}{\sin\theta} \frac{d}{d\theta}\left(\sin\theta \frac{d\Theta_{(\theta)}}{d\theta}\right) - \beta\,\Theta_{(\theta)} = 0$$

This is the desired form of the $\Theta$ equation. The remaining part of the original equation is the $R$ equation,

$$(2\text{–}43) \qquad \frac{1}{r^2} \frac{d}{dr}\left(r^2 \frac{dR_{(r)}}{dr}\right) - \frac{\beta}{r^2} R_{(r)} + \frac{8\pi^2 \mu}{h^2}[E - V_{(r)}] R_{(r)} = 0$$

Thus, the three variables have been successfully separated, and the three independent total differential equations that result are

$$(2\text{-}41) \qquad \frac{d^2\Phi_{(\varphi)}}{d\varphi^2} + m^2\,\Phi_{(\varphi)} = 0$$

$$(2\text{-}42) \qquad \frac{1}{\sin\theta}\frac{d}{d\theta}\left(\sin\theta\,\frac{d\Theta_{(\theta)}}{d\theta}\right) - \frac{m^2\Theta_{(\theta)}}{\sin^2\theta} + \beta\,\Theta_{(\theta)} = 0$$

$$(2\text{-}43) \qquad \frac{1}{r^2}\frac{d}{dr}\left(r^2\frac{dR_{(r)}}{dr}\right) - \frac{\beta}{r^2}R_{(r)} + \frac{8\pi^2\mu}{h^2}\left[E - V_{(r)}\right]R_{(r)} = 0$$

## The $\Phi$ Equation

The first of these equations is the $\Phi$ equation, and it is seen to be of the same form as the wave equation for the particle in a box. In terms of sine and cosine, the solution is

$$(2\text{-}44) \qquad \Phi_m(\varphi) = A\sin m\varphi + B\cos m\varphi$$

In order for a wave function to be acceptable, it must be of the well-behaved class. One of the requirements of such a function is that it be *single-valued*. To meet this restriction, the function $\Phi_m(\varphi)$ must have the same value for $\varphi = 0$ as it does for $\varphi = 2\pi$. For the case of $\varphi = 0$, it is seen that

$$\Phi_m(0) = A\sin(0) + B\cos(0) = B$$

and when $\varphi = 2\pi$, we have

$$\Phi_m(2\pi) = A\sin m2\pi + B\cos m2\pi$$

Since the value of $\Phi_m(\varphi)$ must be the same under both of these conditions, it is necessary that

$$B = A\sin m2\pi + B\cos m2\pi$$

This identity can hold only if $m$ is zero or has a positive or negative integral value. Such characteristics are those that would be expected of a quantum number, and the particular restrictions on $m$ indicate that it is the analog of the magnetic quantum number of the Bohr-Sommerfeld model.

Very often, in the treatment of the hydrogen atom, the exponential solution to the $\Phi$ equation,

$$(2\text{-}45) \qquad \Phi_m(\varphi) = Ce^{\pm im\varphi}$$

is used. In the evaluation of the constant, $C$, it would be most convenient to choose $C$ in such a manner that the wave function, $\Phi(\varphi)$, will be a normalized wave function. This requires that

$$\int_0^{2\pi} \Phi\Phi^*\,d\varphi = 1$$

which leads to

$$\int_0^{2\pi} C^2 e^{\mp im\varphi} e^{\pm im\varphi} \, d\varphi = 1$$

or

$$C^2 \int_0^{2\pi} d\varphi = 2\pi C^2 = 1$$

The value of the constant, $C$, that gives a normalized wave function is thus seen to be $C = 1/\sqrt{2\pi}$, and the final normalized wave function is

(2–46)        $\Phi_m(\varphi) = \dfrac{1}{\sqrt{2\pi}} e^{\pm im\varphi}, \ m = 0, \pm1, \pm2, \ldots$

## The Θ Equation

The solutions to the Θ equation and the radial equation, unfortunately, are not quite so simple as the solution to the Φ equation. However, it happens that the Θ equation can be put into a form that was known by the mathematicians many years before the advent of quantum mechanics. This particular equation is known as *Legendre's equation* and has the normalized solution

(2–47)        $\Theta_{l,m}(\theta) = \sqrt{\dfrac{(2l+1)(l-|m|)!}{2(l+|m|)!}} \, P_l^{|m|}(\cos\theta)$

where $P_l^{|m|}$ is the associated Legendre function of degree $l$ and order $|m|$. The form of the solution is obviously quite complicated. However, as can be seen in Table 2–3, for particular values of the parameters $l$ and $m$ the solution reduces to much simpler forms.

In spite of the complicated nature of the solution, several important features can be observed. Although the mathematics is too complex to be considered here, it can be shown that in Eq. (2–42), $\beta = l(l+1)$, where the allowed values of $l$ are 0, 1, 2, 3, .... This is the source of the new parameter found in Eq. (2–47), and its properties appear to be similar in many ways to those of the azimuthal quantum number of the Bohr-Sommerfeld atom. It can also be seen that there is now a new restriction on the quantum number $m$. In the normalizing factor of the solution to the Θ equation, the term $(l - |m|)!$ occurs. If $|m|$ is allowed to be greater than $l$, the factorial of a negative number results. Since a negative factorial is undefined, the maximum value of $m$ must be $l$.* Thus, the restrictions on the quantum number $m$ now become $m = 0, \pm1, \pm2, \pm3, \ldots, \pm l$. These are the same restrictions that were found to be necessary for the magnetic quantum number in the Bohr-Sommerfeld theory.

---

* It should be mentioned that the method used here to determine the restrictions on the $m$ quantum number is not the ordinary method used to determine these restrictions.

## Spherical Harmonics

Both the solution to the $\Theta$ equation and the solution to the $\Phi$ equation contain trigonometric functions and therefore determine the angular character of the electron wave function. Very often the total wave function can most conveniently be used if it is separated into a *radial* portion and an *angular* portion such that

$$\psi_{(r\theta\varphi)} = R_{n,l}(r)\, Y_{l,m}(\theta,\varphi)$$

The term $Y_{l,m}(\theta,\varphi)$ is referred to as the spherical harmonics, and is given by

(2–48) $\qquad Y_{l,m}(\theta,\varphi) = \Theta_{l,m}(\theta)\, \Phi_m(\varphi)$

It is this portion of the total wave function that will be of primary concern in the treatment of directional bonding.

## The Radial Equation

The remaining equation to be solved is the radial equation

(2–43) $\qquad \dfrac{1}{r^2}\dfrac{d}{dr}\left(r^2\dfrac{dR_{(r)}}{dr}\right) - \dfrac{\beta R_{(r)}}{r^2} + \dfrac{8\pi^2\mu}{h^2}\left[E + \dfrac{Ze^2}{r}\right]R_{(r)} = 0$

This, like the $\Theta$ equation, can be put into a form that has long been known to the mathematicians. This particular equation is the *Laguerre equation*, and its normalized solution is

(2–49) $\qquad R_{n,l}(r) = \sqrt{\left(\dfrac{2Z}{na_0}\right)^3 \dfrac{(n-l-1)!}{2n\,[(n+l)!]^3}}\; e^{-\rho/2}\, \rho^l\, L_{n+l}^{2l+1}(\rho)$

where $\rho = (2Z/na_0)r$, $a_0 = h^2/4\pi^2\mu e^2$, and $L_{n+l}^{2l+1}(\rho)$ represents the associated Laguerre polynomial. As was the case with the $\Theta$ equation, the solution to the radial equation is also rather complex in form. However, again it is possible to make some pertinent observations from the solution. It is to be noted that a new parameter, the quantum number $n$, has been added. Although we have not solved the radial equation, its solution is somewhat analogous to that of the $\Theta$ equation. From this it is found that the new quantum number $n$ is restricted to take on the integral values 1, 2, 3, ... $n$. Both the relation of $n$ to the radial wave function, which is a measure of the position of the electron with respect to the nucleus, and its similar restrictions, indicate that $n$ is the quantum mechanical analog of the principal quantum number of the Bohr theory.

Using the same approach as that applied to the magnetic quantum number, a new restriction can be seen for the quantum number $l$. It is apparent from the normalizing factor of the solution to the radial equation that the term $(n - l - 1)$ requires that the maximum value of $l$ be $(n - 1)$. If $l$ is allowed a value greater than this, the factorial of a negative number would result. Thus, the quantum number is restricted to the values $l = 0$, 1, 2, ..., $(n - 1)$.

## Quantum States

From the solution of the total wave equation, we have arrived at three quantum numbers. These had been postulated out of necessity in the Bohr-Sommerfeld theory, and this was a fundamental weakness of the theory. It is, therefore, gratifying to see that they occur as the result of a few basic postulates in the wave mechanical treatment. However, it might be considered unfortunate that they can no longer be thought of in the pictorial manner of the Bohr theory. We have one quantum number for each degree of freedom, but the idea of precessing orbits is no longer valid.

The quantum numbers with their allowed values may be summarized as follows:

Radial quantum number:    $n = 1, 2, 3, \ldots$,

Azimuthal quantum number:   $l = 0, 1, 2, \ldots, n - 1$ (inclusive)

Magnetic quantum number:  $m = 0, \pm 1, \pm 2, \ldots, \pm l$ (inclusive)

According to these restrictions, there are only certain values of the quantum number $l$ that are permissible for a given value of $n$. The maximum value of $l$ is seen always to be $(n - 1)$. For example, when $n = 4$, $l$ can be any integer up to and including $l = 3$, but no greater. This is illustrated in Table 2–1, where the allowed values of $l$ are shown for the first four radial shells.

**TABLE 2–1. Quantum Numbers With Allowed Values for First Four Radial Shells.**

| Value of $n$: | 1 | 2 | 3 | 4 |
|---|---|---|---|---|
| Allowed Values of $l$: | 0 | 0,1 | 0,1,2 | 0,1,2,3 |

It should be noted that $l = 0$ occurs for every value of $n$, $l = 1$ occurs for every value of $n$ greater than $n = 1$, and so on. These values of the quantum number $l$ play a rather important role in both the geometry and the energy states of the atom. Because of this importance, they are given the following special designations:

$l = 0$        $s$ state
$l = 1$        $p$ state
$l = 2$        $d$ state
$l = 3$        $f$ state

It is much more common to speak in terms of the electron state than the particular value of the $l$ quantum number.

For the first radial shell, the value of the radial quantum number is $n = 1$, and the $l$ quantum number can have only the value $l = 0$. This state is usually represented by (1s) where the 1 represents the value of the $n$ quantum number and $s$ represents $l = 0$. For $n = 2$, the azimuthal quantum number can have the values $l = 0$ and $l = 1$. This gives the two states (2s) and (2p), respectively. For the case of $n = 3$, it can be seen from the allowed values of the $l$ quantum number that there can be the three states (3s), (3p), and (3d) for $l = 0$, $l = 1$, and $l = 2$, respectively. Finally, for the fourth shell for which $n = 4$, it is seen that there can be the four states (4s), (4p), (4d), and (4f). These states determine the possible energies of the electrons, and if the $l$ quantum number contributes to the energy as does the $n$ quantum number, each state will represent a different energy.

## The Electron Spin

In addition to the orbital fine structure which could be explained in terms of the $l$ quantum number, it was experimentally observed that a doublet structure also exists. Spectral lines that had once been thought to be single lines actually turn out to be two lines very close together. An understanding of this doublet character could not be found in the Bohr-Sommerfeld model. In 1925, Uhlenbeck and Goudsmit[6] offered an explanation in which they proposed that the electron has, in addition to its orbital motion, an angular momentum of rotation about its own axis and a corresponding *magnetic moment*. This leads to a new quantum number known as the *spin* quantum number. The magnitude of this spin angular momentum is found to be $\pm\frac{1}{2}$ in units of $h/2\pi$. The plus or minus values of the spin can be thought of as arising from the direction of the spin. For instance, if the electron spin is clockwise it interacts with the orbital magnetic moment of the electron to give a different energy than if it is counterclockwise. The energy difference resulting from these opposite spins is small, but it is sufficient to lead to the observed doublet structure. Actually, there are several serious difficulties that arise from the concept of a physically spinning electron. However, the agreement between theory and experiment is sufficiently good to retain the basic idea.

In 1928, a quantum mechanical answer was found to the problem of electron spin. The wave equation as developed by Schroedinger was non-relativistic. In an attempt to bring wave mechanics into harmony with relativity theory, Dirac succeeded in developing a wave equation that led naturally to the proper spin angular momentum for the electron. According to the Dirac theory, the electron has the same angular momentum and magnetic moment as the spinning electron of Uhlenbeck and Goudsmit. Yet, as was the case for the other three quantum numbers, the quantum mechanical properties of the electron spin are the result of a consistent mathematical treatment and do not offer the problems that result from the physical picture of an electron spinning on its own axis.

## Energy States of the Hydrogen Atom

In the wave equation of the hydrogen-like atom, the energy term occurs only in the radial equation; and it is from its solution that the energy states of the hydrogen atom are obtained. Since the potential energy of the electron with respect to the nucleus is defined as zero when the two particles are infinitely far apart, the potential energy becomes increasingly negative as the electron approaches the nucleus. Consequently, the energy of the electron is defined in such a manner that it is negative, and it is therefore the negative energy states that will be of primary interest. If only these negative states are considered, it can be shown that the allowed values of the energy of the electron are given by

$$(2\text{-}50) \qquad E_n = -\frac{2\pi^2 \mu Z^2 e^4}{n^2 h^2}, n = 1, 2, 3, \ldots$$

Here we see that the expression for the energy of the electron as determined by wave mechanics is identical with that of the Bohr theory.

The quantum number, $n$, that appears in the wave mechanical treatment is related to the principal quantum number of the Bohr theory and can therefore be associated with the electron shell. In the pictorial sense of an electron revolving in a prescribed orbit, this analogy is invalid. However, from the standpoint of an energy state of the electron the two can be thought of in much the same manner. Thus, $n = 1$ corresponds to the first or $K$ shell, and for $n = 2$ we have the $L$ shell, etc.

It is very interesting to note that since the energy of the electron in the hydrogen-like atom is determined only by the $n$ quantum number according to our wave equation, the resulting energy levels must be degenerate. The ground state of the hydrogen atom will be the ($1s$) state, and the subsequent order of energy levels will be

$$1s < 2s = 2p < 3s = 3p = 3d < 4s = 4p = 4d = 4f < 5s\ldots$$

Since the $l$, $m$, and $m_s$ quantum numbers contribute nothing to the energy of the electron state, all of the possible energy states in a given radial shell are of equal energy. This means that only single spectral lines such as those predicted by Bohr will be observed. Yet it is well-known that a fine structure exists in the hydrogen spectrum. This was the incentive for the development of the Bohr-Sommerfeld theory of the hydrogen atom. Obviously, our simple form of the wave equation is inadequate for a completely satisfactory treatment of the hydrogen atom, and thus we are little better off than we were with the Bohr model.

Actually, the sequence of levels obtained is the result of calculations based upon a nonrelativistic hydrogen-like atom in the absence of external electric or magnetic fields. By using a relativistic form of the wave equation, the orbital degeneracy can be removed, thereby leading to the experimentally observed fine structure. Unfortunately, due to the extreme complexity of the mathematics, a relativistic treatment is not practical. For more complex atoms we will see that the orbital degeneracy can be removed by considering the effect of the electron-electron repulsion.

## THE SELF-CONSISTENT FIELD METHOD

It has been noted that an exact solution to a quantum mechanical problem can be obtained in only a very few instances. The hydrogen-like atom is one of these. For any atomic system with more than one electron, various approximation methods must be used in order to obtain a solution. This difficulty arises from the Coulombic repulsion between electrons. As an example, if we consider the helium atom with two electrons, the potential energy term will contain not only the effect of the nucleus on each of the electrons, but also the Coulombic repulsion between the two electrons. The relations between the electrons and the nucleus of the helium atom can be seen in Figure 2–5, where the nucleus is placed at the origin of

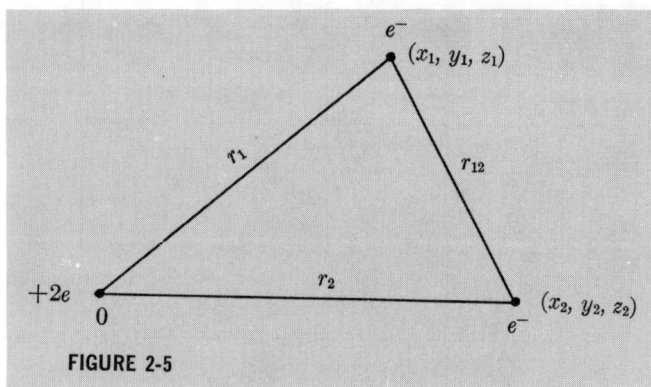

FIGURE 2-5

Coordinates of the Helium Atom.

the coordinate system and the coordinates of the two electrons are $(x_1, y_1, z_1)$ and $(x_2, y_2, z_2)$. The potential energy term is

$$V = -\frac{2e^2}{r_1} - \frac{2e^2}{r_2} + \frac{e^2}{r_{12}}$$

and the resulting wave equation for the helium atom is then

$$(2\text{–}51) \qquad \nabla_1^2 \, \psi_T + \nabla_2^2 \, \psi_T + \frac{8\pi^2 m}{h^2} \left( E + \frac{2e^2}{r_1} + \frac{2e^2}{r_2} - \frac{e^2}{r_{12}} \right) \psi_T = 0$$

where $\nabla_1^2$ is the Laplacian operator for electron 1 and $\nabla_2^2$ is the Laplacian operator for electron 2. Thus, it is seen that for the helium atom, the motion of one electron is dependent on the motion of the other electron, and as we go to more complex atoms the wave equation must contain the coulombic repulsion between all electrons.

The difficulty in the solution arises from the inseparability of the various electron wave functions. This problem can, however, be solved by a method developed by Hartree[7] in which a single electron is treated as if it were moving in the presence of a central electric field resulting from the average charge distribution of the nucleus and the remaining electrons. The general approach is to estimate a potential energy function due to the nucleus and all of the electrons. The wave function of one particular electron is then estimated by considering the chosen electron to be moving in the presence of the average field of the remaining electrons and the nucleus. The solution to the wave equation for the first electron will give a better estimate of the average central field, which can then be used for the wave equation of a second electron, and so on. This general procedure

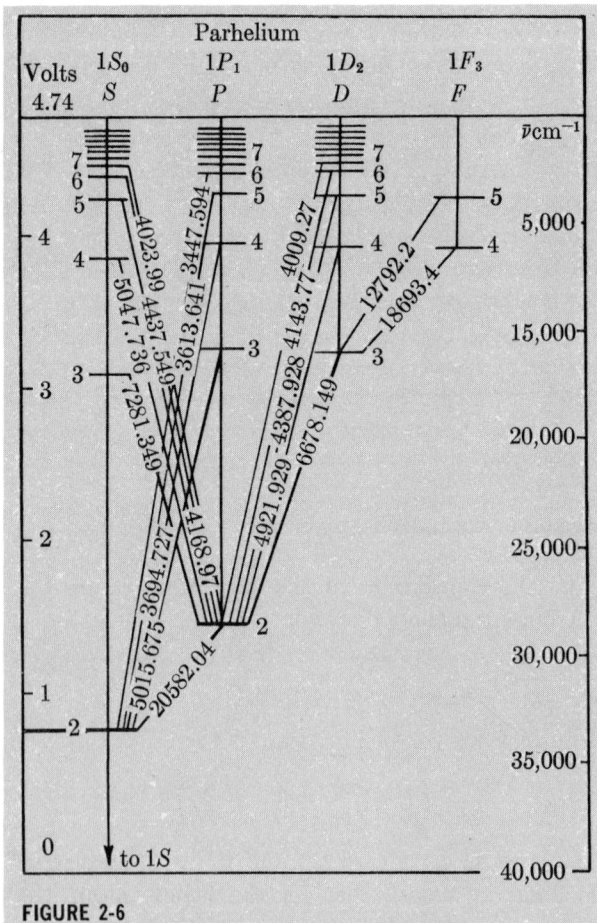

FIGURE 2-6

Energy Level Diagram for Parhelium. Wavelengths of the various transitions are given along the lines in angstroms.

is continued to give successively improved wave functions for the electrons until no appreciable change is noted. At this point the field is said to be *self-consistent*.

One of the important consequences of the electron-electron repulsion is the removal of the orbital degeneracy observed in the solution of the hydrogenic wave equation. Whereas all of the levels in a given radial shell in the hydrogen atom were observed to be degenerate and therefore of the same energy, these levels are found to separate in the more complex atoms. This separation is illustrated in the level scheme for parhelium shown in Figure 2–6, where the order of levels becomes

$$1s < 2s < 2p < 3s < 3p \sim 3d < 4s \ldots$$

On going to yet heavier elements, the order of levels will be altered further, and we find the $ns$ states to be at lower energies than the $(n-1)d$ states.

## WAVE FUNCTIONS OF THE HYDROGEN ATOM

In the Bohr atom, it was postulated that the electron follows a particular circular path in its motion about the nucleus. This particular path was chosen because it is the simplest path that gives agreement between the theoretical model and experiment. Such a circular path led to a symmetrical atom and contributed virtually nothing to an understanding of the geometry of molecules. On the other hand, one of the more outstanding accomplishments of the quantum mechanical approach to atomic structure has been its determination of the general distribution pattern of the electron about the atom, and the relationship of this pattern to molecular structure.

It was postulated that the square of the wave function is a measure of the probability distribution of the electron. This wave function is composed of two parts, an angular portion represented by $Y_{l,m}(\theta,\varphi)$ and a radial portion that is represented by $R_{n,l}(r)$. We shall see that the radial portion of the wave function gives the distribution of the electron with respect to its distance from the nucleus whereas the angular portion gives the geometry of the various electron states.

The normalized solutions of the $\Theta$ equation and also the radial equation are, in general, quite complex. However, they reduce to relatively simple forms on introduction of particular values of the parameters. For the $\Phi$ equation, the allowed values of the parameter $m$ are seen to be $m = 0$, $\pm 1$, $\pm 2, \ldots, \pm l$. These lead to the normalized functions of $\Phi(\varphi)$ shown in Table 2–2, in which both the complex and the real forms are given. Examples of the normalized $\Theta(\theta)$ functions and radial functions are given in Tables 2–3 and 2–4, respectively.

The normalized total wave function for the hydrogen atom is obtained from the relation

$$\psi_{(r\theta\varphi)} = R_{n,l}(r)\, Y_{l,m}(\theta,\varphi)$$

**TABLE 2–2.  Normalized Functions of $\Phi_m(\varphi)$.**

$$\Phi_m(\varphi) = \frac{1}{\sqrt{2\pi}} e^{\pm im\varphi}$$

$$\Phi_0(\varphi) = \frac{1}{\sqrt{2\pi}} \qquad \text{or} \qquad \Phi_0(\varphi) = \frac{1}{\sqrt{2\pi}}$$

$$\Phi_1(\varphi) = \frac{1}{\sqrt{2\pi}} e^{i\varphi} \qquad \text{or} \qquad \Phi_1(\varphi) = \frac{1}{\sqrt{\pi}} \cos \varphi$$

$$\Phi_{-1}(\varphi) = \frac{1}{\sqrt{2\pi}} e^{-i\varphi} \qquad \text{or} \qquad \Phi_{-1}(\varphi) = \frac{1}{\sqrt{\pi}} \sin \varphi$$

**TABLE 2–3.  Normalized Functions of $\Theta_{l,m}(\theta)$.**

$$\Theta_{l,m}(\theta) = \sqrt{\frac{(2l+1)(l-|m|)!}{2(l+|m|)!}} \, P_l^{|m|}(\cos\theta)$$

$$l = 0: \qquad \Theta_{0,0}(\theta) = \frac{\sqrt{2}}{2}$$

$$l = 1: \qquad \Theta_{1,0}(\theta) = \frac{\sqrt{6}}{2} \cos\theta$$

$$\Theta_{1,\pm1}(\theta) = \frac{\sqrt{3}}{2} \sin\theta$$

**TABLE 2–4.  Normalized Functions of $R_{n,l}(r)$.**

$$R_{n,l}(r) = \sqrt{\left(\frac{2Z}{na_0}\right)^3 \frac{(n-l-1)!}{2n\,[(n+l)!]^3}} \, e^{-\rho/2} \, \rho^l \, L_{n+l}^{2l+1}(\rho)$$

$n = 1$, $K$ shell:

$\qquad l = 0: \qquad R_{1,0}(r) = (Z/a_0)^{3/2} \, 2e^{-\rho/2}$

$n = 2$, $L$ shell:

$\qquad l = 0: \qquad R_{2,0}(r) = \dfrac{(Z/a_0)^{3/2}}{2\sqrt{2}} \, (2-\rho) \, e^{-\rho/2}$

$\qquad l = 1: \qquad R_{2,1}(r) = \dfrac{(Z/a_0)^{3/2}}{2\sqrt{6}} \, \rho \, e^{-\rho/2}$

As we have seen, the particular choice of functions of $R_{n,l}(r)$, $\Theta_{l,m}(\theta)$, and $\Phi_m(\varphi)$ are not arbitrary. They are limited by the allowed values of the quantum numbers. By using all possible arrangements of these functions within the limitations of the quantum numbers, we obtain the normalized total wave functions listed in Table 2–5, where $\sigma = \frac{n}{2}\rho$.

---

**TABLE 2–5.   Normalized Hydrogen-like Wave Functions.**

$K$ Shell:

$n = 1, l = 0, m = 0$:     $\psi_{1s} = \dfrac{1}{\sqrt{\pi}} (Z/a_0)^{3/2} e^{-\sigma}$

$L$ Shell:

$n = 2, l = 0, m = 0$:     $\psi_{2s} = \dfrac{1}{4\sqrt{2\pi}} (Z/a_0)^{3/2} (2 - \sigma) e^{-\sigma/2}$

$n = 2, l = 1, m = 0$:     $\psi_{2p_z} = \dfrac{1}{4\sqrt{2\pi}} (Z/a_0)^{3/2} \sigma e^{-\sigma/2} \cos \theta$

$n = 2, l = 1, m = \pm 1$:
$$\begin{cases} \psi_{2p_x} = \dfrac{1}{4\sqrt{2\pi}} (Z/a_0)^{3/2} \sigma e^{-\sigma/2} \sin \theta \cos \varphi \\[2mm] \psi_{2p_y} = \dfrac{1}{4\sqrt{2\pi}} (Z/a_0)^{3/2} \sigma e^{-\sigma/2} \sin \theta \sin \varphi \end{cases}$$

$M$ Shell:

$n = 3, l = 0, m = 0$:     $\psi_{3s} = \dfrac{2}{81\sqrt{3\pi}} (Z/a_0)^{3/2} (27 - 18\sigma + 2\sigma^2) e^{-\sigma/3}$

---

## RADIAL DISTRIBUTION CURVES

In its ground state, the wave function for the hydrogen atom is

$$\psi_{1s} = \frac{1}{\sqrt{\pi a_0^3}} e^{-r/a_0}$$

The square of this function should give the probability distribution for the electron with respect to the nucleus. The fact that a probability distribution exists at all is in conflict with the ideas of the Bohr theory, where the electron was confined to a prescribed orbit at a given distance from the nucleus. Quantum mechanically, however, we see from the plot of the dependence of the radial wave function on the distance, shown in Figure 2–7, that there is a finite probability for the electron to be even at very great distances from the nucleus. This probability, of course, is exceedingly small, but it does exist.

A more indicative means for expressing the electron distribution is in terms of the radial distribution function. This is a measure of the probability of finding the electron in a spherical shell between the distances

$r$ and $r + dr$ from the nucleus. The volume lying between two concentric spheres at a distance of $r$ to $r + dr$ from the origin is obtained by multiplying the surface area of a sphere, $4\pi r^2$, by the distance between it and

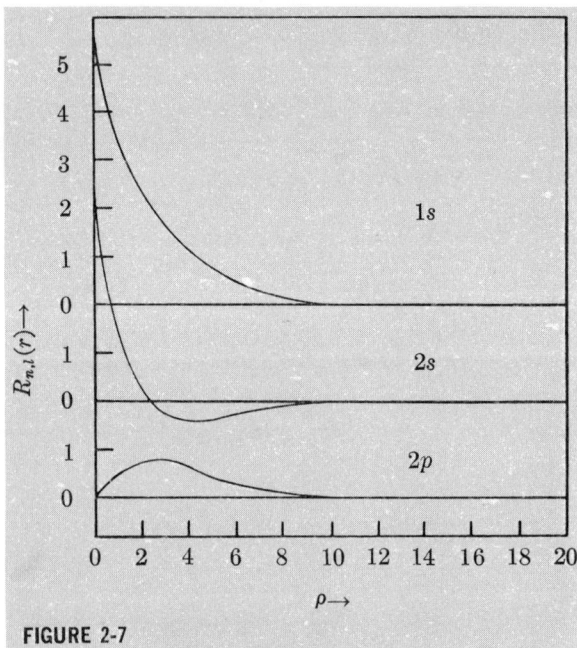

**FIGURE 2-7**

Radial Wave Functions for the $1s$, $2s$, and $2p$ States of the Hydrogen Atom.

the next concentric sphere, $dr$. The probability that the electron is in this volume element should be given by

$$D_{(r)}dr = 4\pi r^2 \,\psi\psi^* \, dr$$

or

$$(2\text{--}52) \qquad D_{(r)}dr = \frac{4}{a_0{}^3} r^2 e^{-2r/a_0} \, dr$$

A plot of the radial distribution function in units of $a_0$ is shown in Figure 2–8. It is interesting to note that, although the electron has a finite probability of existence at great distances from the nucleus, its maximum probability is at $a_0$, exactly the distance calculated by Bohr for the first electron orbit in the Bohr atom.

In addition to the ground state, various excited states of the hydrogen atom also exist. The radial distribution functions for these states can be plotted in the same manner as the ground state function. Examples of a few of these functions are also plotted in Figure 2–8.

## ANGULAR DEPENDENCE OF THE WAVE FUNCTION

The one-electron wave functions of the type we have obtained for the hydrogen atom are often referred to as *atomic orbitals*. The most characteristic feature of these atomic orbitals is their dependence on the angles $\theta$ and $\varphi$, which determine the geometry of the atom. It is generally true that the radial dependence of the wave function is approximately the same for the various $l$ states in a given major shell, $n$. Thus, to a good approximation, an $s$ state and a $p$ state in the same shell can be considered on the basis of their angular dependence alone. This angular dependence is represented by the spherical harmonics

$$Y_{l,m}(\theta,\varphi) = \Theta_{l,m}(\theta)\, \Phi_m(\varphi)$$

In Table 2–6, we see the spherical harmonics for the $s$ state and the three $p$ states for the hydrogen atom, along with their determining $l$ and $m$ quantum numbers.

If we consider the $s$ orbital, the spherical harmonics is seen from Table 2–6 to be

$$Y_{0,0}\,(\theta,\varphi) = \frac{1}{\sqrt{4\pi}}$$

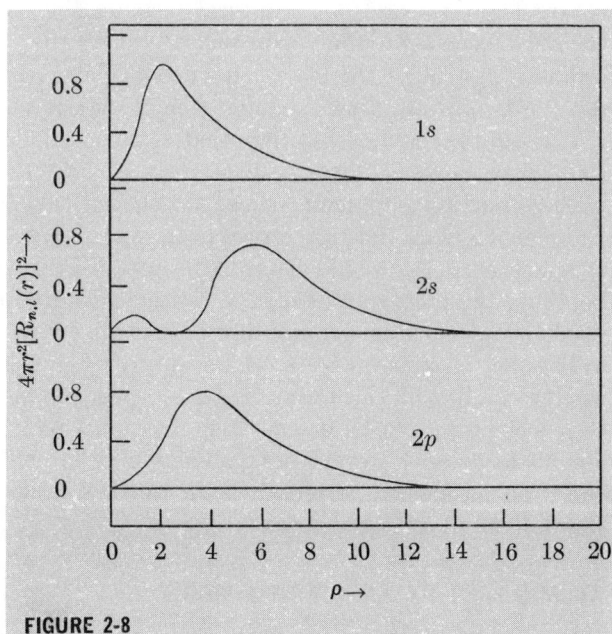

**FIGURE 2-8**

Radial Distribution Functions for the $1s$, $2s$, and $2p$ States of the Hydrogen Atom.

**TABLE 2–6.  Spherical Harmonics for the $s$ State and Three $p$ States of the Hydrogen Atom, with Determining $l$ and $m$ Quantum Numbers.**

$$Y_{l,m}(\theta,\varphi) = \Theta_{l,m}(\theta)\,\Phi_m(\varphi)$$

$$l = 0,\, m = 0: \quad Y_{0,0}(\theta,\varphi) = \frac{1}{\sqrt{4\pi}}$$

$$l = 1,\, m = 0: \quad Y_{1,0}(\theta,\varphi) = \sqrt{\frac{3}{4\pi}}\,\cos\theta$$

$$l = 1,\, m = 1: \quad Y_{1,1}(\theta,\varphi) = \sqrt{\frac{3}{4\pi}}\,\sin\theta\cos\varphi$$

$$l = 1,\, m = -1: \quad Y_{1,-1}(\theta,\varphi) = \sqrt{\frac{3}{4\pi}}\,\sin\theta\sin\varphi$$

Here we see that the $s$ orbital is *independent* of the angles $\theta$ and $\varphi$. Regardless of what values the angles may take, the angular portion of the wave function is a constant and therefore *spherically* symmetrical. This leads to the representation of an $s$ orbital as shown in Figure 2–9. On the other hand, the $p$ orbitals all exhibit the definite geometry shown in Figure 2–9. If a plot is made of each of the three $p$ orbitals, it is found that they form identical *dumbbell*-like patterns lying mutually perpendicular to each other. Thus, we find one $p$ orbital forming the dumbbell pattern along the $x$ axis, one along the $y$ axis, and one along the $z$ axis of a cartesian coordinate system. Although we have not shown the spherical harmonics for the $d$ orbitals, their angular dependence is also given in Fig. 2–9.

It should be pointed out here that since the radial portion of the wave function is omitted, these figures represent only the angular dependence of the $p$ and $d$ wave functions and not their actual geometrical structures. In order to obtain the geometrical patterns for these orbitals, contour lines of constant $\psi_{(r,\theta,\varphi)}$ would have to be used.[8] However, for discussions of bonding, the plots of the angular dependence are quite adequate and, it would seem, more convenient to use than the contour plots.

The fact that the wave mechanical atom has this definite geometry is significant. It is on this character that a quantum mechanical treatment of stereochemistry can be based. Thus, in contrast to the Bohr atom, which had little to offer toward an understanding of molecular geometry, the quantum mechanical approach leads to some quite satisfying results, as we shall see in a later chapter.

## ATOMIC SPECTRA AND TERM SYMBOLS

In expressing the energy of an electron in terms of $s$, $p$, $d$, and $f$ states, we are actually taking into account only two of the four quantum numbers necessary to completely describe the energy of an electron in an atom. In general, such a configuration will be highly degenerate because we are ignoring both *interelectronic repulsion* and *spin-orbit interactions*. While these forces may be relatively small, they nevertheless serve to remove the otherwise high degeneracy of a given electron configuration involving electrons outside a closed shell. In order to see how these additional interactions remove the degeneracy of a given electron configuration, it will be desirable to consider two extreme situations commonly referred to as

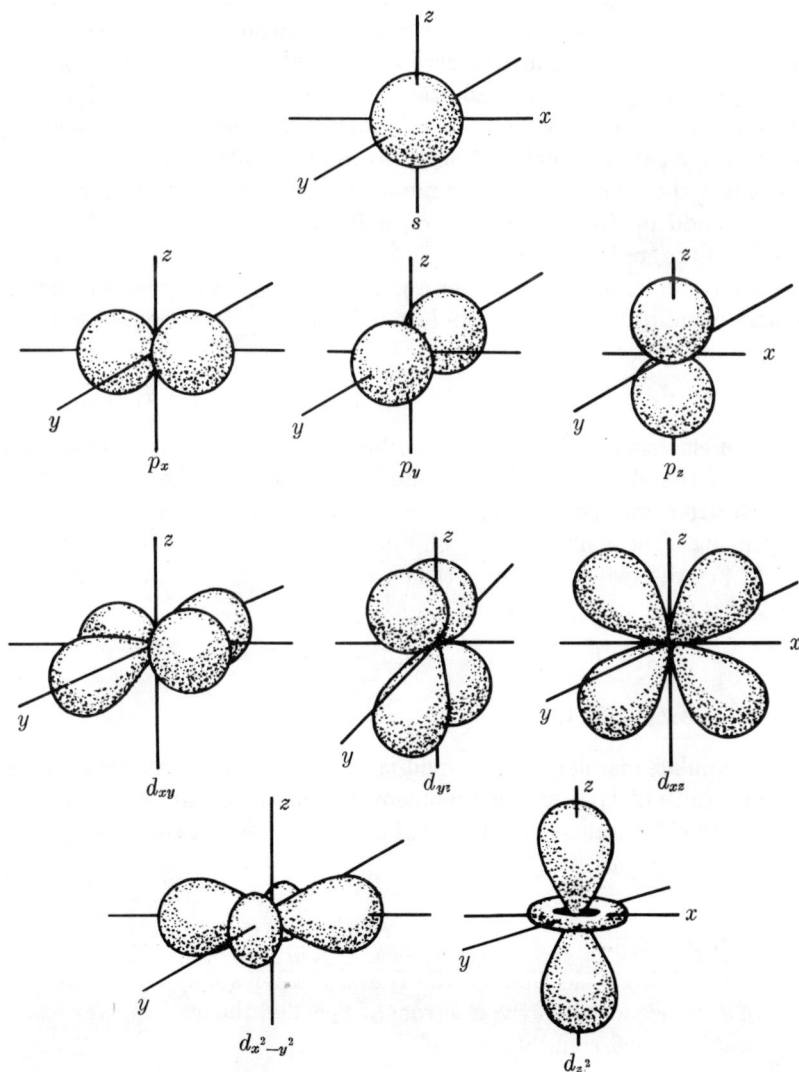

**FIGURE 2-9**

Conventional Boundary Surfaces of the $s$, $p$, and $d$ Atomic Orbitals.

*Russell-Saunders* or *LS coupling* on the one hand and *jj coupling* on the other.

The more common of these is the Russell-Saunders coupling, in which it is assumed that the interaction among the individual orbital moments and among the individual spin moments is stronger than the spin-orbit or *ls* interaction. This assumption appears to be valid for elements lighter than approximately $Z = 30$. In Russell-Saunders coupling, we can assume that all of the angular momenta of the different electrons, $l_i$, in an atom couple together to give a total or resultant orbital angular momentum quantum number $L$. $L$ must be zero or integral according to the quantum restrictions applying to the addition of vector quantities, and it is the vectorial sum of the $l$ values for all of the electrons. The summation is simplified by the fact that the electrons in closed shells do not contribute to $L$ since their orbital angular momenta as well as their spin angular momenta add up to zero. Therefore, only electrons outside a closed shell need be considered.

As an example, we can consider a general case where two electrons have the azimuthal quantum numbers $l_1$ and $l_2$. In this instance, $L$ would take on the values

$$L = l_1 + l_2, l_1 + l_2 - 1, l_1 + l_2 - 2, \ldots, l_1 - l_2$$

For three electrons, the $L$ values are obtained by first finding those values for two of the electrons, and then adding vectorially the $l$ value for the third electron, and so on. To illustrate this point, we can consider three electrons with the configuration $sp^2$. For the $s$ electron, $l = 0$, and therefore $L_1 = 0$. For the two $p$ electrons, the $l$ values are $l_1 = l_2 = 1$, and therefore, $L_2$ is given by

$$L_2 = 1 + 1, 1 + 1 - 1, 1 + 1 - 2(= l - l)$$

or

$$L_2 = 2, 1, 0$$

In a similar manner, the individual spins couple together to give a total or resultant spin angular momentum quantum number, $S$. $S$ is obtained as the algebraic sum of the $s$ values for the separate electrons, that is,

$$S = \sum_i s_i$$

Finally, just as the $l$ and $s$ values may couple (spin-orbit interaction) to give a $j$ for a single electron, so the $L$ and $S$ values may couple to give a series of $J$ values for all of the electrons. $J$ is called the *total angular momentum quantum number* and its possible values are

$$J = L + S, L + S - 1, L + S - 2, \ldots, |L - S|$$

$J$ may only have positive values or be zero, and it will have integral values when $S$ is an integer and half-integral values when $S$ is a half-integer.

An atomic state with given $L$ and $S$ values thus consists of a group of components having energies that are generally relatively close together. The number of components of the group is equal to the number of possible $J$ values. The particular state is then said to be a *multiplet* and to have a *multiplicity* equal to the number of $J$ values. If, for instance, $S = \frac{1}{2}$, then

$$J = L + \tfrac{1}{2}$$

and

$$J = L + \tfrac{1}{2} - 1 = L - \tfrac{1}{2}$$

Therefore, the multiplicity is 2, and we have a *doublet*. If we consider a case where $S = 1$, then

$$J = L + 1$$

$$J = L + 1 - 1 = L$$

and

$$J = L + 1 - 2 = L - 1$$

Here, the multiplicity is 3 and we have a *triplet*. In general, the multiplicity will be $2S + 1$, provided that $L$ is greater than $S$. If $L < S$, there is only one possible value of $J$, although $2S + 1$ may be greater than unity. For instance, a lone $s$ electron outside a closed shell would have $l = 0 = L$ and $s = \frac{1}{2} = S$, so that $J$ can only be $\frac{1}{2}$. Thus, the state is a singlet even though $2S + 1 = 2$.

In order to represent more completely the electronic state of an atom, a scheme based on the use of spectral term symbols was introduced in 1925 by H. N. Russell and F. A. Saunders. The term letters are arrived at as follows:

$$L = 0,\ 1,\ 2,\ 3,\ 4,\ 5, \ldots$$

$$\text{Term letter} = S,\ P,\ D,\ F,\ G,\ H, \ldots$$

The term letter is preceded by a superscript representing the multiplicity of the term, that is, $2S + 1$, and it is followed by a subscript giving the corresponding $J$ value. Thus,

$$^{2S+1}L_J = \text{term symbol}$$

For example, if $L = 2$ and $S = 1$, the term symbol would be $^3D$, and since the possible $J$ values are 3, 2, and 1, the three states of the triplet are $^3D_3$, $^3D_2$, and $^3D_1$.

It is now possible to consider a specific atom and construct the term symbols representing the various energy states in which the atom may be found. If we take the carbon atom with the ground state configuration $1s^2 2s^2 2p^2$, we see that both $L$ and $S$ will be determined only by the two $p$ electrons. $L$ may be 2, 1, or 0 corresponding to $D$, $P$, and $S$ states, respectively. $S$ may be 0 or 1, corresponding to multiplicities of 1 and 3, respectively. Therefore, we can have the following states: $^3D$, $^3P$, $^3S$, $^1D$, $^1P$, and $^1S$. However, application of the Pauli exclusion principle will show that not all of these states are allowed. We may say that some are forbidden, and for the $p^2$ configuration, the only allowed states are $^3P$, $^1D$, and $^1S$. In Table 2–7 a list of the allowed Russell-Saunders states is given

**TABLE 2–7. Allowed Russell-Saunders States for Equivalent $s$, $p$, and $d$ Electrons.**

| Equivalent $s$ Electrons | | | | | |
|---|---|---|---|---|---|
| $s$ | $^2S$ | | | | |
| $s^2$ | $^1S$ | | | | |

| Equivalent $p$ Electrons | | | | | |
|---|---|---|---|---|---|
| $p^1$ or $p^5$ | | $^2P$ | | | |
| $p^2$ or $p^4$ | $^1S$, | | $^1D$, | | $^3P$ |
| $p^3$ | | $^2P$, | | $^2D$, | $^4S$ |
| $p^6$ | $^1S$ | | | | |

| Equivalent $d$ Electrons | | | | | | |
|---|---|---|---|---|---|---|
| $d^1$ or $d^9$ | $^2D$ | | | | | |
| $d^2$ or $d^8$ | $^1(SDG)$, | $^3(PF)$ | | | | |
| $d^3$ or $d^7$ | $^2D$, | $^2(PDFGH)$ | $^4(PF)$ | | | |
| $d^4$ or $d^6$ | $^1(SDG)$, | $^3(PF)$, | $^1(SDFGI)$, | $^3(PDFGH)$, | $^5D$ | |
| $d^5$ | $^2D$, | $^2(PDFGH)$, | $^4(PF)$, | $^4(SDFGI)$, | $^4(DG)$, | $^6S$ |

for equivalent $s$, $p$, and $d$ electrons. It should be noted that an atom or ion consisting of closed shells only is always in a $^1S_0$ state.

For the carbon atom, we will have the states $^3P_0$, $^3P_1$, $^3P_2$, $^1D_2$ and $^1S_0$. Once these terms are known, it is necessary to determine which of the components lies lowest in energy, that is, which represents the ground state. In order to do this, we rely on a set of rules called *Hund's rules*. These are as follows:

(1) Of the Russell-Saunders states arising from a given electron configuration and allowed by the Pauli principle, the most stable state will be the one with the greatest multiplicity.

(2) Of a group of terms with a given value of $S$, the one with the largest value of $L$ lies lowest in energy.

(3) Of the states with given $L$ and $S$ values in a configuration consisting of less than half the electrons in a subgroup, the state with the smallest value of $J$ is usually the most stable. For a configuration consisting of more than half the electrons in a subgroup, the state with the largest $J$

is the most stable. The multiplets of the former are called *normal* multiplets, and those of the latter are called *inverted* multiplets.

In addition to the splitting we have thus far considered, a further splitting of the energy levels can occur from the action of an external magnetic field on the atom. Under these conditions, the total angular momentum quantum number, $J$, is split into $2J + 1$ equally spaced levels corresponding to the number of values that can be assumed by the magnetic quantum number $m$. These values are $-J, \ldots 0, \ldots, +J$.

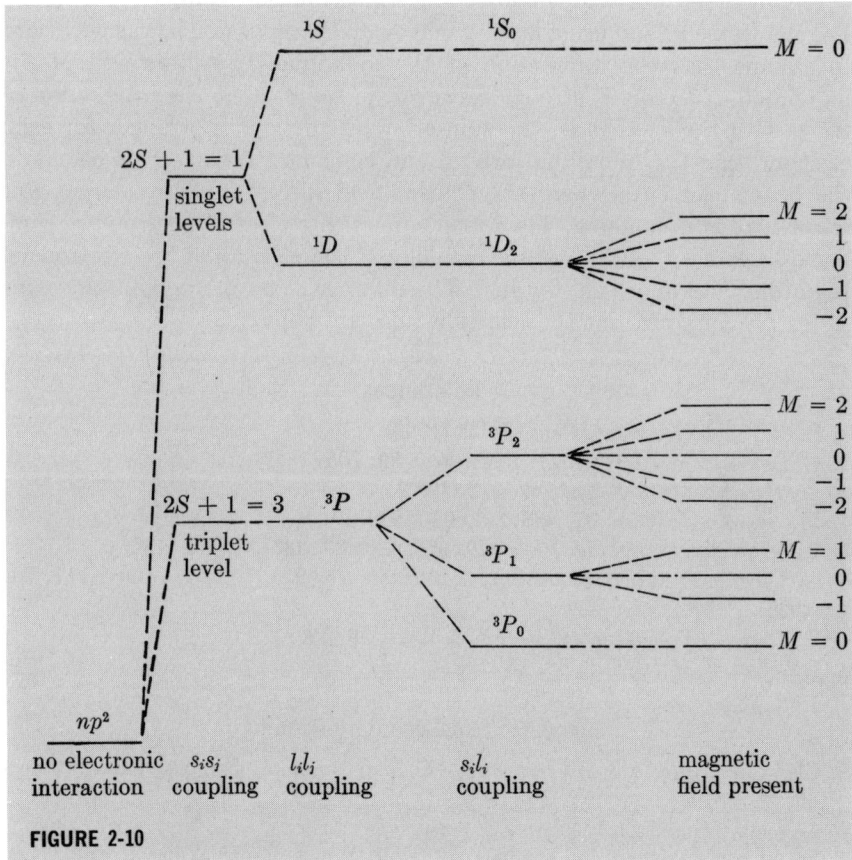

**FIGURE 2-10**

Energy Levels for the Electron Configuration $(np)^2$. (Adapted from H. Eyring, J. Walter, and G. Kimball, "Quantum Chemistry," John Wiley & Sons, Inc., New York, 1944.)

Now we are in a position to write out the various energy levels arising from the $p^2$ configuration, and these are shown in Figure 2–10. It is seen that the $p^2$ configuration is actually 15-fold degenerate. If the $n$ value of the two $p$ electrons is different, then even more energy levels will arise.

For example, the configuration $2p3p$ will split upon application of a magnetic field into 36 energy levels.

The essential importance of the term symbol lies in the fact that in a large number of cases it can be determined from atomic spectra. And from this, it is possible to obtain information regarding electronic configurations. In Table 3–4, the term symbols for the ground state are given for the various elements along with their electron configurations in terms of $s$, $p$, $d$, and $f$ orbitals.

Finally, a few words are in order concerning the more complex $jj$ coupling. For the Russell-Saunders coupling scheme to be applicable it was necessary to assume that the interaction of the individual $l_i$ vectors and of the individual $s_i$ vectors is strong. However, in some cases, the interaction of the $l_i$ with the $s_i$, that is, the spin-orbit interaction, is stronger for each electron than the individual orbital and spin interactions. This occurs in the heavy elements where the central field forces are quite large. The result here is $jj$ coupling. The $l_i$ and $s_i$ moments of each electron are combined to give a $j$ value, and the coupling of these for all of the electrons in the atom determines the total $J$. The number of terms remains the same as for $LS$ coupling, but the selection rules are different.

## References

1. L. de Broglie, *Ann. de. Phys.*, **3**, 22 (1925).
2. C. Davisson and L. Germer, *Phys. Rev.*, **30**, 705 (1927).
3. W. Heisenberg, *Z. Physik*, **43**, 172 (1927).
4. M. Born, *Z. f. Phys.*, **37**, 863; **38**, 803 (1926).
5. E. Schroedinger, *Ann. d. Phys.*, **79**, 361, 489; **80**, 437; **81**, 109 (1926).
6. G. Uhlenbeck and S. Goudsmit, *Naturwiss.*, **13**, 953 (1925); *Nature*, **117**, 264 (1926).
7. D. Hartree, *Proc. Cambridge Phil. Soc.*, **24**, 89 (1928).
8. I. Cohen, *J. Chem. Ed.*, **38**, 20 (1961).

## Suggested Supplementary Reading

S. Glasstone, "Theoretical Chemistry," D. Van Nostrand Company, Inc., Princeton, N. J., 1944.

A. Messiah, "Quantum Mechanics," John Wiley & Sons, Inc., New York, 1965.

C. W. Sherwin, "Quantum Mechanics," Holt, Rinehart and Winston, Inc., New York, 1959.

L. D. Landau, and E. M. Lifshitz, "Quantum Mechanics," Addison-Wesley Publishing Company, Inc., Reading, Mass., 1958.

J. W. Linnett, "Wave Mechanics and Valency," John Wiley & Sons, Inc., New York, 1960.

W. Kauzmann, "Quantum Chemistry," Academic Press, Inc., New York, 1957.

W. Heitler, "Elementary Wave Mechanics," Clarendon Press, Oxford, 1945.

C. A. Coulson, "Valence," 2nd ed., Oxford University Press, Inc., New York, 1961.

G. Herzberg, "Atomic Spectra and Atomic Structure," Dover Publications, Inc., New York, 1944.

## Problems

1. Calculate the energy of a photon if its wavelength is 6000 Å.
2. Calculate the mass of a photon if its energy is $3.2 \times 10^{-12}$ erg.
3. Determine the wavelength of a 100 g ball traveling at 75 mi/hr. What would be the possibility of observing the wave properties of the ball with existing instrumentation?
4. If Planck's constant had a value of unity, what effect would this have on the motion of an automobile?
5. What effect should the uncertainty principle have on the laws of cause and effect?
6. Why is the development of the expression $\lambda = h/p$ valid for a photon but not for an electron?
7. Show that $\psi = A \sin \alpha x \pm B \cos \alpha x \pm C e^{i\alpha x}$ is a solution to the equation

$$\frac{d^2\psi}{dx^2} + \alpha^2\psi = 0$$

8. Use the expression

$$g(t) = A \sin 2\pi\nu t + B \cos 2\pi\nu t + C e^{2\pi i\nu t}$$

to remove the time dependence from Eq. (2–7).

9. If $\psi = A \sin \alpha x$, show that $\dfrac{1}{\psi} \dfrac{\partial^2\psi}{\partial x^2}$ is independent of $x$.

10. Develop the one-dimensional analog of Eq. (2–11) using the time-independent wave function

$$\psi = A \sin \frac{2\pi}{\lambda} x$$

11. Investigate the separation in energy levels for a hydrogen atom confined to a one-dimensional box with $a = 10$ cm. Of what significance is this to the kinetic theory of gases?
12. Can a nondegenerate energy state exist in a cubical box in which the three quantum numbers are not all the same?
13. Consider the ground-state energy of an electron trapped in an infinite well (one-dimensional box) if $a = 10^{-13}$ cm. How does this value compare with conventional energies of beta particles in radioactive decay?
14. Plot $\psi$ and $|\psi|^2$ for the first five energy levels of a particle in a one-dimensional box.
15. Consider the case of a particle in a potential well (region I) that collides with a potential wall of height $V_o$ (region II) and passes through it. Explain how this can happen and set up and solve the wave equation for the particle in each of the three regions.

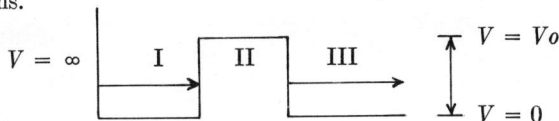

Hint: Note that in regions I and III, $V = 0$, but in region II, $V > E$. This results in an exponential form for $\psi$.

**16.** Rather than 0 and $a$, use $-a/2$ and $+a/2$ for the dimensions of a one-dimensional box. Do not permit the coefficients $A$ and $B$ in the solution to be zero, and determine the character of the quantum number for the *sin* and the *cos* terms.

**17.** Note that the energy of the hydrogen atom is dependent only on the $n$ quantum number. Is this degeneracy a consequence of the actual condition of the hydrogen atom or is it due to a weakness in the theory?

**18.** How can the effect of a magnetic field be introduced into the wave equation? How might this affect the energy and the degeneracy of the electron states?

**19.** From the solution to the $\Phi_m(\varphi)$ equation and the general forms of the normalization constants for the $\Theta_{(\theta)}$ and the $R_{(r)}$ equations, derive the allowed values and restrictions on the three spatial quantum numbers.

**20.** What is the relationship between the four dimensions and the four quantum numbers? Would the necessity of a fifth quantum number imply a fifth dimension?

**21.** By means of a plot, obtain the dumbbell-like pattern of the angular dependence of the $p_z$ orbital.

# The Periodic Table

3

The periodic classification of the elements must be listed among the outstanding contributions to the development of chemistry. Certainly, the importance of such a systemization cannot be questioned, but at times we are prone to forget the difficulties that faced those who were instrumental in the early development of the periodic system. In the light of our present understanding of the relation of electron configurations to the properties of the elements, the correlations seem rather obvious. However, it is necessary to realize that at the time of the development of the periodic system, the quantity and very often the quality of experimental data on which to base such a classification was lacking.

## THE DEVELOPMENT OF THE PERIODIC LAW

At the beginning of the nineteenth century not only was there an insufficient number of elements known on which to base a periodic classification, but equally important, the distinction between atomic weight and equivalent weight had not yet been made. Until this problem was resolved, no significant contribution could be possible. In addition to these difficulties, the accepted values for the atomic weights of many of the elements were questionable. Starting shortly after 1800, this problem was attacked, principally by Berzelius and later by Stas. The accuracy of their atomic weight determinations left little to be desired, and as a result, one of the major barriers to a periodic classification was removed. However, the formulation of a satisfactory periodic law had to wait until after 1860.

An understanding of the relation of atomic weight to equivalent weight did not come until the middle of the nineteenth century. When John Dalton applied the atomic theory to chemistry in 1807, he proposed that when two elements combine to form a compound, one atom of the first element combines with only one atom of the second element. Thus, if hydrogen and oxygen combined to form water, the resultant molecule of water would have the formula HO, rather than $H_2O$ as we know it today. If the atomic weight of hydrogen is taken to be 1, this leads to an atomic weight of 8 for oxygen. This idea was attacked as early as 1809 by Gay-Lussac, but the key to the problem lay in the hypothesis of the physicist Amadeo Avogadro. It was in 1811 that Avogadro published in the *Journal de Physique* the article that contained the basic ideas for what we now refer to as *Avogadro's hypothesis*. In his publication, Avogadro made clear the distinction between atoms (*molecules elementaires*) and molecules (*molecules integrantes*). He pointed out that with the assumption that the elementary gases are actually made up of two or more *molecules elementaires*, the problem of combining volumes with regard to atomic weights can be resolved. Unfortunately, Avogadro's work achieved little recognition at the time of its publication. However, in 1843, it was revived when Gerhardt used Avogadro's hypothesis as a basis for determining molecular weights and volumes. At that time, he came to the conclusion that the water molecule should be written as $H_2O$.

In spite of a growing realization of a difference between the atomic weight and the combining weight of an element, it was not until the first International Chemical Congress held at Karlsruhe in 1860, where Cannizzaro presented a paper based on the hypothesis of Avogadro, that his ideas began to receive the recognition that they deserved. As a result, during the next few years some general agreement on the molecular weights of the more important compounds was possible.

### Dobereiner's Triads

Although a set of atomic weights was not available for all elements, very definite similarities in both chemical and physical properties between a few groups of elements were observed quite early in the nineteenth century. In 1829, Johann Wolfgang Dobereiner made the first truly significant attempt to show a relation between the chemical properties of the elements and their atomic weights. He noted that certain similar elements occur in groups of three which he called *triads*. A particularly interesting feature of these triads is that the atomic weight of the middle member of the triad is very nearly equal to the arithmetic mean of the weights of the other two members of the triad. For instance, such a triad was observed for chlorine, bromine, and iodine where the mean of the atomic weights of chlorine and iodine is 81, which is very nearly the atomic weight of bromine. Other such triads are sulfur, selenium, and tellurium; and lithium, sodium, and potassium. In each of these cases it is seen that the mathematical relationship of the atomic weights holds quite well.

In the years that followed, numerous attempts were made at a more comprehensive classification in which the ideas of Dobereiner were extended and various group similarities were noted. For instance, Dumas noted some analogous trends in the successive molecular weights of a series of hydro-carbon radicals such as shown in Table 3–1.

**TABLE 3–1. Weight-Formula Trends of a Series of Hydrocarbon Radicals.**

| Radical | Formula | Molecular Weight |
|---------|---------|------------------|
| Hydrogen | H | 1 |
| Methyl | $CH_3$ | 15 |
| Ethyl | $C_2H_5$ | 29 |
| Propyl | $C_3H_7$ | 43 |
| Butyl | $C_4H_9$ | 57 |
| Amyl | $C_5H_{11}$ | 71 |

As we pass from one radical to the next, the molecular weight increases by 14 units. If the value of the first term is designated by $a$ and the difference by $d$, the value of any given radical may be represented by the expression $a + nd$. It is also seen that the triad relation of Dobereiner again holds. If any three successive terms are considered, the molecular weight of the middle member is found to be the arithmetic mean of the other two members. Observations such as these point to the existence of some sort of relation between chemical properties and molecular weight, but in themselves are of little value.

## The Telluric Helix

It was in 1862 that a periodic classification of the elements was developed that approached the ideas we have today. At that time, A. E. de Chancourtois, a professor of geology at the Ecole des Mines in Paris, presented an account of his telluric helix in a series of papers before the French Academie des Sciences, in which he indicated a relation between the properties of the elements and their atomic weights. De Chancourtois used a vertical cylinder with 16 equidistant lines on its surface, the lines lying parallel to its axis. He then drew a helix at 45° to the axis and arranged the elements on the spiral in the order of their increasing atomic weights. In this manner, elements that differed from each other in atomic weight by 16 or multiples of 16 fell very nearly on the same vertical line. As can be seen in Figure 3–1, it turns out that in some cases the elements lying directly under each other show a definite similarity. In addition to the 16 vertical lines, de Chancourtois felt that other connecting lines could be drawn, and that all elements lying on such lines were related in some manner. His arrangement resulted in the proposal by de Chancourtois that *the properties of the elements are the properties of numbers*. This was a rather close approach to the basic ideas of the later periodic classifications, but it left much to be desired in the light of the subsequent contributions by Mendeléev and Meyer. Nevertheless, an effort was made by two of his fellow countrymen, de Boisbaudran and Lapparent, to obtain some credit

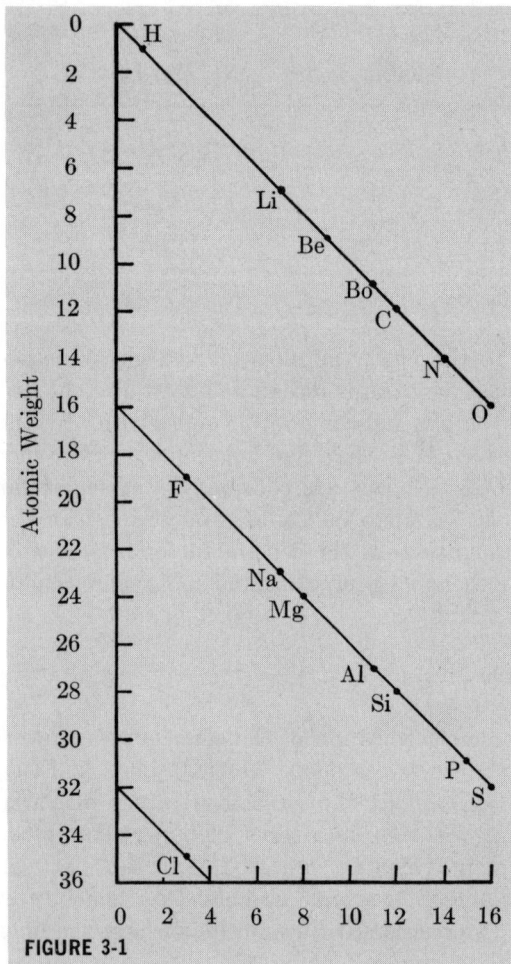

**FIGURE 3-1**

The Telluric Helix of de Chancourtois.

for de Chancourtois for his contributions to the periodic law. However, their opinions were not generally shared. Commenting on the *telluric screw*, around 1900, the British chemist, Dr. W. A. Tilden stated,[1] *"the author seems to have had a dim idea that properties were in some way related to atomic weight, but this idea is so confused by fantastic notions of his own, that it is impossible to be sure that he really recognized anything like periodicity in this relation."*

## The Law of Octaves

Very shortly after the telluric screw of de Chancourtois, John Alexander Reina Newlands in England made the first attempt to correlate the chemical properties of the elements with their atomic weights. This, of course, had not been possible until the difference in atomic weight and equivalent weight was appreciated. In 1864, the first of several papers by Newlands on the subject of periodicity of properties of the elements was published in *Chemical News*. In listing the elements in the consecutive order of their increasing atomic weights, Newlands noted a striking similarity between every eighth element. In the summer of the next year, Newlands published another paper in the *Chemical News* in which he again listed the elements in groups of seven, but found that with a few changes in the order of certain elements, the elements that appeared to belong to the same group would appear on the same horizontal line. This improved table of Newlands is seen in Figure 3–2.

| | No. | | No. | | No. | | No. | | No. | | No. | | No. | | | No. | | No. |
|---|---|---|---|---|---|---|---|---|---|---|---|---|---|---|---|---|---|---|
| H | 1 | F | 8 | Cl | 15 | Co & Ni | 22 | Br | 29 | Pd | 36 | I | | 42 | Pt & Ir | 50 | | |
| Li | 2 | Na | 9 | K | 16 | Cu | 23 | Rb | 30 | Ag | 37 | Cs | | 44 | Tl | 53 | | |
| G | 3 | Mg | 10 | Ca | 17 | Zn | 25 | Sr | 31 | Cd | 38 | Ba & V | 45 | | Pb | 54 | | |
| Bo | 4 | Al | 11 | Cr | 19 | Y | 24 | Ce & La | 33 | U | 40 | Ta | | 46 | Th | 56 | | |
| C | 5 | Si | 12 | Ti | 18 | In | 26 | Zr | 32 | Sn | 39 | W | | 47 | Hg | 52 | | |
| N | 6 | P | 13 | Mn | 20 | As | 27 | Di & Mo | 34 | Sb | 41 | Nb | | 48 | Bi | 55 | | |
| O | 7 | S | 14 | Fe | 21 | Se | 28 | Ro & Ru | 35 | Te | 43 | Au | | 49 | Os | 51 | | |

**FIGURE 3-2**

Periodic Table Proposed by Newlands.

Newlands made the unfortunate mistake of naming his generalization the *law of octaves*, due to its similarity to the musical scale. On the presentation of his periodic system before the Chemical Society in London in 1866, he was met with ridicule to the extent that one chemist asked sarcastically if he had tried arranging the elements in alphabetical order. Nevertheless, there can be little question that Newlands was the first to publish a list of the elements in the order of their increasing atomic weights and to realize that a systematic relationship exists between this order and their chemical and physical properties. In 1887 Newlands finally received the Davy Medal in recognition of his contribution, but even today he is probably given far less credit than he deserves in the development of the periodic law.[2]

Only a very few months after the first paper of Newlands, what would

| | | | Mo  96 | W  184 |
|---|---|---|---|---|
| | | | — | Au  196.5 |
| | | | Pd  106.5 | Pt  197 |
| L  7 | Na  23 | — | Ag  108 | — |
| G  9 | Mg  24 | Zn  65 | Cd  112 | Hg  200 |
| B  11 | Al  27.5 | — | — | Tl  203 |
| C  12 | Si  28 | — | Sn  118 | Pb  207 |
| N  14 | P  31 | As  75 | Sb  122 | Bi  210 |
| O  16 | S  32 | Se  79.5 | Te  129 | — |
| F  19 | Cl  35.5 | Br  80 | I  127 | — |
| | K  39 | Rb  85 | Cs  133 | |
| | Ca  40 | Sr  87.5 | Ba  137 | |
| | Ti  48 | Zr  89.5 | — | Th  231 |
| | Cr  52.5 | — | V  138 | |
| | Mn  55 | — | — | |

**FIGURE 3-3**

Classification of the Elements According to Odling.

appear to be a somewhat superior relation was published in the *Quarterly Journal of Science* by Odling. He pointed out that a purely arithmetical listing of the elements was in very close agreement with their recognized chemical similarities. Odling chose to represent his classification in the form shown in Figure 3–3. It is quite remarkable how similar it is to the table shown in Figure 3–4, which was proposed by Mendeléev in 1869. Never-

| | | | | Ti = 50 | Zr = 90 | ? = 180 |
|---|---|---|---|---|---|---|
| | | | | V = 51 | Nb = 94 | Ta = 182 |
| | | | | Cr = 52 | Mo = 96 | W = 186 |
| | | | | Mn = 55 | Rh = 104.4 | Pt = 197.4 |
| | | | | Fe = 56 | Ru = 104.4 | Ir = 198 |
| | | | Ni = | Co = 59 | Pd = 106.6 | Os = 199 |
| H = 1 | | | | Cu = 63.4 | Ag = 108 | Hg = 200 |
| | Be = 9.4 | Mg = 24 | | Zn = 65.2 | Cd = 112 | |
| | B = 11 | Al = 27.4 | | ? = 68 | Ur = 116 | Au = 197? |
| | C = 12 | Si = 28 | | ? = 70 | Sn = 118 | |
| | N = 14 | P = 31 | | As = 75 | Sb = 122 | Bi = 210 |
| | O = 16 | S = 32 | | Se = 79.4 | Te = 128? | |
| | F = 19 | Cl = 35.5 | | Br = 80 | I = 127 | |
| Li = 7 | Na = 23 | K = 39 | | Rb = 85.4 | Cs = 133 | Tl = 204 |
| | | Ca = 40 | | Sr = 87.6 | Ba = 137 | Pb = 207 |
| | | ? = 45 | | Ce = 92 | | |
| | | ?Er = 56 | | La = 94 | | |
| | | ?Yt = 60 | | Di = 95 | | |
| | | ?In = 75.6 | | Th = 118? | | |

**FIGURE 3-4**

Mendeléev's Original Table of the Elements.

theless, it appears that Odling failed to realize the implications of his table and as a result received virtually no part of the credit for the development of the periodic system of the elements.

## THE PERIODIC LAW

In spite of the importance of the earlier contributions, the major portion of the credit for the development of the periodic system must go to the Russian, Dmitrii Ivanovich Mendeléev, and to the German, Julius Lothar Meyer. Their independent realization that *the properties of the elements can be represented as periodic functions of their atomic weights* made possible a periodic classification that has suffered few significant changes in the subsequent years. Mendeléev published the first account of his periodic system in 1869, a few months before the publication of the table by Meyer. However, there can be no doubt that both men deserve equal credit for the development of the periodic law, regardless of the difference in the publication dates. This fact was recognized by the Royal Society in 1882 by the presentation of the Davy Medal to both Mendeléev and Meyer.

It was in March of 1869 that Mendeléev communicated the first of a series of papers to the Russian Chemical Society in which he set forth the arrangement of the elements in terms of their increasing atomic weights as shown in Figure 3–4. It might be noted that in many respects the Mendeléev arrangement differed little from that proposed five years earlier by Odling. However, Mendeléev appeared to be the first to fully appreciate the significance of this periodicity. In his first paper, Mendeléev emphasized the group similarities of the elements to the extent that he reversed the order of atomic weights where necessary in order to maintain this group similarity. He pointed out that this might be an indication of the correctness of the then accepted value for a given atomic weight and specifically mentioned the relative atomic weights of tellurium and iodine. Of considerable interest and importance is the fact that Mendeléev left vacant positions in his proposed table for yet undiscovered elements and went so far as to express the opinion that the chemical and physical properties of these elements might well be predicted from their positions in the table.

In the summer of 1871, Mendeléev published a much more comprehensive treatment of what he chose to call the *periodic law*. At this time he presented the more familiar form of the periodic table shown in Figure 3–5. Although this particular form of the table differs somewhat from the *short form* that is sometimes used today, it is substantially the same.

It was in his publication of 1871 that Mendeléev utilized the periodic character to predict the properties of the unknown elements lying directly below boron, aluminum, and silicon. These he named *eka-boron, eka-aluminum*, and *eka-silicon*, respectively. Starting in 1875 with the discovery of gallium by Lecog de Boisbaudran, the three blank positions were

| Series | Group I $R_2O$ | Group II $RO$ | Group III $R_2O_3$ | Group IV $RH_4$ $RO_2$ | Group V $RH_3$ $R_2O_5$ | Group VI $RH_2$ $RO_3$ | Group VII $RH$ $R_2O_7$ | Group VIII — $RO_4$ |
|---|---|---|---|---|---|---|---|---|
| 1 | H = 1 | | | | | | | |
| 2 | Li = 7 | Be = 9.4 | B = 11 | C = 12 | N = 14 | O = 16 | F = 19 | |
| 3 | Na = 23 | Mg = 24 | Al = 27.3 | Si = 28 | P = 31 | S = 32 | Cl = 35.5 | |
| 4 | K = 39 | Ca = 40 | — = 44 | Ti = 48 | V = 51 | Cr = 52 | Mn = 55 | Fe = 56 Co = 59 Ni = 59 Cu = 63 |
| 5 | (Cu = 63) | Zn = 65 | — = 68 | — = 72 | As = 75 | Se = 78 | Br = 80 | |
| 6 | Rb = 85 | Sr = 87 | ?Yt = 88 | Zr = 90 | Nb = 94 | Mo = 96 | — = 100 | Ru = 104 Rh = 104 Pd = 106 Ag = 108 |
| 7 | (Ag = 108) | Cd = 112 | In = 113 | Sn = 118 | Sb = 122 | Te = 125 | I = 127 | |
| 8 | Cs = 133 | Ba = 137 | ?Di = 138 | ?Ce = 140 | — | — | — | |
| 9 | (—) | | | | | | | |
| 10 | — | — | ?Er = 178 | ?La = 180 | Ta = 182 | W = 184 | — | Os = 195 Ir = 197 Pt = 198 Au = 199 |
| 11 | (Au = 199) | Hg = 200 | Tl = 204 | Pb = 207 | Bi = 208 | — | — | |
| 12 | — | — | — | Th = 231 | — | U = 240 | — | — |

**FIGURE 3-5**

Mendeléev's Periodic Table of 1871.

soon filled. Eka-boron became scandium, eka-aluminum became gallium, and eka-silicon became germanium. In Table 3–2, we see a comparison of the predictions for the element eka-silicon made by Mendeléev and the observed properties of germanium. The remarkable agreement is certainly a complete justification of Mendeléev's faith in his periodic law.

**TABLE 3–2.  Comparison of Eka-silicon and Germanium.**

| Property | Eka-silicon (1871) | Germanium (1886) |
|---|---|---|
| Atomic weight | 72 | 72.32 |
| Specific gravity | 5.5 | 5.47 |
| Specific heat | 0.073 | 0.076 |
| Atomic volume | 13 cc | 13.22 cc |
| Color | dark gray | grayish white |
| Sp. gr. of dioxide | 4.7 | 4.703 |
| B.p. of tetrachloride | 100° C | 86° C |
| Sp. gr. of tetrachloride | 1.9 | 1.887 |
| B.p. of tetraethyl derivative | 160° C | 160° C |

In December, 1869, a paper was submitted by Lothar Meyer to *Liebig's Annalen* with the title "The Nature of the Chemical Elements as Functions of Their Atomic Weight." This was very shortly after the publication of Mendeléev's first paper on the periodic classification of the elements. In this publication, Meyer proposed the periodic table as seen in Figure 3–6,

| I | II | III | IV | V | VI | VII | VIII | IX |
|---|---|---|---|---|---|---|---|---|
| | B 11 | Al 27.3 | | | | ?In113.4 | | Tl 202.7 |
| | | | — | — | — | | — | |
| | C 12 | Si 28 | | | | Sn 117.8 | | Pb 206.4 |
| | | | Ti 48 | | Zr 89.7 | | — | |
| | N 14 | P 30.9 | | As 74.9 | | Sb 122.1 | | Bi 207.5 |
| | | | V 51.2 | | Nb 93.7 | | Ta 182.2 | |
| | O 16 | S 32 | | Se 78 | | Te 128? | W 183.5 | |
| | | | Cr 52.4 | | Mo 95.6 | | | |
| | F 19.1 | Cl 35.4 | | Br 79.75 | | I 126.5 | | |
| | | | Mn 54.8 | | Ru 103.5 | | Os 198.6? | |
| | | | Fe 55.9 | | Rh 104.1 | | Ir 196.7 | |
| | | Co = | Ni 58.6 | | Rb 106.2 | | Pt 196.7 | |
| Li 7.0 | Na 22.9 | K 39 | | Rb 85.2 | | Cs 132.7 | Au 192.2 | |
| | | | Cu 63.3 | | Ag 107.7 | | | |
| ?Be 9.3 | Mg 23.9 | Ca 39.9 | | Sr 87 | | Ba 136.8 | Hg 199.3 | |
| | | | Zn 64.9 | | Cd 111.6 | | | |

**FIGURE 3-6**

Periodic Table According to Lothar Meyer.

which is very similar to the one published by Mendeléev. Emphasizing the physical properties of the elements, Meyer pointed out that, in general, *the properties of the elements are periodic functions of their atomic weights.* This periodicity was very clearly shown by Meyer by means of the *atomic volume curve.* If the atomic weight of a substance is divided by its density, a quantity known as the atomic volume is obtained. Meyer plotted this quantity against the atomic weight, giving the curve shown in Figure 3–7. In spite of several shortcomings of the concept of atomic volumes as used by Meyer, there can be no question of a general periodic trend. The maximum position in any period is seen to be always occupied by an alkali metal, and each member of a given family consistently holds a given position in its respective period.

## THE LONG FORM OF THE PERIODIC TABLE

Many different forms of a periodic classification of the elements have appeared since the 1871 periodic table by Mendeléev. Each table was designed to point up the various trends and relationships which its author considered most significant. From the literally hundreds of tables that have been proposed, perhaps the most popular and easily reproduced periodic table is the conventional *extended* or *long form,* which is shown in Figure 3–8. First used in a simpler form, by Rang in 1893, it was developed further by Werner in 1905 and evolved into its present form due to the strong support which it received from Bury and many others.

The subgroups of the Mendeléev table are separated with the result that the elements fall into 18 vertical columns, called *families,* representing, as we shall see, the successive filling of *s*, *p*, and *d* orbitals with 2, 6, and 10 electrons, respectively. The elements in each column are true analogs or congeners of each other. The inert gases are placed at the extreme right of the table, reflecting the completion of the *s* and *p* sublevels of their outer or valence shell. There are 7 horizontal rows, called *periods,* the first containing 2 elements, the second and third containing 8 elements, the fourth and fifth containing 18 elements, and the sixth and seventh containing 32 elements. This gives a total of 118 elements, the first 105 of which are now known. We shall see that this particular arrangement becomes quite reasonable when the electronic configurations are considered. To keep the table from being too drawn out, the 14 elements occurring in the sixth and seventh periods belonging in group R3 are placed below the main body of the table and are designated the *lanthanide series* (occurring just beyond lanthanum) and the *actinide series* (occurring just beyond actinium).

There are certainly a great number of advantages to the long form of the periodic table, and there are just as surely some disadvantages. These are discussed in some detail by Foster[3] and Luder,[4] respectively. Due to the shortcomings of the long form, a multitude of more recent periodic

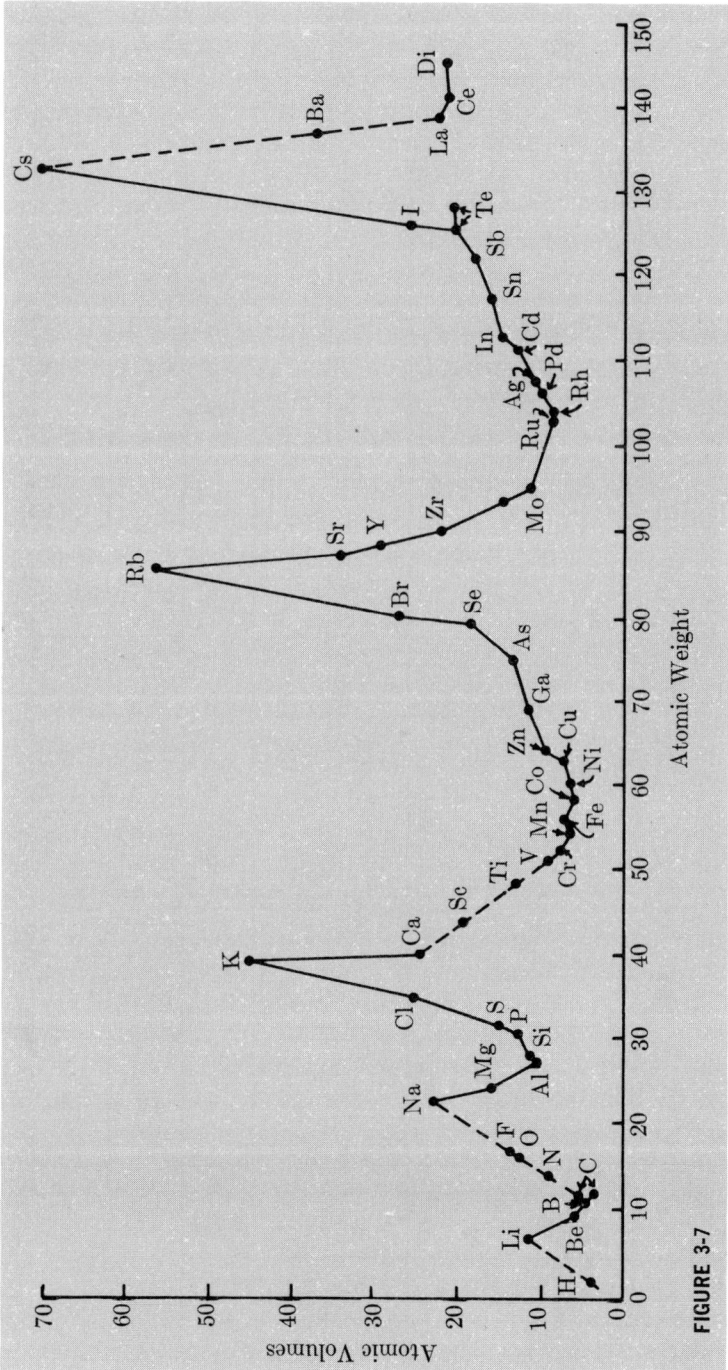

FIGURE 3-7

Atomic Volume Curve of Lothar Meyer.

$$\boxed{\begin{array}{c} 1 \\ \mathrm{H} \\ 1.00797 \end{array}}$$

| R1 | R2 | T3 | T4 | T5 | T6 | T7 | T8 | T9 | T10 | T11 | R2' | R3 | R4 | R5 | R6 | R7 | INERT GASES |
|---|---|---|---|---|---|---|---|---|---|---|---|---|---|---|---|---|---|
| | | | | | | | | | | | | | | | | | 2 He 4.0026 |
| 3 Li 6.939 | 4 Be 9.0122 | | | | | | | | | | | 5 B 10.811 | 6 C 12.01115 | 7 N 14.0067 | 8 O 15.9994 | 9 F 18.9984 | 10 Ne 20.183 |
| 11 Na 22.9898 | 12 Mg 24.312 | | | | | | | | | | | 13 Al 26.9815 | 14 Si 28.086 | 15 P 30.9738 | 16 S 32.064 | 17 Cl 35.453 | 18 Ar 39.948 |
| 19 K 39.102 | 20 Ca 40.08 | 21 Sc 44.956 | 22 Ti 47.90 | 23 V 50.942 | 24 Cr 51.996 | 25 Mn 54.9380 | 26 Fe 55.847 | 27 Co 58.9332 | 28 Ni 58.71 | 29 Cu 63.54 | 30 Zn 65.37 | 31 Ga 69.72 | 32 Ge 72.59 | 33 As 74.9216 | 34 Se 78.96 | 35 Br 79.909 | 36 Kr 83.80 |
| 37 Rb 85.47 | 38 Sr 87.62 | 39 Y 88.905 | 40 Zr 91.22 | 41 Nb 92.906 | 42 Mo 95.94 | 43 Tc (99) | 44 Ru 101.07 | 45 Rh 102.905 | 46 Pd 106.4 | 47 Ag 107.870 | 48 Cd 112.40 | 49 In 114.82 | 50 Sn 118.69 | 51 Sb 121.75 | 52 Te 127.60 | 53 I 126.9044 | 54 Xe 131.30 |
| 55 Cs 132.905 | 56 Ba 137.34 | 57 La 138.91 | 72 Hf 178.49 | 73 Ta 180.948 | 74 W 183.85 | 75 Re 186.2 | 76 Os 190.2 | 77 Ir 192.2 | 78 Pt 195.09 | 79 Au 196.967 | 80 Hg 200.59 | 81 Tl 204.37 | 82 Pb 207.19 | 83 Bi 208.980 | 84 Po (210) | 85 At (210) | 86 Rn (222) |
| 87 Fr (223) | 88 Ra (226) | 89 Ac (227) | | | | | | | | | | | | | | | |

| Lanthanide Series | 58 Ce 140.12 | 59 Pr 140.907 | 60 Nd 144.24 | 61 Pm (147) | 62 Sm 150.35 | 63 Eu 151.96 | 64 Gd 157.25 | 65 Tb 158.924 | 66 Dy 162.50 | 67 Ho 164.930 | 68 Er 167.26 | 69 Tm 168.934 | 70 Yb 173.04 | 71 Lu 174.97 |
|---|---|---|---|---|---|---|---|---|---|---|---|---|---|---|
| Actinide Series | 90 Th 232.038 | 91 Pa (231) | 92 U 238.03 | 93 Np (237) | 94 Pu (242) | 95 Am (243) | 96 Cm (247) | 97 Bk (249) | 98 Cf (251) | 99 Es (254) | 100 Fm (253) | 101 Md (256) | 102 No (253) | 103 Lw (257) |

**FIGURE 3-8**
The Long Form of the Periodic Table.

tables have been proposed, some of which will be discussed. Nevertheless, it is our feeling that the long form is still far superior to any of the presently known periodic tables in that it demands an understanding of the electronic basis of the periodic system and at the same time clearly reflects the similarities, differences, and trends in chemical and physical properties of the elements. Consequently, subsequent discussions on the periodic system will be in terms of the long form of the table.

## BASIS OF A PERIODIC CLASSIFICATION

Although the periodic law of Mendeléev met with considerable success, there were several glaring anomalies. According to the periodic law, the properties of the elements are periodic functions of their atomic weight. If this were strictly true, it would not be possible to have two elements with the same atomic weight and different chemical and physical properties. Yet, this is observed to be very nearly true for the case of cobalt and nickel. It would also be difficult to understand the relation of tellurium to iodine, where the order of increasing atomic weights is reversed. Mendeléev felt that the atomic weight of tellurium must be in error. However, this was shown not to be the case, and we find that tellurium must be placed ahead of iodine in the periodic table even though it has a larger atomic weight. In addition to these problems, there was the questionable location of such groupings of elements as the T8, T9, T10 elements, what we now call the rare earth elements, and, by 1900, the inert gases. It was apparent that there is something in the fundamental structure of the elements that brings about this periodicity, but atomic weight does not appear to be the final answer.

The first major step toward a solution to the problem came with the observation of the characteristic radiation of X rays. If a target is bombarded by high-energy electrons, two different types of X rays are usually observed. One type results in a continuous spectrum similar to that shown in Figure 3–9, in which we see a continuous distribution of wavelengths. The high-energy end of the spectrum is determined by the potential difference through which the electron falls. Superimposed on the continuous spectrum is often found a characteristic spectrum. The wavelengths of the characteristic X rays are found to be a function of the target material but independent of the exciting potential so long as the energy of the incident electrons is greater than some minimum value. In Figure 3–10, the X-ray spectrum for a molybdenum sample is reproduced in which two characteristic lines are seen to be superimposed on the continuous spectrum.

The first serious studies of the characteristic radiation were conducted by Barkla and Sadler, starting around 1908. By studying the absorption properties of these radiations, they were led to the conclusion that the characteristic radiation can be separated into two types, one of which is more penetrating than the other. The more penetrating of the two was called the $K$ radiation and the less penetrating was called the $L$ radiation. Although absorption measurements are not particularly accurate, Barkla and Sadler were able to determine that the energy of either the $K$ or the $L$ radiation increases with an increase in atomic number of the target element.

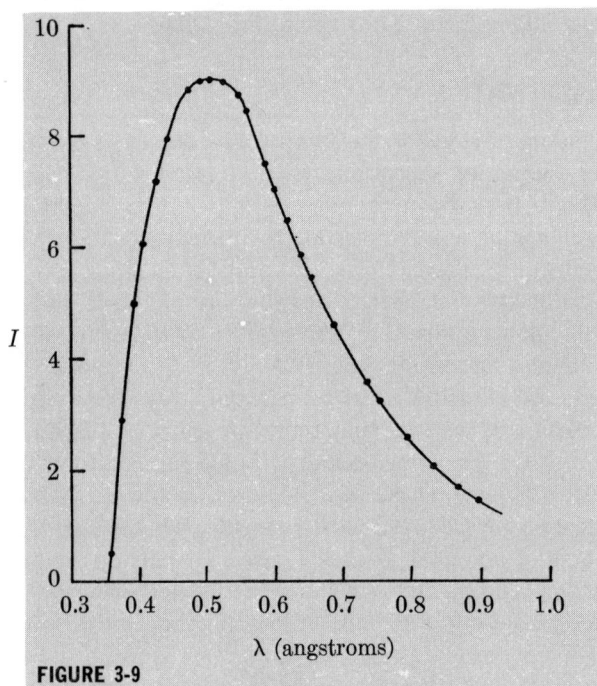

**FIGURE 3-9**

Continuous X-ray Spectrum of Tungsten as a Result of a 35,000-Volt Potential.

In 1913, H. G. J. Moseley made the first detailed study of the characteristic X-ray spectra emitted by different elements.[5] Using the adaptation of the crystal spectrometer developed by Bragg (Figure 3-11), Moseley studied the spectra of 38 elements. On analysis of the characteristic lines, Moseley was able to show that they can be separated into two distinct series which then can be related to the $K$ and $L$ radiations observed earlier by Barkla and Sadler. Figure 3-12 is a reproduction of a photograph of the $K$ series of the elements from calcium to copper. The elements are here arranged in terms of increasing atomic number. It can be seen that for each element, the $K$ radiation consists of two lines, a $K_\alpha$ and a $K_\beta$ line. The apparent discrepancy in the case of cobalt, where four lines are observed, can be attributed to impurities of iron and nickel in the cobalt sample.

It is readily seen that the wavelength of the characteristic $K$ radiation

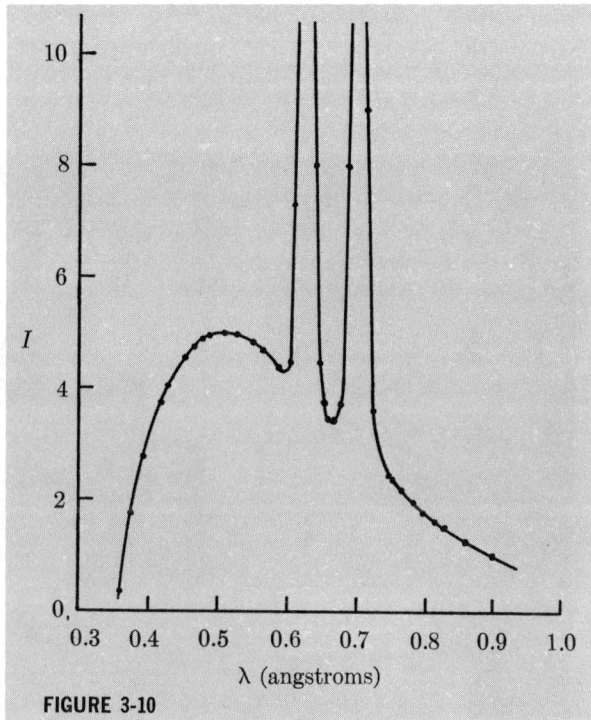

**FIGURE 3-10**

Characteristic X-ray Spectrum of Molybdenum Super-imposed on the Continuous Spectrum.

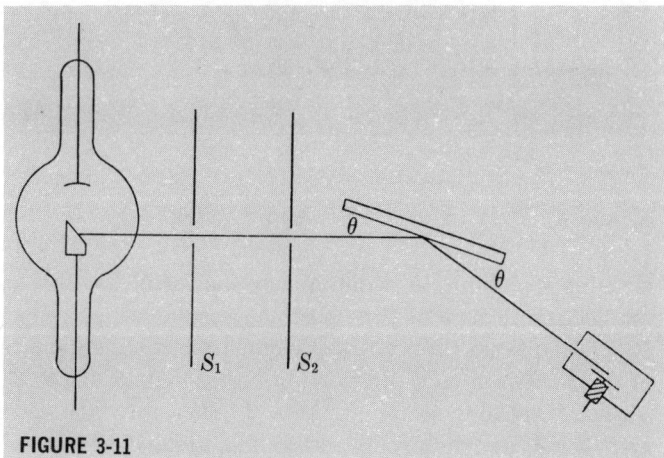

**FIGURE 3-11**

Schematic Diagram of the Crystal Spectrometer Developed by Bragg and Adapted by Moseley for his Study of the Characteristic X-ray Spectra.

decreases in a regular manner with an increase in atomic number. The only irregularity is observed between calcium and titanium, thereby indicating that an element has been omitted. This position, of course, belongs to the

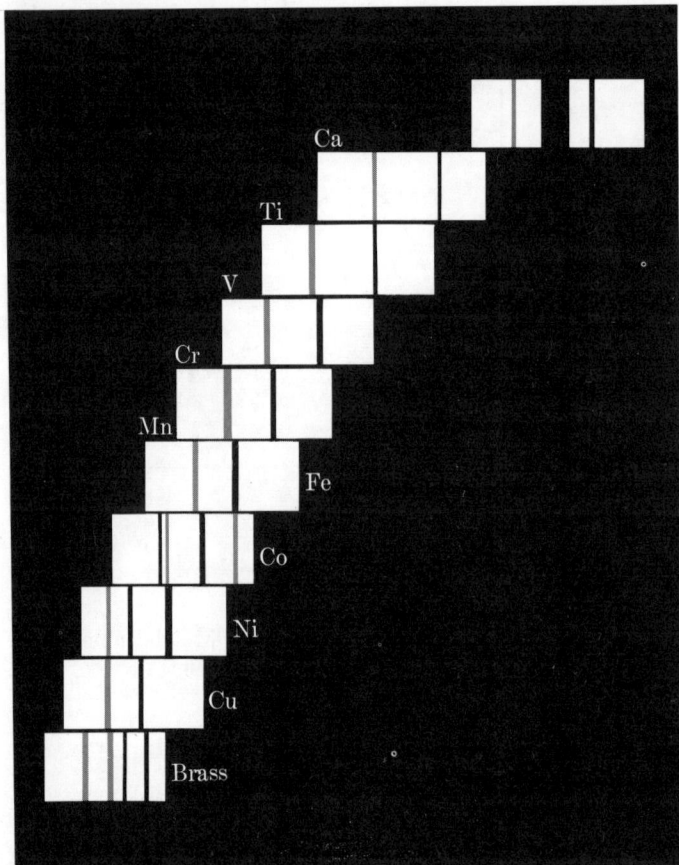

**FIGURE 3-12**

$K_\alpha$ and $K_\beta$ X-ray Lines as Observed by Moseley.

element scandium. A technique so powerful as this immediately found application to several then pertinent problems in the periodic system. By 1913, there were three pairs of elements in the periodic table that, because of their chemical and physical properties, could not be placed in the order of their increasing atomic weights. These were argon and potassium, cobalt and nickel, and tellurium and iodine. By means of the X-ray spectra, it was found that although their order was inverted in terms of atomic weight, it was in the proper sequence with regard to atomic number. This, then, would indicate that the atomic number is a more fundamental quantity than the atomic weight, and is possibly the key to an understanding of the periodic system of the elements.

Empirically, Moseley found that the frequency, $\nu$, of the characteristic X-ray line can be related to the atomic number by means of the relation

$$(3\text{-}1) \qquad \sqrt{\nu} = K(Z - k)$$

where $K$ and $k$ are both constants, and $Z$ is the atomic number of the element. The accuracy of this relation is readily determined by plotting $\sqrt{\nu}$ against the atomic number. As can be seen in Figure 3-13, an excellent

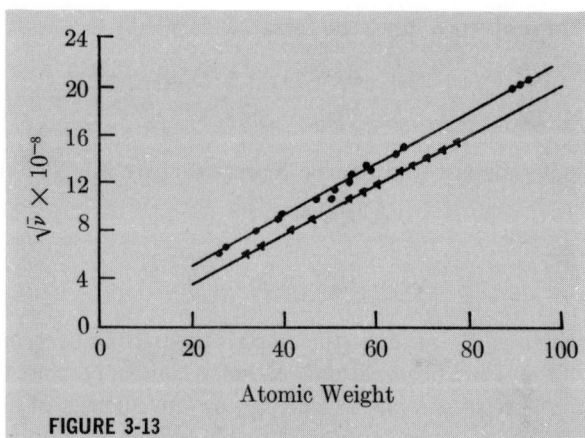

**FIGURE 3-13**

Relationship Between the Frequencies of the Characteristic X-ray Lines and (●) Atomic Weights and (▼) Atomic Numbers as Observed by Moseley.

straight line is obtained. For comparison, a plot of $\sqrt{\nu}$ against the atomic weight is shown on the same graph.

In an attempt to explain the relation between the frequency and the atomic number, Moseley utilized the treatment developed by Bohr for the hydrogen atom. From the Bohr model, it appeared reasonable that an X ray results from an electron transition to the $K$ or $L$ shell. Bohr had shown that the frequency, $\nu$, in the emission spectrum of the hydrogen atom can be represented by the formula

$$(3\text{-}2) \qquad \nu = \frac{2\pi^2 \mu e^4}{h^3} Z^2 \left( \frac{1}{n_f{}^2} - \frac{1}{n_i{}^2} \right)$$

where $\mu$ is the reduced mass, $h$ is Planck's constant, $Z$ is the atomic number, and $n_f$ and $n_i$ are the quantum numbers of the final and initial states, respectively. If this basic formula is applied to a heavy atom in which there are a large number of electrons, it would be expected that the effective nuclear charge seen by a given electron will be less than the nuclear charge, $Ze$, due to the shielding effect of the electrons lying between the nucleus and the given electron. It should more accurately be represented by $(Z - S)e$, where $S$ is a constant known as the screening constant and measures the degree to which other electrons in the atom are able to screen or shield the nucleus from a given electron. If we now use this effective nuclear charge in the Bohr equation we obtain the expression

$$(3\text{-}3) \qquad \nu = \frac{2\pi^2\mu e^4}{h^3} (Z - S)^2 \left( \frac{1}{n_f^2} - \frac{1}{n_i^2} \right)$$

For a given $K$ line, the term

$$\frac{2\pi^2\mu e^4}{h^3} \left( \frac{1}{n_f^2} - \frac{1}{n_i^2} \right)$$

is a constant and can be represented by $K^2$. This leads to the expression

$$\nu = K^2(Z - S)^2$$

or

$$\sqrt{\nu} = K(Z - S)$$

which is the same as the empirical equation determined by Moseley.

From his observations, Moseley concluded that the atomic number is a measure of the positive charge on the nucleus of an atom. This property, for some reason, appeared to be a more fundamental quantity than the atomic weight. The positions in the periodic table of unknown elements can be predicted with a much greater degree of certainty than was possible in terms of atomic weights, and the atomic weight anomalies such as the reversed order of iodine and tellurium can be reconciled. Thus, we are led to a new statement of the periodic law: *The properties of the elements are periodic functions of their atomic numbers.* The realization that the properties of the elements are periodic with respect to the atomic number leads, of course, to a much deeper understanding of the basis of this periodicity. We now recognize that the periodic character of the elements is actually due to the number and arrangement of the electrons which, in turn, are dependent upon the atomic number. Thus, in the light of our present understanding, a more appropriate statement of the periodic law might be expressed in terms of electron configurations.

## ELECTRONIC BASIS FOR THE PERIODIC CLASSIFICATION

With a better understanding of the part that the electron plays in the properties of the elements, a corresponding understanding of the periodic system came about. By 1923, it had been realized that an explanation of the line spectrum of an atom demands the use of four quantum numbers to determine the various energy states in the atom. Of these various states, one might expect that the electrons will exist in the lowest possible energy state available. Thus, if at all possible, an electron should try to be in the state where $n = 1$, $l = 0$, $m = 0$, and $m_s = \frac{1}{2}$. However, it is quite obvious from atomic spectra that it is not reasonable that all of the electrons in a heavy atom are in this particular state. Therefore, before any real advance could be made, it was necessary to find a rule by which the electrons can be fed into these possible energy states. Such a rule came in 1925 in the form of the *Pauli exclusion principle*, which states that in a given atom no two electrons can have the same values for all four quantum numbers.

In essence, the Pauli exclusion principle states that there can be only one electron in each possible energy state of the atom. If a given set of four quantum numbers is nondegenerate, every possible different arrangement of the four quantum numbers will define a new energy state, and in each of these states it will be possible to place only one electron. Thus, the key to the electronic structure of the elements and therefore the properties of the elements is found in the four quantum numbers with their restrictions along with the Pauli exclusion principle.

If we consider the four quantum numbers with their restrictions,

$$n = 1, 2, 3, 4, \ldots,$$
$$l = 0, 1, 2, 3, \ldots, (n-1), \text{ inclusive,}$$
$$m = 0, \pm 1, \pm 2, \pm 3, \ldots, \pm l, \text{ inclusive,}$$
$$m_s = \pm \tfrac{1}{2}$$

we find that only certain combinations are possible. If, for instance, we consider the first or $K$ shell, the principal quantum number equals 1. If $n = 1$, the angular momentum quantum number, $l$, can be only zero since its maximum value is $(n-1)$. Also, if $l = 0$, the magnetic quantum number, $m$, can be only zero. However, the spin quantum number, $m_s$, can be either $+\frac{1}{2}$ or $-\frac{1}{2}$. Thus, we find that in the first shell there are only two possible arrangements of the four quantum numbers within the given restrictions. This means that in the first shell there can be only two electrons, and since $l = 0$, the electrons must be in an $s$ state. For the second, third, and fourth shells, all of the possible arrangements of the four quantum numbers are shown in Table 3–3. It should be noted that the $s$ state occurs in every shell, and it always contains two electrons. In the second shell there is the possibility of both an $s$ and a $p$ state, and the $p$ state occurs in all subsequent shells. The $d$ state occurs first in the third shell and the $f$ state first occurs in the fourth shell.

Since the lower the value of the quantum number, the lower the energy of an electron state, assuming the quantum states are nondegenerate, we should expect the order of increasing energy states to be

$$1s < 2s < 2p < 3s < 3p < 3d < 4s < 4p < 4d < 4f < 5s \ldots$$

## TABLE 3-3. Possible Arrangements of the Electrons.

| $n$ | $l$ | $m$ | $m_s$ | Number of Electrons in a Subgroup | Electron State |
|---|---|---|---|---|---|
| 2 | 0 | 0 | $\pm\frac{1}{2}$ | 2 | $s$ |
|   | 1 | +1 | $\pm\frac{1}{2}$ |   |   |
|   |   | 0 | $\pm\frac{1}{2}$ | 6 | $p$ |
|   |   | −1 | $\pm\frac{1}{2}$ |   |   |
| 3 | 0 | 0 | $\pm\frac{1}{2}$ | 2 | $s$ |
|   | 1 | +1 | $\pm\frac{1}{2}$ |   |   |
|   |   | 0 | $\pm\frac{1}{2}$ | 6 | $p$ |
|   |   | −1 | $\pm\frac{1}{2}$ |   |   |
|   | 2 | +2 | $\pm\frac{1}{2}$ |   |   |
|   |   | +1 | $\pm\frac{1}{2}$ |   |   |
|   |   | 0 | $\pm\frac{1}{2}$ | 10 | $d$ |
|   |   | −1 | $\pm\frac{1}{2}$ |   |   |
|   |   | −2 | $\pm\frac{1}{2}$ |   |   |
| 4 | 0 | 0 | $\pm\frac{1}{2}$ | 2 | $s$ |
|   | 1 | +1 | $\pm\frac{1}{2}$ |   |   |
|   |   | 0 | $\pm\frac{1}{2}$ | 6 | $p$ |
|   |   | −1 | $\pm\frac{1}{2}$ |   |   |
|   | 2 | +2 | $\pm\frac{1}{2}$ |   |   |
|   |   | +1 | $\pm\frac{1}{2}$ |   |   |
|   |   | 0 | $\pm\frac{1}{2}$ | 10 | $d$ |
|   |   | −1 | $\pm\frac{1}{2}$ |   |   |
|   |   | −2 | $\pm\frac{1}{2}$ |   |   |
|   | 3 | +3 | $\pm\frac{1}{2}$ |   |   |
|   |   | +2 | $\pm\frac{1}{2}$ |   |   |
|   |   | +1 | $\pm\frac{1}{2}$ |   |   |
|   |   | 0 | $\pm\frac{1}{2}$ | 14 | $f$ |
|   |   | −1 | $\pm\frac{1}{2}$ |   |   |
|   |   | −2 | $\pm\frac{1}{2}$ |   |   |
|   |   | −3 | $\pm\frac{1}{2}$ |   |   |

According to the *aufbau* principle, we should expect to fill in the 1s state first, then proceed to the 2s, then to the 2p, and so on. Thus, for hydrogen, which has one electron, we should expect the electron to go into the s state of the first shell giving the electron configuration $1s^1$. For helium, with two electrons, the electron configuration would be expected to be $1s^2$, and for lithium, the electron configuration would be $1s^2 2s^1$. In a like manner, it should be possible to write the electron configuration for each of the known elements.

In an attempt to relate the electron configuration of the elements to

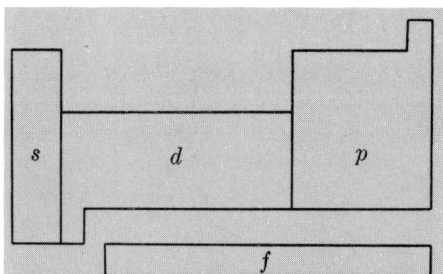

**FIGURE 3-14**

Division of the Long Form of the Periodic Table in Terms of Blocks of $s$, $p$, $d$, and $f$ Electrons.

the periodic table, it is seen that the long form of the table can be divided into the four general sections shown in Figure 3–14. The region represented by $s$ is found to contain a block of two elements in each period, whereas the $p$ block contains six elements, the $d$ block 10 elements, and the $f$ block 14 elements. This, of course, leads one to suspect a link between the periodic table and the $s$, $p$, $d$, and $f$ electron states since they contain 2, 6, 10, and 14 electrons, respectively. In fact, since the periodic table is based on experiment, it should be a guide to the order in which the electrons fill in the successive elements.

If we assume that these blocks are more than coincidence, we can begin

**TABLE 3–4.    Electron Configurations and Term Symbols of Atoms in Their Normal States.**

| Atomic Number | Element | Electron Configuration | Ground State | Atomic Number | Element | Electron Configuration | Ground State |
|---|---|---|---|---|---|---|---|
| 1 | H | $1s$ | $^2S_{1/2}$ | 16 | S | — $3s^23p^4$ | $^3P_2$ |
| 2 | He | $1s^2$ | $^1S_0$ | 17 | Cl | — $3s^23p^5$ | $^2P_{3/2}$ |
| 3 | Li | [He] $2s$ | $^2S_{1/2}$ | 18 | Ar | — $3s^23p^6$ | $^1S_0$ |
| 4 | Be | — $2s^2$ | $^1S_0$ | 19 | K | [Ar] $4s$ | $^2S_{1/2}$ |
| 5 | B | — $2s^22p$ | $^2P_{1/2}$ | 20 | Ca | — $4s^2$ | $^1S_0$ |
| 6 | C | — $2s^22p^2$ | $^3P_0$ | 21 | Sc | — $3d4s^2$ | $^2D_{3/2}$ |
| 7 | N | — $2s^22p^3$ | $^4S_{3/2}$ | 22 | Ti | — $3d^24s^2$ | $^3F_2$ |
| 8 | O | — $2s^22p^4$ | $^3P_2$ | 23 | V | — $3d^34s^2$ | $^4F_{3/2}$ |
| 9 | F | — $2s^22p^5$ | $^2P_{3/2}$ | 24 | Cr | — $3d^54s$ | $^7S_3$ |
| 10 | Ne | — $2s^22p^6$ | $^1S_0$ | 25 | Mn | — $3d^54s^2$ | $^6S_{5/2}$ |
| 11 | Na | [Ne] $3s$ | $^2S_{1/2}$ | 26 | Fe | — $3d^64s^2$ | $^5D_4$ |
| 12 | Mg | — $3s^2$ | $^1S_0$ | 27 | Co | — $3d^74s^2$ | $^4F_{9/2}$ |
| 13 | Al | — $3s^23p$ | $^2P_{1/2}$ | 28 | Ni | — $3d^84s^2$ | $^3F_4$ |
| 14 | Si | — $3s^23p^2$ | $^3P_0$ | 29 | Cu | — $3d^{10}4s$ | $^2S_{1/2}$ |
| 15 | P | — $3s^23p^3$ | $^4S_{3/2}$ | 30 | Zn | — $3d^{10}4s^2$ | $^1S_0$ |

## TABLE 3–4 (cont.)

| Atomic Number | Element | Electronic Configuration | Ground State | Atomic Number | Element | Electronic Configuration | Ground State |
|---|---|---|---|---|---|---|---|
| 31 | Ga | — $3d^{10}4s^24p$ | $^2P_{1/2}$ | 69 | Tm | — $4f^{13}6s^2$ | $^2F_{7/2}$ |
| 32 | Ge | — $3d^{10}4s^24p^2$ | $^3P_0$ | 70 | Yb | — $4f^{14}6s^2$ | $^1S_0$ |
| 33 | As | — $3d^{10}4s^24p^3$ | $^4S_{3/2}$ | 71 | Lu | — $4f^{14}5d6s^2$ | $^2D_{3/2}$ |
| 34 | Se | — $3d^{10}4s^24p^4$ | $^3P_2$ | 72 | Hf | — $4f^{14}5d^26s^2$ | $^3F_2$ |
| 35 | Br | — $3d^{10}4s^24p^5$ | $^2P_{3/2}$ | 73 | Ta | — $4f^{14}5d^36s^2$ | $^4F_{3/2}$ |
| 36 | Kr | — $3d^{10}4s^24p^6$ | $^1S_0$ | 74 | W | — $4f^{14}5d^46s^2$ | $^5D_0$ |
| 37 | Rb | [Kr] $5s$ | $^2S_{1/2}$ | 75 | Re | — $4f^{14}5d^56s^2$ | $^6S_{5/2}$ |
| 38 | Sr | — $5s^2$ | $^1S_0$ | 76 | Os | — $4f^{14}5d^66s^2$ | $^5D_4$ |
| 39 | Y | — $4d5s^2$ | $^2D_{3/2}$ | 77 | Ir | — $4f^{14}5d^76s^2$ | $^4F_{9/2}$ |
| 40 | Zr | — $4d^25s^2$ | $^3F_2$ | 78 | Pt | — $4f^{14}5d^96s$ | $^3D_3$ |
| 41 | Nb | — $4d^45s$ | $^6D_{1/2}$ | 79 | Au | [ ] $6s$ | $^2S_{1/2}$ |
| 42 | Mo | — $4d^55s$ | $^7S_3$ | 80 | Hg | — $6s^2$ | $^1S_0$ |
| 43 | Tc | — $4d^55s^2$ | $^6S_{5/2}$ | 81 | Tl | — $6s^26p$ | $^2P_{1/2}$ |
| 44 | Ru | — $4d^75s$ | $^5F_5$ | 82 | Pb | — $6s^26p^2$ | $^3P_0$ |
| 45 | Rh | — $4d^85s$ | $^4F_{9/2}$ | 83 | Bi | — $6s^26p^3$ | $^4S_{3/2}$ |
| 46 | Pd | — $4d^{10}$ | $^1S_0$ | 84 | Po | — $6s^26p^4$ | $^3P_2$ |
| 47 | Ag | — $4d^{10}5s$ | $^2S_{1/2}$ | 85 | At | — $6s^26p^5$ | $^2P_{3/2}$ |
| 48 | Cd | — $4d^{10}5s^2$ | $^1S_0$ | 86 | Rn | — $6s^26p^6$ | $^1S_0$ |
| 49 | In | — $4d^{10}5s^25p$ | $^2P_{1/2}$ | 87 | Fr | [Rn] $7s$ | $^2S_{1/2}$ |
| 50 | Sn | — $4d^{10}5s^25p^2$ | $^3P_0$ | 88 | Ra | — $7s^2$ | $^1S_0$ |
| 51 | Sb | — $4d^{10}5s^25p^3$ | $^4S_{3/2}$ | 89 | Ac | — $6d7s^2$ | $^2D_{3/2}$ |
| 52 | Te | — $4d^{10}5s^25p^4$ | $^3P_2$ | 90 | Th | — $6d^27s^2$ | $^3F_2$ |
| 53 | I | — $4d^{10}5s^25p^5$ | $^2P_{3/2}$ | 91 | Pa | — $5f^26d7s^2$ | $^4K_{11/2}$ |
| 54 | Xe | — $4d^{10}5s^25p^6$ | $^1S_0$ | 92 | U | — $5f^36d7s^2$ | $^5L_6$ |
| 55 | Cs | [Xe] $6s$ | $^2S_{1/2}$ | 93 | Np | — $5f^57s^2$ | $^6H_{5/2}$ |
| 56 | Ba | — $6s^2$ | $^1S_0$ | 94 | Pu | — $5f^67s^2$ | $^7F_0$ |
| 57 | La | — $5d6s^2$ | $^2D_{3/2}$ | 95 | Am | — $5f^77s^2$ | $^8S_{7/2}$ |
| 58 | Ce | — $4f^26s^2$ | $^3H_4$ | 96 | Cm | — $5f^76d7s^2$ | $^9D_2$ |
| 59 | Pr | — $4f^36s^2$ | $^4I_{9/2}$ | 97 | Bk | — $5f^86d7s^2$ | $^8H_{17/2}$ |
| 60 | Nd | — $4f^46s^2$ | $^5I_4$ | 98 | Cf | — $5f^{10}7s^2$ | $^5I_8$ |
| 61 | Pm | — $4f^56s^2$ | $^6H_{5/2}$ | 99 | Es | — $5f^{11}7s^2$ | $^4I_{15/2}$ |
| 62 | Sm | — $4f^66s^2$ | $^7F_0$ | 100 | Fm | — $5f^{12}7s^2$ | $^3H_6$ |
| 63 | Eu | — $4f^76s^2$ | $^8S_{7/2}$ | 101 | Md | — $5f^{13}7s^2$ | $^2F_{7/2}$ |
| 64 | Gd | — $4f^75d6s^2$ | $^9D_2$ | 102 | No | — $5f^{14}7s^2$ | $^1S_0$ |
| 65 | Tb | — $4f^96s^2$ | $^6H_{15/2}$ | 103 | Lw | — $5f^{14}6d7s^2$ | $^2D_{5/2}$ |
| 66 | Dy | — $4f^{10}6s^2$ | $^5I_8$ | 104 | — | — | — |
| 67 | Ho | — $4f^{11}6s^2$ | $^4I_{15/2}$ | 105 | — | — | — |
| 68 | Er | — $4f^{12}6s^2$ | $^3H_6$ | | | | |

to construct the electron configuration for each of the elements. In Table 3–4, the presently accepted electron configurations for the elements are listed in terms of the $s$, $p$, $d$, and $f$ states. It is interesting to note that the order of filling is not exactly that which one might expect. It appears to be normal as far as element 18, where it is seen to be $1s^2\ 2s^2\ 2p^6\ 3s^2\ 3p^6$. We note that the $s$ state in the first shell and the $s$ and $p$ states of the second shell are completely filled. In the third shell, the $s$ and $p$ states are completed giving a total of 18 electrons. The next state that would be expected to fill is the $3d$ state. However, element number 19 is potassium and lies in the block of 2, which means that the next electron must be an $s$ electron rather than a $d$ electron. This may not be what we would like to see, but it is what nature demands. Thus, the electron configuration for potassium is $1s^2 \, 2s^2 \, 2p^6 \, 3s^2 \, 3p^6 \, 4s^1$. Element number 20 is calcium and it also lies in the $s$ block. Therefore, calcium will complete the $4s$ state giving the configuration $- - - 4s^2$. At this point we find that the next element is the first in a block of 10. This, of course, would indicate the beginning of a $d$ group of electrons. Since the $d$ electrons occur first in the third shell, this must be a $3d$ electron, and the electron configuration for scandium will be $1s^2 \, 2s^2 \, 2p^6 \, 3s^2 \, 3p^6 \, 4s^2 \, 3d^1$. Thus, it is seen that the simple order of filling does not hold and the $4s$ state fills before the $3d$ state. The four transition series are always one shell behind. And for the rare earth series, an even greater overlap exists. The third transition series actually starts with lanthanum, which has an electron configuration $- - - 6s^2 \, 5d^1$. After lanthanum, element 58, cerium, is the first of a group of 14 elements, which indicates that it is the first member of a group of elements filling the $f$ state. Since the $f$ electrons occur first in the fourth shell, the $4f$ state obviously has been skipped until this point. Thus, the electron configuration for cerium would then be expected to be $- - - 6s^2 \, 5d^1 \, 4f^1$. Actually, the $5d$ electron is found to drop down to the $f$ state giving the configuration $- - - 6s^2 \, 4f^2$. Nevertheless, the $f$ state is two shells behind before it begins to fill.

The explanation of this unexpected order of filling of the electron states lies in the splitting of the energy levels as more and more electrons are fed into the atom. This splitting is indicated in Figure 3–15, where it is seen that the $ns$ state overlaps the $(n-1)d$ state, and the $4f$ state is overlapped to an even greater extent.

## SIMILARITIES OF THE ELEMENTS

The original periodic tables were based upon the observed similarities in chemical and physical properties of certain groups of elements. The alkali metals, for instance, were listed in a single column because it was observed that the members of this family had quite surprisingly similar properties. An understanding of the source of this similarity in properties, however, had to wait for a satisfactory model of the atom. In terms of our present model, we can see from the electron configurations of the alkali metals,

Li  $1s^2 \, 2s^1$
Na  $- - \, 2s^2 \, 2p^6 \, 3s^1$
K   $- - - - - - - \, 3s^2 \, 3p^6 \, 4s^1$
Rb  $- - - - - - - - - - - \, 4s^2 \, 3d^{10} \, 4p^6 \, 5s^1$
Cs  $- - - - - - - - - - - - - - - - - - - \, 5s^2 \, 4d^{10} \, 5p^6 \, 6s^1$

that each member of the family has one electron in the *s* state of the outside shell, with a similar internal electron structure. It is this similarity in electron configuration that leads to the similarities in chemical and physical

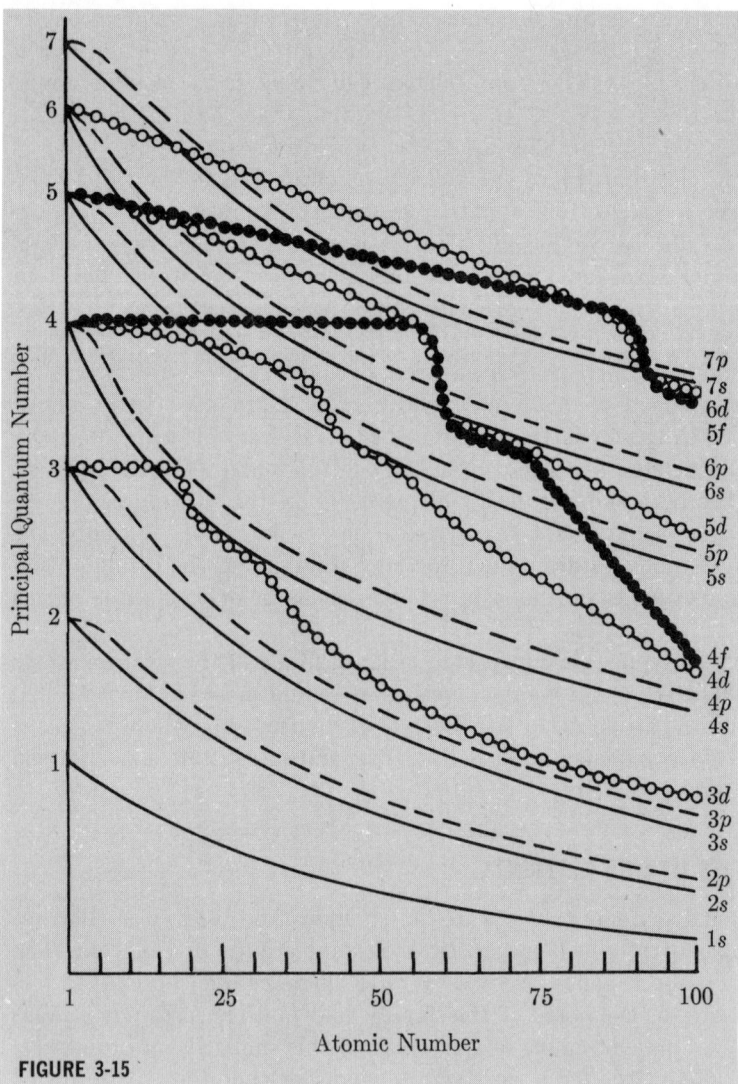

FIGURE 3-15

Energy Dependence of Atomic Orbitals as a Function of Atomic Number.

properties of the alkali metals. Needless to say, this same sort of similarity is found throughout the periodic table.

In addition to the vertical similarities, it is found that horizontal similarities exist for certain groups of elements. This is particularly pronounced for the lanthanide elements, but it is also quite important in the transition elements. For the lanthanides, we note that the electron configurations of the 14 members differ almost exclusively in the $4f$ subshell. If, for example, we consider the following four members with their electron configurations

Ce  $1s^2\ 2s^2\ 2p^6\ 3s^2\ 3p^6\ 4s^2\ 3d^{10}\ 4p^6\ 5s^2\ 4d^{10}\ 5p^6\ 6s^2\ 4f^2$
Pr  - - - - - - - - - - - - - - - - - - - - - - - $6s^2\ 4f^3$
Nd  - - - - - - - - - - - - - - - - - - - - - - - $6s^2\ 4f^4$
.                                                     .
.                                                     .
.                                                     .
Gd  - - - - - - - - - - - - - - - - - - - - - - - $6s^2\ 4f^7\ 5d^1$

the only variation that is noted is the successive increase in the occupancy of the $4f$ orbital and the one electron in the $5d$ state in the case of gadolinium. Inasmuch as these shells are buried somewhat below the valence shell, their contribution to the properties of the elements is quite small, and a strong horizontal similarity results.

## MODERN TRENDS IN PERIODIC TABLES

Realizing that the basis of the periodic system lies in the electron configurations of the elements, many authors have attempted to classify the elements in a manner that better emphasizes this fact.[6] Typical of this modern approach is the table proposed by Longuet-Higgins[7] shown in Figure 3–16. The important feature of the Longuet-Higgins table is the emphasis on the order of filling of electron states in relation to the order of increasing energy states. Visually this is convenient. However, it might be mentioned that, if the $s$, $p$, $d$, $f$ blocks are recognized, the long form of the periodic table tells essentially the same thing. Nevertheless, there are some desirable features in the Longuet-Higgins table just as there are desirable features in most of the hundreds of other tables that have been proposed. The very fact that so many tables have been proposed points out the variety of opinions that exist and at the same time indicates how difficult it would be to prescribe a best periodic table.

## THE TYPES OF ELEMENTS

Using electron configuration as the criterion, we ordinarily recognize four general types of elements; the *inert gas elements*, the *representative* or *main group elements*, the *transition elements*, and the *inner transition elements*. We would like to think that such a classification is definitive, but this is not the case. Based only on electron configuration, the most straightforward manner of classification is in terms of the $s$, $p$, $d$ and $f$ blocks shown in Figure 3–14. Except for the inert gases, which make up the last column

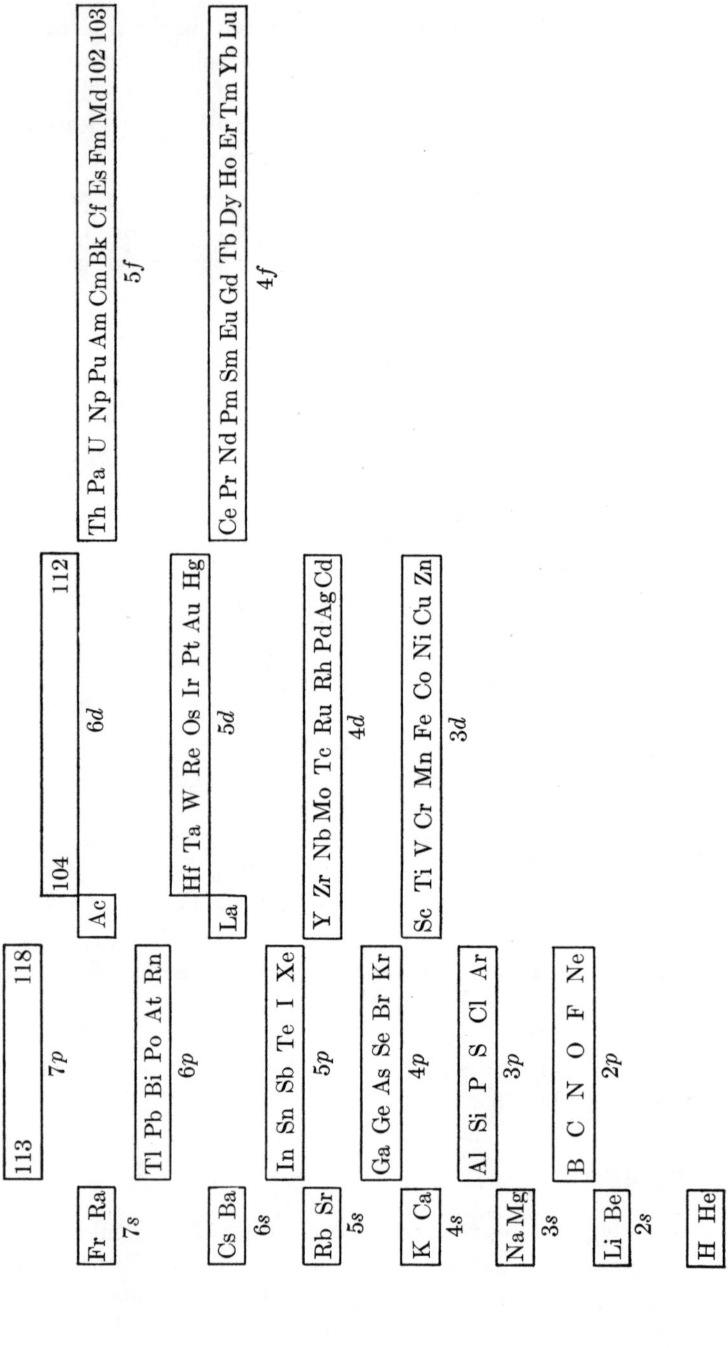

**FIGURE 3-16**

Periodic Table of Longuet-Higgins Based on Electron Configuration.

in the $p$ block, those elements falling in the $s$ or $p$ blocks are classified as representative elements. Those elements falling in the $d$ block are then classified as transition elements, and those falling in the $f$ block, inner transition elements.

Although such a classification is extremely convenient, it overlooks specific chemical characteristics of some of the elements. Here we must recognize that the classification of the elements according to the various types is ultimately for the purpose of emphasizing chemical and physical similarities. Thus the inert gases were originally grouped together and so designated because they are unique among the elements. But according to the classification we have used here, Zn, Cd, and Hg are considered to be transition elements in spite of the fact that chemically they are much closer to the representative elements.

An alternative approach to the classification of the elements into these groups is dependent on the extent to which the $s$, $p$, $d$, and $f$ orbitals are filled. That is, it is dependent on whether the particular subshells are complete or incomplete. By a complete subshell, we refer to the agreement between the number of electrons in the various shells of a given atom and the number in the corresponding shells of the inert gas of next higher or lower atomic number. It is on this basis that we will discuss the individual types of elements.

### Inert Gas Elements

With the exception of helium, which has the configuration $1s^2$, all of the members of this group of elements have completed $s$ and $p$ orbitals. Thus, in general, they may be characterized by the configuration $ns^2np^6$. This is the smallest class of elements, having but six members. Until recently, all of these elements were considered to be inert, but as a consequence of the work of Bartlett[3] reported in 1962, compounds of krypton, xenon, and radon have now been made. In spite of the synthesis of these compounds we must yet recognize the extreme stability of these elements due to the high stability associated with the completed $s$ and $p$ orbitals. As a consequence of the loss of inertness of these heavier inert gases, changes in the family name have been proposed and we find them referred to as the noble gases, rare gases, $M8$ elements, and aerogens. All of these are undoubtedly valid, but relative to the other elements, it would seem that the name inert is still justified.

### Representative Elements

The members of this group of elements have all their occupied subshells filled except their outer electron shell. Thus, any atom with an outer electron configuration from $ns^1$ to $ns^2np^5$ belongs in this class. There are 46 members in this group if strict adherence to the above electronic arrangement is maintained. This includes the copper and zinc families as well as ytterbium and nobelium. Of these, it may reasonably be argued that only the zinc family should be placed in this classification. And this point is emphasized in the long form of the periodic table shown in Figure 3–8, where the zinc family is designated as $R2'$, a modification of the family designations proposed by Sanderson.[9] However, at the same time, Cu, Ag, and Au in their $1+$ oxidation states have a completely filled $d$ subshell and therefore do show representative-type behavior. But this is not true in their higher oxidation states. It might also be argued that ytterbium and nobelium should be included among the representative elements, but in terms of chemical and physical properties they rightly belong with the inner transition elements.

The chemical behavior of the members of this class will be determined to a large extent by the tendency for these atoms to gain, lose, or share electrons in such a manner as to attain the electronic configuration of the inert gas of next higher or lower atomic number, or the so-called *pseudo-inert gas* configuration $(n - 1)s^2p^6d^{10}$. Many of the metals, and all of the nonmetals and metalloids make up the group of representative elements.

## Transition Elements

The transition elements have their two outer principal quantum shells incomplete and are characterized specifically by their incomplete $(n - 1)d$ subshell. Because of the extra stability which is associated with empty, half-filled, and filled subshells, there are some apparent anomalies in electronic arrangements in the transition series. This empirical rule is illustrated by the chromium and copper configurations in the first $d$ series of elements:

|       | Sc | Ti | V | Cr | Mn | Fe | Co | Ni | Cu | Zn |
|-------|----|----|---|----|----|----|----|----|----|----|
| $3d$  | 1  | 2  | 3 | 5  | 5  | 6  | 7  | 8  | 10 | 10 |
| $4s$  | 2  | 2  | 2 | 1  | 2  | 2  | 2  | 2  | 1  | 2  |

There are four transition series corresponding to unfilled $3d$, $4d$, $5d$, and $6d$ orbitals. The series begin with the group T3 elements Sc, Y, La, and Ac, and the first three series end, by our definition, with Ni, Pd, and Pt, respectively. Again, if we adhere strictly to the electron configuration we have defined for this group of elements, lutetium and lawrencium must also be classified as transition elements. It can be argued that chemical similarity justifies keeping lutetium and lawrencium with the inner transition elements. But it can be equally well-argued that chemical similarity justifies their inclusion with the transition elements, and the electron configurations would certainly place them among the transition elements. This point has recently been made by Sanderson in his recommended form of the periodic table.[9]

The elements in this class show striking resemblances to each other, particularly in their physical properties. The oxidation states are very numerous, compounds are highly colored, and coordination compounds are the rule whereas simple compounds are the exception.

## Inner Transition Elements

By definition, these elements are transition series elements, but they have an additional feature in their electronic arrangement which justifies setting them apart from the other transition elements. In this group, the outer three principal quantum levels are incomplete due to incomplete $(n - 2)f$ orbitals. In general, the configuration for this class may be given as $(n - 2)f^{1-14}$ $(n - 1)s^2p^6d^{1 \text{ or } 0}ns^2$. The empty, half-filled, and filled sublevel stability rule holds here also, and indeed this is invoked to explain the oxidation states other than 3 among the lanthanides. The outstanding feature of these elements is the great similarity in chemical and physical properties which they display.

There can be no question that a second inner transition series exists in which $5f$ orbitals are being occupied. However, there is some question as to where this series actually begins, that is, where $5f$-type electrons first appear. The primary difficulty in attempting to assign electrons to specific orbitals in the elements just beyond actinium lies in the closeness in energy between the $5f$ and $6d$ orbitals. The energy released in chemical bond formation is sufficient to transfer electrons between these high levels. By analogy with the $4f$ series, the first $5f$ electron should appear in thorium. However, many properties of this element indicate that it should be placed in family T4, below hafnium, rather than below cerium in family T3. Likewise, on the basis of many of their properties, protactinium and uranium would appear to fit better into families T5 and T6 respectively, rather than below praeseodymium and neodymium. Yet there is now substantial spectral and chemical evidence which supports the contention that the elements beyond actinium do form a second rare-earth series, and that the first appearance of $5f$ electrons is in protactinium.

One fact seems certain with regard to the second inner transition series and that is that, as in the other transition series, the relative energy position of the level which is being filled becomes lower as the successive electrons are added. By the time neptunium, plutonium and the subsequent members of the series are reached, the $5f$ shell seems clearly to be of lower energy than the $6d$ shell.

In conclusion, it should be apparent that a strict classification of the elements in terms of electron configuration is difficult if not impossible. Consequently, several different classifications have arisen. Each has its strengths as well as its weaknesses. It would seem, then, that the adoption of a given mode of classification is not as important as is an appreciation of the basic principles on which these classifications have been constructed.

## References

1. W. Tilden, "A Short History of the Progress of Scientific Chemistry," Longmans, Green and Co., Ltd., London, 1899.
2. W. Taylor, *J. Chem. Ed.*, **26**, 491 (1949).
3. L. Foster, *J. Chem. Ed.*, **16**, 409 (1939).
4. W. Luder, *J. Chem. Ed.*, **20**, 21 (1943).
5. H. Moseley, *Phil. Mag.*, **26**, 1024 (1913), **27**, 703 (1914).
6. G. Quam and M. Quam, *J. Chem. Ed.*, **11**, 27, 217, 288 (1934).
7. H. Longuet-Higgins, *J. Chem. Ed.*, **34**, 30 (1957).
8. N. Bartlett, *Proc. Chem. Soc.*, **218**, (1962).
9. R. T. Sanderson, *J. Chem. Ed.*, **41**, 187 (1964).

## Suggested Supplementary Reading

W. Tilden, "A Short History of the Progress of Scientific Chemistry," Longmans, Green, and Co., Ltd., London, 1899.

E. G. Mazurs, "Types of Graphical Representations of the Periodic System of Chemical Elements," E. Mazurs, 65 Madison Ave., La Grange, Illinois, 1957.

R. Rich, "Periodic Correlation," W. A. Benjamin, Inc., New York, 1965.

J. Main Smith, "Chemistry and Atomic Structure," Ernest Benn Ltd., London, 1924.

R. T. Sanderson, "Chemical Periodicity," Reinhold Publishing Corporation, New York, 1960; and "Inorganic Chemistry," Reinhold Publishing Corporation, New York, 1967.

R. T. Sanderson, *J. Chem. Ed.*, **41**, 187 (1964).

## Problems

1. Write out the electron configurations for:
   (a) Cu     (d) Yb
   (b) Cr     (e) Tl
   (c) Eu     (f) Th

2. The configurations of the first four elements considered in problem 1 are irregular. This is attributed to the particular stability of a filled and a half-filled subshell. Now note the configurations of the elements in the second and the third transition series as well as the remainder of the lanthanide series. Do these support the argument presented for this irregularity?

3. In the light of our present understanding, should the periodic table be constructed in terms of the properties of the elements, in terms of electron configurations, or in terms of a combination of these?

4. Give the arguments for and against classifying copper, silver, and gold with the representative elements.

5. How many electrons can there be in a $g$ state and in an $h$ state? Why are these not necessary to describe the ground states of the known elements?

6. Present three different means of classifying the elements and give a general electron configuration for each type of element in each of these classifications.

7. Compare the following periodic tables and consider the advantages of each:
   (a) Long form
   (b) Longuett-Higgins table — *J. Chem. Ed.*, **34**, 30 (1957)
   (c) Simmons table — *J. Chem. Ed.*, **24**, 588 (1947); 658 (1948)
   (d) Sanderson table — *J. Chem. Ed.*, **41**, 187 (1964)
   (e) Griff spiral representation — *J. Chem. Ed.*, **41**, 191 (1964)

# 4

# Periodic Properties

From the discussion of the periodic table, it is evident that those properties which depend primarily upon the detailed electron configuration of an atom will vary periodically with atomic number. On the other hand, those properties which depend upon the total number of electrons, and these are much fewer in number, will show no such variations. However, it should be pointed out before we discuss some of the more important periodic properties that there are many cases where the more significant similarities among elements occur not within a periodic family (vertically), but either within a period (horizontally) or in a diagonal relationship between elements in adjacent families.

Thus, for example, we find that Fe, Co, and Ni resemble one another both chemically and physically much more than each resembles its own congeners. In fact it is generally observed that among the transition elements *horizontal* similarities are often more pronounced than *vertical* similarities, which is not too surprising when we recall that the differentiating electrons move in the penultimate $(n - 1)$ quantum level. However, this observation must be carefully qualified to indicate that the horizontal similarities predominate in the $3d$ series whereas with the $4d$ and $5d$ series the vertical similarities are paramount. The most striking example of the latter situation may be found with Zr and Hf which, because of their extreme chemical similarity, always occur together in nature and have caused perhaps greater separation problems for inorganic chemists than any other given pair of elements. The $4d$-$5d$ similarity of properties is sustained, but with decreasing degree, all across the transition series of elements. Horizontal similarities are especially evident, indeed reach their zenith, with the inner transition elements, where the differentiating electrons are buried deeply within the atom at the $(n - 2)$ quantum level and, for the most part, exert only secondary effects. These comments do not hold as rigorously for the elements Ac through Am, where the $5f$ and $6d$ energy levels are not only in close proximity but are nearer to the "surface" of the atoms.

The first short-period elements, C, N, and O, share the important capacity for strong $p_\pi$-$p_\pi$ multiple bond formation not found in their congeners, and their relatively high electronegativities and small atomic radii also serve to set them quite apart from the remaining members of their families. Certain *diagonal relationships* stand out in the early non-transition element families. They are of particular prominence with Li and Mg, Be and Al, and B and Si, but exist also to a lesser extent for Na and Ca, K and Sr, etc. Where similarities are most pronounced, they can usually be accounted for by either similar sizes (for example, $Li^+$ and $Mg^{2+}$ radii are 0.60 and 0.65 Å, respectively) or similar charge-to-radius ratios (estimated for $Be^{2+}$ and $Al^{3+}$ at 6.5 and 6.0, respectively).

A great many properties of both elements and their compounds exhibit an unmistakable periodicity based upon electron configuration. Some of these are listed in Table 4–1, a few of which may be singled out as being more im-

## TABLE 4–1.  Properties Dependent upon Electronic Configuration.

| | |
|---|---|
| Atomic radius (and volume) | Heat of solvation of ions |
| Ionic radius | Hardness |
| Density | Malleability |
| Ionization energy | Coefficient of expansion |
| Affinity energy | Optical spectrum |
| Electronegativity | Magnetic behavior |
| Melting point | Thermal conductivity |
| Boiling point | Electrical resistance |
| Standard redox potential | Ion mobility |
| Valence (and oxidation number) | Parachor |
| Compressibility | Refractive index |
| Heats of fusion, vaporization | Heat of formation of a given |
| and sublimation | compound type |
| Bond energy | |

portant and more influencing than the others and thus of great value when it comes to trying to explain and predict chemical behavior of the elements. We suggest that these are the "simple" properties of *ionization energies, affinity energies, oxidation states,* and *atomic and ionic radii;* and the "complex" (or composite) properties of *electronegativities, bond energies,* and *redox potentials.* The remainder of this chapter is devoted to a more or less detailed consideration of several of these properties. However, no special significance should be attached either to the order in which they are discussed or to the length of the discussions. Redox potentials and their periodicity are considered in Chapter 8.

Finally it is perhaps appropriate to point out that there is a current tendency in some advanced-level textbooks and in the teaching of inorganic chemistry toward a de-emphasis of periodic properties *per se*, Rich's[1] and Sanderson's[2] fine monographs notwithstanding. Instead, a separate consideration of the several important periodic properties is presented out of the context of periodicity. We do not question the merits of this trend, but we still feel that there is at least as much merit in the unified approach to periodic properties presented here.

## ATOMIC AND IONIC RADII

The periodicity and fundamental importance of the size property has been recognized for a very long time. In fact it was the strikingly periodic *atomic volume* curve of Lothar Meyer, shown in Figure 3–7, that brought him more fame than his periodic table based on the physical properties of the elements. That the gram-atomic volume, determined by simply dividing the atomic weight by the density, reflects periodicity as well as it does is somewhat surprising when it is realized that the density of an element is a function of such varied factors as physical state, allotropic modification, temperature, and type of crystal structure. For example, to calculate atomic volumes should we take the density of tin to be 7.31 (white form) or 5.75 (gray form), that of carbon to be 3.51 (diamond) or 2.25 (graphite)? Hence it is in terms of *radii* of atoms and ions that the size property is now considered.

Since it is presently impossible to measure the radius of an isolated atom or ion, we must concern ourselves instead with internuclear distances measured in crystals or in gaseous molecules. From these distances we must derive radii in some suitable way. Since all radii are calculated, it is important to know what factors will affect our calculations. These factors include such more or less intangibles as the bond order, the degree of ionic, covalent, or metallic character (that is, bond type), the oxidation states of the bonded neighbors, and the detailed crystal or molecular structure. Thus we find that no single type of atomic radius is completely satisfactory, and we must consequently define and deal with several kinds of radii.

*Bonded* radii may be broadly classed as either *ionic* (crystal) or *atomic*. Atomic radii may then be subdivided into either *metallic*, as in metals, alloys, or intermetallic compounds, or *covalent*, as in the nonmetals and in covalent molecules in general. The covalent radii are sometimes further identified as *tetrahedral* or *octahedral*, etc., and of course, we must distinguish *single-bond* radii from *double-bond* and *triple-bond* radii. However, when multiple bonding is present, the concept of a radius loses much of its significance. Under these conditions the atom can be considered to be grossly distorted from sphericity, and it is more appropriate to deal here only with internuclear distances. In fact this is generally done when we consider molecules that have square planar, trigonal bipyramidal, or any other geometry that does not yield a regular polyhedron. Two additional kinds of radii, both taken to be invariant and related to atomic or covalent radii, are the *Bragg-Slater atomic radii*, to be discussed later, and *orbital* radii, which are defined by the radial distribution functions of the outermost orbitals.

Van der Waals radii can be considered to be nonbonded radii. These arise from the distances between atoms in solids and liquids in which the atoms are in closest proximity to one another but not actually bonded through either ionic, covalent, or metallic bonds. The Van der Waals radii for the atoms in covalent compounds are essentially identical with the univalent or bivalent anionic radii for these atoms.

### Covalent Radii

The *covalent single-bond radius* of an atom is taken as one-half the distance between the nuclei of two like atoms forming a single covalent bond. The postulate of the additivity of these radii holds quite well as long as the bonds considered are mainly covalent and have no appreciable multiple bond character. For example, the C—C bond length in diamond and in a large number of organic molecules is found to be $1.54 \pm 0.01$ Å, leading to a value of 0.77 Å for the covalent radius. Similarly, we find the Si atomic radius to be 1.17 Å, from which we can calculate a C—Si single-bond distance to be 1.94 Å. And this is the experimentally found bond length in SiC or $(CH_3)_4Si$. Although the results of assuming additivity are gratifying here and in many other similar cases (Table 4–2), it is observed that

**TABLE 4–2. Some Calculated and Experimental Bond Lengths, Å.**

| Bond | Atomic Radii | Calculated Sum | Experimental Bond Length | Difference[a] |
|------|------|------|------|------|
| C—I  | 0.77 + 1.35 | 2.12 | 2.14 | −0.02 |
| C—Br | 0.77 + 1.14 | 1.91 | 1.94 | −0.03 |
| C—Cl | 0.77 + 1.00 | 1.77 | 1.76 | +0.01 |
| C—F  | 0.77 + 0.72 | 1.49 | 1.36 | +0.13 |
| Si—I  | 1.17 + 1.35 | 2.52 | 2.44 | +0.08 |
| Si—Br | 1.17 + 1.14 | 2.31 | 2.16 | +0.15 |
| Si—Cl | 1.17 + 1.00 | 2.17 | 2.02 | +0.15 |
| Si—F  | 1.17 + 0.72 | 1.89 | 1.56 | +0.33 |
| B—Br | 0.85 + 1.14 | 1.99 | 1.88 | +0.11 |
| B—Cl | 0.85 + 1.00 | 1.85 | 1.72 | +0.13 |
| B—F  | 0.85 + 0.72 | 1.57 | 1.29 | +0.28 |
| S—Br | 1.04 + 1.14 | 2.18 | 2.27 | −0.09 |
| S—Cl | 1.04 + 1.00 | 2.04 | 1.99 | +0.05 |
| S—F  | 1.04 + 0.72 | 1.76 | 1.56 | +0.20 |
| Cl—Br | 1.00 + 1.14 | 2.14 | 2.14 | 0.0 |

[a] Calculated sum minus experimental bond length.

additivity becomes a poorer approximation in direct proportion to (a) the electronegativity difference between the bonded atoms (or in other words with increasing *ionicity* of the bond) and (b) the multiplicity of the bond. The former dependency has been expressed by an empirical relation proposed in 1941 by Schomaker and Stevenson:

$$(4-1) \qquad r_{AB} = r_A + r_B - 0.09 \, |\chi_A - \chi_B|$$

where $\chi_A$ and $\chi_B$ are the Pauling electronegativity values for elements A and B. There is no similar simple relationship between bond distances and bond multiplicity, although the bond shortening with increasing bond order is well-documented by numerous examples (Table 4–3). It is also

**TABLE 4–3.  Some Multiple Bond Lengths, Force Constants, and Bond Energies.**

| Bond | Bond Length, Å | Stretching Force Constant[a] | Bond Energy, kcal/mole |
|---|---|---|---|
| C—C | 1.54 | ~ 5.0 | 83 |
| C=C | 1.34 | ~ 9.6 | 146 |
| C≡C | 1.20 | ~16 | 200 |
| N—N | 1.45 | — | 38 |
| N=N | 1.24 | — | 100 |
| N≡N | 1.10 | 22.4 | 226 |
| O—O | 1.48 | — | 33 ($H_2O_2$) |
| O=O | 1.21 | 11.4 | 118 ($O_2$) |
| O≡O | 1.12 ($O_2^+$) | — | — |
| C—N | 1.47 | ~ 4.7 | 73 |
| C=N | 1.34 | — | 147 |
| C≡N | 1.14 ($CN^-$) | ~ 18 | 213 |
| C—O | 1.43 | ~ 4.5 | 84 |
| C=O | 1.22 | ~ 12 | ~177 |
| C≡O | 1.13 (CO) | 18.4 | 256 |
| C—S | 1.81 (EtSH) | — | 65 |
| C=S | 1.55 ($CS_2$) | — | 128 |
| N—O | 1.36 | — | — |
| N=O | 1.22 | ~15.5 | — |
| N≡O | 1.06 | — | — |

[a] In dyne/cm × $10^5$. This is also affected by environment.

instructive to compare the bond length in a covalent bond with the *force constant* and the *bond* or *dissociation energy*, as well as with its *bond order* as shown in Table 4–3. In general, the covalent radius for a nonmetal atom coincides with its atomic radius, but the covalent radius for a metal atom is invariably shorter than its atomic (metallic) radius. For example, consider the following atomic and covalent radii:

|                | K     | Ba   | La   | Cr   | Pd   | In   |
|----------------|-------|------|------|------|------|------|
| Metallic radius | 2.31  | 2.17 | 1.88 | 1.59 | 1.38 | 1.62 |
| Covalent radius | 2.025 | 1.98 | 1.69 | 1.45 | 1.28 | 1.50 |

where the consistency of this observation is illustrated for several metals. Periodic trends in covalent radii will be considered after we have examined some of the other kinds of radii.

## Ionic Radii

Although it is a relatively simple experimental matter to measure the internuclear distances in ionic crystals, the determination of that portion of the internuclear distance contributed by the cation and that portion by the anion is not so simple or even obvious. Indeed certain assumptions must be made before the concept of an ionic radius can be meaningful. These assumptions include (a) the existence of ions in the solid state, (b)

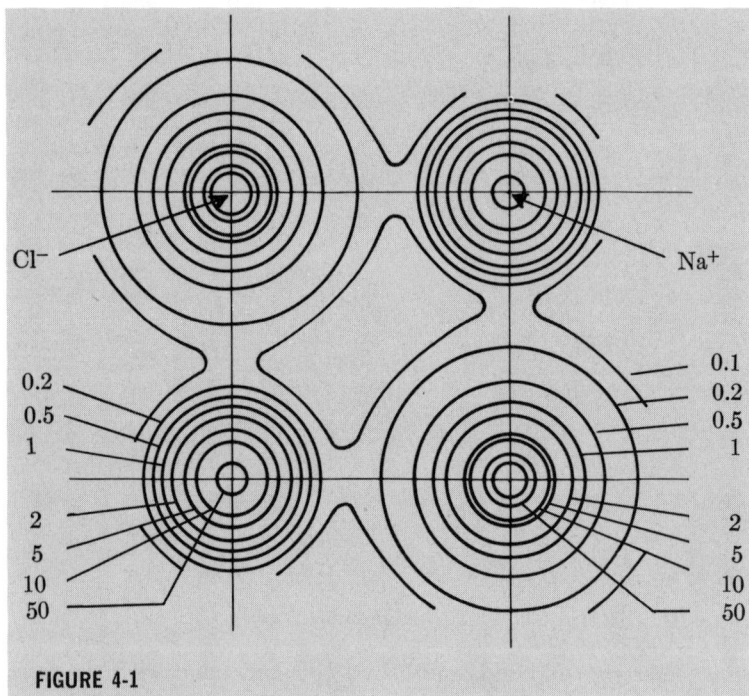

FIGURE 4-1

An electron density map of part of the cubic face of sodium chloride. The electron density, in electrons/$\text{Å}^3$, is constant along each of the contour lines and increases, as shown by the figures, toward the nuclei of the ions.

the possibility of correctly apportioning internuclear distances between ions, and (c) the constancy (the additivity) of the radii.

Although the assumption of the existence of ions in the solid state would probably not be questioned by the experienced chemist, the uninitiated have a right, if not an obligation, to examine this point. Indirect evidence, such as the presence of ions in molten salts and in conducting solutions and the many successful crystal energy calculations based on the presence of ions, is admittedly quite impressive. But the *direct* evidence comes only from *electron density maps*, which are available for only a small number of ionic crystals studied by sophisticated X-ray crystallography. These maps give us not only the relative nuclear positions, but also the *electron charge density* (ED) in the regions surrounding the nuclei. Figure 4–1 shows such an electron density contour map for NaCl, in which each line represents a constant electron density (in electrons/$\text{Å}^3$). The circle nearest the Cl nucleus represents a constant charge density of about 170 electrons/$\text{Å}^3$, and that nearest the Na nucleus represents about 70 electrons/$\text{Å}^3$. It is seen that there is a charge density extending continuously from one atom to the next, but the ED drops off to about 0.2 electrons/$\text{Å}^3$ at the outer "edges," and is of course even lower in the unmarked internuclear regions. If we integrate the ED inward from some arbitrary point of low ED, we find $\sim$10.05 electrons around the sodium nucleus and $\sim$17.70 electrons around the chlorine nucleus, giving the most direct solid state evidence of electron transfer to produce the ions $Na^+$ (10 $e^-$) and $Cl^-$ (18 $e^-$). The $\sim$0.25 electrons unaccounted for must be assumed to be in the internuclear space not included in the somewhat arbitrary integration. The region of minimum ED may be considered to define a boundary between two spherical ions that contain most (but by no means all) of the charge. In this way the following *electron density map radii* have been found:

| Salt | Cation | Anion |
| --- | --- | --- |
| NaCl | 1.18 | 1.64 |
| KCl | 1.45 | 1.70 |
| LiF | 0.92 | 1.09 |
| CaF$_2$ | 1.26 | 1.10 |
| MgO | 1.02 | 1.09 |
| CuCl | 1.10 | 1.25 |
| CuBr | 1.10 | 1.36 |
| NiO | 0.94 | 1.15 |

If these radii are compared with the metallic or covalent radii and the ionic radii determined by the usual procedures, it will be seen that they fall between the two sets of radii but much closer to the apparent ionic radii (Table 4–4). An exception is Br, for which the ED map radius derives from CuBr rather than an R1 metal bromide. In this respect we can note also the smaller ED radius for Cl found in CuCl. A much lower ionicity in the Cu(I) compounds is certainly not unexpected. It is reasonable to

**TABLE 4–4.  Comparison of Various Types of Radii.**

| Atom | Atomic Radius | Covalent Radius | ED Map Radius | Apparent Ionic Radius | Difference Between Last Two Columns |
|------|------|------|------|------|------|
| Li | 1.52 | 1.34 | 0.92 | 0.60 | 0.32 |
| Na | 1.86 | 1.54 | 1.18 | 0.95 | 0.23 |
| K | 2.27 | 1.96 | 1.45 | 1.33 | 0.12 |
| Mg | 1.60 | — | 1.02 | 0.65 | 0.37 |
| Ca | 1.97 | — | 1.26 | 0.99 | 0.27 |
| F | 0.71 | 0.71 | 1.10 | 1.36 | 0.26 |
| Cl | 1.00 | 1.00 | $1.64(Na^+)$ | 1.81 | 0.17 |
|  |  |  | $1.70(K^+)$ |  | 0.09 |
|  |  |  | $1.25(Cu^+)$ |  | 0.56 |
| Br | 1.14 | 1.14 | $1.36(Cu^+)$ | 1.95 | 0.59 |

assume that the nearness of the ED map radii to the apparent ionic radii is some kind of measure of the amount of ionic character or ionicity in the bonds. See, for example, the last column in Table 4–4, where the differences between ED map radii and the Pauling ionic radii reflect the expected ionicity trends, namely, $K^+ > Na^+ > Li^+ \gg Cu^+$; and $Cl^- > Br^-$.

We turn our attention now to the second and third assumptions we made earlier. These are that it is possible to correctly apportion inter-nuclear distances between ions and that because these radii are at least roughly additive, permanent tables of ionic radii may be established. There are essentially only two approaches for obtaining ionic radii, one empirical, the other semiempirical, and we shall very briefly outline these without the details, which may be found elsewhere (for example, Pauling[3]).

The earliest useful values for ionic radii were those obtained in 1920 by Landé from the assumption that in the lithium halide crystals the halogen ions are in mutual contact. Then in 1926, Goldschmidt, using more empirical data, particularly from crystals he considered highly ionic, and Wasastjerne's values of 1.33 Å for $F^-$ and 1.32 Å for $O^{2-}$, revised and greatly extended the radii data to over 80 ions. Pauling later suggested that Goldschmidt's values would be improved if he had taken 1.40 Å for $O^{2-}$, and this is generally accepted today as are Goldschmidt's values for empirical ionic radii. One reason for Pauling's suggestion was that the radii will then agree more closely with the set of semiempirical radii derived by him.

Pauling begins with the observed interionic distances for the five (standard) ionic compounds NaF, KCl, RbBr, CsI, and $Li_2O$. For the first four, most of the factors affecting ionic size are held constant since the ions in each are isoelectronic and univalent and the radius ratios are all roughly 0.75. Pauling then assumes that the size of an ion is inversely proportional to the *effective nuclear charge* operative on its outermost electrons. The effective nuclear charge (see page 139) is taken to be equal to the actual nuclear charge minus the screening constant, $\sigma$, of the other electrons in the ion. A complete set of screening constant values was obtained by Pauling from both theoretical considerations as well as molar refraction and X-ray term values. Thus we may write the equation for the radii of a sequence of isoelectronic ions as

$$(4\text{-}2) \qquad r_{ion} = \frac{C_n}{Z - \sigma}$$

in which $C_n$ is a constant depending only upon the principal quantum number of the outermost electrons.

For each of the five standard crystals chosen by Pauling we then can write two equations with two unknowns. The first of these,

$$(4\text{-}3) \qquad r_c + r_a = \text{observed internuclear distance}$$

is an experimental expression, and the second

$$(4\text{-}4) \qquad \frac{r_c}{r_a} = \frac{Z_a - \sigma}{Z_c - \sigma}$$

is derived from his semitheoretical considerations. Simultaneous solution of these equations leads to the following ionic radii values (in Å):

| Na$^+$ | K$^+$ | Rb$^+$ | Cs$^+$ |
|--------|-------|--------|--------|
| 0.95   | 1.33  | 1.48   | 1.69   |

| F$^-$ | Cl$^-$ | Br$^-$ | I$^-$ |
|-------|--------|--------|-------|
| 1.36  | 1.81   | 1.95   | 2.16  |

A value of 0.60 Å was selected for Li$^+$ to agree with the observed Li$^+ - $O$^{2-}$ distance of 2.00 Å in Li$_2$O and the best empirical value for oxide ion of 1.40 Å.

Using Eq. (4-2) and the values for $C_n$ determined for the standard alkali halides, it is possible to obtain radii for all ions having the inert gas electron configuration. However, the radii determined in this way for all *polyvalent* ions correctly represent only their sizes relative to those for the alkali and halide ions, and are not values whose sums would yield equilibrium interionic distances. These relative values are termed *univalent radii*, and represent the radii that polyvalent ions would possess if they were to retain their electron distributions but enter into ionic interactions as if they were univalent. Fortunately it is possible to obtain the physically significant crystal radii, $r_c$, for polyvalent ions from the univalent radii, $r_u$, by multiplication by a factor derivable from the Born equation (see page 159):

$$(4\text{-}5) \qquad \frac{r_c}{r_u} = z^{-2/(n-1)}$$

where $z$ is the valence of the ion and $n$ is the Born exponent discussed on page 159.

It is either Goldschmidt or Pauling ionic radii which we find tabulated and used most commonly today, and since the values are quite comparable we shall use the Pauling values unless otherwise indicated.

### Bragg-Slater and Orbital Radii

The derivation of empirical atomic and ionic radii dates at least from 1920, when Bragg pointed out that internuclear distances in crystals could be regarded approximately as the sums of radii. He established a set of radii whose sums reproduced observed internuclear distances in several hundred crystals, both ionic and metallic, with an average deviation of

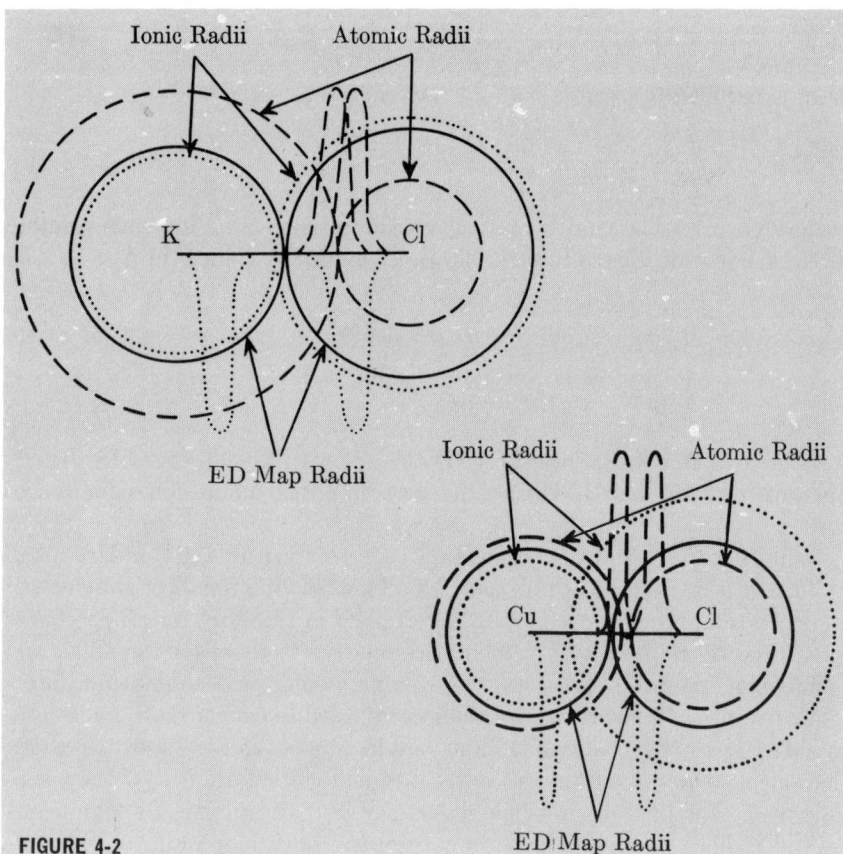

**FIGURE 4-2**

The line connecting metal and nonmetal atom represents the experimental internuclear distance, 3.14 Å for KCl and 2.35 Å for CuCl. The circles represent three different kinds of radii, atomic, ionic, and ED map, as labeled. The principal maxima of the radial charge densities for the outermost electrons of the atoms (pointing upward) and of the ions (pointing downward) are also sketched in at the appropriate positions. Note that the former are half-filled orbitals, hence their close approach, whereas the latter are filled orbitals, hence their large separation.

~0.06 Å. A great many, mostly minor, modifications of the radii have been made since 1920 (see, for example, the excellent general account by Pauling[3]) with the names Born, Lande, Wyckoff, Huggins, Wasastjerne, Goldschmidt, Pauling, Sherman, and Zachariasen standing out among many in this area. However, some of the primary results of the historical development, as we have already seen, have been (a) the establishment of different kinds of radii to describe different sorts of crystals and bonding situations, (b) the detailing of rules for relating one radius to another in order to obtain good agreement with experimental data, and (c) the general introduction of a profusion and confusion of terms and definitions.

In an attempt to simplify and clarify this situation Slater[4] has recently noted that ionic radii for the highly *electropositive* elements are about 0.85 Å *smaller* than the atomic radii, and those for the highly *electronegative* elements are about 0.85 Å *larger* than the atomic radii. Thus, while the differences are not exactly constant, being only close to ±0.85 Å, it is possible to understand how it happens that the sum of ionic radii for an electropositive and electronegative element gives almost the exact same result as the sum of their *atomic radii* (see Table 4–5 and Figure 4–2). Therefore, Slater advocates returning to Bragg's simple scheme of having a unique radius for each type of atom, to be used in every sort of com-

**TABLE 4–5.  Bragg-Slater and Pauling Ionic Radii for Some Ions.**

| Ion | Bragg-Slater Atomic Radius | Pauling Ionic Radius | Difference |
|-----|---------------------------|----------------------|------------|
| $Li^+$ | 1.45 | 0.60 | 0.85 |
| $Na^+$ | 1.80 | .95 | .85 |
| $K^+$ | 2.20 | 1.33 | .87 |
| $Rb^+$ | 2.35 | 1.48 | .87 |
| $Cs^+$ | 2.60 | 1.69 | .91 |
| $Cu^+$ | 1.35 | 0.96 | .39 |
| $Be^{2+}$ | 1.05 | 0.31 | .74 |
| $Mg^{2+}$ | 1.50 | 0.65 | .85 |
| $Ca^{2+}$ | 1.80 | 0.99 | .81 |
| $Sr^{2+}$ | 2.00 | 1.13 | .87 |
| $Ba^{2+}$ | 2.15 | 1.35 | .80 |
| $F^-$ | 0.50 | 1.36 | −0.86 |
| $Cl^-$ | 1.00 | 1.81 | −.81 |
| $Br^-$ | 1.15 | 1.95 | −.80 |
| $I^-$ | 1.40 | 2.16 | −.76 |
| $O^{2-}$ | 0.60 | 1.40 | −.80 |
| $S^{2-}$ | 1.00 | 1.84 | −.84 |
| $Se^{2-}$ | 1.15 | 1.98 | −.83 |
| $Te^{2-}$ | 1.40 | 2.21 | −.81 |

Example: $Rb^+ + Br^- = 2.35 + 1.15 = 3.50$ atomic radii sum
$1.48 + 1.95 = 3.43$ ionic radii sum

pound. He has compared the sums of his atomic radii, as recorded in Table 4–6, with the observed internuclear distances in over 1200 compounds

## TABLE 4–6.   Bragg-Slater Atomic Radii of the Elements.

| H | Li–F col | Na–Cl col | K–Br col | Rb–I col | Cs–At col | Fr–Am col |
|---|---|---|---|---|---|---|
| | | | | | Cs 2.60 | Fr |
| | | | | | Ba 2.15 | Ra 2.15 |
| | | | | | La 1.95 | Ac 1.95 |
| | | | | | Ce 1.85 | Th 1.80 |
| | | | | | Pr 1.85 | Pa 1.80 |
| | | | | | Nd 1.85 | U 1.75 |
| | | | | | Pm 1.85 | Np 1.75 |
| | | | K 2.20 | Rb 2.35 | Sm 1.85 | Pu 1.75 |
| | | | Ca 1.80 | Sr 2.00 | Eu 1.85 | Am 1.75 |
| | | | Sc 1.60 | Y 1.80 | Gd 1.80 | |
| | | | Ti 1.40 | Zr 1.55 | Tb 1.75 | |
| | | | V 1.35 | Nb 1.45 | Dy 1.75 | |
| | Li 1.45 | Na 1.80 | Cr 1.40 | Mo 1.45 | Ho 1.75 | |
| | Be 1.05 | Mg 1.50 | Mn 1.40 | Tc 1.35 | Er 1.75 | |
| | B 0.85 | Al 1.25 | Fe 1.40 | Ru 1.30 | Tu 1.75 | |
| H 0.25 | C 0.70 | Si 1.10 | Co 1.35 | Rh 1.35 | Yb 1.75 | |
| | N 0.65 | P 1.00 | Ni 1.35 | Pd 1.40 | Lu 1.75 | |
| | O 0.60 | S 1.00 | Cu 1.35 | Ag 1.60 | Hf 1.55 | |
| | F 0.50 | Cl 1.00 | Zn 1.35 | Cd 1.55 | Ta 1.45 | |
| | | | Ga 1.30 | In 1.55 | W 1.35 | |
| | | | Ge 1.25 | Sn 1.45 | Re 1.35 | |
| | | | As 1.15 | Sb 1.45 | Os 1.30 | |
| | | | Se 1.15 | Te 1.40 | Ir 1.35 | |
| | | | Br 1.15 | I 1.40 | Pt 1.35 | |
| | | | | | Au 1.35 | |
| | | | | | Hg 1.50 | |
| | | | | | Tl 1.90 | |
| | | | | | Pb 1.80 | |
| | | | | | Bi 1.60 | |
| | | | | | Po 1.90 | |
| | | | | | At | |

of all types of crystals and molecules, and he finds the average deviation to be just 0.12 Å.

Before we attempt to justify the Bragg-Slater radii and to explain why we can get along satisfactorily with only one set of radii if we are only interested in reproducing internuclear distances, it will be instructive to introduce first still another kind of radius. This is a radius defined by the principal maximum in the radial distribution function, $r^2\psi_i^2(r)$, of the outermost orbital and shown in Figure 4–2. The values for these radii given in Table 4–7 are from Waber and Cromer.[5] For comparison purposes the corresponding Bragg-Slater radii are also tabulated. It is perhaps of importance to point out that for cations obtained by removal of all outer quantum shell electrons ($Na^+$, $Mg^{2+}$, $Sc^{3+}$, etc.) the decrease in the distance

**TABLE 4-7.** Radii (in Å) of the principal maxima of the outer orbitals of the elements.[a]

| | 1s | 2s | 2p | 3s | 3p | 3d | 4s | Bragg-Slater Radius |
|---|---|---|---|---|---|---|---|---|
| H | 0.53 | | | | | | | 0.25 |
| He | 0.29 | | | | | | | — |
| Li | 0.19 | 1.59 | | | | | | 1.45 |
| Be | 0.14 | 1.04 | | | | | | 1.05 |
| B | 0.11 | 0.771 | 0.78 | | | | | 0.85 |
| C | 0.090 | 0.62 | 0.60 | | | | | 0.70 |
| N | 0.078 | 0.52 | 0.49 | | | | | 0.65 |
| O | 0.068 | 0.45 | 0.41 | | | | | 0.60 |
| F | 0.060 | 0.40 | 0.36 | | | | | 0.50 |
| Ne | 0.054 | 0.35 | 0.32 | | | | | — |
| Na | | | 0.28 | 1.71 | | | | 1.80 |
| Mg | | | 0.25 | 1.28 | | | | 1.50 |
| Al | | | 0.22 | 1.04 | 1.31 | | | 1.25 |
| Si | | | 0.20 | 0.90 | 1.07 | | | 1.10 |
| P | | | 0.18 | 0.80 | 0.92 | | | 1.00 |
| S | | | 0.17 | 0.72 | 0.81 | | | 1.00 |
| Cl | | | 0.16 | 0.66 | 0.72 | | | 1.00 |
| Ar | | | 0.15 | 0.61 | 0.66 | | | — |
| K | | | | | 0.59 | | 2.16 | 2.20 |
| Ca | | | | | 0.54 | | 1.69 | 1.80 |
| Sc | | | | | 0.50 | 0.54 | 1.57 | 1.60 |
| Ti | | | | | 0.47 | 0.49 | 1.48 | 1.40 |

[a] Reference 5 may be consulted for the values for elements beyond Ti. Underlined values are simply the largest values for the specific atoms.

from the nucleus to the maximum in the radial distribution function is by a factor of approximately 1.5 to 5.0. On the other hand, for an anion formed by putting one or more electrons in the same quantum shell, this distance does not change by any appreciable amount. This may be seen from the following radii for the indicated subshells:

| | Atom | Ion |
|---|---|---|
| Na | 1.71 (3s) | 0.28 (2p) |
| K | 2.16 (4s) | 0.59 (3p) |
| Mg | 1.28 (3s) | 0.25 (2p) |
| Ca | 1.69 (4s) | 0.54 (3p) |
| Sc | 1.57 (4s) | 0.49 (3p) |
| F | 0.40 (2s) | 0.40 (2s) |
| Cl | 0.72 (3p) | 0.74 (3p) |
| Br | 0.85 (4p) | 0.87 (4p) |
| I | 1.04 (5p) | 1.06 (5p) |

Now we are ready to examine more carefully Figure 4–2, in which we have placed the nuclei of the two compounds KCl and CuCl at their respective experimental internuclear distances and drawn circles to represent atomic radii, ionic radii, and ED map radii. In addition, the principal maxima of the radial charge densities for the outermost electrons of the atoms (pointing upward) and of the ions (pointing downward) are sketched in the figure.

We may now try to interpret the difference between atomic and ionic radii and at the same time seek an understanding of why the sums of these two kinds of radii lead to roughly the same internuclear distance for a given pair of atoms. First the following points should be noted. (1) The sum of the radii, whether they be atomic, ionic, or ED map radii, gives the experimental internuclear distance very well for KCl, but only approximately for CuCl. In the latter case, the sum of the atomic radii is a much better approximation than that of the ionic radii. (2) The ionic radii are much closer to the ED map radii for KCl than for CuCl, as already documented earlier in Table 4–4. In each case the map radii fall between the extremes represented by the atomic and ionic radii. (3) The maxima in the radial functions for the outermost electrons nearly coincide for the "combined atoms" (covalent extreme) case, but they are far apart for the "combined ions" (ionic extreme) case.

Thus we discover that atomic radii should be used where atoms are bonded to one another by a covalent bond or a metallic bond, which are essentially of the same nature. The covalent bond depends upon the overlapping of charge in the *incompletely* filled valence shells of the bonded atoms. The overlapping should be a maximum, forming thereby a strong covalent bond, if the maximum charge densities of the valence shells of the two atoms coincide. That is, covalent bonding occurs when the atoms approach each other to the point where each atomic radius is roughly the distance to the radius of maximum radial charge density.

Ionic radii are used when we assume that the outer electron is completely removed from the electropositive atom and placed in the valence shell of the electronegative atom. Using KCl for the convenience of discussion, we have seen that the $4s$ electron of K has an orbital radius of 2.16 Å, whereas the orbital radius of the electron core remaining after its removal to form the $K^+$ is only 0.59 Å. On the other hand, we have seen that the addition of a single electron to the all-but-completed $3p$ subshell of Cl, producing $Cl^-$, causes only a negligibly small increase ($\sim 0.02$ Å) in the orbital radius of this shell. The equilibrium internuclear distance is then determined by the balance of Coulombic attraction between oppositely charged ions and the repulsion due to the closed electronic shells. Thus in the ionic extreme it is necessary only for the outer tails of the radial wave functions to overlap in order to produce sufficient repulsion of the inert gas closed shells to balance the Coulombic attractions. This is far less overlapping than is required for the formation of a covalent bond, where the two radial maxima should nearly coincide. Of course the difference between the two extremes lies in the fact that for the covalent bond there are vacancies in the valence shells of the bonded atoms, and electron sharing can occur, whereas with the filled inert gas shells the exclusion principle strictly forbids overlapping.

In summary, we see that the ionic radius extends well out into the tail of the wave function of the valence electrons, whereas the atomic radius extends out only to the radius of maximum charge density. Thus, the two kinds of radii, atomic and ionic, measure quite different things, and yet they are quite compatible. It would appear that the "best" or "truest" radii are ED map radii, but these are never used as they are tedious to obtain, very few have been measured, and they vary from compound to compound. And, if we are only interested in internuclear distances rather than radii, specific types of radii are not really needed. Indeed, for the latter purpose the Bragg-Slater radii appear to be quite satisfactory.

## PERIODIC TRENDS OF ATOMIC AND IONIC RADII

Vertical size trends in the periodic table are easy to describe, and their explanation is for the most part simple and straightforward. In the non-transition element families, atomic and ionic radii increase with increasing atomic number. The largest percentage of this increase always occurs between the lightest two members, and the smallest percentage generally occurs between the heaviest two members. The addition of electrons to a new quantum shell is simply not compensated for by increased nuclear charge. The addition of increasing numbers of electrons with poorer shielding capacity, such as the $d$ and later the $f$ electrons, appears to be the primary cause of the progressive falling off of the percentage size increase down a family. In the transition element families, size increases are much less pronounced. For example, compare the 44% atomic radius increase from Ge (1.22) to Pb (1.75) with the mere 9% increase from V (1.31) to Ta (1.43). Even this does not tell the whole story since none of the latter percentage increase occurs from Nb to Ta, but rather all of it comes between V and Nb (1.43). Of much greater chemical significance are the analogous trends in ionic radii, for example $Ge^{4+}$ (0.53) to $Pb^{4+}$ (0.84) compared with $Ti^{4+}$ (0.68) to $Hf^{4+}$ (0.78). $Sn^{4+}$ has the intermediate radius of 0.71 whereas the $Zr^{4+}$ value is 0.79 Å! An acceptable explanation for the unexpected near-identical radii for $4d$ and $5d$ congeners is usually given in terms of the nature and number of the intervening electrons and the fact that only one additional quantum shell has been populated. Thus, between $Ti^{4+}$ and $Zr^{4+}$ 18 electrons have been added, 10 of which are poorly shielding $d$ electrons, but nevertheless the addition of a new shell does cause a slight size increase. However, between $Zr^{4+}$ and $Hf^{4+}$ 32 electrons have appeared, and among these are not only an additional 10 $d$-type, but more important 14 $f$-type, which are the poorest-shielding electrons of all. Other analogous vertical trends may be seen from Figure 4-3 and Table 4-8.

Often of more interest and definitely more difficult to account for are

**TABLE 4-8. Isoelectronic Series of Ionic and van der Waals or Nonbonded Radii.**

| Outer electron configuration | | | | | | | | | | | | |
|---|---|---|---|---|---|---|---|---|---|---|---|---|
| $1s^2$ | $H^-$ | $He$ | $Li^+$ | $Be^{2+}$ | $B^{3+}$ | $C^{4+}$ | $N^{5+}$ | | | | | |
|  | >1.5, 1.2[a] | $1.2_8$[a] | 0.68 | 0.35 | 0.23 | 0.16 | 0.13 | | | | | |
| $2s^2 2p^6$ | $N$ | $O^{2-}$ | $F^-$ | $Ne$ | $Na^+$ | $Mg^{2+}$ | $Al^{3+}$ | $Si^{4+}$ | $P^{5+}$ | $S^{6+}$ | $Cl^{7+}$ | |
|  | 1.5[a] | 1.40 | 1.36 | $1.3_9$[a] | 0.97 | 0.67 | 0.52 | 0.42 | 0.35 | 0.30 | 0.27 | |
| $2s^2 2p^6 3s^2$ | | | | | | | $P^{3+}$ | $S^{4+}$ | $Cl^{5+}$ | | | |
|  | | | | | | | 0.44 | 0.37 | 0.33 | | | |
| $3s^2 3p^6$ | $P$ | $S^{2-}$ | $Cl^-$ | $Ar$ | $K^+$ | $Ca^{2+}$ | $Sc^{3+}$ | $Ti^{4+}$ | $V^{5+}$ | $Cr^{6+}$ | $Mn^{7+}$ | |
|  | 1.9[a] | 1.84 | 1.81 | $1.7_1$[a] | 1.33 | 0.99 | 0.81 | 0.68 | 0.59 | 0.52 | 0.46 | |
| $3s^2 3p^6 3d^{10}$ | | | | | $Cu^+$ | $Zn^{2+}$ | $Ga^{3+}$ | $Ge^{4+}$ | $As^{5+}$ | $Se^{6+}$ | | |
|  | | | | | 0.96 | 0.74 | 0.62 | 0.53 | 0.46 | 0.42 | | |
| $3s^2 3p^6 3d^{10} 4s^2$ | | | | | | $Ge^{2+}$ | $As^{3+}$ | $Se^{4+}$ | $Br^{5+}$ | | | |
|  | | | | | | 0.73 | 0.58 | 0.50 | 0.47 | | | |
| $4s^2 4p^6$ | $As$ | $Se^{2-}$ | $Br^-$ | $Kr$ | $Rb^+$ | $Sr^{2+}$ | $Y^{3+}$ | $Zr^{4+}$ | $Nb^{5+}$ | $Mo^{6+}$ | $Tc^{7+}$ | $Rb^{3+}$ |
|  | 2.0[a] | 1.98 | 1.95 | $1.8_6$[a] | 1.47 | 1.12 | 0.92 | 0.79 | 0.69 | 0.62 | 0.57 | 0.5 |
| $4s^2 4p^6 4d^{10}$ | | | | | $Ag^+$ | $Cd^{2+}$ | $In^{3+}$ | $Sn^{4+}$ | $Sb^{5+}$ | $Te^{6+}$ | $I^{7+}$ | |
|  | | | | | 1.26 | 0.97 | 0.81 | 0.71 | 0.62 | 0.56 | 0.50 | |
| $4s^2 4p^6 4d^{10} 5s^2$ | | | | | | $Sn^{2+}$ | $Sb^{3+}$ | $Te^{4+}$ | $I^{5+}$ | | | |
|  | | | | | | 0.93 | 0.76 | 0.70 | 0.62 | | | |
| $4f^0 5s^2 5p^6$ | $Sb$ | $Te^{2-}$ | $I^-$ | $Xe$ | $Cs^+$ | $Ba^{2+}$ | $La^{3+}$ | $Ce^{4+}$ | | | | |
|  | 2.2[a] | 2.21 | 2.16 | 2.0[a] | 1.67 | 1.34 | 1.14 | 0.94 | | | | |
| $4f^7 5s^2 5p^6$ | | | | | | $Eu^{2+}$ | $Gd^{3+}$ | $Tb^{4+}$ | | | | |
|  | | | | | | 1.09 | 0.97 | 0.81 | | | | |
| $4f^{14} 5s^2 5p^6$ | | | | | $Yb^{2+}$ | $Lu^{3+}$ | $Hf^{4+}$ | $Ta^{5+}$ | $W^{6+}$ | $Re^{7+}$ | $Os^{8+}$ | |
|  | | | | | 0.93 | 0.85 | 0.78 | 0.68 | 0.62 | 0.56 | 0.5 | |
| $5s^2 5p^6 5d^{10}$ | | | | | $Au^+$ | $Hg^{2+}$ | $Tl^{3+}$ | $Pb^{4+}$ | $Bi^{5+}$ | | | |
|  | | | | | 1.37 | 1.10 | 0.95 | 0.84 | 0.74 | | | |
| $5s^2 5p^6 5d^{10} 6s^2$ | | | | $Au^-$ | $Tl^+$ | $Pb^{2+}$ | $Bi^{3+}$ | | | | | |
|  | | | | 2.02 | 1.47 | 1.20 | 0.93 | | | | | |
| $5f^0 6s^2 6p^6$ | | $Po^{2-}$ | $At^-$ | $Rn$ | $Fr^+$ | $Ra^{2+}$ | $Ac^{3+}$ | $Th^{4+}$ | $Pa^{5+}$ | $U^{6+}$ | | |
|  | | 2.3 | $2.2_5$ | 2.2[a] | 1.8 | 1.43 | 1.18 | 1.02 | 0.9 | 0.80 | | |

[a] van der Waals radii only. Some radii from miscellaneous sources.
(This table taken from R. Rich, "Periodic Correlations," W. A. Benjamin, Inc., New York, 1965.)

the several horizontal trends which we may consider. However, isoelectronic series, sets of species having the same number and configuration of electrons, are relatively simple to understand and such data as is presented in Table 4–8 requires little explanation. Atomic radii vary horizontally as shown in Figure 4–3, and there are several features of the plots, which are

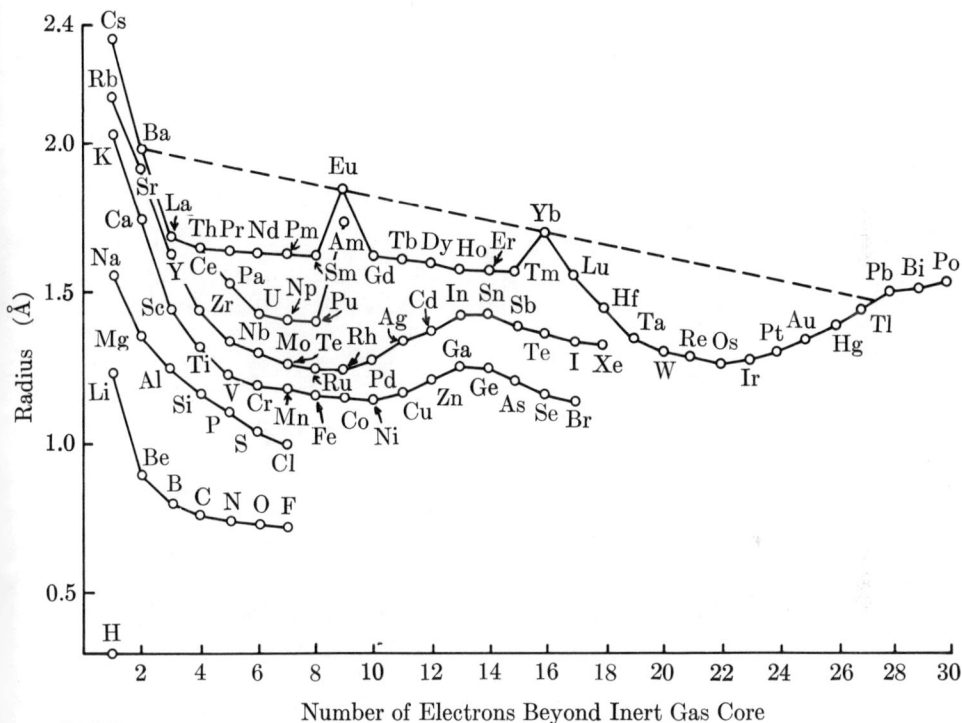

**FIGURE 4-3**

Metallic and covalent radii for most of the elements.

of atomic radii versus the number of electrons beyond the filled inert gas core, which are worthy of comment. The steep drop occurring immediately after the R1 elements is most likely a reflection of the very poor shielding that one $s$ electron affords the other in the $ns^2$ configuration, as well as the large penetrating power of an $s$-type electron. The subtle irregularities observed with the metallic radii of the transition metals, as well as with their ionic radii (see Figure 4–3 and Table 4–9), arise from the particular $d$-orbital occupancy. This is dealt with in more detail in Chapter 11. Ionic radii for *isovalent* ion series may also be compared in Table 4–9, and the trends require no special comment.

The ionic radii for the lanthanide and actinide series also recorded in Table 4–9 are quite regular and we speak of a *lanthanide* (and *actinide*)

**TABLE 4-9. Some Metallic and Ionic Radii.**

Number of Outer $s$, $d$, and $f$ Electrons

| 0 | 1 | 2 | 3 | 4 | 5 | 6 | 7 | 8 | 9 | 10 | 11 | 12 | 13 | 14 |
|---|---|---|---|---|---|---|---|---|---|----|----|----|----|----|
| — — | K 2.27 | Ca 1.97 | Sc 1.61 | Ti 1.45 | V 1.31 | Cr 1.25 | Mn 1.36 | Fe 1.24 | Co 1.25 | Ni 1.25 | | | | |
| — — | — — | Ti²⁺ 0.90 | V²⁺ 0.88 | Cr²⁺ 0.84 | Mn²⁺ 0.80 | Fe²⁺ 0.74 | Co²⁺ 0.73 | Ni²⁺ 0.69 | Cu²⁺ 0.72 | Zn²⁺ 0.74 | — — | Pb²⁺ 1.20 | | |
| Sc³⁺ 0.81 | Ti³⁺ 0.76 | V³⁺ 0.74 | Cr³⁺ 0.63 | Mn³⁺ 0.66 | Fe³⁺ 0.64 | Co³⁺ 0.63 | Ni³⁺ 0.62 | — — | — — | Ga³⁺ 0.62 | | | | |
| Zr⁴⁺ 0.79 | Nb⁴⁺ 0.74 | Mo⁴⁺ 0.69 | — — | Ru⁴⁺ 0.67 | — — | Pd⁴⁺ 0.69 | — — | — — | — — | Sn⁴⁺ 0.71 | | | | |
| Hf⁴⁺ 0.78 | — — | W⁴⁺ 0.70 | Re⁴⁺ 0.72 | Os⁴⁺ 0.69 | Ir⁴⁺ 0.68 | Pt⁴⁺ 0.65 | — — | — — | — — | Pb⁴⁺ 0.84 | | | | |
| La³⁺ 1.14 | Ce³⁺ 1.07 | Pr³⁺ 1.06 | Nd³⁺ 1.04 | Pm³⁺ 1.02 | Sm³⁺ 1.00 | Eu³⁺ 0.98 | Gd³⁺ 0.97 | Tb³⁺ 0.93 | Dy³⁺ 0.92 | Ho³⁺ 0.91 | Er³⁺ 0.89 | Tm³⁺ 0.87 | Yb³⁺ 0.86 | Lu³⁺ 0.85 |
| Ac³⁺ 1.18 | — — | — — | U³⁺ 1.12 | Np³⁺ 1.10 | Pu³⁺ 1.08 | Am³⁺ 1.07 | | | | | | | | |
| Th⁴⁺ 1.02 | Pa⁴⁺ 0.98 | U⁴⁺ 0.97 | Np⁴⁺ 0.95 | Pu⁴⁺ 0.93 | Am⁴⁺ 0.92 | | | | | | | | | |

*contraction* because of the chemical significance of this decrease on the properties of the post-lanthanide elements. The successive addition of $4f$ electrons to the $(n - 2)$ quantum shell does not quite counterbalance the opposing effect of increasing nuclear charge upon the outermost, *radius-determining*, electrons, namely, the $6s^2$ for the atoms and the $5s^2p^6$ for the trivalent ions.

A final point should be made with regard to ionic radii that may be obvious, but is seldom clearly explained. One often finds, as in Table 4–8, species written such as $Ca^{2+}$ and $S^{6+}$, and it should be made quite clear that these do not and chemically cannot mean the same thing. While the former ion may be obtained by the input of approximately 18 eV of ionization energy, the latter represents at the best lower energy approximation an outer electron configuration of $4(sp^3d^2)$, which is some 25 to 31 eV, in *promotion energy*, above the ground state of $4(s^2p^4d^0)$.

## IONIZATION ENERGY

The ionization energy, or ionization potential as it is generally less descriptively termed, is the minimum energy required to remove an electron from a free neutral gaseous atom in its lowest energy or ground state. More precisely, we have defined the *first* ionization energy, $I_1$. The *second*, $I_2$, *third*, $I_3$, etc., ionization energies would remove the most loosely bound electrons from the ground states of singly positive, doubly positive, etc., ions, respectively. Quite clearly, $I_1 < I_2 < I_3 < \cdots I_n$, where $n$ is the total number of electrons in the atom.

The ionization energy is one of the few fundamental properties of an isolated atom that can be measured directly. This fact alone gives it singular significance and stimulates our interest in seeking the major factors that influence its magnitude and the periodic trends, as well as apparent anomalies, it displays.

Each of the following factors, which may be singled out but which are actually strongly interrelated, is expected to contribute in a major way to the ionization energy: (1) the magnitude of the *effective nuclear charge* (see page 139), which is a function of the amount of nuclear shielding supplied by the underlying electrons, (2) the radial distance from the nucleus to the outermost, that is, most loosely bound electron (or more exactly, the radius of its maximum charge density), and (3) the extent to which the outermost electron penetrates the charge cloud set up by the lower-lying electrons.

With regard to the last effect, the degree of penetration of electrons in a given principal quantum level decreases in the order: $s > p > d > f$. This order corresponds also to the extent of binding of the various electrons in a given shell. Thus, an $ns$ electron has more of its charge density nearer the nucleus than an $np$, an $np$ more than an $nd$, etc. For example, the $3s^2 3p^6 3d^{10}$ configuration of the penultimate shell electrons do not screen or shield the nucleus for a $4s$ electron of a copper atom $(Z = 29)$ nearly as well as the $3s^23p^6$ configuration shields that same kind of electron, $4s$, in a potassium atom $(Z = 19)$. Hence the much higher effective nuclear charge of copper effects a sharp decrease in the radial maximum for the copper $4s$ electron, which falls at 1.19 Å, compared to that of the potassium $4s$ electron, which falls at 2.16 Å. This in turn contributes the major portion of the 3.38 eV additional energy required for the removal of the $4s$ electron in Cu compared to that for the $4s$ electron in K. That is, the respective ionization energies are $I_K = 4.34$ and $I_{Cu} = 7.72$ eV. So here we see that factors (1) and (2) are of paramount importance. It may be noticed that in fact it is the first factor, effective nuclear charge, which really controls the second factor, the radius of maximum charge density.

### Trends in Ionization Energies

As we look now at some of the various trends in ionization energies and try to rationalize and understand them, we must keep in mind the foregoing factors and their relative importance in the interplay of forces responsible for a given value of $I$. More subtle and less well understood factors such as the quantum mechanical exchange energy effects and spin (Pauli) and charge (Coulombic) correlation effects appear also to have some influence on ionization energies, but these are more difficult to assess and we shall all but ignore them in our rather qualitative discussion.

The periodicity of the ionization energy can readily be appreciated from the plot of $I_1$ versus atomic number shown in Figure 4–4. Table 4–10 records some ionization energy data for most of the elements. Successive decreases

### TABLE 4–10. Ionization Energies of the Elements.

| Atomic Number | Symbol | Ionization Energies (ev) | | | | | | | |
|---|---|---|---|---|---|---|---|---|---|
| | | I | II | III | IV | V | VI | VII | VIII |
| 1 | H | 13.595 | | | | | | | |
| 2 | He | 24.580 | 54.403 | | | | | | |
| 3 | Li | 5.390 | 75.619 | 122.420 | | | | | |
| 4 | Be | 9.320 | 18.206 | 153.850 | 217.657 | | | | |
| 5 | B | 8.296 | 25.149 | 37.920 | 259.298 | 340.127 | | | |
| 6 | C | 11.264 | 24.376 | 47.864 | 64.476 | 391.986 | 489.84 | | |
| 7 | N | 14.54 | 29.605 | 47.426 | 77.450 | 97.863 | 551.925 | 666.83 | |
| 8 | O | 13.614 | 35.146 | 54.934 | 77.394 | 113.873 | 138.080 | 739.114 | 871.12 |
| 9 | F | 17.418 | 34.98 | 62.646 | 87.23 | 114.214 | 157.117 | 185.139 | 953.60 |
| 10 | Ne | 21.559 | 41.07 | 63.5 | 97.16 | 126.4 | 157.91 | | |

| Atomic Number | Symbol | Ionization Energies | | Atomic Number | Symbol | Ionization Energies | |
|---|---|---|---|---|---|---|---|
| | | I | II | | | I | II |
| 11 | Na | 5.138 | 47.29 | 54 | Xe | 12.127 | 21.2 |
| 12 | Mg | 7.644 | 15.03 | 55 | Cs | 3.893 | 25.1 |
| 13 | Al | 5.984 | 18.823 | 56 | Ba | 5.210 | 10.001 |
| 14 | Si | 8.149 | 16.34 | 57 | La | 5.58 | 11.06 |
| 15 | P | 10.55 | 19.65 | 58 | Ce | 5.54 | 10.85 |
| 16 | S | 10.357 | 23.4 | 59 | Pr | 5.40 | 10.55 |
| 17 | Cl | 13.01 | 23.80 | 60 | Nd | 5.49 | 10.73 |
| 18 | Ar | 15.755 | 27.62 | 61 | Pm | 5.55 | 10.90 |
| 19 | K | 4.339 | 31.81 | 62 | Sm | 5.61 | 11.07 |
| 20 | Ca | 6.111 | 11.87 | 63 | Eu | 5.64 | 11.25 |
| 21 | Sc | 6.54 | 12.80 | 64 | Gd | — | 12.1 |
| 22 | Ti | 6.82 | 13.57 | 65 | Tb | 5.89 | 11.52 |
| 23 | V | 6.74 | 14.65 | 66 | Dy | 5.82 | 11.67 |
| 24 | Cr | 6.764 | 16.49 | 67 | Ho | 5.89 | 11.80 |
| 25 | Mn | 7.432 | 15.64 | 68 | Er | 5.95 | 11.93 |
| 26 | Fe | 7.90 | 16.18 | 69 | Tm | 6.03 | 12.05 |
| 27 | Co | 7.86 | 17.05 | 70 | Yb | 6.04 | 12.17 |
| 28 | Ni | 7.633 | 18.15 | 71 | Lu | 5.32 | 13.9 |
| 29 | Cu | 7.724 | 20.29 | 72 | Hf | 5.5 | 14.9 |
| 30 | Zn | 9.391 | 17.96 | 73 | Ta | 7.88 | 16.2 |
| 31 | Ga | 6.00 | 20.51 | 74 | W | 7.98 | 17.7 |
| 32 | Ge | 7.88 | 15.93 | 75 | Re | 7.87 | 16.6 |
| 33 | As | 9.81 | 20.2 | 76 | Os | 8.7 | 17 |
| 34 | Se | 9.75 | 21.5 | 77 | Ir | 9 | 17 |
| 35 | Br | 11.84 | 21.6 | 78 | Pt | 9.0 | 18.56 |
| 36 | Kr | 13.996 | 24.56 | 79 | Au | 9.22 | 20.5 |
| 37 | Rb | 4.176 | 27.5 | 80 | Hg | 10.434 | 18.751 |
| 38 | Sr | 5.692 | 11.027 | 81 | Tl | 6.106 | 20.42 |
| 39 | Y | 6.377 | 12.233 | 82 | Pb | 7.415 | 15.028 |
| 40 | Zr | 6.84 | 13.13 | 83 | Bi | 7.287 | 16.68 |
| 41 | Nb | 6.88 | 14.32 | 84 | Po | 8.43 | 19.4 |
| 42 | Mo | 7.10 | 16.15 | 85 | At | 9.5 | 20.1 |
| 43 | Tc | 7.28 | 15.26 | 86 | Rn | 10.745 | 21.4 |
| 44 | Ru | 7.36 | 16.76 | 87 | Fr | 3.83 | 22.5 |
| 45 | Rh | 7.46 | 18.07 | 88 | Ra | 5.277 | 10.144 |
| 46 | Pd | 8.33 | 19.42 | 89 | Ac | 6.9 | 12.1 |
| 47 | Ag | 7.574 | 21.48 | 90 | Th | 6.95 | 11.5 |
| 48 | Cd | 8.991 | 16.904 | 91 | Pa | — | — |
| 49 | In | 5.785 | 18.86 | 92 | U | 6.08 | — |
| 50 | Sn | 7.34 | 14.63 | 93 | Np | — | — |
| 51 | Sb | 8.639 | 16.5 | 94 | Pu | 5.8 | — |
| 52 | Te | 9.01 | 18.6 | 95 | Am | 6.0 | — |
| 53 | I | 10.454 | 19.09 | | | | |

down a given family roughly parallel the increasing atomic radii, as expected. However, just as the radius increase becomes more shallow down a family due to the increasing effective nuclear charge in this direction, so also is this behavior found with ionization energies. In fact, it even happens that the $I_1$ value may be slightly *greater* for the heaviest family member than for the next-lighter member, for example, Ba 5.21 and Ra

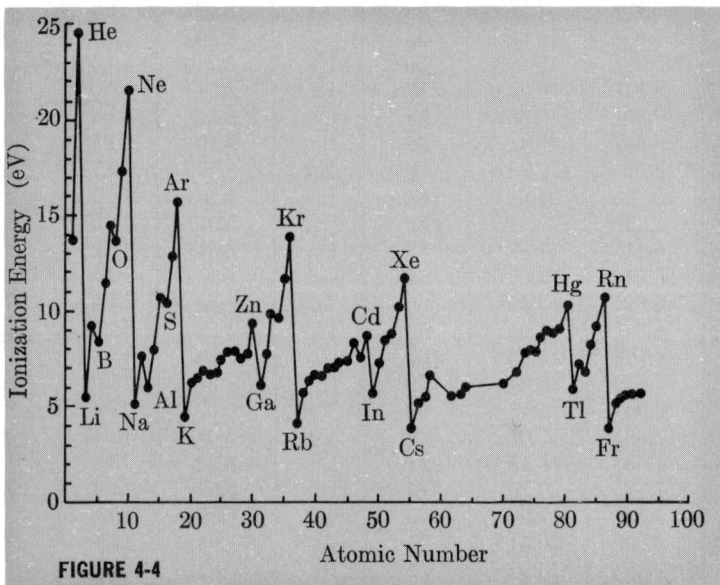

**FIGURE 4-4**

First ionization energies of the elements as a function of atomic number.

5.28; Sn 7.33 and Pb 7.42; Rh 7.5 and Ir 9; or all such values in R2′, R3, T6, T7, T8, T9, T10 and T11 families.

In the transition series of elements there are several vertical trends in ionization energies that are of relatively great importance to the understanding of the properties of these metals. However, before we can examine these trends intelligently we must first consider an important difference between the ionization process for a non-transition element and that for a transition element. When the former loses one electron the ground state of the resultant positive ion always has the (gross) electronic configuration of the ground state of the neutral atom of the preceding element, that is,

$$A, ns^{1 \; or \; 2} \rightarrow A^+, ns^{0 \; or \; 1} + s \text{ electron}$$

or

$$A, ns^2np^x \rightarrow A^+, ns^2p^{x-1} + p \text{ electron}$$

However, when a transition element atom loses an electron, the positive ion produced has an electron configuration *different from that of any neutral atom.* For example, consider the following:

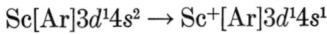

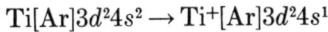

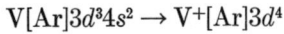

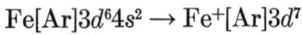

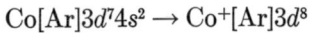

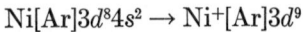

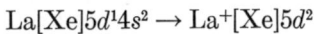

$$Sc[Ar]3d^14s^2 \rightarrow Sc^+[Ar]3d^14s^1$$

$$Ti[Ar]3d^24s^2 \rightarrow Ti^+[Ar]3d^24s^1$$

$$V[Ar]3d^34s^2 \rightarrow V^+[Ar]3d^4$$

$$Fe[Ar]3d^64s^2 \rightarrow Fe^+[Ar]3d^7$$

$$Co[Ar]3d^74s^2 \rightarrow Co^+[Ar]3d^8$$

$$Ni[Ar]3d^84s^2 \rightarrow Ni^+[Ar]3d^9$$

$$La[Xe]5d^14s^2 \rightarrow La^+[Xe]5d^2$$

For *all* three series of transition elements the loss of a *second* electron *always* produces a dipositive ion with a ground state configuration having only outer $d$ electrons remaining. These facts and indeed all other known data for the transition elements suggest the following two generalizations: (1) the rate of increase of binding energy and therefore ionization energy is greater for $(n-1)d$ electrons than for $ns$ electrons and (2) the $(n-1)d$ electrons become even more stable relative to $ns$ electrons with increasing nuclear charge.

The changes in binding energies of various types of electrons in several critical regions of the periodic table are illustrated in Figure 4–5. We see

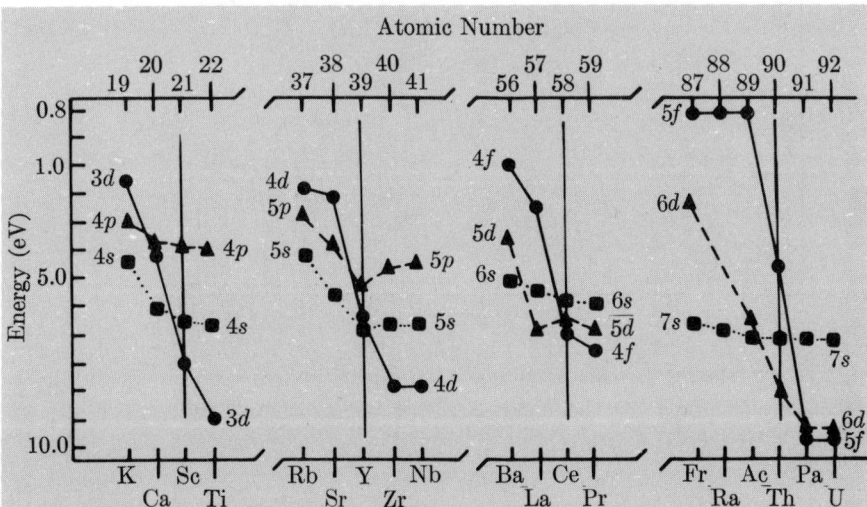

**FIGURE 4-5**

The change in binding energy of an electron with atomic number.

that the first electron to ionize from an atom of scandium is a $4s$ electron. (It should be obvious that we do not need to consider the $p$ orbitals for the ground state.) In the hypothetical process of "building the elements," by simultaneously adding a proton to the nucleus and an electron to the atom, it appears that the electron added between Ca ($4s^2$) and Sc ($3d^14s^2$) is a $d$ electron. But if we carry the process out stepwise as follows this is not really correct:

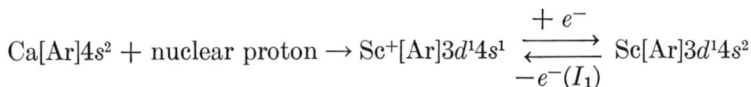

$$\text{Ca[Ar]}4s^2 + \text{nuclear proton} \rightarrow \text{Sc}^+\text{[Ar]}3d^14s^1 \underset{-e^-(I_1)}{\overset{+e^-}{\rightleftarrows}} \text{Sc[Ar]}3d^14s^2$$

The first indicated "reaction" is, of course, not possible to carry out in the laboratory. But the next, the reversible reaction, is experimentally accessible.

Of perhaps more chemical importance for the transition elements than $I_1$ alone is the sum $(I_1 + I_2)$, or the energy required to remove both $ns$ electrons. A plot of this sum versus $x$, the total number of electrons outside the rare gas core, is shown in Figure 4–6. Values for the $5d$ series elements

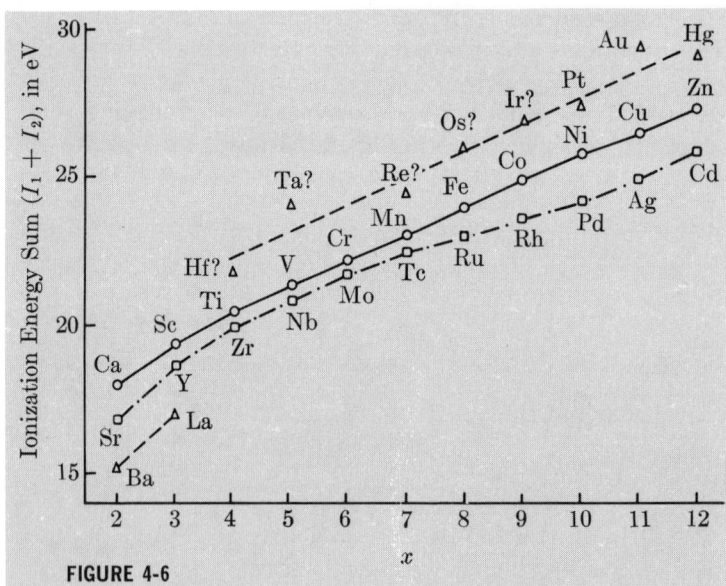

**FIGURE 4-6**

The energy required for the removal of the two $s$ electrons from the configuration $d^{x-2}s^2$ as $x$ goes from 2 to 12 in the three transition series.

are approximate only; and for the $3d$ series all values correspond exactly to $(I_1 + I_2)$ except for Cr and Cu, where the plotted values are less by an amount equivalent to the promotion energies required for electron promotion from $3d^{x-1}4s^1$ to $3d^{x-2}4s^2$. The former represents the ground state configuration for Cr ($x = 6$) and Cu ($x = 11$). Even more corrections of this type have been made for the $4d$ series elements since $4d^{x-1}5s^1$ is the ground state configuration for Nb, Mo, Ru, Rh and Ag, whereas for Pd it is simply $4d^{10}$. From the figure we see that there is the expected gradual increase in binding energy of the $s$ electrons across each series. Perhaps unexpected is the order of stability of the *pair* of $s$ electrons, namely, $5d > 3d > 4d$. This order apparently arises because of the delicate interplay of two important and opposing forces which affect ionization energies, namely, increasing radial magnitude ($5d > 4d > 3d$) and increasing effective nuclear charge ($5d \gg 4d \sim 3d$).

It is of some value to examine the energies required to remove $d$ electrons across the three transition series, that is, the process $(n-1)d^{x-1}$ to $(n-1)d^{x-2}$, in which an electron is ionized from the *singly positive* transition metal ion. In many cases this will simply be equal to $I_2$; in other cases it will involve a promotion energy as well. This has been done for the $3d$ and $4d$ series in Figure 4–7, where it is seen that the binding energy of the

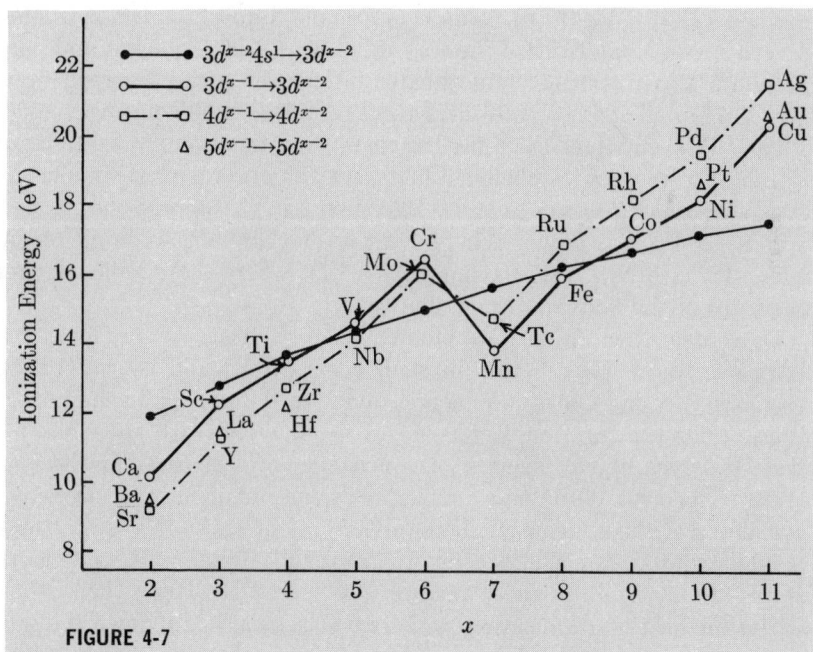

**FIGURE 4-7**

Ionization energies corresponding to the indicated changes in electronic configuration.

$d$ electrons increases steadily across each series, with a sharp break following the half-filled $d$ subshell, in which the electrons are now being removed from doubly occupied orbitals. Because of the extra repulsion energy between electrons occupying the same orbital, which forces them into the same region of space, and because the $d$ orbitals are highly directed spatially, the repulsion between $d$ electrons in the same orbital is unusually high. This is primarily the cause of the observed sharp drop from Cr to Mn seen in the figure. The more shallow drop from Mo to Tc may then be taken as a reflection of the increased size of a $4d$ orbital over that of a $3d$ orbital, the former larger orbital being able to accommodate electron pairing with less lowering of the energy. Although the data are incomplete for the $5d$ series, this same argument should apply, and thus a shallow drop between W and Re would be predicted.

Although there is as yet little data on $d$-electron removal from the $5d$ series, there does appear to be a relative lowering of the $5d$ binding energies compared to $4d$ as can be seen in Figure 4–5. Thus, for example, $d$ electrons in Hf, Pt, and Au are more loosely bound than in their respective congeners, Zr, Pd, and Ag. The relative increase of the binding energies of the $d$ electrons in, for example, Zr and Pd and of the $s$ electrons in Hf and Pt results in the sums of the first and second ionization energies being nearly identical for the respective congeners. Thus we find Zr (19.75), Hf (20.4); and Pd (27.75), Pt (27.6). This is surely one important factor in the observed strong similarities of the respective $4d$ and $5d$ congeners, and perhaps accounts also in part for the greater differences found between divalent Ni (25.78) and its two congeners than between those congeners, divalent Pd and Pt themselves. As a further example of the use of ionization potential data to rationalize chemical facts, we can observe that whereas $PtCl_2$ is expected *and found to be* less stable than $NiCl_2$ (compare the above sum values), $K_2PtCl_6$ is very stable and yet has no counterpart in nickel chemistry; see also $Ni^{4+}$ (117.0) and $Pt^{4+}$ (97.1), where these values represent the sums of the first four ionization energies.

Of course other important factors, such as heats of atomization and lattice energies, play a role in such comparisons and so care must be exercised in using ionization energies alone to try to explain such observations. Thus the sum for Ti (20.4), nearly identical with the sums for Zr and Hf (given above) suggests the existence of a great similarity among these congeners. While this is roughly borne out in fact in much of the chemistry of these elements in the quadrivalent state, it is also true that bivalent and trivalent states for Ti are readily obtainable whereas these states are uncertain or most unstable for Zr and Hf. Here, then, is a case where the heat of atomization, which increases markedly down the family, has apparently forced the heavier two members to adopt higher oxidation states. In this manner they may form stronger covalent bonds, thereby releasing the extra energy necessary to achieve the more difficult breakup of their solid metal lattices. Explanations such as these should become clearer after the Born-Haber cycle has been studied (Chapter 5). But it should be remembered that when explanations are based upon such a thermodynamic cycle they must also examine the possible importance of entropy effects since only enthalpies are normally considered. Generally comparisons of like systems are made, so the entropy effects are of negligible importance.

Finally we shall look briefly at the horizontal trends in the non-transition elements across the periodic table. The first ionization energies for the third-period elements are:

| | Na | Mg | | Al | Si | P | S | Cl | Ar |
|---|---|---|---|---|---|---|---|---|---|
| $I_1$ | 5.14 | 7.64 | — | 6.00 | 8.15 | 10.5 — | 10.36 | 13.01 | 15.76 |
| $I_2$ | 47.29 | 15.03 | | 18.82 — | 16.34 | 19.7 | 23.4 — | 23.80 | 27.62 |

Gradual increases from the R1 to the inert gas family occur and are due primarily to decreasing size and increasing effective nuclear charge, both of which are caused by an increasingly insufficient nuclear shielding by increasing numbers of electrons in the same sublevels. The minor irregularities that appear within the general trend are associated with the more subtle effects of filled, half-filled, and empty orbitals (extra quantum mechanical exchange energy), the added repulsion between electrons which are forced to occupy the same orbital, and the very inefficient mutual shielding by electrons occupying the same orbital. Explanations for the irregularities in the second ionization energies across the series follow the same pattern. The horizontal bars in the above series separate values which without the foregoing considerations would *appear* to be anomalous.

## AFFINITY ENERGY

Almost all atoms appear to have the capacity to accept at least one more electron than is present in the neutral atom. This process occurs with the release of energy. Therefore the *electron affinity* of an atom has been defined to be the energy liberated when a neutral gaseous atom in its ground state accepts an electron to form the gaseous negative ion in its ground state. That is, it is the energy of the reaction

$$X_{(g)} + e^-_{(g)} \rightarrow X^-_{(g)}$$

Since electron affinity represents an energy, we suggest that a better expression to employ is *affinity energy* $(A)$, to parallel the expression *ionization energy* $(I)$. More precisely we have defined the *first affinity energy*, $A_1$. A second, third, etc., $A_2$, $A_3$, ... affinity energy may be defined analogously, although the addition of more than one electron *always* requires an input of energy. Thus the values of $A_n$ (when $n > 1$) are always negative. Furthermore, the value of $A$ for some atom X is identical in magnitude with the value of $I$ for $X^-$, and more generally this $A$ and $I$ relationship holds for the species $X^{n-}$ and $X^{(n+1)-}$.

Although affinity energy, like ionization energy, is a simple property of an isolated atom or ion, it is considerably more difficult to obtain experimentally. As a result, the direct determination of affinity energies has been made for relatively few elements and the most accurate values are available only for the halogens. (See Table 4–11.) Most values are obtained by one of several available calculational procedures, the oldest being the use of a Born-Haber cycle (Chapter 5). More recently a quantum mechanical (Hartree-Fock type) calculation has been used.

### TABLE 4–11. Affinity Energy Values[a] of Some of the Elements (in eV units).

| H | Li | Be | B | C | N | O | F | | | |
|---|----|----|---|---|---|---|---|---|---|---|
| 0.75 | 0.58 | −0.19 | 0.33 | 1.12 | −0.27 | 1.47 | 3.45 | | | |
| | | | | | | (−7.3) | | | | |
| | Na | Mg | Al | Si | P | S | Cl | | | |
| | 0.78 | −0.32 | 0.52 | 1.39 | 0.78 | 2.07 | 3.61 | | | |
| | | | | | | (−3.4) | | | | |
| | Cu | Ag | Au | | | Se | Br | I | | |
| | 1.5 | 2.0 | 2.8 | | | (−4.2)[b] | 3.36 | 3.06 | | |

| K | Ca | Sc | Ti | V | Cr | Mn | Fe | Co | Ni | Cu |
|------|------|-------|------|------|------|-------|------|------|------|--------|
| 0.92 | — | −0.14 | 0.40 | 0.94 | 0.98 | −1.07 | 0.58 | 0.94 | 1.28 | 1.80[c] |
| 0.50 | −1.6 | −0.4 | 0.15 | 0.65 | 0.85 | −1.2 | 0.1 | 0.7 | 1.1 | 0.9[d] |

| Rb | Sr | Y | Zr | Nb | Mo | Tc | Ru | Rh | Pd | Ag |
|-----|------|-----|-----|-----|-----|-----|------|------|-----|--------|
| 0.6 | −0.5 | 0.3 | 1.0 | 1.3 | 1.3 | 1.0 | 1.45 | 1.35 | 1.4 | 0.9[d] |

[a] Recent experimental values are underlined; all other values have been calculated by various methods. Values in parentheses refer to total $A$ for two electrons.
[b] The $A_1$ value for Se is not known.
[c] All data in this line taken from E. Clementi, *Phys. Rev.*, **135A**, 981 (1964); calculated by Hartree-Fock method.
[d] All data in this line taken from O. P. Charkin and M. E. Dyatkina, *Zhur. Strukt. Khim.*, **6**, 422 (1965); calculated by semiempirical Glockler method.

It is seen from Table 4–11 that chlorine possesses the highest affinity energy of all the elements, suggesting that chlorine should be the strongest elemental oxidizing agent. However, fluorine has the latter distinction. The unexpectedly low value for the affinity energy of fluorine may be associated with strong interelectron repulsion forces operating within the relatively compact $2p$ sublevel. These same forces operating between *nonbonding* $2p$ electrons on bonded fluorine atoms are perhaps responsible in part for the greater, and unexpected, ease of dissociation of the fluorine molecule into atoms (1.64 eV), compared to that of chlorine (2.48 eV). There is also very possibly some $\pi$-bonding involving $d$ orbitals and/or hybridization of $p$ and $d$ orbitals contributing to the greater strength of the Cl—Cl bond. But the important point chemically is that the 0.16 eV lower $A$ value for fluorine is more than compensated for by the 0.84 eV lower dissociation energy. From the Born-Haber cycle (page 160) it is seen that these energies contribute oppositely to the total energy of a reaction.

From Table 4–11 it can be seen that the affinity energies for all first members of the non-transition element families are lower than those for the respective second members with the exception of Be and Mg. As in the case of F and Cl, this can be attributed primarily to the much smaller sizes of the first-row elements. This leads to very much higher electron densities for the respective negative ions. For example, representing the charge density as $1e^-/\frac{4}{3}\pi r^3$, the charge density ratio $F^-/Cl^- = 2.35$. But for comparison consider $Cl^-/Br^- = 1.25$, and $Br^-/I^- = 1.36$. High electron density is of course opposed by interelectron repulsion forces.

Very effective nuclear shielding by $s$-type electrons, coupled with the necessity of the added electron entering the next higher sublevel may explain the negative $A$ values for the R2 family. The high positive value for gold is noteworthy since this element displays some halogen-like character in its ability to form ionic *auride* compounds such as CsAu. High effective nuclear charge coupled with the relatively poor nuclear screening of one $ns$ electron for its partner in $Au^-$ are surely important factors here.

## ELECTRONEGATIVITY

Pauling[6] introduced the concept of electronegativity of atoms in 1932 and described its quantitative aspect as a *measure of the tendency of an atom in a molecule to attract electrons to itself*. He used the idea to account for the fact that the energy of a heteropolar bond A—B, which we may symbolize by $D(AB)$, is generally higher than either the average arithmetic or average geometric mean value of the homopolar bond energies of the molecules A—A and B—B. For the case of the arithmetic mean we can write

(4–6)        $D(AB) = \frac{1}{2}[D(AA) + D(BB)] + \Delta_{AB}$

Careful examination of the term $\Delta_{AB}$ for many diatomic molecules showed that it may be given fairly accurately by the equation

(4–7)        $\Delta_{AB} = 23.06(\chi_A - \chi_B)^2$

This may be rearranged to

(4–8)        $|\chi_A - \chi_B| = 0.208\sqrt{\Delta_{AB}}$

where $\chi_A$ and $\chi_B$ are constants. According to Pauling, $\chi_A$ and $\chi_B$ are invariant from molecule to molecule and are characteristic of the atoms A and B. These are referred to as the electronegativities of the respective atoms and are defined by Eq. (4–8). Thus, using the available appropriate thermochemical data and arbitrarily fixing one $\chi$ value (2.1 for hydrogen), Pauling set up the first and most widely used scale of *relative atomic electronegativities*. Through all of the dozens of succeeding different methods of obtaining electronegativities, comparisons are most often made with the original values of Pauling.[3] The Pauling values, derived from more recent thermochemical data, along with values from two other scales are given in Table 4–12.

**TABLE 4‑12.**[a]  Electronegativities of the Elements (Values in bold type are calculated using the Allred‑Rochow formula; those in italics are estimated by Pauling's method and those in Roman type are calculated by Mulliken's method.)[b]

| I | II | III | IV | V | VI | VII | VIII | VIII | VIII | I | II | III | IV | V | VI | Iᶜ |
|---|---|---|---|---|---|---|---|---|---|---|---|---|---|---|---|---|
| H **2.20** | | | | | | | | | | | | | | | | |
| Li **0.97** *0.98* | Be **1.47** *1.57* | | | | | | | | | | | B **2.01** *2.04* 2.01 | C **2.50** *2.55* | N **3.07** *3.04* | O **3.50** *3.44* 3.17 | F **4.10** *3.98* 3.91 |
| Na **1.01** *0.93* | Mg **1.23** *1.31* | | | | | | | | | | | Al **1.47** *1.61* 1.81 | Si **1.74** *1.90* 2.44 | P **2.06** *2.19* 1.81 | S **2.44** *2.58* 2.41 | Cl **2.83** *3.16* 3.00 |
| K **0.91** *0.82* | Ca **1.04** *1.00* | Sc **1.20** *1.36* | Ti **1.32** *1.54* | V **1.45** *1.63* | Cr **1.56** *1.66* | Mn **1.60** *1.55* | Fe **1.64** *1.83* | Co **1.70** *1.88* | Ni **1.75** *1.91* | Cu **1.75** *1.90* 1.36 | Zn **1.66** *1.65* 1.49 | Ga **1.82** *1.81* 1.95 | Ge **2.02** *2.01* | As **2.20** *2.18* 1.75 | Se **2.48** *2.55* 2.23 | Br **2.74** *2.96* 2.76 |
| Rb **0.89** *0.82* | Sr **0.99** *0.95* | Y **1.11** *1.22* | Zr **1.22** *1.33* | Nb **1.23** | Mo **1.30** *2.16* | Tc **1.36** | Ru **1.42** | Rh **1.45** *2.28* | Pd **1.35** *2.20* | Ag **1.42** *1.93* 1.36 | Cd **1.46** *1.69* 1.4 | In **1.49** *1.78* 1.80 | Sn **1.72** *1.96* | Sb **1.82** *2.05* 1.65 | Te **2.01** 2.10 | I **2.21** *2.66* 2.56 |
| Cs **0.86** *0.79* | Ba **0.97** *0.89* | *La | Hf **1.23** | Ta **1.33** | W **1.40** *2.36* | Re **1.46** | Os **1.52** | Ir **1.55** *2.20* | Pt **1.44** *2.28* | Au **1.42** *2.54* | Hg **1.44** *2.00* | Tl **1.44** *2.04* | Pb **1.55** *2.33* | Bi **1.67** *2.02* | Po **1.76** | At **1.96** |
| Fr **0.86** | Ra **0.97** | **Ac | | | | | | | | | | | | | | |

*Lanthanide series:

| III | IV | V | VI | VII | VIII | | | | | | | | | |
|---|---|---|---|---|---|---|---|---|---|---|---|---|---|---|
| *La **1.08** *1.10* | Ce **1.06** *1.12* | Pr **1.07** *1.13* | Nd **1.07** *1.14* | Pm **1.07** | Sm **1.07** *1.17* | Eu **1.01** | Gd **1.11** *1.20* | Tb **1.10** | Dy **1.10** *1.22* | Ho **1.10** *1.23* | Er **1.11** *1.24* | Tm **1.11** *1.25* | Yb **1.06** | Lu **1.14** *1.27* |

**Actinide series:

| III | IV | V | VI | VII | VIII | | | | | | | | |
|---|---|---|---|---|---|---|---|---|---|---|---|---|---|
| **Ac **1.00** | Th **1.11** | Pa **1.13** *1.14* | U **1.22** *1.38* | Np **1.22** *1.36* | Pu **1.22** *1.28* | Am | Cm | Bk | Cf | Es | Fm | Md | |

← **~1.2** (estimated) →

[a] From F. A. Cotton and G. Wilkinson, "Advanced Inorganic Chemistry, Second Edition," Interscience‑Wiley, New York, 1966.

[b] Alfred‑Rochow values from J. Inorg. Nucl. Chem., **5**, 264 (1958); Pauling‑type values from A. L. Allred, J. Inorg. Nucl. Chem., **17**, 215 (1961); Mulliken‑type values from H. O. Pritchard and H. A. Skinner, Chem. Rev., **55**, 745 (1955).

As we have just implied, there have been a great many discussions and controversies over the years since 1932 concerning electronegativity. Within the past decade most of these have been concerned with the importance, value, and real meaning of the electronegativity concept and how best to calculate or measure it. Since the electronegativity concept is not a very precisely defined concept, it should be no surprise that it cannot be measured precisely. It is even difficult to get universal agreement as to what, exactly, is to be measured or calculated and hence what units the numerical values should have. Thus it is understandable why we are faced with many electronegativity scales and methods of computation. However, aside from some relatively minor differences, most scales are in good agreement with each other and are at least internally consistent.

We have space to do little more than briefly outline some of the major contributions that have been made and will refer the reader to the original literature for details. In this connection a fine early review by Pritchard and Skinner[7] should be mentioned as a starting point. Questions of recent concern are:

(1) Is there a theoretical basis for an electronegativity scale?

(2) Is the electronegativity an invariant property of an atom or must it necessarily vary with the atom's environment (oxidation state, hybridization, coordination number, etc.)?

(3) Is electronegativity perhaps better considered a property of an orbital (*orbital electronegativity*), or of a bond (*bond electronegativity*), rather than of an atom?

(4) What should be the units of electronegativity, or of what physical property is it really a measure?

(5) Can we measure or calculate the electronegativity of groups of atoms?

(6) What is meant by *electronegativity neutralization* and of what use is it?

(7) Of what use are electronegativity values anyway, once we agree on how best to obtain them?

With full realization that a long monograph might well be devoted to supplying the answers to these questions, we shall nevertheless proceed to consider them briefly.

## Theoretical Electronegativity Scales

The earliest and perhaps best theoretical definition of electronegativity was supplied by Mulliken.[8] By introducing the idea that the electronegativity of an atom, A, represents an average of the binding energy of the outermost electrons over a range of valence-state ionizations (for example, from $A^+$ to $A^-$ in the A—B molecule), he proposed the very simple relationship:

$$(4\text{--}9) \qquad \chi_A = \tfrac{1}{2}(I_{V^A} + A_{V^A})$$

where $I_V{}^A$ and $A_V{}^A$ are the appropriate *valence-state ionization energy* and *affinity energy*, respectively (*vide infra*). To a good approximation the Mulliken electronegativities, $\chi_M$, calculated from ground-state ionization and affinity energies, are proportional to the Pauling values, $\chi_P$; the relation being,

$$(4\text{-}10) \qquad \chi_P = 0.336 \, (\chi_M - 0.615)$$

Table 4–12 presents the Mulliken values along with the Pauling values for comparison.

Perhaps the most significant advance in recent years in the understanding of electronegativities has been the recognition that the electronegativity of an atom must fluctuate in response to its environment. Using the Mulliken approach, but employing the variable *valence-state* energies rather than the constant *ground-state* energies, it has been possible to obtain such variable electronegativities. But first let us examine the meaning of valence-state energies.

A "valence state" is neither a stationary state nor even a nonstationary state but a statistical average of stationary states chosen so as to have as nearly as possible the same interactions among the electrons of the given atom as they actually have when the atom is part of a molecule. Thus the valence state can be considered to be formed from a molecule by adiabatically removing from a given atom all of the other atoms with their electrons, thereby allowing no electronic rearrangement. If we define $P^0$, $P^+$ and $P^-$ as the *valence-state promotion energies* for A, $A^+$ and $A^-$, respectively, then we can write, for atom A,

$$(4\text{-}11) \qquad I_V = I_1 + P^+ - P^0$$

$$(4\text{-}12) \qquad A_V = A_1 + P^0 - P^-$$

These relationships can be seen more easily by the study of Figure 4–8, which uses the tetrahedral $sp^3$-hybridized carbon atom and its positive and negative ions for the illustration.

Hinze and Jaffe,[9] utilizing Mulliken's exact definition, calculated the promotion energies for a large variety of states of the atoms and ions of the first and second periods[9] and the $3d$ transition series.[10] Then using Eqs. (4–11), (4–12), and (4–9), they obtained electronegativities of a number of valence states and termed these "orbital electronegativities." They discovered that orbital electronegativities for $\sigma$ orbitals are always higher than for $\pi$ orbitals and are linearly related to the amount of $s$ character assumed to be in the hybrid orbitals. As might be expected, the electronegativity was found to increase with an increase in the amount of $s$ character in the hybrid orbital.

Subsequently Hinze and Jaffe[11] put forth a new definition of orbital electronegativity, defining it as the derivative of the energy of the atom with respect to the charge in the orbital,

$$(4\text{--}13) \qquad \chi_j = \frac{\partial E}{\partial n_j}$$

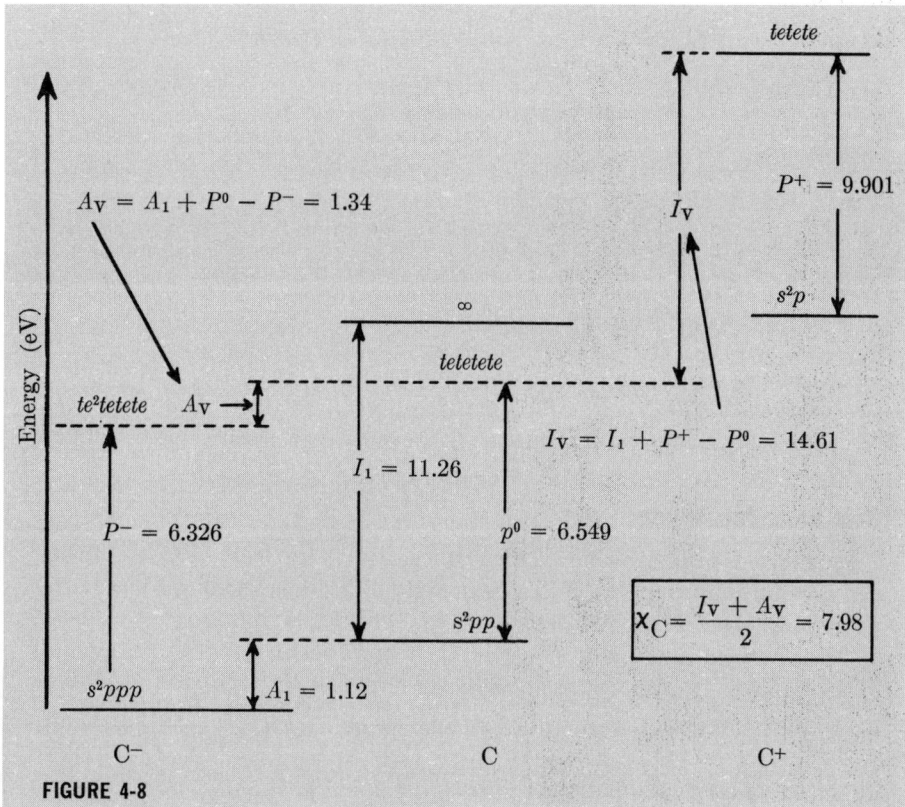

$A_V = A_1 + P^0 - P^- = 1.34$

$te^2tetete \quad A_V \rightarrow$

$P^- = 6.326$

$s^2ppp$

$\infty$

$teletete$

$I_1 = 11.26$

$p^0 = 6.549$

$s^2pp$

$A_1 = 1.12$

$C^-$

$C$

$tetete$

$P^+ = 9.901$

$I_V$

$s^2p$

$I_V = I_1 + P^+ - P^0 = 14.61$

$\boxed{\chi_C = \dfrac{I_V + A_V}{2} = 7.98}$

$C^+$

**FIGURE 4-8**

Valence state term system for C⁻, C, and C⁺ showing only the $sp^3$ hybrid state (*each $sp^3$ hybrid orbital being abbreviated as te*), the evaluation of valence state ionization energy and affinity energy, and finally, in the box, the determination of the Mulliken electronegativity for carbon in the $sp^3$-hybridized state. Using Eq. (4-10) we can calculate the electronegativity on the Pauling scale as 2.47, in excellent agreement with the Pauling value for carbon of 2.5. Reference (9) should be consulted for further details.

where $n_j$ is the occupation number of the $j$'th orbital, $(0 \le n_j \le 2)$, which has electronegativity $\chi_j$. They assume the energy can be written as

$$(4\text{--}14) \qquad E(n_j) = a + bn_j + cn_j^2$$

and they then arbitrarily define the energy scale such that

$$E(0) = 0, \text{ so that}$$

(4–15)
$$E(1) = I_V, \text{ and}$$

$$E(2) = I_V + A_V$$

Therefore, solving for the coefficients $a$, $b$, and $c$ by the simultaneous solution of Eqs. (4–14) and (4–15), we obtain

(4–16)
$$E(n_j) = \tfrac{1}{2}(3I_V - A_V)n_j - \tfrac{1}{2}(I_V - A_V)n_j^2$$

and the derivative

(4–17)
$$\chi(n_j) = \frac{dE(n_j)}{dn_j} = \tfrac{1}{2}(3I_V - A_V) - (I_V - A_V)n_j$$

from which it follows that

$$\chi(0) = \tfrac{1}{2}(3I_V - A_V)$$

(4–18)
$$\chi(1) = \tfrac{1}{2}(I_V + A_V)$$

$$\chi(2) = \tfrac{1}{2}(3A_V - I_V)$$

Thus the new definition permits the computation of orbital electronegativities of *vacant*, $\chi(0)$, and *doubly occupied*, $\chi(2)$, *orbitals*. And the *singly occupied orbital* value, $\chi(1)$, is seen to be just that defined by Mulliken.

## Bond Electronegativity

Furthermore, a new kind of electronegativity, the *bond electronegativity*, can now be defined as the electronegativity of orbitals forming a bond after charge has been exchanged between them. This leads to fractional values of $n_j$. Thus in the formation of molecule AB, transfer of an infinitesimal amount of charge from A to B (or vice-versa) is accompanied by a reduction of charge on A. This requires an expenditure of energy given by $\left(\dfrac{dE_A(n_A)}{dn_A}\right)dn_A$, while at the same time an amount of energy $\left(\dfrac{dE_B(n_B)}{dn_B}\right)dn_B$ is released. A stable bond is obtained when the transfer involves no further energy change. This can be thought of as an equilibrium state where $dn_A = -dn_B$ and $\dfrac{dE_B(n_B)}{dn_B} = \dfrac{dE_A(n_A)}{dn_A}$ or $\chi_{eq}(n_A) = \chi_{eq}(n_B)$. Of course $(n_A + n_B)$ must equal 2. A plot of $\chi_A(n_A)$ versus $n_A$ and a corresponding plot of $\chi_B(n_B)$ versus $n_B$ will yield two straight lines whose intersection gives the equilibrium values of $n_A$ and $n_B$. It should be clear that $n_B$ varies from 2 to 0 as $n_A$ varies from 0 to 2.

## Other Electronegativity Scales

As we have already noted, there have been many attempts to obtain better quantitative measures of atom electronegativities, but we shall take brief notice of only a few of these here. Gordy[12] proposed that electronegativity be defined as

$$(4\text{--}19) \qquad \chi_A = \frac{eZ_{eff}}{r_{cov}}$$

Here electronegativity can be pictured as a *potential* due to the partially screened nuclear charge ($Z_{eff}$) measured at the covalent radius. The electronegativity units are therefore energy/electron.

More recently Iczkowski and Margrave[13] proposed the expression

$$(4\text{--}20) \qquad \chi_A = \left(-\frac{dE}{dN}\right)_{N=0}$$

where $dE$ is the energy change which accompanies a charge change $dN$. This also gives $\chi_A$ in units of energy/electron. It is assumed here that the energy of an atom is a continuous and single-valued function of its charge. For example, a reasonably successful energy equation which has been used is

$$(4\text{--}21) \qquad E(N) = aN + bN^2 + cN^3 + dN^4$$

where $E$ represents the total energy of all the electrons around a nucleus of atomic number $Z$, the neutral atom energy being taken as zero, and $N$ represents the number of electrons present around the nucleus minus the atomic number. That is, $N = n - Z$, where $n$ is the number of extranuclear electrons at any particular state of ionization. This method gives values that correlate well with Mulliken's scale, and indeed the two methods become the same if only the first two terms of the energy equation are considered. It does not, however, consider the orbital dependence of the electronegativity. A recent improvement which obviates this deficiency has been made by Klopman,[14] who differentiates between atomic and molecular electronegativity. He uses a more elaborate Rydberg equation for atom energies which allows the calculation of the electronegativity of atoms in any valency state.

A valuable and widely adopted empirical approach to electronegativity, proposed by Allred and Rochow,[15] pictures the electronegativity as a *force* acting on the atomic electrons at the distance of the covalent radius, so that,

$$(4\text{--}22) \qquad \chi_A = \frac{e^2 Z_{eff}}{r_{cov}^2}$$

This has proven to be a highly successful method for obtaining electronegativity values that appear to reflect chemical trends more accurately than either the Pauling or Mulliken values in those cases where the scales are in poor agreement, such as in the R3, R4, and R5 families. Yet it correlates well with most values obtained by the older methods. The relationship which gives Allred-Rochow electronegativity values on the Pauling scale is

$$(4\text{-}23) \qquad \chi_{AR} = 0.359 \frac{Z_{eff}}{r_{cov}^2} + 0.744$$

For comparison the Allred-Rochow values are recorded in Table 4–12. In line with other recent attempts to allow the electronegativity of an atom to fluctuate in response to its environment, Huheey[16] has proposed a method of modifying the Allred-Rochow expression so as to allow the electronegativity to vary as a function of the partial charge, $\delta$, on the atom. However, even this extension allows no distinction among the various possible valence states that an atom may assume.

Electronegativity values of groups of atoms, such as $CH_3$, $CF_3$, $C_6H_5$, etc., have been calculated by Huheey[17] assuming a variable electronegativity of the central atom in the group and equalization of electronegativity in all the bonds. Others have calculated or estimated group electronegativities by other procedures, but we shall not pursue this point further.

### Electronegativity Equalization

This brings us finally to one of the most interesting and useful contributions to the concept of electronegativity, namely, the *equalization of electronegativity* during the act of stable bond formation. This idea, which we mentioned earlier in connection with bond electronegativity, was put forth by Sanderson[18,2] as the *principle of electronegativity equalization*. Very simply, it postulates that when two or more atoms initially different in electronegativity unite, their electronegativities become equalized at some intermediate value in the molecule. This intermediate electronegativity in the molecule is taken as the geometric mean of the electronegativities of all of the atoms before combination. The driving force for the equalization is easily pictured as follows. Electrons in a stable covalent bond must be equally attracted to both nuclei. If they were not they would move until this equilibrium condition is met. If the two atoms are initially of different electronegativities, their bonding orbitals are necessarily of different energies. Therefore the process of bond formation must provide a pathway by which these energies can become equalized. Such a pathway can be based on the fact that the electronegativity of an atom must decrease as that atom acquires an electron or must increase as it loses an electron. Thus an atom of beryllium has little attraction for electrons, but its ion, $Be^{2+}$, attracts them strongly. Likewise, an atom of oxygen has a very high electronegativity, but the oxide ion has none. Charge exchange in bond formation thus appears to result in a state of *uneven electron sharing* but *even electron attraction*. Some of the uses as well as the limitations of the principle of equalization of electronegativity have been discussed by Sanderson.[2]

The electronegativity concept has been found to be very useful in understanding, explaining, and even predicting many properties related to the energy and charge distribution in chemical bonds. These properties include ionic character of bonds, bond polarity, bond dissociation energies, bond moments, force constants, and the like. It is not our purpose here to pursue the relationships existent between these properties and electronegativity, but only to point out that electronegativity considerations can contribute to a sounder understanding of all such properties.

Finally, the periodic trends in atom electronegativities may be seen from Table 4–12. It will be left for the reader to rationalize these trends in light of all of the foregoing discussion of the electronegativity concept.

## EFFECTIVE NUCLEAR CHARGE

The *effective nuclear charge*, $Z_{eff}$, enters into many calculations as well as qualitative explanations of many properties and trends in properties, and yet there is no thoroughly satisfactory method for evaluating it. The definition appears simple enough, $Z_{eff}$ for a given electron is simply $(Z - \sigma)$, where $Z$ is the actual charge on the nucleus and $\sigma$ is the *screening constant* for that electron. Thus an electron in an atom is acted upon by the nucleus and all of the other electrons. The latter act to reduce the effect of the former. How to determine a good value for $\sigma$ is obviously the not-so-simple problem. In 1930 Slater[19] proposed a set of rules for computing $Z_{eff}$, and these have been widely used in the past and, in fact, are still used.[20]

Slater divides the electrons into groups such as $(1s)$, $(2sp)$, $(3sp)$, $(3d)$, $(4sp)$, etc. Each group is defined as external or internal with respect to the others, with the nucleus taken as the origin. It is then assumed that for a given electron there will be no screening due to electrons of *external* groups. Furthermore it is assumed that the $Z_{eff}$ for a given $s$ or $p$ electron is independent of the $l$ quantum number. Thus, in general, Slater assumes that $\sigma$ is a function of $n$ and $N_i$, where $n$ is the principal quantum number and $N_i$ is the total number of electrons minus those belonging to groups outside the electron under consideration. The rules hold reasonably well for the first-row elements, less well for the second row and rather poorly for those beyond. Although the Slater rules have been generally accepted for a number of years, it would appear from their lack of success in modern theoretical interpretations that their acceptance is no longer justified.

Recently a set of rules for the evaluation of screening constants has been formulated by Clementi and Raimondi.[21] These rules are found to hold quite well for the elements 2 through 36. Clementi and Raimondi have assumed $\sigma$ to be a function of the two quantum numbers $n$ and $l$, the total number of electrons, $N$, the atomic number, $Z$, and for the valency electrons, the total angular momentum and the spin multiplicity of the atom, $L$ and $S$. Denoting the *number* of electrons of principal quantum number $n$, and angular momentum $l$, of a given configuration by $nl$, they propose the following equations for calculating the screening constants for neutral atoms:

$$\sigma(1s) = 0.3(1s - 1) + 0.0072(2s + 2p) \\ + 0.0158(3s + 3p + 4s + 3d + 4p)$$

$$\sigma(2s) = 1.7208 + 0.3601(2s - 1 + 2p) \\ + 0.2062(3s + 3p + 4s + 3d + 4p)$$

$$\sigma(2p) = 2.5787 + 0.3326(2p - 1) - 0.0773(3s) \\ - 0.0161(3p + 4s) - 0.0048(3d) + 0.0085(4p)$$

$$\sigma(3s) = 8.4927 + 0.2501(3s - 1 + 3p) + 0.0778(4s) \\ + 0.3382(3d) + 0.1978(4p)$$

$$\sigma(3p) = 9.3345 + 0.3803(3p - 1) + 0.0526(4s) \\ + 0.3289(3d) + 0.1558(4p)$$

$$\sigma(4s) = 15.505 + 0.0971(4s - 1) + 0.8433(3d) \\ + 0.0687(4p)$$

$$\sigma(3d) = 13.5894 + 0.2693(3d - 1) - 0.1065(4p)$$

$$\sigma(4p) = 24.7782 + 0.2905(4p - 1)$$

## OXIDATION STATES OF THE ELEMENTS

The combining capacity of an element is called its *valence*, and originally the valence was defined as *the number of atoms of hydrogen that the element's atom can combine with or replace*. Since the hydrogen atom can combine with but one other atom at a time, it is *univalent*. More generally we may consider the valence of an element to be defined as the number of atoms of a univalent element with which one atom of the given element combines. Thus in $H_2O$ the oxygen is bivalent, in $NH_3$ the nitrogen is trivalent, in $CH_4$ the carbon is quadrivalent, in gaseous $PF_5$ the phosphorus is quinquevalent, in $SF_6$ the sulfur is sexivalent, in $ReF_7$ the rhenium is septivalent and in $OsO_4$ the osmium is octavalent (O being bivalent in all of its binary oxide compounds with metals). Valence, *per se*, has never been as useful a concept, despite its physical reality, as the more commonly used and related artificial concept of *oxidation state*. The latter is a positive or negative number, not necessarily an integer, which represents the charge that an atom would have if the electrons in a molecule were assigned to the atoms in a certain way. Since this assignment is often somewhat arbitrary, the oxidation state is not always numerically equal to the valence and is not always a useful quantity to know. Rules for assigning oxidation states to each atom in a substance may be formulated as follows, with the understanding that they are not always entirely unambiguous:

(1) The oxidation number of an atom of a free element is taken to be zero. (Note that the *combining capacity* or valence of an *atom in an element*

is most often 1 or greater; it is 0 only in the inert gas elements, $1$ in the R7 elements, $N_2$, $O_2$ and $H_2$, etc., $2$ in the rest of the R6 family, $3$ in the rest of the R5 family, and $4$ in most of the R4 family.)

(2) The oxidation number of a monatomic ion is its charge, including the sign of the charge.

(3) The oxidation number of each atom in a covalent molecule of known structure is the charge that would remain on the atom if each shared electron pair were assigned completely to the more electronegative of the two atoms sharing it. An electron pair shared by like atoms is split equally between them, and in all such cases the valence and oxidation state will differ numerically.

(4) The oxidation number of an element in a compound of uncertain structure may usually be obtained if reasonable oxidation states of all the other elements in the compound can be unambiguously assigned first.

Rules for predicting oxidation states of elements from their positions in the periodic table or even from their known electronic configurations are not quite as clear-cut as the above rules. However, some generalizations are possible which lead to a deeper understanding of some of the basic principles of chemical compound formation.

### Representative Elements, Families R1–R7

The *common* oxidation states of these elements are readily predicted from their electron configurations providing two general principles are observed. (1) *In forming molecules or ions, the atoms will seek to reach the most stable arrangement for their outer electrons.* Such stable arrangements predicted and observed are the *closed shell* outer configurations with two [$s^2$] or eight [$s^2p^6$] electrons, eighteen [$s^2p^6d^{10}$] electrons, or eighteen plus two [$ns^2p^6d^{10}(n+1)s^2$] electrons. (2) *Paired atomic electrons will tend to be lost or shared in pairs.* Consequently, where several positive oxidation states occur within a family, they will differ by two units. However, it should be mentioned that unusual or unexpected oxidation states such as Al(0), Si(0), Cl(IV), etc., are now fairly numerous among these elements and the aforementioned generalizations are to be considered as guidelines only. We shall now consider oxidation states by periodic family.

**Family R1.** The $ns^1$ outer configuration and the very low electronegativities allow only a 1+ oxidation state for members of this family, and ions having the closed shell configurations $s^2$ or $s^2p^6$ are formed readily due to the very low binding energy ($I_1$) of the lone electron. The very large second ionization energy values (see Table 4–13), the highest found in any family, explain the total absence in chemical systems of any higher oxidation states for these elements.

**Family R2.** The outer $ns^2$ configuration suggests that the common oxidation state in this family is 2+. Consideration of ionization energies alone might suggest a 1+ state; but this state, although indicated for Be and

**TABLE 4-13.  Some Energies for R1, R2, T11 and R2′ Metals in eV/atom.**

| Element | $I_1$ | $I_2$ | $I_1 + I_2$ | $-\Delta H_{hyd}$ | $-\Delta H_U{}^a$ | $-\Delta H_{subl}$ |
|---|---|---|---|---|---|---|
| Li | 5.39 | 75.6 | 81.0 | 5.34 | 8.63 | 1.61 |
| Na | 5.14 | 47.3 | 52.4 | 4.21 | 7.94 | 1.13 |
| K | 4.34 | 31.8 | 36.1 | 3.34 | 7.17 | 0.93 |
| Rb | 4.18 | 27.4 | 31.6 | 3.04 | 6.95 | 0.89 |
| Cs | 3.89 | 23.4 | 27.3 | 2.74 | 6.51 | 0.82 |
| Be | 9.32 | 18.2 | 27.5 | 24.8 | — | 3.33 |
| Mg | 7.64 | 15.0 | 22.6 | 20.2 | 40.8 | 1.56 |
| Ca | 6.11 | 11.9 | 18.0 | 16.6 | 36.6 | 2.00 |
| Sr | 5.69 | 11.0 | 16.7 | 15.2 | 34.4 | 1.70 |
| Ba | 5.21 | 10.0 | 15.2 | 13.7 | 32.5 | 1.82 |
| Cu | 7.72 | 20.29 | 28.0 | 6.04; 22.0 (Cu²⁺) | | 3.54 |
| Ag | 7.57 | 21.48 | 29.0 | 5.04 | | 3.00 |
| Au | 9.22 | 20.5 | 29.7 | 6.69 | | 3.57 |
| Zn | 9.39 | 17.96 | 27.3 | 21.4 | | 1.35[b] |
| Cd | 8.99 | 16.90 | 25.9 | 19.0 | | 1.17[b] |
| Hg | 10.43 | 18.75 | 29.2 | 19.1 | | 0.63[b] |

[a] For the chlorides for the R1 elements and for the oxides of the R2 elements, crystal lattice energies.
[b] Heats of vaporization, not sublimation.

Mg under most unusual and forcing conditions, is chemically unimportant for the family. Thus while it is true that the $I_2$ value is roughly double the $I_1$ value, making $I_1 + I_2 \simeq 3I_1$, other energy values, such as hydration or crystal lattice energies, increase in magnitude in going from 1+ to 2+ ions by more than enough to compensate for the extra ionization energy. (See Table 4-13.)

Although it is doubtful that simple $Be^{2+}$ and $Mg^{2+}$ ions form in any compounds, with the possible exception of the fluorides and oxides, the 2+ state is still the only common state. This is true even in the more covalent of their compounds where it is undoubtedly the stronger covalent bonding energy which brings out the 2+ oxidation state in preference to the 1+ state. In fact, the formation of covalent bonds requires the uncoupling of the paired atomic $s$ electrons and their promotion to higher energy states, valence states that are higher relative to the ground state of the metal atom. In these states they can both partake in chemical bond formation, an energy-releasing process.

**Family R2′.**  Atoms of these elements have the $(n - 1)d^{10}ns^2$ configuration and would be predicted to exhibit a 2+ oxidation state. An examination of ionization energies only (Table 4-13) might again suggest the 1+ state. However, except in compounds of the well-established species $Hg_2^{2+}$, and possibly the Cd analog, the predicted 2+ compounds are all that are found. In these two instances there is a metal-metal bond, and the element is therefore *bivalent* even though it must be assigned a 1+ oxidation state.

**Family R3.** The outer electron configuration $ns^2np^1$ suggests the two oxidation states 1+ and 3+, corresponding to the involvement of only the $p$ or the $p$ and both $s$ electrons in chemical bonding. Although there are many reports of 1+ compounds for Al, Ga, and In, the 3+ state is far more important for these elements. On the other hand, the heaviest family member, thallium, is more stable in the 1+ state. For example, in aqueous solution, the standard redox potential is quite high:

$$Tl^{3+} + 2e^- \rightleftharpoons Tl^+ \qquad E^\circ = +1.25 \text{ v}$$

The greater stability of the lower oxidation state in the heaviest family member is found not only with Tl in R3, but also with Pb and Bi in the R4 and R5 families, respectively. In 1933, Sidgwick proposed an unfortunately misleading explanation of this observation by considering the pair of outer $s$ electrons in these elements as well as in the element Hg as being an *inert pair*. However, ionization potential data do not support the early idea that the pair of $s$ electrons is intrinsically inert. Thus, for example, the $(I_2 + I_3)$ sums for Ga, In, and Tl are 51.0, 46.7, and 50.0 eV, respectively. A much more reasonable explanation attributes most of the greater stability of the lower oxidation state of the heaviest family member to the fact that in a given family the strength of covalent bonds in the higher oxidation states formed by these elements decreases as a family is descended. For example, consider the mean bond energies (in kcal/mole) among the trichlorides of Ga (57.8), In (49.2), and Tl (36.5). Drago[22] has attributed the decreased covalent bond-forming energy down a family to (1) a decrease in the amount of overlap of atomic orbitals resulting from the spreading of the valence electrons over a larger volume, and (2) the increased repulsion between the inner (nonbonded) electrons of the bonded atoms. However, the latter cannot be too important a factor inasmuch as we find (*vide infra*) that in descending a transition element family the heaviest family member prefers or is increasingly stable in the *higher* oxidation states.

Compounds that appear from their empirical formulas to be 2+ compounds, for example $GaX_2$ (X = halide) and GaY (Y = S, Se, Te) have been shown not to contain $Ga^{2+}$. The expected paramagnetic behavior of $Ga^{2+}$ is not observed in these compounds. Indeed, the halides are known to have salt-like structures $Ga^I[Ga^{III}X_4]$, and the $Ga^I$ may be complexed with four S, Se or As donor ligands. The chalcogenides have been shown to have Ga-Ga units in a layer lattice with each gallium atom tetrahedrally surrounded by three Y atoms and one gallium atom.

**Family R4.** The members of this family share the $ns^2np^2$ outer configuration leading to the observed stable oxidation states of 4+ and 2+. Again, with the lighter family members the maximum oxidation state is predominant. In fact neither C nor Si occurs in a stable 2+ state. Although carbon in CO would appear to show a 2+ state, all of the valence-shell orbitals and electrons are involved in bonding. And, in addition, the high-temperature silicon species such as SiO, $SiCl_2$, and SiS are apparently thermodynamically quite unstable. However the 2+ state increases progressively in stability in the order Ge < Sn ≪ Pb, which is also the order of decreasing stability of the 4+ state. For example, consider the aqueous solution standard electrode potentials:

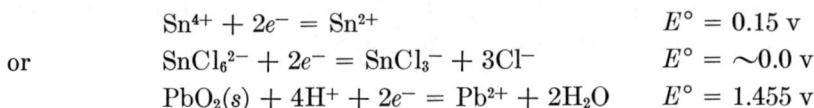

$$Sn^{4+} + 2e^- = Sn^{2+} \qquad\qquad E° = 0.15 \text{ v}$$

or $\qquad$ $SnCl_6^{2-} + 2e^- = SnCl_3^- + 3Cl^- \qquad E° = {\sim}0.0 \text{ v}$

$$PbO_2(s) + 4H^+ + 2e^- = Pb^{2+} + 2H_2O \qquad E° = 1.455 \text{ v}$$

or the decomposition temperatures of the hydrides:

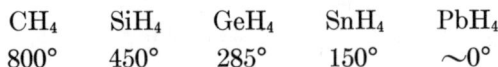

| $CH_4$ | $SiH_4$ | $GeH_4$ | $SnH_4$ | $PbH_4$ |
|---|---|---|---|---|
| 800° | 450° | 285° | 150° | ${\sim}0°$ |

or the addition of chlorine to the dichlorides:

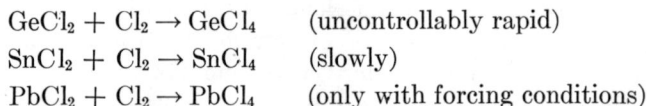

$$GeCl_2 + Cl_2 \rightarrow GeCl_4 \qquad \text{(uncontrollably rapid)}$$
$$SnCl_2 + Cl_2 \rightarrow SnCl_4 \qquad \text{(slowly)}$$
$$PbCl_2 + Cl_2 \rightarrow PbCl_4 \qquad \text{(only with forcing conditions)}$$

or the fact that tetrahalides exist for all R4 elements except for $PbBr_4$ and $PbI_4$.

The formal oxidation state of 4− may be said to exist in carbon, silicon and the few germanium compounds in which these elements are present in binary compounds with more electropositive elements.

**Family R5.** From the electron configuration $ns^2np^3$, which is characteristic of this family, one would readily predict three major oxidation states, 3−, 3+ and 5+. Nitrogen displays all integer oxidation states from 3− to 5+ and, in addition, some fractional states, such as $\frac{1}{3}$− in azides, $N_3^-$. This array of oxidation states arises because of the ability of nitrogen to undergo catenation (self-linkage) and to form $p\pi$ bonds with itself as well as other elements, particularly oxygen. The latter makes it possible for nitrogen to display the 5+ oxidation state even though the presence of only *four* bonding orbitals, one $s$ and three $p$, limits it to quadrivalency.

The 3− oxidation state is characteristic of the lighter family members but becomes progressively unstable down the family just as the *3+ state becomes increasingly stable* and the *5+ state decreasingly stable*. This is particularly abrupt between Sb and Bi.

**Family R6.** The $ns^2np^4$ outer electron configuration characteristic of this family gives rise to oxidation states of 2−, 2+, 4+ and 6+.

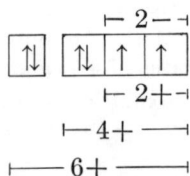

Being just two electrons shy of the very stable inert gas closed shell configuration, these elements, particularly the lighter family members, form the $2-$ state quite readily. Oxygen stands out in its family in this regard and forms very stable binary *oxides* with all elements except the lighter inert gas elements. Many of these occur in several different oxidation states. Except for several unusual negative states such as peroxides $(1-)$, superoxides $(\frac{1}{2}-)$, and ozonides $(\frac{1}{3}-)$, all of which contain O—O bonds, and the $1+$ and $2+$ states which must be ascribed to it in the very reactive compounds with fluorine, $O_2F_2$ and $OF_2$, oxygen will always be found in the $2-$ state. It does not share the $4+$ and $6+$ states with its family members, although it is capable of forming three and even four $\sigma$ bonds in certain compounds. But in these cases its oxidation state nevertheless is still $2-$. The $2-$ state decreases in stability and importance down the family, so that, for example, $H_2Te$ and $H_2Po$ are apparently thermodynamically unstable with respect to their constituent elements. On the other hand both the $2+$ and the $4+$ states grow steadily in importance and stability down the family. The $6+$ oxidation state appears to be unstable, as expected, only with the heaviest member, Po. But with the exception of $MF_6$ compounds, the remaining members of the family are good oxidants in this oxidation state.

**Family R7.** The nearly closed shell configuration, $ns^2np^5$, which characterizes this family, gives rise to a very stable $1-$ state and the predictable positive states of $1+$, $3+$, $5+$ and $7+$, for each member except fluorine. Since fluorine is the most electronegative of all elements it can have no oxidation state but $1-$. However, it is found occasionally to behave as a bridging atom, for example in $(SbF_5)_n$, $(BeF_2)_n$, etc., and is thus capable of simultaneously forming at least two sigma bonds. Positive oxidation states are found when the atoms bonded to the halogen are either oxygen (in oxides, oxyacids or oxysalts) or other halogen atoms (in $XX'$, $XX_3'$, $XX_5'$ and $IF_7$). Generally the stability and occurrence of the positive states falls off in the order $5+ > 7+ > 1+ > 3+$, although exceptions are numerous. Thus a $7+$ bromine compound has just been prepared even though this state is well-known and stable for both Cl and I. Certain compounds having *even* oxidation states are known, particularly with chlorine (for example, $ClO_2$ and $Cl_2O_6$) but these are rare and relatively unimportant.

**Inert Gas Elements.** The closed shell configuration $ns^2np^6$ for the inert gases beyond He was believed to be inviolate to chemical reactions until 1962 when the first chemical compounds of xenon were characterized. Since that time many compounds of xenon have been prepared[23] in the oxidation states of $4+$, $6+$, $2+$ and $8+$, in roughly that order of stability. Compounds have also been prepared with krypton ($2+$ and $4+$) and radon. Only the most highly electronegative elements, fluorine and oxygen, have been able to induce these very stable inert gas elements to form chemical bonds. The field is new, exciting, and rapidly growing, and may have already produced a chloride, $XeCl_4$, to add to the increasing list of new compounds.

### Coordination Numbers and Hybridizations of Representative Elements

More important than a classification by oxidation states for those elements that form covalent bonds are the related classifications by (1) the arbitrary and artificial but useful concept of hybridization of orbitals and (2) the real concept of coordination number or number of sigma bonds formed by the atom in a molecule. This approach is detailed in Table 4–14 for some of the representative elements in which the coordination numbers 2 through 6 are listed along with the corresponding stereochemical configuration of the *valence-shell electron pairs* that are either engaged in sigma bonding or are lone pairs (LP). The coordination number is the number of sigma bonds. The pi-bonding electrons do not affect the gross stereochemistry, as is discussed further in Chapter 7. The hybridized orbitals used to form sigma bonds are not listed whenever there are both bonding and lone pair electrons present in the valence shell since an accurate designation of the content of $s$, $p$ and $d$ atomic orbitals in the hybrid set is quite uncertain under these circumstances.

A few generalizations from Table 4–14 are in order. The coordination number, the number of sigma bonds, for a first short-period (Li through Ne) atom is limited by size and by availability of orbitals to 4. This limitation obviously holds also for Li and Be, which are not included in the table. The coordination number for second short-period atoms is limited to 6 even though it would appear that enough orbitals (one $s$, three $p$, and five $d$, or a total of nine) are available for higher coordination numbers. The limitation is probably to be ascribed to both size requirements, which includes ligand-ligand repulsions, and to the much higher symmetry possible with six-coordination as opposed, for example, to seven-coordination. Higher coordination numbers may occur with third- and later-period elements, but outside of the transition series elements these are very rare.

### Transition Elements

Excluding from consideration the T3 elements, and this family may be taken to include all of those which are filling $4f$ and $5f$ levels (the so-called lanthanide and actinide series), we can make a few generalizations concerning the oxidation states of the transition elements. Reference should be made to Table 4–15 when reading this section.

(1) The lowest common state is 2+. This is suggested by the general external electron configurations $(n - 1)d^x ns^2$ or $(n - 1)d^{x+1} ns^1$ or $(n - 1)d^{x+2} ns^0$, where $x + 2$ is the family number. However, with the $4d$ and $5d$ series the 2+ state is not very stable across the periods until Pd and Pt are reached, and then it is again unstable with Ag and Au, more so with the latter. In the $3d$ series the 2+ state is very unstable and strongly reducing with Ti, and the stability increases with a corresponding decrease

in reducing strength as the series is traversed left to right. Indeed the $2+$ state is quite stable and common from Mn to Cu.

(2) The maximum oxidation state never exceeds the family number. It is $4+$ in T4, $5+$ in T5, $6+$ in T6, etc. Furthermore, in the $3d$ series the oxidizing strength of the maximum oxidation state increases progressively from Ti(IV) to Mn(VII). Beyond Mn, which has a half-filled $d$ subshell, where electron pairing has had to occur in the $d$ sublevel, no state higher than $6+$ is found. In fact, the lower oxidation states have become far more stable. The latter occurrence perhaps reflects the greater increase in binding energy of $(n - 1)d$ electrons relative to $ns$ electrons as the transition series is traversed left to right.

(3) Contrary to the situation that prevails in the representative element families, in the transition element families the heavier members show increasing stability in *higher* oxidation states. Thus Nb(V) and Ta(V) have no comparable oxidizing strength to that of V(V). And the same could be said about the $6+$ state for Cr, Mo, and W, in which only Cr(VI) is a strong oxidizing agent. $Mn_2O_7$ decomposes at $0°C$, whereas $Tc_2O_7$ melts at $119.5°C$ and $Re_2O_7$ melts at $220°C$ and sublimes without decomposition. The increasing stability of higher oxidation states in the order $3d \ll 4d < 5d$ probably results from at least two important factors that change in this order. First of all, the sublimation energy of the metal increases down the family, requiring the formation of stronger and/or more bonds for energy compensation. Secondly, the possibilities for very strong covalent bonds, enhanced by the favorable $\pi$-bond forming abilities of the $(n - 1)d$ orbitals (here $\pi$-acceptor tendencies, from ligands such as $F^-$ and $O^{2-}$), grows rather than diminishes down the family. Thus both $\sigma$- and $\pi$-electron donation from ligands to central metal ions is enhanced by the increase in effective nuclear charge down a family where very little size increase occurs from $3d$ to $4d$ and almost none from $4d$ to $5d$ (see discussion of this point on page 117). Actually the increase in the value of the quotient $Z_{eff}/r_{cov}^2$ is probably much more significant than simply the increase in the value of $Z_{eff}$ by itself. To illustrate how $F^-$ and $O^{2-}$ *bring out* the highest oxidation state of an element, we have listed in Table 4–16 the highest oxidation states for 10 metals in their binary compounds with $F^-$, $Cl^-$, $I^-$, $O^{2-}$, and $S^{2-}$.

(4) As may be seen from Table 4–15 the number of unusual or uncommon oxidation states is much greater than the number of common states. Even zero and formal negative oxidation states are widespread. However, except for one or two most intriguing compounds assumed to derive from Ni($-$I), all negative states may be predicted by not allowing the valence shell to exceed the $d^{10}$ configuration. The latter situation, for example, occurs with Mn(-III), Fe(-II), Co(-I) and Ni(0).

(5) Finally, it is a good time to point out the *great prevalence and preference for the coordination number of 6*. This is followed by coordination number 4 as a not-too-close second. Other coordination numbers are far

TABLE 4-14. Coordination Numbers, Stereochemistry, and Hybrid Orbital Types Found for R3 through R7 Elements.[a]

| Elements | 2, Digonal | 3, Trigonal | 4, Tetrahedral | 5, Trigonal bipyramidal | 6, Octahedral |
|---|---|---|---|---|---|
| B | — | $3(sp^2)\sigma$ | $4(sp^3)\sigma$ | — | — |
| Al | — | $3(sp^2)\sigma + p\pi$ | $4(sp^3)\sigma$ | — | $6(sp^2d^2)\sigma$ |
| Ga, In, Tl | $2(sp)\sigma$ | — | $4(sp^3)\sigma$ | — | $6(sp^2d^2)\sigma$ |
| C | $2(sp)\sigma + 2p\pi$ | $3(sp^2)\sigma + p\pi$ | $4(sp^3)\sigma$ | — | — |
| Si | — | — | $4(sp^3)\sigma^b$ | — | $6(sp^3d^2)\sigma$ |
| Ge, Sn, Pb | — | — | $4(sp^3)\sigma^b$ | $5(sp^3d)\sigma^c$ | $6(sp^3d^2)\sigma$ |
| N | $2(sp)\sigma + 2p\pi$ | $3(sp^2)\sigma + p\pi$ | $4(sp^3)\sigma$<br>$3\sigma + LP$<br>$2\sigma + 2LP$<br>$1\sigma + 3LP$ | — | — |
| P, As, Sb | — | — | $3\sigma + LP$ | $5(sp^3d)\sigma$ | $6(sp^3d^2)\sigma$ |
| O | $(sp)\sigma + 2p\pi + LP$ | $2\sigma + p\pi + LP$<br>$\sigma + p\pi + 2LP$ | $4(sp^3)\sigma$<br>$3\sigma + LP$<br>$2\sigma + 2LP$<br>$1\sigma + 3LP$ | — | — |
| S, Se, Te | $(sp)\sigma + 2\pi + LP$ | $d$ | $4\sigma + 2d\pi$<br>$3\sigma + d\pi + LP$<br>$2\sigma + 2LP$<br>$3\sigma + LP$ | $4\sigma + LP$ | $6(sp^3d^2)\sigma$ |

## TABLE 4-14. (Continued)

| | | | | |
|---|---|---|---|---|
| F | — | — | $2\sigma + 2\text{LP}$ <br> $\sigma + 3\text{LP}$ | — | — |
| Cl, Br, I | — | — | $2\sigma + 2\text{LP}$ <br> $\sigma + 3\text{LP}$ <br> $^e$ | $3\sigma + 2\text{LP}$ <br> $2\sigma + 3\text{LP}$ <br> $5\sigma + 2d\pi^f$ | $5\sigma + \text{LP}$ <br> $5\sigma + 2d\pi^f$ <br> $_g$ |

$^a$ The number in the heading is the number of sigma pairs plus lone pairs, and the geometry is the arrangement of all of these pairs about the central atom.
$^b$ Or $3\sigma + \text{LP}$.
$^c$ Only for Sn.
$^d$ For S there is $\sigma + 2\text{LP} + d\pi$.
$^e$ Also, for Cl and I: $2\sigma + 2\text{LP} + d\pi$; $3\sigma + \text{LP} + d\pi$; $4\sigma + 3d\pi$; for Cl: $2\sigma + 2\text{LP} + 2d\pi$; $3\sigma + \text{LP} + 3d\pi$.
$^f$ Only for I.
$^g$ I has a $7\sigma$ compound, $IF_7$.

TABLE 4-15. Oxidation States and Coordination Numbers of the Transition Elements.[a,b]

| Element | $d^0$ | $d^1$ | $d^2$ | $d^3$ | $d^4$ | $d^5$ | $d^6$ | $d^7$ | $d^8$ | $d^9$ | $d^{10}$ |
|---|---|---|---|---|---|---|---|---|---|---|---|
| Ti | 4+ (4,5,6,7,8) | 3+ (6) | 2+ (6) | — | 0 (6) | 1− (6) | | | | | |
| Zr,Hf | 4+ (4,6,7,8) | 3+ (?) | 2+ (?) | — | 0 (6) | | | | | | |
| V | 5+ (4,5,6) | 4+ (4,5,6,8) | 3+ (4,5,6) | 2+ (6) | 1+ (6) | 0 (6) | 1− (6) | | | | |
| Nb,Ta | 5+ (π,5?,6,7,8,) | 4+ (6,8) | 3+ (?) | 2+ (?) | 1+ (π) | — | 1− (6) | | | | |
| Cr | 6+ (4) | 5+ (4,6,8) | 4+ (4,6) | 3+ (4?,6) | 2+ (6,7) | 1+ (6) | 0 (6) | 1− (6) | 2− (?) | | |
| Mo,W | 6+ (4,5,6,8?) | 5+ (5,6,8) | 4+ (π,6,8) | 3+ (6,7?,8) | 2+ (π,5,6,7,9) | 1+ (π) | 1− (6) | — | 2− (5) | | |
| Mn | 7+ (3,4) | 6+ (4) | 5+ (4) | 4+ (6) | 3+ (5?,6) | 2+ (4,5?,6,7) | 1+ (6) | 0 (6) | 1− (4,5,6) | 2− (4,6) | 3− (4) |
| Tc,Re | 7+ (4,6,7,8,9) | 6+ (6,7,8) | 5+ (5,6,7,8) | 4+ (4?,6,7) | 3+ (π,5,6) | 2+ (5,6) | 1+ (π,6) | 0 (6) | 1− (5) | — | |
| Fe | | 7+ (4,6) | 6+ (4) | 5+ (4) | 4+ (6) | 3+ (4,6,7) | 2+ (4,5?,6) | 1+ (6) | 0 (5,6) | — | 2− (4) |
| Ru,Os[c] | 8+ (4,5,6) | 7+ (4,6) | 6+ (4,5,6) | 5+ (5,6) | 4+ (6,8) | 3+ (6) | 2+ (5,6) | 1+ (?) | 0 (5) | — | 2− (4) |

## TABLE 4-15. (Continued)

| | | | | | | | | |
|---|---|---|---|---|---|---|---|---|
| Co | 1− (4) | 0 (4) | 1+ (4,5,6) | 2+ (4,5,6) | 3+ (4,6) | 4+ (6) | | |
| Rh,Ir[d] | 1− (4,5?) | 0 (?) | 1+ (4,5) | 2+ (4,5,6) | 3+ (5,6) | 4+ (6) | 5+ (6) | 6+ (6) |
| Ni | 0 | 1+ (4?) | 2+ (4,5,6) | 3+ (4?,5,6) | 4+ (6) | | | |
| Pd,Pt[e] | 0 (4) | — | 2+ (4,5,6) | — | 4+ (6) | 5+ (6) | 6+ (6) | |
| Cu | 1+ (2,3,4) | 2+ (4,5,6) | 3+ (4?,6) | | | | | |
| Ag,Au[f] | 1+ (2,3,4) | 2+ (4) | 3+ (4,5,6) | | | | | |

[a] The numbers in parentheses are the known or suspected (?) coordination numbers, with the underlined ones being the more common for the given oxidation state.

[b] The $\pi$ appears in parentheses where a $\pi$-complex is known, in which case coordination number loses its usual significance.

[c] The 2− and 1+ states are not yet known with Os.

[d] The 5+ state is not yet known for Rh.

[e] The 5+ and the 6+ states are not yet known for Pd.

[f] The stability order is Ag: 1+ ≫ 2+ > 3+ and Au: 3+ ∼ 1+ ≫ 2+.

**TABLE 4–16.  Highest Oxidation States Obtainable in Certain Binary Compounds.**

|     | F⁻ | Cl⁻ | I⁻ | O²⁻ | S²⁻ |
|-----|-----|-----|-----|-----|-----|
| V   | 5   | 4   | 3   | 5   | 4   |
| Ta  | 5   | 5   | 5   | 5   | 5   |
| Cr  | 5   | 3   | 3   | 6   | 3   |
| W   | 6   | 6   | 4   | 6   | 6   |
| Mn  | 4   | 4   | 2   | 7   | 3   |
| Re  | 7   | 6   | 4   | 7   | 7   |
| Co  | (4) | 2   | 2   | (4) | 2   |
| Ir  | 6   | 4   | 3   | 6   | 6   |
| Ni  | (4) | 2   | 2   | (4) | 2   |
| Pt  | 6   | 4   | 4   | 4   | 4   |

less common even though they are being discovered with ever-increasing frequency.

### Inner Transition Elements

For the first inner transition series (lanthanide or $4f$ series) the 3+ oxidation state is not only common to all members, but it is the most stable state. The only other states ever found are the 2+ and 4+ states, and these are limited, as shown in Table 4–17. The most stable 4+ state is found with Ce, which has a $4f^2 6s^2$ outer configuration and thus reaches a closed shell inert gas configuration in this state. The 4+ state for Tb might have been predicted since in this state its configuration becomes identical with that of 3+ Gd with a half-filled and hence spherically symmetrical $4f$ sublevel. The most stable 2+ state occurs with Eu, which in this state likewise attains the $4f^7$ configuration. The 2+ state for Yb could then be predicted as its configuration in this state becomes $4f^{14}$, a closed subshell.

The oxidation states for the second inner transition series elements (Table 4–17) show a greater variation, at least in the first six members of the series. Indeed the most stable states found for Th, Pa, and U suggest that they might better be considered the heaviest members of the T4, T5, and T6 families, respectively. The 3+ state is not the most stable one until Am is reached, beyond which it is by far the most stable or only state known. The existence of 4+ compounds of Cm, for example, $CmF_4$ and $CmO_2$, indicates that the stability of the $5f^7$ configuration is not as great as that of the $4f^7$ configuration. The existence of 4+ Bk and the possible existence of 2+ Am are understandable in terms of these ions possessing a half-filled $5f$ subshell.

TABLE 4–17. Oxidation States of the Inner Transition Elements and Sc, Y, La, and Ac.[a,b,c]

| Sc,Y,La | Ce | Pr | Nd | Pm | Sm | Eu | Gd | Tb | Dy | Ho | Er | Tm | Yb | Lu |
|---|---|---|---|---|---|---|---|---|---|---|---|---|---|---|
| — | — | — | 2 | — | 2 | 2 | — | — | — | — | — | 2 | 2 | — |
| 3 | $\underline{3}$ | $\underline{3}$ | $\underline{3}$ | 3 | $\underline{3}$ | $\underline{3}$ | $\underline{3}$ | $\underline{3}$ | $\underline{3}$ | 3 | 3 | $\underline{3}$ | $\underline{3}$ | 3 |
| — | 4 | 4 | 4? | — | — | — | — | 4 | 4? | — | — | — | — | — |

| Ac | Th | Pa | U | Np | Pu | Am[d] | Cm | Bk | Cf | Es | Fm | Md | No | Lw |
|---|---|---|---|---|---|---|---|---|---|---|---|---|---|---|
| 3 | 3 | 3 | 3 | 3 | 3 | $\underline{3}$ | $\underline{3}$ | $\underline{3}$ | $\underline{3}$ | 3 | 3 | 3 | 3 | 3 |
| — | $\underline{4}$ | 4 | $\underline{4}$ | 4 | $\underline{4}$ | 4 | 4 | 4 | 4 | | | | | |
| | | $\underline{5}$ | 5 | $\underline{5}$ | 5 | 5 | | | | | | | | |
| | | | 6 | 6 | 6 | 6 | | | | | | | | |

[a] Stability order of 2+ state Eu > Yb ≫ Sm > Tm ~ Nd. $MI_2$ solids (M = La, Ce, Pr, and Gd) do not contain $M^{2+}$ ions, but are metallic in nature.
[b] Stability order of 4+ state Ce ≫ Tb ~ Pr > (Nd ~ Dy)? (last two doubtful).
[c] The most stable states are underlined.
[d] A 2+ state may also exist and would be predicted from the $5f^7$ configuration.

## References

1. R. Rich, "Periodic Correlations," W. A. Benjamin, Inc., New York, 1965.
2. R. T. Sanderson, "Chemical Periodicity," Reinhold Publishing Corporation, New York, 1960; and "Inorganic Chemistry," Reinhold Publishing Corporation, New York, 1967.
3. L. Pauling, "The Nature of the Chemical Bond," 3rd ed., Cornell University Press, Ithaca, New York, 1960.
4. J. C. Slater, *J. Chem. Phys.*, **41**, 3199 (1964).
5. J. T. Waber and D. T. Cromer, *J. Chem. Phys.*, **42**, 4116 (1965).
6. L. Pauling and D. M. Yost, *Proc. Natl. Acad. Sci. U. S.*, **14**, 414 (1932); L. Pauling, *J. Am. Chem. Soc.*, **54**, 3570 (1932).
7. H. O. Pritchard and H. A. Skinner, *Chem. Revs.*, **55**, 745 (1955).
8. R. S. Mulliken, *J. Chem. Phys.*, **2**, 782 (1934), **3**, 573 (1935); **46**, 497 (1949).
9. J. Hinze and H. H. Jaffe, *J. Am. Chem. Soc.*, **84**, 540 (1962).
10. J. Hinze, M. A. Whitehead and H. H. Jaffe, *J. Am. Chem. Soc.*, **85**, 148 (1963).
11. J. Hinze and H. H. Jaffe, *Can. J. Chem.*, **41**, 1315 (1963); *J. Phys. Chem.*, **67**, 1501 (1963).
12. W. Gordy, *Phys. Rev.*, **69**, 604 (1946).
13. R. P. Iczkowski and J. L. Margrave, *J. Am. Chem. Soc.*, **83**, 3547 (1961).
14. G. Klopman, *J. Chem. Phys.*, **43**, S124 (1965).
15. A. L. Allred and E. G. Rochow, *J. Inorg. Nucl. Chem.*, **5**, 264, 269 (1958).
16. J. E. Huheey, *J. Inorg. Nucl. Chem.*, **27**, 2127 (1965).
17. J. E. Huheey, *J. Phys. Chem.*, **69**, 3284 (1965).
18. R. T. Sanderson, *Science*, **114**, 670 (1951).
19. J. C. Slater, *Phys. Rev.*, **36**, 57 (1930).
20. J. C. Slater, "Theory of Atomic Structure," Vol. I, McGraw-Hill, Inc., New York, 1960.
21. E. Clementi and D. L. Raimondi, *J. Chem. Phys.*, **38**, 2686 (1963).
22. R. Drago, *J. Phys. Chem.*, **62**, 353 (1958).
23. J. G. Malm, H. Selig, J. Jortner, and S. A. Rice, *Chem. Revs.*, **65**, 199 (1965).

## Problems

1. Using the new set of rules for calculating effective nuclear charges, redetermine electronegativity values for elements 3 through 35 using the Allred-Rochow definition, and compare these new values with the old.

2. In light of the recent elucidations of a chemistry for at least half of the inert gas elements, discuss the chemical significance of electronegativity for these elements. In particular, what would you expect to happen to the trend established with the horizontally preceding elements when this is extended to include the inert gases? (See R. T. Sanderson, *J. Inorg. Nucl. Chem.*, **27**, 989 (1965).)

3. In 1955, Sanderson (*J. Chem. Phys.*, **23**, 2467 (1955)) proposed an interesting electronegativity scale based upon a quantity he defined as a *stability ratio*, SR. This is defined for a given atom as the ratio of the average electron density, ED, around the atom to the hypothetical electron density, $ED_i$, the atom would have were it to be an inert gas atom. ED is obtained by dividing the number of electrons around the atom by $4/3 \pi r^3$, where $r$ is the covalent radius of the atom. The $ED_i$ is to be calculated for a given atom by linear interpolation between the electron density values of the preceding and succeeding inert gas atoms.

Calculate SR values for the atoms in the first two short periods and relate these values to both the Pauling and the Allred-Rochow electronegativity values for the same elements. What do you find to recommend the SR values over other electronegativity scales? What are the serious disadvantages to the SR values?

4. The great importance and prevalence of coordination number 6 is obvious from Table 4–15. Suggest as many reasons as you can for this. You may find that you can add some additional reasons after reading Chapters 7, 10, and 11.

5. How is the electronegativity concept useful in the determination *of*, or the determination *from* (a) percent ionic character in a bond, (b) bond polarity, and (c) bond stretching force constants? (See J. K. Wilmshurst, *J. Chem. Ed.*, **39,** 132 (1962), and W. Gordy, *J. Chem. Phys.*, **14,** 304 (1946).)

6. Document the chemical similarities found for (a) Li and Mg, (b) Be and Al, and (c) B and Si. Suggest explanations (more detailed than found in the text here) in each case for the observed similarities.

7. Show how one might set up a procedure for calculating percent ionic character utilizing ED map, ionic and atomic radii. What are the advantages and the disadvantages to such a procedure?

8. Notice from Figure 4–3 that the atomic radii for europium and ytterbium are well out of line; that is, they are much larger than those of their immediate neighbors, and in fact fall on an almost straight line drawn from barium to lead. Explain.

9. In the R11, R2′, R3, and R4 families, as well as in most of the transition element families, the first ionization energy for the heaviest member of the family is *higher* than that of the next-lightest family member. Explain.

10. Why have chemists expected the chemical behavior of the transition elements, in particular the $3d$ series, to roughly parallel the $(I_1 + I_2)$ sum?

11. In trying to rationalize the peculiar trend in the $I_1$ values across the period from Na to Ar, different explanations have been offered. For example, consider 5.14, 7.64, 6.00 eV for Na, Mg, and Al. One explanation is that the Mg value is *abnormally high* because of the very poor shielding one $s$ electron affords its partner, the other $s$ electron. Another explanation is that the Al value is *abnormally low* because a $p$ electron is being removed, it is less penetrating than an $s$ electron, and furthermore it is being well-shielded by a pair of $s$ electrons. How would *you* explain the observed trend? Offer explanations for the other apparently anomalous values in both the $I_1$ and $I_2$ set of values.

12. Why do we not find univalent ions of the transition elements?

13. Find a dozen examples of inorganic compounds in which the oxidation state and the valence of a given element are numerically different. Explain why this is so.

14. Why should the 1− oxidation state for Ni, for example in the carbonylate anion $[Ni_2(CO)_6]^{2-}$, be considered far more unusual than certain other 1− oxidation states, for example with Mn, Co, Cr, V, or Ti?

15. What are some of the properties of the elements or their compounds which parallel the Allred-Rochow electronegativity trends better than the Pauling or Mulliken trends, for example in R3, R4, and R5?

# 5

# Chemical Bonding

Before the discovery of the electron, an understanding of the nature of a chemical bond was not possible. It is true that the idea of valence existed as early as 1852, and some appreciation of molecular geometry existed shortly thereafter. In this respect, the tetrahedral structure of the carbon atom was recognized by van't Hoff and by le Bel, and the stereochemistry of complex ions was deduced by Werner. For such structures to exist, it was apparent that some type of bonding force must be present. For the lack of something better, the chemical bond was represented by a straight line between the symbols of the bonded atoms. This indicates the existence of a bond, but it, of course, fails to give any description of the nature of the bond.

Even before the discovery of the electron, the independent existence of ions was proposed by Arrhenius. In terms of this concept, several attempts were made to offer an explanation of the bonding forces between atoms. Although they were unsuccessful, they contributed to the idea of electrical charge as a basis for bond formation. Then, with the discovery of the electron, our present-day advances were made possible. During the first few years, a variety of explanations of bond formation were offered in terms of positive and negative ions, but little or no attempt was made to relate the charges to the structure of the atom. Then, in 1916, G. N. Lewis presented his theory of valence.[1] Since that time, we have made great strides in the mathematical formulation of the valency theory, but the basic ideas of the chemical bond can still be traced to the original ideas of Lewis.

According to Pauling,[2] *a chemical bond exists between two atoms when the bonding force between them is of such strength as to lead to an aggregate of sufficient stability to warrant their consideration as an independent molecular species.* Although it would appear that this definition permits some freedom of choice, we ordinarily find it convenient to consider five types of chemical bonds. These are *ionic bonds, covalent bonds, metallic bonds, hydrogen bonds,* and *van der Waals forces.* All of these are important, and the first three are quite strong. However, in spite of its importance, we will not consider the metallic bond in this discussion. There is a brief account of the metallic bond in Chapter 6, and references are given there for additional study.

## THE IONIC BOND

From a mathematical standpoint, the simplest type of chemical bond is one that can be considered strictly electrostatic in character. Such a treatment has proved successful for the alkali halides where the bonding occurs between the cation of a highly electropositive atom and the anion of a highly electronegative atom. Although it is possible to consider a bond to be partially covalent and partially ionic, the extent of ionic character in the bond is dependent on the difference in electronegativity between the combining atoms. In the instance of the alkali halides, it would be safe to consider the bonds to be almost exclusively ionic. However, the test of this postulate will actually rest on the success we have in quantitatively evaluating various properties of the resultant compounds. In general, we can define a bond as purely ionic in terms of the success of the electrostatic model.

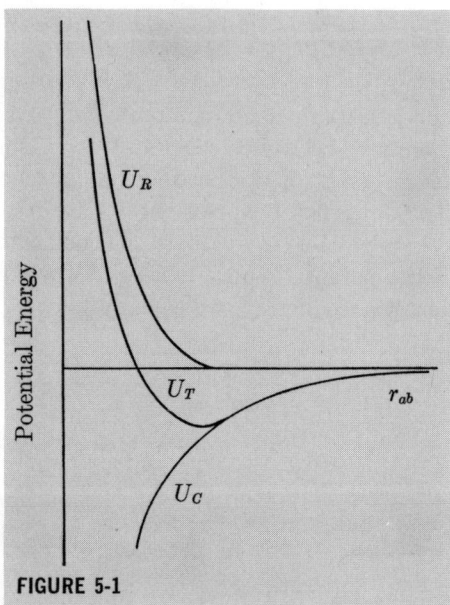

**FIGURE 5-1**

Potential Energy Diagram for the Formation of an Ionic Bond as a Function of Ion Separation.

As a first test of the electrostatic model, it should be possible to calculate the energy of a crystalline lattice. And, as a first-order approximation, we can consider the ions to be point charges. For the formation of a single uni-univalent molecule, we can imagine a positive and a negative point charge to be separated by an infinite distance and then allow them to approach each other. From our general knowledge of electrostatics, we are aware that an attraction will exist between the two ions that can be described by Coulomb's law. At the same time, we also know that two ions would not be expected to approach each other sufficiently close to coalesce. Consequently, there must be a *repulsion* term that comes into play at some place apparently rather close to the equilibrium distance between the two ions. Such a situation can be described by the potential energy diagram shown in Figure 5–1. The potential energy of the system is thus seen to be expressible as the sum of the attractive potential, given by Coulomb's law, and a repulsive potential that must be represented in some alternative manner.

It should be apparent that the repulsive term is due to a short-range type of force; that is, it is not effective except when the ions are extremely close to each other. It was proposed by Born that such a term could be represented by $U_R = B/r_{ij}^n$, where $r_{ij}$ is the distance between the two ions, and $n$ is a parameter that is usually found to be of the order of 9. This, then, gives the mathematical description of the potential energy diagram

$$(5–1) \qquad U_T = U_C + U_R = -\frac{z_i z_j e^2}{r_{ij}} + \frac{B}{r_{ij}^n}$$

For a uni-univalent compound, $z_i = z_j = 1$, and in this case they can, therefore, be neglected. For this reason $z_i$ and $z_j$ will be omitted in the development of the pertinent equations, but they will be introduced into the final general expression.

It is possible to evaluate the coefficient, $B$, by recognizing that the slope of the potential curve is zero at the equilibrium distance, $r_0$, between the ions. Consequently, we can say that

$$\left(\frac{\partial U_T}{\partial r}\right)_{r=r_0} = \frac{e^2}{r_0^2} - \frac{Bnr_0^{n-1}}{r_0^n r_0^n} = 0$$

or

$$(5–2) \qquad \frac{e^2}{r_0^2} - \frac{nB}{r_0^{n+1}} = 0$$

Solving Eq. (5–2) for $B$, we obtain

$$B = \frac{e^2 r_0^{n-1}}{n}$$

and by substituting for $B$ in Eq. (5–1), we obtain the expression for the potential at the equilibrium position

$$U_T = -\frac{e^2}{r_0} + \frac{e^2 r_0^{n-1}}{n r_0^n} = -\frac{e^2}{r_0} + \frac{e^2}{n r_0}$$

or

(5–3)      $$U_T = -\frac{e^2}{r_0}\left(1 - \frac{1}{n}\right)$$

If we now consider the energy of an entire crystal, it is necessary to multiply by the number of ions present, and in addition, we must take into consideration higher-order attractive and repulsive terms. These arise due to the effects of second, third, and so on nearest neighbors. The first factor can readily be determined by considering a crystal containing 1 mole of ions of each type. The second factor is found to be dependent on the particular crystalline structure. It has been evaluated for numerous crystal lattices and is referred to as the *Madelung constant*. For general comparison, a few examples of the Madelung constant for uni-univalent salts are given in Table 5-1. If these terms are now incorporated into our basic equation,

**TABLE 5-1.  Values of the Madelung Constant for Several Different Crystal Lattices.**

| Lattice | Madelung Constant |
| --- | --- |
| Rock salt | 1.74756 |
| Cesium chloride | 1.76267 |
| Zinc blende | 1.63806 |
| Wurtzite | 1.641 |

we can obtain the general expression for the lattice energy of the crystal[3]

(5–4)      $$U = -\frac{z_i z_j \, A N e^2}{r_0}\left(1 - \frac{1}{n}\right)$$

where $A$ is the Madelung constant and $N$ is Avogadro's number.

Finally, the parameter, $n$, can be evaluated from the compressibility of the crystal by virtue of the relationship

(5–5)      $$\kappa = \frac{18 r_0^4}{A e^2 (n - 1)}$$

where $\kappa$ is the compressibility.

Although the lattice energies calculated in terms of this model are quite good, they can be improved by considering two additional factors: the repulsion arising from electron-electron interactions and the zero point energy of the ions. These have been incorporated along with a different form of the repulsion term variously by Mayer, Born, and Helmholz in the form

(5–6)        $$U = -\frac{z_i z_j A N e^2}{r} + N B e^{-kr} - \frac{NC}{r^6} + N E_0$$

where $k$ and $C$ are two new parameters and $E_0$ is the zero point energy.[4]

An experimental test of the electrostatic model can be made either by a direct determination of the vapor phase equilibrium

$$MX \rightleftharpoons M^+ + X^-$$

in conjunction with the $MX$ sublimation pressure[5] or by means of an indirect method using a cyclic process known as the *Born-Haber cycle*.[6] As it turns out, a direct measurement of a lattice energy is rather difficult to make and, for that reason, the Born-Haber cycle is more commonly used.

The Born-Haber cycle for sodium chloride can be represented as

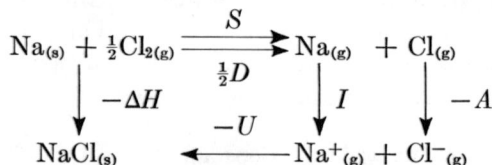

$$\begin{array}{ccc}
\text{Na}_{(s)} + \tfrac{1}{2}\text{Cl}_{2(g)} & \xrightarrow[\tfrac{1}{2}D]{S} & \text{Na}_{(g)} + \text{Cl}_{(g)} \\
\Big\downarrow {\scriptstyle -\Delta H} & & \Big\downarrow {\scriptstyle I} \quad \Big\downarrow {\scriptstyle -A} \\
\text{NaCl}_{(s)} & \xleftarrow{\;-U\;} & \text{Na}^+_{(g)} + \text{Cl}^-_{(g)}
\end{array}$$

Here,

$S$ = heat of sublimation
$D$ = dissociation energy
$I$ = ionization energy
$A$ = affinity energy
$\Delta H$ = heat of formation
$U$ = lattice energy

Those quantities that require the expenditure of energy are represented as positive and those that take place with the release of energy are represented as negative. It is apparent from the cycle that

(5–7)        $$U = \Delta H + S + \tfrac{1}{2}D + I - A$$

Thus, if all the necessary quantities are known, we can evaluate the crystal lattice energy. Such calculations and measurements have been carried out according to the two theoretical equations as well as by direct measurement and by the Born-Haber cycle. Some idea of the success of the electrostatic model can be obtained from a comparison of these determinations as seen in Table 5–2.

## THE VARIATION METHOD

Although a simple electrostatic model is adequate for the treatment of an ionic bond, it is necessary to use a quantum mechanical approach to the covalent bond. This requires a solution of the Schroedinger wave

**TABLE 5-2.**[a] **Comparison of the Theoretical and the Experimental Determinations of the Lattice Energies of the Alkali Halides in kcal/mole.**

| Compound | $U_{calc}$ | | $U_{exp}$ | |
|---|---|---|---|---|
| | Eq. (5-4) | Eq. (5-6) | Born-Haber Cycle | Direct Measurement |
| LiF | 238.9 | 240.1 | 242.8 | |
| NaF | 213.8 | 213.4 | 216.6 | |
| KF | 189.2 | 189.7 | 191.8 | |
| RbF | 180.6 | 181.6 | 184.6 | |
| CsF | 171.6 | 173.7 | 176.0 | |
| LiCl | 192.1 | 199.2 | 201.7 | |
| NaCl | 179.2 | 184.3 | 183.9 | 181.3 |
| KCl | 163.2 | 165.4 | 168.3 | |
| RbCl | 157.7 | 160.7 | 162.8 | |
| CsCl | 147.7 | 152.2 | 157.2 | |
| LiBr | 181.9 | 188.3 | 191.0 | |
| NaBr | 170.5 | 174.6 | 175.5 | 176 |
| KBr | 156.6 | 159.3 | 160.7 | 160 |
| RbBr | 151.3 | 153.5 | 157.1 | 151 |
| CsBr | 142.3 | 146.3 | 151.2 | |
| LiI | 169.5 | 174.1 | 178.4 | |
| NaI | 159.6 | 163.9 | 164.8 | 166 |
| KI | 147.8 | 150.8 | 151.5 | 153 |
| RbI | 143.0 | 145.3 | 147.9 | 146 |
| CsI | 134.9 | 139.1 | 143.7 | 141.5 |

[a] From J. A. A. Ketelaar, "Chemical Constitution," 2nd ed., Elsevier Publishing Co., Amsterdam, 1958.

equation. When we recall that an exact solution of the wave equation is possible only for a one-electron system, it is apparent that we must resort to approximation methods when treating a many-electron system such as we would expect to encounter in a molecule.

There are two rather common approximation methods used in quantum mechanics. These are the *perturbation method* and the *variation method*.[7] The variation method is by far the most useful in the treatment of chemical bonding, and for that reason, it will be the only method that we shall consider here. The development of the variation principle is given in appendix A. Here we shall only illustrate its application to the problem of chemical bonding.

It can readily be seen that the time-independent wave equation

$$(2-13) \qquad \nabla^2 \psi + \frac{8\pi^2 m}{h^2} (E - V) \psi = 0$$

can be rearranged to the form

$$\left(-\frac{h^2}{8\pi^2 m}\nabla^2 + V\right)\psi = E\psi$$

The term on the left side of the equation contains the differential operator, $\nabla^2$, and some multiplying constants as well as the wave function, $\psi$. The quantities within the parentheses can be considered, together, to be an operator acting on $\psi$. Consequently, the equation may be expressed in the more convenient form

(5–8) $\qquad H\psi = E\psi$

where $H$ represents this operator and is known as the *Hamiltonian operator*.

Now, if we multiply both sides of Eq. (5–8) by $\psi^*$ and integrate over the configuration space, we will obtain the expression

$$\int \psi^* H\psi \, d\tau = E \int \psi\psi^* \, d\tau$$

or

(5–9) $\qquad E = \dfrac{\displaystyle\int \psi^* H\psi \, d\tau}{\displaystyle\int \psi\psi^* \, d\tau}$

Assuming that we know $\psi$ as well as the form of the Hamiltonian operator, and assuming that we can solve the equations, we obviously have an expression that will permit the calculation of the energy of the system. As it turns out, we can ordinarily set up the Hamiltonian operator for a given problem, but it would be unusual if we happened to know the form of $\psi$. Nevertheless, it is still possible to use the variation principle to obtain an approximate wave function.

To begin with, we can guess a wave function and calculate the energy in terms of this proposed function. If we were so fortunate as to guess the correct one, we would obtain the correct energy of the system. On the other hand, if we happened to choose a poor wave function, we would expect to get a rather poor value for the energy. Thus, it would appear that unless we had some means of making a particularly good choice of wave functions, Eq. (5–9) would be of little value. However, this is not the case. According to the variation principle, if $\psi_0$ is the correct wave function for a system, that is, the one that will give the correct energy of the system, $E_0$, then any other acceptable wave function, $\psi_i$, will give an energy greater than $E_0$. Therefore, any energy, $E_i$, obtained from a wave function, $\psi_i$, will be greater than $E_0$. And although this will not give us the correct wave function, it will, at least, tell us which of several trial wave functions is closest to the true wave function in the sense of giving the best energy.

The problem is really not as bad as it may at first appear. In choosing a reasonable wave function, one is often aided by his chemical intuition, and a reasonable choice is often quite obvious. In addition, it is possible to consider a general group or family of wave functions at one time. This can be done by introducing one or more variable parameters in the trial function and then minimizing the energy with respect to the parameters. Of course, the more parameters that are introduced in a given trial function, the closer one comes to the actual wave function, but at the same time, the more parameters used, the greater will be the work involved in the computation.

## Ground-State Energy of the Hydrogen Atom

As a simple example of the use of the variation method, we can consider the calculation of the ground-state energy of the hydrogen atom. We are already aware of the fact that the wave function is of the general form

$$\psi = e^{-ar}$$

and to illustrate the method, we shall choose our trial function to be of this same form. The problem now becomes one of evaluating the two integrals $\int \psi^* H \psi \, d\tau$ and $\int \psi \psi^* d\tau$. For the ground state of the hydrogen atom, the potential energy term will be $-e^2/r$, and this leads to the expression for the Hamiltonian operator

$$(5\text{-}10) \qquad H = -\frac{h^2}{8\pi^2 m} \nabla^2 - \frac{e^2}{r}$$

From our discussion of the hydrogen atom, we should recall that the energy term occurs only in the radial equation, and for this reason we shall need only to consider the radial portion of the Laplacian operator. In spherical coordinates, this can be seen from Eq. (2–43) to be

$$\nabla^2 = \frac{1}{r^2} \frac{\partial}{\partial r} \left( r^2 \frac{\partial}{\partial r} \right)$$

Applying the indicated operation to $\psi$ gives

$$\nabla^2 \psi = \left[ \frac{1}{r^2} \frac{\partial}{\partial r} \left( r^2 \frac{\partial}{\partial r} \right) \right] e^{-ar}$$

$$(5\text{-}11) \qquad = \frac{1}{r^2} [-r^2 (-a^2 e^{-ar}) + (ae^{-ar})(-2r)]$$

$$= \left( a^2 - \frac{2a}{r} \right) e^{-ar}$$

The limits of integration are determined by considering the electron to be in a spherical shell at a distance of $r$ to $r + dr$ from the nucleus as $r$ goes from zero to infinity. The volume of such a spherical shell at a distance, $r$, from the nucleus is given by $4\pi r^2 dr$, and this, when substituted into Eq. (5–9) along with the value for $\nabla^2\psi$, leads to

$$E = \frac{\displaystyle\int \psi^* H\psi \, d\tau}{\displaystyle\int \psi\psi^* \, d\tau}$$

$$= \frac{\displaystyle\int_0^\infty e^{-ar}\left[-\frac{h^2}{8\pi^2 m}\left(a^2 - \frac{2a}{r}\right)e^{-ar} - \frac{e^2}{r}e^{-ar}\right]4\pi r^2 dr}{\displaystyle\int_0^\infty e^{-2ar}\,4\pi r^2 dr}$$

$$= \frac{\displaystyle\int_0^\infty -\frac{h^2 a^2}{8\pi^2 m}r^2 e^{-2ar}\,dr + \int_0^\infty \frac{2h^2 a}{8\pi^2 m}re^{-2ar}dr - \int_0^\infty e^2 re^{-2ar}\,dr}{\displaystyle\int_0^\infty r^2 e^{-2ar}\,dr}$$

The solutions to these integrals are readily obtained from the relation

$$\int_0^\infty x^n e^{-mx}\,dx = \frac{\Gamma(n+1)}{m^{(n+1)}}$$

where, if any number $p$ is an integer then $\Gamma(p) = (p - 1)!$ Thus, $\Gamma(n + 1) = (n + 1 - 1)!$ Solving these equations and substituting the solutions to the integrals, we can obtain the expression for the energy

$$E = \frac{-\dfrac{h^2 a^2}{8\pi^2 m}\left(\dfrac{2}{8a^3}\right) + \dfrac{h^2 a}{4\pi^2 m}\left(\dfrac{1}{4a^2}\right) - e^2\left(\dfrac{1}{4a^2}\right)}{\left(\dfrac{2}{8a^3}\right)}$$

$$= \frac{h^2 a^2}{8\pi^2 m} - e^2 a$$

The best energy obtainable from this wave function will be the minimum energy, and this will depend on the magnitude of the parameter, $a$. Therefore, it will be necessary to choose $a$ in such a manner as to give a minimum energy value. This can be done by minimizing $E$ with respect to $a$:

$$\frac{\partial E}{\partial a} = \frac{2h^2}{8\pi^2 m}a - e^2 = 0$$

or

$$a = \frac{4\pi^2 m e^2}{h^2}$$

If we now substitute the value for $a$ into the general expression for the energy, we obtain

$$E_{min} = -\frac{2\pi^2 m e^4}{h^2}$$

which is the same as the ground-state energy of the hydrogen atom as obtained in Chapter 2 by the closed form calculation. Of course, the exact energy of the ground state of the hydrogen atom would not have been obtained if we had not used the correct form of the wave function. However, a variety of trial functions could have been tried, and the closer we came to the proper function, the better would have been our energy calculation.

## The Secular Equations

In most instances, it is desirable to express the variation function in terms of a set of functions, $\varphi_i$. These functions will be of the class $Q$ and will frequently, but not necessarily, be normalized. Accordingly,

$$(5\text{-}12) \qquad \psi = a_1\varphi_1 + a_2\varphi_2 + \cdots + a_n\varphi_n$$

where the $a_1, a_2, \ldots, a_n$ are arbitrary parameters that can be varied to give a minimum in the energy. Actually, it is often found that the wave function can be adequately represented simply as

$$\psi = a_1\varphi_1 + a_2\varphi_2$$

and for convenience, we shall consider the variation function in terms of this representation. In essence, this is an application of the principle of superposition.

If we now substitute for $\psi$ in Eq. (5–9), we obtain

$$E = \frac{\int (a_1\varphi_1{}^* + a_2\varphi_2{}^*) \, H \, (a_1\varphi_1 + a_2\varphi_2) \, d\tau}{\int (a_1\varphi_1{}^* + a_2\varphi_2{}^*)(a_1\varphi_1 + a_2\varphi_2) \, d\tau}$$

or

$$E \int (a_1\varphi_1{}^* + a_2\varphi_2{}^*)(a_1\varphi_1 + a_2\varphi_2)d\tau = \int (a_1\varphi_1{}^* + a_2\varphi_2{}^*) \, H \, (a_1\varphi_1 + a_2\varphi_2)d\tau$$

which leads to

$$E\left(a_1{}^2 \int \varphi_1{}^*\varphi_1 d\tau + 2a_1a_2 \int \varphi_1{}^*\varphi_2 d\tau + a_2{}^2 \int \varphi_2{}^*\varphi_2 d\tau\right) = a_1{}^2 \int \varphi_1{}^*H\varphi_1 d\tau$$

$$(5\text{-}13) \qquad + 2a_1a_2 \int \varphi_1{}^*H\varphi_2 d\tau + a_2{}^2 \int \varphi_2{}^*H\varphi_2 d\tau$$

If we consider an operator $\alpha$ and the two wave functions $\varphi_1$ and $\varphi_2$, it would not be expected in general that

$$\int \varphi_1{}^*\alpha\varphi_2 \, d\tau = \int \varphi_2\alpha \, \varphi_1{}^* \, d\tau$$

Nevertheless, we assumed this to be the case in Eq. (5-13) when we combined $\int \varphi_1{}^*H\varphi_2 d\tau$ and $\int \varphi_2 H\varphi_1{}^*d\tau$ to give $2a_1a_2\int \varphi_1{}^*H\varphi_2 d\tau$. However, if $\varphi_1$ and $\varphi_2$ are class $Q$ functions, this will be valid for a particular class of operators known as *Hermitean* operators. The unique character of these operators is that their eigenvalues for functions in class $Q$ are always real. In our quantum mechanical treatment, it is assumed that we are dealing with real quantities, and we find, then, that the operators must be of the Hermitean variety. Proof that the eigenvalues for Hermitean operators are real, along with proof that the Hamiltonian operator is Hermitean, is given in appendices B and C.

Now, returning to Eq. (5–13), we can see that since the minimum value of $E$ is desired, it is necessary to minimize $E$ with respect to both $a_1$ and $a_2$. Using the differentiation with respect to $a_1$ as an example we find that

$$E\left[2a_1\int \varphi_1^*\varphi_1 d\tau + 2a_2\int \varphi_1^*\varphi_2 d\tau\right] + \frac{\partial E}{\partial a_1}\left[a_1^2\int \varphi_1^*\varphi_1 d\tau + 2a_1a_2\int \varphi_1^*\varphi_2 d\tau\right.$$

$$(5–14) \qquad \left.+ a_2^2\int \varphi_2^*\varphi_2 d\tau\right] = 2a_1\int \varphi_1^*H\varphi_1 d\tau + 2a_2\int \varphi_1^*H\varphi_2 d\tau$$

and, of course, an equivalent equation would be obtained for $a_2$. Now, in order to minimize $E$ with respect to $a_1$ and $a_2$ it is necessary that

$$\left(\frac{\partial E}{\partial a_1}\right)_{a_2} = \left(\frac{\partial E}{\partial a_2}\right)_{a_1} = 0$$

For the sake of simplicity, it is desirable to introduce the symbolism

$$(5–15) \qquad H_{ij} = \int \varphi_i^*H\varphi_j d\tau \qquad \text{and} \qquad S_{ij} = \int \varphi_i^*\varphi_j d\tau$$

If this is done, and we apply the condition $\partial E/\partial a_1 = 0$ to Eq. (5–14), we can obtain, on rearrangement,

$$(5–16a) \qquad (H_{11} - ES_{11})\, a_1 + (H_{12} - ES_{12})\, a_2 = 0$$

The equivalent equation obtained by applying the condition $\partial E/\partial a_2 = 0$ will be

$$(5–16b) \qquad (H_{21} - ES_{21})\, a_1 + (H_{22} - ES_{22})\, a_2 = 0$$

These two equations are known together as the *secular equations*.

It is to be noted that the secular equations are of the form

$$ax + by = 0$$
$$cx + dy = 0$$

and if we solve this set of linear homogeneous equations, we see that

$$(ad - bc)\, y = 0$$

In order for this equality to be valid, it is apparent that either $y$ is identically zero or else the coefficient of $y$ is zero. If $y$ is zero, no problem really exists; therefore, a nontrivial solution requires that the coefficient of $y$ be zero. This can be expressed in determinental form as

$$\begin{vmatrix} a & b \\ c & d \end{vmatrix} = 0$$

Obviously, the same condition applies to our secular equations, and they can therefore be expressed as

$$(5\text{-}17) \quad \begin{vmatrix} H_{11} - ES_{11} & H_{12} - ES_{12} \\ H_{21} - ES_{21} & H_{22} - ES_{22} \end{vmatrix} = 0$$

To be more general, if $\psi$ is expressed as $n$ independent terms, then the secular determinant becomes

$$(5\text{-}18) \quad \begin{vmatrix} H_{11} - ES_{11} & H_{12} - ES_{12} \cdots H_{1n} - ES_{1n} \\ H_{21} - ES_{21} & H_{22} - ES_{22} \cdots H_{2n} - ES_{2n} \\ \cdot & \cdot \qquad\qquad \cdot \\ \cdot & \cdot \qquad\qquad \cdot \\ \cdot & \cdot \qquad\qquad \cdot \\ H_{n1} - ES_{n1} & H_{n2} - ES_{n2} \cdots H_{nn} - ES_{nn} \end{vmatrix} = 0$$

## MOLECULAR ORBITAL THEORY

In the variation method we have a means of approximating the energy of a system, but it is still necessary to choose a trial wave function. This is not always an easy task, and for the calculation of molecular energy levels two basic approaches to the problem have evolved. These are referred to as the *molecular orbital theory* and the *valence bond theory*. The two theories lead to quite different means of constructing a trial function, but equally important, they reflect quite different conceptual approaches to the basic structural model of a molecule.

The valence bond theory was developed several years prior to the molecular orbital theory, but it would appear that in recent years the latter has generally become the more popular of the two. This can be attributed to the greater conceptual and mathematical simplicity of the molecular orbital approach. However, with respect to certain aspects of molecular structure, the valence bond theory offers a simpler pictorial representation and therefore has been more frequently used for qualitative considerations.

Whereas the valence bond theory retains the individuality of the atoms composing a molecule, the molecular orbital theory attempts to construct a molecule in terms of the basic concepts of atomic structure. Just as there are atomic orbitals in an atom, there are *molecular orbitals* in a molecule. The most obvious difference is that the molecular orbitals are *polycentric*. Nevertheless, the molecular orbital theory attempts to construct a wave function for an electron in a molecule that lends itself to the same interpretations that proved valid for the wave function of an electron in an atom. Thus, the probability that the electron will be confined to a volume element, $d\tau$, will be proportional to $\int \psi \psi^* d\tau$, and, as with atomic orbitals, each molecular orbital will be dependent on a set of quantum numbers that determine its energy and spatial distribution. It is also assumed that the building-up or *aufbau* principle holds for the molecule just as it does for the atom. That is, each molecular orbital can accommodate two electrons with opposite spins, and starting with the lowest-energy orbital, the electrons must go into the available molecular orbitals one at a time.

According to the molecular orbital theory, if we have a simple covalent bond composed of two electrons described by one-electron molecular orbital wave functions, $\psi_1$ and $\psi_2$, respectively, these can be combined to give the configurational wave function for the system

$$\psi_{MO} = \psi_1 \psi_2$$

Or, in general, if there are $n$ electrons, the configurational wave function will be

$$(5\text{--}19) \qquad \psi_{MO} = \psi_1 \psi_2 \cdots \psi_n$$

where the $\psi_i$ are one-electron molecular orbitals. In determining the individual one-electron molecular orbitals such as $\psi_1$ and $\psi_2$, a linear combination of the bonding atomic orbitals of the atoms composing the molecule can be used.

### The Hydrogen Molecule Ion

As an example, we can consider the simplest possible molecular species, the hydrogen molecule ion.[8] If we are interested in determining the ground-state energy of this species, we can construct our trial wave function from a linear combination of the $1s$ orbitals of the hydrogen atoms. We can begin by imagining the nuclei to be separated to an infinite distance, thereby leading to the arrangement

$$\begin{array}{cc} \text{H} \cdot & \text{H}^+ \\ (a) & (b) \end{array}$$

In this instance we have the electron on atom $a$, and in the ground state the molecular orbital will be represented by the atomic orbital $\psi_a = \psi_{1s_a}$. If the other extreme is considered, the ground-state molecular orbital will be the atomic orbital $\psi_b = \psi_{1s_b}$, indicating that the electron is associated with atom $b$. If the two nuclei are now allowed to come together, it would be reasonable to consider that the resultant one-electron molecular orbital will be characteristic of the two atomic orbitals. This leads to the approximation of *linear combination of atomic orbitals* (*LCAO*). Thus, we can say that the one-electron molecular orbital for the first electron is

$$(5\text{--}20) \qquad \psi_1 = a_1 \psi_a + a_2 \psi_b$$

For this particular problem, there is only one electron, and the total wave function is $\psi = \psi_1$. However, if we were considering the hydrogen molecule, the configurational wave function would be

$$\psi_{MO} = \psi_1 \psi_2 = \{a_1 \psi_a(1) + a_2 \psi_b(1)\}\{a_3 \psi_a(2) + a_4 \psi_b(2)\}$$

where $\psi_a(1)$ is the wave function for electron 1 on atom $a$, $\psi_a(2)$ is the wave function for electron 2 on atom $a$, and so on. Of course, due to the symmetry of the hydrogen molecule, the absolute magnitudes of all of the coefficients $a_i$ are the same.

Since the molecular orbital of the one electron in the hydrogen molecule ion is here represented as a linear combination of two independent terms, the secular determinant will take the same form we have already developed in Eq. (5–17)

$$\begin{vmatrix} H_{aa} - ES_{aa} & H_{ab} - ES_{ab} \\ H_{ba} - ES_{ba} & H_{bb} - ES_{bb} \end{vmatrix} = 0$$

This can be simplified considerably by noting the symmetry of this particular molecule. Since the hydrogen atoms, and therefore, the ground-state atomic orbitals, are identical, it should be apparent that

$$H_{aa} = H_{bb}$$
$$H_{ba} = H_{ab}$$

and

$$S_{ba} = S_{ab}$$

Further, if we use normalized wave functions, $S_{aa} = S_{bb} = 1$. Consequently, the secular determinant can now be reduced to

(5–21) $$\begin{vmatrix} H_{aa} - E & H_{ba} - ES \\ H_{ba} - ES & H_{aa} - E \end{vmatrix} = 0$$

and this leads to the expression*

(5–22) $$(H_{aa} - E)^2 - (H_{ba} - SE)^2 = 0$$

If this expression is now rearranged and solved for $E$ by means of the quadratic equation, two roots are obtained,

(5–23a) $$E_S = \frac{H_{aa} + H_{ba}}{1 + S}$$

and

(5–23b) $$E_A = \frac{H_{aa} - H_{ba}}{1 - S}$$

where $E_S$ and $E_A$ denote *symmetric* and *antisymmetric* energy states, respectively.

**Bonding and antibonding orbitals.** Here we note that the original $1s$ energy states of the two hydrogen atoms are degenerate, but, on combination, they split into two new energy states, one of lower energy and one of higher energy than the original atomic $1s$ states. This is represented schematically in Figure 5–2. In terms of the molecular orbital theory, the lower-energy orbital is considered to be a *bonding orbital* and the higher-

---

* Very often, this equation is seen in terms of the symbolism $H_{aa} = E_a$, $H_{bb} = E_b$, and $H_{ab} = \beta$.

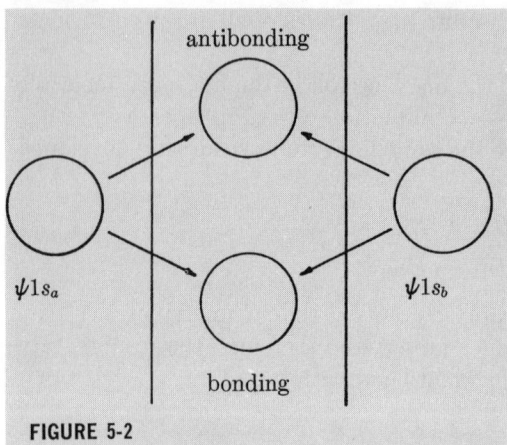

**FIGURE 5-2**

Combination of Two Atomic Orbitals to Form
Two Molecular Orbitals.

energy orbital is considered to be an *antibonding orbital*. Both orbitals can accommodate a pair of electrons, but the bonding orbital, since it is of lower energy, will be filled first. Thus, for the hydrogen molecule ion, the electron will go into the bonding orbital. However, for more complex systems, we shall see that the occupancy of an antibonding orbital is extremely important in determining bond character.

**Electron distribution in the hydrogen molecule ion.** Along with the energy states, it is also interesting to consider the electron distribution in the hydrogen molecule ion for both the symmetric and the antisymmetric states. From Eq. (5–20) we have seen that the wave function for the hydrogen molecule ion is

$$\psi_{MO} = a_1 \psi_a + a_2 \psi_b$$

Now, from the fact that two energy states were obtained, we can conclude that two wave functions must exist: one for the symmetric state and one for the antisymmetric state. These can be determined from the general expression relating $a_1$ to $a_2$,

(5–16a)        $(H_{11} - ES_{11})a_1 + (H_{12} - ES_{12})a_2 = 0$

In terms of our present symbolism, this can be represented as

$$(H_{aa} - E)a_1 + (H_{ba} - ES_{ba})a_2 = 0$$

In order to obtain the symmetric solution, it is necessary to substitute $E_S$ for $E$ in this equation, and for the antisymmetric solution, it is necessary to substitute $E_A$ for $E$. If this is done, and the resultant expression is solved, it is found, respectively, that

$$a_1 = a_2$$

and

$$a_1 = -a_2$$

This, then, leads to the two molecular orbital wave functions

(5–24a) $\qquad \psi_S = N_S(\psi_a + \psi_b)$

(5–24b) $\qquad \psi_A = N_A(\psi_a - \psi_b)$

where $N$ is a normalizing constant.

In order to obtain the final wave functions, it is necessary to carry out the normalization of $\psi_S$ and $\psi_A$. This can be done quite simply for both the symmetric and the antisymmetric functions. Using the symmetric function to illustrate the procedure, we can recognize that for a normalized function

$$\int \psi_S^* \psi_S d\tau = 1$$

or

$$\int N_S^2 \, (\psi_a + \psi_b)^* \, (\psi_a + \psi_b) d\tau = 1$$

On expansion, this becomes

(5–25) $\qquad N_S^2 \left[ \int \psi_a{}^2 d\tau + 2 \int \psi_a \psi_b d\tau + \int \psi_b{}^2 d\tau \right] = 1$

The complex conjugate forms have been dropped here since both $\psi_a$ and $\psi_b$ are real. Now, if we have originally chosen $\psi_a$ and $\psi_b$ to be normalized wave functions, it follows that

$$\int \psi_a{}^2 d\tau = \int \psi_b{}^2 d\tau = 1$$

and by definition

$$\int \psi_a \psi_b d\tau = S_{ab}$$

Thus, we see that Eq. (5–25) becomes

$$N_S^2(1 + 2S_{ab} + 1) = 1$$

or

(5–26a) $\qquad N_S = \dfrac{1}{\sqrt{2 + 2S_{ab}}}$

In the same manner, it can be shown that

(5–26b) $\qquad N_A = \dfrac{1}{\sqrt{2 - 2S_{ab}}}$

This now gives us the normalized wave functions

(5–27a) $\qquad \psi_S = \dfrac{1}{\sqrt{2 + 2S_{ab}}} \, (\psi_a + \psi_b)$

and

(5–27b) $\qquad \psi_A = \dfrac{1}{\sqrt{2 - 2S_{ab}}} \, (\psi_a - \psi_b)$

From the wave functions we can determine the distribution of the electron charge in the molecule, and from the expressions for the energy states, we can calculate the molecular energy levels. Considering the charge distribution first, we can see that if $S_{ab}$ is sufficiently near to zero, then

$$(5\text{--}28a) \qquad \psi_S{}^2 = \frac{1}{2}\{\psi_a{}^2 + \psi_b{}^2 + 2\,\psi_a\,\psi_b\}$$

and

$$(5\text{--}28b) \qquad \psi_A{}^2 = \frac{1}{2}\{\psi_a{}^2 + \psi_b{}^2 - 2\,\psi_a\,\psi_b\}$$

Here we see that the symmetric function leads to an increase in electron charge density in the region of overlap between the two atoms over that of the individual atoms as described by the functions $\psi_a{}^2$ and $\psi_b{}^2$. On the other hand, the antisymmetric function leads to a decrease in the charge density. This is represented graphically in Figure 5–3. The dotted lines represent the respective charge densities on the individual atoms when separated to infinity, and the heavy line represents the electron charge distribution in the hydrogen molecule ion along a line through the nuclei.

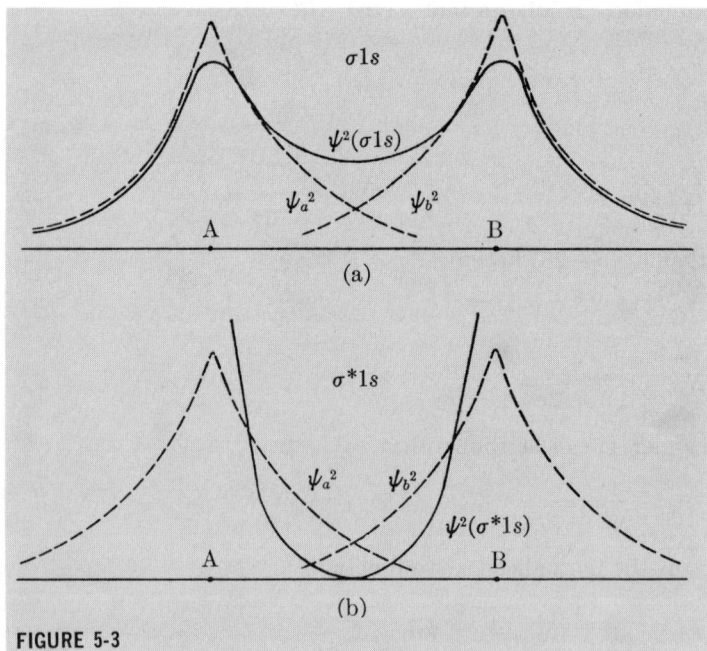

FIGURE 5-3

Electron Densities Along the Nuclear Axis for (a) the Symmetrical and (b) the Antisymmetrical States of the Hydrogen Molecule Ion.

It is apparent that the bonding orbital favors a charge distribution that is concentrated between the nuclei, whereas the antibonding orbital tends to decrease the charge density in this region and to concentrate it around the individual atoms. Since we consider the formation of a covalent bond to be associated with an electron charge buildup in the region of the bond, it would seem that only the symmetric function should lead to the formation of a stable molecule.

**Stability of the hydrogen molecule ion.** If the hydrogen molecule ion actually does form a stable species, we should expect a diagram of the potential energy as a function of the distance of separation of the two nuclei to show a minimum at some equilibrium separation of the two atoms, $a$ and $b$. It should be possible to plot such a curve if we can evaluate the expression for the energy of the molecule as a function of the internuclear separation. Immediately, we should recognize that two potential energy diagrams will be obtained: one for the bonding orbital and one for the antibonding orbital. In both $E_S$ and $E_A$ the same integrals will appear, but the energies will be different due to the signs of the various terms. From Eq. (5–23), it can be seen that these integrals are

$$H_{aa} = \int \psi_a H \psi_a d\tau$$

$$H_{ba} = \int \psi_b H \psi_a d\tau$$

and

$$S_{ba} = \int \psi_b \psi_a d\tau$$

Now, according to Figure 5–4, the Hamiltonian operator for the hydrogen molecule ion is

(5–29) $$H = -\frac{h^2}{8\pi^2 m} \nabla^2 - \frac{e^2}{r_a} - \frac{e^2}{r_b} + \frac{e^2}{r_{ab}}$$

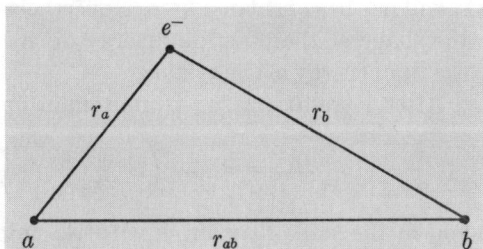

**FIGURE 5-4**

Coordinates for the Hydrogen Molecule Ion.

and $H_{aa}$ will therefore become

$$(5\text{-}30) \qquad H_{aa} = \int \psi_{1s_a} \left[ -\frac{h^2}{8\pi^2 m} \nabla^2 - \frac{e^2}{r_a} - \frac{e^2}{r_b} + \frac{e^2}{r_{ab}} \right] \psi_{1s_a} \, d\tau$$

This integral can be simplified by recognizing that

$$-\frac{h^2}{8\pi^2 m} \nabla^2 - \frac{e^2}{r_a}$$

is the Hamiltonian operator for the hydrogen atom with the electron around atom $a$, and since, in general,

$$H\psi = E\psi$$

the Hamiltonian operator for the hydrogen molecule ion can be expressed as

$$(5\text{-}31) \qquad H = E_0 - \frac{e^2}{r_b} + \frac{e^2}{r_{ab}}$$

where $E_0$ is the ground-state energy of the hydrogen atom. This then gives

$$H_{aa} = \int \psi_{1s_a} \left[ E_0 - \frac{e^2}{r_b} + \frac{e^2}{r_{ab}} \right] \psi_{1s_a} \, d\tau$$

Now $E_0$ and $r_{ab}$ are both constants, and for this reason it is possible to remove them from under the integral sign, giving

$$H_{aa} = E_0 \int \psi_{1s_a} \psi_{1s_a} \, d\tau + \frac{e^2}{r_{ab}} \int \psi_{1s_a} \psi_{1s_a} \, d\tau - e^2 \int \frac{1}{r_b} \psi_{1s_a} \psi_{1s_a} \, d\tau$$

But, since the $1s$ wave functions are normalized, it follows that

$$(5\text{-}32) \qquad H_{aa} = E_0 + \frac{e^2}{r_{ab}} - J$$

where $J$ denotes the integral

$$(5\text{-}33) \qquad J = e^2 \int \frac{1}{r_b} \psi_{1s_a} \psi_{1s_a} \, d\tau$$

The evaluation of this integral is not a simple matter, and for that reason it will not be considered here. Nevertheless, it will yet be possible to discuss the shape of the potential energy diagram in terms of its contribution to the total energy of the system.

After introducing the Hamiltonian operator, the integral $H_{ba}$ becomes

$$H_{ba} = \int \psi_{1s_a} \left( E_0 + \frac{e^2}{r_{ab}} - \frac{e^2}{r_b} \right) \psi_{1s_b} \, d\tau$$

and in the same manner as with the integral $H_{aa}$, the integral $H_{ba}$ can be represented as

$$H_{ba} = E_0 \int \psi_{1s_a} \psi_{1s_b} \, d\tau + \frac{e^2}{r_{ab}} \int \psi_{1s_a} \psi_{1s_b} \, d\tau - e^2 \int \frac{1}{r_b} \psi_{1s_a} \psi_{1s_b} \, d\tau$$

Since $\int \psi_{1s_a} \psi_{1s_b} d\tau$ is defined as $S_{ab}$, then $H_{ba}$ can be expressed as

(5–34)           $H_{ba} = E_0 S_{ab} + \dfrac{e^2}{r_{ab}} S_{ab} - K$

where $K$ denotes the integral

(5–35)           $K = e^2 \int \dfrac{1}{r_b} \psi_{1s_a} \psi_{1s_b} \, d\tau$

Just as with the integral $J$, the integral $K$ is rather difficult to evaluate, but it can be helpful to see how it will, in general, affect the energy of the molecule.

When the solutions of the various integrals are substituted into the equations for the energy states, we obtain for the symmetric state,

$$E_S = \frac{H_{aa} + H_{ba}}{1 + S_{ba}}$$

$$= \frac{E_0 + \dfrac{e^2}{r_{ab}} - J + E_0 S_{ba} + \dfrac{e^2}{r_{ab}} S_{ba} - K}{1 + S_{ba}}$$

or

(5–36a)           $E_S - E_0 = \dfrac{e^2}{r_{ab}} - \dfrac{J + K}{1 + S_{ba}}$

For the antisymmetric state, it is found that the corresponding equation is

(5–36b)           $E_A - E_0 = \dfrac{e^2}{r_{ab}} - \dfrac{J - K}{1 - S_{ba}}$

Although the arithmetic is rather complex, it is possible to evaluate the integrals $J$ and $K$ as a function of the internuclear separation of the hydrogen nuclei. This result can be represented by means of a potential energy diagram such as that shown in Figure 5–5. Here the antisymmetric state is seen to correspond to an unstable energy state, and if the electron were in the antisymmetrical orbital, we could conclude that the hydrogen molecule ion would be an unstable species. On the other hand, the symmetric energy state leads to a potential minimum and, therefore, a stable molecular species.

Using the particular wave function we have chosen, the potential energy minimum is equivalent to a dissociation energy of 1.76 eV and an equilibrium separation of 1.32 Å. Experimentally, it is found that the dissociation energy is 2.791 eV, and the equilibrium separation is 1.06 Å. Thus, our calculations indicate that the hydrogen molecule ion is a stable species, but they also indicate that the trial wave function we have used can be improved. A first step in this direction would be to introduce a variable parameter. Such a trial function that has proven successful is

$$\psi_a = \left( \frac{\alpha^3}{\pi} \right)^{1/2} e^{-\alpha r_a}$$

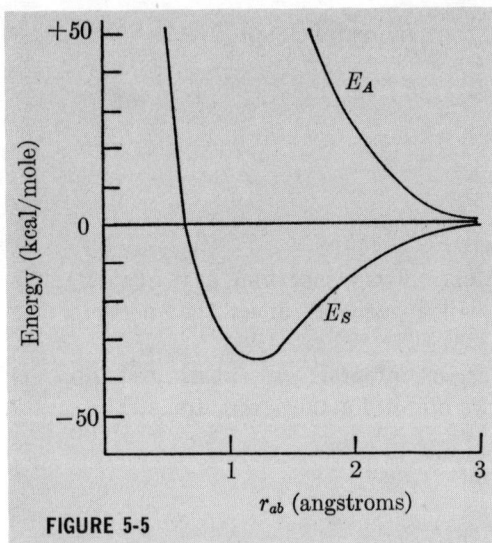

**FIGURE 5-5**

Potential Energy Diagram for the Hydrogen Molecule Ion Showing the Symmetrical ($E_S$) and the Antisymmetrical ($E_A$) Energy States.

If the energy is minimized with respect to the parameter, $\alpha$, somewhat closer agreement between theory and experiment is obtained than with our original wave function, but some difference yet exists. Finally, James used the function

$$\psi = e^{-c_1(r_a + r_b)} \{ 1 + c_2(r_a + r_b)^2 \}$$

and obtained very good agreement between theory and experiment. This function is of particular interest in that it is not based on the $LCAO$ method, and the concept of resonance does not appear.

## THE VALENCE BOND THEORY

The natural extension of the Lewis concept of an electron-pair bond finds its quantum mechanical expression in the valence bond theory. Just as with the molecular orbital theory, the valence bond theory is also an approximation method. However, its basic approach to the structure of a molecule is closely related to our conventional ideas of localized chemical bonds. Here, the atoms are considered to maintain their individuality, and the bond arises from the interaction of the valence electrons as the atoms come together. Such a view is more readily adapted to qualitative arguments than that presented by the molecular orbital theory, and for this reason the language of the valence bond theory is frequently the more familiar. In spite of this desirable feature, the valence bond method often leads to more complicated mathematics than does the molecular orbital theory. And, in addition, the determination of a trial function for a complex molecule is not so straightforward as it is in the molecular orbital theory.

## The Hydrogen Molecule

To some extent, the ideas of the valence bond method can be appreciated by an analysis of the quantum mechanical treatment of the hydrogen molecule.[9] This problem was first treated successfully by Heitler and London in 1927, and their approach was basically that of the valence bond method. The trial wave function will be somewhat different in the simple valence bond theory than it is in the simple molecular orbital theory. According to the molecular orbital theory, the wave function for electron (1) is made up of a linear combination of the hydrogen $1s$ orbitals just as it is in the hydrogen molecule ion. Thus,

$$\psi_1 = \psi_a(1) + \psi_b(1)$$

But, in the hydrogen molecule there are two electrons, and a one-electron molecular orbital wave function for the second electron will be

$$\psi_2 = \psi_a(2) + \psi_b(2)$$

This leads to the configurational wave function

(5–37) $$\psi_{MO} = \psi_1 \psi_2 = \{\psi_a(1) + \psi_b(1)\}\{\psi_a(2) + \psi_b(2)\}$$

In this case it is seen that the configurational wave function is expressed as a product of one-electron molecular orbital wave functions.

A quite different construction is used in the valence bond method. Here it is proposed that if two unconnected systems can be separately described by wave functions $\psi_a$ and $\psi_b$, respectively, then the wave function for the combined system will be

(5–38) $$\psi_I = \psi_a \psi_b$$

Consequently, for the hydrogen molecule

(5–39) $$\psi_I = \psi_{1s_a}(1) \psi_{1s_b}(2)$$

This wave function places electron (1) on atom $a$ and electron (2) on atom $b$.

Although the potential energy diagram that results from this wave function has a minimum, thereby indicating a stable molecule, the agreement with experiment is not good. However, a much better trial function can be obtained by introducing what has now become a fundamental principle of quantum mechanics. In the wave function $\psi_I$ we have arbitrarily placed electron (1) on atom $a$ and electron (2) on atom $b$. There is no basis for such a choice, and it is equally valid to reverse this arrangement, giving

(5–40) $$\psi_{II} = \psi_{1s_a}(2) \psi_{1s_b}(1)$$

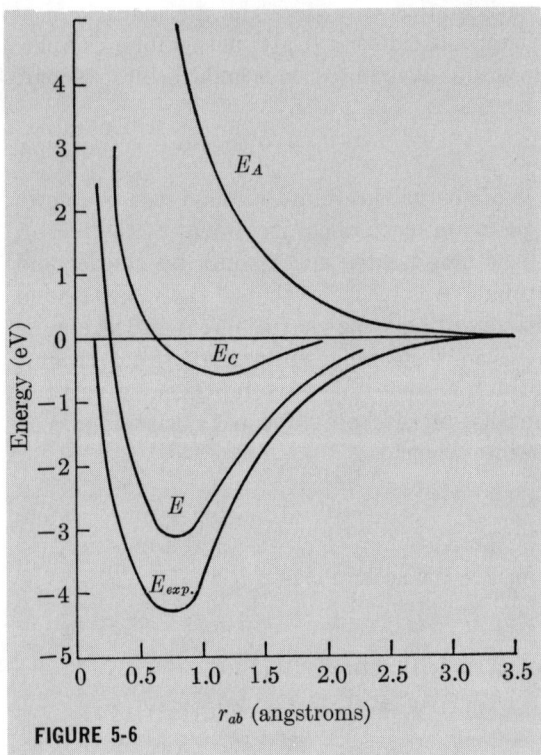

**FIGURE 5-6**

Potential Energy Diagram for the Hydrogen Molecule Showing the Antisymmetrical State $(E_A)$, the Classical Coulombic Interaction $(E_C)$, the Simple Valence Bond Energy $(E)$, and the Experimental Curve $(E_{exp})$.

This emphasizes the fact that the electrons are *indistinguishable*, and it is significant that this recognition leads to a much better trial function. This is evident from Figure 5–6, where the potential energy diagram for the hydrogen molecule is plotted for various trial functions. This new effect can be introduced into the total wave function by taking the usual linear combination according to the principle of superposition, thereby giving

$$(5\text{–}41) \qquad \psi_{VB} = \psi_{1s_a}(1)\,\psi_{1s_b}(2) + \psi_{1s_a}(2)\,\psi_{1s_b}(1)$$

**Symmetric and antisymmetric energy states.** Now that we have a trial wave function, the problem is essentially the same as that of the hydrogen molecule ion. Using the symbolism of Figure 5–7, the Hamiltonian operator for the hydrogen molecule will be

$$(5\text{–}42) \qquad H = \left[ -\frac{h^2}{8\pi^2 m}\,(\nabla_1{}^2 + \nabla_2{}^2) - \frac{e^2}{r_{a1}} - \frac{e^2}{r_{b1}} - \frac{e^2}{r_{a2}} - \frac{e^2}{r_{b2}} + \frac{e^2}{r_{12}} + \frac{e^2}{r_{ab}} \right]$$

where $\nabla_1{}^2$ is the Laplacian operator for electron (1) and $\nabla_2{}^2$ is the Laplacian

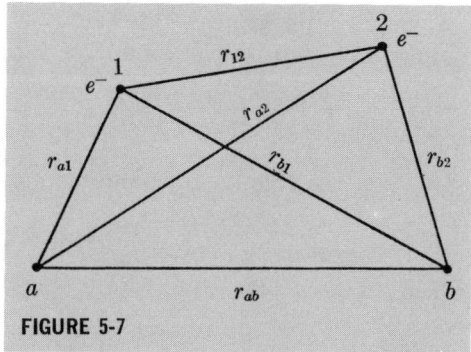

**FIGURE 5-7**

Coordinates of the Hydrogen Molecule.

operator for electron (2). The wave function can be expressed in the general form

$$(5\text{-}43) \qquad \psi_{VB} = a_1 \psi_I + a_2 \psi_{II}$$

and this leads to the same secular determinant that was used for the hydrogen molecule ion. Additionally, we will also obtain the symmetric and antisymmetric energy terms

$$E_S = \frac{H_{11} + H_{12}}{1 + S_{12}}$$

and

$$E_A = \frac{H_{11} - H_{12}}{1 - S_{12}}$$

The resultant integrals, however, will be more complex than those encountered in the hydrogen molecule ion.

The integral $H_{11}$ is now defined as

$$H_{11} = \int \int \psi_I H \psi_I d\tau_1 d\tau_2$$

where the double integral results from a consideration of the coordinates of both electron (1) and electron (2). On introducing the Hamiltonian operator, the integral becomes

$$H_{11} = \int \int \psi_a(1) \, \psi_b(2) \left\{ -\frac{h^2}{8\pi^2 m} (\nabla_1^2 + \nabla_2^2) \right.$$
$$\left. - \frac{e^2}{r_{a1}} - \frac{e^2}{r_{b1}} - \frac{e^2}{r_{a2}} - \frac{e^2}{r_{b2}} + \frac{e^2}{r_{12}} + \frac{e^2}{r_{ab}} \right\} \psi_a(1) \, \psi_b(2) \, d\tau_1 d\tau_2$$

Now,

$$\left( -\frac{h^2}{8\pi^2 m} \nabla_1^2 - \frac{e^2}{r_{a1}} \right) \psi_a(1) = E_0 \psi_a(1)$$

and

$$\left( -\frac{h^2}{8\pi^2 m} \nabla_2^2 - \frac{e^2}{r_{b2}} \right) \psi_b(2) = E_0 \psi_b(2)$$

where $E_0$ is the ground-state energy of the hydrogen atom. Therefore, the integral $H_{11}$ can be expressed more simply as

$$H_{11} = \int\int \psi_a(1)\psi_b(2) \left(2E_0 + \frac{e^2}{r_{ab}} + \frac{e^2}{r_{12}} - \frac{e^2}{r_{a2}} - \frac{e^2}{r_{b1}}\right) \psi_a(1)\,\psi_b(2)\,d\tau_1 d\tau_2$$

Or, since the 1s wave functions are assumed to be normalized,

$$(5\text{-}44) \qquad H_{11} = 2E_0 + \frac{e^2}{r_{ab}} + J_1 - 2J_2$$

Here,

$$(5\text{-}45) \qquad J_1 = e^2 \int\int \frac{1}{r_{12}} [\psi_a(1)\,\psi_b(2)]^2\,d\tau_1 d\tau_2$$

and, because of the equivalence of the two electrons,

$$(5\text{-}46a) \qquad J_2 = e^2 \int\int \frac{1}{r_{a2}} [\psi_a(1)\,\psi_b(2)]^2\,d\tau_1 d\tau_2$$

$$(5\text{-}46b) \qquad = e^2 \int\int \frac{1}{r_{b1}} [\psi_a(1)\,\psi_b(2)]^2\,d\tau_1 d\tau_2$$

The solution of these integrals is even more complicated than that for the corresponding integrals of the hydrogen molecule ion, and for this reason they will not be treated further here. However, it will again be useful to consider the basic form of the energy expression.

By the same general argument as was used for $H_{11}$, it can be shown that

$$H_{12} = 2E_0 S_{12} + \frac{e^2}{r_{ab}} + K_1 - 2K_2$$

where

$$(5\text{-}47) \qquad K_1 = e^2 \int\int \frac{1}{r_{12}} [\psi_a(1)\,\psi_b(2)\,\psi_a(2)\,\psi_b(1)]\,d\tau_1 d\tau_2$$

and, again because of the equivalence of the electrons,

$$(5\text{-}48a) \qquad K_2 = e^2 \int\int \frac{1}{r_{a1}} [\psi_a(1)\,\psi_b(2)\,\psi_a(2)\,\psi_b(1)]\,d\tau_1 d\tau_2$$

$$(5\text{-}48b) \qquad = e^2 \int\int \frac{1}{r_{b2}} [\psi_a(1)\,\psi_b(2)\,\psi_a(2)\,\psi_b(1)]\,d\tau_1 d\tau_2$$

If these results are now substituted into the expressions for the energy states of the hydrogen molecule and are rearranged, we can obtain

$$(5\text{-}49a) \qquad E_S - 2E_0 = \frac{e^2}{r_{ab}} + \frac{J_1 - 2J_2 + K_1 - 2K_2}{1 + S_{12}}$$

and

$$(5\text{-}49b) \qquad E_A - 2E_0 = \frac{e^2}{r_{ab}} + \frac{J_1 - 2J_2 - K_1 + 2K_2}{1 - S_{12}}$$

which correspond to the symmetric and the antisymmetric wave functions, respectively.

The potential energy curves resulting from both the symmetric and the antisymmetric states are seen in Figure 5–6. By setting the ground-state energies of the isolated hydrogen atoms equal to zero, that is, $E_0 = 0$, then the resultant potential energy curve represents the *interaction energy* between the two hydrogen atoms as they form the hydrogen molecule. The antisymmetric state fails to show a minimum, and as would be expected, it represents an unstable state. On the other hand, the symmetric state shows a relatively deep minimum in the curve, thereby indicating a stable molecular species. It is significant that the potential minimum using this trial function is considerably closer to the experimental potential minimum than that obtained by ignoring the indistinguishability of the electrons. However, we should also note that better agreement can be obtained by using still other wave functions.

**The classical interaction energy.** Acknowledging the indistinguishability of the electrons in the hydrogen molecule is comparable to admitting the existence of the two structures

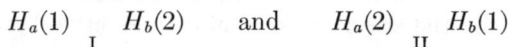

$$H_a(1) \underset{\text{I}}{\quad} H_b(2) \quad \text{and} \quad H_a(2) \underset{\text{II}}{\quad} H_b(1)$$

Because of the identity of any two isolated hydrogen atoms, the wave functions of the two structures are identical as are the energies of the two structures. Now, in order to appreciate the significance of the indistinguishability of the two electrons, it is necessary to interpret the molecular energy in terms of either one or the other of these structures. If, for example, we consider structure I, we can recall that the corresponding wave function is

$$\psi_I = \psi_a(1)\,\psi_b(2)$$

and the energy of the system will be given by

$$E = \frac{\int \psi H\,\psi d\tau}{\int \psi\,\psi d\tau}$$

$$= \frac{\int \psi_a(1)\,\psi_b(2)\,H\,\psi_a(1)\,\psi_b(2)d\tau}{\int \psi_a(1)\,\psi_b(2)\,\psi_a(1)\,\psi_b(2)\,d\tau}$$

or simply

$$E_c = \frac{H_{11}}{S_{11}}$$

This can be simplified further by using normalized 1s wave functions. If this is done, $S_{11} = 1$, and

(5–50) $\qquad E_c = H_{11}$

Earlier, it was shown that

(5-44)        $H_{11} = 2E_0 + \dfrac{e^2}{r_{ab}} + J_1 - 2J_2$

Therefore, we can say that

(5-51)        $E_c = 2E_0 + \dfrac{e^2}{r_{ab}} + J_1 - 2J_2$

The interaction energy of the two atoms can be determined by equating the energy of the isolated atoms, $E_0$, to zero. The resultant potential energy diagram then gives the decrease in energy of the system over that of the isolated atoms as the two hydrogen atoms are allowed to approach each other. If a potential minimum as a function of $r_{ab}$ is observed, then the theory would predict a stable molecule. In Figure 5–6, the second line represents $E_c$, and it is seen to show a slight minimum in the potential diagram. This indicates that the molecule is stable, but the calculated dissociation energy is seen to be only about one-tenth of the experimental value.

If we consider the various terms in the energy expression, we can conclude that $E_c$ is essentially the classical Coulombic interaction between the two hydrogen atoms. The integral $J_1$ was seen to be

$$J_1 = e^2 \int\int \frac{1}{r_{12}} \psi_a{}^2(1)\, \psi_b{}^2(2)\; d\tau_1 d\tau_2$$

The term $e^2/r_{12}$ represents the electrostatic repulsion between two unit charges and $\psi_a{}^2(1)$ and $\psi_b{}^2(2)$ represent the charge densities of electron (1) and electron (2). We can, therefore, conclude that $J_1$ represents the electrostatic repulsion between the charge clouds of electron (1) and electron (2).

On the other hand, the integral $J_2$ was seen to be

$$J_2 = e^2 \int\int \frac{1}{r_{a2}} \psi_a(1)\psi_b(2)\psi_a(1)\psi_b(2)d\tau_1 d\tau_2$$

$$= e^2 \int\int \frac{1}{r_{b1}} \psi_a(1)\, \psi_b(2)\, \psi_a(1)\, \psi_b(2)d\tau_1 d\tau_2$$

and these reduce effectively to

$$J_2 = e^2 \int \frac{1}{r_{a2}} \psi_b{}^2(2)\; d\tau_2$$

$$= e^2 \int \frac{1}{r_{b1}} \psi_a{}^2(1)\; d\tau_1$$

This can be seen to represent the attraction between the electron charge around each individual atom and the opposite nucleus. For instance, it is the attraction between the positively charged nucleus $b$ and the negatively charged electron cloud around nucleus $a$.

Finally, $e^2/r_{ab}$ represents the nuclear repulsion between the two positively charged nuclei. Thus, we have the two electrostatic repulsion terms, $e^2/r_{ab}$ and $J_1$, and we have the one electrostatic attraction term $2J_2$. $E_c$ should then represent the classical electrostatic attraction between the two hydrogen atoms. This leads to a stable hydrogen molecule, but it is not nearly so stable according to this picture as it is experimentally known to be.

**Resonance.** It is here that the significance of the indistinguishability of the electrons can be appreciated. By recognizing that the two structures I and II can exist, a new wave function is seen to result that leads to a dissociation energy that is about 70% of the experimental value. When we recall that the simple electrostatic interaction gives a dissociation energy of only about 10% of the experimental value, it is understandable why the effect is considered a quantum mechanical principle. There is no classical analog of this effect, but in terms of our quantum mechanical model it has meaning. In valence bond language, this effect is referred to as *resonance*, and it plays a critical role in the valence bond theory.

Mathematically, one might conclude that electron (1) exists on atom $a$ and electron (2) exists on atom $b$ at a given instant, and at another instant the reverse arrangement exists. Thus, the electrons could be considered to resonate between the two structures. This conclusion, however, is not the presently accepted conclusion. It is more realistic to say that the actual state of the system is represented by neither structure I nor II at any instant of time, but rather the actual structure is some intermediate that has some of the character of each of the independent structures. Thus, we are constructing a wave function to describe the actual structure by introducing the various common characteristics expressed by each individual structure. Carrying out this construction results in a much better wave function than that of either structure by itself and correspondingly, a significant increase in stability of the system with respect to any single structure. This difference in energy is referred to as the *resonance energy*.

It is important to recognize that the concept of resonance is, in a sense, a *fictitious* concept. It has its existence by virtue of the mode of construction of the trial wave function, and its reality is thus a product of the quantum mechanical model used to describe the system. As long as we use the valence bond theory, resonance has a reality, for it is basic to the language and concepts of this theory. However, in terms of another equally good model, it may have no meaning and, therefore, no existence. The same argument, of course, applies to any scientific concept, but because of the success and use of some concepts, they are often given more credence than they deserve. Resonance falls into this category. Thus, in terms of the valence bond theory, resonance will have an existence, and we will feel justified in speaking as if it were a fact. But in terms of other models, we may tend to deny its existence.

**Contribution of ionic terms.** Using the language of the valence bond theory, we can improve the trial function by recognizing that two additional structures may contribute to the total wave function of the hydrogen molecule. These are the ionic structures

$$:H_a^- \quad H_b^+ \quad \text{and} \quad H_a^+ \quad :H_b^-$$
$$\text{III} \qquad\qquad\qquad \text{IV}$$

in which both of the electrons are placed on either atom $a$ or atom $b$. These structures obviously cannot be of equal weight with the homopolar structures I and II. Consequently, the total wave function can be represented as

$$\psi_{VB} = a\,\psi_{I} + a\,\psi_{II} + b\,\psi_{III} + b\,\psi_{IV}$$
$$= a\,\psi_a(1)\,\psi_b(2) + a\,\psi_a(2)\,\psi_b(1)$$
$$+ b\,\psi_a(1)\,\psi_a(2) + b\,\psi_b(1)\,\psi_b(2)$$

Here we note that the two homopolar terms are of equal weight and the two ionic terms are also of equal weight. If the atoms were not the same, this would not be true. Thus, in general, it can be said that

(5–52)    $$\psi_{VB} = a\,\psi_{I} + b\,\psi_{II} + c\,\psi_{III} + d\,\psi_{IV} + \cdots$$

The values of the parameters $a$, $b$, $c$, $d$, etc., can be determined by minimizing the energy of the system with respect to each of the parameters. Such a calculation for the hydrogen molecule implies that the bond has about 17% ionic character. It is of interest to note that the simple molecular orbital wave function for the hydrogen molecule contains the ionic terms represented by structures III and IV, but they are given the same weight as the homopolar terms. This can be seen by multiplying out the function

(5–37)    $$\psi_{MO} = \{\psi_a(1) + \psi_b(1)\}\{\psi_a(2) + \psi_b(2)\}$$

to give

$$\psi_{MO} = \psi_a(1)\,\psi_b(2) + \psi_a(2)\,\psi_b(1) + \psi_a(1)\,\psi_a(2) + \psi_b(1)\,\psi_b(2)$$

Thus, it would appear that the simple molecular orbital theory places too much emphasis on the ionic terms. This, of course, can be remedied by arbitrarily introducing parameters to give the proper weight to the various terms.

In addition to the resonance effect, the valence bond function can be further improved by introducing other parameters. These are usually attributed to various physical aspects of the molecular structure. For instance, we can represent the $1s$ wave function as $\psi = (1/\pi a_0^3)e^{-\alpha r/a_0}$ rather than $\psi = (1/\pi a_0^3)e^{-r/a_0}$. The parameter, $\alpha$, can then be used to minimize the energy, and at the same time it can be justified in terms of a contraction of the electron orbit due to an attraction to two nuclear centers. Of course, no justification is really necessary. The more parameters that are used in the trial function, the more adaptable it should be, if we ignore the effort required to solve the resultant equations. In fact, essentially exact agreement between theory and experiment was obtained by James and Coolidge using a wave function with 13 terms.[10]

## DIATOMIC MOLECULES

In the construction of the one-electron molecular orbitals for the hydrogen molecule, we used linear combinations of the atomic $1s$ orbitals of the separated hydrogen atoms. In this instance, the isolated atoms are identical, and the ground state in each is the same. If, on the other hand, a molecule is to be constructed from two different atoms, it is very unlikely that the same orbitals would be used from both atoms. For instance, if we consider the molecule HCl, it is obvious that, although the $1s$ orbital of the hydrogen atom is used, the $1s$ orbital of the chlorine atom will contribute nothing to the formation of the bond. This emphasizes an important point with regard to bond formation. In order for two orbitals to make an effective molecular orbital, it is necessary that they be of comparable energies. In this particular instance, the $1s$ orbital of the chlorine atom is at a much lower energy than the $1s$ orbital of the hydrogen atom, and the two will, therefore, not combine.

It is also important to consider the extent of *overlap* between the combining orbitals. Although the overlap is, by itself, an insufficient criterion for bond formation, it nevertheless is important. Mathematically, the overlap is expressible in terms of the overlap or orthogonality integral, $S_{ab} = \int \psi_A \psi_B d\tau$. When $S_{ab}$ is large, the overlap of the orbitals $\psi_A$ and $\psi_B$ is large. In directional bonding, the overlap criterion will be of extreme importance. However, in general, we shall say that it is one of the factors that must be considered in the choice of atomic orbitals to be used in the construction of the molecular orbital.

Along the same lines, it is important to consider the symmetry of the combining orbitals. We can recall that a $p$ orbital has a positive and a negative lobe whereas a $1s$ orbital is everywhere positive. Now, as can be seen in Figure 5–8, two $s$ orbitals overlap effectively as do an $s$ and a $p_x$

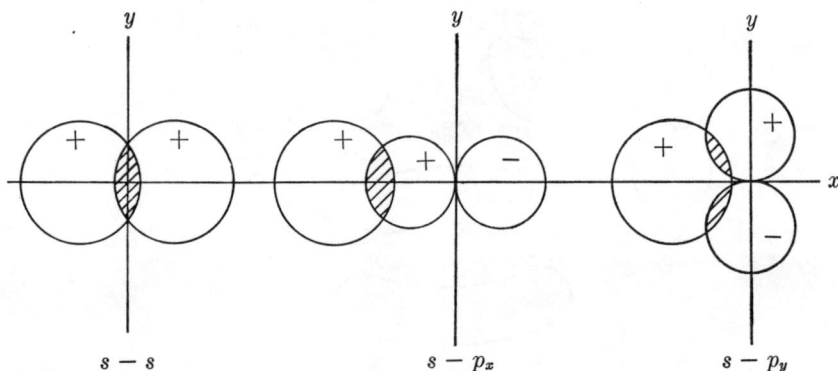

**FIGURE 5-8**

Overlap of $s$- and $p$-Type Orbitals with Respect to Axial Symmetry.

orbital. However, it is apparent that the symmetry of the $p_y$ and $p_z$ orbitals prevent an effective overlap with the $s$ orbital. Here we see that both the positive and the negative lobes of the $p_y$ orbital overlap the $s$ orbital and each exactly cancels the effect of the other. The $p_z$ orbital would, of course, give the same result as the $p_y$ orbital. Just as no overlap at all causes $S_{ab} = 0$, the type of overlap observed with the $p_y$ and $p_z$ orbitals also leads to $S_{ab} = 0$. Thus, it is evident that we must consider the relative energies, the extent of overlap, and the symmetry of the combining orbitals in the construction of a molecular orbital wave function.

In the case of homonuclear diatomics, we can give a simple pictorial representation of the molecular orbitals as they are formed from the individual atomic orbitals. For the $s$ and $p$ states, we can obtain two distinct types of molecular orbitals, as seen in Figure 5–9. The atomic orbitals combine to form two molecular orbitals, the bonding and the antibonding molecular orbitals corresponding to the symmetric and the antisymmetric functions, respectively. The molecular orbitals formed from the $1s$ atomic orbitals have cylindrical symmetry about the line of centers between the two

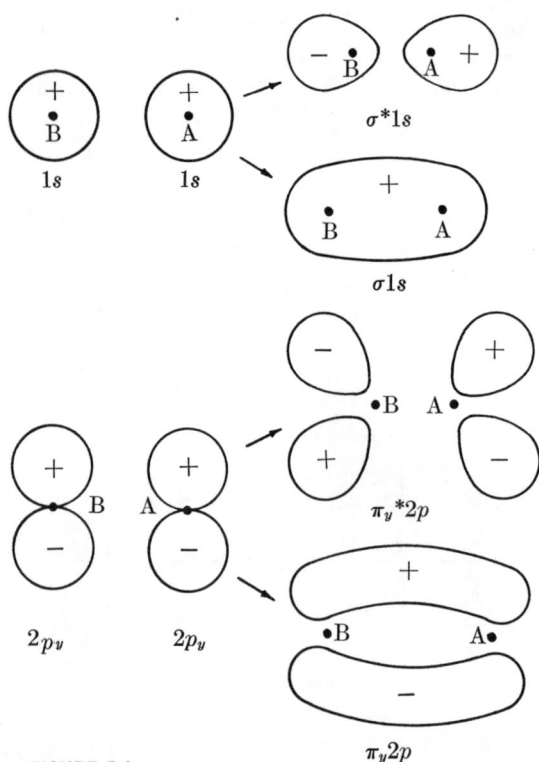

**FIGURE 5-9**

Formation of $\sigma$ and $\pi$-Type Molecular Orbitals from Atomic Orbitals.

atoms, **A** and **B**. Such bonds are referred to as $\sigma$ bonds and would also be expected from the combination of two $p_x$ orbitals or an $s$ orbital with a $p_x$ orbital. To the contrary, the $p_y$ orbitals shown in Figure 5–9 show a quite different symmetry with respect to the bond axis. These bonds are called $\pi$ bonds, and since they can be formed from either the $p_y$ or the $p_z$ atomic orbitals, it is necessary to designate them as $\pi_y$ and $\pi_z$, respectively.

In addition to the source of the orbitals, it is also necessary to differentiate between the bonding and the antibonding orbitals. This can be done by designating an antibonding $\sigma$ orbital as $\sigma^*$ and an antibonding $\pi$ orbital as $\pi^*$. The same symbolism can be used for $\delta$ orbitals, which originate from atomic $d$ orbitals, and so on.

Just as there is an order of filling of atomic orbitals, there is also an order of filling of molecular orbitals. The latter has not been determined strictly from theoretical considerations because of mathematical complexity, but in conjunction with spectroscopic evidence, it has been concluded that for homonuclear diatomic molecules of the first-row elements the order is generally

$$\sigma_{1s} < \sigma^*_{1s} < \sigma_{2s} < \sigma^*_{2s} < \pi_{y2p} = \pi_{z2p} < \sigma_{2p} < \pi^*_{y2p} = \pi^*_{z2p} < \sigma^*_{2p}$$

with the positions of the $\pi_{2p}$ and $\sigma_{2p}$ changing in the heavier members. Once this order is established, it is possible to consider the formation of a homonuclear diatomic molecule between the first-row elements. This can be done in much the same manner as that used to build up the electron levels in an atom. Considering first the hydrogen molecule, we can recognize that there will be two electrons to place in the molecular orbitals, and since each orbital will accommodate two electrons, we can represent the molecular formation as

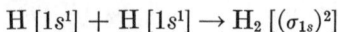

$$H\,[1s^1] + H\,[1s^1] \rightarrow H_2\,[(\sigma_{1s})^2]$$

This tells us that the hydrogen molecule has two electrons in the bonding $\sigma_{1s}$ orbital that was constructed from the atomic $1s$ states of the isolated hydrogen atoms. Since both electrons are in the bonding orbital, the resultant molecule should be stable. We should also recognize that the hydrogen molecule ion should be stable. Its configuration is $H_2^+\,[(\sigma_{1s})]$, and the one electron in the bonding orbital should lead to a stable species.

The next normal homonuclear diatomic molecule is $He_2$, and with four electrons, it can be represented as

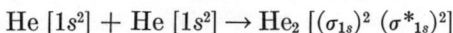

$$He\,[1s^2] + He\,[1s^2] \rightarrow He_2\,[(\sigma_{1s})^2\,(\sigma^*_{1s})^2]$$

Here, we see that there are two electrons in the bonding orbital and two electrons in the antibonding orbital. The latter electrons cancel the effect of the bonding electrons, and a stable $He_2$ molecule should not exist. This, of course, is what is observed. However, the helium molecule ion, $He_2^+$, has been detected. Its stability can be attributed to an excess of bonding over antibonding electrons. But it should be mentioned that an anti-

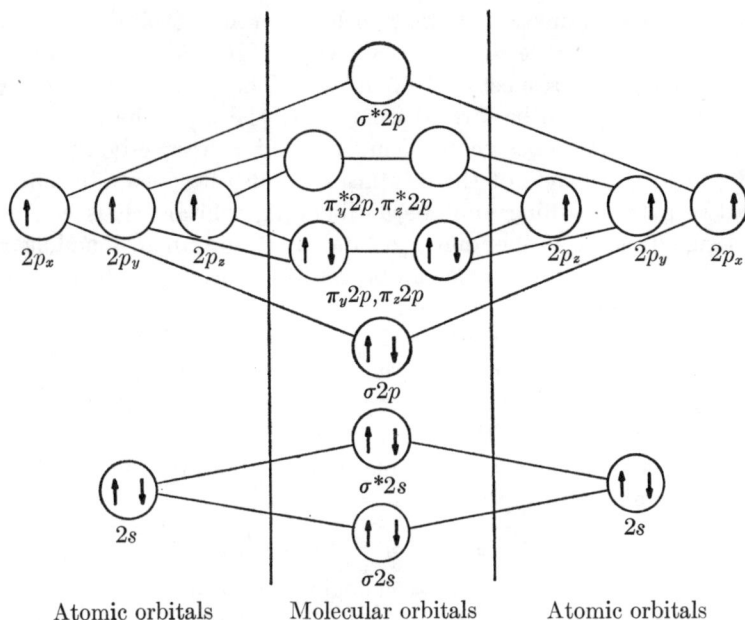

**FIGURE 5-10**

Molecular Orbitals and Their Occupancy for the Nitrogen Molecule.

bonding electron more than compensates for the bonding effect of a single bonding electron. Thus, the helium molecule ion would not be expected to be as stable as a molecule having a simple one-electron bond.

Two molecules that are of particular interest are nitrogen and oxygen. The bond formation in nitrogen can be considered to be

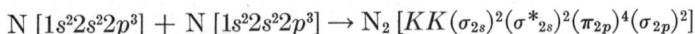

$$N\,[1s^2 2s^2 2p^3] + N\,[1s^2 2s^2 2p^3] \rightarrow N_2\,[KK(\sigma_{2s})^2(\sigma^*_{2s})^2(\pi_{2p})^4(\sigma_{2p})^2]$$

The $KK$ indicates the completion of the $K$ shell, namely $(\sigma_{1s})^2(\sigma^*_{1s})^2$, and $(\pi_{2p})^4$ represents the occupancy of the $\pi_{y_{2p}}$ and the $\pi_{z_{2p}}$ orbitals. Here, we see a total of six bonding electrons distributed in three molecular orbitals. We can consider this to represent a triple bond, composed of a $\sigma$-type bond and two $\pi$-type bonds. It is of interest to consider the formation and the occupancy of the molecular orbitals for this molecule. As can be seen in Figure 5–10, the bonding results from the $\sigma_{2p}$ and the two equivalent bonding $\pi$ orbitals.

The oxygen molecule has the unusual property of being paramagnetic in its ground state, and any valency theory must adequately explain this property. If we consider the bond formation for this molecule, we see that it is

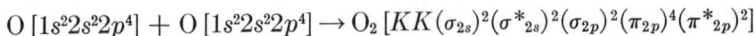

$$O\,[1s^2 2s^2 2p^4] + O\,[1s^2 2s^2 2p^4] \rightarrow O_2\,[KK(\sigma_{2s})^2(\sigma^*_{2s})^2(\sigma_{2p})^2(\pi_{2p})^4(\pi^*_{2p})^2]$$

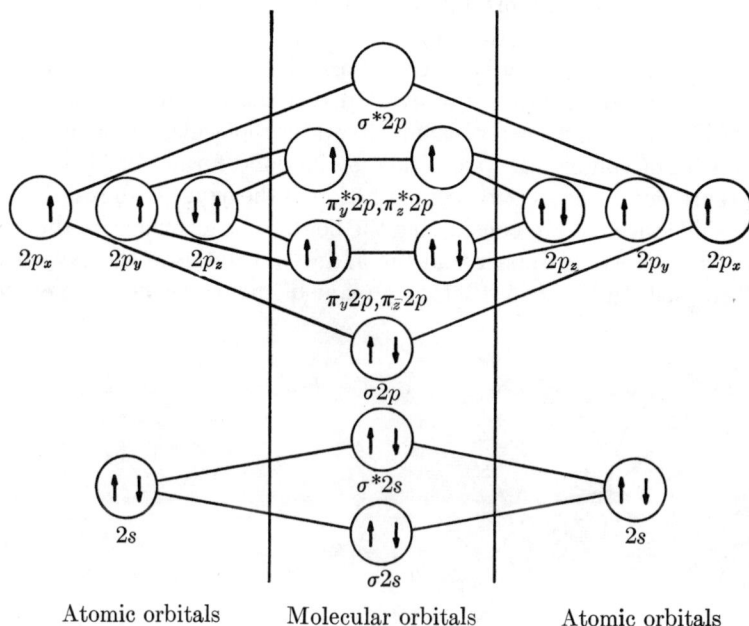

**FIGURE 5-11**

Molecular Orbitals and Their Occupancy for the Oxygen Molecule.

Two of the electrons go into the antibonding $\pi$ orbitals leading effectively to a double bond. Now, if we look at the diagram of the oxygen molecular orbitals shown in Figure 5-11, we see that rather than pair up, the electrons located in the antibonding orbitals go into different orbitals of equal energy according to Hund's rule. This still gives the double-bond character demanded, but at the same time, it gives two unpaired electrons. This, then, explains the resultant paramagnetic susceptibility of the oxygen molecule.

## DIRECTIONAL BONDING

One of the more outstanding accomplishments of the quantum mechanical approach to molecular structure has been its success in the treatment of molecular geometry. If we consider the bonds between the atoms in a molecule to be associated with the overlap of atomic orbitals, we will then expect a definite geometry to be associated with the molecule. There are a variety of ways to handle molecular structure, and the approach used in the treatment of stereochemistry in Chapter 7 will be different from the one used here. However, the use of localized atomic orbitals in terms of the valence bond theory has proved so successful that it is worthy of discussion.

In terms of the valence bond approach to molecular structure, we can imagine a covalent bond to be formed as the result of the pairing of two electrons in the atomic orbitals of two different atoms. The bond, then, should lie along the directions of the overlapping atomic orbitals, and we would expect the strongest bond to be formed in the position that permitted the maximum possible extent of overlap between the two orbitals. For the sake of illustration, we can consider the water molecule. The two hydrogen atoms will, of course, use $1s$ orbitals in the bond formation. However, we have represented the electron configuration of the oxygen atom as $1s^2 2s^2 2p_x{}^1 p_y{}^1 p_z{}^2$, and since we are considering the bond to be formed from the pairing of electrons, it would appear that the $p_x$ and $p_y$ orbitals of the oxygen atom must be used. In Figure 5–12, the $xy$ plane of the water molecule is shown,

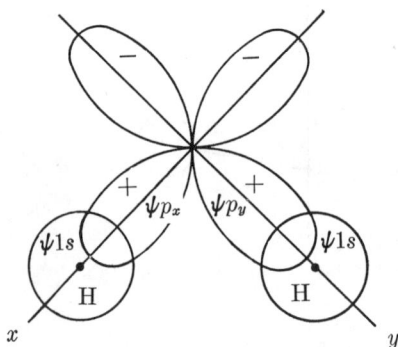

**FIGURE 5-12**

Overlap of Bonding Orbitals in the Water Molecule.

and we see that the $p_x$ and $p_y$ orbitals lie mutually perpendicular to each other in this plane. In order to obtain a maximum possible overlap between the $1s$ orbitals of the hydrogen atoms and the $p_x$ and $p_y$ orbitals of the oxygen atom, it is necessary for the hydrogen atoms to approach along the $x$ and the $y$ axes. This leads to the overlap shown, and the HOH bond angle should amount to 90°. This is not exactly in agreement with experiment. The angle has been shown to be 104°31′. However, the deviation from 90° can be attributed to a repulsion between the two hydrogen atoms and to the effect of some contribution to the bond of the $2s$ electrons of the oxygen atom. Subsequently, we shall see that the bond may also be treated as an $sp^3$ hybrid bond.

The same structure would be expected for $H_2S$, $H_2Se$, and $H_2Te$ as is predicted for $H_2O$, and, in fact, the agreement between experiment and theory is found to be quite good for these molecules. The HXH angle for $H_2S$ is observed to be 92.2°, for $H_2Se$ it is 91.0°, and for $H_2Te$ it is 89.5°. The same general arguments may also be applied to the series $NH_3$, $PH_3$, $AsH_3$, and $SbH_3$. Here, all three of the $p$ orbitals are involved in the bonding and the HXH bond angles should again be 90°. Experimentally, the angles are found for the series to be 107.3°, 93.3°, 91.8°, and 91.3°, respectively. Various explanations can be given for these deviations within the structure of the valence bond picture as well as explanations in terms of different theoretical models. However, we can see from this simple picture that a basis for molecular geometry exists in terms of the overlap of atomic orbitals.

## Hybridization

Although the simple approach based on the overlap of $s$ and $p$ atomic orbitals works well for a variety of molecules, it fails completely for the organic compounds of carbon. The occupancy of the atomic orbitals of the carbon atom is

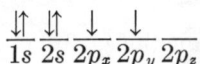

$$\frac{\uparrow\downarrow}{1s}\ \frac{\uparrow\downarrow}{2s}\ \frac{\downarrow}{2p_x}\ \frac{\downarrow}{2p_y}\ \frac{}{2p_z}$$

From this configuration we might expect a molecular formation similar to water, where the carbon atom has a valence of 2 with the bonds being mutually perpendicular. This, of course, is not what is observed for carbon. The quadrivalence and tetrahedral structure of carbon are well-established.

It is possible to justify the quadrivalence of carbon by the promotion of one of the $2s$ electrons to the empty $2p_z$ orbital. This would lead to the configuration

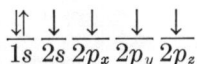

$$\frac{\uparrow\downarrow}{1s}\ \frac{\downarrow}{2s}\ \frac{\downarrow}{2p_x}\ \frac{\downarrow}{2p_y}\ \frac{\downarrow}{2p_z}$$

Although this gives the known quadrivalence, it still does not agree with observation. According to the valence bond picture, such a configuration would predict that three of the hydrogen atoms in a molecule such as $CH_4$ are equivalent in energy and are also mutually perpendicular, but the fourth hydrogen atom is held by a weaker bond at an angle of about 125° with respect to the other bonds. Again, this result is contrary to the known geometry of the organic carbon atom.

In spite of the apparent difficulties, it is possible to resolve the problem within the framework of the valence bond approach.[11] In our previous application of the valence bond theory, it was found that the wave function was improved by taking linear combinations of several reasonable functions which described different representations or resonance structures of the system. We might likewise imagine that the actual wave functions for the four carbon bonds are represented by a mixture of the $s$ and $p$ orbitals available for bond formation. This character would be introduced into the various total wave functions by taking linear combinations of the $s$ and $p$ wave functions. The best combination should be the one that would lead to the strongest bond.

Although it is quite reasonable to assume that the best wave function will be the one that gives the strongest bond, it is not obvious what basis we can use to measure the strength of the bond. It would seem reasonable, however, to assume that the strongest bond would be the one that permitted the maximum possible overlap between the bonding orbitals. This is referred to as *the criterion of maximum overlap*, and it is fundamental to the valence bond treatment of directional bonding.

**sp³ hybridization.**   In the construction of the hybrid orbitals, it is of importance to note that the radial portions of the orbitals in a given shell are approximately the same. This is apparent from a comparison of the radial portions of the 2s and the 2p orbitals shown in Figure 2–8. On this basis, it is assumed that the hybridized orbitals can be constructed from the angular parts of the individual wave functions. Thus, for the quadrivalent carbon atom, these should be four bonds of the form

$$(5\text{--}53) \qquad \psi_i = a_i\,\psi_s + b_i\,\psi_{p_x} + c_i\,\psi_{p_y} + d_i\,\psi_{p_z}$$

where the $\psi_i$ are now purely angular wave functions.

If we restrict ourselves to normalized functions, we can say that

$$\int \psi_i\,\psi_i{}^*\,d\tau = 1$$

Now, using the $s$ orbital for convenience, the charge can be imagined to be spread over a spherical distribution pattern. Introducing the element of volume for a sphere then gives

$$\int\!\!\int\!\!\int \psi_i\,\psi_i{}^*\,r^2 \sin\theta\,d\theta\,d\varphi\,dr = 1$$

Inasmuch as we are assuming the radial portion of the wave function to be constant for all orbitals in the valence shell, our concern is with the angular portion of the wave function. Thus we see that integration over the angular portion of the wave function results in

$$\int_0^{2\pi}\!\!\int_0^{\pi} \sin\theta\,d\theta\,d\varphi = 4\pi$$

From this, we say that the angular portion of the wave function is normalized to $4\pi$.

It has been stated that we shall neglect the radial portions of the wave functions. Consequently, the individual angular wave functions can be expressed in terms of spherical harmonics as seen in Table 2–6. Here, we note that $\psi_s = 1/\sqrt{4\pi}$ and the $p$ orbitals all contain the term $\sqrt{3/4\pi}$. The result of the normalization is effectively to set $\psi_s$ equal to unity. This results in the following equations for the individual angular wave functions

$$(5\text{--}54a) \qquad \psi_s = 1$$

$$(5\text{--}54b) \qquad \psi_{p_x} = \sqrt{3}\,\sin\theta\,\cos\varphi$$

$$(5\text{--}54c) \qquad \psi_{p_y} = \sqrt{3}\,\sin\theta\,\sin\varphi$$

$$(5\text{--}54d) \qquad \psi_{p_z} = \sqrt{3}\,\cos\theta$$

Because of the orthogonality of the various atomic orbitals, this leads to the restriction of the coefficients in Eq. (5–53)

(5–55)          $a_i^2 + b_i^2 + c_i^2 + d_i^2 = 1$

In determining the angular relationship between the bonds, it is convenient to let the first bond lie along one of the axes, say the $x$ axis. If this is done, the $p_y$ and $p_z$ orbitals should make no contribution to the bond. For this reason, $c_i$ and $d_i$ will both be zero and

(5–56)          $\psi_1 = a_1 \psi_s + b_1 \psi_{p_x}$

Since

$$a_1^2 + b_1^2 = 1$$

we can say that

$$\psi_1 = a_1 \psi_s + \sqrt{1 - a_1^2}\, \psi_{p_x}$$

If we now substitute the wave functions expressed in terms of $\psi_s$ equal to unity, we obtain,

(5–57)          $\psi_1 = a_1 + \sqrt{1 - a_1^2}\, \sqrt{3} \sin\theta \cos\varphi$

We have decided that the bond will be formed in such a manner as to give the strongest bond possible, and if we are to use the principle of maximum overlap, then we are interested in choosing $a_1$ so as to make $\psi_1$ as large as possible in the direction of the bond. This is based on the assumption that the larger the extension of the orbital in space, the stronger the bond it will be able to form. Now, the value of $\psi_1$ along the $x$ axis is given by

$$\psi_1 = a_1 + \sqrt{3}\, \sqrt{1 - a_1^2}$$

The $\sin\theta \cos\varphi$ term is unity here because $\theta = \pi/2$ and $\varphi = 0$. This, of course, is the maximum possible value for this term. If we differentiate with respect to $a_1$ and set the result equal to zero, we obtain

(5–58)          $1 - a_1 \sqrt{3}\, (1 - a_1^2)^{-1/2} = 0$

from which it can be seen that $a_1 = \frac{1}{2}$ and $b_1 = \sqrt{3}/2$. This, then, gives us the wave function for the first hybrid bond,

$$\psi_1 = \tfrac{1}{2} + (\sqrt{3}/2)\sqrt{3} \sin\theta \cos\varphi$$

which, in terms of the orbital symbols, can be expressed as

(5–59)          $\psi_1 = \tfrac{1}{2}\psi_s + (\sqrt{3}/2)\, \psi_{p_x}$

Previously, we have set the value of $\psi_s$ at unity, and in terms of this value, the $p$ orbitals assume the form shown in Eq. (5–54). If we evaluate their maximum magnitude in space on this basis by setting $\sin\theta$ and $\cos\varphi$ each equal to unity, we obtain a value of 1.732. This would indicate that a $p$ orbital is capable of a greater amount of overlap than a pure $s$ orbital, and according to the principle of maximum overlap, it should form a stronger bond. On the same basis, we can evaluate the magnitude of the hybrid bond $\psi_1$. By setting $\sin \theta$ and $\cos \varphi$ both equal to unity, a value of 2 is obtained for $\psi_1$ in the direction of the bond. This is considerably greater than either a pure $s$ or a pure $p$ orbital, and we would, therefore, expect the hybrid orbital to form a stronger bond than either a pure $s$ or a pure $p$ orbital.

The formation of the hydrid bond can be pictured in the manner shown in Figure 5–13. The $s$ orbital is everywhere positive and the $p$ orbital has

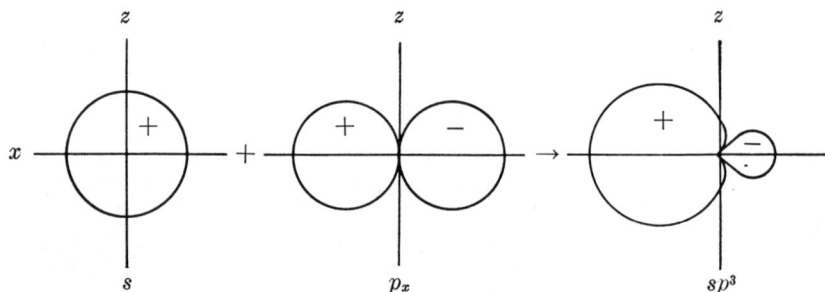

**FIGURE 5-13**

Formation and Structure of an $sp$ Hybrid Orbital.

a positive and a negative lobe. The positive lobe of the $p$ orbital combines with the $s$ orbital and the negative lobe of the $p$ orbital is reduced by the mixing.

The second of the hybrid bonds can be chosen to lie in the $xz$ plane. Here, the $p_y$ orbital will make no contribution, and the new wave function will be

$$(5\text{–}60) \qquad \psi_2 = a_2 \psi_s + b_2 \psi_{p_x} + d_2 \psi_{p_z}$$

It can be seen from Figure 2–4 that since the orbital will lie in the $xz$ plane, the angle $\varphi$ will be $0°$ or $180°$ for the bond direction. However, for $\varphi = 0°$, we are in the region of $\psi_1$. Therefore, $\psi_2$ will have its maximum value when $\varphi$ is $180°$, and $\cos \varphi$ will be $-1$. Using this fact and substituting the expressions for the wave functions into Eq. (5–60), we obtain

$$\psi_2 = a_2 - b_2 \sqrt{3} \sin \theta + d_2 \sqrt{3} \cos \theta$$

This equation can be expressed in terms of the parameter $b_2$ and the angle $\theta$ by using the two relations

$$a_2{}^2 + b_2{}^2 + d_2{}^2 = 1$$

and

$$a_1 a_2 + b_1 b_2 = 0$$

The latter expression arises from the orthogonality condition. Since $d_1$ is zero, the $d$ parameters vanish. Knowing the values for $a_1$ and $b_1$, we can evaluate $a_2$ and $d_2$ in terms of $b_2$. Thus

$$a_2 = -(b_1/a_1)b_2 = -\frac{\sqrt{3}/2}{1/2} b_2 = -\sqrt{3}\, b_2$$

and

$$d_2 = \pm\sqrt{1 - a_2{}^2 - b_2{}^2} = \pm\sqrt{1 - 3b_2{}^2 - b_2{}^2} = \pm\sqrt{1 - 4b_2{}^2}$$

It is immaterial which sign is chosen for $|\,d_2\,|$, so for convenience we will use the positive sign. Now we can express $\psi_2$ as

(5-61) $\qquad \psi_2 = -b_2 \sqrt{3}\, (1 + \sin\theta) + \sqrt{3(1 - 4b_2{}^2)}\, \cos\theta$

Just as with the first hybrid orbital, it is again necessary to maximize the orbital in the bond direction in order to evaluate the parameter, $b_2$. However, we find that in this case, the angle, $\theta$, is also a variable and it determines the bond direction, $\theta_2$. Consequently, it is necessary to maximize the function with respect to both $b_2$ and $\theta$. This leads to two equations which can be solved simultaneously. The first of these can be obtained by differentiating $\psi_2$ with respect to $b_2$ and setting the differential equal to zero. This results in the following equation

$$\left(\frac{\partial\psi}{\partial b_2}\right)_\theta = -\sqrt{3}\,(1 + \sin\theta_2) + (\tfrac{1}{2})(3 - 12b_2{}^2)^{-1/2}\,(-24b_2)\cos\theta_2 = 0$$

or

(5-62) $\qquad -\sqrt{3}\,(1 + \sin\theta_2) - \dfrac{12b_2}{\sqrt{3 - 12b_2}}\cos\theta_2 = 0$

Similarly,

$$\left(\frac{\partial\psi}{\partial\theta}\right)_{b_2} = -b_2\sqrt{3}\cos\theta_2 - \sqrt{3 - 12b_2{}^2}\sin\theta_2 = 0$$

or

(5-63) $\qquad -b_2\sqrt{3}\cos\theta_2 - \sqrt{3 - 12b_2{}^2}\sin\theta_2 = 0$

These two equations can now be used to determine the angle, $\theta_2$, and the parameter, $b_2$. If we rearrange them to the form

$$-\frac{12b_2}{\sqrt{3 - 12b_2{}^2}} = \frac{\sqrt{3}\,(1 + \sin\theta_2)}{\cos\theta_2}$$

and

$$-\frac{\sqrt{3}b_2}{\sqrt{3 - 12b_2{}^2}} = \frac{\sin\theta_2}{\cos\theta_2}$$

it can be seen that

$$\frac{12\sin\theta_2}{\sqrt{3}\cos\theta_2} = \frac{\sqrt{3}\,(1 + \sin\theta_2)}{\cos\theta_2}$$

And this equation can be solved for sin $\theta_2$ to give sin $\theta_2 = \frac{1}{3}$. Once sin $\theta_2$ is known we can evaluate $b_2$, and this is found to be $b_2 = -\frac{1}{2}\sqrt{3}$. If these two quantities are both known, we can calculate the magnitude of the new hybrid orbital and we can determine its angle with respect to the first hybrid orbital. In terms of Eq. (5–61), the magnitude is

$$\psi_2 = -\left(-\frac{1}{2\sqrt{3}}\right)\sqrt{3}\left(1 + \frac{1}{3}\right) + \sqrt{3 - 12\left(\frac{1}{2\sqrt{3}}\right)^2}\left(\frac{2\sqrt{2}}{3}\right) = 2$$

This is the same value that was obtained for $\psi_1$. Therefore, we can conclude that the two bonds are of equal strength.

In considering the angle $\theta_2$, it is seen that for the sin $\theta_2$ to equal $\frac{1}{3}$, $\theta_{\max}$ will equal 19° 28' or 160° 32'. The former corresponds to choosing the positive sign for $d_2$. Since it was necessary to consider the angle $\varphi$ to be 180°, it is apparent that, with respect to the first bond, the angle $\theta_2$ is 90° + 19° 28', or 109° 28'.

This same general procedure can be continued to determine the angles and magnitudes of the remaining two bonds. The general forms of the wave functions of these orbitals along with those of the first two orbitals are

$$\psi_1 = \frac{1}{2}\psi_s + \sqrt{3}/2\,\psi_{p_x}$$

$$\psi_2 = \frac{1}{2}\psi_s - \frac{1}{2\sqrt{3}}\psi_{p_x} + \frac{\sqrt{2}}{\sqrt{3}}\psi_{p_z}$$

$$\psi_3 = \frac{1}{2}\psi_s - \frac{1}{2\sqrt{3}}\psi_{p_x} + \frac{1}{\sqrt{2}}\psi_{p_y} - \frac{1}{\sqrt{6}}\psi_{p_z}$$

$$\psi_4 = \frac{1}{2}\psi_s - \frac{1}{2\sqrt{3}}\psi_{p_x} - \frac{1}{\sqrt{2}}\psi_{p_y} - \frac{1}{\sqrt{6}}\psi_{p_z}$$

Needless to say, the angles all turn out to be 109° 28', and the relative magnitudes are all 2. In the methane molecule, we know that all four of the bonds are equivalent and the HCH angles are all 109° 28'. Thus, it would appear that the valence bond theory is capable of giving a satisfactory treatment of the tetrahedral carbon atom. The tetrahedral hybrid bonds are constructed from one $s$ orbital and three $p$ orbitals, and for this reason they are referred to as $sp^3$ *hybrids*.

**$sp^2$ and $sp$ hybridization.** The treatment of double-bonded structures is in terms of $sp^2$ hybrid orbitals. For a molecule such as ethylene, we can imagine the *skeletal* structure to be

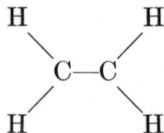

It can be seen that there are three sigma bonds from each carbon atom. These can be considered to arise from the $2s$ and two of the three $2p$ electrons of the carbon atom in the form of an $sp^2$ hybridization. This leaves one unused $2p$ electron on each carbon atom. By hybridizing these three orbitals, three equivalent bonds are obtained that lie in a plane at an angle of 120° with respect to one another. This is in general agreement with experiment.[2] Further, their magnitude compared to unity for an $s$ orbital is 1.991. This is less than that of an $sp^3$ hybrid, but it is considerably greater than that of a pure $p$ orbital or a pure $s$ orbital.

The double-bond character of the carbon—carbon bond can be attributed to the overlap of the $p$ orbitals occupied by the unused $p$ electrons. This is illustrated in Figure 5–14. The overlap of the $sp^2$ hybrid orbitals

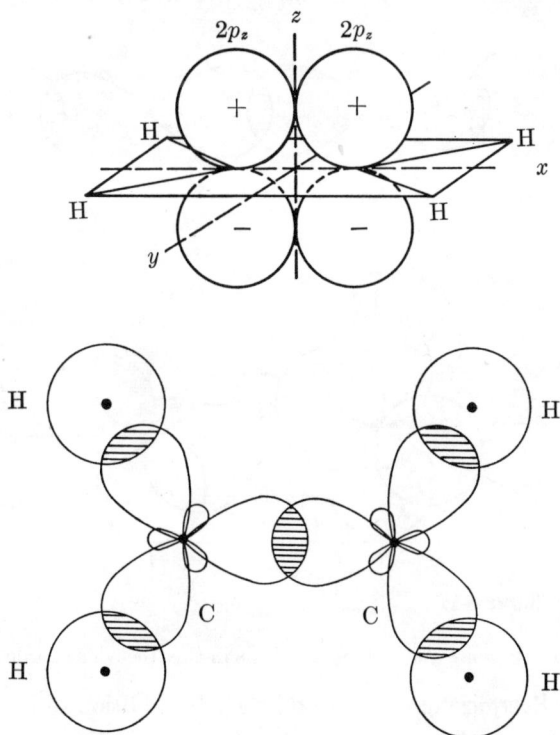

FIGURE 5-14

Overlap of $\sigma$ and $\pi$-Type Orbitals in the Ethylene Molecule.

leads to the sigma bonds and the overlap of the $p$ orbitals leads to a pi bond. From the figure, it is apparent why restricted rotation should exist around a double bond. In order for one of the carbon atoms to rotate around an axis through the line of centers of the two carbon atoms, it is necessary to break the overlap of the $\pi$ orbitals.

Finally, for the acetylenic-type linkage, the skeletal structure is

$$H—C—C—H$$

This structure requires the use of only one of the $2p$ orbitals along with the $2s$ orbital to form the hybrid bond. Here, we have two $p$ electrons on each carbon atom that are unused in the formation of the sigma bond. The calculations of the geometry and relative strengths of two equivalent $sp$ hybrids are again found to be in agreement with the experimentally determined structure of the molecule. Both bonds are of equal strength and they are linear. The triple-bond character of the acetylenic linkage can then be attributed to the overlap of the two $p$ orbitals that are left on each carbon atom, as can be seen in Figure 5–15.

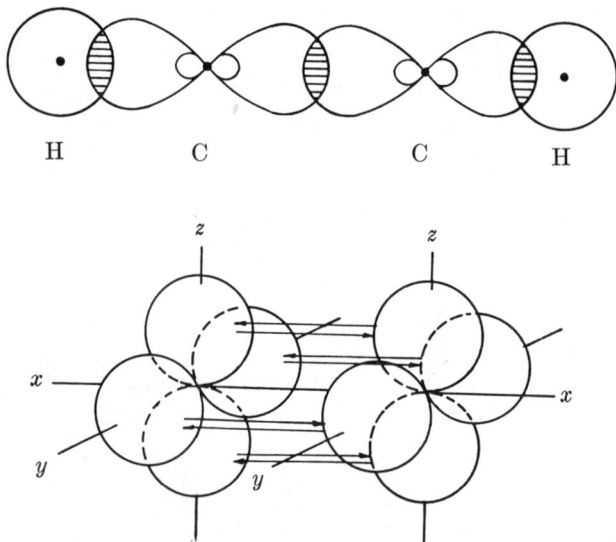

FIGURE 5-15

Overlap of $\sigma$ and $\pi$-Type Orbitals in the Acetylene Molecule.

**Hybridization with $d$ orbitals.** In addition to its success in the treatment of the stereochemistry of the carbon compounds, the valence bond theory has also proved moderately successful in its treatment of the structure of coordination compounds. Here, it is usually necessary to consider the effects of the $d$ orbitals. The most common structures that result from the hybridization of $d$ orbitals are the $dsp^2$ hybrid bonds, which lead to square planar structures, and the $d^2sp^3$ hybrid bonds which lead to octahedral structures. These are discussed in some detail in Chapter 10, and for that reason they will not be considered further here. However, a summary of the more common bonds, along with their geometry and their relative strengths, are listed in Table 5–3. In all cases the strength of the bonds is based on a value of unity for the pure $s$ orbital.

TABLE 5–3.  **Relative Strengths and Angular Distributions of Bonding Orbitals.**

| Coord. No. | Orbitals Used | Angular Distribution | Relative Strength |
|:---:|:---:|:---:|:---:|
| 1 | $s$ | none | 1.000 |
| 3 | $p$ | mutually $\perp$ | 1.732 |
| 2 | $sp$ | linear | 1.932 |
| 3 | $sp^2$ | planar 120° | 1.991 |
| 4 | $sp^3$ | tetrahedral | 2.000 |
| 4 | $dsp^2$ | square planar | 2.694 |
| 6 | $d^2sp^3$ | octahedral | 2.923 |

**Hybridization in water and ammonia.** Previously, we considered the structures of water and ammonia as well as the analogous compounds in their families in terms of pure $p$ bonding. This would predict HXH bond angles of 90°, and these were seen to differ considerably from the experimental values for both water and ammonia. Another means of considering the structures of these compounds is in terms of hybridization.[12] If we take the three molecules $CH_4$, $NH_3$, and $H_2O$, we note that they are isoelectronic, each having 10 electrons. Now, the methane molecule is considered to be using $sp^3$ hybrid orbitals, and its structure, therefore, is tetrahedral. If we consider an imaginary process whereby a proton from one of the hydrogen atoms coalesces with the carbon nucleus, we have the ammonia molecule. Further, if we imagine a proton from one of the hydrogen atoms in the ammonia molecule coalescing with the nitrogen nucleus, we have a water molecule. The only difference between the three structures is the exchange of a **M—H** bond for a *lone pair* of electrons on the $(M + 1)$ nucleus. We can now wonder whether a lone pair of electrons on an atom can occupy a tetrahedral position just as can a *bonding pair* of electrons. If this is possible, and it is certainly reasonable, then we can consider the water molecule as well as the ammonia molecule to be a tetrahedral structure based on $sp^3$ hybridization, and our problem is no longer one of understanding a deviation from 90°, but rather one from 109.5°. This idea has been developed with some success, but it will not be considered further here. The details of stereochemistry will be discussed in Chapter 7.

## THE THREE-CENTERED BOND

Because of the prejudices we carry over from our early ideas of chemical bonding, we are prone to think of a covalent bond only in terms of one or more electron pairs. The classical example of a violation of this principle is found among the boron hydrides. Based on the conventional electron pair bond, we would expect to find the compound $BH_3$. But this compound does not occur. Rather, the first member of the series is $B_2H_6$. If the electron distribution is drawn, it is evident that there are not enough electrons available to meet the demands of electron pairing between each pair of bonded atoms. That is, the compound is *electron-deficient*.

The structure of diborane has now been shown to be that given in Figure 5–16. At present, the nature of the bonding in such electron-deficient

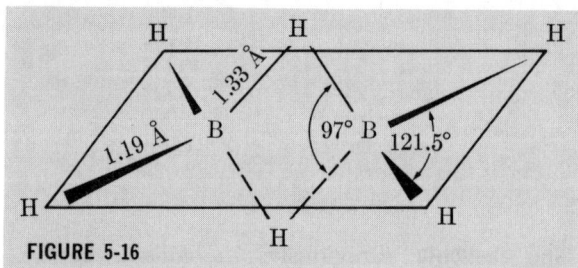

**FIGURE 5-16**

Structure of $B_2H_6$.

compounds can best be treated in terms of molecular orbital theory. In terms of the M.O. approach, a conventional covalent bond is formed by the combination of one atomic orbital on a bonded atom with one atomic orbital on the second of the bonded atoms. These result in two new orbitals, a bonding and an antibonding orbital as was shown in Figure 5–2. If instead we consider the combination of three atoms, two of one kind, $X$ and $X'$, and a third $Y$, with one atomic orbital from each, three molecular orbitals would be expected to occur. One of these will be a bonding orbital, $\psi_b$, one will be an antibonding orbital, $\psi_a$, and intermediate in energy between these two will be a third orbital which is, in essence, nonbonding, $\psi_n$. These are represented in the energy level diagram shown in Figure 5–17.

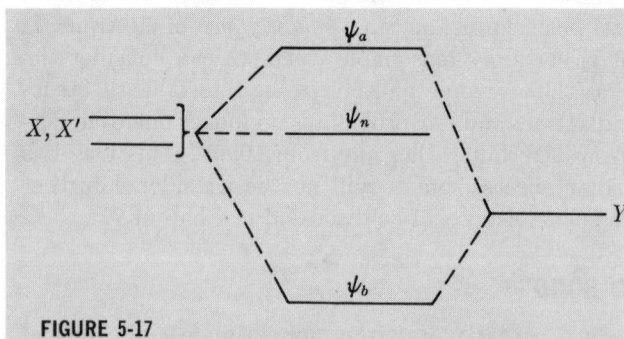

**FIGURE 5-17**

Energy Level Diagram for a Three-Centered Molecular Orbital.

If we now apply this to the molecule $B_2H_6$, we can imagine two $BH_2$ groups with all the atoms lying in a plane as shown in the Figure 5–16. In each $BH_2$ fragment, there are two electron-pair bonds and one electron left over,

$$\begin{array}{c} \overset{..}{\phantom{}}\text{H} \\ \cdot\text{B}\overset{.}{\phantom{.}} \\ \overset{.}{\phantom{.}}\text{H} \end{array}$$

These two fragments are then linked together by means of hydrogen bridging. It is here that we find the deficiency of electrons; a given BHB linkage has only two electrons available. We can now construct two three-centered bonds, each having two boron atoms and one hydrogen atom contributing to a molecular orbital. This requires the use of hybridized orbitals on the boron atoms that approach $sp^3$ hybridization combined with the $1s$ orbital on each hydrogen atom. This gives us two sets of three-centered molecular orbitals and, therefore, two bonding three-centered molecular orbitals. Since there are four electrons available, we can put two into each of the bonding orbitals, thereby leading to a stable structure.

In general, molecular orbitals can be constructed from linear combinations of any number of bonded atoms. We might, therefore, consider the conventional two-centered or electron-pair bond to be a simple example of a *multicentered* bond. And we could expect not only three-centered bonds, but four-, five-, etc. centered bonds.

There is, of course, no reason why the three-centered bond should be limited to diborane. In fact, it is observed in the higher boranes where additionally a three-centered bond involving three boron atoms is also found. It can also be used to rationalize the structures of $FHF^-$, the dimers of the $Al(alkyls)_3$ such as $[Al(CH_3)_3]_2$, and the polymer $[Be(CH_3)_2]_n$.

## LINNETT'S DOUBLE-QUARTET APPROACH[13]

Fundamental to our ideas of a covalent bond is the concept of a shared electron pair. In the treatment of the normal sigma bond, basically we use a quantum mechanical extension of the classical ideas developed before 1920 primarily by Lewis, Langmuir, and Kossel. It is in terms of this classical view that we find the *Lewis Structures* and the *Lewis-Langmuir octet rule*. According to the octet rule, the tendency of an atom to have an electron arrangement similar to that of an inert gas results in eight electrons in its valence shell arranged as four electron pairs. These classical ideas are common tools to the beginning chemistry student. But in 1961 Linnett proposed a very interesting modification of the Lewis-Langmuir Octet Rule. He suggested that the importance of the *pairing* of electrons has been overemphasized. Accordingly, he has proposed that the eight valence electrons, for example those present around first short-period atoms in their molecules, be treated as *two groups of four*, the *double-quartet*, rather than as *four pairs.* It is generally accepted that the stable octet is made up of four electrons of one spin and four of the opposite spin. Because of the effects of like charge, *charge correlation*, and like spin, *spin correlation*, each set of four will have a high probability of being arranged in an approximately regular tetrahedral pattern around the nucleus. However, Linnett suggests that the two sets of four electrons around a given nucleus, which individually have an approximately tetrahedral arrangement, may be treated as essentially uncorrelated spatially *relative to one another.* This results from the opposing effects of charge correlation, which tends to keep *all* electrons apart, and of the Pauli principle or spin correlation effect that tends to keep all electrons of opposite spins together (see Chapter 7). It is supposed that the two opposing effects roughly cancel one another as far as the spatial correlation of the two tetrahedral sets is concerned. The two tetrahedra, relatively uncorrelated in species such as $Na^+$, Ne, $F^-$, $O^{2-}$, will of course become strongly correlated by the localization of electrons in covalent *bonding pairs.*

The assumption of different dispositions around a nucleus by the two tetrahedral sets should lead to a lowering of the mean interelectron repulsion energy. Thus, if the adoption of spatially separated dispositions by the two spin sets can be achieved with no reduction in the *number* of electrons in regions of space influenced by two nuclei, that is, in the *bond regions*, a structure involving two separated dispositions will have a lower energy than one having the spin sets identically disposed with the same number of electrons in bond regions. However, in molecules such as $CH_4$, $NH_3$, $H_2O$, etc., and more generally for the great majority of diamagnetic molecules which can be described by a *single* valence bond structure, it is only possible to put two electrons into each bond region if the two spin sets adopt essentially the same spatial disposition. That is, this must occur in spite of some resulting increase in the interelectron repulsion energy. For multiple bonds, a *bent bond* representation must be used. Thus a double bond is represented by two tetrahedra (on the separate bonded atoms) sharing an edge and a triple bond by the two tetrahedra sharing a face.

It is obvious, then, that for most molecules the double-quartet representation is no different from the four-pair representation usually envisioned. However, for certain other molecules, particularly paramagnetic ones, and stable radicals only the double-quartet treatment is capable of representing the molecule or ion with a single structure, and hence, in this sense, it is preferable to the conventional representations.

## Some Applications of the Double-Quartet Approach

We have already seen that the valence bond approach is not satisfactory for representing the paramagnetic ground state of the oxygen molecule. The molecular orbital approach does succeed in accounting for the paramagnetism in terms of a molecular orbital energy level scheme. The double-quartet approach allows us to write one structural configuration for the ground state of the oxygen molecule that satisfies all requirements, such as two unpaired electrons, eight electrons around each oxygen, and four electrons in the internuclear region. Additionally, it minimizes the inter-electron repulsion energy. To do this we assign the 12 valence-shell electrons in $O_2$ such that there are five of one spin set and seven of the other, as shown diagrammatically below.

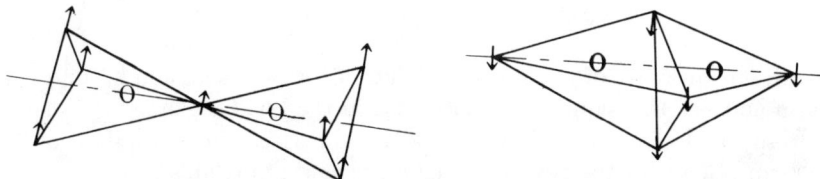

On the left is the minimum energy arrangement of seven electrons of like spin and on the right is the minimum energy arrangement of five electrons of like spin. When the two are superimposed we have the desired ground-state ($^3\Sigma_g^-$) structure:

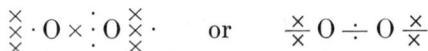

$$ \overset{\times}{\underset{\times}{\times}} \cdot O \times \vdots O \overset{\times}{\underset{\times}{\times}} \cdot \qquad \text{or} \qquad \overset{\times}{\underset{\times}{}} O \div O \overset{\times}{\underset{\times}{}} $$

where a bar represents an electron pair having opposite spins. The excited diamagnetic state $^1\Sigma_g^+$, which is 38 kcal/mole above the ground state, would have two sets of six electrons arranged as six pairs at the corners of two tetrahedra having a common edge between the nuclei; that is, the two sets have the same spatial distribution:

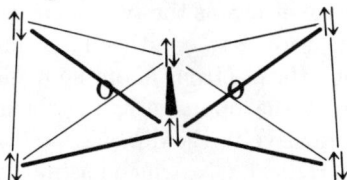

But there is a lower-energy excited diamagnetic state ($^1\Delta_g$) just 25 kcal/mole above the ground state in which it may be postulated that the lower energy arises from *no correlation* between the two sets of six electrons. Note that because all three states have four electrons in the bond region, the bond lengths are approximately equal: $^3\Sigma$:1.207; $^1\Delta$:1.216; $^1\Sigma$:1.227 Å.

The nitric oxide molecule has 11 valence-shell electrons and hence is paramagnetic. The equilibrium bond length is 1.15 Å, intermediate between that in triply bonded $N_2$(1.06 Å) and doubly bonded $O_2$(1.21 Å). The vibration frequency of NO in its ground state is 1876 cm$^{-1}$, which is again intermediate between that of $N_2$(2330 cm$^{-1}$) and that of $O_2$(1556 cm$^{-1}$), and the dissociation energy of 150 kcal/mole is likewise intermediate between the $N_2$(225 kcal/mole) and $O_2$(117 kcal/mole) values. These data clearly indicate that the bond in NO is intermediate in order between that in $N_2$, which consists of six electrons and that of $O_2$, which consists of four electrons. Thus we deduce that in the ground state of the NO molecule there should be five electrons in the internuclear region. The double-quartet model very nicely accommodates these facts:

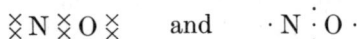

$$\overset{\times}{\underset{\times}{N}} \overset{\times}{\underset{\times}{O}} \overset{\times}{\underset{\times}{}} \quad \text{and} \quad \cdot N \overset{\cdot}{\cdot} O \cdot$$

yield the single structure

$$\cdot \overset{\times}{\underset{\times}{N}} \overset{\cdot}{\cdot} \overset{\times}{\underset{\times}{O}} \overset{\times}{\underset{\times}{}} \cdot \quad \text{or} \quad \overset{\times}{-}N \doteq O \overset{\times}{-}$$

which has an octet of electrons made up of four of each spin around each atom and has the required five electrons in the bond region.

Many odd electron molecules tend to dimerize, but the failure of NO to form $N_2O_2$ with the electronic structure containing an

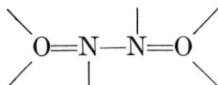

$$\overset{\displaystyle \diagdown \quad\quad | \quad\quad \diagup}{\underset{\displaystyle \diagup \quad\quad | \quad\quad \diagdown}{O=N-N=O}}$$

even number of electrons, five bond pairs and six lone pairs, has long been a puzzling problem. However, it should be noted that two NO molecules taken together *also* contain 10 bonding electrons and 12 unshared electrons. Therefore dimerization would not yield an increase in the number of bonding electrons, and this means that there would be little, if any, gain energetically. Indeed, the entropy factor would be unfavorable. But even more important, perhaps, is that in the proposed dimer the electrons of one spin must adopt the same spatial pattern as the electrons of opposite spin. Thus for each of 11 pairs of electrons the two electrons are forced to occupy the same spatial region, leading to a rather large interelectron repulsion. However, in the monomer the electrons of one spin adopt a different spatial pattern from electrons of the opposite spin, which is what apparently stabilizes the monomer relative to the dimer.

Note that the cyanide radical, CN, which has nine valence electrons that may be distributed either as

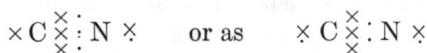

$$\times C \overset{\times}{\underset{\times}{\overset{\cdot}{\cdot}}} \overset{\cdot}{\cdot} N \times \quad \text{or as} \quad \times C \overset{\times}{\underset{\times}{\overset{\cdot}{\cdot}}} \overset{\cdot}{\cdot} N \times$$

readily dimerizes to cyanogen, $(CN)_2$, of electronic structure

$$-N \equiv C - C \equiv N-$$

Here we find *fourteen* bonding electrons as opposed to a maximum of 12 for two CN radicals. (For another approach see reference 14.)

Because Lewis formulas fail for many polyatomic molecules, Pauling introduced the concept of resonance. Molecules and ions such as $O_3$, $NO_2^-$, $C_2O_4^{2-}$, $C_6H_6$, $N_2O$, $NO_2^+$, $N_3^-$, $NCO^-$, $CN_2^{2-}$, $CO_2$, $NO_2$, etc., have long been written using so-called resonance-hybrid structures. Without going into further details here, we shall depict the electronic structures for a few of these molecules in terms of the double-quartet approach:

$O_3$

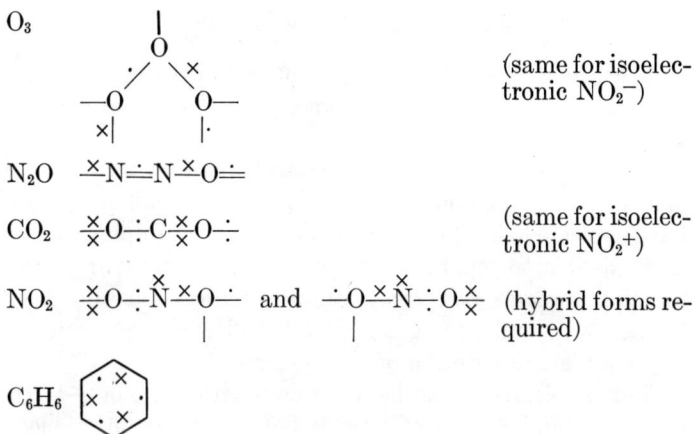

(same for isoelectronic $NO_2^-$)

$N_2O$   $\overset{\times}{N}=N\overset{\times}{O}=$

$CO_2$   $\overset{\times}{\times}O \div C\overset{\times}{\times}O\div$    (same for isoelectronic $NO_2^+$)

$NO_2$   $\overset{\times}{\times}O \div \overset{\times}{N}\overset{\times}{O}\div$   and   $\div O\overset{\times}{N}\div O\overset{\times}{\times}$   (hybrid forms required)

$C_6H_6$

The heavy bar in the $O_3$ structure represents a pair of electrons that are forced into the *same spatial orbital*. Analogous structures for $F_2$ and $N_2$ would be

$-\text{F}-\text{F}-$          $-N\equiv N-$

Extensions of this approach to more complicated molecules and radicals may be found in both Linnett's and Luder's book.[13] A quantitative, that is, mathematical, application of these ideas is in the formative stage by Linnett and others and is referred to as the *non-paired-spatial-orbital* (n.p.s.o.) model.

## THE PAULI EXCLUSION PRINCIPLE

In our previous discussion of the hydrogen molecule, it was seen that both a symmetrical and an antisymmetrical function arose. In terms of the valence bond language, these can be represented respectively as

$$\psi_S = \psi_a(1) \, \psi_b(2) + \psi_a(2) \, \psi_b(1)$$

and

$$\psi_A = \psi_a(1) \, \psi_b(2) - \psi_a(2) \, \psi_b(1)$$

Now, if we take the electrons in the system described by these two wave functions and exchange the coordinates of electron (1) with the coordinates of electron (2), we will obtain for the symmetric function

$$\psi_S' = \psi_a(2) \, \psi_b(1) + \psi_a(1) \, \psi_b(2)$$

But, this is the same as $\psi_S$. Therefore, on exchange of coordinates, we can say that $\psi_S = \psi_S{}'$. On the other hand, if we exchange the coordinates of the electrons in the antisymmetrical function, we obtain

$$\psi_A{}' = \psi_a(2)\,\psi_b(1) - \psi_a(1)\,\psi_b(2)$$

This expression is not the same, but rather it is opposite in sign from $\psi_A$. Consequently, in this case, we can say that for the exchange of coordinates, $\psi_A = -\psi_A{}'$.

Up until this point, we have considered only the *orbital wave functions*, and these are based on the spatial quantum numbers $n$, $l$, and $m$. Thus, we have neglected the *spin wave functions*. For a simple discussion of valency, it is possible to separate the spin wave functions from the orbital wave functions because spin-orbit interactions are generally negligible. Nevertheless, we should recognize that the electron spin must be included in any complete description of the system.

Since an electron must have a spin of either $+\frac{1}{2}$ or $-\frac{1}{2}$, we can represent these as $\alpha$ and $\beta$, respectively. If both electrons in an electron pair have spins of $+\frac{1}{2}$, then their combined wave function will be $\alpha(1)\,\alpha(2)$. If both of the electrons have a spin of $-\frac{1}{2}$, then their combined wave function will be $\beta(1)\,\beta(2)$. Finally, if one of the electrons has a spin of $+\frac{1}{2}$ and the other has a spin of $-\frac{1}{2}$, the combined wave function will be

$$\alpha(1)\,\beta(2) \pm \alpha(2)\,\beta(1)$$

The linear combination of wave functions arises here because of the indistinguishability of the electrons just as it did in the treatment of the hydrogen molecule. Further, if we exchange the electrons we can see that the function

$$\alpha(1)\,\beta(2) + \alpha(2)\,\beta(1)$$

is a symmetrical function, but the function

$$\alpha(1)\,\beta(2) - \alpha(2)\,\beta(1)$$

is an antisymmetrical function.

Now that we have both the spatial portion and the spin portion of the system, it is necessary to combine these to give the total wave function including spin. In Chapter 2, it was seen that the total wave function for the hydrogen atom could be expressed as a product of the individual wave functions, such that

$$\psi_T = R_{(r)}\Theta_{(\theta)}\Phi_{(\varphi)}$$

If we introduce a fourth degree of freedom, we expect the product relationship to still hold. Thus, the total wave function for the hydrogen molecule, including spin, should be formed from the product of the orbital function and the spin function. That is, for this system

$$\psi_T = \{\psi_a(1)\,\psi_b(2) \pm \psi_a(2)\,\psi_b(1)\}\,\{\text{spin functions}\}$$

Immediately, we can see that this leads to the eight possible combinations

$$\{\psi_a(1)\,\psi_b(2) + \psi_a(2)\,\psi_b(1)\}\begin{cases}\alpha(1)\,\beta(2) - \alpha(2)\,\beta(1)\\ \alpha(1)\,\alpha(2)\\ \beta(1)\,\beta(2)\\ \alpha(1)\,\beta(2) + \alpha(2)\,\beta(1)\end{cases}$$

$$\{\psi_a(1)\,\psi_b(2) - \psi_a(2)\,\psi_b(1)\}\begin{cases}\alpha(1)\,\alpha(2)\\ \beta(1)\,\beta(2)\\ \alpha(1)\,\beta(2) + \alpha(2)\,\beta(1)\\ \alpha(1)\,\beta(2) - \alpha(2)\,\beta(1)\end{cases}$$

The question now arises as to whether all eight of these combinations are significant. The answer lies in the Pauli exclusion principle. As it was first enunciated in 1925, it was stated, in essence, that *no two electrons in the same system can have the same values for all four quantum numbers.* In terms of quantum mechanics, however, it takes on the more subtle form: *every total wave function must be antisymmetric in the exchange of the coordinates of every pair of electrons.* In terms of this restriction, only four of the eight combinations will be acceptable. These are

$$\{\psi_a(1)\,\psi_b(2) + \psi_a(2)\,\psi_b(1)\}\ \{\alpha(1)\,\beta(2) - \alpha(2)\,\beta(1)\}$$

$$\{\psi_a(1)\,\psi_b(2) - \psi_a(2)\,\psi_b(1)\}\begin{cases}\alpha(1)\,\alpha(2)\\ \beta(1)\,\beta(2)\\ \alpha(1)\,\beta(2) + \alpha(2)\,\beta(1)\end{cases}$$

Thus, it would appear that there are three states corresponding to the antisymmetric spatial function, whereas there is only one state corresponding to the symmetric spatial function. Accordingly, we refer to the ground state as a *singlet state* and to the repulsive or antisymmetric state as a *triplet state*.

On the surface, the relationship between the quantum mechanical statement of the exclusion principle and the more familiar statement in terms of the four quantum numbers may not be evident. However, it can readily be shown that the original statement of the exclusion principle is a necessary feature of the more general statement of the principle in terms of symmetry. If we take the total ground-state function

$$\psi_T = \{\psi_a(1)\,\psi_b(2) + \psi_a(2)\,\psi_b(1)\}\{\alpha(1)\,\beta(2) - \alpha(2)\,\beta(1)\}$$

we obtain on multiplication

$$\begin{aligned}\psi_T = &\ \psi_a(1)\alpha(1)\,\psi_b(2)\beta(2) - \psi_a(2)\alpha(2)\,\psi_b(1)\beta(1)\\ &+ \psi_a(2)\beta(2)\,\psi_b(1)\alpha(1) - \psi_a(1)\beta(1)\,\psi_b(2)\alpha(2)\end{aligned}$$

This can be expressed more conveniently in the determinantal form

$$\psi_T = \begin{vmatrix}\psi_a(1)\alpha(1) & \psi_b(1)\beta(1)\\ \psi_a(2)\alpha(2) & \psi_b(2)\beta(2)\end{vmatrix} - \begin{vmatrix}\psi_a(1)\beta(1) & \psi_b(1)\alpha(1)\\ \psi_a(2)\beta(2) & \psi_b(2)\alpha(2)\end{vmatrix}$$

If we consider the three quantum numbers $n$, $l$, and $m$ to be identical for both electrons, then $\psi_a = \psi_b$, and the determinant can be expressed as

$$\psi_T = \begin{vmatrix} \psi_a(1)\alpha(1) & \psi_a(1)\beta(1) \\ \psi_a(2)\alpha(2) & \psi_a(2)\beta(2) \end{vmatrix} - \begin{vmatrix} \psi_a(1)\beta(1) & \psi_a(1)\alpha(1) \\ \psi_a(2)\beta(2) & \psi_a(2)\alpha(2) \end{vmatrix}$$

or

$$\psi_T = 2 \begin{vmatrix} \psi_a(1)\alpha(1) & \psi_a(1)\beta(1) \\ \psi_a(2)\alpha(2) & \psi_a(2)\beta(2) \end{vmatrix}$$

Now, if we permit $\alpha$ to equal $\beta$, the two rows of the determinant become identical and the determinant vanishes. Thus, the wave function is seen to be equal to zero when all four quantum numbers are the same for two different electrons in the same system. And this, of course, is what is demanded by the Pauli exclusion principle.

## VAN DER WAALS FORCES

Although the covalent, ionic, as well as the metallic bond can be used to explain the structural characteristics and the solid, liquid, and gaseous states of many substances, there is yet a large number of systems that do not fit into these categories. We can look at the inert gases as the most obvious example. These atoms are spherically symmetrical and incapable of forming any of the above-mentioned bonds. Yet there must be some force active between adjacent atoms if a liquid or solid state exists above absolute zero. We can go further and recognize that the same necessity exists for the multitude of saturated molecules such as $H_2$, $N_2$, $CH_4$, etc. These species have used their available electrons in the formation of the molecule and thus have no apparent means to bond to adjacent molecules.

There is one particularly significant characteristic of these forces. They are extremely weak in comparison to ionic or covalent forces. This is evident from the observable properties of substances that are predominantly dependent on such forces. In an ionic crystal where each ion is held by a number of ionic interactions, the boiling point of the substance will be quite high. However, substances in which the adjacent molecules are held together by these apparently very weak forces are frequently gases at room temperature, and in many instances their boiling points are extremely low. This is particularly true for the inert gases. As a comparison with a covalent bond, we can consider the sublimation of $Cl_2$. Here, the heat of sublimation of $Cl_2$ is of the order of 5 kcal/mole, whereas the Cl—Cl bond dissociation energy is 57 kcal/mole. Obviously, the forces holding one $Cl_2$ molecule to another $Cl_2$ molecule are extremely weak in comparison to the covalent bond holding one chlorine atom to the other chlorine atom in a $Cl_2$ molecule.

The existence of such weak attractive forces was first recognized by van der Waals as early as 1813. At that time, he introduced the $a/V^2$ term in his equation of state to allow for such interactions. It is for this reason that these forces are referred to as *van der Waals forces*.

If we carefully consider the systems involved, we can find two obvious sources of the van der Waals forces. First, if the molecules have a permanent dipole, it is easy to see that a weak attractive force should exist simply from the electrostatic interactions of two *dipoles*. This would be the case for any polar molecule. In molecules like HCl and $H_2O$ the effect should be rather large. However, in a molecule such as CO, which has an anomalously low dipole moment, the effect will be present, but much smaller than in the former instances. The interaction energy of this particular effect is found to be expressible as

$$(5\text{--}64) \qquad E_K = -\frac{2\,\mu^4}{3r^6kT}$$

where $r$ is the distance between centers of the dipoles, $\mu$ is the permanent dipole moment, and $k$ and $T$ have their usual significance.

A second source of the van der Waals interaction is that of a *dipole-induced dipole* type. This results from the polarization of one molecule by the dipoles of the surrounding molecules. This effect can be superimposed upon the dipole-dipole interaction and thus lead to a slight increase in the attraction. This particular aspect of the van der Waals interaction can be expressed as

$$(5\text{--}65) \qquad E_D = -\frac{2\alpha\mu^2}{r^6}$$

where $\alpha$ represents the polarizability.

Although these two factors are important in the evaluation of the van der Waals interaction, they are obviously not the only factors to be considered. Neither of these effects can be used to explain the solid or liquid states of such species as the inert gases, $Cl_2$, $H_2$, $N_2$, $CH_4$, etc. In addition, it is also found that neither the individual equations nor their combination gives any agreement with experiment.

A third contributing term to the total van der Waals energy was developed in 1930. This particular term represents the *London dispersion forces*. Using a quantum mechanical treatment, London arrived at the equation

$$(5\text{--}66) \qquad E_L = -\frac{3\,\alpha^2 h\nu_0}{4r^6}$$

where $h\nu_0$ is the zero point energy that is inherently present in any atom or molecule. The London dispersion forces are existent between any atoms or molecules regardless of their geometry. Pictorially, the dispersion forces can be thought to arise from the synchronization of instantaneous dipoles in the interacting species. For the sake of argument, we can consider a helium atom. This atom can certainly be considered to be spherically symmetrical, and consequently, to have no dipole moment. However, this is based on a time average. If we were to take an instantaneous photograph of a helium atom, we should find a quite unsymmetrical distribution of the electrons around the nucleus at that moment. Consequently, a temporary dipole should exist. This instantaneous dipole can act to induce an instantaneous dipole in an adjacent atom and this then leads to a synchronized field throughout the system. Such an effect will lead to a lowering of the energy of the system, but it will necessarily be a very weak interaction.

In terms of the three contributing effects to the total van der Waals interaction, quite satisfactory agreement can be obtained between theory and experiment. For a symmetrical atom or molecule, the London term is the only term operative. However, the more polar a molecule becomes and the more polarizable it becomes, the more significant will be the first two effects. This can be appreciated from Table 5–4, where the contributions

**TABLE 5–4.   Contributions to the Total van der Waals Lattice Energy in kcal/mole.**

| Molecule | Dipole-Dipole | Dipole-induced Dipole | Dispersion | Total | B. Pt., °K |
|----------|---------------|-----------------------|------------|-------|------------|
| Ar | 0.000 | 0.000 | 2.03 | 2.03 | 76 |
| CO | 0.000 | 0.002 | 2.09 | 2.09 | 81 |
| HCl | 0.79 | 0.24 | 4.02 | 5.05 | 188 |
| $NH_3$ | 3.18 | 0.37 | 3.52 | 7.07 | 239.6 |
| $H_2O$ | 8.69 | 0.46 | 2.15 | 11.30 | 373.1 |

of the three terms to the total van der Waals attraction are given. *Dipole-quadrapole* and *quadrapole-quadrapole* interactions can also be considered, but they result in $1/r^8$ and $1/r^{10}$ effects respectively, and these can generally be neglected.

## THE HYDROGEN BOND

It has long been recognized that a hydrogen atom can be attracted simultaneously to two different atoms, thereby showing a coordination number of 2. Under these circumstances, the hydrogen atom serves as a bridge between the two species and can be considered the basis of a bond between the two atoms. The resultant bond is much weaker than a covalent bond. Nevertheless, the effects of the hydrogen bond are of considerable significance. This is evident in many physical and structural properties of matter, and it is known to be of particular significance in physiological processes.[15]

We can imagine the hydrogen bond to be formed by a process of the type

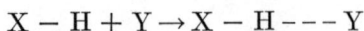

$$X - H + Y \rightarrow X - H --- Y$$

where X and Y can be the same kind or different atoms. This reaction implies that the hydrogen atom is bonded simultaneously to both atom X and atom Y. In addition, it implies that the bond to Y is different in some respect from the bond to X. This is found to be the case in every instance except for the ion FHF⁻. The structure of the hydrogen bond can be appreciated from a consideration of the dimer of formic acid. The geometry of the dimer, including the bond distances and bond angles, is shown in Figure 5–18. Here we see that the hydrogen atom is bonded to two different oxygen atoms. However, from the bond distances, it is apparent that the bonding is not identical. The O—H bond still retains its identity, but it is seen to be elongated from its normal distance of 0.97 Å in the monomer to a value of 1.07 Å in the dimer.

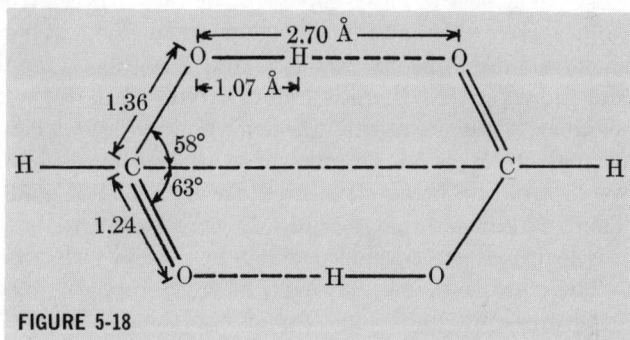

**FIGURE 5-18**

Hydrogen Bonding in the Dimer of Formic Acid.

In an attempt to determine the nature of the hydrogen bond, we can consider several different approaches. In terms of purely covalent bonding, it can be assumed that the hydrogen atom can be divalent. This would require the use of the 1s orbital for the formation of one bond and either the 2s or the 2p orbital for the second bond. Because of the extreme difference in energy between the 1s state and the 2s or 2p states of the hydrogen atom, a divalent hydrogen is not very likely. A second possibility involves the existence of resonance structures, and a third possibility is that the bond can be treated as if it were purely electrostatic in origin. All three of these possibilities have been seriously considered, and it is now generally assumed that we can satisfactorily treat the hydrogen bond in terms of electrostatic interactions.

If we can consider the hydrogen bond to be solely electrostatic in origin, it would appear that the van der Waals-type forces would be adequate to describe its character. The major contribution to the bond energy arises from the dipole-dipole interaction with the dipole-induced dipole and the dispersion forces making lesser contributions. This basic structure is supported by the fact that the bond occurs only when the atoms X and Y are the highly electronegative atoms F, O, N, and Cl. The surprisingly large magnitude of the dipole-dipole interaction can be attributed to the small size of the hydrogen atom and the absence of inner electron shells. This combination permits the close approach of a second atom that is not possible with other positive ions such as those of lithium or sodium.

Although the hydrogen bond can generally be treated solely in terms of electrostatic interactions, in valence bond language, there is evidence of resonance contributions in terms of the structures

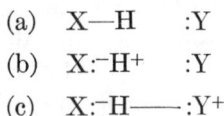

$$(a) \quad X\text{—}H \quad :Y$$
$$(b) \quad X\text{:}^-H^+ \quad :Y$$
$$(c) \quad X\text{:}^-H\text{——}:Y^+$$

It would be expected that structure (a), where the normal X—H bond exists, will be the most important of these three structures. In structure (c), the H——Y bond is a long bond, and a covalent bond over this distance should be insignificant. In ice, X and Y will both represent oxygen atoms, and knowing that there is only a trivial shift in the normal O—H bond distance, we can say that the long bond is about 0.80 Å longer than the normal O—H bond. On this basis, it has been shown that the relative weights of the three structures are 61%, 34%, and 5%, respectively.[16] Thus, we can see that, because of the small contribution of structure (c), it is legitimate to consider the bond essentially electrostatic in origin.

There are instances, however, where it appears that the three resonance structures contribute more equally to the final state of the system. This occurs when there is a decrease in the $X\cdots X$ distance over what is normally observed. Such is found to be the case in the $\beta$-form of oxalic acid where the $O\cdots O$ distance is approximately 2.5 Å, but it is particularly true in the extreme case of $FHF^-$ as found in $KHF_2$. In this instance, the hydrogen atom is located at the midpoint between the two fluorine atoms. Further, it is observed that the H—F distance is only 0.2 Å greater than in the isolated HF molecule. Both of these facts indicate that all three of the resonance structures

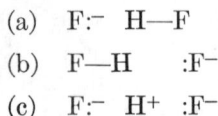

$$(a) \quad F\text{:}^- \quad H\text{—}F$$
$$(b) \quad F\text{—}H \quad :F^-$$
$$(c) \quad F\text{:}^- \quad H^+ \quad :F^-$$

probably contribute about equally to the bond. In this particular case, the hydrogen bond energy is of the order of five times greater than is usually observed, being of the order of 27 kcal/mole.

The relative ability of atoms in a given family to form hydrogen bonds is evident from a consideration of the boiling points of the nonmetallic hydrides. In Figure 5–19 these are given for the four nonmetal families. According to molecular weight considerations and van der Waals forces, we observe the expected trend for $CH_4$, $SiH_4$, $GeH_4$, and $SnH_4$ as well as for the three heavier members of each of the other families. However,

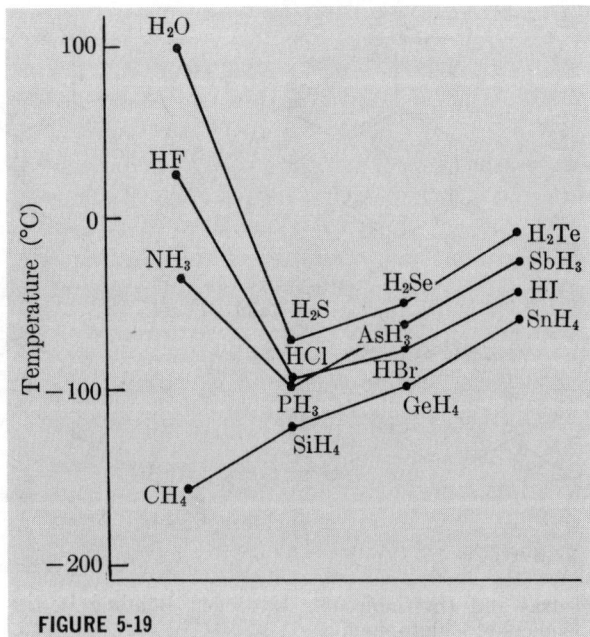

**FIGURE 5-19**

The Effect of Hydrogen Bonding on the Boiling Points of Nonmetal Hydrides.

we note the anomalous behavior of the lightest member of each of the three latter families. The unusually high boiling points for HF, $H_2O$, and $NH_3$ must be attributed to the existence of hydrogen bonding. It is interesting to note that a stronger hydrogen bond exists in HF than in $H_2O$, yet the boiling point of water is the higher. This can be attributed to the statistical fact that water can form two such bonds to one for hydrogen fluoride per molecule.

Thus far, we have considered only intermolecular hydrogen bonding. However, there is also the possibility of the formation of intramolecular hydrogen bonds where the hydrogen bond exists between two atoms of the same molecule. This type of hydrogen bonding is quite common, and it can also have a marked effect on the chemical and physical properties of the molecule. As an example, we can consider the three isomers of chlorophenol. For the *ortho* isomer, we should expect to find both the *cis* and the *trans* form of the molecule, as shown in Figure 5–20. From the proximity of the chlorine atom to the hydrogen atom, the *cis* form should be stabilized over the *trans* form as a result of the formation of an intramolecular hydrogen bond between the oxygen atom and the chlorine atom. This is, indeed, observed. From spectral studies of *o*-chlorophenol in carbon tetrachloride, the distribution is approximately 91% *cis* and 9% *trans*. The possibility of hydrogen bond formation is also reflected in the physical properties of the

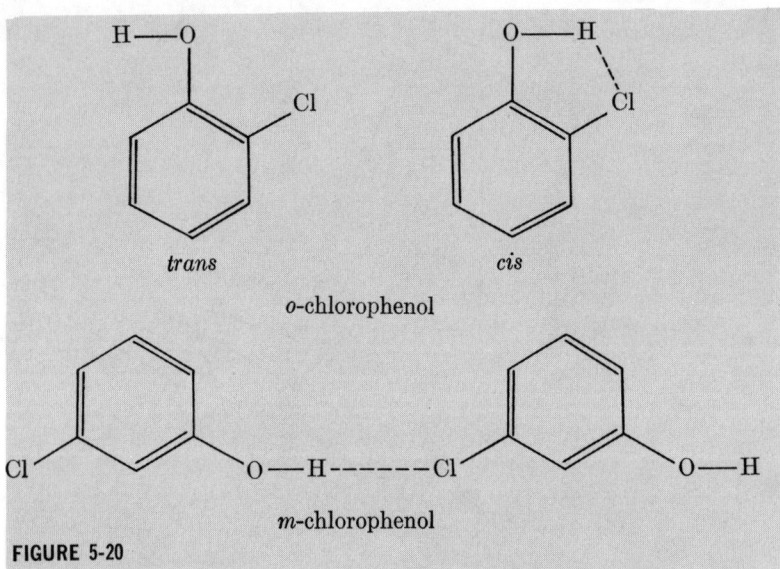

FIGURE 5-20

Intra- and Intermolecular Hydrogen Bonding in the *Ortho* and *Meta* Isomers of Chlorophenol.

chlorophenols. For instance, the boiling point of the *o*-chlorophenol is 176°C, whereas it is considerably greater for both the *meta* and the *para* isomers. In the latter two structures intermolecular hydrogen bonding plays a much more important part than it does in the *ortho* isomer.

## References

1. G. Lewis, *J. Am. Chem. Soc.*, **38,** 762 (1916).
2. L. Pauling, "The Nature of the Chemical Bond," 3rd ed., Cornell University Press, Ithaca, N. Y., 1960.
3. M. Born and A. Lande, *Preuss. Akad. Wiss. Berlin, Ber.*, **45,** 1048 (1918).
4. J. Mayer and L. Helmholz, *Z. f. Physik,* **75,** 1 (1932).
5. L. Helmholz and J. Mayer, *J. Chem. Phys.,* **2,** 245 (1934).
6. M. Born, *Verhandl. deut. physik. Ges.,* **21,** 13 (1919); Haber, F., *Verhandl. deut. physik. Ges.,* **21,** 750 (1919).
7. S. Glasstone, "Theoretical Chemistry," D. Van Nostrand Company, Inc., New York, 1944.
8. L. Pauling, *Chem. Revs.,* **5,** 173 (1928).
9. W. Heitler and F. London, *Z. f. Phys.,* **44,** 455 (1927).
10. H. James and A. Coolidge, *J. Chim. phys.,* **1,** 825 (1933).
11. L. Pauling, *J. Am. Chem. Soc.,* **53,** 1367 (1931).
12. H. Bent, *J. Chem. Ed.,* **37,** 616 (1960).
13. J. W. Linnett, "The Electronic Structure of Molecules," Methuen & Co., Ltd., London, 1964; W. F. Luder, "The Electron-Repulsion Theory of the Chemical Bond," Reinhold Book Corporation, New York, 1967.
14. E. A. Guggenheim, *J. Chem. Ed.,* **43,** 474 (1966).

15. G. Pimentel and A. McClellan, "The Hydrogen Bond," W. H. Freeman Co., San Francisco, 1959.
16. L. Pauling, *J. Chim. phys.*, **46**, 435 (1949).

## Suggested Supplementary Reading

C. S. G. Phillips and R. J. P. Williams, "Inorganic Chemistry, Vol. 1," Oxford University Press, Inc., New York, 1965.

C. A. Coulson, "Valence," 2nd ed., Oxford University Press, Inc., New York, 1961.

J. W. Linnett, "The Electronic Structure of Molecules," Methuen & Co., Ltd., London, 1964.

E. Cartmell and G. W. A. Fowles, "Valency and Molecular Structure," 2nd ed., Butterworth & Co. (Publishers), Ltd., London, 1961.

J. A. A. Ketelaar, "Chemical Constitution," 2nd ed., Elsevier Publishing Co., Amsterdam, 1958.

H. B. Gray, "Electrons and Chemical Bonding," W. A. Benjamin, Inc., New York, 1964.

L. Pauling, "The Nature of the Chemical Bond," 3rd ed., Cornell University Press, Ithaca, N. Y., 1960.

## Problems

1. Use Eq. (5–1) to obtain the relation $\left(\dfrac{\partial U}{\partial r}\right) = 0$, and determine the ionic radii for $z_+ = z_- = 1$ and for $z_+ = z_- = z$. From this show that $r_z = r_1 z^{-2/(n-1)}$. (See p. 111).

2. Use the wave function $\psi = e^{-ar^2}$ to calculate the ground-state energy of the hydrogen atom, and compare the energy with that obtained with $\psi = e^{-ar}$.

3. Show why

$$E = \frac{\displaystyle\int \psi H \psi^* \, d\tau}{\displaystyle\int \psi\psi^* \, d\tau}$$

is used to evaluate $E$ rather than simply $E = \dfrac{H\psi}{\psi}$.

4. Determine the Hamiltonian operator for a harmonic oscillation.

5. Show why the Hamiltonian operator is Hermetian.

6. Give the valence bond wave function $\psi_{VB}$ for the hydrogen molecule ion.

7. Show that in a given atom an $s$ orbital is orthogonal to a $p$ orbital.

8. In terms of the quantum mechanical model, can we consider the electrons to revolve about the two atoms in a hydrogen molecule in the path of a figure eight?

9. Determine the stability the $H_2^+$ ion will have due to electron delocalization if the hydrogen atoms are treated as one-dimensional boxes and the hydrogen molecule ion is treated as a larger one-dimensional box.

10. Show that a total wave function $\psi_{(r\theta\varphi s)} = R_{(r)}\Theta_{(\theta)}\Phi_{(\varphi)}S_{(s)}$ that is antisymmetric

in the exchange of coordinates of every pair of electrons does not violate the Pauli exclusion principle.

11. Draw the Linnett structures for

(a) CO    (d) FCN
(b) $CN^-$    (e) COS
(c) $NO^-$    (f) $N_2O_4$

# The Crystalline
State

If we use the conventional definition that a crystal is a homogeneous solid bounded by naturally formed plane faces, it is true that most of the solids we come in contact with in our everyday experiences do not appear to be crystalline. This can generally be attributed to one of two causes. On the one hand many solids are extremely complex, being composed of a mixture of compounds usually having extremely large molecules of variable size. If, however, these are separated to give pure compounds, crystalline structures tend to occur. Thus, for instance, certain proteins and cellulose, both of which are constituents of common naturally occurring solids, have been obtained in crystalline form in spite of the fact that they are not found in this manner in nature.

On the other hand, many substances that appear to be amorphous are actually composed of microcrystalline units. Thus we find that although such common items as soil, various powders, concrete, etc. appear to be noncrystalline to the unaided eye, microscopy and X-ray techniques show that this is not the case. Rather they consist of aggregates of small crystalline units.

Yet, along with this preponderance of amorphous solids, there are many substances that do occur naturally in the crystalline state. From the standpoint of beauty alone, an interest in the crystalline state might be justified. In fact, this has undoubtedly been the initial cause of interest of many of those who have devoted their professional efforts to a study of crystallography. But in a more subtle vein, one cannot help but be impressed with the unique development of specific faces and the obvious symmetry that occurs in a well-formed crystal. Yet such aesthetic reasons can hardly be the sole justification for the study of crystallography. To the present-day inorganic chemist, the multitude of structures of inorganic compounds, along with the diversity of types of chemical bonds, makes crystal structure analysis an indispensable research tool.

The inorganic chemist is usually concerned with structure on a molecular level. Very often this can be determined by means of X-ray studies on the corresponding crystal, a classic example being the coordination of the uranyl ion $UO_2^{2+}$. This is a linear ion, and complexation about the equatorial plane was generally assumed to be either 4-fold or 6-fold. In many instances, these are observed to occur. However, crystal structure analysis has now confirmed that the five-coordinated uranyl ion is equally common, if not the predominant form.[1] In general, we can say that strong indications of molecular geometry can be obtained by a variety of means, but for a solid, crystal analysis is capable of giving the definitive answer.

A solid, of course, occurs only because the forces between the adjacent structural units are sufficiently strong to hold them in a rigid position. In a pure compound there will invariably be a regular distribution of these units that will permit a maximum interaction. Consequently, if the physical conditions will permit, the structural units will pack in this regular manner with the resultant ordered arrangement giving the crystalline form. Thus, the crystalline state is the natural state of a pure solid.

In Figure 6–1, well-formed crystals of fluorite ($CaF_2$), pyrite (FeS), galena (PbS), and quartz ($SiO_2$) are shown. A tendency to form cubic structures is apparent among the first three of these. In both the fluorite and the pyrite almost perfect cubes can be seen to protrude from the body of the corresponding crystal. An additional crystal of galena is shown in which the cube has been modified by octahedral faces. In all of these crystals the regularity displayed by the external structure is striking, and we might go further to state that this regularity implies some sort of regularity of internal structure. At the same time, the obvious difference between the quartz and the other crystals indicates that the structural arrangements here differ. The untrained individual might conclude that the two differently appearing galena crystals also result from a difference in internal structure, but in this instance, we shall see that such a conclusion is incorrect.

It is a rare experience to find a crystal that even approaches a perfect external structure. The specific conditions that exist during the period of growth will strongly affect the shape of the resulting crystal. If NaCl crystals are very carefully grown in a stirred solution, they will approach a cube in shape. Generally, however, they are more prone to be shaped like a wafer due to the low growth rate on the side in contact with the container. Yet, if they happen to be grown from a well-stirred solution containing a small amount of urea, they turn out to be octahedral. Thus, in Figure 6–2, we see three apparently different structures that can occur for the same substance. With this in mind, one can imagine the complexities that face the crystallographer, who must base his studies on the myriad of structures and distortions of structures that occur in nature. Here we see one reason why any understanding of crystal structure was so late in coming, in

**FIGURE 6-1**

Photographs of Some Well-Formed Crystals: (a) fluorite ($CaF_2$) — cubic; (b) pyrite (FeS) — cubic; (c) galena (PbS) — cubic; (d) a galena crystal showing the development of octahedral faces; and (e) quartz ($SiO_2$) — hexagonal.

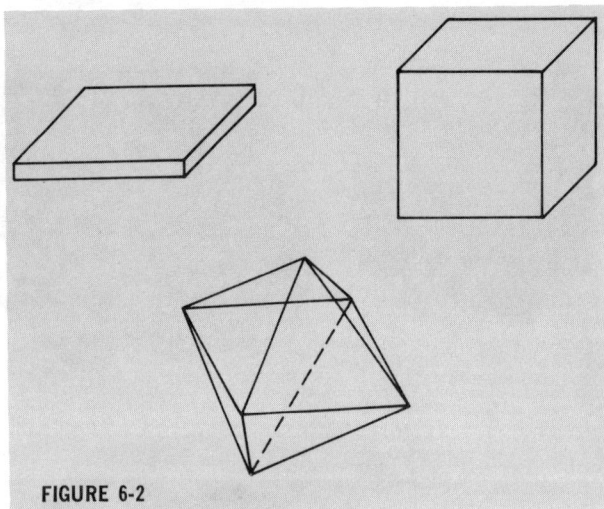

FIGURE 6-2

Some Natural Habits of NaCl.

spite of the fact that crystals have been available for investigation through-out history.

## Constancy of Interfacial Angles

The first truly significant contribution to crystallography came in 1669 when Niels Stenson reported his observations of the angles between equivalent faces of quartz crystals. By cutting sections from different quartz crystals and tracing their outline on paper, Stenson laid the foundation for what later became known as the law of constancy of interfacial angles. In Figure 6-3(a) some examples of such crystal tracings are shown. If normals are drawn to any two faces, they will intersect to form an inter-facial angle as shown in Figure 6-3(b). Using the tracings from various sections of different quartz crystals, Stenson showed that corresponding interfacial angles are the same. This does not mean that adjacent faces necessarily have the same interfacial angle. As can be seen in Figure 6-3(b), the comparison must always be between equivalent faces.

The ideas of Stenson were extended by Domenico Guglielimini and finally confirmed by Jean Baptiste Louis Rome Delisle. The culmination of this work came about 1780, more than a century after Niels Stenson first published his findings. The definitiveness of the work by Delisle can be attributed to the invention of the contact goniometer by his assistant, Carangeot. This permitted quite accurate measurements, by their stan-dards, of the interfacial angles. A diagram of a contact goniometer is shown in Figure 6-4. Although it has long since been replaced for accurate measurements by the reflecting goniometer, its original impact on the study of crystal angles should be apparent. As a consequence of observations

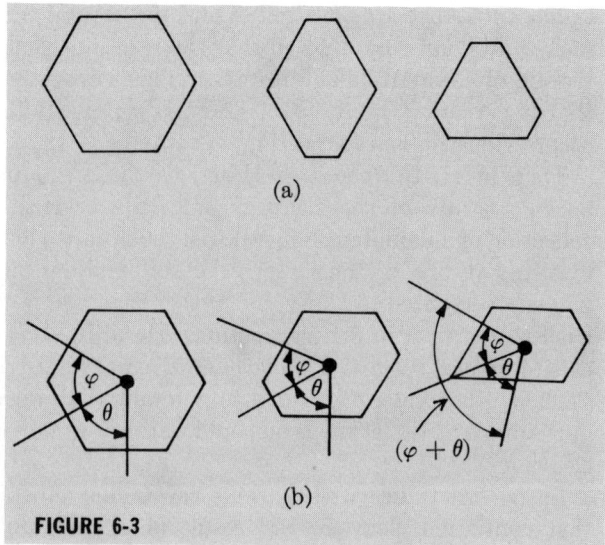

FIGURE 6-3

Crystal Tracings to Illustrate the Law of Constancy of Interfacial Angles.

on a large number of crystals, the constancy of interfacial angles was established. In general, the law of constancy of interfacial angles can be stated as: *In all crystals of the same substance, the angles between corresponding faces have the same value.*

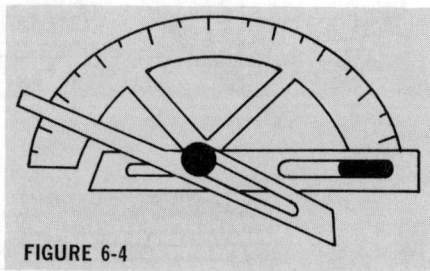

FIGURE 6-4

A Simple Contact Goniometer.

## CRYSTAL SYMMETRY

The early studies on the structure of crystals were of necessity *morphological*, that is, based on the external appearance of the crystal. However, it was soon recognized that this regularity of external structure implies a regularity of internal structure. It was mentioned earlier that Domenico Guglielmini extended the ideas of Niels Stenson on the constancy of interfacial angles. He had noted that cleavage directions in a given crystalline substance are constant. Thus, if a crystal is split, it may form plane faces, and Guglielmini noted that the resultant cleavage planes are characteristic of the particular substance. This led Guglielmini to the conclusion that a crystal is built up of microscopic crystalline units, and cleavage occurs along planes connecting these units.

Then in 1784, Rene Just Haüy, as the result of a fortunate accident, began a study of the cleavage of calcite crystals. While inspecting the collection of an amateur mineralogist, he dropped a group of calcite crystals, resulting in one of the larger crystals breaking off. The crystal was forgivingly presented to Haüy, but rather than add it to his own collection, he tried to cleave it in other directions. He ultimately was able to cleave the calcite crystal to give a *rhombohedral* unit. As a result of this success, he studied other calcite crystals and found that regardless of the external appearance of the crystal, he could obtain the same rhombohedral unit by cleavage.

Based on studies with various crystalline substances, Haüy concluded that continued cleavage will result in a basic unit, repetition of which gives the macroscopic crystal. Some examples of various structures built up from cubic units by Haüy are shown in Figure 6–5. Although the views

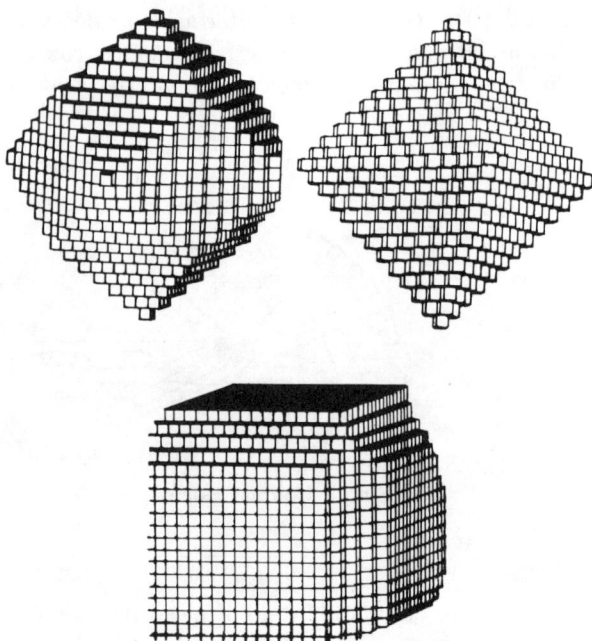

**FIGURE 6-5**

Different Geometrical Structures Formed by the Stacking of Cubelets.

proposed by Haüy are not accepted as such today, our present idea of the basic structural unit defined by the positions of atoms in the unit cell is very similar to Haüy's *molecule integrante*.

Rather than consider a macroscopic crystal to be composed of solid parallelepipeds, as did Haüy, we now consider it to result from a regular internal arrangement of atoms or ions in a crystalline lattice. And the morphology of a crystal will reflect this regularity of internal structure. We can thus choose a section of this lattice structure which will permit us to recreate a crystal, like Haüy, by stacking these units one upon the other. Operationally, it may be more correct to imagine this unit to be translated successively from one position in the lattice to an equivalent adjacent position, but the net result is the same as that obtained by the stacking procedure. Thus, if we wish to construct a crystal of the type Haüy would have built with perfect cubes, we might use a cubical arrangement of atoms or ions, as shown in Figure 6–6.

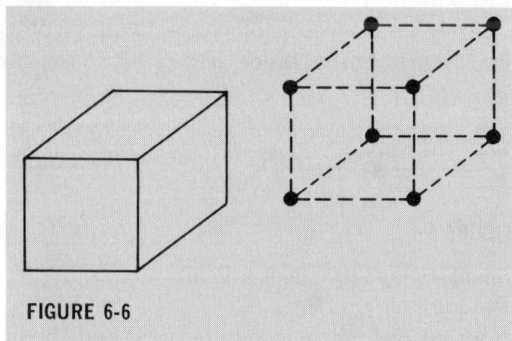

FIGURE 6-6

Analogy between the Structural Unit Envisioned by Haüy and a Unit Cell from a Crystal Lattice.

In attempting to differentiate between various types of crystals, we can note that there are only a limited number of geometrical shapes a parallelepiped representing the *unit cell* can have. And it is tempting to say that the macroscopic crystal will in some way reflect this shape. Now it is possible to differentiate between these simple structures in terms of their symmetry. For instance, based on our general feeling for balance, it is apparent that a perfect cube is more symmetrical than a rhombohedron. Thus we might conclude that a crystal synthesized from cubic unit cells would show a different symmetry from one synthesized from rhombohedral unit cells. It should, therefore, be possible to devise a system of crystal classification in terms of the *crystal symmetry*, and this, of course, is what has been done.

A consideration of the symmetry of the unit cell is a rather convenient means to represent pictorially the approach to crystal classification in terms of symmetry. However, we shall later see in some detail that the approach is actually the reverse. Rather than base the symmetry of a crystal on that of the unit cell, the unit cell is chosen to conform to the properties of the macroscopic crystal. This, then, re-emphasizes the point that the early studies of crystallography were morphological in character.

## Symmetry Elements and Operations

There are numerous ways to describe the symmetry of a system. The chemist normally is concerned with molecules. In this respect, an origin or point in the molecule is assigned and the symmetry with respect to lines and planes through this point is considered. Thus, we are here concerned with *point symmetry*. In crystals point symmetry is also found to be of value. But we must realize that in a crystal, symmetry is not restricted to a point. Rather, it extends in space throughout the crystal. Consequently, the point symmetry permitted in a crystal is restricted in that it cannot violate any *translation symmetry* requirements.

An inherent feeling of symmetry in an object is a product of our everyday experience. For instance, if we look at the two bars in Figure 6–7, we would

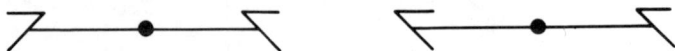

**FIGURE 6-7**

Simple Symmetry about a Point.

say that one figure is symmetrical and the other is not. Actually, there is a degree of symmetry associated with both figures, but our feeling for symmetry here generally revolves around the center point. It might be said that there is *a sense of balance* around this center point.

To give a quantitative description of the symmetry of a figure, more than a nebulous feeling of balance is needed. It becomes necessary to define certain symmetry elements that can give an unambiguous expression of the symmetry. In this respect, the elements that are generally used to describe point symmetry in a crystal are: (a) *proper rotation axis*, (b) *mirror plane*, and (c) *rotation-inversion axis*. These are not the only choices that can be made. However, all of the point symmetry of a crystal can be expressed in terms of these three elements. For example, one recognizes that a *center of symmetry* exists in many geometrical structures, yet it is not included here. But we shall soon show that a center of symmetry can be represented by a rotation-inversion axis.

## Proper Rotation Axis

If it is possible to rotate an object about an imaginary axis to produce an orientation that is *indistinguishable* from the original, the object is said to have a *rotation axis*. The rotation axis is an example of a *symmetry element*, and the operation of rotation around this axis is an example of a *symmetry operation*. A familiar example of such an axis can be seen for the water molecule in Figure 6–8(a). Since the two hydrogen atoms are indistinguishable, the two orientations are equivalent. The angle through which the molecule is rotated to give an indistinguishable orientation can be represented by $\theta$, and the molecule or whatever object is being studied will have a $360°/\theta$-*fold* rotation axis. Thus, the axis shown for the water molecule is a $360°/180° = 2$-fold rotation axis.

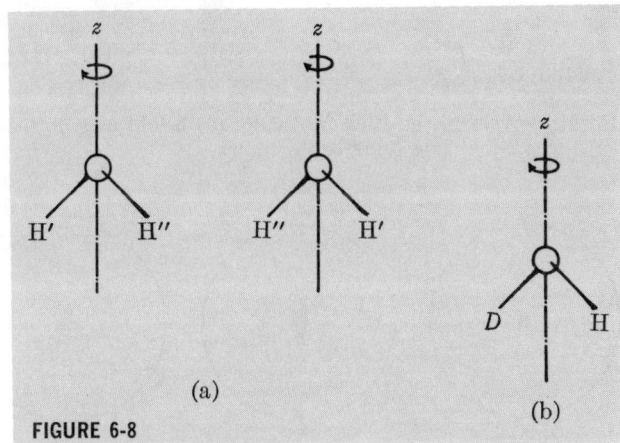

(a)

(b)

**FIGURE 6-8**

Rotation Axis in a Water Molecule: (a) 2-fold axis in $H_2O$ and (b) 1-fold axis in DHO.

Now, if one of the atoms in the water molecule is replaced by a deuterium atom as shown in Figure 6–8(b), a 2-fold rotation axis no longer exists. It is now necessary to rotate through a full 360° to obtain an indistinguishable structure. That is, there are no equivalent orientations around this axis, and we must rotate completely around to the original orientation. This structure is *identical* rather than *equivalent* to the original, and the axis is, therefore, a 1-fold or *identity* axis.

As a final example we can consider the molecule $BCl_3$. The geometry of this molecule is triangular planar, and if we designate the equivalent chlorine atoms with primes as shown in Figure 6–9(a), it can be seen that rotation about an axis normal to the plane of the molecule and passing through the boron atom will generate an equivalent figure every 120°. Thus, this axis is a 3-fold rotation axis. In addition, it can be seen in Figure 6–9(b) that three 2-fold axes are also present, one lying along each of the three different B—Cl bonds.

The rotation axes of concern to the crystallographer are summarized in Figure 6–10. By means of the circular figures we have shown succes-

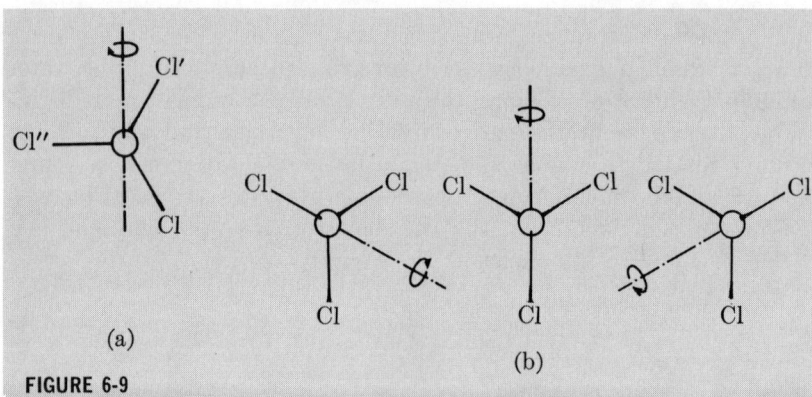

**FIGURE 6-9**

Rotation Axes in the $BCl_3$ Molecule: (a) 3-fold axis normal to the plane and (b) three 2-fold axes lying in the plane.

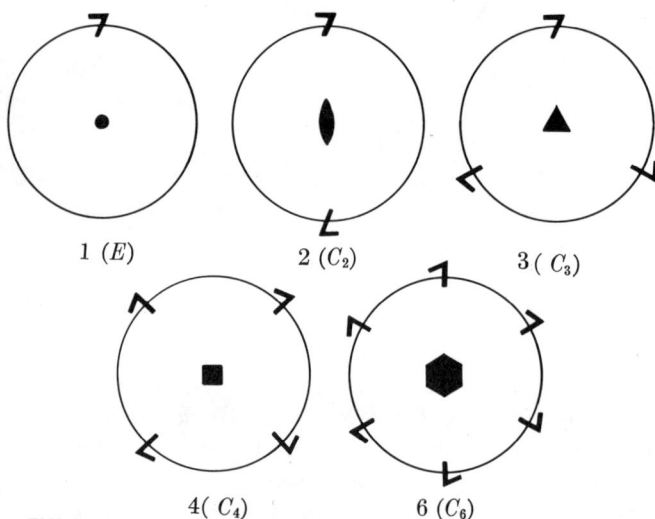

1 (E)　　　　2 ($C_2$)　　　　3 ($C_3$)

4 ($C_4$)　　　　6 ($C_6$)

**FIGURE 6-10**

Symbolism for the Acceptable Crystal Rotation Axes.

sively 1, 2, 3, 4, and 6-fold rotation axes. In each case, the axis is assumed to be normal to the plane of the circle and passing through its center. The geometrical figure in the center of each circle indicates the order of the axis through that point.

## Symmetry Notation

There are two common conventions for symbolizing point symmetry. The chemist dealing with molecules would most likely use the Schoenflies notation. Except for the identity axis, which is symbolized by $E$, the symbol $C_n$ is used to represent a rotation axis, where $n$ is the order or fold of the axis. On the other hand, the crystallographer ordinarily uses the Hermann-Mauguin notation. Here the order of the axis is represented by the integer corresponding to the order of the rotation axis. Both of these notations are given in Figure 6–10.

We have spoken of the rotation axis as an element of symmetry and the operation of rotation about this axis as a symmetry operation. If $n$ equivalent positions exist about this axis, the axis is of order $n$ or is an $n$-fold rotation axis. Thus, in summary:

| Order of Axis $n$ | Type of Axis | Degrees Rotated to Equivalent Orientation | Symbol |
|---|---|---|---|
| 1 | identity | 360 | 1 or $E$ |
| 2 | diad | 180 | 2 or $C_2$ |
| 3 | triad | 120 | 3 or $C_3$ |
| 4 | tetrad | 90 | 4 or $C_4$ |
| 6 | hexad | 60 | 6 or $C_6$ |

## Allowed Rotation Axes

It is significant that we have restricted our considerations to the above rotation axes. In the description of a molecule, a 5-fold or even higher axis would be expected to occur. For instance, $IF_7$ is considered to exist as a pentagonal bipyramid and will, therefore, have a 5-fold axis passing through the polar positions. However, in a crystal we can readily show that the axes given in the tabular summary are the only ones consistent with translational periodicity.

Using Figure 6–11, consider that the points $p_n$ represent points in a crystal lattice with points $p_1$ and $p_2$ being adjacent points on a given row.

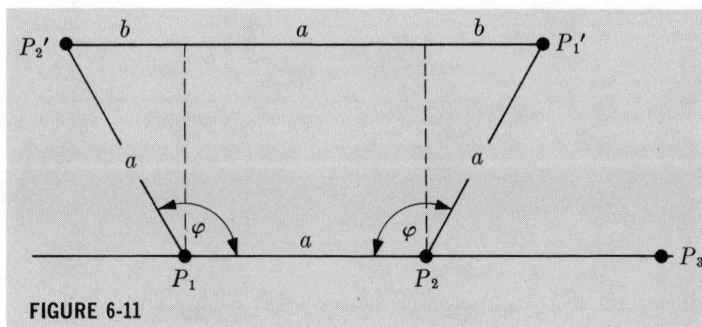

**FIGURE 6-11**

Geometrical Demonstration of the Allowed Crystal Rotation Axes.

Since the sites in a lattice must all be equivalent, the surroundings of point $p_2$, for instance, are the same as those of points $p_1$, $p_1'$, $p_2'$, $p_3$ and so on. Consequently, if a point is translated from one location to another, an equivalent orientation is obtained. Imagine a rotation axis perpendicular to the plane of the paper and passing through point $p_1$. If we rotate through an angle $\varphi$, point $p_2$ will be carried into point $p_2'$. Likewise, a rotation through the same angle but in the opposite direction around an axis through $p_2$ will carry point $p_1$ into point $p_1'$. Points $p_1'$ and $p_2'$ lie on a new row which, from the geometry of the system, can be seen to be parallel to the original row.

Since all sites in the lattice are equivalent, the distance between two adjacent points in any parallel row must be the same. This requires that the distance between any two points on a row must be an integral multiple of the distance between adjacent points. Inasmuch as the distance between the adjacent points has been designated as $a$, the distance between $p_2'$ and $p_1'$ will be $na$, where $n$ is an integer $0, 1, 2, 3, \ldots n$. Now it can be seen that

$$\overline{p_1'p_2'} = a + 2b$$
$$= a + 2a \sin (\varphi - 90°)$$
$$= a - 2a \cos \varphi$$

But since

$$\overline{p_1'p_2'} = na \text{ where } n = 0, 1, 2, \ldots, n$$

then

$$na = a - 2a \cos \varphi$$

and

$$\cos \varphi = \frac{1 - n}{2}$$

The $\cos \varphi$ must, of course, lie between $-1$ and $+1$. This places a restriction on the values the angle $\varphi$ can have. The consequence of this is shown in Table 6–1, where it can be seen that only 2, 3, 4, and 6-fold rotations are

**TABLE 6–1.**

| $n$ | Cos $\varphi$ | $\varphi$ (deg) | Order of Rotation |
|:---:|:---:|:---:|:---:|
| 0 | $\frac{1}{2}$ | 60 | 6 |
| 1 | 0 | 90 | 4 |
| 2 | $-\frac{1}{2}$ | 120 | 3 |
| 3 | $-1$ | 180 | 2 |

consistent with translational periodicity. Although it does not occur in the table, one can readily appreciate that the special case of an identity or 1-fold rotation axis is also acceptable.

In the discussion used here, we have shown the rotation of only two points. But it must be realized that a rotation axis will rotate all the points around the axis. From Figure 6–11, one might conclude that a 2-fold rotation axis must always be present. Certainly a 2-fold rotation around point $p_2$ will take point $p_1$ into the equivalent point $p_3$. However, we must also consider the other points around the rotation axis. These must also be carried into equivalent positions by the 2-fold rotation before we can say that the axis does, in fact, exist. What this calculation does show is that if there is a rotation axis about a point, it cannot possibly translate all points about the axis into equivalent points in the lattice unless the axis is one of the four types shown in Table 6–1.

## Mirror Plane

If a plane can be drawn through an object such that one side is a mirror image of the other, a plane of symmetry exists for that object. Using the water molecule as an example, two planes of symmetry can be observed. Both of these contain the $z$ axis, the 2-fold rotation axis, as shown in Figure 6–12. One of these is the $xz$ plane and the other is the $yz$ plane. Using the Schoenflies symbolism, both of these planes are denoted as $\sigma_v$. A plane that encompasses the principal rotation axis is, by convention, considered a

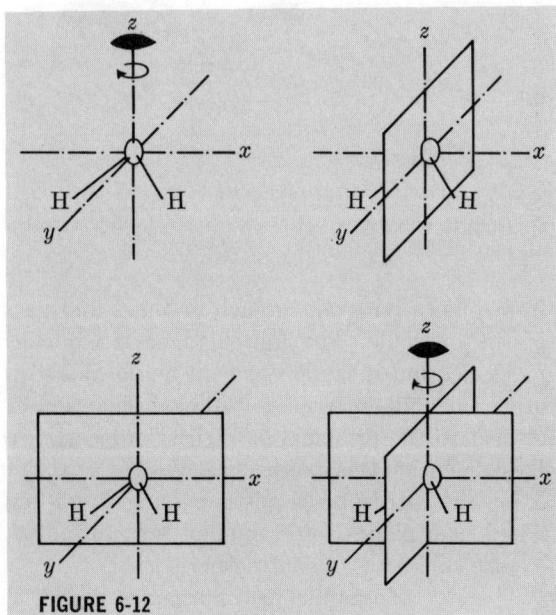

**FIGURE 6-12**

Symmetry Planes in a Water Molecule.

vertical plane with respect to that axis, hence the subscript $v$. The $\sigma$ is the symbol for a mirror plane.

Previously we considered the $BCl_3$ molecule. It was shown to have one 3-fold rotation axis and three 2-fold rotation axes. As can be seen in Figure 6–13. There are four mirror planes in the molecule. Three of these result

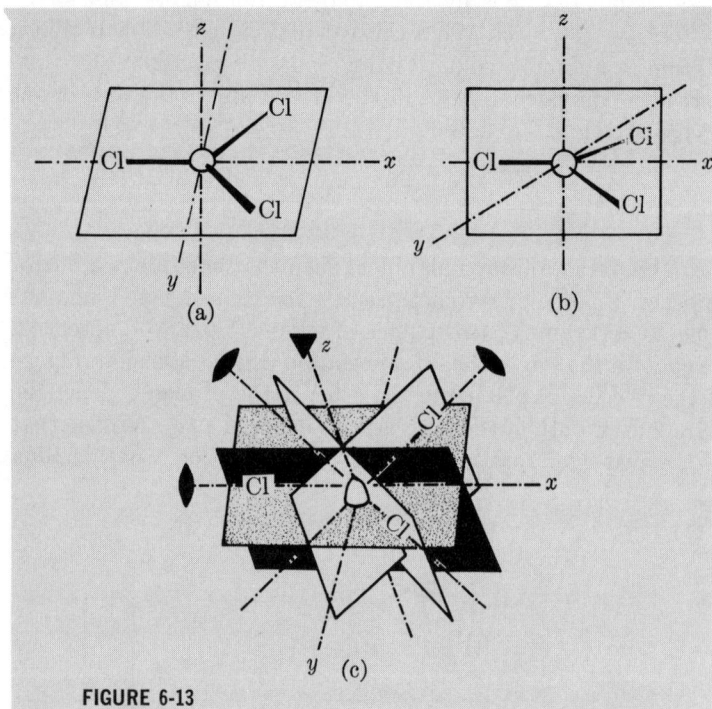

**FIGURE 6-13**

Symmetry Planes in the $BCl_3$ Molecule: (a) horizontal plane (normal to the $z$ axis), (b) one of the three vertical planes, and (c) all four planes combined.

from planes lying along each of the 2-fold axes and normal to the plane of the molecule. The fourth plane is coincident with the plane of the molecule, and it is also seen to be normal to the 3-fold axis whereas the other three all encompass it. The designation of these planes is made with respect to the principal or highest-order axis; the 3-fold axis in this case. By convention this axis is chosen as the $z$ axis and is considered the vertical axis. The three planes encompassing the 3-fold axis are, therefore, designated as $\sigma_v$ planes, and the plane normal to the 3-fold axis is designated as $\sigma_h$, the subscript $h$ denoting horizontal.

The crystallographer will not generally use this symbolism for a mirror plane. Rather, he will represent all mirror planes merely with the letter $m$, and the vertical or horizontal nature will be indicated by the manner in which the $m$ is combined with the symbol for the rotation axis. The combination $2m$ symbolizes a 2-fold axis with a mirror plane encompassing it, whereas $2/m$ represents a 2-fold axis normal to the mirror plane.

## Rotary-Inversion Axis

The third symmetry element we shall consider is a compound element arising from a rotation followed by an inversion across a center. This is referred to as an *improper* rotation axis in contrast to the proper axes we have previously discussed.

This particular symmetry element is not so readily visualized as the previous two. In Figure 6–14, the numeral 7 is used to indicate the nature

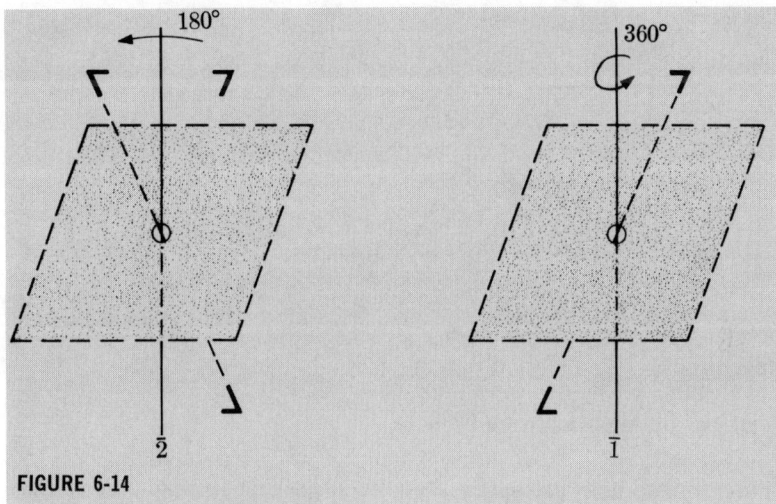

**FIGURE 6-14**

Illustration of Rotation-Inversion Axes.

of the operation. The identification of a particular rotary-inversion axis involves the numeral corresponding to the rotation axis with a bar drawn over the numeral. Thus, we see a 2-fold rotary-inversion ($\bar{2}$) and a 1-fold rotary-inversion ($\bar{1}$) in Figure 6–14. In the 2-fold rotary-inversion, the 7 has first been rotated through 180° and then inverted through the center. From the figure one can see that this gives the same result as a mirror plane normal to the axis. Consequently, an object that shows a 2-fold rotary-inversion axis would normally have this symmetry represented by a mirror plane in preference to the improper rotation.

The 1-fold rotary-inversion axis shown in the figure can be seen to be equivalent to a center of symmetry. In crystallography, this particular element of symmetry is used in preference to the center of symmetry. However, the chemist dealing with molecular symmetry would be prone to speak in terms of the symmetry center. From here on we shall use the Hermann-Mauguin symbolism, but a detailed comparison of the two systems is given in appendix D.

## Symmetry of a Cube

Now that we have developed the necessary symmetry elements, it will be helpful to apply them to a specific geometrical structure. A highly symmetrical structure that can also be conveniently related to a crystalline material is the cube. If we consider first the proper rotation axes, we find that there are three different types. These are shown in Figure 6–15.

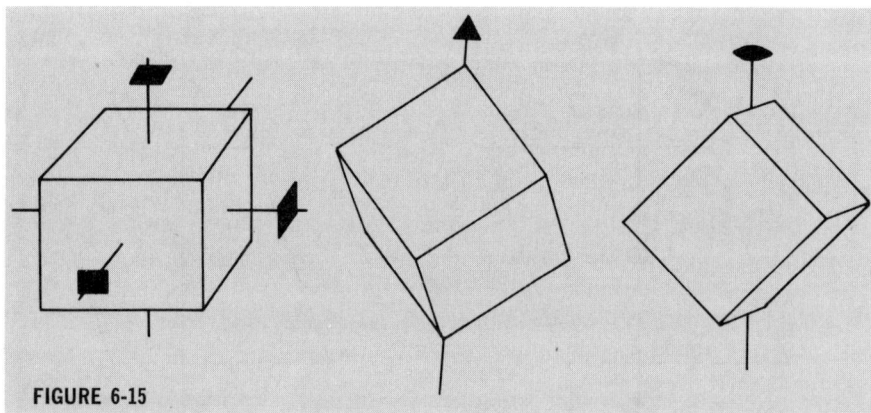

**FIGURE 6-15**

Types of Rotation Axes of a Cube.

Three tetrad axes are easily seen, each passing through the centers of the opposite faces of the cube. The triad axes are less easily seen unless one takes a cube and carries out the operation. Four of these axes exist, one passing through each of the four pairs of opposite corners of a cube. Finally, six diad axes exist. Again, it is almost essential to have a model of a cube in order to appreciate these. There will be six such axes, one passing through each of the six pairs of opposite edges of the cube. In summary, the 13 proper rotation axes of a cube are shown in Figure 6–16.

The symmetry planes found in a cube are much simpler to visualize than the rotation axes. There are two types of mirror planes as shown in Figure 6–17. The first type is composed of three planes parallel to the cube faces, and the second type is composed of six diagonal planes. This gives a total of nine mirror planes. The latter six have been separated in order to more clearly illustrate their positions on a stereographic projection in the next section.

Finally, we can see that there is a center of symmetry in a cube. Operationally we can prove this by inverting every point of the cube to an

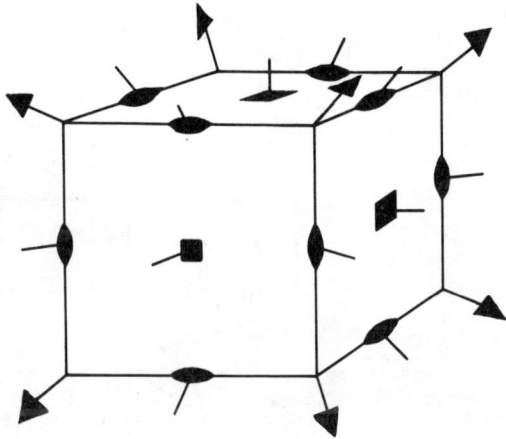

**FIGURE 6-16**

Rotation Axes of a Cube.

equivalent point on the opposite side of the center. Or, we can use the 1-fold rotary-inversion axis to demonstrate an equivalent structure. However, our general feeling for symmetry certainly is sufficient to tell us that a center of symmetry does exist in the cube.

There are means of describing symmetry with elements other than those we have considered here, and in fact, we have not even carried out all the operations within the three different symmetry elements we have discussed. The center of symmetry is equivalent to a 1-fold rotary-inversion axis, but we have not considered the higher-fold rotary-inversion axes. This does not mean we have overlooked any symmetry elements for the cube. It can readily be shown that these can all be expressed in terms of the elements and operations we have already used. For instance, the 2-fold rotary-inversion is equivalent to a mirror plane, and we have already determined all the mirror planes for the cube. Thus, it would be superfluous to describe the same symmetry in terms of two different operations. For a cube, then, we have for the full symmetry the following 23 elements:

1 center of symmetry

3 parallel planes ⎫
6 diagonal planes ⎭ 9 planes

3 tetrad axes ⎫
4 triad axes ⎬ 13 proper rotation axes
6 diad axes ⎭

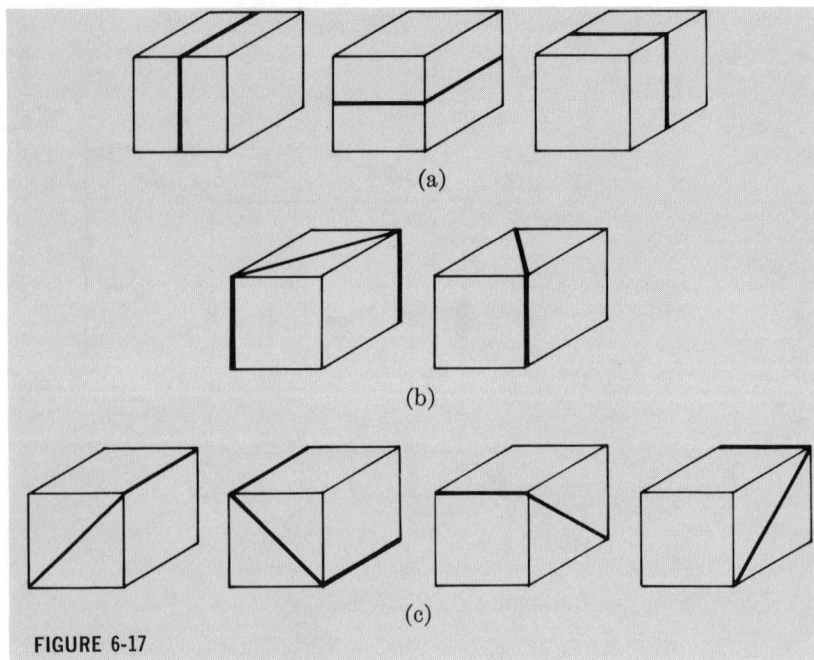

FIGURE 6-17

Symmetry Planes of a Cube.

## Stereographic Projections

The normal development of a crystal is such as to obscure its symmetry. But since the symmetry is our primary concern here, it is desirable to have a means of representing the symmetry of a crystal without having to deal with its specific idiosyncrasies of growth. It is for this reason that the method of *stereographic projection* is utilized. By this means we can clearly illustrate the planes and axes of rotation characteristic of a given crystal type.

The general procedure is to imagine the crystal to be circumscribed by a sphere. Normals to the crystal faces are then drawn and permitted to intersect the surface of the sphere as shown in Figure 6–18(a). We now imagine a plane to pass through the center of the sphere as seen in Figure 6–18(b). The intersections of the normals with the surface of the sphere can now be projected onto the horizontal circle.

The figure we have chosen for this representation shows the full symmetry of the cube and is particularly useful for consideration here. It can be seen that the normals to the faces each coincides with one of the rotation axes of a cube, whereas, if we had chosen the cube itself, only the normals coinciding with the three 4-fold rotation axes would appear.

If only those normals that lie in the horizontal plane are considered first,

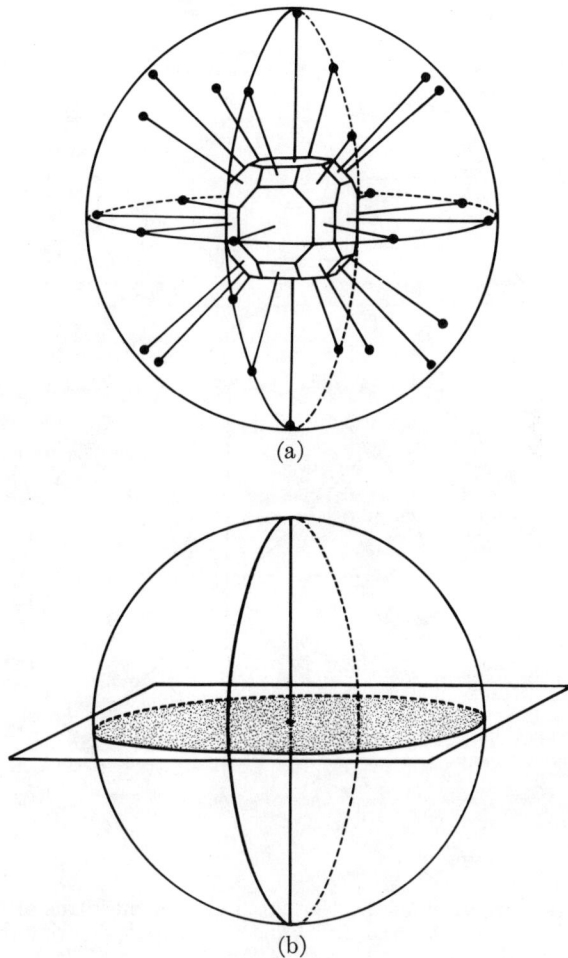

FIGURE 6-18

Stereographic Projection: (a) spherical projection of face normals, and (b) position of the projection circle.

we shall obtain the points of intersection that lie on the circumference of the *projection circle*. This is more clearly shown in Figure 6–19, where only these normals are drawn along with their projections.

With reference to our previous discussion of the symmetry elements found in a cube (Figure 6–16), it is apparent that those normals lying in the horizontal plane correspond to two of the cubic tetrad axes and two of the cubic diad axes. These are so indicated in the projection. The third tetrad axis results from the normal to the horizontal plane and is represented by a square in the center of the projection circle. The remaining intercepts fall somewhere inside the circle, and their exact positions can be calculated.

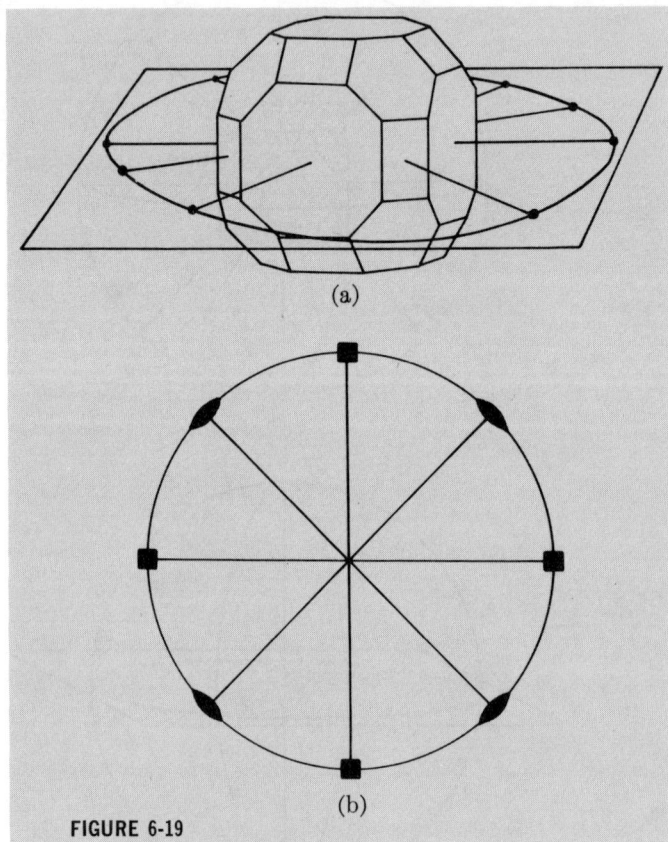

**FIGURE 6-19**

(a) Projection of the Normals Lying in the Plane of the Projection Circle, and (b) Projection Circle Showing Rotation Axes Lying in the Projection Plane.

However, we shall see that they can also be located by means of the mirror planes.

It is, of course, necessary that the projection circle be drawn in our stereographic projection. However, it may or may not coincide with a mirror plane. We have chosen a structure for illustrative purposes that happens to show the full cubic symmetry, and we can use our previous detailed discussion of cubic symmetry as a guide. Thus, from the position of the horizontal plane as shown in Figure 6–18, we can see that the projection circle does, in fact, coincide with a mirror plane for this structure. To indicate this, the projection circle is drawn with a solid line. If, in a given projection, the projection circle does not coincide with a mirror plane, it will be represented by a dashed line.

Earlier it was seen that there are nine mirror planes in a perfect cube. The projection plane represents one of the three parallel planes. In Figure

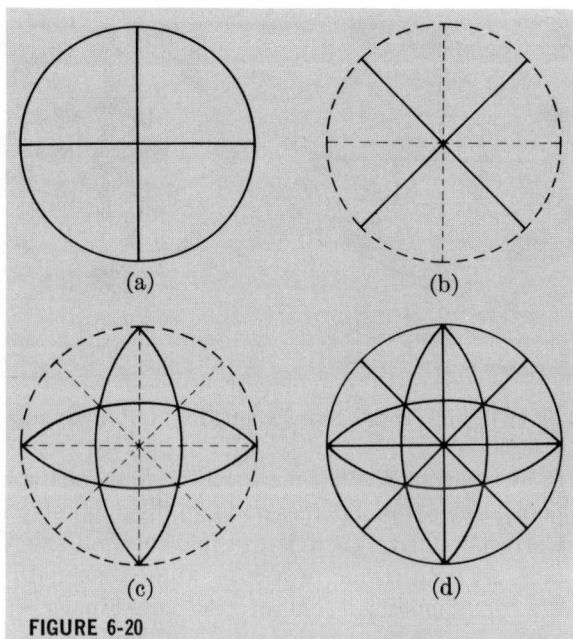

FIGURE 6-20

Representation on the Projection Circle of the Symmetry Planes of a Cube Corresponding to Figure 6-17.

6–20, all nine of the planes are shown in projection. These are grouped in such a manner that they can be compared to the corresponding planes for a perfect cube shown earlier in Figure 6–17. Thus, the three mirror planes parallel to the cube faces shown in Figure 6–17(a) are shown in projection in Figure 6–20(a). The diagonal planes shown in Figure 6–17(b) are shown in projection in Figure 6–20(b), and the second type of diagonal plane shown in Figure 6–17(c) is shown in projection in Figure 6–20(c).

The mirror planes shown in Figure 6–20(a) and (b) are seen to coincide with the tetrad and diad axes lying in the horizontal plane as seen in Figure 6–19. We can now combine the nine mirror planes and these four rotation axes to give the partial stereogram shown in Figure 6–21(a). In addition, we have included the tetrad axis normal to the projection circle. Thus, this figure shows all nine of the mirror planes, three tetrad axes, and two diad axes. We have yet to consider the four triad axes, four additional diad axes, and a center of symmetry.

Although a rigorous proof is possible, it should be sufficient to point out that from a study of our diagrams of the planes and rotation axes observed for a cube, it can be seen that a rotation axis occurs at the intersection of two or more mirror planes. The order of the axis will be the same as the number of planes that intersect at that point. For instance, if we choose the center

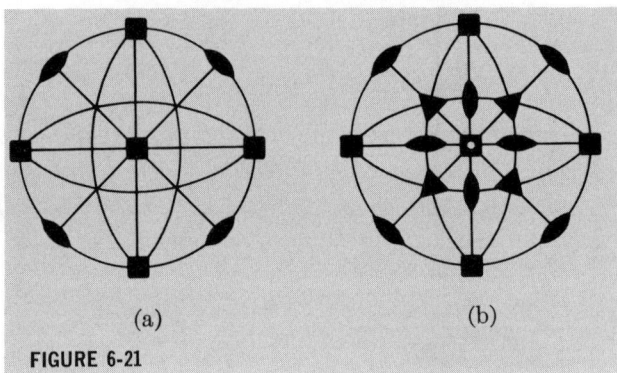

FIGURE 6-21

Partial and Total Stereogram for Cubic Symmetry.

of any face of a cube, there will be four mirror planes intersecting at that point, two parallel planes and two diagonal planes, as seen in Figure 6–22. Correspondingly, there is a 4-fold rotation axis through the center of each face. At each corner of a cube, three diagonal planes intersect, and there is a corresponding 3-fold rotation axis. Finally, at the center of each edge of a cube, it can be seen that one parallel plane and one diagonal plane intersect, and here we find a 2-fold rotation axis.

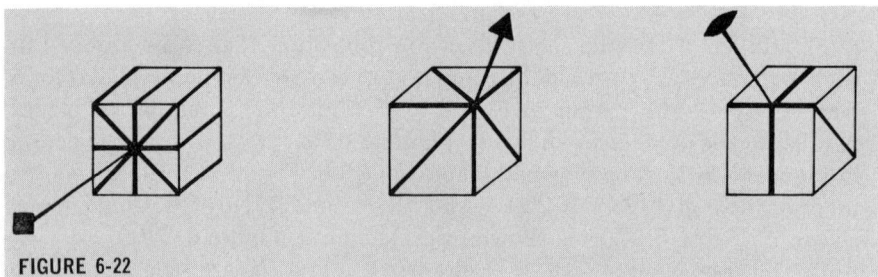

FIGURE 6-22

Correspondence between the Intersections of Symmetry Planes and the Order of Rotation Axes.

Looking at the partially completed stereogram in Figure 6–21(a), it can be seen that the 4-fold and 2-fold axes occur at the intersections of the corresponding number of mirror planes. This principle can now be carried further to locate the remaining rotation axes. There are four positions where three mirror planes intersect. These determine the locations of the four triad axes. There are four additional positions where two planes intersect. These determine the locations of the four remaining diad axes. Finally, the center of symmetry is indicated by a white dot in the center of the stereogram. This, then, gives us the completed stereographic projection of the full cubic symmetry shown in Figure 6–21(b).

It should be pointed out that those rotation axes that lie in the horizontal plane have the symbol for the particular order of the respective axes shown on each end of the axis. This practice cannot be followed for the remaining axes. Thus, the number of symbols for rotation axes shown on the stereogram will be greater than the actual number of rotation axes.

## The Crystal Systems

Now that the various symmetry elements have been considered, it is possible to apply them to the problem of crystal classification. Again, it is helpful to use the symmetry of a parallelepiped to pictorially illustrate the presence of the symmetry elements. We can imagine at one extreme a parallelepiped with the minimum possible symmetry. Thus it would have no center of symmetry, no symmetry planes, and no rotation axes other than a 1-fold axis which, of course, is common to any structure. At the other extreme we might consider a cube with its 23 elements of symmetry. Between these two, it is possible to place all crystals.

On the basis of symmetry, seven crystal systems have been defined. These are listed in Table 6–2, where it can be seen that each system is

**TABLE 6–2.  The Seven Crystal Systems.**

| Crystal System | Minimum Symmetry | Parallelepiped Dimensions |
|---|---|---|
| triclinic | 1 (or $\bar{1}$) | $a \neq b \neq c$ <br> $\alpha \neq \beta \neq \gamma$ |
| monoclinic | 2 (or $\bar{2}$) | $a \neq b \neq c$ <br> $\alpha = \beta = 90° \neq \gamma$ |
| orthorhombic | 222 or ($\bar{2}\bar{2}\bar{2}$) | $a \neq b \neq c$ <br> $\alpha = \beta = \gamma = 90°$ |
| tetragonal | 4 (or $\bar{4}$) | $a = b \neq c$ <br> $\alpha = \beta = \gamma = 90°$ |
| cubic (isometric) | four 3 (or four $\bar{3}$) | $a = b = c$ <br> $\alpha = \beta = \gamma = 90°$ |
| trigonal | 3 (or $\bar{3}$) | |
| hexagonal | 6 (or $\bar{6}$) | $a = b \neq c$ <br> $\alpha = \beta = 90°; \gamma = 120°$ |

dependent upon the presence of certain rotation axes. In the last column of Table 6–2 are given the dimensions of parallelepipeds having the maximum possible symmetry of their respective systems. The lengths of the edges and the angles between the axes of the parallelepiped are defined in Figure 6–23. In line with our previous discussion, it can be seen from the

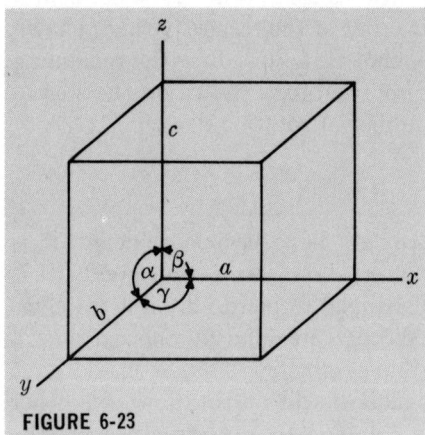

**FIGURE 6-23**

General Representation of Crystallographic Axes.

dimensions given in the table that the corresponding parallelepiped in the cubic system is a cube. And we see that a member of the cubic system must have four 3-fold rotation axes. Of the various parallelepipeds one can draw, only a cube will have this set of axes. Thus, this uniquely defines the cubic system.

Whereas a cube is distinguished by four triad axes, it also contains most of the symmetry elements observed in the other systems. We can conclude, therefore, that the presence of these triad axes are indicative of a rather high degree of symmetry. Just as these axes set the cube apart from the other parallelepipeds, the presence of only a limited symmetry can likewise set a system apart. Thus we find at the other extreme the triclinic system, which is distinguished by the lack of any symmetry other than a possible center.

In Table 6-2, seven crystal systems have been included. However, no dimensions for the corresponding parallelepiped have been given for the trigonal system. A problem arises here because of the possibility of more than one set of dimensions. Frequently this difficulty is resolved by considering the trigonal system a special case of the hexagonal system, thereby giving a total of six crystal systems.

## Stereographic Projection of Crystal Systems

In stereographic projections the crystal systems, excluding the trigonal system, can be represented as shown in Figure 6-24. The stereograms are each of the *holosymmetric* or *normal class* of their respective system. That is, they show the maximum possible symmetry for that system.

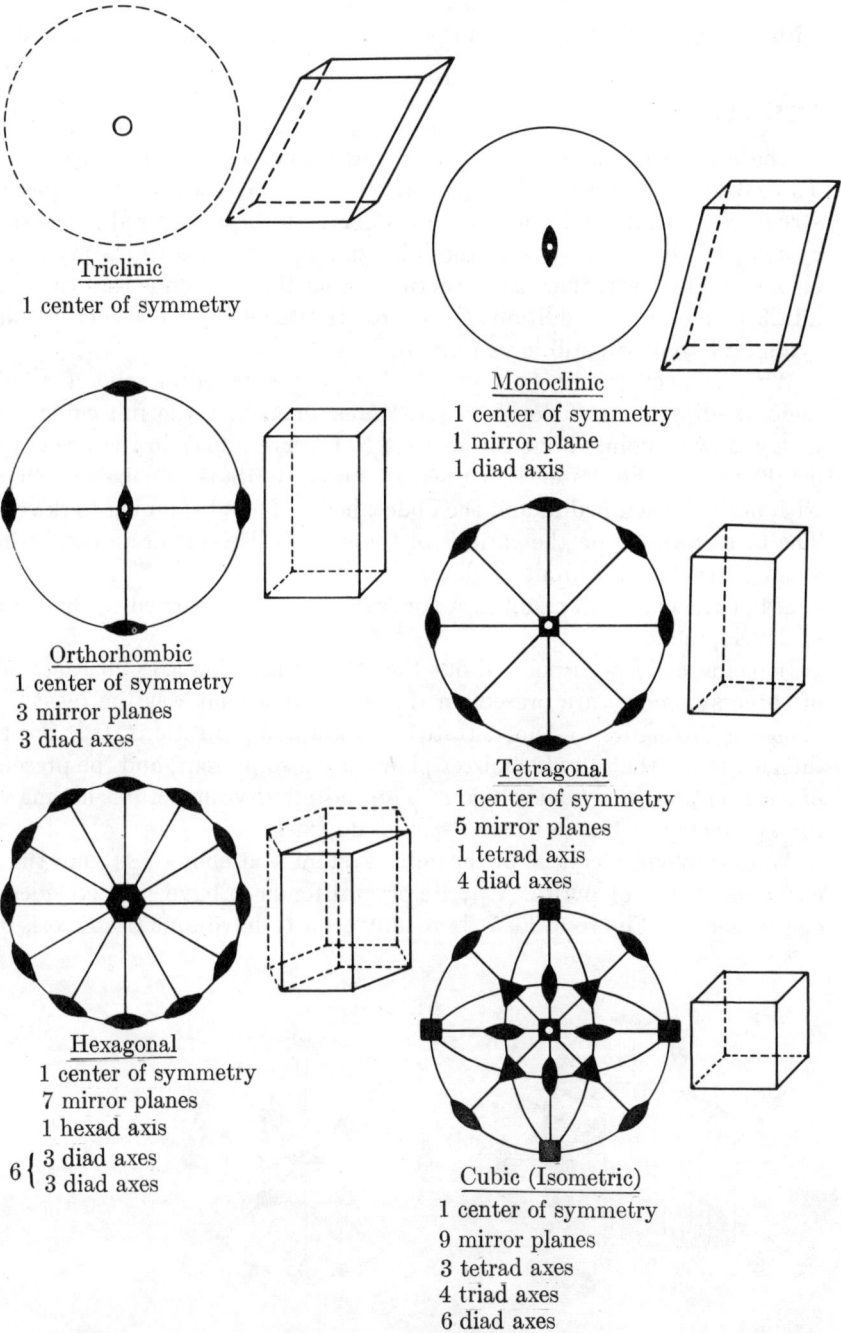

Triclinic
1 center of symmetry

Monoclinic
1 center of symmetry
1 mirror plane
1 diad axis

Orthorhombic
1 center of symmetry
3 mirror planes
3 diad axes

Tetragonal
1 center of symmetry
5 mirror planes
1 tetrad axis
4 diad axes

Hexagonal
1 center of symmetry
7 mirror planes
1 hexad axis
$6 \begin{cases} 3 \text{ diad axes} \\ 3 \text{ diad axes} \end{cases}$

Cubic (Isometric)
1 center of symmetry
9 mirror planes
3 tetrad axes
4 triad axes
6 diad axes

**FIGURE 6-24**

The Crystal Systems in Stereographic Projection.

Included with each stereogram is a list of the symmetry elements, along with a parallelepiped which shows the complete symmetry of the system.

## Crystal Classes

The arrangement of crystals in the seven crystal systems was seen in Table 6–2 to be based on certain rotation axes. Such a restriction permits structures with different degrees of symmetry to be considered in the same system. In the cubic system, the full symmetry was shown to involve 23 elements. However, there are structures other than the cube that can show all 23 elements. In addition, there are structures with considerably less symmetry that will still have four triad axes.

With respect to the first point, a regular octahedron and a regular dodecahedron are two common structures that show the full cubic symmetry. It was pointed out earlier that NaCl crystallizes in the form of an octahedron in the presence of urea. In Figure 6–18(a), a cube was shown with both the octahedral and the dodecahedral faces beginning to develop. The faces forming on the corners of the cube will result in an octahedron when carried to the limit of development, and the faces forming on the edges of the cube will result in a dodecahedron when carried to their limit of development.

It has already been pointed out that the 13 rotation axes characteristic of cubic symmetry are present in this structure. That was the reason for choosing this figure to demonstrate stereographic projections. It is not too difficult to see that all nine mirror planes are also present, and the presence of a center of symmetry is obvious. Thus, alternative structures having the full symmetry of this particular system do exist.

An example of a crystal in the cubic system that shows less than the 23 elements is that of pyrite. A pyrite crystal tends to have striated sides as can be seen in Figure 6–25. It is readily seen to have four triad axes, but

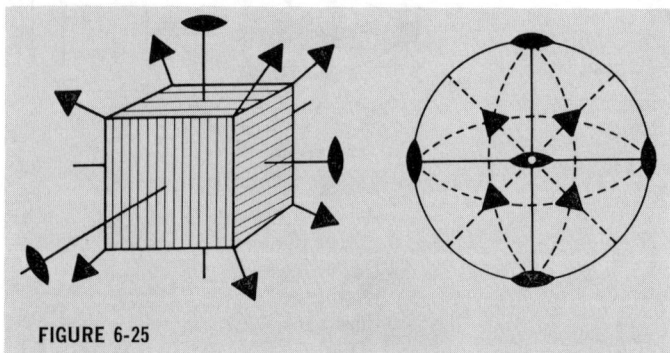

**FIGURE 6-25**

A Pyrite Crystal with Its Stereographic Projection Illustrating a Lower Class of the Cubic System.

because of the striations its full symmetry is much less than the 23 elements possible for cubic symmetry, being

> 1 center of symmetry
>
> 3 planes of symmetry
>
> 3 diad axes
>
> 4 triad axes

The mirror planes are those we have previously considered parallel planes, and the diad axes are the three normals to the cube faces.

The tetrahedron also falls into the cubic system by virtue of four triad axes, but, again, it has a symmetry less than that of the normal class. In this case there are

> 4 triad axes
>
> 3 tetrad rotary inversion axes
>
> 6 diagonal planes

Rather than three tetrad rotary inversion axes, three diad axes are sometimes shown. It is true that these are present, but the symmetry is actually greater than this. For this reason the rotary inversion axes are more appropriate. One example of each of these elements is shown in Figure 6–26. If the four legs of the tetrahedron used to illustrate the mirror plane

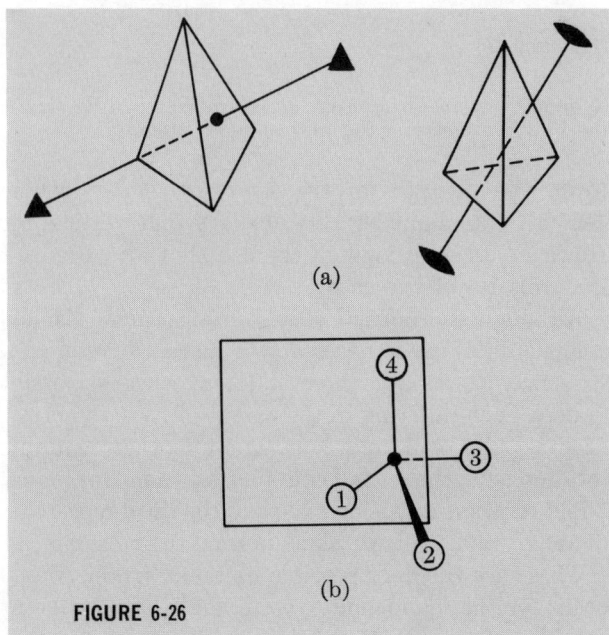

**FIGURE 6-26**

The Symmetry Elements of a Tetrahedron.

are labeled 1, 2, 3, and 4, it can be seen that there are six combinations that can be made of the type shown in the figure, where in this particular example the combination 1, 4 is used.

As a final example, we can consider the monoclinic system. From Table 6–2, it can be seen that the characteristic parallelepiped has the dimensions: $a \neq b \neq c$ and $\alpha = \beta = 90° \neq \gamma$. The full symmetry of this system is shown in Figure 6–27 to be simply

> 1 center of symmetry
>
> 1 diad axis
>
> 1 mirror plane normal to the diad axis

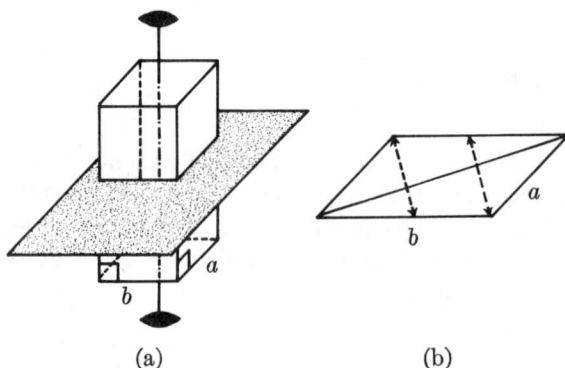

(a)  (b)

FIGURE 6-27

Symmetry Elements Present in a Parallelepiped Having the Full Symmetry of the Monoclinic System.

From the diagram of the appropriate parallelepiped shown in Figure 6–27(a), one might be led to believe that vertical mirror planes also exist. However, the top view of the cell shown in Figure 6–27(b) clearly shows that this is not the case.

Although the normal class contains three elements of symmetry, only a single diad axis is necessary to place a crystal in this system. As a consequence, there are three different classes possible in the monoclinic system, each differing in the extent of its symmetry. These can be symbolized as 2, $m$ or $\bar{2}$, and $2/m$. The first class contains only a proper 2-fold rotation axis; the second contains only a mirror plane, or its equivalent — a 2-fold rotation-inversion axis; and the third represents the normal class — a diad axis and a mirror plane normal to the axis.

These are the only possible different symmetries that geometrically can occur for the monoclinic system, and they can be arrived at without any consideration of whether actual crystals with these symmetries exist or not. In a similar manner, it can be shown that 32 different symmetry classes

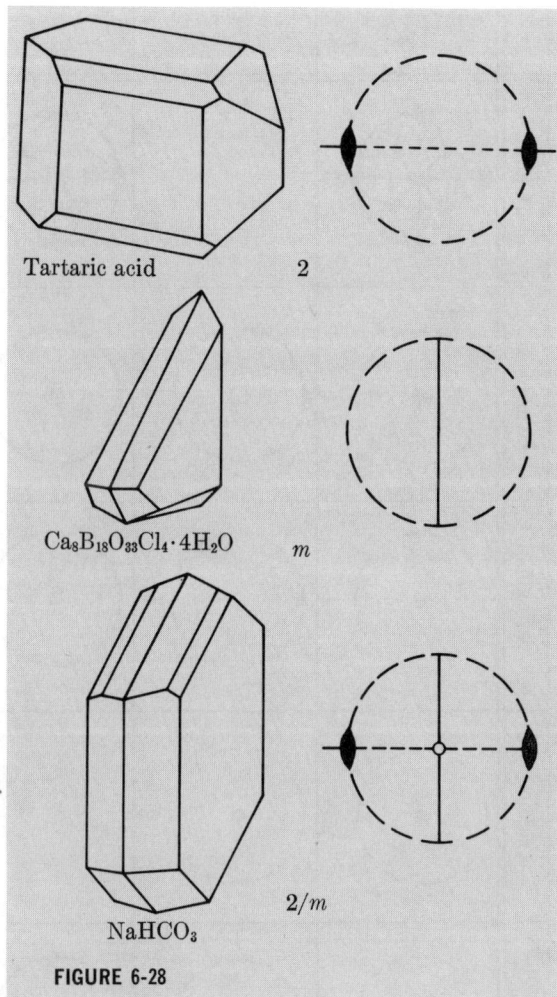

**FIGURE 6-28**

Examples of the Three Classes of the Monoclinic
System with Their Stereographic Projections.

exist throughout the seven crystal systems, and actual crystals have now
been found for each of these. For the monoclinic system, an example of each
of the three classes is shown in Figure 6–28, along with the appropriate
stereogram.

## Thirty-two Crystal Classes

The 32 crystal classes are shown in stereographic projection in Figure
6–29. The plan that has been followed is to list the classes within a crystal
system in the respective column. Except for the cubic system, the rows

| | $X$ | $\bar{X}$ | $X/m$ | $Xm$ |
|---|---|---|---|---|
| Triclinic | $1$ | $\bar{1}$ | $1/m = \bar{2}$ | $1/m = \bar{2}$ |
| Monoclinic and Orthorhombic | $2$ | $\bar{2} = m$ | $2/m$ | $2m = mmm$ |
| Trigonal | $3$ | $\bar{3}$ | $3/m = \bar{6}$ | $3m$ |
| Tetragonal | $4$ | $\bar{4}$ | $4/m$ | $4m = 4mm$ |
| Hexagonal | $6$ | $\bar{6}$ | $6/m$ | $6m = 6mm$ |
| Cubic | $23$ | $\bar{23} = 2/m3$ | $2/m\bar{3} = m\bar{3}$ | $2m\bar{3} = \bar{4}3m$ |

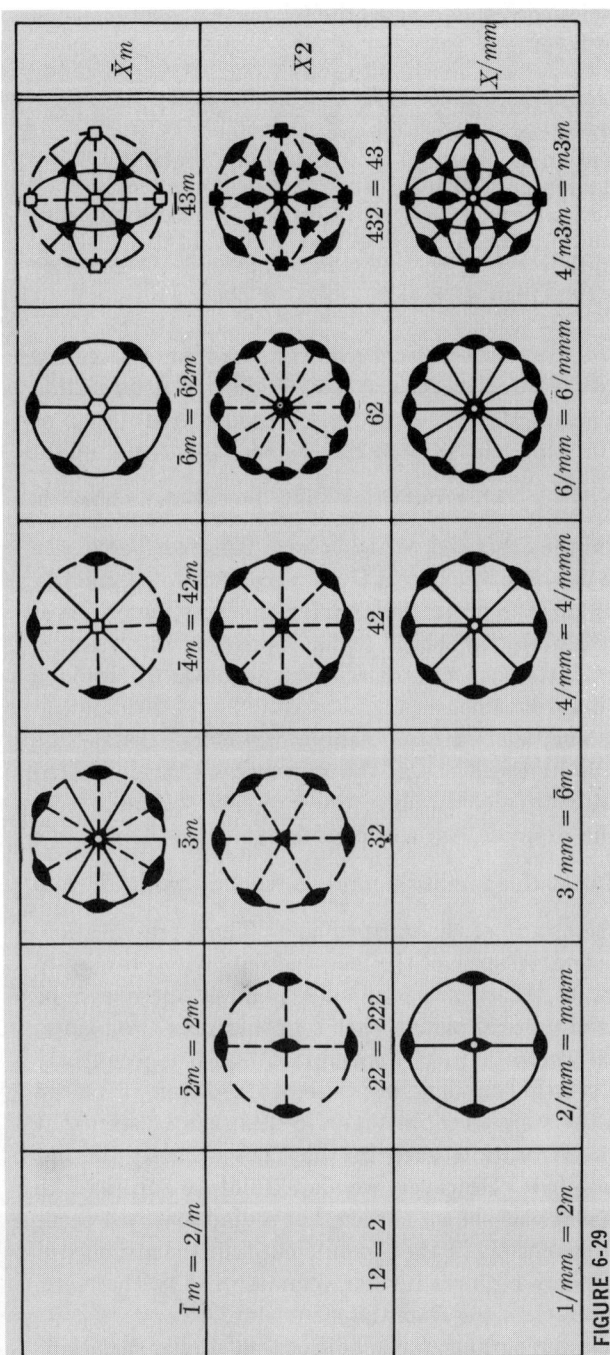

**FIGURE 6-29** Stereographic Projections of the 32 Crystal Classes.

are composed of symmetries corresponding to the following elements, using $X$ to denote the order of rotation

$X$      rotation axis only
$\overline{X}$      rotation-inversion axis only
$X/m$    rotation axis normal to a plane of symmetry
$Xm$     rotation axis with a vertical plane of symmetry
$\overline{X}m$     rotary inversion axis with a vertical plane of symmetry
$X2$      rotation axis with a diad axis normal to it
$X/mm$   rotation axis with a normal plane and one or more vertical planes

By convention, the order of the *principal* axis is listed first. Except for the cubic system, this will always be the axis with the highest rotation order. Again, by convention, the principal axis is considered the $z$ or vertical axis. With this in mind, the symmetries denoted by $X/mm$ might better be represented as $\dfrac{X}{m} m$. This indicates that one of the planes is normal to the axis, $X$, and the other is a vertical plane. However, since only one plane can exist normal to a given axis, it is not necessary to specifically indicate that the second plane is a vertical plane. Inasmuch as the principal axis is chosen normal to the projection circle, a plane normal to the principal axis will be coincident with the projection circle and is shown by drawing the projection circle with a solid line.

In many instances the symbols used to denote a particular crystal class appear to be incomplete. In these cases, the presence of the indicated symmetry elements automatically implies the presence of additional elements. Using a specific example, we can consider the class $4/mm$. We have already mentioned that this might be represented as $\dfrac{4}{m} m$ in order to more clearly point out the nature of the mirror planes. The representation, however, is still not complete. In spite of the fact that only one $m$ is used to indicate the vertical planes, there are four such planes. These can be considered to be two sets of vertical planes, each set containing two planes. Thus the symbol $4/mmm$ would be more appropriate, and it is frequently used. But even here we see that a single $m$ is used to represent each set of vertical planes. From the stereogram in Figure 6–29 it can be seen that there are also four 2-fold rotation axes in this class. These are not indicated by the symbol $4/mm$. However, we should not be surprised to find the diad axes because it was shown earlier that a diad axis will occur at the intersection of two planes. This, then, points out that the presence of certain symmetries necessarily requires further symmetry. It is, therefore, not necessary to show more than the essential symmetry.

By studying the various stereograms, it can be seen that similar situations arise in a number of instances. A particularly obvious example is the class $3m$, where three vertical mirror planes are present in spite of the fact that the class symbol specifically shows only one. Thus the 3-fold axis generates the other two parallel planes.

In Figure 6–29, several spaces are noted where the appropriate combinations of symmetry elements are shown, but no stereogram is given. In each of these instances, the combination is equivalent to one elsewhere in the table. In most cases, it will be a member of a different crystal system. For instance, the combination $1m$ is equivalent to $\bar{2}$ or $m$, which is a class in the monoclinic system. It should also be noted that the combination $\bar{X}/m$ and $\bar{X}2$ are not considered at all. In every case these can be shown to be equivalent to elements considered elsewhere in the table. For instance, $\bar{2}/m = m$ and $\bar{3}2 = \bar{3}m$.

The cubic system is unique. It is the only system based on a secondary set of axes, the four triad axes. Thus every class in the cubic system contains the symbol 3. Since a secondary axis is always written after the principal axis, the 3 is never written first in the cubic system.

To offer a limited familiarity with the multitude of crystal shapes, some simple examples are shown in Figure 6–30, along with their respective crystal class. An example of each of the three monoclinic classes was shown in Figure 6–28. For the examples we have chosen, the symmetry elements can readily be seen, and it is worthwhile to compare these with the appropriate stereographic projections.

## Methods of Crystal Classification

It is not too difficult to look at a stereographic projection and determine the elements of symmetry expressed by the projection. It is quite another matter to see how this symmetry may apply to an actual crystal. Early in our discussion, three different habits of NaCl were shown. In spite of the fact that they appear to be quite different, all three belong to the normal class of the cubic system, $m3m$. In the case of the cube and the octahedron, all 23 of the elements of symmetry characteristic of this system can be seen to be present. However, this is certainly not true for the wafer-like structure shown in Figure 6–2. This geometrical shape is compatible with several crystal systems as well as various classes in these systems.

It should be recalled that the purpose of using stereographic projections is to permit one to study the crystal symmetry without encountering the confusion that might arise from the various irregularities of crystal growth. The procedure used in our specific discussion suffers, in a sense, from the fact that we chose a geometrical structure that shows the full symmetry of the cubic system. This is an advantage that one does not have when working with an unknown crystal. In dealing with an actual crystal, all that can generally be done is to draw normals to the developed faces and note the positions on a projection circle. It is not usually possible to say that the axis is of a certain order or that so many mirror planes are necessarily present, although these may be clearly suggested. Thus,

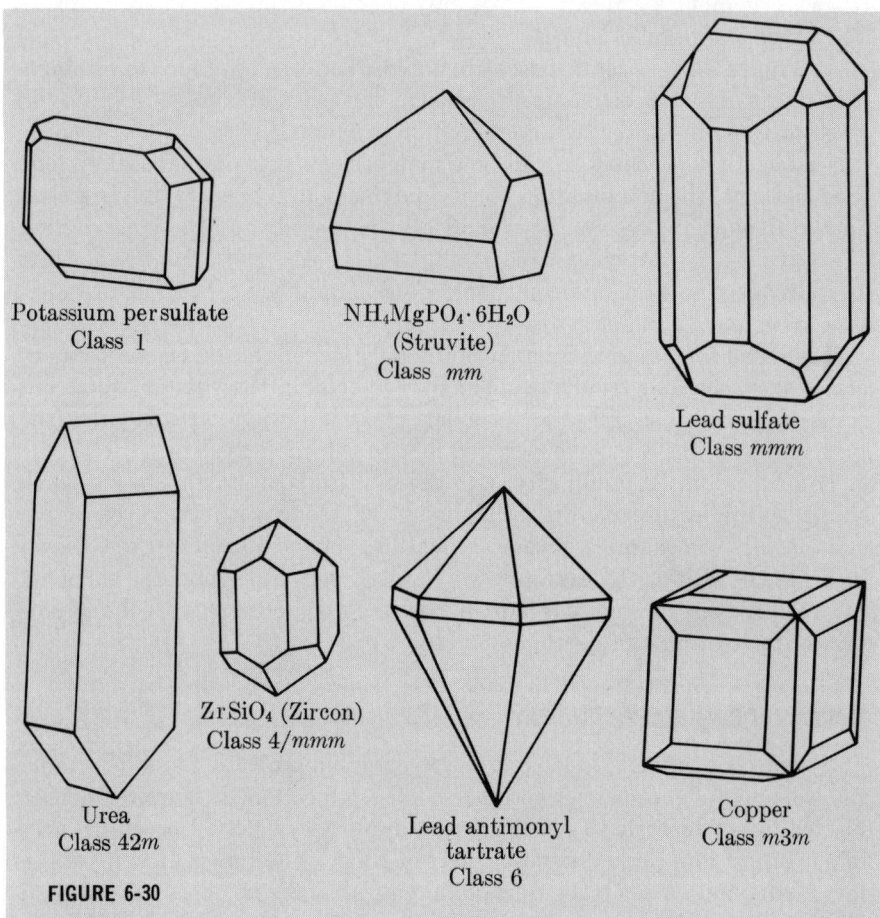

Potassium per sulfate
Class 1

NH₄MgPO₄·6H₂O
(Struvite)
Class mm

Lead sulfate
Class mmm

ZrSiO₄ (Zircon)
Class 4/mmm

Urea
Class 42m

Lead antimonyl
tartrate
Class 6

Copper
Class m3m

**FIGURE 6-30**

Examples of Crystal Shapes.

for a crystal that has the general shape of a cube, the stereographic projection is simply that shown in Figure 6–31. The four solid dots on the circumference of the circle represent the intersections of the normals to the four vertical faces. The solid dot in the center of the circle represents the intersection of the normal from the top face, and the coincident open circle represents the intersection of the normal from the bottom face.

From this stereogram, one can state only that there are six faces that are parallel to the faces of a cube. Now, if we draw the stereogram for a crystal having the wafer-like form shown in Figure 6–2, it should be apparent that the same stereogram will result as that just seen for the cube. Such a stereogram is consistent with more than one class in the cubic system, and it is also compatible with structures that might occur within the orthorhombic and the tetragonal systems. This is apparent from the parallelepiped characteristic of these systems.

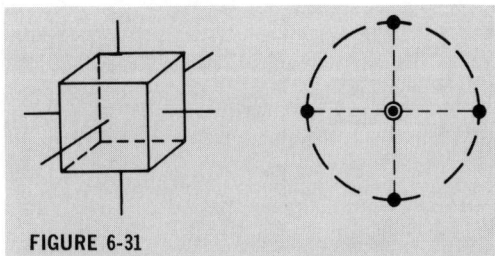

FIGURE 6-31

Simple Stereographic Projection of a Cube.

This, then, points up the difficulty of classifying a crystal on the basis of only morphological considerations. It is true that by careful growth of crystals additional faces may be developed that will give a more definitive indication of the crystal class. But only in rare instances can all the essential faces be grown. Consequently, unequivocal classification of a crystal is generally not possible by means of morphological considerations alone. The complete stereograms which we have shown for each crystal class give the unique symmetry of that class. But these stereograms are not the result of morphological studies alone. They generally depend on an accumulation of structural information from a number of different sources.

In spite of this fact, we have emphasized that the development of the faces of a crystal is determined by the underlying symmetry characteristic of that particular crystal. In addition, it is essential to recognize that not only crystal faces, but *all parts and properties of a crystal are repeated by the symmetry elements characteristic of that particular crystal.* Thus, other observable properties of a crystal will also reflect its symmetry. This can be seen in the optical, electrical, and thermal properties, etch figures, X-ray patterns, etc., of a crystal. All of these are dependent on the internal symmetry of the crystal, and it should be apparent that they can also be used for the determination of the appropriate class of a crystal.

Etch figures are among the oldest of these techniques. If a solvent is placed on a crystal face, the dissolution of the crystal occurs in a manner characteristic of the crystal symmetry. For instance, etch figures occurring on the cube faces of crystals in the classes $m3m$ and $m3$ are shown in Figure 6–32. Here we observe the respective 4-fold and 2-fold symmetries of these faces.

A great amount of study has been conducted on the optical properties of crystals. A crystal in the cubic system has the same atomic arrangement along all three axial directions, and we would, therefore, expect the properties of such a crystal to correspondingly be the same along these equivalent axes. It is not necessarily obvious why, but a crystal in this system is optically isotropic. Thus, regardless of the orientation of a light source, only one refractive index is observed. This, however, is not the case with crystals in other systems. If, for the sake of convenience, only the directions

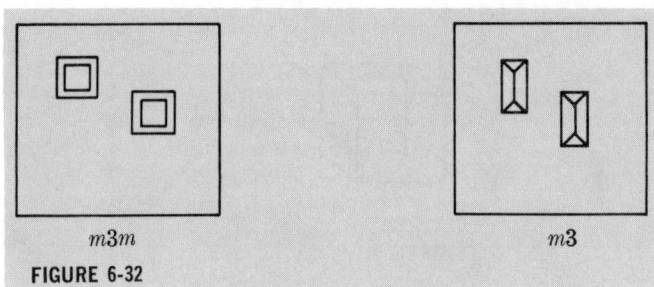

$m3m$    $m3$

**FIGURE 6-32**

Etch Figures of Classes $m3m$ and $m3$.

parallel to the crystallographic axes are considered, the environment seen by an incident beam of light will thus depend upon which axis it parallels. This leads to different optical properties, and the resultant difference in behavior can be used to help classify the crystal in terms of its symmetry.

An interesting phenomenon found in some crystals is the tendency to be pyro- and piezoelectric. When a pyroelectric crystal is heated or cooled, a separation of charge occurs and the crystal becomes positive at one end and negative at the other. An analogous type of behavior is noted for a piezo-electric crystal when it is subjected to a physical strain such as compression. These effects are observed only in crystals that do not have a center of symmetry. Consequently, an observation of either of these effects auto-matically rules out any crystal class containing a symmetry center.

It should be apparent that any one of these methods by itself is inade-quate for crystal classification. However, with utilization of all the various tools available, a crystal can be unequivocally placed in its proper class. The magnitude of the classification procedure can be appreciated when one considers that the number of classified crystalline materials, both natural and synthetic, is in the neighborhood of 20,000. Approximately 90% of these, however, fall into only three systems: monoclinic, orthorhombic, and triclinic, and about 50% of these belong to the monoclinic system.

## INTERNAL STRUCTURE

### Space Lattices

In studying the morphology of a crystal we have insisted that the regu-larity of external appearance of a crystal in some way reflects a regularity of internal arrangement. Thus we should conclude that the angles between the various faces of a crystal depend on the arrangement of the atoms, ions, or molecules in the crystal. Yet, by considering different crystals of the same substance, one often finds different interfacial angles. A crystal in the cubic system, for instance, would likely show cubic faces, but it may also show octahedral and dodecahedral faces as seen in Figure 6–18. Or it may even develop faces other than these. And, of course, crystals in different systems will develop faces that are characteristic of their particular system.

The development of the particular faces on a crystal depends very critically on the conditions under which the crystal is grown. Temperature, rate of growth, extent of stirring, and the presence of impurities all play an important role in the determination of the crystal shape. However, under normal conditions of growth, only a small number of faces tend to develop on a given crystal, and these are quite characteristic of that particular crystal class. The reasons for this can best be appreciated by representing the crystal in terms of a *lattice*.

A crystal can be imagined to be generated from the repetition of some basic *unit of pattern*. But rather than draw out the entire unit of pattern, it is much more convenient to represent the unit of pattern by a point. Each point then represents the position of an atom, ion, molecule, or group of ions or molecules, resulting in a three-dimensional orderly array of points called a *space lattice*. An example of such a lattice is shown in Figure 6–33.

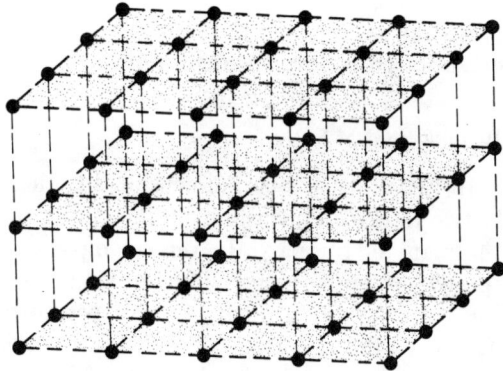

**FIGURE 6-33**

Example of a Simple Space Lattice.

We can define a space lattice as *a regular three-dimensional array of identical points in space, arranged such that a straight line passing through any two points will pass at equal intervals through a succession of similar points*. This definition is strictly geometrical in character and represents the three-dimensional translational repetition of the centers of gravity of the units of pattern in the crystal. From this, it is apparent that the lattice points do not represent the actual atoms in the crystal. Rather *they reflect the spatial arrangement of the units of pattern*.

Fundamental to a space lattice is the fact that the environment around any point is the same as that around any other point in the lattice. Keeping this in mind, we can see that a lattice point may represent something quite different depending on the nature of the specific crystal. For instance, in a crystal composed of simple spheres packed together, such as might be observed with a metal or inert gas, the position of each atom would probably be represented by a lattice point. On the other hand, a lattice point in a crystal of methane would most reasonably represent the center of a methane molecule. In an ionic crystal, it is often found convenient to consider a lattice of positive ions independently of a lattice of negative ions. Then the two can be interlocked to give an adequate representation of the crystal. In this case, each ion in its respective lattice would be represented by a lattice point. However, we might in an alternative approach choose a point equidistant between a positive and a negative ion. In this case, a lattice point would represent an ion pair, resulting in a single lattice to represent the ionic crystal. In addition to these examples, it is not uncommon for a lattice point to represent a grouping of two or more molecules.

It should be appreciated, then, that the distribution of atoms around a lattice point is somewhat arbitrary, and more than one possible choice may be made, although only one may be reasonable. But any choice will have to be such that every lattice site is identical to every other lattice site in the particular lattice, and the translation of the lattice sites must generate the crystal according to the definition.

## Development of Faces

In Figure 6–34, a two-dimensional lattice is shown. We might extend this lattice to three dimensions by imagining the $z$ axis to be perpendicular to the plane of the page. In this example, the lattice points are all equally spaced and can be imagined to lie along an orthogonal cartesian axis. Herein the basis for the growth of only specific faces in a crystal can be seen. There is a direct correlation between the planes one can draw through the lattice points and the development of faces on a corresponding crystal. It is apparent from the figure that essentially an infinite number of planes might be drawn through the various lattice points. Theoretically any of these planes could represent a crystal face. In spite of this, only a few faces are actually observed to develop. This is understandable when we realize that the reason a regular pattern occurs at all is because the atoms, ions, or molecules in the crystal tend to seek positions of minimum energy. Growth would thus be expected to occur on the faces having a large surface density. Or more correctly, we might say that a face with a large surface density would be expected to occur where the atoms, ions, or molecules can find a position of minimum energy. From the planes shown in the figure, it is apparent that the number of lattice sites falling on the various planes may differ greatly. The planes parallel to the cartesian axes have the highest

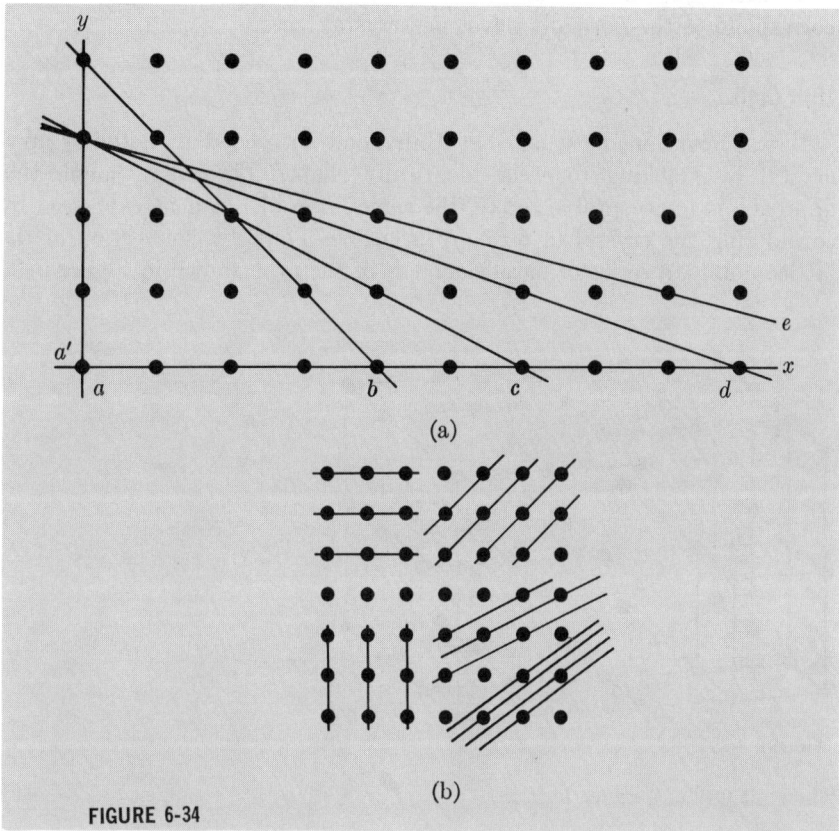

**FIGURE 6-34**

Two-Dimensional Lattice Characteristic of the Cubic System: (a) possible crystal faces showing the relative densities of lattice sites, and (b) distance of separation of various planes.

density of points, and the density decreases in the order of planes $a$ to $e$. Thus, we can imagine the development of crystal faces from a statistical point of view, and we find that only a few faces show a high probability of development. More complex planes are generally observed only as small faces occurring as modifications of the predominant faces.

## Crystal Cleavage

In terms of the crystal lattice, it is also possible to understand why crystals tend to show characteristic cleavage planes. From Figure 6-34(b) it can be seen that the planes with the highest site density are also the planes with the greatest distance of separation. Because of this separation, it is here that the interatomic interaction between planes should be the least. It is expected, then, that the natural fractures will occur parallel to faces with high surface density. Consequently, the cleavage planes should correspond to the normally developed crystal faces.

## Unit Cells

It was mentioned earlier that Haüy had succeeded in building up a crystal by stacking together identically shaped blocks. Although this approach is unacceptable today, the same basic idea can be expressed by considering the crystal in terms of a lattice. Thus, by connection of the lattice points, a series of parallelepipeds of the type shown in Figure 6–35

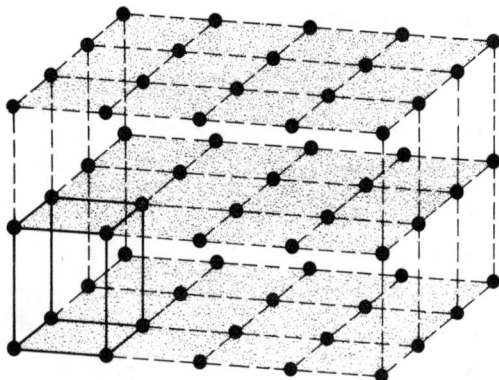

**FIGURE 6-35**

The Unit Cell of a Space Lattice.

can be obtained. Each of these parallelepipeds contains a complete unit of pattern of the crystal. By translation or stacking of the parallelepipeds, the entire crystal structure can be obtained. Such a parallelepiped can be drawn for any crystal lattice, and it is called a *unit cell*.

The choice of a unit cell in a crystal is somewhat arbitrary. If we use the two-dimensional lattice shown in Figure 6–36, it can be seen that a number of different parallelograms will generate the surface by repetition. However, for the sake of convenience it is desirable, although not necessary, to have the lattice points at the corners of the unit cell. Four different cells are shown in the figure, and each one contains one unit of pattern. This is most easily appreciated from the cell ABCD. Point B is shared by four unit cells in this two-dimensional representation. Thus, only one-fourth of the point can be considered to belong to cell ABCD. This, of course, is also true for the other corner points. If this lattice is extended to three dimensions, the corresponding points will each contribute only one-eighth of a lattice point to a given cell.

Referring back to the two-dimensional lattice shown in Figure 6–36, the cell JKLM obviously contains one lattice point. This is also true of cell

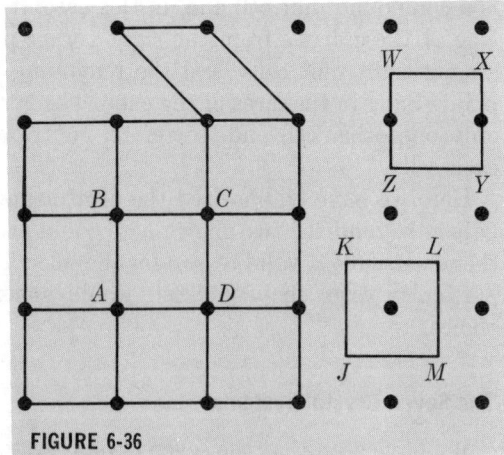

**FIGURE 6-36**

Illustration of Different Possible Choices of a Unit Cell.

WXYZ, where each point is shared with one other cell. Consequently, in this case each point will contribute one-half of a unit of pattern to the cell.

A cell containing only one unit of pattern is referred to as a *primitive* cell and is symbolized by a *P*. A primitive unit cell can be drawn for any space lattice, but frequently, such a cell is not the most convenient one for the purpose of calculation. Also, as we shall later see, it does not always best represent the properties of the crystal. To illustrate this point, two different representations of the unit cell of a copper crystal are shown in Figure 6–37. The first of these is a primitive cell having a rhombohedral structure. In the second of these, a cubic arrangement is maintained. However, the second cell obviously contains more than one unit of pattern. This

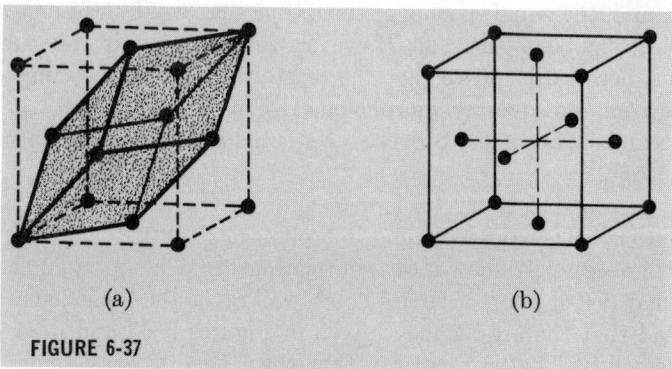

(a)          (b)

**FIGURE 6-37**

Two Different Unit Cells for a Copper Crystal: (a) rhombohedral and (b) face-centered cubic.

is a compound unit cell and, in this case, it contains four units of pattern. One of these arises from the eight corner points, each of which is shared among eight unit cells, and the remaining three are the result of the six points lying in the faces of the cube. The latter points are each shared with only one other cell and, therefore, contribute one-half a lattice point to each cell.

Here we have emphasized the arbitrariness of the choice of a unit cell. Others beyond the two shown here could also have been chosen. However, there is usually a valid reason for the selection, as we shall soon see. In this particular case, the cubic cell would generally be the most reasonable choice.

### The Seven Crystal Systems

We have seen that the seven crystal systems are defined in terms of the crystal morphology. This, of course, is a consequence of the historical development of our understanding of crystal structures. The internal arrangement of a crystal could not be directly observed until the advent of X-ray diffraction techniques. However, in spite of the limitations of morphological measurements along with etch figures, optical properties, etc., a proper representation of internal structure was determined for many crystals before X-ray crystallography was developed.

In both Figures 6–33 and 6–34, a lattice is drawn that is consistent with the cubic system. From a morphological standpoint, the general interfacial angles observed in a crystal are unique to a given system. It is true that major faces may be common even among different systems, but secondary faces can uniquely determine the system to which a crystal belongs and, in many instances, even a crystal class. Now, from the lattice shown in Figure 6–33, we would expect that the most prominent faces, those with the greatest site density, should be parallel to the orthogonal cartesian axes. If a crystal is carefully grown from a solution containing this compound, the equal spacing of the lattice sites would lead us to expect that a perfect cube might result. However, the occurrence of only cubic faces would not necessarily prove that the crystal belongs to the cubic system. If we restrict ourselves to morphological considerations, only in terms of the secondary and more obscure faces could it be uniquely placed in the cubic system.

In practice, a crystal is determined to be cubic as a consequence of a variety of tests, including morphological tests. The interfacial angles will necessarily show a certain relationship. The question then arises as to what distribution of lattice points will favor these particular faces, and we find that the arrangement given in Figures 6–33 and 6–34 satisfies these demands. Thus we can say that the lattice is consistent with the cubic system. Now, if a unit cell is drawn with the sides parallel to the orthogonal set of cartesian axes, we obtain a perfect cube. We can then imagine the crystal to be generated by stacking these cubes together. It is tempting to say that the symmetry of the unit cell determines the crystal symmetry. However, it must be remembered that alternative unit cells can be drawn, some of which do not show cubic symmetry. From this we should see that *the crystal symmetry generates the unit cell*, not the reverse. It is usually desirable to maintain a correlation between the crystal symmetry and that of the unit cell. Thus it is because a copper crystal shows the properties of the cubic system that the cubic unit cell is generally chosen. But at the same time, it must be recognized that the unit cell does not have to be cubic, even though the crystal symmetry is cubic.

In Figure 6–38, a two-dimensional representation of a tetragonal lattice is shown. Again it is seen that the planes parallel to the $x$ and $y$ axes have

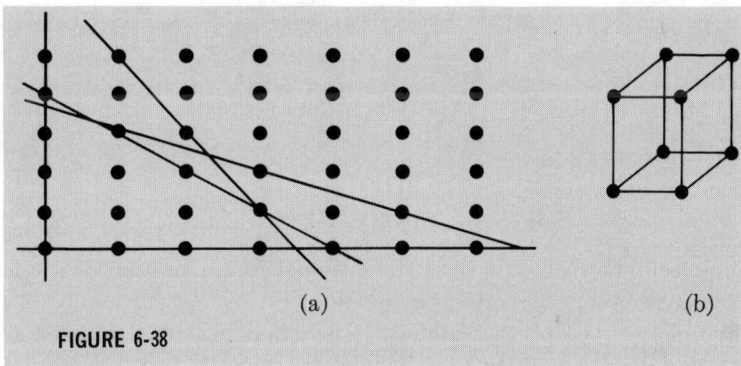

(a)            (b)

FIGURE 6-38

Two-Dimensional Tetragonal Lattice and Unit Cell.

the highest site densities, and if the lattice is extended to three dimensions, the same can be said with regard to the planes parallel to the $z$ axis. Thus, we might expect faces to develop that are mutually perpendicular. These would be the same faces expected in the cubic system, thereby emphasizing the point made earlier concerning the uniqueness of crystal faces. It is, then, the secondary faces that will define the tetragonal system, and from the figure these are seen to form different interfacial angles from those observed in the cubic system.

In Figure 6–38(b), a unit cell is drawn in which the sides of the parallelepiped are parallel to the cartesian axes. Here a unit cell is obtained that shows the full symmetry of the tetragonal system, one tetrad axis, four diad axes, five mirror planes, and a center of symmetry.

## Crystallographic Axes

In designating the crystal faces, it is necessary to define a set of crystal axes. These can be any three straight lines that do not lie in the same plane. In general, it is desirable to define them in such a manner that they give the simplest possible relationship between the faces. This requires that they be parallel to actual or possible edges of the crystal. Since the angular relation between the edges of a crystal is a consequence of the internal arrangement of the lattice points, we can see that the points of the more common faces should lie parallel to the axes.

In our first two examples, an orthogonal set of axes was appropriate. But in the lattice shown in Figure 6–39, this is not the case. In fact, the

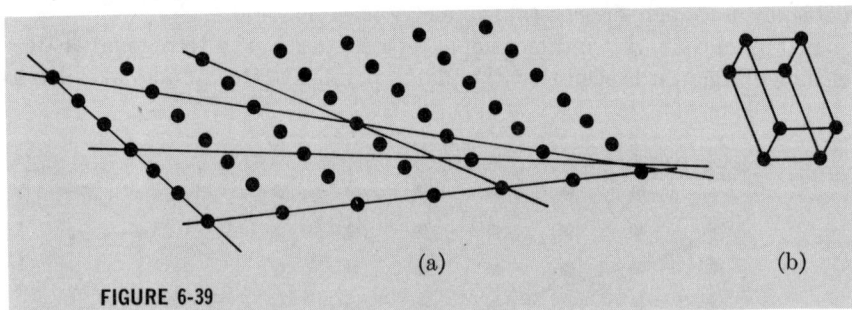

FIGURE 6-39

Two-Dimensional Triclinic Lattice and Unit Cell.

unit cell shown in the three-dimensional representation of the lattice has the symmetry characteristic of the triclinic system; only a center of symmetry. Using the same approach that has been followed with these three lattices, one can draw a lattice compatible with each of the seven crystal systems. Further, it is possible to represent a unit cell in each of these lattices that shows the symmetry of the system. The edges of the parallelepiped defining the unit cell will lie parallel to the respective crystallographic axes. In fact, we now find that the parallelepipeds we have previously used to express the crystal symmetry can become the unit cells.

### The Fourteen Bravais Lattices

Thus far we have considered two types of cubic space lattices. In one instance it was seen that the lattice can be generated by the repetition of a simple unit cube. In the second instance, a compound unit cell was chosen for copper in which a lattice point exists in the center of each face as well as at the corners of the cube. Both of these unit cells show four triad axes and are, therefore, compatible with the cubic system. But there is yet one other cubic space lattice. This is shown along with the resultant cubic unit cell in Figure 6–40. Here there is a lattice point at the center as well as at each corner of the cell. The simple unit cell that was first considered contains only one unit of pattern and is referred to as a primitive unit cell ($P$). The cell with lattice points in each face is called a face-centered cube

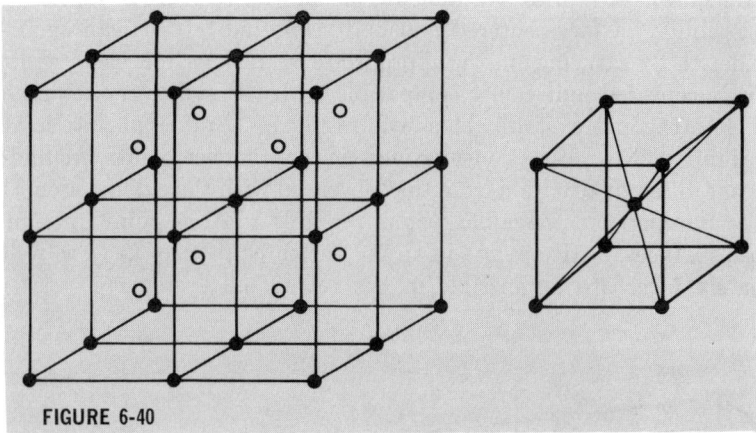

FIGURE 6-40

Space Lattice Based on a Body-Centered Cubic Unit Cell.

(*F*), and the last example is called a body-centered cube (*I*). The three different cubic lattices can be differentiated in terms of their respective unit cells as shown in Figure 6–41.

In each of these lattices, the number of nearest neighbors to any lattice point is different from that of the others. Referring back to Figure 6–35, we can see that any given point in the primitive space lattice is surrounded by six nearest neighbors, and from Figure 6–40 it is seen that each point in the body-centered cubic lattice is surrounded by eight nearest neighbors. Here we should recall that the environment of all the points in a given lattice must be the same. Thus, we note that a point at the corner of a unit cell in a body-centered cubic lattice represents the point at the center of another unit cell in the lattice. Finally, a lattice point in a face-centered cubic lattice has 12 nearest neighbors.

In the tetragonal system, the simplest lattice can be generated by the

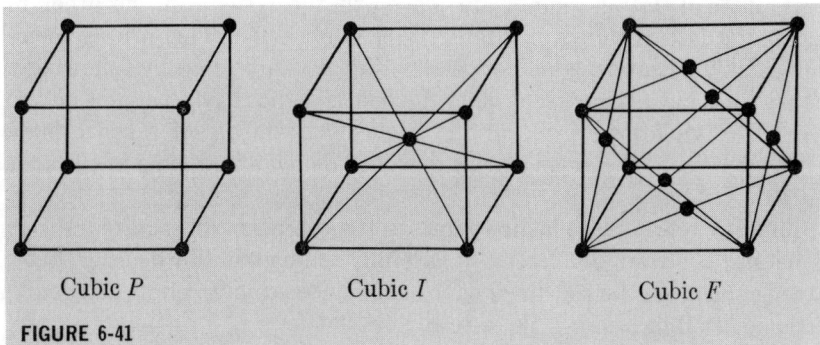

Cubic *P*        Cubic *I*        Cubic *F*

FIGURE 6-41

Three Types of Cubic Unit Cells.

translation of a primitive unit cell as shown in Figure 6–42. However, the possibility of face-centered and body-centered lattices can again be considered. By carrying out the appropriate operations, it is easily shown that a body-centered unit cell is compatible with the symmetry requirements of the tetragonal system. This will result in a different lattice from the primitive. Now, if we consider face-centered lattice points, we find a somewhat different situation from that observed with the cubic system. Whereas the symmetry of the cubic system requires a lattice point to be in each of the six faces, lattice points in any of the three pairs of faces as well as four or six faces are permissible in the tetragonal system.

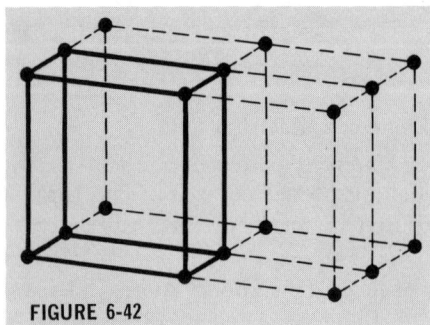

FIGURE 6-42

Primitive Unit Cell in a Tetragonal Space Lattice.

In Figure 6–43 a tetragonal lattice is drawn in which only the end pair of faces is centered. This arrangement is represented by the dotted lines. However, by rotating the $x$ and $y$ axes by 45°, a new unit cell can be drawn that is primitive but smaller than the primitive tetragonal cell shown in Figure 6–42. Consequently, it must be concluded that this lattice can still be described by the repetition of a primitive unit cell and is, therefore, not a new type of lattice.

If, instead, lattice points are placed in the centers of the other four faces, the lattice shown in Figure 6–44 is obtained. This arrangement is compatible with tetragonal symmetry, but it can be seen by following the heavy lines in the figure that all of the points do not have the same environment — a violation of one of the basic requirements of a space lattice. For this reason, the arrangement does not define a new type of tetragonal lattice.

Finally, we can place lattice points in the centers of all six faces. If this is done and the resultant lattice is carefully studied, it can be seen that the arrangement can be redefined as a body-centered unit cell. So, again, this does not lead to a new type of tetragonal lattice.

With these we have exhausted the possible arrangements of lattice points in a tetragonal lattice, and we find that only two distinct lattices

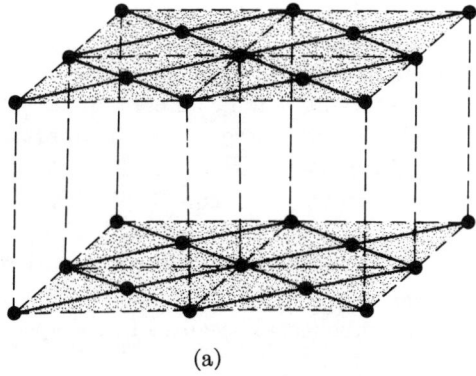

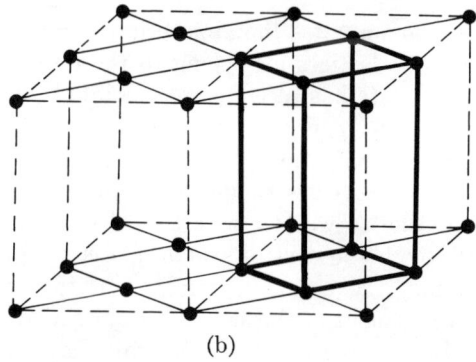

(a)

(b)

**FIGURE 6-43**

Representation of a Tetragonal Space Lattice by either an End-Centered or a Smaller Primitive Unit Cell.

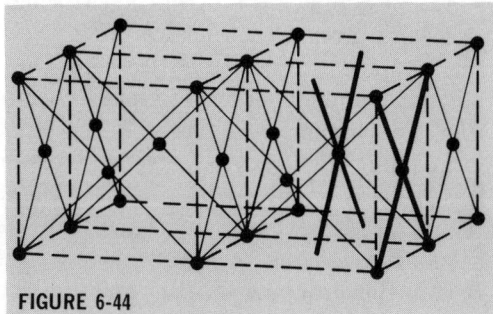

**FIGURE 6-44**

Illustration of the Different Environments of Two Points in a Tetragonal Lattice Based on the Face-Centering of the Four Equivalent Faces of the Unit Cell.

exist in this system. These are a tetragonal $P$ space lattice and a tetragonal $I$ space lattice, corresponding respectively to a primitive tetragonal unit cell and a body-centered tetragonal unit cell.

The same type of reasoning can be extended to the other crystal systems with the result that 14 different space lattices can be drawn for the seven crystal systems. These are called *Bravais lattices* after Auguste Bravais who determined their existence in 1848. The unit cell characteristic of each of these is shown in Figure 6–45. The symbols for the primitive, body-centered, and face-centered lattices have been given earlier. The $C$ space lattice represents a lattice in which only one set of the faces in the unit cell is faced-centered. Although the trigonal unit cell is primitive, it is referred to as rhombohedral because of its shape, and is designated by the letter $R$.

## The Crystal Classes

If one applies the appropriate symmetry operations, it is seen that the Bravais lattices all belong to the normal class of their respective crystal system. It has already been pointed out that the arrangement of lattice points determines which faces in a crystal can develop, and since the arrangements must conform to the Bravais lattices, it would seem that only the normal class should occur in each crystal system. This, then, should lead one to wonder why the 32 crystal classes exist. The answer is that although the crystal system is determined by the arrangement of lattice points, the overall symmetry of the crystal is limited by the specific arrangement of the structural units about a lattice point.

Here it is important to recall that a lattice point does not necessarily represent a single molecule, ion pair, or atom. Rather, we imagine a crystal structure to result from the repetition at regular intervals in space of groupings of molecules, ions, or atoms in a fixed orientation. The smallest such group from which we can generate the crystal is chosen as the unit of pattern, and it is this unit that is represented by a lattice point.

As an example, a 2-fold rotation axis might arise from the grouping of two molecules around each lattice point. It is important to realize that it is the grouping around the lattice point that gives rise to the 2-fold axis and not the symmetry of the molecules themselves. Individually they may be totally asymmetric. Of course, if a molecule does have a 2-fold axis, it can form a unit of pattern in a crystal demanding this symmetry provided that it is properly oriented with respect to the edges of the unit cell. However, it should be clear that a single molecule cannot form a unit of pattern in a lattice belonging to a crystal system of higher symmetry than itself. It is in this grouping of structural units around the lattice points that we find the basis for the 32 crystal classes. Depending on the structural units and how they are arranged, symmetries less than the normal class can occur. These can favor the development of specific faces characteristic of the particular class as well as displaying characteristic symmetry in etch figures, optical properties, etc.

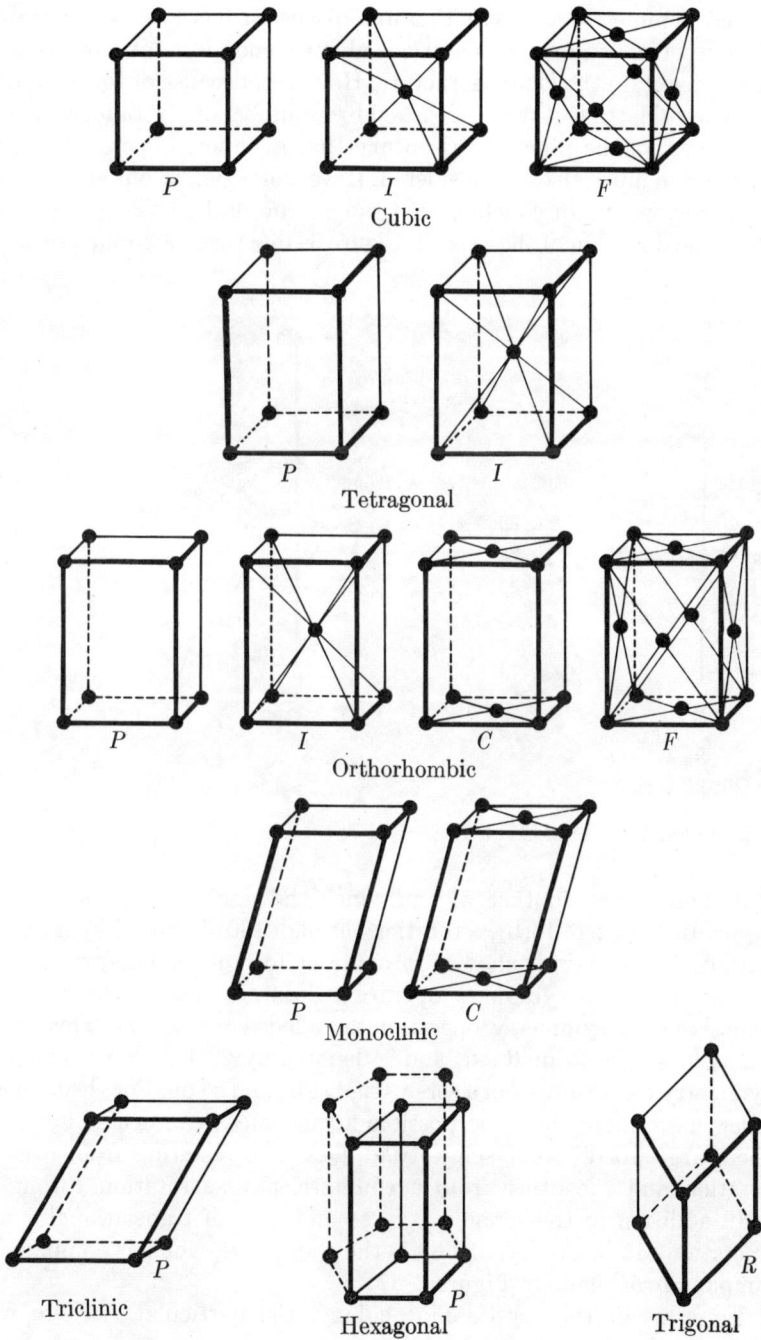

FIGURE 6-45

The 14 Bravais Lattices.

## Space Groups

The symmetry we have thus far discussed has been point symmetry. This involves symmetry operations about a chosen point, not necessarily a lattice point, in the lattice structure. However, because of the spatial nature of a crystal lattice, it was recognized by a number of crystallographers in the latter part of the nineteenth century that symmetry operations involving translation must also be considered. If we look again at our crystal having a diad axis, we might conclude that some structural unit is arranged in pairs along the direction of the axis. In Figure 6–46(a) such a 2-fold axis is shown.

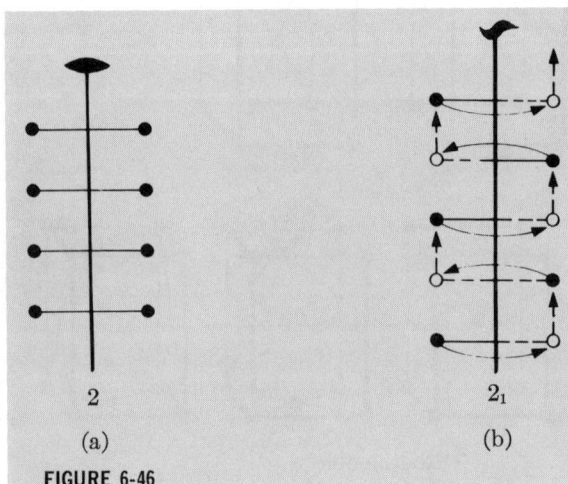

2          $2_1$

(a)          (b)

**FIGURE 6-46**

A 2-Fold and a 2-Fold Screw Axis.

But there is yet another way in which the same external symmetry can occur. In Figure 6–46(b), a rotation through 180° followed by a translation parallel to the axis is shown. In contrast to point symmetry, continuous repetition of this symmetry operation does not return us to the original point. Such a symmetry operation is called a *screw axis*. This particular example is a screw diad axis and is denoted by $2_1$. In a like manner, 3-fold symmetry can result from a screw triad axis, and so on. For the higher-order screw axes, more than one possible arrangement can occur. For instance, there are two different screw triad axes, one resulting from a clockwise rotation and the other from a counterclockwise rotation through 120°.

In addition to the screw axis, a second type of translational symmetry operation can be observed. This is the *glide plane*, which is compared to the simple mirror plane in Figure 6–47.

In terms of the crystal morphology, the particular class to which a crystal belongs is determined by the symmetry of the arrangement of structural units about each point in the appropriate Bravais lattice. As an

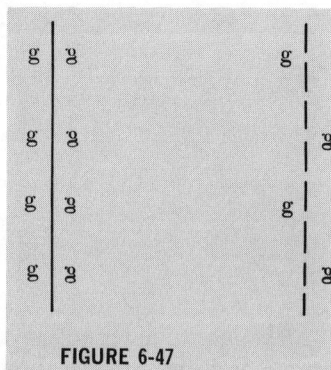

FIGURE 6-47

A Simple Symmetry Plane and
a Glide Plane.

example, in the monoclinic system there are two Bravais lattices, the primitive and the $C$ face-centered. As do all Bravais lattices, these show the full symmetry of the corresponding normal class. Now if the symmetry of the grouping of structural units around each lattice point in either of these lattices is such that a mirror plane and a perpendicular diad axis are present, the crystal will be in the normal class of the monoclinic system, $2/m$. On the other hand, if the point symmetry is such that only a diad axis is present, the crystal will belong to class 2. And, of course, if the point symmetry is such that only a mirror plane is present, the crystal class will be $m$.

Carrying this treatment to all seven crystal systems, we can see the source of the 32 crystal classes. But at the same time, we find that this external symmetry can result from a number of different internal arrangements. In the case of the monoclinic system, two different Bravais lattices were seen to give the same crystal morphology. Going a step further, we note that class $m$ may result from either a true mirror plane or a glide plane. Likewise, class 2 may result from either a true diad axis or a screw diad axis. Finally, in the normal class the various combinations of diad and screw diad axes with mirror and glide planes must be considered.

We are now in a position to determine all of the internal arrangements of structural units that can occur. This is done by associating the symmetry elements of the various crystal classes with each point in the appropriate Bravais lattice, and recognizing the existence of screw axes and glide planes. When this is done 230 different arrangements are found. These are the 230 *space groups*. Relative to the 32 point groups we have discussed in such detail, the greater complexity in the space groups arises primarily from the extension to a space lattice by means of the screw axis and the glide plane.

**TABLE 6–3.  Space Groups in the Monoclinic System.**

| Monoclinic | | | | | |
|---|---|---|---|---|---|
| Class 2 | | Class $m$ | | Class $2/m$ | |
| $P2$ | $C2$ | $Pm$ | $Cm$ | $P2/m$<br>$P2_1/m$ | $C2/m$ |
| $P2_1$ | | $Pc$ | $Cc$ | $P2/c$<br>$P2_1/c$ | $C2/c$ |

Although we shall not go through the development of the space groups here, it is helpful to see the number of combinations that can occur. The 13 space groups in the monoclinic system are given in Table 6–3. The first letter in the symbol refers to the type of Bravais lattice. In this case it is either primitive or $C$ face-centered. The symbols 2 and $m$ are the usual Hermann-Maugin symbols for a diad axis and a mirror plane respectively. The $2_1$ symbolizes a screw diad axis, and the lower case $c$ symbolizes a glide plane.

Of course, crystal morphology, of itself, does not tell us the space group to which a crystal belongs. Yet the space group is necessary for a complete description of the crystal structure. In a few limited cases, the early crystallographers were able to guess at a probable space lattice, but it was not until the development of X-ray techniques that unambiguous assignments were possible.

## Miller Indices

Although the importance of the planes in a crystal lattice have been emphasized, we have not yet described the means of representing them. Historically, there have been a number of systems devised for this purpose, but presently only the method devised by Miller is commonly used, and the planes are designated in terms of *Miller indices*. The representation of the crystal planes requires first that a set of crystallographic axes be chosen. These were defined earlier as any three straight lines that do not lie in the same plane, but for convenience, they should lie parallel to possible edges of the crystal. Secondly, it is necessary to choose a *parametral plane*. This is any plane parallel to one crystal face and intersecting each of the crystallographic axes. This will be the reference plane. From the definition, it is obvious that the choice of a parametral plane is arbitrary. Consequently, more than one set of Miller indices may be given for a particular plane, depending on the choice of a parametral plane.

In Figure 6–48(a), the intercepts of the parametral plane with the crystallographic axes are all equal, and the intercepts are arbitrarily chosen to be $a$, $b$, and $c$, along the $x$, $y$, and $z$ axes respectively. We can

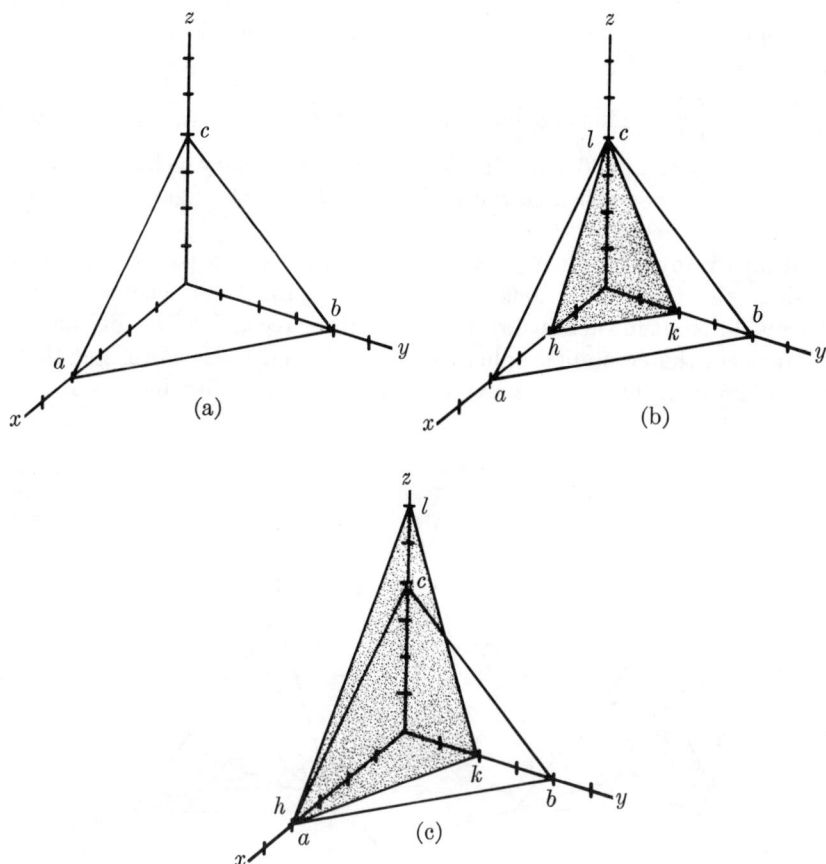

**FIGURE 6-48**

Miller Indices Based on a Chosen Parametral Plane.

define the Miller indices of any plane in terms of our chosen parametral plane as $\dfrac{a}{h}, \dfrac{b}{k}, \dfrac{c}{l}$ reduced to the simplest whole numbers. The symbols $h$, $k$, and $l$ represent the intercepts of the plane in question with the respective $x$, $y$, and $z$ axes relative to the intercepts of the parametral plane. The parametral plane will thus have the Miller indices $\dfrac{a}{a}\dfrac{b}{b}\dfrac{c}{c}$ or 111. These, of course, will always be the indices of the parametral plane because the intercepts are always chosen to be $a$, $b$, $c$.

In Figure 6–48(b), a plane $hkl$ is shown along with the same parametral plane we have just described. If, for convenience, the magnitudes of $a$, $b$, and $c$ are each set equal to unity, the Miller indices will be $\dfrac{a}{h}\dfrac{b}{k}\dfrac{c}{l}$ or $\dfrac{1}{\frac{1}{2}}\dfrac{1}{\frac{1}{2}}\dfrac{1}{1}$.

On clearing fractions, this becomes 221. It is here that we can see why the intercepts $hkl$ are referred to as reciprocal intercepts. Finally, as a third example, we can consider the plane shown in Figure 6–48(c). Its indices will be $\frac{a}{h} \frac{b}{k} \frac{c}{l}$ or $\frac{1}{1} \frac{1}{\frac{1}{2}} \frac{1}{\frac{3}{2}}$. This can be rearranged to give 1, 2, $\frac{2}{3}$, and on clearing fractions, it becomes 362. In practice, it is not customary to have indices as large as 6, but nevertheless, the plane indicated in the figure would be so designated.

To illustrate the arbitrary nature of our choice of a parametral plane, consider the three situations depicted in Figure 6–49. The parametral plane is different in each instance, but it can be recognized readily because it intersects the crystallographic axes at $a$, $b$, and $c$. Here we find that a particular plane $hkl$ will be defined by different Miller indices in each

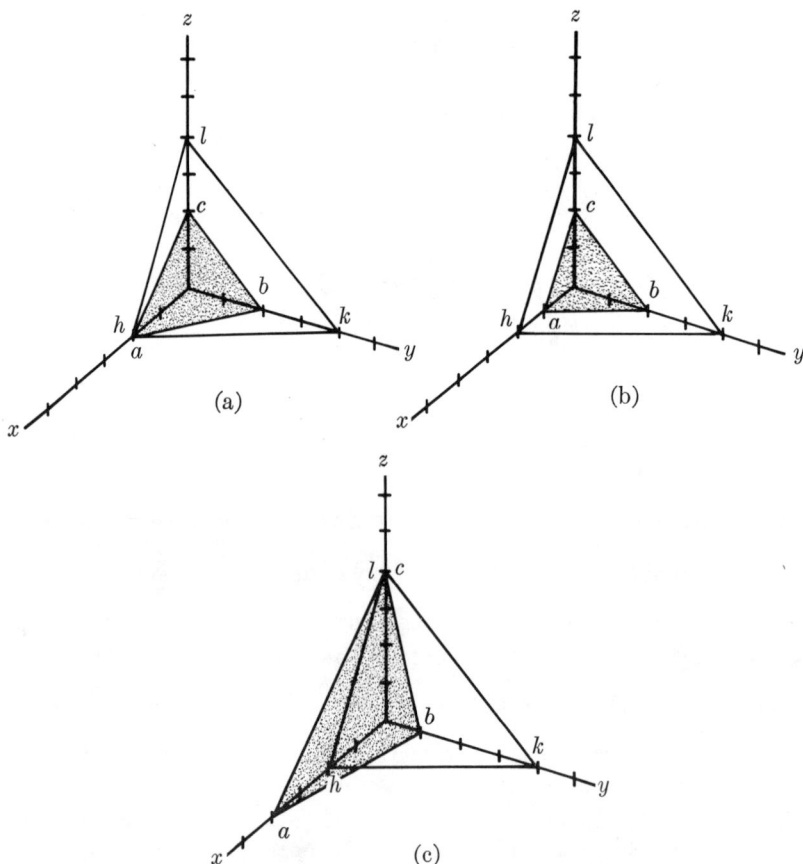

FIGURE 6-49

Miller Indices for a Given Plane Based on Different Parametral Planes.

instance as a consequence of a different choice of a parametral plane. Thus, for the three planes we have,

(a) $\dfrac{a}{h}\dfrac{b}{k}\dfrac{c}{l}$   or   $\dfrac{1}{1}\dfrac{1}{2}\dfrac{1}{2} \equiv 211$

(b) $\dfrac{a}{h}\dfrac{b}{k}\dfrac{c}{l}$   or   $\dfrac{1}{2}\dfrac{2}{4}\dfrac{2}{4} \equiv 222$  Reduced to the simplest whole numbers, this becomes 111.

(c) $\dfrac{a}{h}\dfrac{b}{k}\dfrac{c}{l}$   or   $\dfrac{1}{\frac{1}{2}}\dfrac{1}{4}\dfrac{1}{1} \equiv 814$

Of particular importance, the plane in Figure 6–49(b) is parallel to the parametral plane and will, therefore, have the same indices as the parametral plane, 111.

It is not uncommon that a face is parallel to one or two of the crystallographic axes. For instance, this will generally occur when we consider the cubic faces of a crystal in this system, as shown in Figure 6–50. The para-

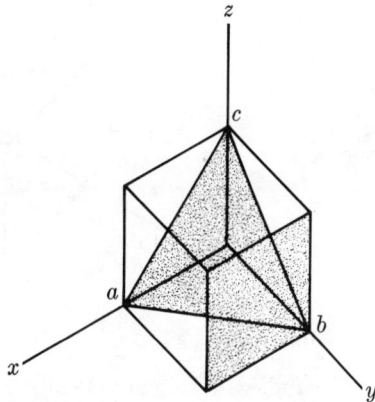

FIGURE 6-50

Miller Indices of the Cubic Faces.

metral plane is shown here in the interior of the cube. Now consider the indices of the plane parallel to the $xz$ plane. It can be seen that the plane intersects the $y$ axis at $b$, coincident with the intercept of the parametral plane on this axis. However, the plane is parallel to both the $x$ axis and the $z$ axis and, therefore, does not intercept these axes at all. Thus, the intercept is considered to be at infinity. Representing the indices in the usual manner, we thus obtain

$\dfrac{a}{h}\dfrac{b}{k}\dfrac{c}{l}$   or   $\dfrac{1}{\infty}\dfrac{1}{1}\dfrac{1}{\infty} \equiv 010$

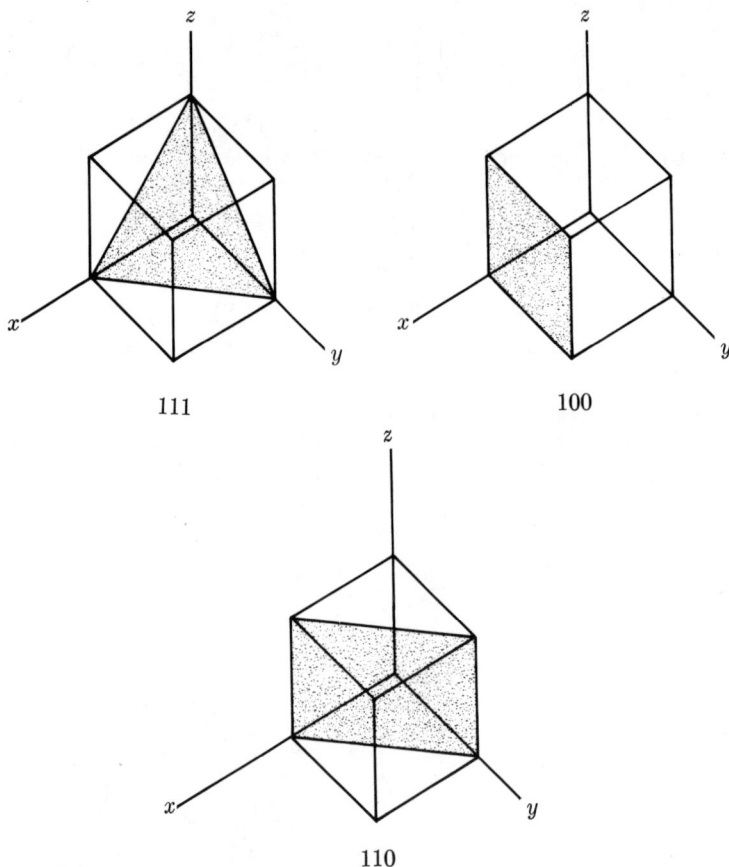

FIGURE 6-51

Miller Indices of Common Planes in a Cube.

To further illustrate the point, several common planes in the cubic system are shown in Figure 6–51.

We are now in a position to represent the specific faces that might show up in an actual or ideal crystal. In Figure 6–52, a geometrical structure is presented which has been studied in some detail earlier. The appropriate indices are placed on the faces, with the $(-)$ above an index merely indicating the sign of the axis in that region. Thus, we find the cubic faces being of the form $\{100\}$, the octahedral faces to be of the form $\{111\}$, and the dodecahedral faces to be of the form $\{110\}$. It might be pointed out, however, that these particular indices are a consequence of the most reasonable choice of a parametral plane. If an alternative plane were chosen, the indices would also be different.

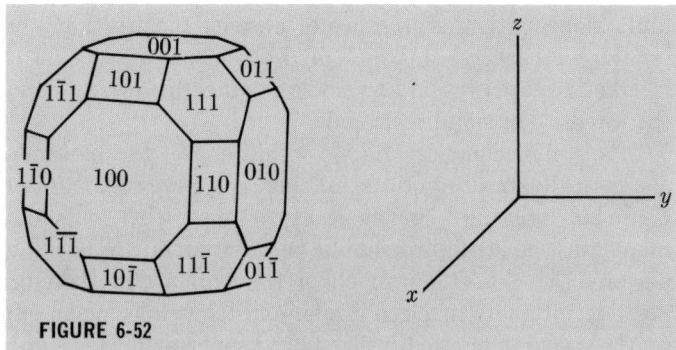

FIGURE 6-52

Miller Indices of the Cubic, Octahedral, and Dodecahedral Faces of a Crystal in the Cubic System.

## CLASSIFICATION BASED ON BOND TYPE

Although it is possible to classify crystals in terms of the crystal symmetry, this tells us very little about the chemical and physical properties of a given crystalline material. Such properties as solubility, crystal energy, heat of fusion, conductance, etc., are of particular concern, and a classification that permits us to understand these properties is needed. Although there are limitations to the method, a classification based on bond type seems to be most useful.

The difficulties with this means of classification are twofold. First, we have seen that a given bond is rarely, if ever, of a single type. If it is covalent, it has some ionic character, and if it is ionic it has some covalent character. Superimposed upon either of these is the van der Waals interaction. The second problem is due to the fact that two or more different types of forces may simultaneously be present in the crystal. For instance, $Na_2SO_4$ may be considered to form an ionic crystal, but covalent bonding is certainly present in the sulfate ion. Or in graphite, the carbon atoms in each plane are held together by covalent bonds, but the planes are held to each other by van der Waals forces.

However, in spite of the difficulties, the properties of most crystals are found to conform to one of the four general types of chemical bonds, in terms of which it is possible to give a sufficiently unique classification of crystals to justify the approach. On this basis we can then consider four crystal types: ionic, covalent, molecular, and metallic.

### Molecular Crystals

Molecular crystals are those in which the crystalline state is composed of an aggregate of discrete molecules held together by van der Waals forces. The uniqueness in properties and structure of these crystals can be understood in terms of the weak nature of the interactions between structural units along with the presence or absence of directional character. Except for those structures in which hydrogen bonding occurs and where dipole interactions are present at very low temperatures, we can generally assume the forces to be nondirectional.

In any molecular crystal we would expect the molecules to pack in the energetically most favorable manner. But if we restrict our considerations to nonpolar molecules, or at least to ones with sufficiently small dipole moments, the structure should be determined by packing efficiency. This permits the closest approach of the atoms and therefore the maximum possible interaction. Here the simplest possible model can be represented by the stacking of equally sized hard spheres. This is a good representation of the solid state of the inert gases, many of the metals, and some molecules, such as $CH_4$, that approximate a spherical structure. Additionally, it is found that nonpolar molecules that are not spherical tend to stack in a similar manner, but the structures distort in accordance with their geometries.

Using spheres for illustrative purposes, the most efficient packing gives an arrangement that is referred to as a *closest-packed* structure. In such an arrangement, each sphere is in contact with the maximum possible number of nearest neighbors. In Figure 6–53 a closest-packed layer of spheres is

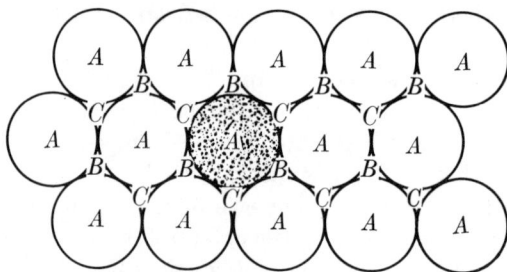

**FIGURE 6-53**

Two-Dimensional Closest Packing of Spheres.

shown. Each sphere is surrounded by six nearest neighbors lying in the plane. The symmetry around any sphere in the plane is seen to be hexagonal and in the symmetry class $6/mm$. Further, it can be seen that two different types of *voids* exist, each having a symmetry $3/mm$ but pointing in opposite directions.

We can now define three distinct locations in our layer of spheres. These have been designated $A$, $B$, and $C$; the $B$ and $C$ designating the two different types of voids. If a second layer of closest-packed spheres is placed upon the first layer, the spheres of the second layer can occupy the region above either the $B$ voids or the $C$ voids. But because of the size of the spheres, both types of voids cannot be occupied simultaneously.

No immediate significance can be placed on the arrangement of the second layer of spheres. It is in the third layer that different stacking patterns can be realized. In the third layer, the regions above either one or the other of two different sets of voids can still be occupied, but they now have different surroundings. If, for the sake of illustration, we choose to place the spheres of the second layer in $B$ sites, one of the available sets of voids for the third layer will be directly above the spheres in the original layer. These are $A$ sites. The other set of voids will be directly above the voids designated by $C$ in the original layer. If the spheres of the third layer are placed in the $A$ sites, the order will be $ABA$ .... If, on the other hand, they are placed in the $C$ sites, the order will be $ABC$ .... Thus we see that different packing arrangements can occur in a closest-packed structure. Quite obviously, the addition of more layers can lead to a variety of patterns such as $ABCACB...$, $ABAC...$, etc. But the most symmetrical arrangements, and the ones in which we shall be most interested, are $ABABAB...$, and $ABCABCABC...$, where the original stacking pattern is continued. For many of the more common substances that form closest-packed structures, one or the other of these arrangements is observed. However, examples of the more complex packings are well-known.

If we consider the stacking of layers in the order $ABABAB...$, it should readily be apparent that our simple 6-fold symmetry has been destroyed. At worst we are limited to a simple 3-fold axis. This can be seen from Figure 6–54 to result from the fact that an axis through a given sphere in one layer must pass through a void in an adjacent layer. We can see from the geometry

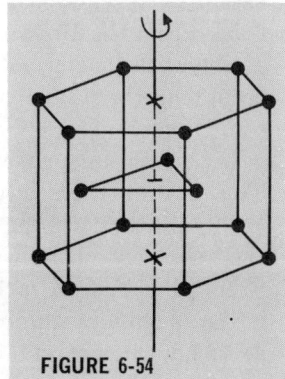

**FIGURE 6-54**

Symmetry about an Axis Passing through Three Planes of Closest-Packed Spheres.

of the voids that the arrangement of spheres around each void will have trigonal symmetry along this axis. However, the scheme $ABAB\ldots$ will have a symmetry considerably greater than a simple triad axis. If we consider the presence of a hexad screw axis and a glide plane, the arrangement falls into the space group $P\dfrac{6_3}{m}mc$, where the $6_3$ represents the particular hexad screw axis that occurs here. This is one that rotates through $180°$ before translation. Because of its symmetry, this particular type of closest-packed arrangement is referred to as a *hexagonal closest-packed* structure (*hcp*) and is illustrated in Figure 6–55.

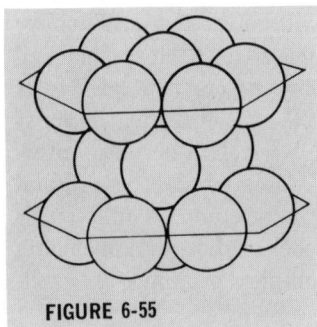

**FIGURE 6-55**

Stacking    Arrangement    in Hexagonal Closest Packing of Spheres.

The packing scheme $ABCABC\ldots$ is unique among the closest-packed structures in that it shows cubic symmetry. Consequently, it is referred to as a *cubic closest-packed* (*ccp*) structure. By shifting the crystallographic axes, a face-centered cubic unit cell can be obtained as shown in Figure 6–56. From the Bravais lattices, we have seen that this structure is compatible with the normal class of the cubic system. Thus, with single spheres occupying the lattice sites, we would expect and we find the space group to be $Fm3m$.

In both closest-packed structures, a given sphere has 12 nearest neighbors. This is best appreciated by considering the six nearest neighbors that are in physical contact with the reference sphere in the plane. We can then see that there will be three spheres in physical contact with the reference sphere in each of the adjacent layers. Twelve nearest neighbors is the maximum number that can exist in an ordered arrangement such as we have in a crystal. This represents the most efficient means of packing equal-sized spheres, resulting in $74\%$ of the available space being filled.

Because of the spherical geometry of the inert gas atoms, one would expect a closest-packed structure for the solid states of these elements, and this is found to be the case. Helium forms a hexagonal closest-packed struc-

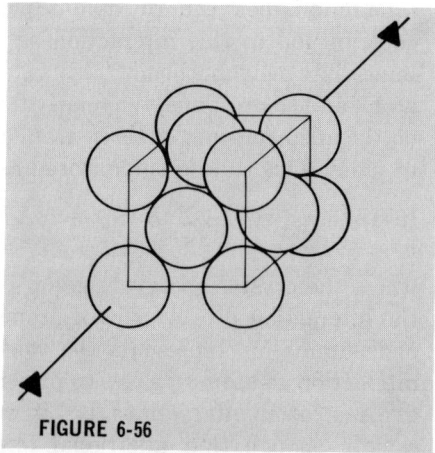

FIGURE 6-56

Stacking Arrangement in Cubic Closest
Packing of Spheres.

ture, and the remaining inert gases form cubic closest-packed structures. In
order to understand this preference by the inert gases, a number of theoretical calculations of the relative stability of the *ccp* to the *hcp* structure
have been made. One of the more conventional approaches is to sum up the
pair-wise interactions of nearest neighbors. This is analogous to the approach we used in the calculation of an ionic crystal lattice energy. There
we determined the interaction between one cation and one anion and then
multiplied by a geometrical factor called the Madelung constant. Using
this approach for the inert gas crystals, the *hcp* structure is consistently
found to be more stable — in contradiction to the experimental facts.

Inasmuch as the prevalence of the *ccp* structure for the inert gases
cannot be understood in terms of pair-wise interactions of spherically
symmetric particles, two alternative approaches have been considered.
In the first of these, it is assumed that the charge distribution is not
spherically symmetric. This approach has been used by Cuthbert and
Linnett[2] where they propose that the eight electrons in the outer shell
of an inert gas atom are distributed in pairs in an $sp^3$ hybrid set of orbitals.
This, then, gives a tetrahedral distribution of charge.

The treatment by Cuthbert and Linnett has been attacked by Jansen[3]
for a number of reasons, including the lack of quantitative calculations.
At the same time, Jansen has approached the problem along quite different
lines. In the previous treatments only two-atom interactions were considered, but Jansen has attempted to show than an additional contribution
to the crystal energy arises from three-atom interactions, and, in principle,
from *n*-atom interactions. That is, the electron distribution in one atom will
affect and be affected by the electron distributions in the surrounding
atoms. Certainly the major contribution to the energy arises from the

interactions between any two adjacent atoms, but there is an additional contribution to this interaction as a consequence of a third atom in the immediate neighborhood. This is, in fact, true for a fourth atom, fifth atom, and so on. Thus, the relative stabilities of two crystal lattices depend on the arrangements of three-atom clusters as well as on the arrangements of pairs. This can be summarized as

$$(6\text{-}1) \qquad U = \sum_{i<j} E\,(r_{ij}) + \sum_{i<j<k} E(r_{ij}, r_{ik}, r_{jk}) + \cdots$$

where the crystal energy is given by $U$, the first summation represents the interaction of any pair of atoms $i$ and $j$ separated by $r_{ij}$, and the second summation represents the contribution to the crystal energy of the interaction of atom triplets. In principle, the series should continue to four, five, etc., atom interactions, but it would appear that the series converges rapidly enough that additional terms will not be too important. Using quantum mechanical calculations based on his model of three-body interactions, Jansen has shown, at least qualitatively, that the *ccp* structure is energetically favored for the higher inert gas crystals.

It is often found that nonspherical molecular compounds also tend to crystallize in a closest-packed structure, but due to their geometries, the crystals will be distorted from the simple hexagonal or cubic symmetry. At the same time, some of these substances will also show a modification, frequently at higher temperatures, in which the distortion is removed. As seen in Figure 6–57, the crystal structure of both $Br_2$ and $I_2$ can be understood in terms of a distortion of a *fcc* structure. In these cases, the crystals show orthorhombic symmetry. Now $N_2$ might be expected to show similar behavior, but it is found that a cubic modification of $N_2$ exists below 35°K which transforms to a *hcp* form above this temperature. The low-temperature modification of $N_2$ is not a closest-packed structure, but the

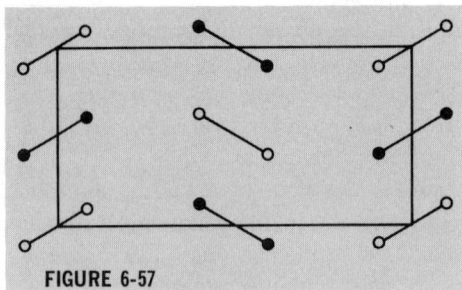

FIGURE 6-57

Distorted Closest-Packed Arrangement of $Br_2$ and $I_2$ Molecules in the Solid State. The open circles represent atoms in the front face and the solid circles represent atoms halfway through the face-centered structural unit.

$N_2$ molecules are located at the same positions as one would find in a *fcc* cell. The unit cell is considered primitive, but there are four $N_2$ molecules per unit cell.[4] Thus we must conclude that each lattice point in the primitive cell represents an arrangement of several $N_2$ molecules about it.

At higher temperatures, the $N_2$ crystal structure changes from the low-temperature $\alpha$-form to the higher-temperature $\beta$-form, which is *hcp*. Being closest-packed, the latter form requires effectively a spherical shape for the molecule and is attributed to a rotation of the $N_2$ molecule. The presence of rotating molecules is also used to explain the crystal structures of HCl, HBr, HI, $H_2S$, $H_2Se$, $CH_4$, and $SiH_4$. The low-temperature structure of $\alpha$-$N_2$ is attributed to quadrupole interactions between adjacent $N_2$ molecules.[5] Thermal motion at the higher temperatures tends to break up the quadrupole interaction, permitting the free rotation. Interpretations of these structures is yet very qualitative, and it is likely that more sophisticated treatments will also consider three-body type interactions.

We have considered only extremely simple examples of molecular crystals. More complex crystals may still tend to pack in an analogous manner. However, a point must be reached where molecular geometry will not permit even a distortion of a closest-packed structure. Benzene is geometrically a more complex molecule than those we have considered. Yet it tends to pack in a *fcc* structure as shown in Figure 6–58. It is easy

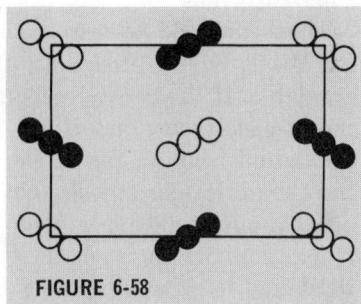

**FIGURE 6-58**

Distorted Closest-Packed Arrangement of Benzene Molecules in the Solid State. The open circles represent atoms in the front face.

to see here why crystalline benzene shows orthorhombic symmetry rather than the hexagonal symmetry of the individual molecule.

Proceeding to an example of a very complicated molecule in which molecular forces are important, we can consider the structure of graphite. Here the carbon atoms in a given plane are held by $sp^2$ hybrid bonds. Since an $sp^2$ hybrid set of orbitals is triangular planar, the covalently bonded portions of the crystal form large sheets, which are held together by van der Waals forces. This is clearly indicated in Figure 6–59, where the C--C

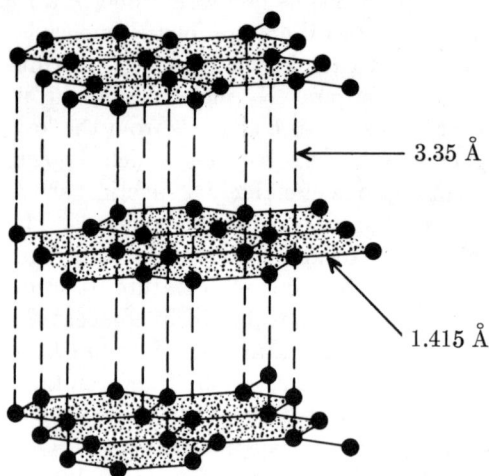

**FIGURE 6-59**

Layer Structure of Graphite Showing the Presence of Different Bond Types.

distance in a given layer is seen to be 1.415 Å, whereas it is 3.35 Å between adjacent layers.

Thus, it should be reasonable to say that in terms of the nature of van der Waals-type interactions, the properties of these crystals are understandable. If the geometry permits, the crystal structures will tend to be closest-packed or distortions of closest-packed arrangements, melting points and heats of fusion should be low, crystals will be relatively soft, and the optical spectra are changed very little from the gas phase. Additionally, one would expect molecular crystals to be electrical nonconductors. Where conduction is observed, as is the case with graphite, it can be attributed to something other than the molecular forces.

## Ionic Crystals

By definition we would consider an ionic crystal one in which the structural units are held in position by electrostatic forces. However, more realistically, the nature of the crystal is considered in terms of the success of the model. That is, if an ionic model permits one to calculate the properties of a crystal, then the crystal can be considered ionic. In Chapter 5, crystal energies were discussed in terms of an electrostatic approach. Here it was seen that an ionic model is reasonably adequate for the alkali halides. Of course, this is where we should expect the best possible results, but even with the alkali halides our general feeling for chemical bonding tells us that an ionic model should not be completely satisfactory. Certainly van der Waals interactions must be present and possibly quite significant in a compound such as LiI, where a very small cation and a highly polarizable anion are present.

In addition to the ionic and van der Waals terms that must be present, one cannot help but wonder at the importance of covalent contributions in supposedly ionic crystals. From a pragmatic point of view, we can see that crystal energies can be calculated for the alkali halides without introduction of covalent terms. However, the question is certainly meaningful in the cases of many other less obviously "ionic" crystals. Assuming the validity of the ionic model as expressed by Eq. (5–6), this point is illustrated by the data of Brackett and Brackett[6] shown in Table 6–4. Here they have

**TABLE 6–4. Comparison of Calculated and Experimental Lattice Energies of the Alkaline Earth Halides.[a]**

| Salt | $U_{0\,calc}$ (kcal/mole) | $U_{0\,exp}$ (kcal/mole) | $\Delta$ (kcal/mole) |
|---|---|---|---|
| $MgF_2$ | 696.4 | 702.3 | 5.9 |
| $MgCl_2$ | crystal structure questionable | | |
| $MgBr_2$ | 501.2 | 573 | 71.8 |
| $MgI_2$ | 464.6 | 547.1 | 82.5 |
| $CaF_2$ | 623.7 | 623.4 | −0.3 |
| $CaCl_2$ | 531.4 | 532.1 | 0.7 |
| $CaBr_2$ | 509.6 | 509.4 | −0.2 |
| $CaI_2$ | 455.4 | 487.4 | 32.0 |
| $SrF_2$ | 592.0 | 591.6 | −0.4 |
| $SrCl_2$ | 508.5 | 508 3 | −0.2 |
| $SrBr_2$ | 480.0 | 487.4 | 7.4 |
| $SrI_2$ | | 463.6 | |
| $BaF_2$ | 559.6 | 557.1 | −2.5 |
| $BaCl_2$ | 488.5 | 484.8 | −3.7 |
| $BaBr_2$ | 466.1 | 465.6 | −0.5 |
| $BaI_2$ | 437.7 | 441.0 | 3.3 |

[a] Taken from T. E. Brackett and E. B. Brackett, *J. Phys. Chem.*, **69**, 3611 (1965).

calculated the lattice energies of the alkaline earth halides and compared the results with the corresponding lattice energies determined by the Born-Haber cycle. If the experimental data is considered accurate, the majority of the alkaline earth halides also conform to an ionic representation. However, the rather large deviations observed for $MgBr_2$, $MgI_2$, and $CaI_2$ can most reasonably be attributed to the presence of covalent character in the bonds.

In spite of the success of the ionic model for the alkali and most of the alkaline earth halides, the presence of covalent character in these bonds has been investigated. In general, if a chemical bond can be considered to have covalent character, it should be treated quantum mechanically. Thus, the wave function should be expressed in the form

$$\psi = \psi_{ionic} + \lambda\psi_{cov}$$

And if the crystal is primarily ionic, the ionic term will make the major contribution to the wave function. This is the approach that has recently been discussed by Slater.[7] Slater's emphasis has been on the meaning of atomic and ionic radii, but his interpretation depends on a meaningful contribution of the covalent term to the crystal structure of even the alkali halides.

In Chapter 4, Slater's atomic radii were presented along with his quantum mechanical interpretation of atomic and ionic radii. If we accept the validity of these atomic radii, then a covalent mechanism can be proposed for the formation of a so-called ionic crystal. To contrast the two mechanisms, the conventional view of the formation of an ionic crystal permits the ions at infinite distance of separation having zero kinetic energy to come together to their equilibrium positions in the crystal as a consequence of the electrostatic interactions. This, of course, is expressed in terms of the conventional potential energy diagram shown in Figure 5–1. In Slater's view, the neutral atoms are envisioned as coming together from an infinite distance of separation. However, in this case the energy will not show the characteristic decrease on approach until the charge distributions actually begin to overlap. Using KCl as an example, with the atomic radius of potassium equal to 2.20 Å and that of chlorine equal to 1.00 Å, Slater points out that the electron distribution between the atoms will be in the region where the outer wave functions of both atoms are large. Since this region will correspond to the atomic radii, it will be much closer to the chlorine atom than to the potassium atom. Inasmuch as each potassium atom is surrounded by six chlorine atoms, each potassium atom can be considered to donate one-sixth of an electron to each chlorine atom. But since each chlorine atom is surrounded by six potassium atoms, each chlorine atom receives a total average of one electron. Thus, by considering this strictly atomic-type model, an electron has been effectively placed on each chlorine atom and removed from each potassium atom.

Slater maintains that the ionic contribution to the wave function still is of primary importance, but he considers that the covalent part is of sufficient importance to determine the distance of separation of the atoms in the crystal. This is an interesting approach to bonding in crystals that are conventionally considered ionic. Its primary justification is that it rationalizes the success Slater has had with the additivity of atomic radii. But it also emphasizes the versatility of the quantum mechanical approach, whereby any extent of ionic character can be introduced into the wave function. However, at the same time it leads us into the potential trap of using the mixing parameter $\lambda$ as nothing more than an adjustable parameter capable of giving a correct answer once the answer has already been experimentally determined. Thus, in addition to the ease of calculation, there is still something to be said for the simple electrostatic approach.

## Rock Salt and CsCl Structures

Of the almost infinite variety of possible crystal structures, only four are of particular importance for simple ionic compounds, although a much greater variability can be expected if we include the borderline crystals. The simplest example of this type of crystal arises from compounds with the general formula MX. Here there are equal numbers of positive and negative ions, and the coordination number of the cation is the same as that of the anion. In crystals characterized by the formula $MX_2$, the coordination number of the anion is one-half that of the cation, and the reverse is true for compounds of the type $M_2X$. Most ionic compounds fall into one of these categories, and the crystal structures, of course, have to be consistent with the coordination numbers.

The conventional representatives of the MX-type ionic compounds are the alkali halides. Because of the spherical nature of the ions and the nondirectional character of the ionic bond, we would expect quite regular structures to occur. Thus, we might expect the coordination number to be determined strictly by the geometrical consideration of how many ions of one kind can be packed around one of the other kind. The crystal structure would then be expected to be the regular geometrical arrangement characteristic of the particular coordination number. We shall see that other factors must be considered, but nevertheless this is a reasonable starting point for this type of crystal.

If we consider only the first layer of a closest-packed arrangement, it is apparent that a balance of charge will not permit closest packing in an ionic crystal. Thus, if a given positive ion is surrounded by six negative ions as nearest neighbors, a negative ion cannot then have six positive ions as nearest neighbors. Rather, in order to maintain a balance of charge, the coordination number will be limited by the number of larger ions that can pack around the smaller ion. Except for RbF and CsF, the anion of an alkali halide will always be larger than the cation. Thus, it is the number of negative ions that can surround the positive ion that determines the coordination number. This, then, implies that the relative sizes of the ions tend to determine the coordination number and, therefore, the crystal structure.

The alkali halides tend to crystallize in either the *rock salt* (NaCl) or the CsCl structure, with most showing the rock salt structure. A section of the NaCl structure is shown in Figure 6–60, where each ion is seen to be surrounded octahedrally by six ions of the opposite type. Either set of ions forms a face-centered cubic lattice, and the crystal can be thought of as being composed of two interlocking face-centered lattices. The structural unit then contains four cations and four anions. Recalling that a face-

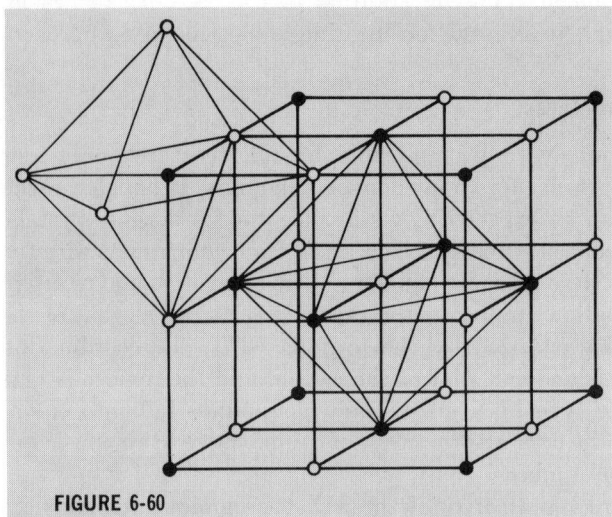

FIGURE 6-60

Rock Salt Structure Showing the Octahedral Distribution
of Opposite Charges about Each Ion Type.

centered cubic lattice is consistent with the normal class of the cubic system
and noting that each ion is a simple spherical structure, we would expect
the simple ionic compounds crystallizing in the rock salt lattice to be in
the symmetry class $m3m$. Typical examples are most of the alkali halides,
the oxides, sulfides, selenides, and tellurides of the alkaline earth metals,
and the fluoride and chloride of silver.

The only alkali halides that show the CsCl structure at normal conditions
of temperature and pressure are the chloride, bromide, and iodide of
cesium. The corresponding halides of rubidium also show this structure
under high pressures, as do a number of thallium compounds at normal
conditions. From the structural unit shown in Figure 6–61, each ion is
seen to be surrounded by eight of the opposite type. This, then, gives a
simple cubic lattice in each ion type. Again, for simple ionic compounds
showing this type of lattice, the symmetry class $m3m$ would be expected.

Herein we frequently find a problem in terminology. From the structural
unit shown in Figure 6–61, we may be prone to imagine a lattice site at
each corner of the cube and one in the center and say that this represents
a body-centered cubic unit cell. But this is not the case, because the identi-
ties of the lattice points are not the same. More appropriately, this type of
structure can be derived from the interlocking of two simple cubic lattices;
one composed of the cations and one composed of the anions. From this
standpoint, each lattice in the CsCl structure has a primitive unit cell.
However, once the symmetry of the unit cell has been determined, it is
desirable to determine the arrangement of atoms within some arbitrarily
chosen structural unit which may or may not be the unit cell. We might,

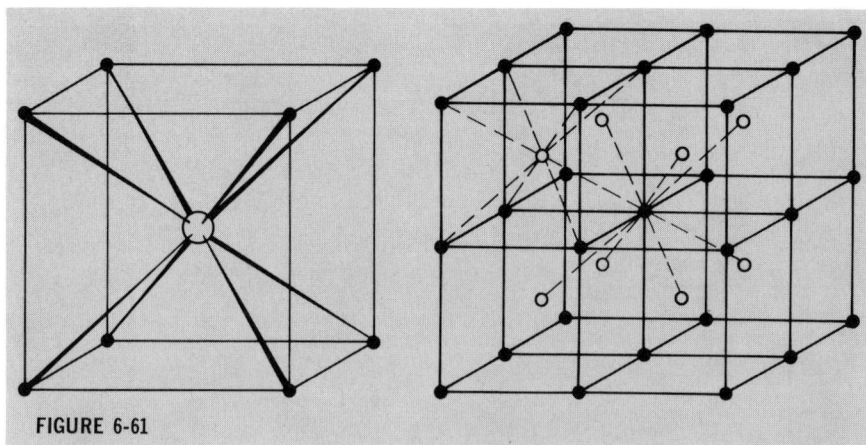

FIGURE 6-61

Cesium Chloride Structure. The solid circles represent the positive ions.

then, consider the CsCl structure shown in Figure 6-61 to represent the arrangement of atoms in a unit cell of positive ions. But again it should be emphasized that this is a primitive and not a body-centered unit cell.

## Radius Ratio

The question of why a substance might favor the NaCl structure over the CsCl structure, or the reverse, can be partially explained in terms of solely geometrical considerations. It was pointed out that the coordination number is determined, to some extent, by the packing of the larger ion around the smaller. In this respect, the three possibilities shown in Figure 6-62 can occur. If it is assumed that a stable arrangement will exist when the ions are just in contact as illustrated in Figure 6-62(b), we can calculate the ratios of the ionic radii $(r_+/r_-)$ necessary to give a stable distribution for the various coordination numbers.

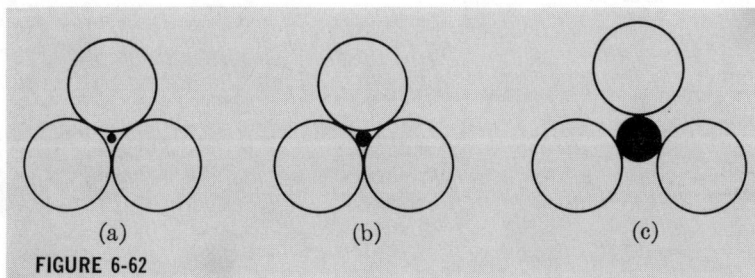

(a)                    (b)                    (c)

FIGURE 6-62

Effect of the Ion Size on the Distribution of Negative Ions about a Positive Ion.

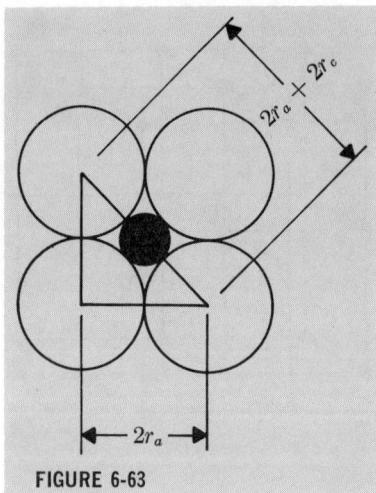

**FIGURE 6-63**

Radius Ratio for an Octahedral
Structure.

For the point of illustration, we can consider an octahedral distribution.
Here it is necessary to consider only the four anions lying in the plane as
shown in Figure 6–63. Since the triangle drawn in the figure is a right
triangle, we can say that:

$$a^2 + b^2 = c^2$$

or

$$(2r_a)^2 + (2r_a)^2 = (2r_a + 2r_c)^2$$

Solving for $r_c/r_a$ by the quadratic equation, we obtain

$$r_c/r_a = 0.414$$

This same type of calculation can be made for the other coordination
numbers giving the ranges shown in Table 6–5.

**TABLE 6–5.**

| CN | Structure | $r_c/r_a$ Limits |
|----|-----------|------------------|
| 2 | linear | <0.155 |
| 3 | triangular | 0.155–0.225 |
| 4 | tetrahedral | 0.225–0.414 |
| 6 | octahedral | 0.414–0.732 |
| 8 | cubic | >0.732 |

If only geometrical factors were involved, a direct correlation should be
observed between the crystal structure and the radius ratio. However, this

is not generally the case, although it does appear to be the deciding factor in many instances. The alkali halides are a case in point. The radius ratios of the alkali halides are tabulated below.

|    | Li | Na | K | Rb | Cs |
|----|------|------|------|------|------|
| F  | 0.44 | 0.70 | 0.98 | 0.92 | 0.81 |
| Cl | 0.33 | 0.52 | 0.73 | 0.82 | 0.93 |
| Br | 0.31 | 0.49 | 0.68 | 0.76 | 0.87 |
| I  | 0.28 | 0.44 | 0.62 | 0.69 | 0.78 |

The enclosed salts show crystal structures consistent with the radius ratio. Only those crystals with a radius ratio between 0.414 and 0.732 should crystallize in the rock salt lattice. Yet all except the chloride, bromide, and iodide of cesium show this structure, clearly pointing to the existence of other factors.

A problem somewhat analogous to that of the relative stability of the *ccp* and *hcp* structures in the inert gases now arises here, and, in a similar manner, attempts have been made to understand the distribution of the NaCl and CsCl structures among the alkali halides. The conventional calculations in terms of pairwise interactions shown in Eqs. (5–4) and (5–6) favor the NaCl structure. And the superposition of multipole interactions does not contribute enough stability to the crystal lattice to alter the results. As a consequence of this failure, Tosi and Fumi[8] have rationalized the experimental results by assuming that the parameters $B$ and $k$ in Eq. (5–6) are structure-dependent. However, Jansen[3] takes issue with this interpretation and again relies on three-body interactions. In this manner he has been able to show that the CsCl structure is stabilized to a sufficient extent for the heavier cesium halides to favor this structure over the NaCl structure. Thus, at the moment, it would appear that this is the most reasonable approach.

## Fluorite and Rutile Structures

Compounds of the form $MX_2$ and $M_2X$ will, of course, have different coordination numbers for the cation and the anion. This permits quite a large number of possible arrangements of structural units, but only two of these structures turn out to be particularly common for ionic species. These are the *fluorite* structure, characterized by $CaF_2$, and the *rutile* structure, characterized by $TiO_2$. In terms of radius ratios, the fluorite structure is favored for values greater than 0.732, and the rutile structure is favored in the range of 0.414 and 0.732. Although exceptions occur as usual, the radius ratio rule is generally followed, and we observe the fluorite structure primarily with the small fluoride anion. Most notably, it is observed for the alkaline earth fluorides of calcium, strontium, and barium as well as those of cadmium, mercury, and lead.

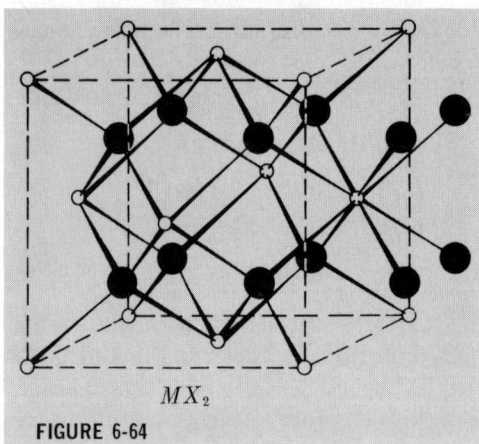

$MX_2$

**FIGURE 6-64**

Fluorite ($CaF_2$) Structure.

The fluorite structure is shown in Figure 6–64, where the tetrahedral coordination of the anion is apparent. The structure can be seen to be face-centered cubic with respect to the cation, and each of the face-centered ions is seen to have four ions of the opposite type associated with it inside the cell. It should be obvious that an additional four anions will be in the equivalent positions in the adjacent cell. The cation then has a cubic distribution of anions as nearest neighbors giving a coordination number of 8. An antifluorite structure is observed for the chalconides of lithium, sodium, and potassium, where the only difference from the fluorite structure is the interchange of positive and negative ions.

Whereas coordination numbers of 8 and 4 are noted in the fluorite structure, the rutile structure results in coordination numbers of 6 and 3. Additionally, in contrast to the previous three structures we have considered, the rutile structure falls into the tetragonal system. In Figure 6–65,

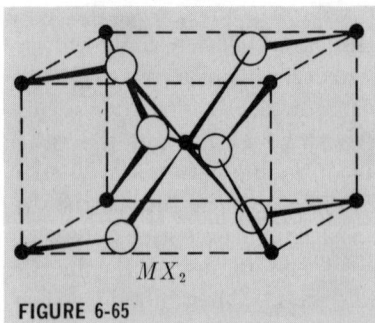

$MX_2$

**FIGURE 6-65**

Rutile ($TiO_2$) Structure.

the cation is seen to be octahedrally surrounded by the anions, and the three cations lie in a triangular plane around each anion. The general shape of the unit cell is tetragonal with the cations occupying the positions of a body-centered tetragonal lattice.

## Polyatomic Ions

Ionic crystals having a polyatomic ion still tend to crystallize in one of the simple lattices. However, the geometry of the ion generally causes a distortion of the crystalline lattice. This point is well-illustrated in the structure of $CaC_2$, which crystallizes in a distortion of the rock salt lattice. Here the carbide ions are arranged parallel to each other as shown in Figure 6–66. As a consequence, the $CaC_2$ crystal shows tetragonal rather than cubic symmetry.

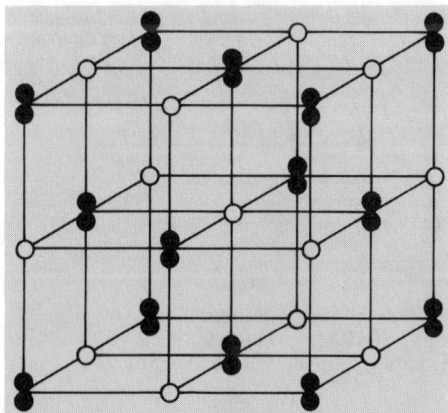

**FIGURE 6-66**

Structure of $CaC_2$ Showing the Tetragonal Distortion of a Face-Centered Cubic Pattern Due to the Linear Nature of the Carbide Ion.

Again we should recognize that the structure of $CaC_2$ shown in Figure 6–66 is not a space lattice. Rather it shows the relative locations of the calcium and carbide ions. However, because of its similarity to the NaCl structure, we might wonder why it does not fall into the cubic system. If the ions are replaced by lattice points in both the $CaC_2$ and the NaCl structures, we might expect both to form a cubic lattice with the $CaC_2$ falling into a lower class of the cubic system because of the 2-fold symmetry of the carbide ion. However, we have already pointed out that this is not the case. The reason, of course, is that the charge distribution in the $C_2^{2-}$ ion distorts the lattice from the expected cubic symmetry, and because of the parallel alignment of the ions, the effect will be in only one direction. Thus, it is quite easy to understand the resultant tetragonal symmetry.

Along the same lines, we find that the rhombohedral symmetry of calcite is also the result of a distortion of the rock salt lattice. In Figure 6–67, the distribution of calcium and carbonate ions in the front faces of the calcite rhombohedron are shown, and it can be seen that their relative positions are the same as those in NaCl. However, the trigonal symmetry of the carbonate ion, along with its orientation in the crystal, leads to a rhombohedral distortion.

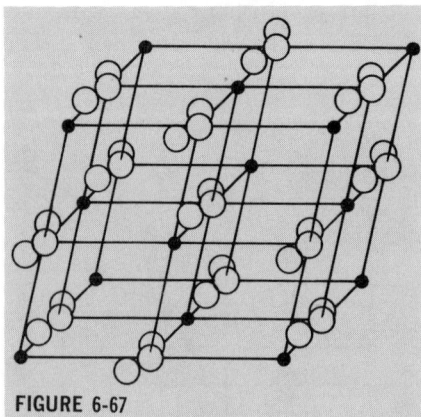

**FIGURE 6-67**

Rhombohedral Distortion of the Rock Salt Structure Due to the Trigonal Symmetry of the Carbonate Ion.

## Properties of Ionic Crystals

Because of our familiarity with many ionic crystals, we are naturally prone to draw some rather general conclusions concerning their properties. Interestingly enough, these do not always hold if some of the more exotic ionic substances are considered. Relatively high melting points and heats of fusion, solubility in polar solvents, and ionic conductance in both the fused state and in solution are typical of the properties we ascribe to ionic compounds. These certainly seem reasonable in terms of the model we have used for the ionic bond.

Inasmuch as covalent character, as well as polarization effects, is undoubtedly present in most compounds, an interpretative problem can always arise in a discussion of variations from the expected ionic behavior. Nevertheless, there are many compounds that can be more reasonably considered ionic that show radically different behavior from what one might normally expect. At the same time, however, the behavior can be rationalized in terms of a strictly electrostatic model.

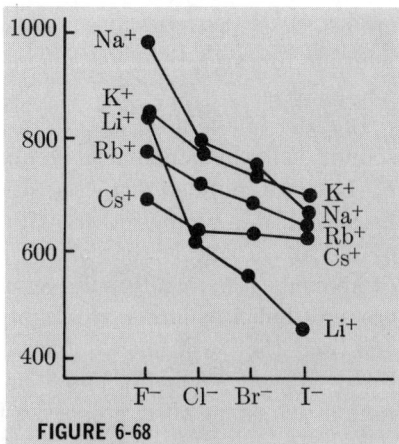

**FIGURE 6-68**

Melting    Points    of    the    Alkali
Halides.

To minimize the effect of covalent character we can consider only univalent salts of the alkali metals. In Figure 6–68, plots of the melting points of the alkali halides are given. Obviously, the melting points are not dependent simply on the sizes of the ions. If this were the case, a consistent series of parallel straight lines would be obtained. Nevertheless, the trends clearly indicate that larger ion sizes favor lower melting points. Thus, if the trends are continued, we should expect alkali metal compounds with very large anions to have quite low melting points. In many instances this is found to be the case. In Table 6–6, the melting points of several aluminum

**TABLE 6–6.    Melting Points of Sodium Salts as a Function of Anion Size.**

| Na Salt | $F^-$ | $Cl^-$ | $Br^-$ | $I^-$ | $AlMe_4^-$ | $AlEt_4^-$ | $Al(nBu)_4^-$ | $Al(n\text{-octyl})_4^-$ |
|---------|-------|--------|--------|-------|------------|------------|---------------|--------------------------|
| Mp, °C  | 992   | 801    | 755    | 651   | 240 d      | 128        | 64            | liquid at room temperature |

alkyl salts of sodium are given, along with those of the halide salts. Here we see that the general trend continues. But of most interest is the fact that $NaAl(octyl)_4$ is a liquid at room temperature. Yet, based on conductance studies, it is a compound that appears to be ionic.

The rule of thumb that ionic substances dissolve in polar solvents is quite generally used, and it certainly seems to be valid in most instances. Correspondingly, ionic compounds are assumed to be insoluble in nonpolar solvents. However, again we must be careful with such generalizations. A great amount of study has been carried out on ionic conductance in relatively nonpolar solvents. The most commonly used compounds for such studies have been the alkyl ammonium salts such as $N(i\text{-amyl})_4NO_3$. In addition to water, these salts have been studied in such solvents as dioxane and benzene.[9]

In order to avoid the limitation to a large cation, a number of studies of sodium salts in nonpolar solvents have been made in recent years. One of the more common of these salts is sodium boron tetraphenyl which has been studied in various ethers. However, the extreme example of a violation of the above rule is that of the sodium aluminum alkyls.[10] All of these compounds are readily soluble in ethers, and salts in which the alkyl group is butyl or larger are highly soluble in saturated hydrocarbons. In addition, they will show ionic conductance in these solvents. Although these salts are soluble in nonpolar solvents, because of their reactivity we cannot determine their properties in aqueous or alcoholic solvents.

## Covalent Crystals

In both molecular and ionic crystals we have seen examples of crystals in which covalent bonding is involved. These, however, are not examples of *covalent crystals*. Rather, we are here concerned with three-dimensional arrays in which covalent bonding exists between all the structural units. It turns out that simple crystals falling into this category are not overly common, examples being diamond, silicon, germanium, grey tin, silica, and silicon carbide. In contrast to ionic and most molecular crystals, we should expect to find the arrangement of atoms in a covalent crystal to be determined by the directional character of the atomic orbitals used in the bonding. This will exclude closest-packed arrangements.

The existence of directional character of equivalent bonds is probably the most unique aspect of a covalent crystal. In the examples of covalent crystals given above, except for the oxygen in silica, the atoms all show a tetrahedral coordination which implies the use of $sp^3$ hybrid orbitals. The characteristic structure is that of diamond, shown in Figure 6–69, where each atom is tetrahedrally surrounded by four covalently bonded nearest neighbors.

In addition to these clear-cut examples of covalent crystals, we may attribute the structures of some borderline ionic crystals to the presence of covalent character. This point is well-illustrated by the sulfide and oxide of zinc in which a tetrahedral coordination occurs in defiance of the radius ratio, and the zinc blende and wurtzite structures shown in Figure 6–70 are observed.

In silica, oxygen shows its conventional valence of two. The disposition of oxygen atoms about each silicon atom is tetrahedral or distorted tetrahedral, and it is reasonable to assume the use of $sp^3$ hybrid orbitals on the silicon. There are three main crystalline modifications of silica: quartz, tridymite, and cristobalite, with quartz being the most common form at

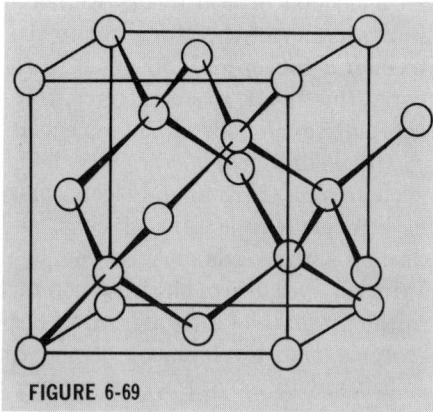

FIGURE 6-69

Structure of Diamond.

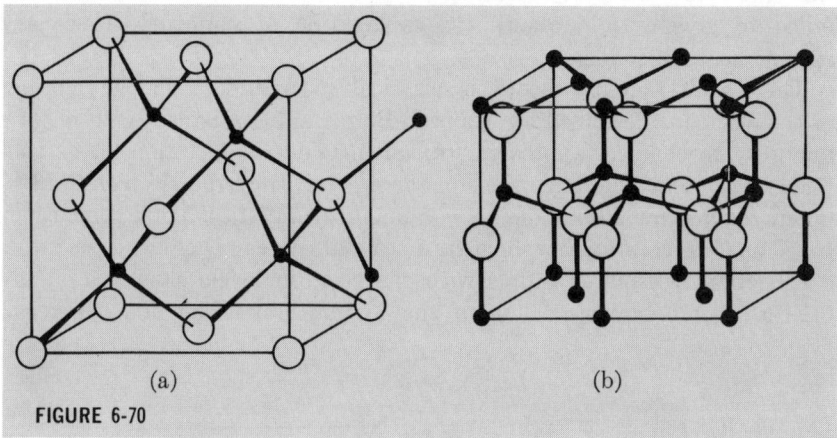

FIGURE 6-70

(a) Zinc Blende Structure and (b) Wurtzite Structure.

normal temperatures. In all of these modifications, the crystals are built up of $SiO_4$ tetrahedral units linked in such a manner that each oxygen atom is common to two of these units.

## Metallic Crystals

In the three types of crystals we have thus far discussed, the bonding principles were developed for interactions on an atom-to-atom basis. These same principles were then applied to the solid state. In contrast, the theoretical treatment of metals begins directly with the crystal. The first approach seems to appeal more to the prejudice of the human mind, which has concluded that matter is discrete and continues to see it in this light.

In molecular orbital theory we make a limited break with the idea of the individuality of the atom, but to the unconditioned mind this often seems to offer a conceptual difficulty. In the theoretical treatment of a metal we carry this break to an extreme, but in this respect the treatment is conceptually analogous to the molecular orbital approach.

Metals are most successfully defined in terms of their unique properties such as high thermal and electrical conductivities, metallic luster, opaqueness to all wavelengths of light, and closest-packed structures. Theories dealing with metals must, of course, be able to rationalize these properties. Primarily because of electrical conductivity, a model based on free electrons in the presence of a regular array of positive metal ions was proposed very early in the development of the theory of metals. The electrons were considered to move freely according to classical statistics throughout the solid in a manner analogous to a gas. Thus the stability of the metal is a consequence of the attraction between the positive ions and the electron gas. This approach was first proposed by Drude and shortly thereafter was extended by Lorentz. Although it resulted in some success, the model failed to give even a qualitative explanation of semiconductance and specific heats of solids.

The Drude-Lorentz theory is classical in nature, with the motions of the electrons described by Maxwell-Boltzmann statistics. In terms of a quantum mechanical approach, we can imagine the electrons to be confined to a box of the dimensions of the crystal. Here the electrons will be quantized and must obey the laws of quantum statistics. Thus, conclusions based on classical statistics should not be valid. We can begin by considering a collection of isolated atoms with the electron levels shown in Figure 6–71(a). From our discussion of the hydrogen molecule ion, we recall

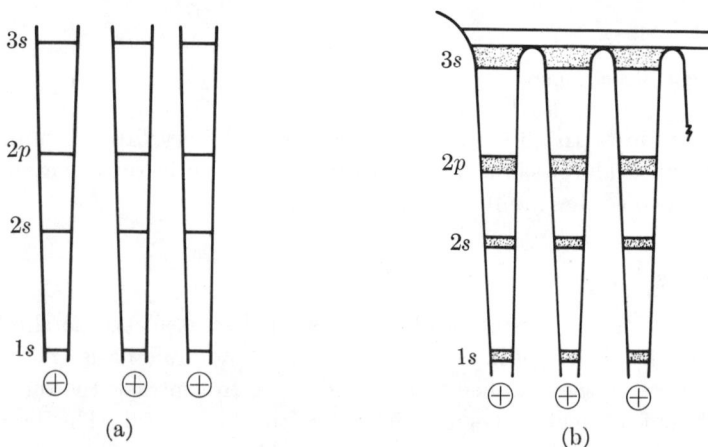

**FIGURE 6-71**

Energy Bands in a Metallic Crystal.

that the overlap of the atomic orbitals leads to two new orbitals, one of higher and one of lower energy. In a metal we can imagine $n$ atoms combining with each set of orbitals, forming $2n$ new sets of orbitals in the solid. Further, from the effect of the box size on the energy of a particle confined to the box,

$$(2\text{--}19) \qquad E = \frac{n^2 h^2}{8ma^2}$$

We know that the difference between energy levels $(E_2 - E_1)$ decreases with an increase in box size. Thus, levels that are close together in energy will tend to merge. And if there are enough energy levels they will form an essentially continuous distribution in a limited energy range which can be represented as a band.

In a metallic crystal we are dealing with a very large number of atoms, of the order of $10^{20}$. Thus, there will be an exceedingly large number of energy states. Correspondingly, because of the volume of a macroscopic crystal, the energy separation between these levels will be exceedingly small, thereby leading to the energy bands shown in Figure 6–71(b). The width of the bands is small for tightly held orbitals because of the very limited interaction with corresponding orbitals on neighboring atoms, but as can be seen in Figure 6–72, the width at the equilibrium separation is much more significant with the outer shell electrons.

It is in terms of this basic picture that many of the properties of metals can be understood. In Figure 6–73, two distinctly different arrangements of bands are shown. At the equilibrium separation in Figure 6–73(a), a break exists between the upper filled level and the lowest empty level. On the other hand, this separation does not exist in the diagram shown in Figure

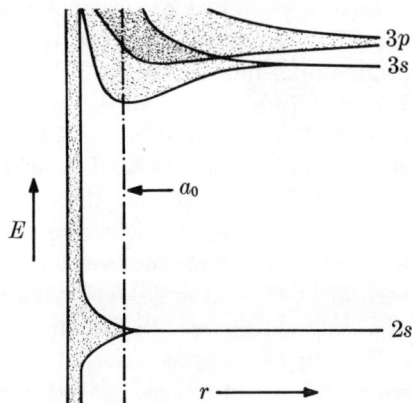

**FIGURE 6-72**

Effect of Distance of Separation on Band Width.

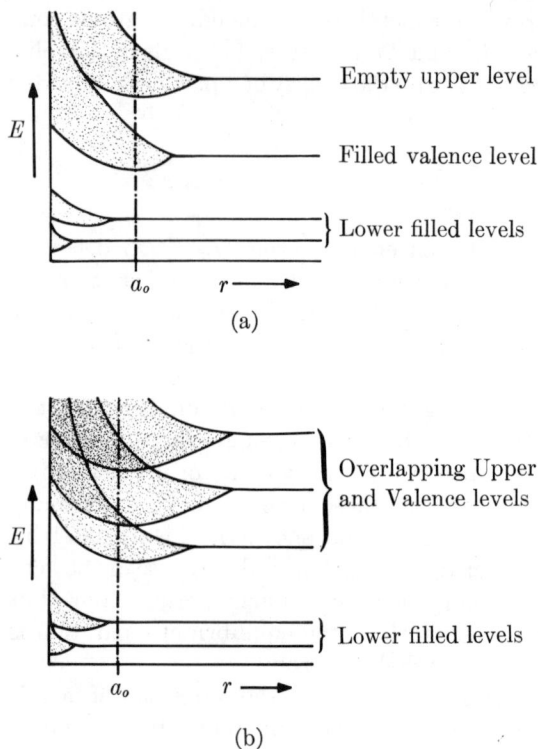

Empty upper level

Filled valence level

} Lower filled levels

(a)

Overlapping Upper and Valence levels

} Lower filled levels

(b)

**FIGURE 6-73**

Distribution of Electron Bands Leading to: (a) an insulator or semiconductor and (b) a conductor.

6-73(b). The ability of electrons to move from the highest filled or valence band to the adjacent unfilled band distinguishes metal conductors, semiconductors, and insulators. This can be appreciated by noting that in Figure 6-73(a), a *forbidden* region of energy states exists between these bands. If this separation is sufficiently great, an electron cannot absorb enough energy from an applied field to move from its bound state, and the substance is an insulator. If, on the other hand, the bands merge as shown in Figure 6-73(b), the electron can readily be raised to this conduction band and the substance is a conductor. Semiconduction can then be attributed to a relatively small forbidden region that will permit the promotion of a limited number of electrons from the valence band.

We can also appreciate that the metallic bond is nondirectional. Thus we would expect closest-packed structures for pure metals. Metal crystals are generally found to be *ccp*, *hcp*, or *bcc* with various modifications frequently being observed. The body-centered cubic structure is not a closest-packed structure, but the occupied space is very close to that of a true closest-packed structure. As yet there has been no explanation for the various crystal modifications in metals, but with the success the three-body interaction has had with the inert gases and the alkali halides, it would appear that this approach holds the greatest promise.

## Crystals with Mixed Bonding

It was mentioned at the outset that a classification of crystals in terms of bond type faces difficulties. In spite of this we have discussed crystals in terms of the four types of bonds. Yet there are many crystals in which the properties are clearly dependent upon the simultaneous presence of two of these bond types. We have already seen that the structure of graphite can be interpreted in terms of sheets of covalently bonded carbon atoms, the sheets being connected by van der Waals forces. Because of the presence of infinite sheets, the crystal cannot be classified simply as a molecular crystal. To further emphasize this point, the silicates are a particularly useful group of crystals.

**Silicates.** The variety of silicates that occur in nature is almost limitless, but it is possible to place them in four general groupings depending on whether the silicate complex is finite or whether it combines to form infinite chains, sheets, or three-dimensional framework structures. In all of these the silicon is tetrahedrally surrounded by four oxygen atoms. The various groupings, then, are a consequence of the different ways the $SiO_4$ tetrahedra can link. The resultant negatively charged $Si_nO_m$ complexes are then bonded together in the crystal by metal cations. The great variety of silicates can largely be attributed to the isomorphous replacement of one cation by another. Additionally, silicon in an $SiO_4$ tetrahedron is frequently replaced by aluminum. In order to maintain a balance of charge, this requires the addition of another cation or substitution of one with a higher positive charge.

**Finite anions.** Orthosilicates such as zircon, $ZrSiO_4$, willemite, $Zn_2SiO_4$, and phenacite, $Be_2SiO_4$, have relatively simple structures in which the $SiO_4^{4-}$ grouping exists as a discrete entity in the crystal lattice. More complicated orthosilicates are characterized by the garnets, which have the general stoichiometry $M_3^{2+}M_2^{3+}(SiO_4)_3$ where $M^{2+}$ is Mg, Ca, or Fe and $M^{3+}$ is Al, Cr, or Fe.

The linkage of two $SiO_4$ tetrahedra through a common oxygen atom to form a finite $Si_2O_7^{6-}$ anion is observed in the pyrosilicates. The only common simple example of this type of silicate is $Sc_2Si_2O_7$. A common, but more complicated example is found in the zinc ore hemimorphite, $Zn_4(OH)_2Si_2O_7$.

Finally, by the sharing of two oxygen atoms of each $SiO_4$ tetrahedron, discrete cyclic silicate anions can be formed. In general, these are represented as $(SiO_3^{2-})_n$, but only the $Si_3O_9^{6-}$ and $Si_6O_{18}^{12-}$ ions are observed.

**Infinite chain silicates.** There are two general types of infinite chains among the silicates, as shown in Figure 6–74. In the *pyroxenes*, simple

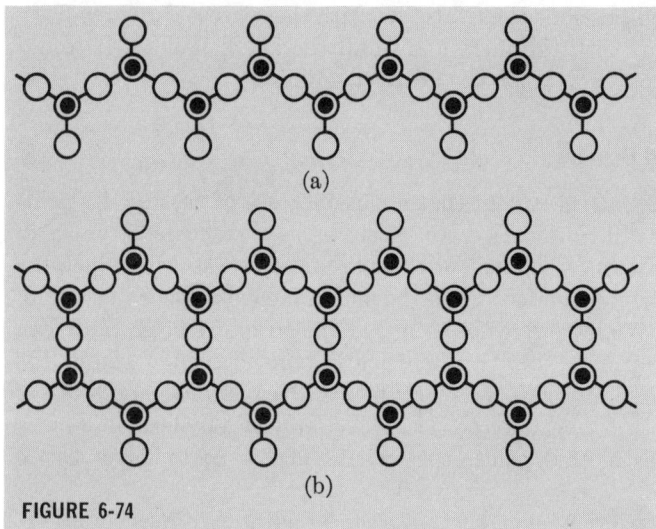

FIGURE 6-74

Silicate Chain Structures: (a) pyroxenes and (b) amphiboles. Solid circles represent the silicon atoms. Four oxygen atoms are tetrahedrally positioned about each silicon atom.

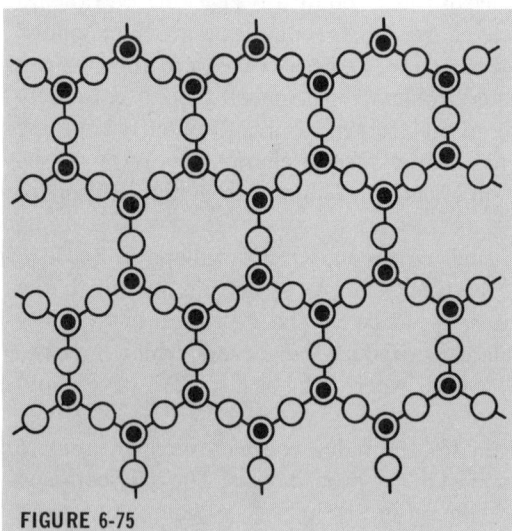

FIGURE 6-75

Silicate Sheet-Type Structure. The solid circles represent silicon atoms.

chains are formed having the composition $(SiO_3^{2-})_n$, whereas in the *amphiboles*, double chains with the composition $(Si_4O_{11}^{6-})_n$ are found. Within the chain, bonding is covalent, but the chains are held together by sharing of metal ions. For instance, in diopside, $CaMg(SiO_3)_2$, parallel pyroxene chains are held together by metal ions lying between the chains. The structural principles are quite similar between the pyroxenes and the amphiboles; in both types of structures the chains are held together by metal ions. However, in the latter case, half of the silicon atoms share two oxygen atoms with adjacent silicon atoms, whereas the other half share three oxygen atoms.

**Infinite sheet silicates.** If each silicon atom in an amphibole-type chain shares three oxygen atoms with the adjacent silicon atoms, an infinite sheet with the composition $(Si_2O_5^{2-})_n$ is formed. From the diagram in Figure 6–75 rings containing six silicon atoms are obtained. In the kaolins and micas this type of linkage occurs with the unshared oxygen atoms pointing in the same direction. Two such sheets lie parallel to each other with the unshared oxygen atoms on the adjacent sheets pointing towards each other. These sheets are then connected by metal ions. Although the structures are not quite so simple, the cleavage characteristic of the micas can be appreciated in terms of the weak interaction between adjacent pairs of silicate sheets.

**Framework silicates.** If all four of the oxygen atoms in the $SiO_4$ tetrahedra are shared, a three-dimensional lattice is obtained. In quartz this leads to a covalent crystal. However, in silicate structures a silicon atom is often replaced by aluminum and charge balance requires an additional positive ion. This then results in three-dimensional silicate structures in which the crystal properties may be dependent on ionic as well as covalent forces. Among the framework silicates we can distinguish three groupings: *feldspars, zeolites,* and *ultramarines.*

The feldspars are extremely common, being the most important of the rock-forming minerals. In general, they can be represented by the formula

**Table 6–7.   Crystal Classification of Some Silicates.**

| Mineral | Formula | Type | Crystal System |
|---------|---------|------|----------------|
| zircon | $ZrSiO_4$ | finite $SiO_4^{4-}$ | tetragonal |
| thortveitite | $Sc_2Si_2O_7$ | finite $Si_2O_7^{6-}$ | monoclinic |
| beryl | $Be_3Al_2Si_6O_{18}$ | finite $Si_6O_{18}^{12-}$ | hexagonal |
| benitoite | $BaTiSi_3O_9$ | finite $Si_3O_9^{6-}$ | trigonal |
| enstatite | $MgSiO_3$ | chain $(SiO_3^{2-})_n$ | orthorhombic |
| tremolite | $Ca_2Mg_5(Si_4O_{11})(OH)_2$ | chain $(Si_4O_{11}^{6-})_n$ | monoclinic |
| muscovite | $KAl_2(Si_3AlO_{10})(OH)_2$ | sheet $(Si_2O_5^{2-})_n$ | monoclinic |
| orthoclase | $KAlSi_3O_8$ | framework $(SiO_2)_n$ | monoclinic |
| leucite | $KAlSi_2O_6$ | framework $(SiO_2)_n$ | cubic |

$M(Al,Si)_4O_8$. If one quarter of the silicon atoms are replaced by aluminum, M will represent an alkali metal, whereas the replacement of two silicon atoms requires that M be a bipositive cation. Typical of each of these is orthoclase, $KAlSi_3O_8$, and anorthite, $CaAl_2Si_2O_8$.

The zeolites are of particular interest here because they clearly show the presence of ions in the crystal lattice. The zeolite structure is more open than that of the feldspars and consequently it has the ability to absorb water and to exchange cations. For this reason certain of the zeolites are commonly used as water softeners. The *permutites*, which are sodium-containing zeolites, are used to exchange sodium ions for calcium ions in hard water.

In summary, a few common examples of the various types of silicates are given in Table 6–7.

## References

1. H. T. Evans, Jr., *Science*, **141**, 154 (1963).
2. J. Cuthbert and J. W. Linnett, *Trans. Faraday Soc.*, **54**, 617 (1958).
3. L. Jansen, "Advances in Quantum Mechanics," Vol. 2, Academic Press, New York, 1965; L. Jansen and E. Lombardi, *Discussions of the Faraday Soc.*, Intermolecular Forces, **40**, 78 (1965).
4. L. H. Bolz, M. E. Boyd, F. A. Mauer, and H. S. Peiser, *Acta. Cryst.*, **12**, 247 (1959).
5. C. S. Barrett and L. J. Meyer, *J. Chem. Phys.*, **42**, 107 (1965).
6. T. E. Brackett and E. B. Brackett, *J. Phys. Chem.*, **69**, 3611 (1965).
7. J. C. Slater, *J. Chem. Phys.*, **41**, 3199 (1964).
8. M. P. Tosi and F. G. Fumi, *J. Phys. Chem. Solids*, **23**, 359 (1962).
9. C. A. Kraus, *J. Phys. Chem.*, **60**, 129 (1956).
10. M. C. Day, H. M. Barnes, and A. J. Cox, *J. Phys. Chem.*, **68**, 2595 (1964).

## Suggested Supplementary Reading

F. C. Phillips, "An Introduction to Crystallography," 2nd ed., Longmans, Green and Co., Ltd., London, 1956.

C. W. Bunn, "Chemical Crystallography," 2nd ed., Oxford University Press, London, 1961.

A. F. Wells, "Structural Inorganic Chemistry," 3rd ed., Oxford University Press, Inc., New York, 1961.

L. V. Azaroff, "Introduction to Solids," McGraw-Hill, Inc., New York, 1960.

W. E. Addison, "Structural Principles in Inorganic Compounds," John Wiley & Sons, Inc., New York, 1961.

K. B. Harvey and G. B. Porter, "Introduction to Physical Inorganic Chemistry," Addison-Wesley Publishing Company, Inc., Reading, Mass., 1963.

L. Pauling, "Nature of the Chemical Bond," 3rd ed., Cornell University Press, Ithaca, New York, 1960.

J. A. A. Ketelaar, "Chemical Constitution, Elsevier, New York, 1958.

S. Raimes, "The Wave Mechanics of Electrons in Metals," Interscience, New York, 1961.

C. Kittel, "Introduction to Solid State Physics," John Wiley & Sons, Inc., New York, 1953.

H. H. Jaffe and M. Orchin, "Symmetry in Chemistry," John Wiley & Sons, Inc., New York, 1965.

## Problems

1. Determine the rotation axes and symmetry planes and give both the Hermann-Mauguin and the Schoenflies symbols for the point group symmetry classification of
   (a) $H_2$ (linear)           (d) $CH_4$ (tetrahedral)
   (b) $NH_3$ (pyramidal)        (e) $PtCl_4^{2-}$ (square planar)
   (c) $CoF_6^{3-}$ (octahedral) (f) $PCl_5$ (trigonal bipyramid)
   (See Appendix D)

2. For a cube, find the equivalent representations in terms of rotation axes and symmetry planes for $\bar{1}$, $\bar{2}$, $\bar{3}$, and $\bar{4}$.

3. Draw the complete stereograms, showing rotation axes, symmetry planes, and centers of symmetry for
   (a) square plane      (d) plane hexagon
   (b) octahedron        (e) trigonal bipyramid
   (c) tetrahedron       (f) bipyramid (distorted octahedron)

4. Classify the following molecules in the proper point groups using Hermann-Mauguin and Schoenflies symbolism

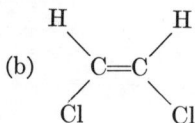

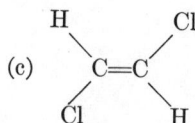

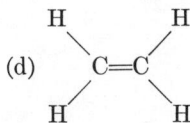

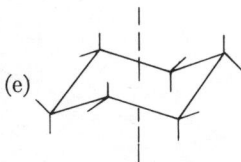

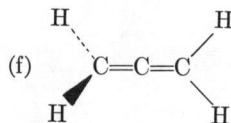

(See Appendix D)

5. Without studying the appropriate stereograms shown in the text, and remembering that the presence of certain symmetry elements requires the presence of others, determine all of the symmetry present in class
   (a) $2/m$         (e) 23
   (b) $2/mm$        (f) $2m3$
   (c) $4m$          (g) $\bar{4}m$
   (d) 62            (h) 432

6. What would be the Schoenflies symbols for the classes given in Problem 5?

7. What symmetry element is absent in the molecule $NH_3$ that is present in the $D_{3h}$ point group?

8. Using Schoenflies symbolism, and carrying out the indicated operations from right to left, determine the single operation that is equivalent to the following for a methane molecule.
   (a) $S_4 \times S_4 =$
   (b) $C_3 \times C_2 =$
   (c) $\sigma_d \times C_2 =$
   (d) $C_2 \times E =$

9. Show that the symmetry expressed by $2m$ must place a crystal in the orthorhombic system.

10. By considering all possible symmetries, show that there can be only three classes in the orthorhombic system.

11. Show why there cannot be an orthorhombic space lattice based on a cell with only two sides centered.

12. Illustrate all of the possible hexad screw axes, $6_1$, $6_2$, . . . .

13. In a face-centered cubic lattice, which plane will have the maximum density of lattice sites?

14. Calculate the limiting radius ratio for a tetrahedral structure.

15. Using the melting points in Table 6–5, and assuming a linear relationship, estimate the ion size of the three aluminum alkyl anions listed in the table. Are these ion sizes reasonable?

# Inorganic
# Stereochemistry

7

Theories of valence and stereochemistry were developed in the last century side by side, usually the one being the result of the other. In 1852 Frankland proposed the concept of valence. He stated that the elements form compounds by combining with a definite number of what we now call equivalents of other elements. Kekule (1858) and Kolbe (1859) extended the valence idea and postulated the quadrivalence of the carbon atom. In 1858, Kekule suggested that carbon atoms may bond to one another in an indefinite number to form chains, and in the same year Couper introduced the concept of a valence bond and drew the first structural formulas. The term *chemical structure* was introduced by Butlerov in 1861. He stressed the importance of writing a single formula for a compound showing how the atoms are linked together in one of its molecules. He also stated that the properties of compounds are determined by their molecular structure, and that a knowledge of the latter will permit one to predict the former. However, it was not until 1874 that the first major step was taken toward a visualization and assignment of molecular structures in three dimensions. In that year J. H. van't Hoff and J. A. le Bel independently postulated the tetrahedral disposition of the four valence bonds of the carbon atom and thereby placed classical organic stereochemistry twenty years ahead of inorganic stereochemistry.

The *coordination theory* of A. Werner (1893) (see Chapter 10) may be considered the beginning of inorganic stereochemistry, for it is concerned with the spatial arrangement of molecules and ions about a central metal atom. For many years after the Werner theory, inorganic chemists were occupied with the preparation and characterization of a great number of new complex compounds, particularly of the transition metals, and the deduction of their stereochemistries. Such deductions were based primarily on chemical methods of study, and these are necessarily limited.

The great advances that were subsequently made in the field of stereochemistry can be attributed to two completely independent developments, both originating in the second decade of this century. The first of these was the development, beginning in 1912, of physical methods for the determination of structure. This began with the application of the X-ray diffraction method of structural analysis and was followed much later by such physical techniques as the diffraction of electron and neutron waves, the measurement of electric and magnetic dipole moments, and the interpretation of infrared, Raman, visible-ultraviolet, microwave and nuclear magnetic resonance spectra. All of these methods are capable of revealing, in more or less detail, three-dimensional geometry and have served to stimulate the interest in, and rapid growth of, inorganic stereochemistry. At the same time, the electronic theory of valence, beginning in 1916, stimulated the later development of quantum chemistry and its subsequent application to more sophisticated theories of the chemical bond and molecular structure. The application of these powerful experimental and theoretical methods for the past forty years has produced a great wealth of structural information as well as uncovered new structural principles which have produced a greater understanding and a higher degree of predictability of the properties of compounds. It will be a primary purpose of this chapter to point out the principles which underlie some of the modern approaches to stereochemistry and to show how they may be applied in specific cases.

It is apparent at the outset that we shall have to limit somewhat sharply the scope of the presentation. For example, we shall not consider the structures of crystalline solids. This subject has been dealt with already in Chapter 6. Furthermore, no attempt will be made to discuss the very complex and ever-increasing amount of theoretical and experimental information concerning the structure of liquids as opposed to the structure of solute *molecules* in solution. What we shall primarily be concerned with is the stereochemistry of *molecules* and *molecule-ions* in the solid, liquid, or gaseous phase which are derived from multivalent atoms. *Bond lengths* and particularly *bond angles* will be the relevant experimental quantities that we shall want to explain and to predict. Actually, many of the basic ideas necessary for a discussion of stereochemistry have already been introduced in Chapters 5 and 6; and in Chapters 10 and 11 the stereochemistry of transition metal complexes has been considered, so that subject will not be covered here. Therefore, we shall further limit our discussion to simple inorganic molecules and to complex molecules and ions of non-transition elements. But first it would seem appropriate to present a brief review of the modern experimental techniques available to the stereochemist.

## EXPERIMENTAL METHODS

The stereochemistry of a molecule will be considered defined by a statement of the spatial positions of the atoms or groups of atoms relative to a given multivalent atom to which they are attached by bonds that have some substantial degree of covalent character. Thus, for an elucidation of the stereochemistry of a molecule, we require precise internuclear distances and the angles between imaginary lines joining the centers of the bonded atoms. The main experimental methods used for obtaining such structural information may be conveniently grouped into four categories: (1) diffraction methods; (2) spectroscopic methods; (3) resonance methods; (4) other physical methods. Because chemical methods are in general fewer and less informative — for example, no bond lengths or angles may ever be obtained — they will not be considered here.

## Diffraction Methods

In this category are included *X-ray, electron*, and *neutron diffraction*. The most direct and accurate method for the determination of internuclear distances in individual molecules is one that utilizes the diffraction of radiation having wavelengths comparable to molecular dimensions. For example, the wavelengths for bombarding X rays and neutrons range from about 0.7 to about 2.5 Å, and those for electrons range from 0.05 to 0.07 Å. X-ray diffraction has been a powerful, perhaps the most powerful, tool for structure determination. It had its experimental beginning in 1912, when Bragg determined the structures of NaCl, KCl, and ZnS, by directing a beam of monochromatic X rays on crystals of these compounds.

As was discussed in Chapter 1, any particle moving with a momentum $mv$ has associated with it a deBroglie wavelength, $\lambda = h/mv$, and a beam of such particles can give rise to diffraction phenomena under appropriate conditions. Beams of monochromatic electrons are used primarily for investigations of molecular structure in the gaseous state, although in more recent years they have also been used for studying the surface and internal structures of crystals. Neutrons, unlike X rays and electrons which interact with the extranuclear electrons of the atoms they encounter, are scattered by the atomic nuclei. Neutron diffraction is particularly valuable in that it provides a means of locating hydrogen atoms in molecules. Thus hydrogen atoms scatter neutrons as well as do the much heavier atoms. This is in marked contrast to the behavior of X rays, where the scattering increases smoothly with an increase in the number of orbital electrons in the scattering atom. Another advantage of neutron diffraction over X-ray diffraction is that two chemically dissimilar atoms having nearly (or exactly) the same number of orbital electrons cannot be distinguished by the latter method, yet neutron diffraction may easily distinguish between them. For example, it has been shown by neutron diffraction that in spinel, $MgAl_2O_4$, the magnesium ions occupy tetrahedral sites in the crystal and the aluminum ions are in octahedral sites. But since $Mg^{2+}$ and $Al^{3+}$ are isoelectronic, X-ray scattering cannot differentiate between the two elements.

## Spectroscopic Methods

Spectroscopic methods of structural analysis are concerned primarily with the absorption of radiant energy by molecules. Although the divisions are not sharply defined, we can state that molecules may absorb energy in the four general regions of the electromagnetic spectrum illustrated in Figure 7-1. These give rise to spectra which may be labeled *rotational*,

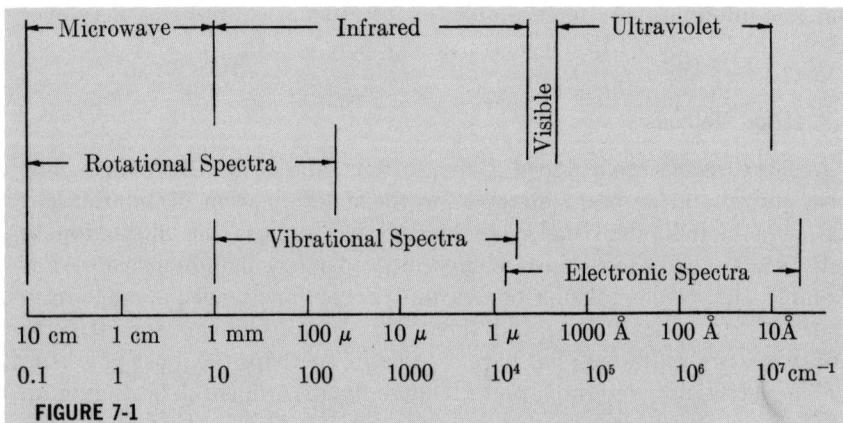

**FIGURE 7-1**

Part of the Electromagnetic Spectrum Showing the Approximate Regions Which Give Rise to Rotational, Vibrational, and Electronic Spectra.

*vibration-rotational*, *vibrational* and *electronic*. Excitation of electrons generally involves energies in the range of 1 to 8 electron volts (eV). This requires radiation in the *near-infrared*, *visible*, and *near-ultraviolet* regions of the spectrum, that is, wavelengths in the 14,000 to 1500 Å range. In the vacuum ultraviolet (below 1500 Å) the energies are high enough to cause the dissociation of molecules, in other words, the breaking of chemical bonds. Electronic spectra may yield structural information about both the ground and the lower excited states of molecules, but interpretations are quite difficult and not nearly as straightforward and revealing of structure as with other methods. However, the emphasis in recent years being placed upon electronic spectra of transition metal complexes (Chapter 11), and the attempts at the correlation and determination of the structures of these ions from their electronic spectra, indicates an increasing interest in this method.

The energy *differences* existing between vibrational energy levels within a molecule generally fall in the range from about 0.05 to 1.2 eV or 4000 to 200 $cm^{-1}$. This corresponds to absorptions of light of wavelengths in a vibrational excitation of 2.0 to 50 $\mu$ ($1\mu = 10^4$ Å). Absorptions due to the fundamental stretching of most chemical bonds, as well as many other more complicated vibrational modes, occur in this spectral region. Comparison of calculated frequencies (assuming a particular structural model for a molecule) with measured frequencies is the most important use of IR data in structure elucidations.

Often the use of infrared spectral data for structure determinations must be supplemented by Raman spectral information. The *Raman effect*, which gives rise to Raman spectra, may be pictured briefly as arising in the following manner. A beam of monochromatic light from any convenient source and with almost any desired frequency, passing through a gas, liquid, or transparent solid, is partially scattered. Most of the scattered light has the same frequency as the incident light, but a small fraction has slightly altered frequencies, called *Raman frequencies*. The alteration in frequency arises from the partial removal of energy by a molecule from an incident photon. The absorbed photon results in some rotational or vibrational energy of the molecule being increased by a discrete amount, and the remainder of the energy is re-emitted as a photon of lower frequency. It is also possible for a molecule in an excited energy state to transfer energy to the incident photon and thereby increase the frequency of the scattered photon. The frequencies of the photons with altered energy are recorded as lines in a spectrogram relative to those of the unaltered incident light.

The selection rules that prescribe the number of lines obtained in the IR or Raman spectra are strongly dependent upon the molecular symmetry. For example, a linear triatomic molecule XAX shows two strong lines in its IR spectrum and only one strong line in its Raman spectrum. The two spectra of a given molecule are said to be *complementary* or *mutually exclusive* if a strong line in one does not appear in the other. This is the theoretical situation that must occur if a molecule has a center of symmetry. If the *same* vibration gives rise to lines in both the IR and Raman spectra, the molecule has no symmetry center. So, for example, the appearance of three lines in both types of spectra of $SO_2$ rules out linearity for this molecule. In general, it is necessary that a vibrational mode involve a change in the dipole moment of the molecule in order for an IR absorption by that mode to occur, whereas any vibration that causes a change in the polarizability of a molecule should produce a Raman shift.

The rotation spectra of polar gas molecules lie in the *far-infrared* and *microwave* regions of the electromagnetic spectrum. The study of rotation spectra, which record the energy differences existing between rotational energy levels, yields the moments of inertia of molecules. The latter may be used to derive very precise bond lengths and angles when combined with similar data for isotopically substituted molecules that can reasonably be expected to have the same lengths and angles. Excellent precision is possible since resolution to about $10^{-8}$ $cm^{-1}$ can be realized in the microwave region, compared, for example, to about 1 $cm^{-1}$ in the IR region. Limitations on microwave studies include the necessity of gaseous samples and the less restrictive requirement that the molecule have a permanent dipole moment in the ground state. That is, mere rotation cannot create a dipole moment in a molecule.

Finally in this section we should point out that the excellent book by Drago, "Physical Methods in Inorganic Chemistry,"[1] may profitably be used as a starting point for pursuing the subjects mentioned in this section and, indeed, for the subjects of the next two sections. Extensive references to other literature may be found in this book, but one additional reference worthy of special note here is the book by Nakamoto, "Infrared Spectra of Inorganic and Coordination Compounds."[2]

## Resonance Methods

Structural information is attainable by other modern methods which employ energies in the *radio-frequency* range. These include nuclear magnetic resonance (NMR), nuclear quadrupole resonance (NQR), and electron paramagnetic resonance (EPR). Nuclei that possess magnetic moments may exist in various quantum states in the presence of a magnetic field. The nuclear resonance effect arises from transitions among the energy levels corresponding to different orientations of nuclear magnetic moments with respect to the applied external field. The energy separation of the quantum states is very small, falling in the frequency range 10 to 60 megacycles (the 30- to 5-meter wavelength range). Since the field that determines the energy gaps depends partially upon the exact electron distribution about the nucleus, variations in the latter caused by bonding or molecular environment produce shifts, termed *chemical shifts*, in the position of the resonance absorption peaks. These shifts therefore lead, in principle, to structure information.

If a nucleus possesses an electric quadrupole moment, it will interact with the inhomogeneous electrical field resulting from the asymmetry in the electron distribution surrounding it. Thus, in a nuclear quadrupole resonance experiment, radio frequency radiation is used to effect transitions among the various orientations of a quadrupolar nucleus in an asymmetric electrical field. Structural information may be obtained by considering how different structural and electronic effects influence the asymmetry of the electronic environment of the nucleus. However, the number of structures determined by NQR is very small indeed compared to the exponentially increasing number determined by high-resolution NMR.

Electron paramagnetic resonance absorption spectroscopy is a technique which may be applied to atoms, ions, molecules, or molecular fragments that have unpaired electrons. Magnetic moments here are approximately 2000 times greater than those observed for nuclei. Consequently this results in the absorption of energy in the microwave region (usually at 4- to 1-cm wavelength range). Energy absorption results in a change in the orientation of the magnetic moment from one allowed position to another. The actual frequency absorbed is dependent upon the magnetic field, and thus, by varying the field, the absorption may be detected at some appropriate microwave frequency. Little direct structural information is forthcoming from EPR studies, but indirectly certain subtle effects may support or refute structural information that has been guessed or obtained in other ways.

### Other Physical Methods

Finally, it should be mentioned that a certain amount of general structural information may also be determined from (1) dipole moment data, (2) detailed magnetic susceptibility data, (3) electronic spectral intensities, (4) Mössbauer spectroscopy, and (5) circular dichroism and optical rotatory dispersion studies. All of these physical methods suffer from severe restrictions as to the number and kind of molecules that can be investigated and other drawbacks. And they do not yield bond lengths or angles. Furthermore, in many cases because of interpretation problems they may be misleading about the structure and hence are seldom employed for this kind of information.

## MODERN STEREOCHEMICAL THEORIES

In the years following the introduction of the valence bond method and its application to the hydrogen molecule by Heitler and London, Pauling,[3] Slater,[4] and later, others extended the theory and were able to describe the shapes of simple molecules formed by multicovalent atoms. We have already seen in Chapter 5 that the valence bond approach, because of its emphasis on localized hybrid orbitals having well-defined directional properties, is particularly well-suited to stereochemical descriptions. Thus, it is possible to predict the stereochemistry of a molecule with very nearly the same degree of certainty that we have of the orbitals which are hybridized on the central multivalent atom and of the relative contribution of each of the separate atomic orbitals in the hybrid set. The number of examples of linear $(sp)$, trigonal planar $(sp^2)$, tetrahedral $(sp^3)$, square planar $(dsp^2)$, trigonal bipyramidal $(sp^3d)$ and octahedral $(d^2sp^3$ or $sp^3d^2)$ molecules whose geometry is very neatly described in terms of hybrid orbitals certainly is a significant testimony to the theoretical value, range, and simplicity of this approach. Nevertheless there is a sufficient element of artificiality in the familiar hybrid orbital model, as there is in such concepts as *resonance* and *exchange* and related mathematical *descriptions*, that we should remember that they should not in themselves be considered *explanations* for the actual cause of the phenomena they are used to describe.

Indeed there are some relatively simple molecules for which the hybridized orbital picture appears to be inadequate. For example, we describe the ethylene molecule in terms of the formation of $\sigma$ and $\pi$ bonds between the carbon atoms, and we account for the near-120° bond angle by assuming that each carbon atom is hybridized $sp^2$. But consider the unsymmetrical dihalo derivatives, $H_2C{=}CX_2$, where the expected *opening* of the XCX angle, due to the electrostatic repulsion of the more electronegative halogens, does not occur. In fact there is actually a *closing* to about 114°. Or consider the 104.5° angle in $H_2O$. A current picture using hybrid orbitals assumes that the two lone pairs of electrons also occupy hybrid orbitals, but since the angle is less than the predicted 109°28′ there must be more "$p$-character" in the bond orbitals and more "$s$-character" in the lone pair orbitals, whatever all that may really mean *physically*. It is not surprising that some chemists find such descriptions employing the *mathematically* sound variation of $s$- or $p$- or $d$- or $f$-character not very satisfying.

Finally we might note that when there are five electron pairs to be accommodated, a choice must be made between $sp^3d_{z^2}$ and $sp^3d_{x^2-y^2}$ (or $sp^3d_{xy}$). These lead to trigonal bipyramidal and square pyramidal arrangements, respectively. Furthermore, without making additional unjustifiable assumptions, the valence bond theory cannot account for the difference in equatorial and axial bond lengths in the trigonal bipyramid. Similar and additional problems arise with six and seven electron pairs in the valence shell of the central atom. For example, why does the bromine atom in the square pyramidal $BrF_5$ molecule lie *below* the plane of the four fluorine atoms?

## A New Theory

In 1940 Sidgwick and Powell[5] laid the basis for a new theory when they concluded that the arrangements of bonds around multicovalent atoms are related simply to the *total number of valence shell electrons*, the unshared as well as the shared pairs. From 1950 onward Lennard-Jones, Pople, Linnett, Mellish, Walsh, and particularly Gillespie, to name only the most active contributors, have been stressing the importance of the lone pairs of electrons and the Pauli principle in stereochemistry. In reality they have been building a new theory. In 1954, Lennard-Jones wrote the following statement concerning the Pauli principle: "Its all-pervading influence does not seem hitherto to have been fully realized by chemists, but it is safe to say that ultimately it will be regarded as the most important property to be learned by those concerned with molecular structure."

The new theory, which we shall call, following Gillespie's[6] suggestion, the *Valence-Shell-Electron-Pair-Repulsion Theory* (VSEPR), proposes that *the stereochemistry of an atom in a molecule is determined primarily by the repulsive interactions among all the electron pairs in its valence shell*. The theory assumes that the valence shell electrons occupy essentially localized orbitals spatially oriented so as to maximize their average distance apart. This happens as a consequence of three very basic scientific assumptions: (1) the indistinguishability of electrons; (2) the operation of the familiar $1/r^2$

Coulombic force; (3) the operation of a *Pauli repulsion force*. The Pauli repulsion force may be considered to arise from the subtle operation of the Pauli principle which, in essence, keeps electrons of like spin as far apart as possible while allowing electrons of opposite spin to be drawn together. If we can properly think of this as a force, it may be approximated by a potential function involving a $1/r^m$ term, where $m$ is large.

It is a relatively simple mathematical exercise to demonstrate how the Pauli antisymmetry principle operates to keep electrons maximally separate. Consider first the two-electron system of a helium atom in the excited state $1s^1 2s^1$. The electrons in the *singlet state*, spins opposed, may be shown to have a tendency to be drawn together, whereas in the triplet state the opposite must occur, not as a result of any electrostatic force but rather as a consequence solely of the form required by quantum mechanics of the wave functions if assumptions (1) and (3) above are to hold. For this particular case, in which both electrons have been placed in nondirectional $s$ orbitals, the spatial distributions of the electrons are found specifically to be as follows, with only the results but not the proofs given here:

(a) The symmetric or singlet state has *three* configurations with high probability: two in which one electron is near the nucleus and the other far out from it and one in which both electrons are simultaneously near the nucleus.

(b) The antisymmetric or triplet state has only *two* configurations with high probability and both result in one electron near the nucleus and one electron far from the nucleus. Since $s$ orbitals have no angular dependence, the *electron correlation*, the correlation between electron positions, is radial only. However, from the standpoint of stereochemistry we are interested only in wave functions that involve angular terms. Therefore it will be of far more interest to consider the next-simplest excited state in the helium atom, $1s^1 2p^1$.

For this particular case, we can designate the two occupied orbitals as $\psi_a$ and $\psi_b$. Each of these is assumed to represent a solution of the one-electron problem and therefore is an unperturbed wave function. That is, we shall neglect interelectronic repulsion effects that arise from electrostatics. Satisfactory solutions to the unperturbed Schroedinger equation for electrons 1 and 2 will be

(7–1) $\qquad \psi_{(1,2)} = \psi_a(1)\psi_b(2)$

and

(7–2) $\qquad \psi_{(2,1)} = \psi_a(2)\psi_b(1)$

where these are orbital wave functions rather than total wave functions. Since electrons are indistinguishable, the probability, $|\psi|^2$, for a configuration in which electron 1 is at position $a$ and electron 2 is at position $b$ must always equal the probability of the configuration for which the electrons are interchanged. That is,

(7–3) $\qquad |\psi_{(1,2)}|^2 = |\psi_{(2,1)}|^2$

Therefore

$$(7-4) \qquad \psi_{(1,2)} = +\psi_{(2,1)}$$

and

$$(7-5) \qquad \psi_{(1,2)} = -\psi_{(2,1)}$$

Now two solutions that satisfy the conditions set forth in Eqs. (7–4) and (7–5) are

$$(7-6) \qquad \psi_S = \frac{1}{\sqrt{2}} [\psi_a(1)\psi_b(2) + \psi_a(2)\psi_b(1)]$$

and

$$(7-7) \qquad \psi_A = \frac{1}{\sqrt{2}} [\psi_a(1)\psi_b(2) - \psi_a(2)\psi_b(1)]$$

where Eq. (7–6) is *symmetrical to electron interchange*, that is, it does not change sign if we interchange electrons 1 and 2, displaying therefore the property of Eq. (7–4); and Eq. (7–7) is *antisymmetrical to electron interchange*, showing the property of Eq. (7–5). The factor $1/\sqrt{2}$ is necessary to ensure a probability of unity of finding the electrons somewhere in the whole of space.

Now for the specific configuration of $1s^1 2p^1$, satisfactory wave functions are:

$$(7-8) \qquad \psi_a = \psi_{1s} = N_{1s}(e^{-\alpha r_1}) = f_{1s}(r_1)$$

and

$$(7-9) \qquad \psi_b = \psi_{2p} = N_{2p} r_2 (e^{-\alpha r_2}) \cos \theta = f_{2p}(r_2) \cos \theta$$

where $N_{1s}$, $N_{2p}$, and $\alpha$ are constants; $r_1$ and $r_2$ are the respective distances of electrons 1 and 2 from the nucleus; $\theta$ is the angle between a line joining the $p$ electron and the nucleus and a fixed axis (for example, the $z$ axis for a $p_z$ orbital) about which the wave function is symmetric; and $f_{1s}(r_1)$ and $f_{2p}(r_2)$ are the radial portions of the wave functions for the $s$ and $p$ states, respectively.

Next we may rewrite Eq. (7–6) and Eq. (7–7) as

$$(7-10) \qquad \psi_S = \frac{1}{\sqrt{2}} [\psi_{1s}(1)\psi_{2p}(2) + \psi_{1s}(2)\psi_{2p}(1)]$$

and

$$(7-11) \qquad \psi_A = \frac{1}{\sqrt{2}} [\psi_{1s}(1)\psi_{2p}(2) - \psi_{1s}(2)\psi_{2p}(1)]$$

and then using Eqs. (7–8) and (7–9) we can rewrite Eqs. (7–10) and (7–11) as

$$(7-14) \qquad \psi_S = [f_{1s}(r)f_{2p}(r)][\cos \theta_2 + \cos \theta_1]$$

and

$$(7-15) \qquad \psi_A = [f_{1s}(r)f_{2p}(r)][\cos \theta_2 - \cos \theta_1]$$

It is readily seen that $|\psi_S|^2$ will be a maximum when $\theta_1 = \theta_2 = 0$ or $\pi$; and $|\psi_A|^2$ will be a maximum when $\theta_1 = 0$, $\theta_2 = \pi$ (or vice versa).

From the above results we reach the following conclusions. For $\psi_S$ the electrons must have opposed spins, for $\psi_A$ they may have opposed or parallel spins. The important point here is that if the electrons have the same spin, then they can have only the $\psi_A$ space function and hence *must* move apart with a maximum probability of being found at $\pi$, that is, 180° from each other. Neglecting electrostatic forces, physically we interpret this result to mean that electrons in the singlet state, $\psi_S$ (spins opposed), will tend to be on the same side of the nucleus, whereas in the triplet state, $\psi_A$ (spins parallel), they will tend to be on opposite sides of the nucleus.

We may construct *contour diagrams* to show the dependence of these wave functions on the positions, $z_1$ and $z_2$, of the two electrons, where $z$ is the chosen fixed axis and the subscripts refer to electrons 1 and 2. These diagrams are shown in Figure 7–2(a) and (b). The former shows the contour of $\psi_S$, the singlet state, with electrons 1 and 2 located on the $z$ axis, and the latter shows the contour of the triplet state $\psi_A$ on that axis. It can be seen that in both cases there are maxima for the configurations in which one electron is at the nucleus ($z_i = 0$) and one removed from the nucleus by the radius at which the $2p$ orbital has its maximum ($z_j = 1$). More significantly, it can be observed from the contour diagram for the singlet state that the configurations having both electrons on the same side of the nucleus (shaded areas) are favored, whereas for the triplet state, configurations having the electrons on opposite sides of the nucleus (clear areas) are favored. In fact, electrons with the same spin will have a high probability of being located on opposite sides of the nucleus on a straight line passing through the nucleus, thus at an angle of 180° with respect to each other. This is seen to be the same result that is obtained by hybridization of one $s$ and one $p$ atomic orbital to yield the two equivalent $sp$ hybrid orbitals. The only difference here is that we have considered the atomic configuration $1s^1 2p^1$, and assumed a constant radial portion of the wave function for simplicity of presentation, rather than $ns^1 np^1$ which is necessary for forming equivalent hybrid orbitals. But the *angular correlation*, which we were seeking, turns out to be the same regardless of whether or not the $s$ and $p$ orbitals have the same principal quantum number.

We cannot verify our predictions of the radial or angular electron distribution for isolated excited or even ground state atoms, but we can study the molecules which they form. So we are ready now to try to relate these results to molecular stereochemistry. Atoms, such as Be, Zn, Cd, and Hg, which can adopt the $ns^1 np^1$ configuration with the input of a suitable amount of promotion energy will therefore be expected to form bonds at an angle of 180°. In their triatomic molecules they do. Analogously, but not so readily demonstrated mathematically, the atoms that can reach the configuration $ns^1 np^1 np^1$ must adopt bond angles of $\frac{2}{3}\pi$ or 120°, and so forth. Thus, even though the unpaired electrons of our idealized systems just discussed be-

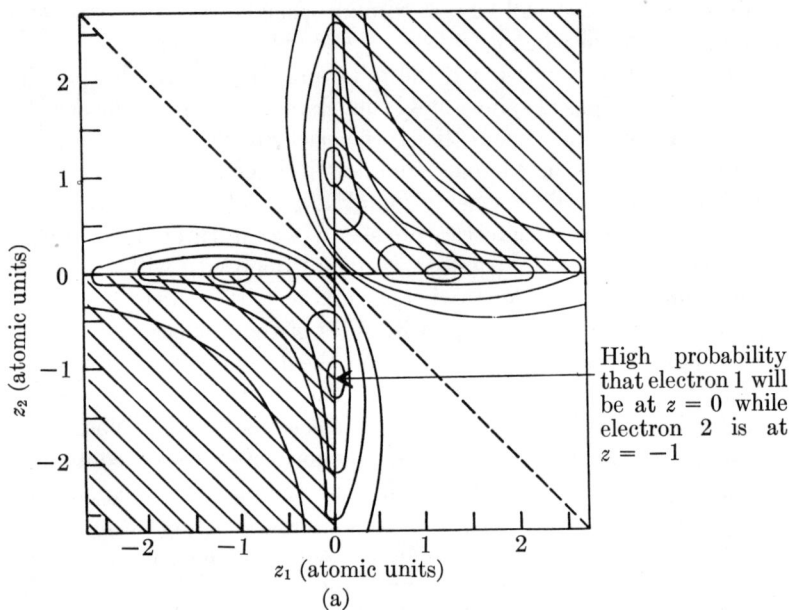

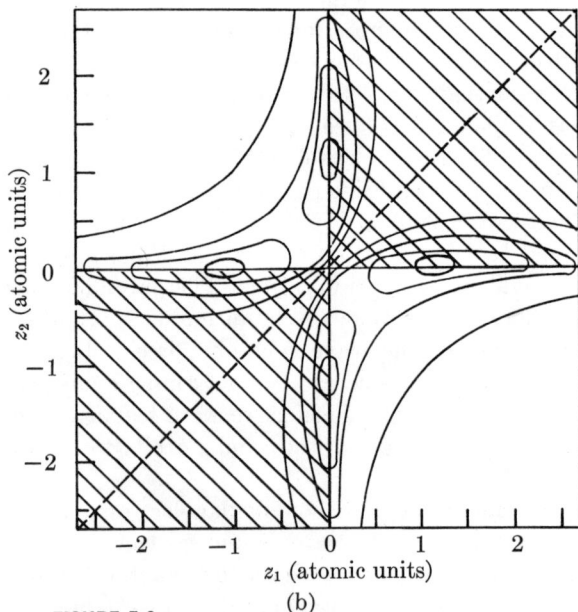

High probability
that electron 1 will
be at $z = 0$ while
electron 2 is at
$z = -1$

FIGURE 7-2

Contour Diagrams of the Wave Function Dependence on Electron Positions $z_1$
and $z_2$ along the Chosen Fixed $z$ Axis, for the Helium Atom in the $1s^1 2p^1$
Configuration: (a) the singlet, $\psi_S$ state; (b) the triplet, $\psi_A$, state. The shaded
areas are regions in which both electrons are on the *same* side of the nucleus.
Both states are seen to include the high probability state in which one electron is
at the nucleus, $z = 0$, and the other at $z = \pm 1$.

come paired with electrons from the surrounding atoms in a completed molecule, we still predict that the electrons with parallel spin must distribute themselves as far apart as possible, consistent with their being radially constrained. We draw the very important, simple conclusion that *the most probable arrangement for* n *electrons of the same spin will, in a molecule, be the most probable arrangement for* n *pairs of electrons.* This conclusion must be slightly revised for certain molecules if we adopt Linnett's double quartet approach (page 201). In these cases electrons of opposing spin are assigned to the same spatial orbital only if there is no alternative lower-energy arrangement available.

We can postulate that in the Ne atom, or $F^-$, $O^{2-}$, $N^{3-}$, and $C^{4-}$ ions, there will be a set of four electrons of the same spin having a maximum probability of being found at the vertices of a regular tetrahedron and another set of four electrons having spin opposite to that of the first set and having maximum probability of being at the corners of a second, not totally independent, tetrahedron. The two tetrahedral sets of electrons will have a tendency to be drawn into coincidence by the operation of the Pauli principle, but this will be strongly opposed by the Coulombic repulsion operating between electrons. However, the tetrahedra will be brought into increasing approximate coincidence by the formation of localized bonds. We may justify this reasoning by considering the following hypothetical steps, which should lead to an even greater aligning of the two tetrahedra: (1) take a proton from the Ne nucleus and form HF; (2) take a proton from the F nucleus and form $H_2O$; (3) take a proton from the O nucleus and form $H_3N$; (4) take a proton from the N nucleus and form $H_4C$. Now note the increasing order of protonic or Lewis base strength, $HF < H_2O < H_3N$ or better, $F^- < HO^- < H_2N^-$ $< H_3C^-$, which may be attributed to the increasing alignment of the remaining nonbonding electrons, bringing pairs into closer coincidence.

In this connection, it is interesting to note that Ne, Ar, Kr, and Xe, all with four outer electron pairs or two antiparallel sets of four, each has the cubic closest packing arrangement in its solid state, just as $CH_4$ does, rather than the hexagonal closest packing found for solid He and theoretically predicted for all spherically symmetrical species.

Now let us assume that the valence shell electron pairs in a multivalent atom, A, are all at the same average distance from the nucleus. This will only be strictly true for $AX_n$ molecules where $n = 2, 3, 4$, and 6, and there are no multiple bonds or lone pairs on A. Then, from what we have already stated, we may reason that the most probable distribution of the $n$ electron pairs is the same as the most probable distribution of a set of $n$ particles on the surface of a sphere under the influence of an appropriate force law. Both the Coulombic force law, $1/r^2$, and what we may call the *Pauli force law*, $1/r^m$ with $m$ large (for example, $> 6$), lead to the same arrangements of particles if we use these laws to perform the mathematical operation of *maximizing the least distance between particles* or, equivalently, minimizing the interaction between particles. (See Table 7–1.) An exception is the case

**TABLE 7-1. Equilibrium Arrangements of Like Particles on the Surface of a Sphere Obeying a Force Law $1/r^2$ or $1/r^m (m > 2)$.**

| No. of Particles (or Electron Pairs) | Arrangement |
|:---:|:---|
| 2 | linear |
| 3 | equilateral triangle |
| 4 | tetrahedron |
| 5 | trigonal bipyramid |
| 6 | octahedron |
| 7 | (see text) |
| 8 | square (Archimedes) antiprism |
| 9 | tripyramid[a] |

[a] A trigonal prism plus an extra particle opposite each of the three rectangular faces.

of seven particles or electron pairs,[7] in which we have the following results for different values of $m$ (in $1/r^m$), in order of decreasing symmetry:

if $m = 1$, $D_{5h}$, pentagonal bipyramid, 1:5:1,
e.g., $UF_7^{3-}$, $UO_2F_5^{3-}$, $IF_7(?)$, $Zr(HF)F_7^{3-}$

if $m > 6$, $C_{3v}$, irregular octahedron (7th particle outside a face of a distorted octahedron), 1:3:3, e.g., $Nb(Ta)F_7^{2-}$, $NbOF_6^{3-}$

if $m > 4$, $C_{2v}$, irregular trigonal prism (7th particle outside a rectangular face of a distorted prism), 1:4:2, e.g., $A$-modification oxides, $M_2O_3$, of La,Ce,Pr,Nd

if $2 < m < 5.6$, $C_2$, figure has no name, 1:2:2:2, $IF_7$ and perhaps most others?

There is some reason to believe that in most of the molecules having seven electron pairs on the central atom the electron pairs are arranged in the low-symmetry $C_2$ structure. It also seems very likely that for molecules of the type $AX_n$ for $n \geq 6$ the shape may be determined primarily or solely by the repulsions operating between the ligands. Therefore the predictions of molecular structure derived from calculations for particles on a sphere must be considered less reliable in the case of seven particles than when smaller numbers are considered.

It is a significant fact that the predicted shapes given in Table 7-1 for $AX_n$ molecules are found to hold for all values of $n$ from 2 to 6, whether or not the electron pairs are equivalently involved, for *all* known molecules of the non-transition elements, with the possible exception of $Sb(C_6H_5)_5$, which appears to have near-tetragonal pyramidal symmetry in the solid state.

Perhaps the most important feature of the VSEPR theory is its emphasis upon valence shell lone pairs as well as the bond pairs in the explanation and prediction of molecular stereochemistry. In Table 7-2 the various possible

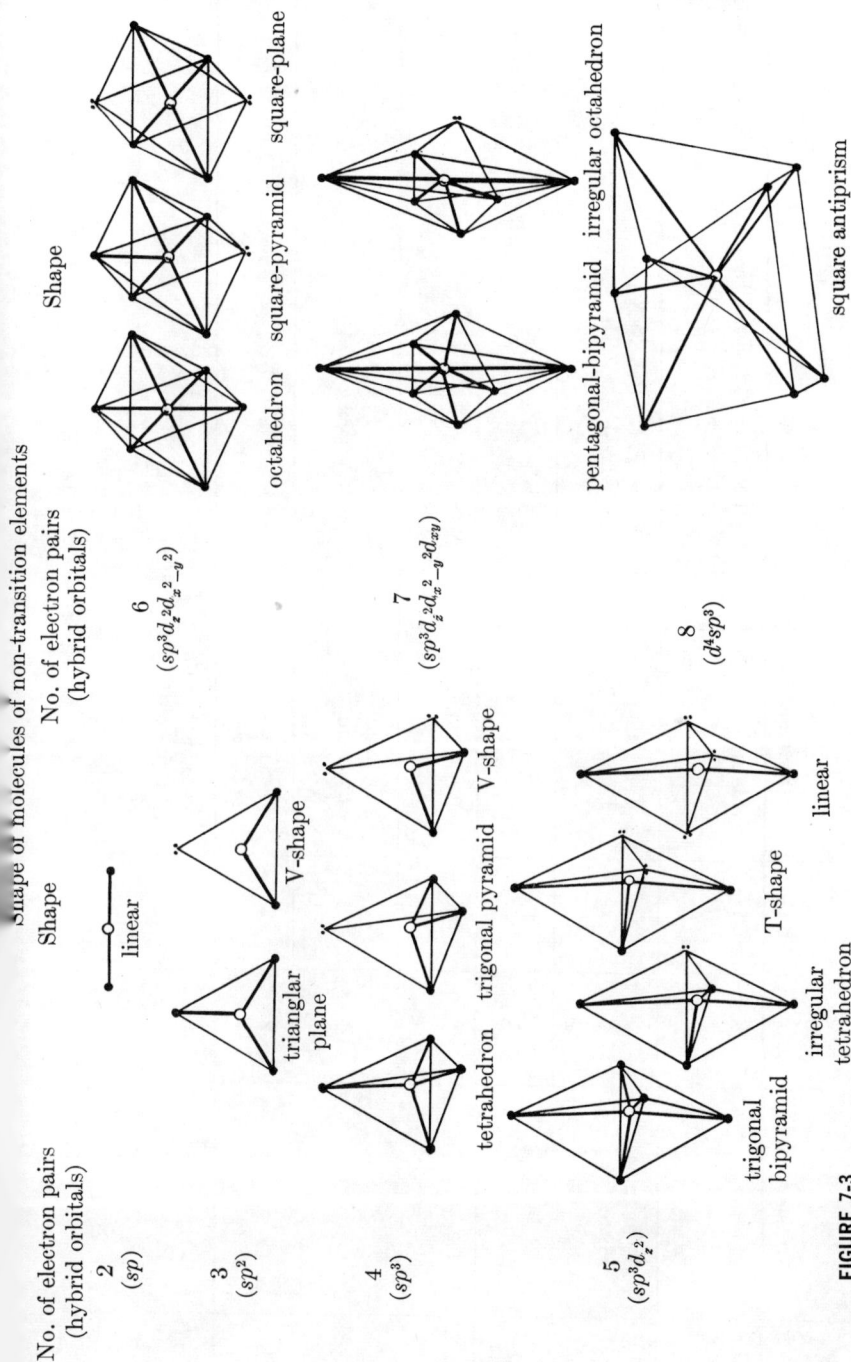

Shape of molecules of non-transition elements

No. of electron pairs (hybrid orbitals)

Shape

6
$(sp^3d_{z^2}d_{x^2-y^2})$

square-plane

square-pyramid

octahedron

7
$(sp^3d_{z^2}d_{x^2-y^2}d_{xy})$

irregular octahedron

pentagonal-bipyramid

square antiprism

8
$(d^4sp^3)$

No. of electron pairs (hybrid orbitals)

Shape

2
$(sp)$

linear

3
$(sp^2)$

triangular plane

V-shape

4
$(sp^3)$

tetrahedron

trigonal pyramid

V-shape

5
$(sp^3d_{z^2})$

trigonal bipyramid

irregular tetrahedron

T-shape

linear

**FIGURE 7-3**

Shapes of Molecules of Non-transition Elements. (See text for discussion of the case of seven electron pairs.)

**TABLE 7–2.   Arrangements of Electron Pairs in Valence Shells and the Shapes of Molecules.**[a]

| No. Electron Pairs | Electron Pairs Arrangement | No. B.P. | No. L.P. | Molecular Shape (formula) | Examples |
|---|---|---|---|---|---|
| 2 | Linear | 2 | 0 | Linear $AX_2$ | $Ag(NH_3)_2^+$, $(Zn, Cd, Hg)(CH_3)_2$, $(Au, Ag)(CN)_2^-$, $AgCl_2^-$ $UO_2^{2+}$, $(HgO)_2$, $(AgCN)_x$ |
| 3 | Triangular Plane | 3 | 0 | Triangular Plane $AX_3$ | $BX_3(X = F, Cl, Br)$, $GaX_3$, $InX_3$, $B(CH_3)_2F$, $In (CH_3)_3$ |
|  |  | 2 | 1 | V-Shape $AX_2E$ | $SnX_2(gas)$, $PbX_2$ $(X = Cl, Br, I)$ |
| 4 | Tetrahedron | 4 | 0 | Tetrahedron $AX_4$ | $BeX_4^{2-}$, $BX_4^-$, $CX_4$, $NH_4^+$, $BeO$, $ZnO$, $AsX_4^+$, $GeF_4$, $Al_2Cl_6$ |
|  |  | 3 | 1 | Trigonal Pyramid $AX_3E$ | $NX_3(X = H, F, Cl)$, $PF_3$, $AsX_3$ $SbX_3$, $P_4O_6$, $As_4O_6$, $Sb_2O_3$ $H_3O^+$ |
|  |  | 2 | 2 | V-Shape $AX_2E_2$ | $H_2O$, $F_2O$, $SCl_2$, $SeX_2$, $TeBr_2$, $NH_2^-$ |
| 5 | Trigonal Bipyramid | 5 | 0 | Trigonal Bipyramid $AX_5$ | $PF_5$, $PCl_5$ (gas), $PF_3Cl_2$ $(Nb, Ta)Cl_5$, $(Nb, Ta)Br_5$, $V_2O_5$, $Sb(CH_3)_3Cl_2$, $Zn(terpy)Cl_2$,[c] $Zn(aca)_2H_2O$[c] |

**Table 7-2.** (Continued)

| | | | | | |
|---|---|---|---|---|---|
| 5 | Trigonal Bipyramid (continued) | 4 | 1 | Irregular Tetrahedron $AX_4E$ | $TeCl_4$, $(S,Se)F_4$, $R_2(Se,Te)X_2$ |
| | | 3 | 2 | T-Shape* $AX_2E_2$ | $ClF_3$, $BrF_3$, $C_6H_5ICl_2$ |
| | | 2 | 3 | Linear* $AX_2E_3$ | $ICl_2^-$, $I_3^-$, $XeF_2$ |
| 6 | Octahedron | 6 | 0 | Octahedron $AX_6$ | $AlF_6^{3-}$, $SiF_6^{2-}$, $PF_6^-$, $PCl_6^-$, $(S,Se, Te)F_6$, $S_2F_{10}$, $(Sn, Pb)Cl_6^{2-}$ $(Sn, Pb)(OH)_6^{2-}$, $SbF_6^-$, $Te(OH)_6^-$, $(Ta, Nb, V, Fe)F_6^{n-}$ |
| | | 5 | 1 | Square Pyramid $AX_5E$ | $IF_5$, $BrF_5$, $ClF_5$ $SbF_5^{2-}$, $SbCl_5^{2-}$ |
| | | 4 | 2 | Square Plane* $AX_4E_2$ | $ICl_4^-$, $L_2Cl_6$, $BrF_4^-$, $XeF_4$ |
| 7 | Pentagonal Bipyramid | 7 | 0 | Pentagonal Bipyramid[b] $AX_7$ | $IF_7$ |
| | | 6 | 1 | Irregular* Octahedron $AX_6E$ | $SbBr_6^{3-}$, $SeBr_6^{2-}$, $XeF_6$ |

[a] The starred molecular configurations are the ones actually observed, although other molecular geometries are theoretically possible for these particular numbers of electron pairs.

[b] This structure may not be that for $IF_7$; see the text, page 316.

[c] Terpy = terpyridyl; aca = acetylacetonate ion.

arrangements of electron pairs in the valence shells of atoms and the resultant molecular shapes are given along with examples drawn from non-transition elements, or $d^0$, $d^5$ (spin-free), and $d^{10}$ transition elements, since the latter have a spherically symmetrical nonbonding core. In Figure 7–3 are illustrated the electron and molecular shapes for 2 to 8 electron pairs. It may be noted that for the molecules $AX_4E$, $AX_3E_3$, $AX_2E_3$, $AX_4E_2$, and $AX_6E$, where A = central atom, X = ligand, E = lone electron pair, there are alternative positions for the lone electron pairs, but all of the observed shapes are as shown in Figure 7–3 and fit a stereochemical rule to be elaborated later. It should also be noted that the shapes *predicted* for the new molecules $XeF_2$, $XeF_4$, and $XeF_6$ *have* been experimentally verified, although the *extent* of distortion of $XeF_6$ from a regular octahedron is as yet uncertain. If $XeF_8$ (eight bond pairs) really exists, as reported by at least one laboratory, and if $XeF_7^-$ (one lone and seven bond pairs) and $XeF_8^{2-}$ (one lone and eight bond pairs) ions really exist, they would be predicted to have the antiprism, distorted antiprism, and distorted tripyramid structures, respectively.

It is, of course, no mere accident that it requires the most highly electronegative of elements, that is, the most electronegative in $\sigma$-bond formation, to attain very high coordination numbers as well as very high oxidation states. This fact may be traced to the necessity (*vide infra*) for the $\sigma$-bonding electron pairs to be drawn out sufficiently from the central atom to effectively lower their repulsive interactions. In this manner we can understand many similar observations such as: $AlCl_4^-$ but $AlF_6^{3-}$; $SiCl_4$ but $SiF_6^{2-}$; $PCl_4^+$ but $PF_6^-$, which is more stable than $PCl_6^-$; $SCl_4$ but $SF_6$; $BrCl_3$ but $BrF_5$; $ICl_5$ but $IF_7$. Hybridization theory would propose that the fluoride ion permits a more ready use of the outer $d$ orbitals of the central ion for additional hybridization. The mechanism would involve the withdrawing of charge from the central atom, causing its effective nuclear charge to increase thereby resulting in a contraction of the radial size of the $d$ orbitals. While this contraction may or may not actually occur, its assumption is unnecessary in the new theory. We shall generalize some of the foregoing observations later in a stereochemical rule.

## STEREOCHEMICAL RULES FOR NON-TRANSITION ELEMENTS

Adopting in part the method of presentation by Gillespie,[6] we shall formulate and briefly discuss a set of rules that appear to be generally valid for nontransition elements and indeed for all elements when their atoms possess a spherically symmetrical nonbonding core. This generally arises from a closed or half-filled penultimate shell or subshell. These rules are actually stereochemical generalizations. Although we are well aware of their qualitative nature, because of their success in logically correlating molecular structure we feel there is great value in their detailed discussion.

*Rule 1: Lone pair electrons (E) repel adjacent electron pairs more strongly* *than do bonding electron pairs $(X)$; furthermore, the repulsions are expected to increase in the order:* $X$—$X < X$—$E < E$—$E$.

This rule is readily understandable since the lone pair electrons are under the influence of but a single positive center. They would therefore be expected to occupy a broader orbital with a greater electron density radially distributed closer to the central atom than the bond pair electrons which are drawn out between two positive centers. This might be pictured diagrammatically as in Figure 7–4.

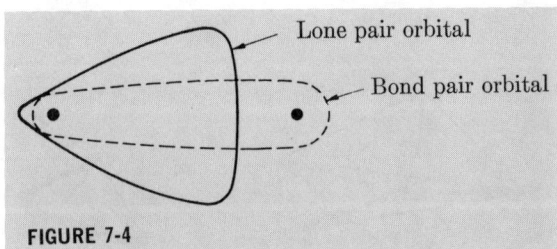

**FIGURE 7-4**

Diagrammatic Representation of the Spatial Differences between a Lone Pair and a Bond Pair Orbital.

This rule may be used to explain the well-known bond angle decrease in the series $CH_4$, 109.5° > $NH_3$, 107.3° > $H_2O$, 104.5°. Furthermore, it is expected that the lone pair will exert a greater effect than a lone electron, and the bond angles in the following series bear this out:

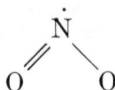          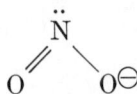

180°                    134°                    115°

Likewise this rule offers a reasonable explanation of why the lone pairs occupy equatorial rather than axial positions in the trigonal bipyramidal electron arrangements of $AX_4E$, $AX_3E_2$, $AX_2E_3$ molecules; and occupy *trans* rather than *cis* positions in the octahedral electron arrangement of $AX_4E_2$ molecules. In the trigonal bipyramid a group in an equatorial position has *two nearest neighbors at 90°* and two next-nearest neighbors at 120°, whereas an axial group has *three nearest neighbors at 90°* and one next-nearest neighbor at 180°. If we accept the postulate that repulsions due to Pauli forces decrease rapidly with distance, or with the angle between electron pairs, then repulsions only become large at 90° and are far less significant at 120° or greater. The structures shown in Figure 7–5 are readily explained in these terms. The detailed matter of bond lengths and bond angles will be dealt with later under another rule which applies more specifically to molecules of this type.

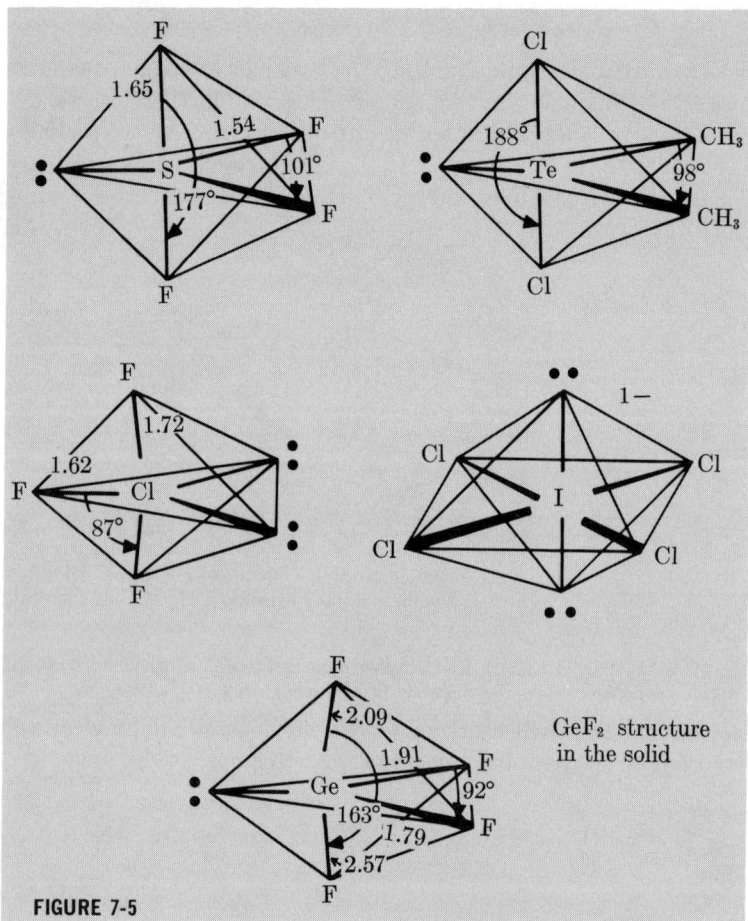

**FIGURE 7-5**

Structures of Some Molecules Having Five and Six Valence Shell Electron Pairs on the Central Atom.

*Rule 2: Repulsions exerted by bond pairs decrease as the electronegativity of the bonded atom increases.*

Here we recognize that a bonded atom draws a sigma-bonded electron pair away from the central atom nucleus, thereby contracting and thinning out the orbital. The more electronegative the bonded atom, therefore, the more the electron density is displaced toward it in the $\sigma$ bond. This is a plausible explanation for the following observed decreases in bond angle:

$$OH_2, 104.5° > OF_2, 103.2°$$

$$NH_3, 107.3° > NF_3, 102°$$

$$PI_3, 102° > PBr_3, 101.5° > PCl_3, 100.3° > PF_3, 97.8°$$

$$AsI_3, 100° > AsBr_3, 99.7° > AsCl_3, 98.4° > AsF_3, 96°$$

However, since $PF_3$ has the angle 104° and $AsF_3$, 102°, we have yet to account for these apparent exceptions to this rule. An important factor not yet discussed takes precedence in these cases, and we shall delay the explanation until we come to a later rule which will cover it.

*Rule 3: Multiple bonds do not affect the gross stereochemistry of a molecule. Rather, the geometry is determined primarily by the number of sigma bond pairs and the number of lone pairs.*

This rule is easy to understand when we recognize that a sigma bond pair and a lone pair each occupies a distinct stereochemical site on the central atom, whereas the pi bonds occupy roughly the same site as the sigma bond they accompany. That is, the difference between a sigma bond and its accompanying one or two pi bonds is solely in the symmetry and strength of the bond itself, not in its direction. In support of this rule we list, in Table 7–3, examples of molecules containing multiple bonds, mainly involving oxygen, in which the central atoms have varying numbers of sigma and lone pairs of electrons.

Now, of course, there are going to be some stereochemical effects of multiple bonds. The most obvious effect of a multiple bond will be upon the bond lengths and, at least indirectly, these have some influence upon bond angles and hence stereochemistry. The length of a bond varies not only with the nature and size of the bonded atoms, but with the *order* of the bond formed between them. The stereochemical theory we have been discussing is of no value to us in this matter, but it will be instructive, nevertheless, to briefly examine variations in internuclear distances. Ideally, we can assign a covalent radius to an atom which will remain nearly constant in all covalent molecules formed by that atom (see Chapter 4). Furthermore, if the bond order changes for a given atom from molecule to molecule, we can estimate the corresponding change in its radius, and in certain cases, notably with carbon compounds, we can describe the change in terms of the degree and kind of hybridization which is occurring. The latter is illustrated in Table 7–4, where the characteristic carbon-carbon bond lengths are listed for various types of compounds, reproducing reported bond lengths for virtually all representatives of that type to within ±0.02 Å. This table illustrates the usefulness of the hybridization theory in the classification of bond types and the estimation of bond lengths.[8] From the table it is seen that the superposition of one or two $\pi$ bonds on a $\sigma$ bond effects a marked reduction in the bond length. This observation may be verified for many other atoms besides carbon which are known to form $\pi$ bonds. Values for multiple bond covalent radii are listed for several elements in Table 7–5. A general conclusion that can be drawn from the values listed in Table 7–5 is that a second $\pi$ bond causes a further, but smaller, decrease in a given radius.

We now look briefly at the effects that multiple bonding will have upon bond angles. In Table 7–6, we have recorded some bond angles for molecules possessing multiple bonds. It is seen that for all of the phosphorus compounds, $POX_3$ and $PSX_3$, the bond angles XPX fall not far from, but

**TABLE 7-3. Shapes of Molecules Containing Multiple Bonds.**

| No. σ Bonds + L.P. | Arrangement of Electron Pairs | No. σ Bonds | No. L.P. | Molecular Shape | Examples |
|---|---|---|---|---|---|
| 2 | Linear | 2 | 0 | Linear | $O=C=C=C=O$, $H-N=C=S$, $O=C=O$, $H-C\equiv N$, $H-N=C=O$, $O=N=O$, $H_2C=C=CH_2$ |
| 3 | Triangular Plane | 3 | 0 | Triangular Plane | $Cl_2C=O$, $H-\overset{\overset{\displaystyle O-}{\displaystyle \vert}}{C}=O$, $SO_3, NO_3^-$, $NO_2X$, $X_2C=CX_2$ |
|  |  | 2 | 1 | V-Shape | $:SO_2$, $:OO_2$, $:NOCl$, $:NO_2^-$, $:N_2F_2$, |
| 4 | Tetrahedron | 4 | 0 | Tetrahedron | $SO_2Cl_2, SO_4^{2-}$, $(P, As)OCl_3$, $ClO_3F$, $PO_2F_2$, $PO_3F^{2-}$ |
|  |  | 3 | 1 | Triangular Pyramid | $:SOCl_2$ $:SO_3^{2-}$, $:ClO_3^-$, $:SeOCl_2$, $:IClO_3^-$, $:XeO_3$ |
|  |  | 2 | 2 | V-Shape | $BrO_2^-$, $ClO_2^-$ |

**TABLE 7-3.** (Continued)

| | | | | | |
|---|---|---|---|---|---|
| 5 | Trigonal Bipyramid | 5 | 0 | Trigonal Bipyramid | $O{=}SF_4$ |
| | | 4 | 1 | Irregular Tetrahedron | $:IO_2F_2^-$, $:IOF_3$ |
| 6 | Octahedron | 6 | 0 | Octahedron | $O{=}I(OH)_5$, $O{=}IF_5$ |
| | | 5 | 1 | Square Pyramid | $O{=}\ddot{Se}(py)_2Cl_2$<br>$O{=}\ddot{X}eF_4$ |

**TABLE 7–4.   Effect of Atom Hybridization on the C—C Bond Length.[8]**

| C—C Hybridization[a] | % of $s$ Character in the $\sigma$ Bond | Valence-Bond Structure | Characteristic C—C Bond length, Å |
|---|---|---|---|
| te-te | 25 | $-\overset{|}{\underset{|}{C}}-\overset{|}{\underset{|}{C}}-$ | 1.54 |
| te-tr | 29 | $-\overset{|}{\underset{|}{C}}-\overset{|}{C}\!\!=$ | 1.50 |
| te-di | 33 | $-\overset{|}{\underset{|}{C}}-C\!\equiv$ | 1.46 |
| tr-tr | 33 | $\overset{|}{C}\!\!=\!\!\overset{|}{C}$ | 1.47 |
| tr-di | 40 | $\overset{|}{C}\!-\!C\!\equiv$ | 1.42–1.47 |
| di-di | 50 | $\equiv\!C\!-\!C\!\equiv$ | 1.38 |
| tr-tr + $\pi$ | 33 | $\overset{|}{C}\!=\!\overset{|}{C}$ | 1.34 |
| tr-di + $\pi$ | 40 | $\overset{|}{C}\!=\!C\!=$ | 1.31 |
| di-di + $\pi$ | 50 | $=\!C\!=\!C\!=$ | 1.28 |
| di-di + $2\pi$ | 50 | $-C\!\equiv\!C-$ | 1.20 |
| tr-tr + $\frac{1}{2}\pi$ | 33 | benzene | 1.40[b] |
| tr-tr + $\frac{1}{3}\pi$ | 33 | graphite | 1.42[b] |

[a] te stands for $sp^3$ hybrid and tr and di stand for $sp^2$ and $sp$ hybrids, respectively.
[b] These values, obtained by linear interpolation between tr-tr and tr-tr + $\pi$ values are to be compared with the actual values for benzene and graphite of 1.397 and 1.421 Å, respectively.

**TABLE 7–5.   Multiple Bond Covalent Radii (Å).**

| Atom | Single Bond | Double Bond | Triple Bond | Atom | Single Bond | Double Bond |
|---|---|---|---|---|---|---|
| C | 0.771 | 0.665 | 0.602 | Ge | 1.223 | 1.12 |
| N | 0.74 | 0.60 | 0.55 | As | 1.21 | 1.11 |
| O | 0.74 | 0.55 | 0.50 | Se | 1.17 | 1.07 |
| Si | 1.173 | 1.07 | 1.00 | Sn | 1.412 | 1.30 |
| P | 1.10 | 1.00 | 0.93 | Sb | 1.41 | 1.31 |
| S | 1.04 | 0.94 | 0.87 | Te | 1.37 | 1.27 |

**TABLE 7-6.   Bond Angles in Molecules Containing Multiple Bonds.**

| Molecule | XAX Angle | Molecule[a] | XSX Angle | XSO Angle |
|---|---|---|---|---|
| $O{=}PF_3$ | 102.5° | $O{=}SF_2$ | 92.3° | 106.8° |
| $O{=}PCl_3$ | 103.5° | $O{=}SBr_2$ | 96° | 108° |
| $O{=}PBr_3$ | 108° | | | |
| $S{=}PF_3$ | 100.3° | $O{=}S(CH_3)_2$ | 100° | 107° |
| $S{=}PCl_3$ | 100.5° | $O{=}S(C_6H_5)_2$ | 97.3° | 106.2° |
| $S{=}PBr_3$ | 106° | | | |
| $O{=}CH_2$ | 118° | $O_2SF_2$ | 96.1° | 124°[a] |
| $O{=}CF_2$ | 112.5° | $O_2SCl_2$ | 112.2° | 110.8° |
| $O{=}CCl_2$ | 111.3° | | | |
| $O{=}C(CH_3)_2$ | 119.6° | $O_2S(CH_3)_2$ | 115° | 110.4° |
| $O{=}C(NH_2)_2$ | 118° | $O_2S(NH_2)_2$ | 112.1° | 125° |

[a] Note that the difference in angles is greatest with the F compounds because F draws out the bond pairs to greatest extent.

definitely below, the tetrahedral angle. Since there are four $\sigma$-bond pairs and no lone pairs, a tetrahedral angle would be predicted for the gross stereochemistry. However, one of the bonds is multiple, that is, it consists of four electrons rather than just two, and the added electron density of the multiple bond exerts a greater repulsive effect than that exerted by a single bond; hence, the reduction in bond angle. We see the same effect in the $COX_2$ compounds, where instead of the 120° angle predicted for three sigma and no lone pairs, we obtain smaller XCX angles due to the additional repulsion of the C=O bond electrons. The same effect must be responsible for the observed bond angles in the sulfur compounds, $SOX_2$ and $SO_2X_2$, given in the Table. Thus, it would appear that a safe generalization would be that *multiple-bond orbitals repel other orbitals more strongly than single-bond orbitals.*

*Rule 4: Repulsions between electron pairs in filled shells are greater than those between electron pairs in incompleted shells.*

We may reasonably assume that in a filled shell the orbitals effectively fill all of the available space around an atom. Therefore, anything that attempts to reduce an angle between such filled orbitals will be strongly resisted by Pauli repulsion forces, preventing appreciable orbital overlap.

The valence shells of the first-row atoms, Li to Ne, are completely filled by four electron pairs. Therefore, for the molecules $AX_4$, $AX_3E$, and $AX_2E_2$, where *both* A and X are first-row atoms, the bond angles will not deviate very greatly from 109.5°. In fact, they are always found within several degrees of this value. The largest deviation we know about occurs in $NF_3$, which has an angle of 102.1°, or a 7.4° deviation. Note that it requires the three bonds in the $AX_3E$ molecule to be to the very electronegative fluorine atoms, which are able to draw out the bond pairs to a point that allows such a large reduction in angle from the preferred tetrahedral value.

The atoms of the next period, Na to Ar, can, in theory, hold up to nine electron pairs in their valence shells. For reasons not well understood, they appear to be completely filled by less than that number. Thus, if they are all lone pairs, as in argon, the maximum number is four, but if they are bond pairs, it appears to be six. That is, it seems that 90° bond angles are allowed for these larger atoms, but that great resistance to a decrease below 90° exists. This is similar to the resistance to decreases below 109.5° with the members of the preceding period. It is interesting, however, that the bond pairs must be drawn out strongly from the central atom before a stable species with six electron pairs can form. Consider the known species, $SF_6$, $PF_6^-$, $SiF_6^{2-}$, and $AlF_6^{3-}$, a series in which the lability of the fluoride increases rapidly in the order given due to the increasing charge accumulating on the species. The important point to be made in connection with this rule is that when an element, either from this period or from a later period, has only four valence shell electron pairs, the orbitals may be compressed, presumably because of the larger volume available, to make 90° angles with each other at the nucleus before strong Pauli repulsion sets in.

Now we can use the foregoing discussion to explain the following observed bond angle decreases:

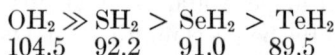

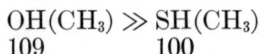

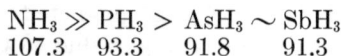

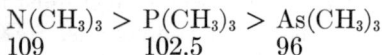

$$OH_2 \gg SH_2 > SeH_2 > TeH_2$$
$$104.5 \quad 92.2 \quad \;\; 91.0 \quad \;\; 89.5$$

$$OH(CH_3) \gg SH(CH_3)$$
$$109 \qquad \quad\;\; 100$$

$$NH_3 \gg PH_3 > AsH_3 \sim SbH_3$$
$$107.3 \quad 93.3 \quad\;\; 91.8 \quad\;\; 91.3$$

$$N(CH_3)_3 > P(CH_3)_3 > As(CH_3)_3$$
$$109 \qquad\;\; 102.5 \qquad\; 96$$

For example, since the four valence shell pairs in $PH_3$, three bond pairs and one lone pair, do not constitute a filled shell, the lone pair orbital can force the bond pair orbitals to close to an angle near 90°. At this point the Pauli repulsions begin to become large. Thus it is expected that the lone pair in $PH_3$ occupies a much larger volume of space than the lone pair in $NH_3$. This may well account for the marked decrease in basicity and metal coordinative ability observed for $PH_3$ compared to $NH_3$. Note also that the existence of two lone pairs in $H_2S$ causes the bond angle to approach even closer to 90°. In contrast, the valence bond "explanation" that is usually offered for the large decrease in the HXH bond angle going from $H_2O$ to $H_2S$ or $NH_3$ to $PH_3$ is that the hybrid orbitals have "mostly $p$ character" whereas the lone pair orbital has "mostly $s$ character." This, it would seem is not really an explanation. Actually, the $3s$ and $3p$ orbitals are closer in energy to each other than are the $2s$ and $2p$ orbitals and these should be expected to hybridize with each other even better. Accordingly this would lead one to expect an order that is the reverse of that which is experimentally observed.

*Rule 5: When an atom with a filled valence shell and one or more lone pairs is bonded to an atom with an incomplete valence shell, or a valence shell that can become incomplete by electron shifts, there is a tendency for the lone pairs to be partially transferred from the filled to the unfilled shell.*

This generalization is reasonable if we may assume that the repulsions between electron pairs in the filled shell are much larger than the repulsions in the incomplete shell. This effect probably is the major cause of bond shortening found in $BF_3$ and $SiF_4$, for example, where important contributions must arise from forms such as:

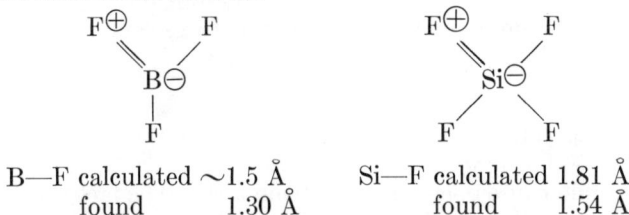

$$F^{\oplus} \diagdown \quad \diagup F$$
$$B^{\ominus}$$
$$| $$
$$F$$

$$F^{\oplus} \diagdown \quad \diagup F$$
$$Si^{\ominus}$$
$$\diagup \quad \diagdown$$
$$F \qquad F$$

B—F calculated $\sim 1.5$ Å     Si—F calculated 1.81 Å
    found         1.30 Å             found         1.54 Å

The high charge density that must reside on the relatively small fluoride can be lowered if the atom to which the fluoride is bonded has a much lower charge density, and the formal mechanism for this delocalization of charge away from the fluoride is through formation of a $p\pi \rightarrow p\pi$ or $p\pi \rightarrow d\pi$ bond. Thus while the fluorine has a very high, perhaps the highest, "sigma-electronegativity," its "pi-electronegativity" must be very much lower, and in fact we perhaps should say that fluorine possesses some "pi-electropositive" character. In accord with the proposed structure, $BF_3$ is known to be a poorer Lewis acid than $BCl_3$ and $BBr_3$.

More important for stereochemistry are the effects on bond angles of this mechanism for lowering high charge densities. Consider the following angles:

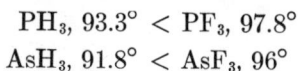

$$PH_3, \; 93.3° \; < \; PF_3, \; 97.8°$$
$$AsH_3, \; 91.8° \; < \; AsF_3, \; 96°$$

and recall that for $NH_3$, 107.3° and $NF_3$, 102.1°, there is an angle decrease rather than an increase. Now from what we have already stated in Rule 2, it would seem that since fluorine is surely more electronegative than hydrogen, the angles should all decrease. However, we can account for the observed trend here by assuming considerable delocalization of fluorine electrons toward the central atoms, P and As, which serves not only to strengthen the bond, but also to reduce the intravalence-shell repulsions in the highly compact fluorine atom. Then partial multiple bonding such as

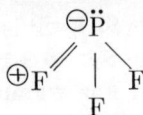

$$\ominus \ddot{P}$$
$$\oplus F \diagdown \quad | \quad \diagdown F$$
$$F$$

serves to increase the FPF bond angles. This may also help explain why $PF_3$ can coordinate so much better to metals than can either $PCl_3$ or $PBr_3$. With the heavier halogens there will not exist the same necessity for delocalization of charge that seems to exist for fluorine. Hence bond angles decrease with increasing ligand electronegativity as we have already observed.

So far our examples have been ones in which the central atom has the incompleted shell. We find examples also of this effect when the central atom has the filled shell and the bonded atoms have incompleted shells. Consider the angles:

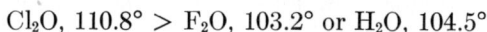

$$Cl_2O, 110.8° > F_2O, 103.2° \text{ or } H_2O, 104.5°$$

This order tempts us to postulate structures such as

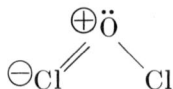

which would cause opening of the bond angle. The fact that the bond angle in $SCl_2$ is only 102° lends support to this proposal. Here the sulfur atom will not have the same need to delocalize its nonbonding electrons.

There exists much more evidence of this phenomenon and we shall simply record some of it here. (a) The molecule $N(SiH_3)_3$ is planar with 120° bond angles, even though the nitrogen ostensibly has a lone pair. (b) The molecules $[Cl_5RuORuCl_5]$ and $[TiCl_2(C_2H_5)]_2O$ are linear, suggesting species such as $\ominus M{=}O{=}M\ominus$ contributing to the electronic structures. (c) The ion $O(HgCl)_3^+$ is triangular planar rather than pyramidal which would have been expected if the oxygen lone pair remains entirely on the oxygen atom.

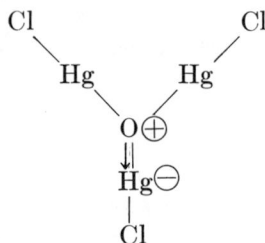

(d) SiOSi bond angles fall in the range 130 to 150°, whereas SiSSi angles are much smaller. For example, in $(SiS_2)_n$ the angle is 80°, in $[(CH_3)_2SiS]_2$ it is 75°, and in $[(CH_3)_2SiS]_3$ it is 110°; but in $\alpha$-quartz it is 142°, in $(SiO_3{}^{2-})_n$ it is 137.5°, in $[(CH_3)_3Si]_2O$ it is 137°, and in $(Cl_3Si)_2O$ it is 130°.

Likewise SOS angles are invariably larger than SSS angles. For example the angle in $(SO_3)_3$ is 114°, in $S_2O_{10}{}^{2-}$ it is 122° and in $S_2O_7{}^{2-}$ it is 124°; but in $S_8$ it is 105°, in $S_4O_6{}^{2-}$ the SSS angle is 103°, and in $S_4{}^{2-}$ it is 104.5°.

Similarly POP angles are larger than PSP angles. For example the angle in $P_4O_{10}$ is 123.5°, but in $P_4S_{10}$ it is 109.5°, and all other known PSP angles fall below the tetrahedral angle whereas all other known POP angles fall into the 120 to 140° range.

(e) COC angles in aromatic compounds are generally in the 120 to 125° range, whereas in aliphatic compounds the angles are very near the tetrahedral angle. For example, the angles in the aliphatic compounds $(C_2H_5)_2O$, $(CH_3)_2O$, 1,4-dioxane, and paraldehyde are 108°, 110°, 108°, and 109.5°, respectively, whereas in $(p\text{-}IC_6H_4)_2O$, $(p\text{-}BrC_6H_4)_2O$, $(C_6H_5)_2O$, and $p\text{-}(CH_3O)_2C_6H_4$ they are 123°, 123°, 124°, and 121°, respectively. We may picture the delocalization as occurring in the following way:

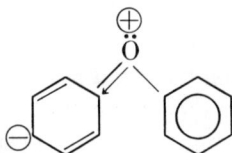

(f) The planar arrangement with 120° bond angles around the nitrogen atoms in urea and formamide may be explained similarly:

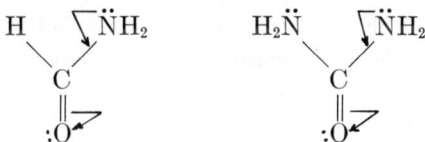

*Rule 6: In a valence shell containing five or seven electron pairs, in which all electron pairs cannot have the same number of nearest neighbors, those pairs with the largest number of nearest neighbors will be located at a greater average distance from the nucleus than the other electron pairs.*

As we have already noted, the axial groups in the trigonal bipyramid have one more nearest neighbor than do the equatorial groups. Therefore it is not surprising that we find axial bond lengths longer than equatorial bond lengths in molecules of the types $AX_5$, $AX_4E$ and $AX_3E_2$. The pertinent data for some examples of molecules of these types is collected in Table 7-7.

**TABLE 7-7.  Bond Lengths (Å) in Some $AX_5$, $AX_4E$, and $AX_3E_2$ Molecules.**

| Molecule | Axial Length | Equatorial Length |
|---|---|---|
| $PCl_5(g)$ | 2.10 | 2.04 |
| $PF_5(g)$ | 1.577 | 1.534 |
| $P(C_6H_5)_5(s)$ | 1.987 | 1.850 |
| $SbCl_5(s)$ | 2.34 | 2.29 |
| $CH_3PF_4(g)$ | 1.612 | 1.543 |
| $SF_4(g)$ | 1.646 | 1.545 |
| $ClF_3(s)$ | 1.716 | 1.621 |
| $BrF_3$ | 1.81 | 1.72 |

Experimentally it is found in $AX_4E$ molecules that the equatorial bond angles are reduced well below the ideal 120° to the range of 96° to 110°. However, the angle between axial groups is not altered significantly from the ideal

180°. This is not surprising since the angle between the equatorial electron pairs (one lone pair and two bond pairs) and the axial groups is already 90°. In light of the earlier discussions we would not expect this angle to deviate to any large extent. For $AX_3E_2$ molecules, in which both lone pairs are in equatorial positions, the added repulsions caused by two lone pairs might be expected to decrease the angles between axial and equatorial bond pairs to slightly below the ideal 90°. This, in fact, is observed with, for example, $ClF_3$ (87.5°), $BrF_3$ (86.2°), and $C_6H_5ICl_2$ (86°).

Finally, it is an empirical fact that *in the trigonal bipyramidal configuration the more electronegative ligands tend to assume the axial positions.*[9] Thus in all of the following trigonal bipyramidal compounds, it is fluorine, or the other halogen if fluorine is not present, that is found in the axial positions: $PF_3Cl_2$, $PF_4(R_2N)$, $PFCl_4$, $R_2PF_3$, $R_3PF_2$, $RPF_4$, $Sb(CH_3)_3X_2(X = Cl, Br, I)$, $Sb(C_6H_5)_3Cl_2$, $Sb(C_6H_5)_2Cl_3$, etc.[9] Since it is possible for the more electronegative group to draw out a bond pair further than a less electronegative group, it is not unexpected that the former groups will be forced or will find their way into the position where the greater interelectron repulsion effects reside.

The possibility that certain molecules or ions in which the central atom has seven valence shell pairs, one lone and six bond pairs, $AX_6E$, are actually regular octahedral structures raises the question of whether the lone pair may become *stereochemically inert* in these systems. Urch[10] has suggested that if the $ns$ orbital ($a_{1g}$) of the central atom plays little or no part in bonding then the extra electron pair may be accommodated in the $a_{1g}$ antibonding molecular orbital without distorting the $O_h$ symmetry of the molecule. It is certainly possible that in the elements which form these species, namely, all later-period elements in high oxidation states with relatively very high effective nuclear charges, the $ns$ electron pair is indeed pulled into the core. Being spherically symmetrical this pair then loses its stereochemical effects. It is also possible that the $O_h$ molecular symmetry is conferred by strong crystal forces. It is interesting in this connection that $XeF_6E$ is not a regular octahedron, but neither is the distortion as large as VSEPR theory would predict.

## References

1. R. S. Drago, "Physical Methods in Inorganic Chemistry," Reinhold Publishing Corporation, New York, 1965.
2. K. Nakamoto, "Infrared Spectra of Inorganic and Coordination Compounds," John Wiley & Sons, Inc., New York, 1963.
3. L. Pauling, *J. Am. Chem. Soc.*, **53**, 1367 (1931).
4. J. Slater, *Phys. Rev.*, **37**, 481 (1931).
5. N. Sidgwick and H. Powell, *Proc. Roy. Soc.*, **A, 176,** 153 (1940).
6. R. Gillespie, *J. Chem. Ed.*, **40**, 295 (1963). See also R. Gillespie and R. S. Nyholm, *Quart. Rev.*, **9**, 339 (1957); R. Gillespie, *Can. J. Chem.*, **38**, 818 (1961);

*J. Chem. Soc.*, **1963**, 4672, 4679, *Angew. Chem.* (*Internat. Ed.*), **6**, 819 (1967); H. B. Thompson and L. S. Bartell, *Inorg. Chem.*, **7**, 488 (1968).

7. T. A. Claxton and G. C. Benson, *Can. J. Chem.*, **44**, 157 1730 (1966).
8. H. Bent, *J. Chem. Ed.*, **37**, 616 (1960), *Chem Revs.*, **61**, 275 (1961); D. R. Lide, *Tetrahedron*, **17**, 125 (1962).
9. E. L. Meutterties and R. A. Schunn, *Quart. Revs.*, **20**, 245 (1966).
10. D. S. Urch, *J. Chem. Soc.*, **1964**, 5775.

## Problems

1. On what basis is the Pauli antisymmetry principle accepted by scientists, that is, how is it justified? Can it be derived from anything more fundamental? Write out its mathematical formulation for electrons.

2. Predict the molecular structures and estimate all of the bond angles in the following molecules and ions:

   (a) $BrNO$              (h) $BO_2^-$
   (b) $TeCS$              (i) $TeBr_4$
   (c) $SeCl_2$            (j) $N_2F_2$
   (d) $NO_2F$             (k) $COBr_2$
   (e) $PBr_2F_3$          (l) $PCl_4^+$
   (f) $SF_3(CF_3)$        (m) $TeBr_6^{2-}$
   (g) $SOF_6$             (n) $(CH_3)_2SnF_2$

3. R. J. Gillespie (*J. Chem. Soc.*, **1963**, 4679) has attempted to extend the VSEPR theory to the transition elements, applying it specifically to the stereochemistry of five-coordination in the molecules of these elements. He proposes that when the interaction between the ligand electron pairs is relatively more important than their interaction with the nonbonding $d$ electrons, which is probably the case in complexes having a large amount of "covalent" character, a trigonal-bipyramidal structure is to be expected. But when the interaction between the bonding electron pairs and the $d$ shell predominates, as is probably the case in complexes with essentially "ionic" bonding, then a square-pyramidal structure is expected. When all the interactions are comparable an intermediate structure might be expected. What reasoning was used to arrive at the foregoing predictions? (*Hint:* Consider ligand field stabilization energies for $d^1$ to $d^9$ configurations in both the square-pyramidal and trigonal-bipyramidal geometries.) What known structures can be cited in support of his suggestions?

4. The $PF_5$ structure is known [K. W. Hansen and L. S. Bartell, *Inorg. Chem.*, **4**, 1775 (1965)] to be trigonal-bipyramidal with different axial (1.577) and equatorial (1.534) bond lengths. Yet the $^{19}F$ NMR spectrum for this compound consists of a widely separated doublet due to P-F spin-spin coupling that shows no indication of separate signals or spin-spin coupling due to the nonequivalent axial and equatorial fluorine atoms. How might these separate experimental results be reconciled? (R. J. Gillespie, *J. Chem. Soc.*, **1963**, 4672.)

5. The very unstable radical $SiF_2$ was recently prepared and studied. Predict its electronic as well as molecular structure and estimate the magnitude of the bond angle. [J. C. Thompson and J. L. Margrave, *Science*, **155**, 669 (1967).]

6. The compound $AlH_3[N(CH_3)_3]_2$ is monomeric. Predict its structure from the knowledge that the electronegativity of H is 2.1 and that of the group $-N(CH_3)_3$

is about 2.4. Check your prediction against the single crystal X-ray study by
C. W. Keitsch, C. E. Nordman and R. W. Parry, *Inorg. Chem.* **2,** 508 (1963).

7. In a recent paper, M. M. Rochkind and G. C. Pimentel, *J. Chem. Phys.*, **42,**
(1965), it is concluded from infrared data and the resulting vibrational assignments and bond stretching force constant that the Cl—O bond in $Cl_2O$ is a single
bond. If this is the case, then how might you rationalize the bond angle data:
$Cl_2O$, 110° and $F_2O$, 103.2°, in some other manner than was done by the VSEPR
model on page 330?

8. Predict the bond angle in the cation $ClF_2^+$. Check your prediction against that of
Christe and Sawodny, *Inorg. Chem.*, **6,** 313 (1967), which is based upon vibrational spectra (IR and Raman) and force constant information.

9. Discuss the problem of the structure of such species as $SeX_6^{2-}$, $TeX_6^{2-}$, $PoX_6^{2-}$,
$AsX_6^{3-}$, $SbX_6^{3-}$, $BiX_6^{3-}$, $IF_6^-$, and $XF_6$. Under what circumstances might the lone
pair become stereochemically inert? (See Ref. 10.)

# Electromotive Force

The driving force of a chemical reaction results from a tendency for the system to approach equilibrium, and from thermodynamics we recall that this tendency is expressed by means of the free energy change, $\Delta G$, in going from the initial to the final state of the system. For this reason it is of interest to determine $\Delta G$, and as it turns out, one way of determining it for a redox system is by means of the relation between the free energy change and the electromotive force of a galvanic cell.

To appreciate this relationship between the free energy change and the cell emf, we can consider a reaction such as that of zinc metal with dilute sulfuric acid,

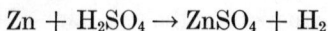

$$Zn + H_2SO_4 \rightarrow ZnSO_4 + H_2$$

If this reaction is allowed, for instance, to take place in a calorimeter, an amount of heat, $q'$, will be evolved and an amount of work, $w'$, will be done by the expansion of the hydrogen gas along with other volume changes. According to the first law of thermodynamics, the change in internal energy of the system will be given by

$$(8\text{--}1) \qquad \Delta E = E_f - E_i = q' - w'$$

If a galvanic cell, such as that shown schematically in Figure 8–1, is now set up, the same net reaction can be made to take place in such a manner that the electrons will be forced to flow through an external circuit and the resultant electrical energy can be harnessed to do useful work. In this particular cell, one electrode is made of zinc metal and the other is of some inert metallic conductor such as platinum. These are placed in solutions containing $ZnSO_4$ and $H_2SO_4$ and are then connected externally by means of an electrical conductor. The same basic reaction will take place as before, but we now note that the hydrogen gas is evolved from the surface of the plati-

**FIGURE 8-1**

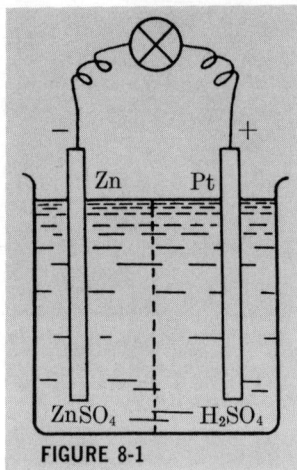

Schematic Diagram of a Galvanic Cell.

num electrode rather than from the zinc. The current flow in the external circuit results from the oxidation reaction

$$Zn = Zn^{2+} + 2e^-$$

which occurs at the zinc electrode-solution interface, and the reduction reaction

$$2\,H^+ + 2e^- = H_2$$

which occurs at the platinum electrode-solution interface. The electrons released at the zinc electrode proceed through the external circuit to the platinum electrode where they participate in the reduction of the $H^+$. This, then, gives the total cell reaction

$$Zn + 2H^+ = Zn^{2+} + H_2$$

which can be considered to be made up of the two half-cell reactions shown.

Again, it can be said that

$$\Delta E = q - w$$

but the $w$ we now have is made up of two parts, the work done against the atmosphere and the electrical work done in the external circuit. Consequently, $\Delta E$ can be expressed as

(8–2) $$\Delta E = q - (w_e + P\Delta V)$$

Although the change in internal energy is independent of the path, the same is not true of the heat, $q$, and the work, $w$. If a large current flow is permitted, considerable heating effects will be observed as a result of electrical resistance in the cell. On the other hand, if the current flow can be made imperceptibly small, the heating effects will become negligible. The great advantage of such cells in the study of free energy changes lies in the fact that they can be made to operate very nearly reversibly. This is accomplished by placing an electrical potential in the external circuit in such a manner that an emf occurs in opposition to that of the galvanic cell. The opposing emf is varied by means of a potentiometer until the current flow from the cell is essentially zero. Under these conditions, the cell may very well approach reversibility. This is readily tested by changing the direction of the current and allowing an infinitesimally small current flow in the opposite direction. If the cell is reversible, the cell reaction will proceed in the reverse direction with the same efficiency as it did in the forward direction.

For a reversible reaction

$$(8\text{-}3) \qquad \Delta G = -w_{net} = -w_{max} + P\Delta V$$

and for our system, $w_{net}$ is the same as $w_e$, the electrical work done by the cell. Thus, it is seen that a relationship does exist between the free energy change of the system and the electrical work done by the reversible cell. This relationship becomes more useful when it is recognized that the available electrical energy for a mole of reactants is

$$(8\text{-}4) \qquad w_e = n_{FE}$$

where $\varepsilon$ is the emf of the cell, $F$ is Faraday's constant, and $n$ is the number of electrons changed per atom or ion. This, then, leads to the fundamental relation

$$(8\text{-}5) \qquad \Delta G = -n_{FE}$$

and we now find that it is possible to express the thermodynamic driving force of a reaction in terms of the cell emf as well as in terms of the free energy change.

## THE NERNST EQUATION

To give a thermodynamic description of a system in which a chemical reaction is taking place, it is usually not sufficient that the temperature, pressure, and volume be specified. It is also necessary to specify the composition of the system in terms of the concentrations of the various components present. This leads to the free energy expression

$$(8\text{-}6) \qquad dG = -SdT + VdP + \sum_i \mu_i \, dn_i$$

where $S$, $V$, $T$, and $P$ have their usual significance, $n_i$ is the number of moles of the $i$th component, and $\mu_i$ is the chemical potential of the $i$th component and may be defined as

$$(8\text{-}7) \qquad \mu_i = \left(\frac{\partial G}{\partial n_i}\right)_{T,P,n_j}$$

If the conditions of constant temperature and pressure are imposed, the free energy expression will then become

$$(8\text{-}8) \qquad (dG)_{T,P} = \mu_1 dn_1 + \mu_2 dn_2 + \cdots$$

If we consider the reaction in the galvanic cell to be

$$n_a A + n_b B = n_c C + n_d D$$

we can be more explicit and say that

(8–9) $\quad (dG)_{T,P} = \mu_C dn_c + \mu_D dn_d - \mu_A dn_a - \mu_B dn_b$

which on integration gives

(8–10) $\quad \Delta G = \mu_C n_c + \mu_D n_d - \mu_A n_a - \mu_B n_b$

assuming that the chemical potentials of the reactants and products remain constant. Now, since the chemical potential in terms of the activity is given by

(8–11) $\quad \mu_i = \mu_i{}^\circ + RT \ln a_i$

it is possible to combine Eq. (8–10) and Eq. (8–11) to give

$$\Delta G = n_c \left( \mu_C{}^\circ + RT \ln a_C \right) + n_d \left( \mu_D{}^\circ + RT \ln a_D \right)$$

(8–12) $\qquad - n_a \left( \mu_A{}^\circ + RT \ln a_A \right) - n_b \left( \mu_B{}^\circ + RT \ln a_B \right)$

or on rearrangement

(8–13) $\quad \Delta G = n_c \mu_C{}^\circ + n_d \mu_D{}^\circ - n_a \mu_A{}^\circ - n_b \mu_B{}^\circ$

$$+ RT \ln \frac{(a_C)^{n_c} (a_D)^{n_d}}{(a_A)^{n_a} (a_B)^{n_b}}$$

Finally, we can say that

(8–14) $\quad \Delta G = \Delta G^\circ + RT \ln \frac{(a_C)^{n_c} (a_D)^{n_d}}{(a_A)^{n_a} (a_B)^{n_b}}$

where $\Delta G^\circ$ is the *standard state* free energy change.

The transition to the reversible emf of a galvanic cell is now quite straightforward. Combining Eq. (8–14) with Eq. (8–5), we obtain the Nernst equation

(8–15) $\quad \mathbf{E} = \mathbf{E}^\circ - \frac{RT}{n\mathbf{F}} \ln \frac{(a_C)^{n_c} (a_D)^{n_d}}{(a_A)^{n_a} (a_B)^{n_b}}$

This now permits us to evaluate the thermodynamic driving force of a redox reaction in terms of a measurable cell emf. And going further, it is possible to utilize the relationship between the standard state potential and the standard state free energy to derive an expression for the equilibrium constant of a redox reaction in terms of the emf. Thus,

(8–16) $\quad \mathbf{E}^\circ = \frac{RT}{n\mathbf{F}} \ln K_{eq}$

## STANDARD ELECTRODE POTENTIALS

The relationship between a standard half-cell emf and the corresponding standard free energy change is given by the equation

(8–17) $\quad \Delta G^\circ = -n\mathbf{F}\mathbf{E}^\circ$

Just as is conventional with other thermodynamic functions, the choice of standard states used to determine $\varepsilon^\circ$ is completely arbitrary and is left to the discretion of the experimenter. For pure substances, this is usually quite routine. Ordinarily, gases at pressures of 1 atm or less are chosen to have an activity equal to their partial pressures, and for pure solids or liquids, it is customary to choose the standard state such that the activity is unity at any specified temperature. On the other hand, when we consider ionic solutions, it is most advantageous to choose the standard state of the solute in such a manner that the activity is equal to the concentration in regions where the concentration is very low. That is, as the concentration approaches zero, the activity approaches the concentration.

Such a choice of the standard state of an electrolyte has several advantages. To begin with, it permits the determination of the standard state potential by means of a rather straightforward extrapolation procedure. Secondly, it allows one to interpret the activity coefficients in terms of a reasonable theoretical model. With respect to this latter point, it can be recognized that at very high dilution the only interactions of the ions are those with the solvent. However, as the ionic concentration is increased, the average distance of separation of the ions becomes less and ion-ion interactions become significant. By defining the standard state in such a manner that the activity equals the concentration at infinite dilution, we are essentially saying that the solution is ideal when the only interactions of the ions are those with the solvent. Therefore, it is then possible to attribute to ion-ion interactions any deviations from this ideal behavior that are observed with an increase in ionic concentration.

## The Activity Coefficient

Deviations from this ideal state where the activity of the solute is equal to the concentration of the solute are expressed in terms of the *activity coefficient*. This is defined as

$$(8\text{–}18) \qquad \gamma = \frac{a}{m}$$

where $a$ represents the activity and $m$ the molal concentration of the solute.* Now, it can be seen that as the system deviates from the defined ideality, the activity coefficient reflects this deviation by showing a corresponding deviation from unity. Thus, according to our chosen standard state, as $m \to 0$, $a \to m$ and $\gamma \to 1$. Furthermore, in the case of electrolytes, the activity coefficient will be a measure of the nonideality of the system due primarily to ion-ion interactions.

---

* It is not necessary that the concentration be expressed in terms of the molal scale; mole fraction or molarity could be used equally well. Accordingly, the activity coefficient on these latter two scales is, respectively, $f = a/N$ and $y = a/c$.

## Determination of the Standard Electrode Potential

In studying thermodynamic properties of electrolytic solutions, both the standard state potential and the activity coefficients of the solute are of fundamental importance. From Eq. (8–15) it should be evident that if the value of the standard electrode potential is known, it is possible to determine the activities and, therefore, the activity coefficients of the electrolyte used in the cell. Consequently, our first endeavor will be to determine the standard potential, and to illustrate the general procedure we can consider a simple type of cell without liquid junction. In order to study the thermodynamic properties of a solute such as hydrochloric acid, the cell

$$\text{Pt, H}_{2(1 \text{ atm})} \,|\text{HCl}_{(m)}|\, \text{AgCl,Ag}$$

can be used. This involves simply the immersion of a hydrogen electrode and a silver-silver chloride electrode in a solution of hydrochloric acid. The half-cell reactions will be

$$\tfrac{1}{2}\,\text{H}_2 = \text{H}^+ + e^-$$

and

$$\text{AgCl} + e^- = \text{Ag} + \text{Cl}^-$$

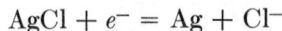

This gives the overall cell reaction

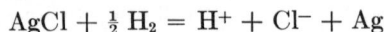

$$\text{AgCl} + \tfrac{1}{2}\,\text{H}_2 = \text{H}^+ + \text{Cl}^- + \text{Ag}$$

With this cell reaction, the emf is given by

$$(8\text{–}19) \qquad \text{E} = \text{E}^\circ - \frac{RT}{\text{F}} \ln \frac{(a_{\text{Ag}})(a_{\text{H}+})(a_{\text{Cl}-})}{(a_{\text{AgCl}})(a_{\text{H}_2})^{1/2}}$$

Now, if we use the conventional standard states, the activities of the silver, silver chloride, and the hydrogen at a pressure of 1 atm can be set equal to unity. The cell emf can then be expressed as

$$(8\text{–}20) \qquad \text{E} = \text{E}^\circ - \frac{RT}{\text{F}} \ln a_{\text{H}+}\, a_{\text{Cl}-}$$

As it turns out, there is no thermodynamically valid way to measure the activity of a single ion. We must always measure the two together. Consequently, it is necessary to speak of a *mean* ionic activity, and this leads to the expression

$$(8\text{–}21) \qquad \text{E} = \text{E}^\circ - \frac{RT}{\text{F}} \ln a_{\pm}{}^2$$

and this can be changed further to give

$$\text{E} = \text{E}^\circ - \frac{2RT}{\text{F}} \ln a_{\pm}$$

In terms of the activity coefficient, this becomes

$$(8\text{–}22) \qquad \text{E} = \text{E}^\circ - \frac{2RT}{\text{F}} \ln \gamma_{\pm} m$$

At this point, several different approaches can be used to determine the $E°$ value.[1] The most common of these utilize some form of the Debye-Huckel theory. If we have confidence in the theory, it is possible to supplement our experimental data with the theoretical treatment in regions of very great dilution. However, an experimental method for the determination of $E°$ was developed by Lewis and Randall[2] before the advent of the Debye-Huckel theory, and it will suffice to show the basic principles involved in addition to giving, in many instances, quite acceptable values for the standard emf.

In order to illustrate the method, we can rearrange Eq. (8–22) to the form

$$(8\text{–}23) \qquad E = E° - \frac{2RT}{F} \ln \gamma_{\pm} - \frac{2RT}{F} \ln m$$

or

$$(8\text{–}24) \qquad E + \frac{2RT}{F} \ln m = E° - \frac{2RT}{F} \ln \gamma_{\pm}$$

Since the standard state has been defined in such a manner that $\gamma_{\pm} \to 1$ as $m \to 0$, it can be seen that the term containing the activity coefficient disappears as we approach infinite dilution. Thus, if a plot is made of the left side of Eq. (8–24) against some function of the concentration, the left side of the equation approaches $E°$ as $m \to 0$. Such a plot is shown in Figure 8–2, where potential measurements have been made with the cell

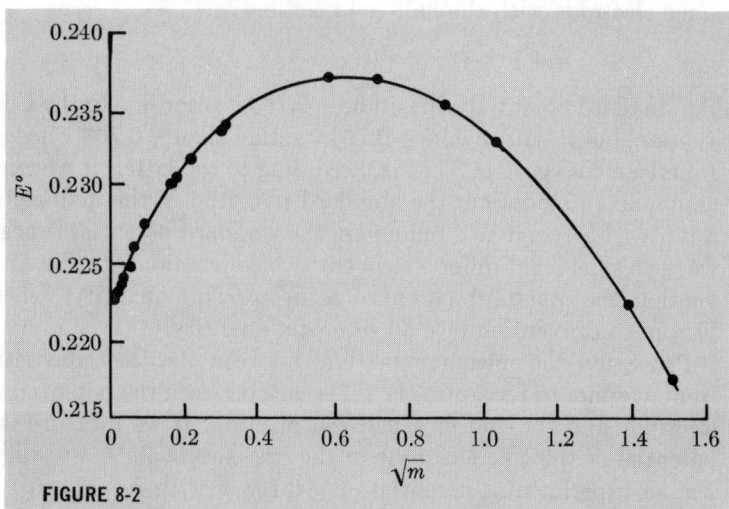

$$\text{Pt, } H_{2(1\text{ atm})} | HCl_{(m)} | AgCl, Ag$$

FIGURE 8-2

Determination of the Standard Electrode Potential of the Ag,AgCl Electrode by the Extrapolation Procedure of Lewis and Randall. (After G. N. Lewis and M. Randall, "Thermodynamics," McGraw-Hill Book Co., New York, 1923.)

at various concentrations of hydrochloric acid. The extrapolation to infinite dilution yields a standard half-cell emf for the AgCl,Ag half-cell of $-0.2234$ v when measured with respect to a defined potential of zero for the standard hydrogen electrode. This is in fair agreement with the value of $-0.2225$ v determined by the more modern methods of extrapolation.

## SIGN CONVENTIONS

For years, two different sign conventions for electrode potentials have been in common use. Without real justification, they are usually referred to as the *European convention* and the *American convention*. The latter of these was originated by Lewis and Randall and has found quite general favor among the physical chemists. Unfortunately, the difference in the two conventions has been quite generally misunderstood, for the difference is not merely that of a sign as usually supposed, but rather lies in a basic difference in meaning.[3,4]

To illustrate the two conventions, we can consider the Zn,Zn²⁺ electrode at which the reaction

$$Zn = Zn^{2+} + 2e^-$$

takes place. According to the European convention, we would assign a value of $-0.763$ v for the standard potential of this electrode with respect to the standard hydrogen electrode. On the other hand, the value according to the American convention would be taken as $+0.763$ v. Thus, it would appear that the difference is merely that of a sign. However, if we now consider the same electrode with the half-cell reaction

$$Zn^{2+} + 2e^- = Zn$$

the standard potential will still be $-0.763$ v according to the European convention, but it will now be $-0.763$ v rather than $+0.763$ v according to the American convention. That is, according to the latter, if we express the reaction as an oxidation, the standard potential of this half-cell is positive, but if we express it as a reduction, the standard potential is negative. Here we see a significant difference in the two conventions. In the American convention the standard potential is a *bivariant* quantity, whereas in the European convention it is an *invariant* quantity.

Primarily, the misunderstanding between the two conventions results from a failure to recognize the difference between the potential of an actual electrode and the emf of a half-cell reaction. If we measure the standard potential of the zinc electrode in the presence of Zn²⁺, we will find that it has an experimental potential of $-0.763$ v with respect to the standard hydrogen electrode. That is, its actual potential will be negative with respect to the standard hydrogen electrode, and this is true regardless of whether we write the reaction as a reduction or as an oxidation. It is for this reason that the European convention assigns an invariant value of $-0.763$ v to its standard potential. On the other hand, from a thermodynamic stand-point, it is known that the reaction of zinc going to zinc ions tends to take place spontaneously in the presence of acid, and $\Delta G°$ is therefore negative. Since

(7–17)     $\Delta G° = -n_{\text{FE}}°$

it can be seen that $\text{E}°$ should be positive for the $Zn, Zn^{2+}$ half-cell expressed as an oxidation potential. It is for this reason that the American convention assigns a positive sign to the emf of the cell if it is expressed as an oxidation, but a negative sign if it is expressed as a reduction. Herein lies the difference between the two conventions. The European convention is referring to an experimentally observed *electrostatic potential* of an electrode with respect to the hydrogen electrode, whereas the American convention is referring to the *thermodynamic tendency* of a particular reaction to take place. And it might be pointed out that the problem is not solved by stating that

$$\Delta G° = +n_{\text{FE}}°$$

This would lead to agreement in sign between the two conventions as long as we continue the usual convention of expressing the cell emf as an oxida-tion potential. Nevertheless, the American convention will still lead to a bivariant potential and the European convention will still lead to an in-variant potential.

In an attempt to reconcile the two conventions, several authors as well as the IUPAC have recommended changes in terminology.[5,6] Basically, it is recommended that the term "electrode potential" be reserved for the European convention, and that the American convention refer to the "electromotive force" of a half-cell.

## OXIDATION POTENTIALS

We have seen that it is possible to consider a cell reaction to be made up of two half-cell reactions, one of which may be the reaction associated with the hydrogen electrode as a primary reference. In the instance of the zinc electrode, the half-cell reactions can be expressed as

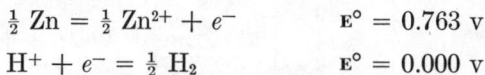

$$\tfrac{1}{2} Zn = \tfrac{1}{2} Zn^{2+} + e^- \qquad \text{E}° = 0.763 \text{ v}$$
$$H^+ + e^- = \tfrac{1}{2} H_2 \qquad \text{E}° = 0.000 \text{ v}$$

Thus it can be said that with respect to the standard hydrogen electrode, the standard zinc electrode potential is $-0.763$ v, or the half-cell emf is $+0.763$ v, expressed as an oxidation. Now, if we compare a large number of half-cell reactions to the standard hydrogen half-cell reaction, it is possible to set up a table of relative oxidation potentials. From the relationship be-tween $\Delta G$ and the cell emf, it follows that the more positive the oxidation potential of a half reaction, the greater will be its thermodynamic tendency to take place. Since the emf is being expressed as an oxidation potential, it

## TABLE 8–1.  Oxidation-Reduction Couples in Acid and Basic Solutions.[a]

### Acid Solutions

| Couple | E° (volts) | Couple | E° (volts) |
|---|---|---|---|
| $HN_3 = \frac{3}{2} N_2 + H^+ + e^-$ | 3.09 | $2Ta + 5H_2O = Ta_2O_5 + 10H^+ + 10\ e^-$ | 0.81 |
| $Li = Li^+ + e^-$ | 3.045 | $Zn = Zn^{2+} + 2e^-$ | 0.763 |
| $K = K^+ + e^-$ | 2.925 | $Tl + I^- = TlI + e^-$ | 0.753 |
| $Rb = Rb^+ + e^-$ | 2.925 | $Cr = Cr^{3+} + 3e^-$ | 0.74 |
| $Cs = Cs^+ + e^-$ | 2.923 | $H_2Te = Te + 2H^+ + 2e^-$ | 0.72 |
| $Ra = Ra^{2+} + 2e^-$ | 2.92 | $Tl + Br^- = TlBr + e^-$ | 0.658 |
| $Ba = Ba^{2+} + 2e^-$ | 2.90 | $2Nb + 5H_2O = Nb_2O_5 + 10H^+ + 10e^-$ | 0.65 |
| $Sr = Sr^{2+} + 2e^-$ | 2.89 | $U^{3+} = U^{4+} + e^-$ | 0.61 |
| $Ca = Ca^{2+} + 2e^-$ | 2.87 | $AsH_3 = As + 3H^+ + 3e^-$ | 0.60 |
| $Na = Na^+ + e^-$ | 2.714 | $Tl + Cl^- = TlCl + e^-$ | 0.557 |
| $La = La^{3+} + 3e^-$ | 2.52 | $Ga = Ga^{3+} + 3e^-$ | 0.53 |
| $Ce = Ce^{3+} + 3e^-$ | 2.48 | $SbH_3(g) = Sb + 3H^+ + 3e^-$ | 0.51 |
| $Nd = Nd^{3+} + 3e^-$ | 2.44 | $P + 2H_2O = H_3PO_2 + H^+ + e^-$ | 0.51 |
| $Sm = Sm^{3+} + 3e^-$ | 2.41 | $H_3PO_2 + H_2O = H_3PO_3 + 2H^+ + 2e^-$ | 0.50 |
| $Gd = Gd^{3+} + 3e^-$ | 2.40 | $Fe = Fe^{2+} + 2e^-$ | 0.440 |
| $Mg = Mg^{2+} + 2e^-$ | 2.37 | $Eu^{2+} = Eu^{3+} + e^-$ | 0.43 |
| $Y = Y^{3+} + 3e^-$ | 2.37 | $Cr^{2+} = Cr^{3+} + e^-$ | 0.41 |
| $Am = Am^{3+} + 3e^-$ | 2.32 | $Cd = Cd^{2+} + 2e^-$ | 0.403 |
| $Lu = Lu^{3+} + 3e^-$ | 2.25 | $H_2Se = Se + 2H^+ + 2e^-$ | 0.40 |
| $H^- = \frac{1}{2} H_2 + e^-$ | 2.25 | $Ti^{2+} = Ti^{3+} + e^-$ | 0.37 ca. |
| $H(g) = H^+ + e^-$ | 2.10 | $Pb + 2I^- = PbI_2 + 2e^-$ | 0.365 |
| $Sc = Sc^{3+} + 3e^-$ | 2.08 | $Pb + SO_4^{2-} = PbSO_4 + 2e^-$ | 0.356 |
| $Pu = Pu^{3+} + 3e^-$ | 2.07 | $In = In^{3+} + 3e^-$ | 0.342 |
| $Al + 6F^- = AlF_6^{3-} + 3e^-$ | 2.07 | $Tl = Tl^+ + e^-$ | 0.3363 |
| $Th = Th^{4+} + 4e^-$ | 1.90 | $\frac{1}{2} C_2N_2 + H_2O = HCNO + H^+ + e^-$ | 0.33 |
| $Np = Np^{3+} + 3e^-$ | 1.86 | $Pt + H_2S = PtS + 2H^+ + 2e^-$ | 0.30 |
| $Be = Be^{2+} + 2e^-$ | 1.85 | $Pb + 2Br^- = PbBr_2 + 2e^-$ | 0.280 |
| $U = U^{3+} + 3e^-$ | 1.80 | $Co = Co^{2+} + 2e^-$ | 0.277 |
| $Hf = Hf^{4+} + 4e^-$ | 1.70 | $H_3PO_3 + H_2O = H_3PO_4 + 2H^+ + 2e^-$ | 0.276 |
| $Al = Al^{3+} + 3e^-$ | 1.66 | $Pb + 2Cl^- = PbCl_2 + 2e^-$ | 0.268 |
| $Ti = Ti^{2+} + 2e^-$ | 1.63 | $V^{2+} = V^{3+} + e^-$ | 0.255 |
| $Zr = Zr^{4+} + 4e^-$ | 1.53 | $V + 4H_2O = V(OH)_4^+ + 4H^+ + 5e^-$ | 0.253 |
| $Si + 6F^- = SiF_6^{2-} + 4e^-$ | 1.2 | $Sn + 6F^- = SnF_6^{2-} + 4e^-$ | 0.25 |
| $Ti + 6F^- = TiF_6^{2-} + 4e^-$ | 1.19 | $Hg + 4Br^- = HgBr_4^{2-}$ | |
| $Mn = Mn^{2+} + 2e^-$ | 1.18 | | |
| $V = V^{2+} + 2e^-$ | 1.18 ca. | | |
| $Nb = Nb^{3+} + 3e^-$ | 1.1  ca. | | |
| $Ti + H_2O = TiO^{2+} + 2H^+ + 4e^-$ | 0.89 | | |
| $B + 3H_2O = H_3BO_3 + 3H^+ + 3e^-$ | 0.87 | | |
| $Si + 2H_2O = SiO_2 + 4H^+ + 4e^-$ | 0.86 | | |
| $Ni = Ni^{2+} + 2e^-$ | 0.250 | | |

## TABLE 8–1. (Continued)

### Acid Solutions — (Continued)

| Couple | E° (volts) | Couple | E° (volts) |
|---|---|---|---|
| $N_2H_5^+ = N_2 + 5H^+ + 4e^-$ | 0.23 | $+ 2e^-$ | $-0.21$ |
| $S_2O_6^{2-} + 2H_2O = 2SO_4^{2-}$ | | $Ag + Cl^- = AgCl + e^-$ | $-0.222$ |
| $+ 4H^+ + 2e^-$ | 0.22 | $(CH_3)_2SO + H_2O$ | |
| $Mo = Mo^{3+} + 3e^-$ | 0.2 $ca.$ | $= (CH_3)_2SO_2$ | |
| $HCOOH(aq) = CO_2$ | | $+ 2H^+ + 2e^-$ | $-0.23$ |
| $+ 2H^+ + 2e^-$ | 0.196 | $As + 2H_2O = HAsO_2(aq)$ | |
| $Cu + I^- = CuI + e^-$ | 0.185 | $+ 3H^+ + 3e^-$ | $-0.247$ |
| $Ag + I^- = AgI + e^-$ | 0.151 | $Re + 2H_2O = ReO_2$ | |
| $Ge + 2H_2O = GeO_2$ | | $+ 4H^+ + 4e^-$ | $-0.252$ |
| $+ 4H^+ + 4e^-$ | 0.15 | $Bi + H_2O = BiO^+ + 2H^+$ | |
| $Sn = Sn^{2+} + 2e^-$ | 0.136 | $+ 3e^-$ | $-0.32$ |
| $HO_2 = O_2 + H^+ + e^-$ | 0.13 | $U^{4+} + 2H_2O = UO_2^{2+}$ | |
| $Pb = Pb^{2+} + 2e^-$ | 0.126 | $+ 4H^+ + 2e^-$ | $-0.334$ |
| $W + 3H_2O = WO_3(c)$ | | $Cu = Cu^{2+} + 2e^-$ | $-0.337$ |
| $+ 6H^+ + 6e^-$ | 0.09 | $Ag + IO_3^- = AgIO_3 + e^-$ | $-0.35$ |
| $HS_2O_4^- + 2H_2O = 2H_2SO_3$ | | $Fe(CN)_6^{4-} = Fe(CN)_6^{3-}$ | |
| $+ H^+ + 2e^-$ | 0.08 | $+ e^-$ | $-0.36$ |
| $Hg + 4I^- = HgI_4^{2-} + 2e^-$ | 0.04 | $V^{3+} + H_2O = VO^{2+}$ | |
| $H_2 = 2H^+ + 2e^-$ | 0.00 | $+ 2H^+ + e^-$ | $-0.361$ |
| $Ag + 2S_2O_3^{2-}$ | | $Re + 4H_2O = ReO_4^-$ | |
| $= Ag(S_2O_3)_2^{3-} + e^-$ | $-0.01$ | $+ 8H^+ + 7e^-$ | $-0.363$ |
| $Cu + Br^- = CuBr + e^-$ | $-0.033$ | $HCN(aq) = \frac{1}{2} C_2N_2 + H^+$ | |
| $UO_2^+ = UO_2^{2+} + e^-$ | $-0.05$ | $+ e^-$ | $-0.37$ |
| $HCHO(aq) + H_2O$ | | $S_2O_3^{2-} + 3H_2O = 2H_2SO_3$ | |
| $= HCOOH(aq) + 2H^+$ | | $+ 2H^+ + 4e^-$ | $-0.40$ |
| $+ 2e^-$ | $-0.056$ | $Rh + 6Cl^- = RhCl_6^{3-}$ | |
| $PH_3(g) = P + 3H^+ + 3e^-$ | $-0.06$ | $+ 3e^-$ | $-0.44$ |
| $Ag + Br^- = AgBr + e^-$ | $-0.095$ | $2Ag + CrO_4^{2-} = Ag_2CrO_4$ | |
| $Ti^{3+} + H_2O = TiO^{2+}$ | | $+ 2e^-$ | $-0.446$ |
| $+ 2H^+ + e^-$ | $-0.1$ | $S + 3H_2O = H_2SO_3 + 4H^+$ | |
| $SiH_4 = Si + 4H^+ + 4e^-$ | $-0.102$ | $+ 4e^-$ | $-0.45$ |
| $CH_4 = C + 4H^+ + 4e^-$ | $-0.13$ | $Sb_2O_4 + H_2O = Sb_2O_5$ | |
| $Cu + Cl^- = CuCl + e^-$ | $-0.137$ | $+ 2H^+ + 2e^-$ | $-0.48$ |
| $H_2S = S + 2H^+ + 2e^-$ | $-0.141$ | $2Ag + MoO_4^{2-}$ | |
| $Np^{3+} = Np^{4+} + e^-$ | $-0.147$ | $= Ag_2MoO_4 + 2e^-$ | $-0.49$ |
| $Sn^{2+} = Sn^{4+} + 2e^-$ | $-0.15$ | $2NH_3OH^+ = H_2N_2O_2$ | |
| $2Sb + 3H_2O = Sb_2O_3$ | | $+ 6H^+ + 4e^-$ | $-0.496$ |
| $+ 6H^+ + 6e^-$ | $-0.152$ | $ReO_2 + 2H_2O = ReO_4^-$ | |
| $Cu^+ = Cu^{2+} + e^-$ | $-0.153$ | $+ 4H^+ + 3e^-$ | $-0.51$ |
| $Bi + H_2O + Cl^- = BiOCl$ | | $S_4O_6^{2-} + 6H_2O = 4H_2SO_3$ | |
| $+ 2H^+ + 3e^-$ | $-0.16$ | $+ 4H^+ + 6e^-$ | $-0.51$ |
| $H_2SO_3 + H_2O = SO_4^{2-}$ | | $C_2H_6 = C_2H_4 + 2H^+$ | |
| $+ 4H^+ + 2e^-$ | $-0.17$ | $+ 2e^-$ | $-0.52$ |
| $CH_3OH(aq) = HCHO(aq)$ | | | |
| $+ 2H^+ + 2e^-$ | $-0.19$ | $Cu = Cu^+ + e^-$ | $-0.521$ |

## TABLE 8-1. (Continued)

### Acid Solutions — (Continued)

| Couple | $E°$ (volts) | Couple | $E°$ (volts) |
|---|---|---|---|
| $Te + 2H_2O = TeO_2(c)$ $+ 4H^+ + 4e^-$ | $-0.529$ | $Pt + 4Cl^- = PtCl_4^{2-} + 2e^-$ | $-0.73$ |
| $2I^- = I_2 + 2e^-$ | $-0.5355$ | $C_2H_4 = C_2H_2 + 2H^+ + 2e^-$ | $-0.73$ |
| $3I^- = I_3^- + 2e^-$ | $-0.536$ | $Se + 3H_2O = H_2SeO_3$ $+ 4H^+ + 4e^-$ | $-0.74$ |
| $CuCl = Cu^{2+} + Cl^- + e^-$ | $-0.538$ | $Np^{4+} + 2H_2O = NpO_2^+$ | |
| $Ag + BrO_3^- = AgBrO_3$ $+ e^-$ | $-0.55$ | $+ 4H^+ + e^-$ | $-0.75$ |
| | | $2CNS^- = (CNS)_2 + 2e^-$ | $-0.77$ |
| $Te + 2H_2O = TeOOH^+$ $+ 3H^+ + 4e^-$ | $-0.559$ | $Ir + 6Cl^- = IrCl_6^{3-} + 3e^-$ | $-0.77$ |
| $HAsO_2 + 2H_2O = H_3AsO_4$ | | $Fe^{2+} = Fe^{3+} + e^-$ | $-0.771$ |
| $+ 2H^+ + 2e^-$ | $-0.559$ | $2Hg = Hg_2^{2+} + 2e^-$ | $-0.789$ |
| $Ag + NO_2^- = AgNO_2 + e^-$ | $-0.564$ | $Ag = Ag^+ + e^-$ | $-0.7991$ |
| $MnO_4^{2-} = MnO_4^- + e^-$ | $-0.564$ | $N_2O_4 + 2H_2O = 2NO_3^-$ | |
| $2H_2SO_3 = S_2O_6^{2-} + 4H^+$ $+ 2e^-$ | $-0.57$ | $+ 4H^+ + 2e^-$ | $-0.80$ |
| | | $Rh = Rh^{3+} + 3e^-$ | $-0.8$ *ca.* |
| $Pt + 4Br^- = PtBr_4^{2-}$ $+ 2e^-$ | $-0.58$ | $Os + 4H_2O = OsO_4(c)$ $+ 8H^+ + 8e^-$ | $-0.85$ |
| $2SbO^+ + 3H_2O = Sb_2O_5$ $+ 6H^+ + 4e^-$ | $-0.581$ | $H_2N_2O_2 + 2H_2O = 2HNO_2$ $+ 4H^+ + 4e^-$ | $-0.86$ |
| $CH_4 + H_2O = CH_3OH(aq)$ $+ 2H^+ + 2e^-$ | $-0.586$ | $CuI = Cu^{2+} + I^- + e^-$ | $-0.86$ |
| | | $Au + 4Br^- = AuBr_4^-$ $+ 3e^-$ | $-0.87$ |
| $Pd + 4Br^- = PdBr_4^{2-}$ $+ 2e^-$ | $-0.6$ | $Hg_2^{2+} = 2Hg^{2+} + 2e^-$ | $-0.920$ |
| $Ru + 5Cl^- = RuCl_5^{2-}$ $+ 3e^-$ | $-0.60$ | $PuO_2^+ = PuO_2^{2+} + e^-$ | $-0.93$ |
| $U^{4+} + 2H_2O = UO_2^+$ $+ 4H^+ + e^-$ | $-0.62$ | $HNO_2 + H_2O = NO_3^-$ $+ 3H^+ + 2e^-$ | $-0.94$ |
| $Pd + 4Cl^- = PdCl_4^{2-}$ $+ 2e^-$ | $-0.62$ | $NO + 2H_2O = NO_3^-$ $+ 4H^+ + 4e^-$ | $-0.96$ |
| $CuBr = Cu^{2+} + Br^- + e^-$ | $-0.640$ | $Au + 2Br^- = AuBr_2^- + e^-$ | $-0.96$ |
| $Ag + C_2H_3O_2^-$ $= AgC_2H_3O_2 + e^-$ | $-0.643$ | $Pu^{3+} + Pu^{4+} + e^-$ | $-0.97$ |
| $2Ag + SO_4^{2-} = Ag_2SO_4$ $+ 2e^-$ | $-0.653$ | $Pt + 2H_2O = Pt(OH)_2$ $+ 2H^+ + 2e^-$ | $-0.98$ |
| $Au + 4CNS^-$ $= Au(CNS)_4^- + 3e^-$ | $-0.66$ | $Pd = Pd^{2+} + 2e^-$ | $-0.987$ |
| | | $IrBr_6^{4-} = IrBr_6^{3-} + e^-$ | $-0.99$ |
| $PtCl_4^{2-} + 2Cl^- = PtCl_6^{2-}$ $+ 2e^-$ | $-0.68$ | $NO + H_2O = HNO_2 + H^+$ $+ e^-$ | $-1.00$ |
| $H_2O_2 = O_2 + 2H^+ + 2e^-$ | $-0.682$ | $Au + 4Cl^- = AuCl_4^-$ $+ 3e^-$ | $-1.00$ |
| $2NH_4^+ = HN_3 + 11H^+$ $+ 8e^-$ | $-0.69$ | $VO^{2+} + 3H_2O = V(OH)_4^+$ $+ 2H^+ + e^-$ | $-1.00$ |
| $H_2Te = Te + 2H^+ + 2e^-$ | $-0.70$ | $IrCl_6^{3-} = IrCl_6^{2-} + e^-$ | $-1.017$ |
| $H_2N_2O_2 = 2NO + 2H^+$ $+ 2e^-$ | $-0.71$ | $TeO_2 + 4H_2O = H_6TeO_6(c)$ $+ 2H^+ + 2e^-$ | $-1.02$ |
| $OH + H_2O = H_2O_2 + H^+$ $+ e^-$ | $-0.72$ | $2NO + 2H_2O = N_2O_4$ $+ 4H^+ + 4e^-$ | $-1.03$ |
| | | $Pu^{4+} + 2H_2O = PuO_2^{2+}$ $+ 4H^+ + 2e^-$ | $-1.04$ |

## TABLE 8-1. (Continued)

### Acid Solutions — (Continued)

| Couple | E° (volts) | Couple | E° (volts) |
|--------|-----------|--------|-----------|
| $2Cl^- + \frac{1}{2} I_2 = ICl_2^- + e^-$ | $-1.06$ | $Au = Au^{3+} + 3e^-$ | $-1.50$ |
| $2Br^- = Br_2(1) + 2e^-$ | $-1.0652$ | $H_2O_2 = HO_2 + H^+ + e^-$ | $-1.5$ |
| $2HNO_2 = N_2O_4 + 2H^+$ | | $Mn^{2+} = Mn^{3+} + e^-$ | $-1.51$ |
| $\quad + 2e^-$ | $-1.07$ | $Mn^{2+} + 4H_2O = MnO_4^-$ | |
| $Cu(CN)_2^- = Cu^{2+}$ | | $\quad + 8H^+ + 5e^-$ | $-1.51$ |
| $\quad + 2CN^- + e^-$ | $-1.12$ | $\frac{1}{2} Br_2 + 3H_2O = BrO_3^-$ | |
| $Pu^{4+} + 2H_2O = PuO_2^+$ | | $\quad + 6H^+ + 5e^-$ | $-1.52$ |
| $\quad + 4H^+ + e^-$ | $-1.15$ | $\frac{1}{2} Br_2 + H_2O = HBrO$ | |
| $H_2SeO_3 + H_2O = SeO_4^{2-}$ | | $\quad + H^+ + e^-$ | $-1.59$ |
| $\quad + 4H^+ + 2e^-$ | $-1.15$ | $2BiO^+ + 2H_2O = Bi_2O_4$ | |
| $NpO_2^+ = NpO_2^{2+} + e^-$ | $-1.15$ | $\quad + 4H^+ + 2e^-$ | $-1.59$ |
| $4Cl^- + C + 4H^+ = CCl_4$ | | $IO_3^- + 3H_2O = H_5IO_6$ | |
| $\quad + 4H^+ + 4e^-$ | $-1.18$ | $\quad + H^+ + 2e^-$ | $-1.6$ |
| $ClO_3^- + H_2O = ClO_4^-$ | | $Bk^{3+} = Bk^{4+} + e^-$ | $-1.6$ |
| $\quad + 2H^+ + 2e^-$ | $-1.19$ | $Ce^{3+} = Ce^{4+} + e^-$ | $-1.61$ |
| $\frac{1}{2} I_2 + 3H_2O = IO_3^-$ | | $\frac{1}{2} Cl_2 + H_2O = HClO$ | |
| $\quad + 6H^+ + 5e^-$ | $-1.195$ | $\quad + H^+ + e^-$ | $-1.63$ |
| $HClO_2 + H_2O = ClO_3^-$ | | $AmO_2^+ = AmO_2^{2+} + e^-$ | $-1.64$ |
| $\quad + 3H^+ + 2e^-$ | $-1.21$ | $HClO + H_2O = HClO_2$ | |
| $2H_2O = O_2 + 4H^+ + 4e^-$ | $-1.229$ | $\quad + 2H^+ + 2e^-$ | $-1.64$ |
| $2S + 2Cl^- = S_2Cl_2 + 2e^-$ | $-1.23$ | $Au = Au^+ + e^-$ | $-1.68$ $ca.$ |
| $Mn^{2+} + 2H_2O = MnO_2$ | | $Ni^{2+} + 2H_2O = NiO_2$ | |
| $\quad + 4H^+ + 2e^-$ | $-1.23$ | $\quad + 4H^+ + 2e^-$ | $-1.68$ |
| $Tl^+ = Tl^{3+} + 2e^-$ | $-1.25$ | $PbSO_4 + 2H_2O = PbO_2$ | |
| $Am^{4+} + 2H_2O = AmO_2^+$ | | $\quad + SO_4^{2-} + 4H^+ + 2e^-$ | $-1.685$ |
| $\quad + 4H^+ + e^-$ | $-1.26$ | $Am^{3+} + 2H_2O = AmO_2^{2+}$ | |
| $2NH_4^+ = N_2H_5^+ + 3H^+$ | | $\quad + 4H^+ + 3e^-$ | $-1.69$ |
| $\quad + 2e^-$ | $-1.275$ | $MnO_2 + 2H_2O = MnO_4^-$ | |
| $HClO_2 = ClO_2 + H^+ + e^-$ | $-1.275$ | $\quad + 4H^+ + 3e^-$ | $-1.695$ |
| $PdCl_4^{2-} + 2Cl^- = PdCl_6^{2-}$ | | $Am^{3+} + 2H_2O = AmO_2^+$ | |
| $\quad + 2e^-$ | $-1.288$ | $\quad + 4H^+ + 2e^-$ | $-1.725$ |
| $N_2O + 3H_2O = 2HNO_2$ | | $2H_2O = H_2O_2 + 2H^+$ | |
| $\quad + 4H^+ + 4e^-$ | $-1.29$ | $\quad + 2e^-$ | $-1.77$ |
| $2Cr^{3+} + 7H_2O = Cr_2O_7^{2-}$ | | $Co^{2+} = Co^{3+} + e^-$ | $-1.82$ |
| $\quad + 14H^+ + 6e^-$ | $-1.33$ | $Fe^{3+} + 4H_2O = FeO_4^{2-}$ | |
| $NH_4^+ + H_2O = NH_3OH^+$ | | $\quad + 8H^+ + 3e^-$ | $-1.9$ |
| $\quad + 2H^+ + 2e^-$ | $-1.35$ | $NH_4^+ + N_2 = HN_3 + 3H^+$ | |
| $2Cl^- = Cl_2 + 2e^-$ | $-1.3595$ | $\quad + 2e^-$ | $-1.96$ |
| $N_2H_5 + 2H_2O$ | | $Ag^+ = Ag^{2+} + e^-$ | $-1.98$ |
| $\quad = 2NH_3OH^+ + H^+ + 2e^-$ | $-1.42$ | $2SO_4^{2-} = S_2O_8^{2-} + 2e^-$ | $-2.01$ |
| $Au + 3H_2O = Au(OH)_3$ | | $O_2 + H_2O = O_3 + 2H^+$ | |
| $\quad + 3H^+ + 3e^-$ | $-1.45$ | $\quad + 2e^-$ | $-2.07$ |
| $\frac{1}{2} I_2 + H_2O = HIO + H^+$ | | $H_2O + 2F^- = F_2O + 2H^+$ | |
| $\quad + e^-$ | $-1.45$ | $\quad + 4e^-$ | $-2.1$ |
| $Pb^{2+} + 2H_2O = PbO_2$ | | $Am^{3+} = Am^{4+} + e^-$ | $-2.18$ |
| $\quad + 4H^+ + 2e^-$ | $-1.455$ | $H_2O = O(g) + 2H^+ + 2e^-$ | $-2.42$ |

## TABLE 8–1.   (Continued)

### Acid Solutions — (Continued)

| Couple | E° (volts) | Couple | E° (volts) |
|---|---|---|---|
| $H_2O = OH + H^+ + e^-$ | $-2.8$ | $2F^- = F_2 + 2e^-$ | $-2.87$ |
| $N_2 + 2H_2O = H_2N_2O_2$ $+ 2H^+ + 2e^-$ | $-2.85$ | $2HF(aq) = F_2 + 2H^+$ $+ 2e^-$ | $-3.06$ |

### Basic Solutions

| Couple | E° (volts) | Couple | E° (volts) |
|---|---|---|---|
| $Ca + 2OH^- = Ca(OH)_2$ $+ 2e^-$ | $3.03$ | $= Na_2UO_4 + 4H_2O + 2e^-$ | $1.61$ |
| $Sr + 2OH^- + 8H_2O$ $= Sr(OH)_2 \cdot 8H_2O + 2e^-$ | $2.99$ | $H_2PO_2^- + 3OH^- = HPO_3^{2-}$ $+ 2H_2O + 2e^-$ | $1.57$ |
| $Ba + 2OH^- + 8H_2O$ $= Ba(OH)_2 \cdot 8H_2O + 2e^-$ | $2.97$ | $Mn + 2OH^- = Mn(OH)_2$ $+ 2e^-$ | $1.55$ |
| $H(g) + OH^- = H_2O + e^-$ | $2.93$ | $Mn + CO_3^{2-} = MnCO_3$ $+ 2e^-$ | $1.48$ |
| $La + 3OH^- = La(OH)_3$ $+ 3e^-$ | $2.90$ | $Zn + S^{2-} = ZnS + 2e^-$ | $1.44$ |
| $Lu + 3OH^- = Lu(OH)_3$ $+ 3e^-$ | $2.72$ | $Cr + 3OH^- = Cr(OH)_3$ $+ 3e^-$ | $1.3$ |
| $Mg + 2OH^- = Mg(OH)_2$ $+ 2e^-$ | $2.69$ | $Zn + 4CN^- = Zn(CN)_4^{2-}$ $+ 2e^-$ | $1.26$ |
| $2Be + 6OH^- = Be_2O_3^{2-}$ $+ 3H_2O + 4e^-$ | $2.62$ | $Zn + 2OH^- = Zn(OH)_2$ $+ 2e^-$ | $1.245$ |
| $Sc + 3OH^- = Sc(OH)_3$ $+ 3e^-$ | $2.6\ \ ca.$ | $Ga + 4OH^- = H_2GaO_3^-$ $+ H_2O + 3e^-$ | $1.22$ |
| $Hf + 4OH^- = HfO(OH)_2$ $+ H_2O + 4e^-$ | $2.50$ | $Zn + 4OH^- = ZnO_2^{2-}$ $+ 2H_2O + 2e^-$ | $1.216$ |
| $Th + 4OH^- = Th(OH)_4$ $+ 4e^-$ | $2.48$ | $Cr + 4OH^- = CrO_2^-$ $+ 2H_2O + 3e^-$ | $1.2$ |
| $Pu + 3OH^- = Pu(OH)_3$ $+ 3e^-$ | $2.42$ | $Cd + S^{2-} = CdS + 2e^-$ | $1.21$ |
| $U + 4OH^- = UO_2 + 2H_2O$ $+ 4e^-$ | $2.39$ | $6V + 33OH^- = 16H_2O$ $+ HV_6O_{17}^{3-} + 30e^-$ | $1.15$ |
| $Zr + 4OH^- = H_2ZrO_3$ $+ H_2O + 4e^-$ | $2.36$ | $Te^{2-} = Te + 2e^-$ | $1.14$ |
| $Al + 4OH^- = H_2AlO_3^-$ $+ H_2O + 3e^-$ | $2.35$ | $HPO_3^{2-} + 3OH^- = PO_4^{3-}$ $+ 2H_2O + 2e^-$ | $1.12$ |
| $U(OH)_3 + OH^- = U(OH)_4$ $+ e^-$ | $2.2$ | $S_2O_4^{2-} + 4OH^- = 2SO_3^{2-}$ $+ 2H_2O + 2e^-$ | $1.12$ |
| $U + 3OH^- = U(OH)_3$ $+ 3e^-$ | $2.17$ | $Zn + CO_3^{2-} = ZnCO_3$ $+ 2e^-$ | $1.06$ |
| $P + 2OH^- = H_2PO_2^- + e^-$ | $2.05$ | $W + 8OH^- = WO_4^{2-}$ $+ 4H_2O + 6e^-$ | $1.05$ |
| $B + 4OH^- = H_2BO_3^-$ $+ 3e^-$ | $1.79$ | $Mo + 8OH^- = MoO_4^{2-}$ $4H_2O + 6e^-$ | $1.05$ |
| $Si + 6OH^- = SiO_3^{2-}$ $+ 3H_2O + 4e^-$ | $1.70$ | $Cd + 4CN^- = Cd(CN)_4^{2-}$ $+ 2e^-$ | $1.03$ |
|  |  | $Zn + 4NH_3 = Zn(NH_3)_4^{2+}$ $+ 2e^-$ | $1.03$ |

## TABLE 8–1. (Continued)

### Basic Solutions — (Continued)

| Couple | E° (volts) | Couple | E° (volts) |
|---|---|---|---|
| $U(OH)_4 = 2Na^+ + 4OH^-$ | | $Fe + S^{2-} = FeS_{(\alpha)}$ | 1.01 |
| $In + 3OH^- = In(OH)_3 + 3e^-$ | 1.0 | $Cd + 4NH_3 = Cd(NH_3)_4^{2+} + 2e^-$ | 0.597 |
| $CN^- + 2OH^- = CNO^- + H_2O + 2e^-$ | 0.97 | $ReO_2 + 4OH^- = ReO_4^- + 2H_2O + 3e^-$ | 0.594 |
| $2Tl + S^{2-} = Tl_2S + 2e^-$ | 0.96 | $Re + 8OH^- = ReO_4^- + 4H_2O + 7e^-$ | 0.584 |
| $Pb + S^{2-} = PbS + 2e^-$ | 0.95 | $S_2O_3^{2-} + 6OH^- = 2SO_3^{2-} + 3H_2O + 4e^-$ | 0.58 |
| $Pu(OH)_3 + OH^- = Pu(OH)_4 + e^-$ | 0.95 | $Re + 4OH^- = ReO_2 + 2H_2O + 4e^-$ | 0.576 |
| $Sn + S^{2-} = SnS + 2e^-$ | 0.94 | $Te + 6OH^- = TeO_3^{2-} + 3H_2O + 4e^-$ | 0.57 |
| $SO_3^{2-} + 2OH^- = SO_4^{2-} + H_2O + 2e^-$ | 0.93 | $Fe(OH)_2 + OH^- = Fe(OH)_3 + e^-$ | 0.56 |
| $Se^{2-} = Se + 2e^-$ | 0.92 | $O_2^- = O_2 + e^-$ | 0.56 |
| $Sn + 3OH^- = HSnO_2^- + H_2O + 2e^-$ | 0.91 | $2Cu + S^{2-} = Cu_2S + 2e^-$ | 0.54 |
| $Ge + 5OH^- = HGeO_3^- + 2H_2O + 4e^-$ | 0.9 | $Pb + 3OH^- = HPbO_2^- + H_2O + 2e^-$ | 0.54 |
| $HSnO_2^- + H_2O + 3OH^- = Sn(OH)_6^{2-} + 2e^-$ | 0.90 | $Pb + CO_3^{2-} = PbCO_3 + 2e^-$ | 0.506 |
| $PH_3 + 3OH^- = P + 3H_2O + 3e^-$ | 0.89 | $S^{2-} = S + 2e^-$ | 0.48 |
| $Fe + 2OH^- = Fe(OH)_2 + 2e^-$ | 0.877 | $Ni + 6NH_3(aq) = Ni(NH_3)_6^{2+} + 2e^-$ | 0.47 |
| $Ni + S^{2-} = NiS_{(\alpha)} + 2e^-$ | 0.83 | $Ni + CO_3^{2-} = NiCO_3 + 2e^-$ | 0.45 |
| $H_2 + 2OH^- = 2H_2O + 2e^-$ | 0.828 | $2Bi + 6OH^- = Bi_2O_3 + 3H_2O + 6e^-$ | 0.44 |
| $Cd + 2OH^- = Cd(OH)_2 + 2e^-$ | 0.809 | $Cu + 2CN^- = Cu(CN)_2^- + e^-$ | 0.43 |
| $Fe + CO_3^{2-} = FeCO_3 + 2e^-$ | 0.756 | $Hg + 4CN^- = Hg(CN)_4^{2-} + 2e^-$ | 0.37 |
| $Cd + CO_3^{2-} = CdCO_3 + 2e^-$ | 0.74 | $Se + 6OH^- = SeO_3^{2-} + 3H_2O + 4e^-$ | 0.366 |
| $Co + 2OH^- = Co(OH)_2 + 2e^-$ | 0.73 | $2Cu + 2OH^- = Cu_2O + H_2O + 2e^-$ | 0.358 |
| $Hg + S^{2-} = HgS + 2e^-$ | 0.72 | $Tl + OH^- = Tl(OH) + e^-$ | 0.3445 |
| $Ni + 2OH^- = Ni(OH)_2 + 2e^-$ | 0.72 | $Ag + 2CN^- = Ag(CN)_2^- + e^-$ | 0.31 |
| $2Ag + S^{2-} = Ag_2S + 2e^-$ | 0.69 | $Cu + CNS^- = Cu(CNS) + e^-$ | 0.27 |
| $As + 4OH^- = AsO_2^- + 2H_2O + 3e^-$ | 0.68 | $OH + 2OH^- = HO_2^- + H_2O + e^-$ | 0.24 |
| $AsO_2^- + 4OH^- = AsO_4^{3-} + 2H_2O + 2e^-$ | 0.67 | $Cr(OH)_3 + 5OH^- = CrO_4^{2-} + 4H_2O + 3e^-$ | 0.13 |
| $2FeS + S^{2-} = Fe_2S_3 + 2e^-$ | 0.67 | $Cu + 2NH_3 = Cu(NH_3)_2^+ + e^-$ | 0.12 |
| $Sb + 4OH^- = SbO_2^- + 2H_2O + 3e^-$ | 0.66 | | |
| $Co + CO_3^{2-} = CoCO_3 + 2e^-$ | 0.64 | | |

## TABLE 8-1. (Continued)

### Basic Solutions — (Continued)

| Couple | E° (volts) | Couple | E° (volts) |
|---|---|---|---|
| $Cu_2O + 2OH^- + H_2O$ $= 2Cu(OH)_2 + 2e^-$ | 0.080 | $2Ag + 2OH^- = Ag_2O$ $+ H_2O + 2e^-$ | −0.344 |
| $HO_2^- + OH^- = O_2 + H_2O$ $+ 2e^-$ | 0.076 | $ClO_3^- + 2OH^- = ClO_4^-$ $+ H_2O + 2e^-$ | −0.36 |
| $TlOH + 2OH^- = Tl(OH)_3$ $+ 2e^-$ | 0.05 | $Ag + 2NH_3 = Ag(NH_3)_2^+$ $+ e^-$ | −0.373 |
| $Ag + CN^- = AgCN + e^-$ | 0.017 | $TeO_3^{2-} + 2OH^- = TeO_4^{2-}$ $+ H_2O + 2e^-$ | −0.4 |
| $Mn(OH)_2 + 2OH^-$ $= MnO_2 + 2H_2O + 2e^-$ | 0.05 | $OH^- + HO_2^- = O_2^-$ $+ H_2O + e^-$ | −0.4 |
| $NO_2^- + 2OH^- = NO_3^-$ $+ H_2O + 2e^-$ | −0.01 | $4OH^- = O_2 + 2H_2O + 4e^-$ | −0.401 |
| $Os + 9OH^- = HOsO_5^-$ $+ 4H_2O + 8e^-$ | −0.02 | $2Ag + CO_3^{2-} = Ag_2CO_3$ $+ 2e^-$ | −0.47 |
| $2Rh + 6OH^- = Rh_2O_3$ $+ 3H_2O + 6e^-$ | −0.04 | $Ni(OH)_2 + 2OH^- = NiO_2$ $+ 2H_2O + 2e^-$ | −0.49 |
| $SeO_3^{2-} + 2OH^- = SeO_4^{2-}$ $+ H_2O + 2e^-$ | −0.05 | $I^- + 2OH^- = IO^- + H_2O$ $+ 2e^-$ | −0.49 |
| $Pd + 2OH^- = Pd(OH)_2$ $+ 2e^-$ | −0.07 | $Ag_2O + 2OH^- = 2AgO$ $+ H_2O + 2e^-$ | −0.57 |
| $2S_2O_3^{2-} = S_4O_6^{2-} + 2e^-$ | −0.08 | $MnO_2 + 4OH^- = MnO_4^{2-}$ $+ 2H_2O + 2e^-$ | −0.60 |
| $Hg + 2OH^- = HgO(r)$ $+ H_2O + 2e^-$ | −0.098 | $RuO_4^{2-} = RuO_4^- + e^-$ | −0.60 |
| $2NH_4OH + 2OH^- = N_2H_4$ $+ 4H_2O + 2e^-$ | −0.1 | $Br^- + 6OH^- = BrO_3^-$ $+ 3H_2O + 6e^-$ | −0.61 |
| $2Ir + 6OH^- = Ir_2O_3$ $+ 3H_2O + 6e^-$ | −0.1 | $ClO^- + 2OH^- = ClO_2^-$ $+ H_2O + 2e^-$ | −0.66 |
| $Co(NH_3)_6^{2+} = Co(NH_3)_6^{3+}$ $+ e^-$ | −0.1 | $IO_3^- + 3OH^- = H_3IO_6^{2-}$ $+ 2e^-$ | −0.7 |
| $Mn(OH)_2 + OH^-$ $= Mn(OH)_3 + e^-$ | −0.1 | $N_2H_4 + 2OH^- = 2NH_2OH$ $+ 2e^-$ | −0.73 |
| $Pt + 2OH^- = Pt(OH)_2$ $+ 2e^-$ | −0.15 | $2AgO + 2OH^- = Ag_2O_3$ $+ H_2O + 2e^-$ | −0.74 |
| $Co(OH)_2 + OH^-$ $= Co(OH)_3 + e^-$ | −0.17 | $Br^- + 2OH^- = BrO^-$ $+ H_2O + 2e^-$ | −0.76 |
| $PbO(r) + 2OH^- = PbO_2$ $+ H_2O + 2e^-$ | −0.248 | $3OH^- = HO_2^- + H_2O$ $+ 2e^-$ | −0.88 |
| $I^- + 6OH^- = IO_3^-$ $+ 3H_2O + 6e^-$ | −0.26 | $Cl^- + 2OH^- = ClO^-$ $+ H_2O + 2e^-$ | −0.89 |
| $PuO_2OH + OH^-$ $= PuO_2(OH)_2 + 2e^-$ | −0.26 | $FeO_2^- + 4OH^- = FeO_4^{2-}$ $+ 2H_2O + 3e^-$ | −0.9 |
| $Ag + 2SO_3^{2-} = Ag(SO_3)_2^{3-}$ $+ e^-$ | −0.30 | $ClO_2^- = ClO_2 + e^-$ | −1.16 |
| $ClO_2^- + 2OH^- = ClO_3^-$ $+ H_2O + 2e^-$ | −0.33 | $O_2 + 2OH^- = O_3 + H_2O$ $+ 2e^-$ | −1.24 |
| | | $OH^- = OH + e^-$ | −2.0 |

[a] The values reported in this table were taken directly from Latimer, "The Oxidation States of the Elements and Their Potentials in Aqueous Solutions," 2nd ed., Prentice-Hall, Inc., Englewood Cliffs, N. J., 1952.

can also be said that the more positive potentials are associated with the better reductants and the more negative potentials are associated with the better oxidants; those with potentials above hydrogen are better reducing agents than hydrogen and those with potentials below hydrogen are better oxidizing agents than the hydrogen ion. The standard potentials of a collection of inorganic couples as given by Latimer[7] are shown in Table 8–1. The most positive potentials, and therefore the best reducing agents in their standard states, are placed at the top of the table.

Now the algebraic difference between two such couples will give the emf for the total cell reaction if all species are at unit activity. If the resultant emf is positive, the reaction will be expected to proceed spontaneously in the direction written, but if the emf is negative, the reverse reaction would be favored. As an example, let us consider the reaction

$$Fe + Cd^{2+} = Cd + Fe^{2+}$$

The two half-cell reactions with their standard state potentials are

$$Fe = Fe^{2+} + 2e^- \qquad E° = 0.440 \text{ v}$$
$$Cd = Cd^{2+} + 2e^- \qquad E° = 0.403 \text{ v}$$

and by combining these we can obtain the standard state emf of the cell

$$E°_{cell} = E°_{Fe} - E°_{Cd}$$

or

$$E°_{cell} = 0.440 \text{ v} - 0.403 \text{ v} = 0.037 \text{ v}$$

The emf for the cell reaction is seen to be positive, so the reaction will be expected to proceed spontaneously as written.

There are two important limitations to this argument. Firstly, we are neglecting the kinetics of the reaction, and secondly, we are using standard state potentials, which are valid only at unit activity. With respect to the first point, it should be recognized that thermodynamics considers the initial and the final states of a system, but it does not consider what takes place between these two states. There are numerous examples of reactions with negative free energy changes that apparently do not occur spontaneously. This can result from such factors as a mechanism involving an intermediate step which has a positive free energy change, the formation of a surface coating, alternative reaction paths that are more probable than those proposed, and so on. With respect to the second point, we should recognize that it is not likely that a reaction would be carried out at such specific activities, and in fact it would be coincidental if such were possible. So it can be seen that the actual experimental situation is only approximated by the use of the standard oxidation potentials. To be more nearly correct, it would be necessary to calculate the cell emf under the specific conditions of the experiment. For this particular reaction,

$$E = E° - \frac{RT}{2F} \ln \frac{a_{Fe^{2+}}}{a_{Cd^{2+}}}$$

or, at 25°C,

$$E = 0.037 - \frac{0.059}{2} \log \frac{a_{Fe^{2+}}}{a_{Cd^{2+}}}$$

It follows from the above that, although from the standard potentials we would predict that iron will reduce cadmium ion, if the ratio of $a_{Fe^{2+}}$ to $a_{Cd^{2+}}$ is sufficiently large, the cell emf may actually be negative, and under these conditions, iron will not reduce cadmium ion. It might be mentioned that the activities are not usually used since the activity coefficients of each ion type are not ordinarily known in a system of this complexity. However, if the concentrations are not too great, it is possible to substitute concentration for activity and still obtain a good first-order approximation of the cell emf.

## PERIODIC TRENDS IN OXIDATION POTENTIALS

It has already been pointed out that an electrode potential is related to the free energy change by the relation

$$\Delta G = -n_{FE} = \Delta H - T\Delta S$$

From this it is apparent that the magnitude of an electrode potential cannot be considered from an energetic viewpoint alone. It is also necessary to consider the effect of an entropy change. Using a metal, metal ion half-cell in aqueous media, $M/M^{n+}_{(aq)}$ as an example, the emf will be proportional to the free energy change for the reaction

$$M_{(s)} \rightarrow M^{n+}_{(aq)} + ne^-$$

The enthalpy change for this reaction can now be expressed in terms of a Born-Haber type cycle involving the three steps: (1) the sublimation of the metal, (2) the ionization of the gaseous metal atom, and (3) the solvation of the metal ion. Thus we have

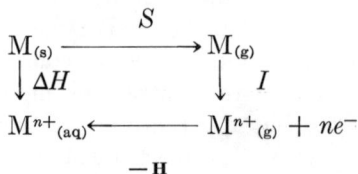

$$
\begin{array}{ccc}
 & S & \\
M_{(s)} & \longrightarrow & M_{(g)} \\
\downarrow \Delta H & & \downarrow I \\
M^{n+}_{(aq)} & \longleftarrow & M^{n+}_{(g)} + ne^- \\
 & -\text{H} &
\end{array}
$$

where        $S$ = heat of sublimation

$I$ = ionization energy

H = hydration energy

$\Delta H$ = enthalpy change

From this we can say that

$$\Delta H = S + I - \text{H}$$

An analogous type of cycle can, of course, be constructed for the cathode.

Now, if the entropy terms are relatively small, then $\Delta G$ will be approximately given by $\Delta H$. Such an approximation may not be generally valid, but it does seem to work in many instances. But even if it is not valid in an absolute sense, it is not unreasonable to expect the entropy terms to be similar for similar solutes. This would permit us to correlate the potentials within a series of half-cell couples and, in this manner, consider the effects on the potential of the various terms in the Born-Haber cycle.

Since an electrode potential is measured relative to the standard hydrogen electrode, the potential can be related to the free energy change of the reaction

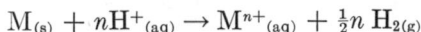

$$M_{(s)} + nH^+_{(aq)} \rightarrow M^{n+}_{(aq)} + \tfrac{1}{2}n\, H_{2(g)}$$

By neglecting the entropy change, the standard oxidation potential for a couple $M/M^{n+}_{(aq)}$ will be approximately given by

$$M/M^{n+}_{(aq)} = \frac{1}{n}[(\tfrac{1}{2}nD_{H_2} + nI_H - n\,\text{H}_{H^+}) - (I_M + S_M - \text{H}_M{}^{n+})]$$

The $\dfrac{1}{n}$ term is introduced because the standard potentials are given for a one-electron change.

If we now use the data in Table 8–2, along with the pertinent data for

**TABLE 8–2.   Oxidation Potentials of R1 Elements.**

| Element | $I$(eV) | $-\text{H}$(eV) | $S$(eV) | $\text{E}$(v) | $M \rightarrow M^+ + e^-$ |
|---------|---------|-----------------|---------|---------------|---------------------------|
| Li | 5.39 | 5.34 | 1.61 | 3.045 | |
| Na | 5.14 | 4.21 | 1.13 | 2.714 | |
| K  | 4.34 | 3.34 | 0.93 | 2.925 | |
| Rb | 4.18 | 3.04 | 0.89 | 2.925 | |
| Cs | 3.89 | 2.74 | 0.82 | 2.923 (3.08) | |

hydrogen ($D_{H_2} = 4.48$ eV; $I_H = 13.60$ eV; $\text{H}_{H^+} = 11.18$ eV), it is possible to approximate the oxidation potentials for the alkali metals. Using lithium as an example and noting that $n = 1$, the volt equivalent is seen to be

$$\text{E}^\circ{}_{Li/Li^+(aq)}$$
$$= [(\tfrac{1}{2} \times 4.48 + 13.60 - 11.18) - (5.39 + 1.61 - 5.34)]$$
$$= 3.0\ v$$

This is in surprisingly good agreement with the experimental value of 3.045 v.

From the table of oxidation potentials, we have taken the value of 2.923 v for cesium as reported by Latimer. A number of other values have been proposed. The value of 3.08 v given in parentheses in Table 8–2 is one of these. If we accept the higher value, a trend is roughly observed for the $\text{E}^\circ$ values among all the alkali metals except lithium. In terms of ionization energy and sublimation energy, lithium would be expected to be the poorest reducing agent among the alkali metals. But according to the $\text{E}^\circ$ value, it is the best reducing agent of the series. The anomolous behavior of lithium is attributed to the high hydration energy of the lithium ion resulting from its very small radius, or more correctly, its relatively high charge density. This point can more clearly be seen if the $\varepsilon°$ values of the remaining alkali metals are calculated.

The existence of a general trend in oxidation potentials is more distinctive among the alkaline earth metals than among the alkali metals, as can be seen from Table 8–3.

### TABLE 8–3.    Oxidation Potentials for R2 Elements.

| Element | $I$(eV) | $-H$(eV) | $S$(eV) | $\varepsilon$(v) | $M \rightarrow M^{2+} + 2e^-$ |
|---------|---------|----------|---------|------------------|-------------------------------|
| Be | 27.5 | 24.8 | 3.33 | 1.85 | |
| Mg | 22.6 | 20.2 | 1.56 | 2.37 | |
| Ca | 18.0 | 16.6 | 2.00 | 2.87 | |
| Sr | 16.7 | 15.2 | 1.70 | 2.89 | |
| Ba | 15.2 | 13.7 | 1.82 | 2.90 | |

Considering the alkali and the alkaline earth metals, it would seem that general trends in oxidation potentials do, in fact, occur. But this should not be surprising because the terms that contribute to the electrode potential show general trends themselves. We might also note that a consistency occurs in the horizontal groupings

$$Na/Na^+_{(aq)} \qquad Mg/Mg^{2+}_{(aq)} \qquad Al/Al^{3+}_{(aq)}$$
$$(2.714\ v) \qquad (2.37\ v) \qquad (1.66\ v)$$

$$K/K^+_{(aq)} \qquad Ca/Ca^{2+}_{(aq)} \qquad Sc/Sc^{3+}_{(aq)}$$
$$(2.925\ v) \qquad (2.87\ v) \qquad (2.08\ v)$$

$$Rb/Rb^+_{(aq)} \qquad Sr/Sr^{2+}_{(aq)} \qquad Y/Y^{3+}_{(aq)}$$
$$(2.925\ v) \qquad (2.89\ v) \qquad (2.37\ v)$$

$$Cs/Cs^+_{(aq)} \qquad Ba/Ba^{2+}_{(aq)} \qquad La/La^{3+}_{(aq)}$$
$$(3.08\ v) \qquad (2.90\ v) \qquad (2.52\ v)$$

These trends might well be attributed to the increase in ionization energy corresponding to the successive increase in the number of electrons that are removed.

In the data above, a consistent trend is also observed in the oxidation potentials of scandium, yttrium, and lanthanum. But in spite of all of these positive observations, we must be very careful in discussing periodic trends. There are a number of interworking properties that contribute to the oxidation potential, and they must be considered individually for each series. For instance, it is most probable that the $\varepsilon°$ value for cesium is 2.923 v rather than 3.08 v. If this value is used, the general trend presented for the alkali metals is not valid. But even more instructive is the trend observed for the R3 family shown in Table 8–4. Here we observe a consistent trend, but it is the reverse of the one found for families R1 and R2. Additionally, the only parallel between the trend in $\varepsilon°$ values and the contributing terms is found with the sublimation energy, but this term is not sufficiently large to establish the trend. It would therefore appear that the regularity of this particular trend is more or less accidental.

**TABLE 8–4. Oxidation Potentials for R3 Elements.**

| Element | $I$(eV) | $-H$(eV) | $S$(eV) | $\varepsilon°_{calc}$(v) | $\varepsilon°_{obs}$(v) |
|---------|---------|----------|---------|--------------------------|-------------------------|
| Al | 53.3 | 48.7 | 3.3 | 2.00 | 1.66 |
| Ga | 57.3 | 48.8 | 2.9 | 0.83 | 0.53 |
| In | 52.7 | 43.2 | 2.5 | 0.63 | 0.342 |
| Tl | 56.4 | 42.7 | 1.8 | −0.53 | −0.720 |

Thus we might conclude that periodic trends in oxidation potentials do exist. But at the same time, we must not forget that the potentials are the result of the interplay of several different factors. And meaningful trends will arise only when the trends in one or more of the contributing terms are sufficient in magnitude and consistency to establish an overall trend.

## SINGLE ELECTRODE POTENTIALS

From the Born-Haber cycle used to predict oxidation potentials, it would seem that we have a means of evaluating single electrode potentials. Yet at the same time we know that in order to have an oxidation there must be a corresponding reduction. In a galvanic cell this requires the use of two electrodes, and the potential that is measured will necessarily be relative to the value assigned to one of these. A closely related question is that of a single ion activity. In this case, the requirement of electrical neutrality prevents us from having an ion in solution without at the same time having one of opposite charge. Thus we find it necessary to look at the overall effect of all the ions rather than that of a single ion type.

Both of these questions have been of general concern for years, but from a thermodynamic standpoint, they were resolved around 1930, when it was shown that they have no thermodynamic validity. This, of course, does not mean they cannot be evaluated or that they have no meaning. It only points out that they cannot be evaluated within the structure of thermodynamics; and although the concept of a single electrode potential and of a single ion activity are thermodynamically inaccessible, attempts to evaluate them are not dead, as can be seen from the recent discussion by H. S. Frank[8] on single ion activities and the proposal of a means to evaluate a single electrode potential by I. Oppenheim.[9]

A number of models have been proposed by which single electrode potentials and single ion activities can be evaluated. The absolute potential of the calomel electrode has been calculated in essentially the manner we have used here, but with the additional consideration of the single ion entropy change. The difficulty in this case, however, is that a method based on a theoretical model and empirical curve fitting is required to assign the single ion values for hydration and entropy terms. In this treatment, Latimer, Pitzer, and Slansky[10] calculate the single ion hydration energies by means of the Born equation (see Chapter 12). Here it is necessary to define an ion radius as well as to distribute the hydration energy of a given salt between the positive and the negative ions. Based on a model of the hydrated ions, they set the radius of a cation equal to the Pauling ionic radius $+0.85$ Å, that is, $r_+ = (r_c + 0.85)$ Å (see Chapter 4). The distribution of hydration energy was then determined by dividing it between the positive and negative ions in CsI such that the values for both ions fall on the Born curve. This, of course, may very well give the correct values for the single ion hydration energies, but there is no thermodynamic means of verification.

## APPLICATIONS OF OXIDATION POTENTIALS

We have seen that, once the standard state emf values for the various half-cell reactions have been determined, it is possible to combine these to determine the overall cell emf and therefore, the thermodynamic tendency for the reaction to proceed as written. In combining half-cell potentials it is not necessary to consider the number of electrons involved in the change since the half-cell potentials are evaluated for a one-electron change. Thus, in adding two half-cell potentials, we can directly add the listed potentials from the tables. This point can be appreciated if we consider a reaction where the half-cells do not involve the same number of electrons, such as in the reaction

$$3 Ag^+ + Al = 3 Ag + Al^{3+}$$

The half-cell reactions are

$$Ag = Ag^+ + e^- \qquad E^\circ = -0.7991 \text{ v}$$
$$Al = Al^{3+} + 3e^- \qquad E^\circ = 1.66 \text{ v}$$

This, then, gives an over-all cell potential of $E^\circ = 2.46$ v, and we therefore conclude that if all species are at unit activity at 25°C, aluminum will reduce silver ion.

The situation is somewhat different when two half-cell reactions are added to obtain a third half-cell reaction. Ordinarily, it is necessary to evaluate the free energy change for each half reaction in order to take into consideration the number of electrons involved. These are then additive. Once the free energy change is evaluated for the desired half-cell reaction, it is possible to convert back to the half-cell emf. As an example, we can consider the half-cell

$$Tl = Tl^{3+} + 3e^-$$

From the table of oxidation potentials, it can be seen that

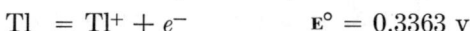

$$\text{Tl} = \text{Tl}^+ + e^- \qquad\qquad \text{E}° = 0.3363 \text{ v}$$

and

$$\text{Tl}^+ = \text{Tl}^{3+} + 2e^- \qquad\qquad \text{E}° = -1.250 \text{ v}$$

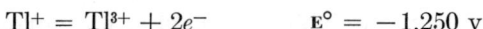

Converting these to standard free energies according to Eq. (8–17), we obtain $\Delta G° = -1 \times (0.3363)$F and $\Delta G° = -2 \times (-1.250)$F $= 2.50$F, respectively. Adding these, we find that $\Delta G°$ for the half-cell reaction

$$\text{Tl} = \text{Tl}^{3+} + 3e^-$$

is equal to $(-0.3363 + 2.50)$F or $2.16$F, from which we obtain the value $-(2.16\text{F}/3\text{F})$ or $-0.720$ v for E°.

## Stabilization of Oxidation States of Metals

One of the important applications of half-cell potentials is that of predicting the relative stabilities of different oxidation states of a given element. As an example of such a calculation, the disproportionation of Au(I) can be considered. For the reaction

$$3 \text{ Au}^+ = 2 \text{ Au} + \text{Au}^{3+}$$

the two half-cell reactions

$$\text{Au}^+ = \text{Au}^{3+} + 2e^- \qquad\qquad \text{E}° = -1.41 \text{ v}$$

and

$$\text{Au}^+ + e^- = \text{Au} \qquad\qquad \text{E}° = 1.68 \text{ v}$$

can be written. This yields a positive emf of 0.27 v for the total cell reaction, and therefore it would be expected that gold(I) will disproportionate to gold(0) and gold(III) if all species are at unity activity.

It should be pointed out that in every reaction we have written thus far and in those that will follow, all ionic species are written as if only the simple ion were involved. This, of course, is not the case, since all ions are hydrated to some extent in an aqueous solution. Thus, each ion may be considered surrounded by a number of water molecules, the number, firmness of attachment, and the distance of closest approach being determined primarily by the size, charge, and electronic nature of the ion. With this in mind, it is interesting to note that the standard potential for either of these half-cell reactions can be changed merely by changing the environment of the gold(I) species, that is, by replacing the water molecules which surround the metal ion by other molecules or ions. We can illustrate this effect of environment on the central ion by considering the ions Br⁻, NCS⁻, I⁻, and CN⁻ with respect to the standard potential of the following half-cells:

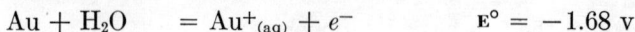

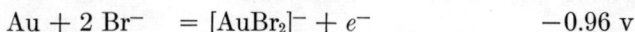

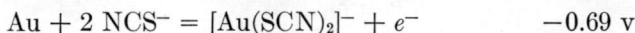

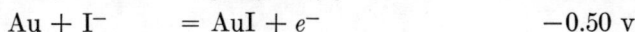

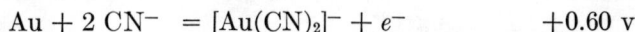

$$\text{Au} + \text{H}_2\text{O} \quad= \text{Au}^+_{(aq)} + e^- \qquad\qquad \text{E}° = -1.68 \text{ v}$$
$$\text{Au} + 2 \text{ Br}^- \quad= [\text{AuBr}_2]^- + e^- \qquad\qquad\qquad -0.96 \text{ v}$$
$$\text{Au} + 2 \text{ NCS}^- = [\text{Au(SCN)}_2]^- + e^- \qquad\qquad -0.69 \text{ v}$$
$$\text{Au} + \text{I}^- \qquad= \text{AuI} + e^- \qquad\qquad\qquad\quad -0.50 \text{ v}$$
$$\text{Au} + 2 \text{ CN}^- \quad= [\text{Au(CN)}_2]^- + e^- \qquad\qquad\quad +0.60 \text{ v}$$

Here we note that the various ions have effected a stabilization of the gold(I) toward reduction, with the $CN^-$ giving the greatest stabilization. This is an example of the stabilization toward reduction of an oxidation state through the formation of either an insoluble compound or a complex ion which is more stable than the simple hydrated ion.

As an example of the stabilization of a given oxidation state toward oxidation, we can consider the iron(II)-iron(III) system for which we have the following half-cell potentials:

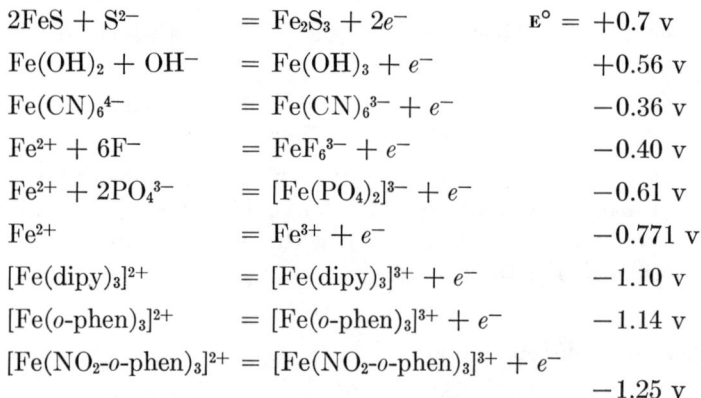

$$2FeS + S^{2-} \qquad = Fe_2S_3 + 2e^- \qquad E^\circ = +0.7 \text{ v}$$

$$Fe(OH)_2 + OH^- = Fe(OH)_3 + e^- \qquad\qquad +0.56 \text{ v}$$

$$Fe(CN)_6^{4-} \qquad = Fe(CN)_6^{3-} + e^- \qquad\qquad -0.36 \text{ v}$$

$$Fe^{2+} + 6F^- \qquad = FeF_6^{3-} + e^- \qquad\qquad -0.40 \text{ v}$$

$$Fe^{2+} + 2PO_4^{3-} \quad = [Fe(PO_4)_2]^{3-} + e^- \qquad -0.61 \text{ v}$$

$$Fe^{2+} \qquad\qquad = Fe^{3+} + e^- \qquad\qquad -0.771 \text{ v}$$

$$[Fe(dipy)_3]^{2+} \qquad = [Fe(dipy)_3]^{3+} + e^- \qquad -1.10 \text{ v}$$

$$[Fe(o\text{-phen})_3]^{2+} = [Fe(o\text{-phen})_3]^{3+} + e^- \qquad -1.14 \text{ v}$$

$$[Fe(NO_2\text{-}o\text{-phen})_3]^{2+} = [Fe(NO_2\text{-}o\text{-phen})_3]^{3+} + e^-$$
$$-1.25 \text{ v}$$

(where dipy $\equiv$ dipyridyl, $o$-phen $\equiv$ *ortho*phenanthroline, and $NO_2$-$o$-phen $\equiv$ nitro-*ortho*phenanthroline). From these values it is concluded that sulfide, hydroxide, cyanide, fluoride, and phosphate ions all stabilize iron(III) against reduction to a greater degree than does water. Likewise, the organic complexing agents stabilize iron(II) against oxidation to a greater degree than water.

Many examples are known of the stabilization of particular oxidation states for the large number of multivalent metals. However, it should again be emphasized that the standard potentials give the thermodynamic tendency of a particular reaction to proceed when all components are at unit activity, and the actual rate of a reaction is dependent on the kinetics of the reaction. So, for example, while the above values predict $Fe(CN)_6^{4-}$ to be less stable toward oxidation than the aquated $Fe^{2+}$, the cyano complex shows greater chemical stability. This is attributed to the slow rate of oxidation under ordinary conditions.

## Equilibrium Constants from Half-Cell Potentials

Recalling the two basic relationships

$$\Delta G^\circ = -RT \ln K_{eq}$$

and

$$\Delta G^\circ = -nF E^\circ$$

it can be seen that the equilibrium constant of a reaction can readily be related to the standard-cell potential by the expression

$$(8\text{–}16) \qquad \varepsilon^\circ = \frac{RT}{n\mathbf{F}} \ln K_{eq}$$

As an example of the calculation of a simple equilibrium constant, let us consider the reduction of selenous acid with stannous chloride. The overall reaction can be represented as

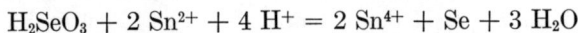

$$H_2SeO_3 + 2\ Sn^{2+} + 4\ H^+ = 2\ Sn^{4+} + Se + 3\ H_2O$$

and this can be broken down into the two half reactions

$$H_2SeO_3 + 4\ H^+ + 4e^- = Se + 3\ H_2O \qquad \varepsilon^\circ = 0.74\ v$$

and

$$Sn^{2+} = Sn^{4+} + 2e^- \qquad\qquad\qquad \varepsilon^\circ = -0.15\ v$$

This leads to an overall cell potential of $\varepsilon^\circ = +0.59$ v. Solving Eq. (8–16) for $\log K$, we obtain for 25°C

$$\log K = \frac{n\varepsilon^\circ}{0.059} = \frac{(4)(0.59)}{0.059} = 40$$

or $K = 10^{40}$.

It is also possible to calculate solubility products for relatively insoluble substances by the same general means. For instance, the solubility of silver chloride can be determined by considering the reaction

$$AgCl = Ag^+ + Cl^-$$

to be made up of the two half-cell reactions

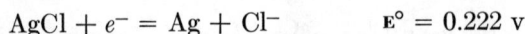

$$AgCl + e^- = Ag + Cl^- \qquad \varepsilon^\circ = 0.222\ v$$

and

$$Ag = Ag^+ + e^- \qquad\qquad \varepsilon^\circ = -0.799\ v$$

The standard potential for a cell composed of these two half reactions is $\varepsilon^\circ = -0.577$ v, and $\log K$ at 25°C is given by the following equation:

$$\log K = \frac{(1)(-0.577)}{0.059} = -9.78$$

or $K = 10^{-9.78} = 1.7 \times 10^{-10}$.

In solutions of this dilution it is permissible to substitute concentrations for activities, and therefore calculation of the solubilities of silver chloride in aqueous solutions having a low ionic strength is straightforward.

## References

1. H. S. Harned and B. B. Owen, "The Physical Chemistry of Electrolytic Solutions," 3rd ed., Reinhold Publishing Corporation, New York, 1958.

2. G. N. Lewis and M. Randall, "Thermodynamics," McGraw-Hill, Inc., New York, 1923.

3. J. J. Lingane, "Electroanalytical Chemistry," Interscience Publishers, New York, 1953.

4. F. C. Anson, *J. Chem. Ed.*, **36**, 394 (1959).

5. J. A. Christiansen and M. Pourbiax, "Conventions Concerning the Signs of Electromotive Forces and Electrode Potentials," *Comptes Rendus of the 17th Conference of the IUPAC*, Maison de la Chimie, Paris, 1954.

6. T. S. Licht and A. J. Bethune, *J. Chem. Ed.*, **34**, 433 (1957).

7. W. M. Latimer, "Oxidation Potentials," 2nd ed., Prentice-Hall, Inc., Englewood Cliffs, N. J., 1952.

8. H. S. Frank, *J. Phys. Chem.*, **67**, 1554 (1963).

9. I. Oppenheim, *J. Phys. Chem.*, **68**, 2959 (1964).

10. W. M. Latimer, K. S. Pitzer, and C. Slansky, *J. Chem. Phys.*, **7**, 108 (1939).

## Suggested Supplementary Reading

C. B. Monk, "Electrolytic Dissociation," Academic Press, Inc., New York, 1961.

G. Lewis and M. Randall, "Thermodynamics," 2nd ed., McGraw-Hill, Inc., New York, 1961.

D. MacInnes, "The Principles of Electrochemistry," Reinhold Publishing Corporation, New York, 1939.

J. Kleinberg, W. Argersinger, Jr., and E. Griswold, "Inorganic Chemistry," D. C. Heath and Company, Boston, 1960.

## Problems

1. Using the data reported by G. A. Linhart, *J. Am. Chem. Soc.*, **41**, 1175 (1919) for the cell

$$Pt, H_{2(1\ atm)}\ |HCl_{(m)}|\ AgCl, Ag$$

| Molal Conc. ($m$) | 0.04826 | 0.00965 | 0.004826 | 0.00100 | 0.000483 | 0.000242 | 0.000136 |
|---|---|---|---|---|---|---|---|
| $E_{(corrected)}$ | 0.3874 | 0.4658 | 0.5002 | 0.5791 | 0.6161 | 0.6514 | 0.6805 |

(a) obtain the $E°$ value by a direct extrapolation of the experimental data according to the method of Lewis and Randall.

(b) rearrange the Nernst equation and use the Debye-Huckel limiting law, $-\log \gamma_{\pm} = A\sqrt{m}$ to obtain the $E°$ value by extrapolation.

2. From the data in Problem 1, determine the activity coefficient for HCl at each concentration.

3. By considering the terms that contribute to the $E°$ value and the manner of evaluating them, why is it not thermodynamically possible to distribute an activity coefficient between the cation and the anion?

4. Determine the standard oxidation-reduction potential for the half-cell

$$Mn^{3+} + 2H_2O = MnO_2 + 4H^+ + e^-$$

5. Calculate the equilibrium constant at 25°C for the reaction

$$H_2 + O_2 \rightleftharpoons H_2O_2$$

**6.** Using standard electrode potentials, predict the chemical reactions that should occur when

(a) an acid solution of $H_2O_2$ is added to $MnO_2$

(b) $Ce(SO_4)_2$ is added to a solution of $KI$

(c) $VO_2^+$ is reduced by $Sn^{2+}$

(d) $Zn$ is added to a solution of $Pb(NO_3)_2$

# 9

# Acids and Bases

The realization that an acid is a unique type of compound existed long before its properties were first systematized by Robert Boyle. Yet, after almost three centuries of experience with definitions and theories of acid behavior, we still lack a unanimity of opinion. Of necessity, the first concepts of an acid were empirical in character. The fact that the particular species of compounds known as acids has certain observable properties characteristic only of that species led to the original setting apart of these compounds. It had very early been observed that calcareous earths such as limestone effervesce when dissolved in certain solvents of which vinegar is typical. In addition, dilute solutions of such solvents were observed to have a characteristic sour taste. Because of this property, they were termed "acids" from the Latin *acetum* for vinegar. And by the end of the seventeenth century, other properties were found that also proved to be characteristic of this class of compounds, such as their general solvent ability and their action on certain plant dyes.

Another class of compounds that have properties quite different from those of acids was early recognized. These compounds were first derived from the ashes of various plants and were, for this reason, given the name *alkali* from the Arabic term for ashes of a plant. Just as acids had certain characteristic properties, the alkalies could be recognized by their unique properties. These consist of the ability to dissolve sulfur and oils, to affect the colors of certain plant dyes, and in particular, to neutralize the effects of acids.

It should be noted that the designation of a substance as an acid or an alkali depended on the unique observable properties of the substance. This idea was emphasized by Robert Boyle when he recognized the properties of an acid to consist of its solvent action on many substances, its ability to precipitate sulfur from alkali solutions, and its ability to turn certain blue plant dyes to red. Along the same lines, Rouelle extended the concept of an alkali in 1744 by introducing the more general class of compounds known as *bases*. Rouelle maintained that a salt results from the combination of an acid and a base, and a base, therefore, can be defined as any species that reacts with an acid to form a salt. According to Rouelle, the group of compounds known as bases includes the alkalies, the alkaline earth equivalents of the alkalies, metals, and certain oils.

The first deviation from a strictly experimental definition of an acid came as a result of Lavoisier's studies on oxidation. During this work, Lavoisier observed that many of the more common acids result from the union of oxygen with nonmetals such as sulfur and phosphorous. This led him to conclude that the peculiar properties of an acid can be attributed to the presence of oxygen. Such a view was not necessarily intended to displace the experimental definition of an acid, but rather, it represented a first attempt to understand this characteristic behavior.

The breakdown of Lavoisier's theory came as a result of studies on *muriatic acid* (HCl). There could be no question that muriatic acid was an acid by all experimental criteria. Thus, according to Lavoisier, it must contain oxygen as all acids must. At that time, an acid was thought to be the oxide of the nonmetal, and the compound we now know to have the formula $H_2SO_4$, for instance, was considered to be the acid plus water of crystallization. In a similar fashion, muriatic acid would be represented as XO,HO where XO is the anhydrous acid and HO is the water of crystallization written in the characteristic form of that time. Such a structure for muriatic acid seemed to receive support from the studies of neutralization reactions. If we consider the reaction of sulfuric acid with lime, we find that it would have been expressed at that time as

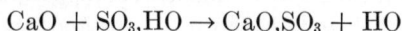

$$CaO + SO_3,HO \rightarrow CaO,SO_3 + HO$$

An analogous reaction for muriatic acid would be expressed as

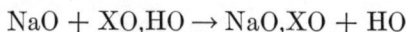

$$NaO + XO,HO \rightarrow NaO,XO + HO$$

Just as the reaction of sulfuric acid results in a salt plus water, the reaction of muriatic acid also results in a salt plus water. Thus, it would appear that the two acids are similar insofar as they both contain the so-called water of crystallization. Consequently, if the anhydrous form of sulfuric acid could be obtained, it seemed reasonable that the anhydrous form of muriatic acid could also be obtained. All attempts to make the anhydrous form, however, either resulted in what appeared to be the hydrated form of the acid or in a substance that apparently contained no oxygen. Many experiments were carried out along these lines by Sir Humphrey Davy in an attempt to show that muriatic acid contains the required oxygen of the Lavoisier theory, but they always met with failure. Finally, Davy concluded that there is no oxygen in muriatic acid, but rather, it contains the *dephlogisticated* muriatic acid gas discovered by Scheele. Davy maintained that this is an element, and he named it chlorine.

Shortly after he had shown that muriatic acid contains no oxygen, Davy proposed that acidity cannot be attributed to any particular element, but rather, it is due to particular arrangements of various substances. However, in 1816, he finally expressed the opinion that hydrogen is peculiar to all acidic species. This idea did not meet with immediate approval. In particular, Gay-Lussac was not convinced that muriatic acid contains no oxygen until he, himself, showed that there is none in HI and HCN. Still refusing to make a complete break with the oxygen theory of Lavoisier, Gay-Lussac proposed that those acids which do not contain oxygen should be considered a new class of compounds known as *hydracides*. This view lasted for a short while but was thrown out when Liebig presented a rather convincing argument that it is much simpler to consider an acid to be a compound containing a hydrogen atom that can be replaced by a metal. This brought all of the then recognized acids into a common classification.

## THE ARRHENIUS DEFINITION

Before the Arrhenius theory of electrolytic dissociation, which was developed between 1880 and 1890, acids were classified either in terms of their observable properties or in terms of some acidifying species such as hydrogen. Actually, these two approaches were not independent. The search for an acidifying species was a search for the source of the particular properties of this class of compounds known as acids. Thus, an acid was ultimately recognized by its properties. This can be appreciated when we recall that the oxygen theory of Lavoisier was discarded because it failed to give complete agreement with experiment. The subsequent hydrogen theory of Davy was acceptable only because all substances that could then be classified as acids by virtue of their properties also contained a replaceable hydrogen atom.

The proposal of an acidifying species to explain the properties of an acid represents a rather crude but first attempt to find a model for acid character. Such a model offered little to the understanding of acid properties and was of absolutely no value from a quantitative point of view. Yet, with the extremely limited knowledge of solutions at that time, little more could reasonably be expected. The door was opened, however, with the Arrhenius theory of electrolytic dissociation. The new concept of ionization processes made possible the Arrhenius model of acid behavior. This was the first model to approach the somewhat more sophisticated views of the present day on acid-base character. According to the Arrhenius model, *an acid is any hydrogen-containing compound which gives hydrogen ions in aqueous solution, and a base is any hydroxyl-containing compound which gives hydroxyl ions in aqueous solution*. The process of neutralization of an acid by a base can be represented by the reaction

$$H^+ + OH^- = H_2O$$

With the Arrhenius definition, many of the aspects of acid-base behavior were now understandable in terms of a mechanistic picture, and for the first time quantitative relationships could be determined. This is the kind of model that leads to great advances in any phase of science. For instance, the constant heat of neutralization of a strong acid by a strong base can readily be understood in terms of the Arrhenius picture. Since the reaction involves only the combination of a hydrogen ion and a hydroxyl ion in all such neutralization reactions, the approximately constant molar heat of neutralization would be expected. The theory also leads to quantitative determinations of acid or base strengths from the evaluation of an equilibrium relation such as

$$(9\text{--}1) \qquad K_i = \frac{a_{H^+} a_{B^-}}{a_{HB}}$$

One of the more significant correlations of the Arrhenius model with experiment lay in the catalytic properties of acids. It had been found quite early that along with the other acid properties, catalytic character is also fundamental, and the Arrhenius theory of electrolytic dissociation offered a means of correlating this character with the concentration of the hydrogen ion. Due to the high mobility of the hydrogen ion and the relatively low mobilities of the various anions, the conductivity of a solution should parallel the catalytic activity of the solution if the hydrogen ion is truly the source of the catalytic properties. One of the more outstanding examples of such a correlation is shown in Table 9–1, where the relative conductivities and catalytic effects are compared for a series of acids in anhydrous ethanol as determined by Goldschmidt.

**TABLE 9–1. Catalytic Effects of Various Acids in Anhydrous Ethanol Relative to HCl = 100.**

| Acid | Relative Conductivity | Catalytic Effect on the Esterification of Formic Acid |
|---|---|---|
| Hydrochloric | 100 | 100 |
| Picric | 10.4 | 10.3 |
| Trichloroacetic | 1.00 | 1.04 |
| Trichlorobutyric | 0.35 | 0.30 |
| Dichloroacetic | 0.22 | 0.18 |

In spite of the obvious successes of the Arrhenius definition, several serious shortcomings were soon apparent. It was only natural that a serious complaint should arise with regard to the limitation of all acid-base reactions to aqueous media. It is certainly true that the large majority of chemical reactions are carried out in aqueous systems, but it must be admitted that other solvents exist and their chemistry is also important.

This was particularly true for liquid ammonia at that time. Unfortunately, the Arrhenius theory would deny the use of acid-base concepts in these nonaqueous solvents. Equally important, reactions in the gas phase where no solvent is present were also excluded from the acid-base concept. Another serious weakness of the definition lay in the restriction of bases to hydroxyl compounds. Although chemists had come to think of acids as hydrogen-containing compounds, they were not yet convinced that a base had to be a hydroxyl compound. Many organic substances as well as ammonia were known to show basic properties insofar as their chemistry is concerned, and it was difficult to ignore these. By analogy, it would seem that if some acidic compounds that do not contain hydrogen had been known at that time, a corresponding feeling would undoubtedly have existed with regard to the restriction of acids to hydrogen-containing compounds. This, however, was not the case, and the hydrogen concept of an acid was greatly strengthened by the Arrhenius definition.

## THE PROTONIC DEFINITION

The climax to the hydrogenic concept of an acid came in 1923, when Bronsted[1] in Denmark and Lowry[2] in England independently proposed that *an acid is a species that tends to give up a proton, and a base is a species that tends to accept a proton.* It is readily apparent that such a definition does not change the status of an Arrhenius acid or base. A hydroxyl-containing compound that yields hydroxide ions in aqueous solution is still a base, and a hydrogen-containing compound that yields hydrogen ions in aqueous solution is still an acid. Yet the scope of acid-base reactions has been greatly extended with the corresponding removal of many of the shortcomings that were found in the Arrhenius definition.

The protonic definition of a base is obviously far different from that proposed by Arrhenius. Whereas the Arrhenius model restricts a base to a hydroxyl compound that gives a hydroxyl ion in aqueous solution, the protonic definition requires no particular type of ion or any particular solvent. Although they are still bases, the hydroxyl bases are no more so than molecules such as pyridine or ammonia or a multitude of other molecules or ions. The protonic definition of an acid, on the other hand, is not quite so radically removed from that of Arrhenius. The most significant difference is the inherent acid character of a Bronsted acid in the absence or presence of any solvent. Thus, HCl is an acid by virtue of the fact that it can donate a hydrogen ion, not because it might have already done so in an aqueous solution. The restriction of an acid-base reaction to an aqueous medium has, therefore, been removed, and we see that it is possible to have such reactions in any medium or in the absence of a medium.

In the light of the protonic definition, an acid-base reaction can be considered to involve a competition for a proton by two different bases.

If we consider the ionization of an acid such as HCl, it is obvious that the resultant chloride ion is a base, inasmuch as it can accept a proton. Such an acid-base pair is referred to as a *conjugate pair*, the chloride ion being the conjugate base of the acid HCl. In order for the acid to donate its proton, it is necessary that there be a base to accept the proton. For the simple ionization of an acid, the solvent will act as that base. For instance,

$$HCl + H_2O = H_3O^+ + Cl^-$$

Here we see that the water accepts the proton from the HCl and is, therefore, a base. The reverse reaction that expresses the competition for the proton is represented by the reaction of the chloride ion with the hydronium ion. Since this particular reaction proceeds essentially 100% to the right, it is quite apparent that the chloride ion is a much weaker base than the water molecule. In general, we can say that *the conjugate base of a strong acid is a weak base, and the conjugate base of a weak acid is a strong base.*

In the instance of an aqueous solution of HCl, we have seen that the water acts as a base. This is not always the case. If a solute that is more basic than water had been used, the water would have donated the proton and would, itself, be the acid. Thus, we can say that water is an *amphiprotic* solvent. Actually, we find that the acidity or basicity of a particular substance may vary with the basicity or acidity of the medium in which it is placed. Towards a sufficiently strong acid, such as anhydrous perchloric acid, even nitric acid behaves as a base.

Although the meaning is somewhat different in the protonic definition, the relative strengths of acids and bases can be determined in much the same manner as in the Arrhenius definition. If we consider the general acid-base reaction to be expressable as

$$acid_1 + base_1 = acid_2 + base_2$$

the thermodynamic equilibrium constant for the reaction will be given by

$$(9-2) \qquad K' = \frac{a_{A_2} a_{B_2}}{a_{A_1} a_{B_1}}$$

Here we have an expression that will give the relative acidity of $acid_1$ with respect to any $base_1$. Ordinarily, $base_1$ will be a solvent, such as water. This is not necessarily the case. However, under such circumstances, the strength of an acid, HA, can be expressed in terms of the ionization constant

$$(9-3) \qquad K = K' a_{H_2O} = \frac{a_{H^+} a_{A^-}}{a_{HA}}$$

which is the same as the expression for the acid strength in the Arrhenius theory.

The success of the protonic definition in handling acid-base systems has been more than impressive, and as a result, we might be prone to forget the original criteria of acid-base character. Yet, if a definition is adequate, it should be consistent with the experimental data. If acid-base character can truly be attributed to the proton, then it should be possible to correlate all acid-base reactions in terms of this definition. If, on the other hand, there are acid-base reactions that cannot be described by the protonic definition, then we must admit that the Bronsted-Lowry approach is not sufficient and seek a new definition.

## THE SOLVENT SYSTEM DEFINITION

One of the more important weaknesses of the Arrhenius model of acid-base behavior is its undue restriction of acid-base reactions to aqueous media. This very early led to some conflicts, primarily as the result of solvent studies in liquid ammonia. On the basis of the experimental criteria, similarities could be pointed out between acid-base reactions in aqueous media and certain types of reactions in ammonia. If we consider the auto-ionization of water to form the hydronium and the hydroxyl ions, we can note an analogous autoionization of ammonia to form the ammonium and the amide ions,

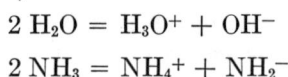

$$2\ H_2O = H_3O^+ + OH^-$$
$$2\ NH_3 = NH_4^+ + NH_2^-$$

If, further, we consider an ordinary neutralization reaction between a strong acid and a strong base in aqueous media, we can again recognize an exact analog in liquid ammonia. This similarity is apparent from the reactions

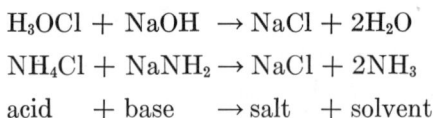

$$H_3OCl + NaOH \rightarrow NaCl + 2H_2O$$
$$NH_4Cl + NaNH_2 \rightarrow NaCl + 2NH_3$$
$$\text{acid} + \text{base} \rightarrow \text{salt} + \text{solvent}$$

Such reactions as these are typical of a wide variety of acid-base type reactions which were recognized by Franklin[3] at the turn of the century.

The ammono and the aquo systems of acids and bases are formally very similar. This becomes apparent when we realize that the ammonium ion and the hydronium ion can both be considered to be solvated protons. Although, historically, a solvent system definition of acid-base behavior existed before the protonic definition, it is apparent that the latter will, in general, suffice for protonic solvents. The protonic definition, however, does not satisfy the needs for nonprotonic solvents. Researches on such solvents as $COCl_2$, $SeOCl_2$, and $SO_2$ indicate that acid-base type behavior can be observed in systems where a proton plays no part in the reactions at all, and it would be desirable to talk in terms of acid-base concepts here as well as in protonic solvents.

In an attempt to apply the acid-base concept to the various nonprotonic solvents, several definitions of acids and bases have been proposed.[4,5] Using the definition of Cady and Elsey, we can say that *an acid is a solute that, either by direct dissociation or by reaction with the solvent, gives the cation characteristic of the solvent, and a base is a solute that, either by direct dissociation or by reaction with the solvent, gives the anion characteristic of the solvent.* If, for example, we consider the solvent $SO_2$, its characteristic cation and anion are seen to be

$$2 SO_2 = SO^{2+} + SO_3{}^{2-}$$

A neutralization reaction would then result from the reaction of a substance such as $SOCl_2$, which should be an acid, and tetramethylammonium sulfite, which should be a base:

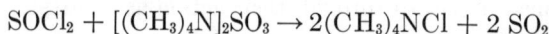

$$SOCl_2 + [(CH_3)_4N]_2SO_3 \rightarrow 2(CH_3)_4NCl + 2 SO_2$$

Just as with the Arrhenius definition, the product of neutralization in the solvent system definition is a salt plus the solvent. This can be seen from Table 9–2, where examples of typical neutralization reactions along with the characteristic cations and anions are given for several solvents.

**TABLE 9–2.  Neutralization Reactions in Various Solvents.**

| Solvent | Cation | Anion | Neutralization Reaction |
|---------|--------|-------|-------------------------|
| $H_2O$ | $H_3O^+(H^+)$ | $OH^-$ | $HCl + NaOH \rightarrow NaCl + H_2O$ |
| $NH_3$ | $NH_4{}^+(H^+)$ | $NH_2{}^-$ | $NH_4Cl + NaNH_2 \rightarrow NaCl + 2 NH_3$ |
| $HC_2H_3O_2$ | $H_2C_2H_3O_2{}^+(H^+)$ | $C_2H_3O_2{}^-$ | $HCl + NaC_2H_3O_2 \rightarrow NaCl + HC_2H_3O_2$ |
| $SO_2$ | $SO^{2+}$ | $SO_3{}^{2-}$ | $SOCl_2 + Na_2SO_3 \rightarrow 2NaCl + 2 SO_2$ |

There can be no question that an acid-base definition in terms of the parent solvent extends the realm of acid-base reactions to systems that previously had been ignored. But it omits a large segment of typical acid-base reactions that had been included in the protonic definition, and for this reason, it can in no way be considered to displace the protonic definition. In actuality, the solvent system approach differs little from a series of Arrhenius-type models, one for each solvent. Consequently, virtually all of the complaints that were leveled against the Arrhenius definition can be repeated against the solvent system approach, except for the Arrhenius restriction of all acid-base reactions to aqueous media. Nevertheless, we cannot help but see the desirability of devising an acid-base definition to cover these systems, and at the same time appreciate the need for a definition of sufficient generality to include all possible systems.

## THE LEWIS DEFINITION

In the same year that the Bronsted-Lowry definition was published, G. N. Lewis[6] attacked the problem from a direction that has now led to the unification of virtually all of the existent acid-base definitions. Re-emphasizing the experimental basis, Lewis proposed the following four necessary and sufficient criteria for acid-base classification:

(1) "When an acid and a base can combine, the process of combination, or neutralization, is a rapid one.

(2) "An acid or a base will replace a weaker acid or base from its compounds.

(3) "Acids and bases may be titrated against one another by the use of substances, usually colored, known as indicators.

(4) "Both acids and bases play an extremely important part in promoting chemical processes through their action as catalysts."

Going a step further, Lewis attributed these characteristic properties to the electronic structure of the acid or base along with the formation of a coordinate covalent bond. According to Lewis, *a base is any species that is capable of donating a pair of electrons to the formation of a covalent bond, and an acid is any species that is capable of accepting a pair of electrons to form a covalent bond*. Neutralization can then be attributed to the formation of the covalent bond. For instance, in the reaction

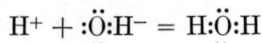

$$H^+ + :\ddot{O}:H^- = H:\ddot{O}:H$$

it is seen that the $H^+$ is capable of accepting a pair of electrons to form a covalent bond and is therefore an acid, whereas, the $OH^-$ has an electron pair to donate to the formation of the covalent bond and is, therefore, a base.

Invariably, the most thorough understanding is obtained when a system is viewed in terms of the most fundamental concepts possible. In the instance of a base, it is found that, although a variety of compounds and ions are admitted under this classification, they all have one fundamental property in common. This, of course, is a pair of available electrons. In like manner, the properties of an acid might be attributed to the availability of an empty orbital for the acceptance of a pair of electrons. The hydrogen ion has this fundamental property. However, this property is not limited to the hydrogen ion, and we might, therefore, expect other species with this same fundamental property to also behave as acids. Here we see that the Lewis definition does not attribute acidity to any particular element, but rather to a unique electronic arrangement. This is apparent when we consider the electron arrangements for the Lewis acid-base reaction

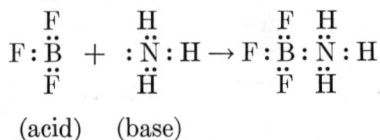

$$\begin{array}{ccc} F & H & F\ H \\ F:\ddot{B} & + & :\ddot{N}:H \rightarrow F:\ddot{B}:\ddot{N}:H \\ \ddot{F} & \ddot{H} & \ddot{F}\ \ddot{H} \end{array}$$
$$\text{(acid)} \quad \text{(base)}$$

The classification of a reaction such as this as an acid-base type reaction might appear to be rather extreme in the light of our experience with protonic acids. Yet it is found that Lewis acid-base reactions of this type can actually be titrated to an end point using a colored indicator in the same manner as can be done with a protonic acid.

Among the most extreme examples of a Lewis acid-base reaction is the formation of a coordination compound. As an example, the formation of the copper ammine complex involves the donation of a pair of electrons by each ammonia molecule to the copper ion, forming the complex $Cu(NH_3)_4^{2+}$. Such a broad applicability might be considered a weakness of the Lewis definition. This, however, will depend on one's own opinion of the purpose of the acid-base concept. If the purpose is to offer a means of correlating data, then it must be admitted that the Lewis definition has a definite value in the study of coordination compounds. This can be seen from Figure 9–1, where the relation between the basicity of coordinating ammines and the stability constants of the corresponding silver ammine complexes is given.

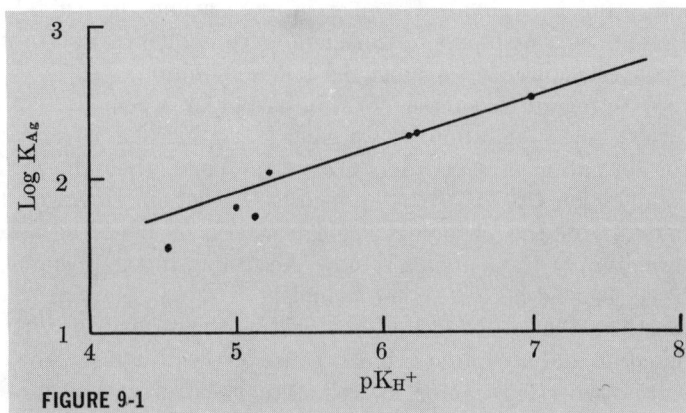

**FIGURE 9-1**

Stability of Silver Ammine Complexes as a Function of the Basicity of the Coordinating Ammines: Aniline, Quinoline, Hexamethylenetetramine, Pyridine, $\alpha$-Picoline, $\gamma$-Picoline, and 2, 4 Lutidine. (Based on data from J. Bjerrum, *Chem. Rev.*, **46**, 38 (1950).)

Along with a more fundamental understanding of the mechanism of acid-base reactions, the Lewis approach obviously offers a much greater generality than any previous approach. However, we do not obtain these without making some serious sacrifices. In fact, these sacrifices are deemed sufficiently serious by some that a definite opposition to the Lewis definition[7] has arisen. One of the more obvious weaknesses of this definition lies in its treatment of the conventional protonic acids such as $H_2SO_4$ and $HCl$. According to the Lewis approach, an acid must be able to accept a pair of electrons to form a covalent bond. This, of course, does not take place with the protonic acids. An attempt has been made to overcome this difficulty by a rather indirect approach. It is considered that the acid-base reaction between an acid, HX, and a base, B, is initiated by the formation of a hydrogen bond, thereby giving an intermediate species represented by X—H$\cdots$B, which then breaks down to give the final neutralization products. It must be admitted that the necessity of using such an indirect approach for the most common of the known acids is undesirable.

Although the above weakness might be passed off as more or less trivial, the question of relative acid strengths is not so trivial. One of the strong points of the protonic definition lies in the quantitative determination of relative acid and base strengths that it permits. In terms of the Lewis definition, this is no longer possible. The strength of an acid or base is found to depend on the particular reaction. This can be seen by comparing the stabilities of complexes of a pair of positive ions with different ligands. For instance, we find that the fluoride complex of beryllium is considerably more stable than that of copper, thereby indicating that divalent beryllium is more acidic than divalent copper. Yet, on the other hand, we find that the ammine complex of copper is more stable than that of beryllium indicating that divalent copper is more acidic than divalent beryllium. Herein lies a weakness of considerable consequence, and in it is found the strongest indictment against the Lewis definition.

With regard to complex ions, in particular, we can note a contradiction in the Lewis definition. If we consider an acid-base reaction in terms of the formation of a coordinate covalent bond, then the formation of the complex ion $Cr(NH_3)_6^{3+}$ is an acid-base reaction. However, that is not so in terms of the phenomenological criteria of Lewis because the rate of formation of the complex is slow. According to the phenomenological criteria, an acid-base reaction should be a rapid reaction. There are many reactions that will fall into the acid-base category by virtue of electron donation and acceptance, but because of the kinetics of the reaction, they will be slow. This factor, as well as the arbitrariness of our definition of a covalent bond, makes quantitative comparisons among Lewis acids and bases rather difficult.

A third instance in which some question has arisen concerning the validity of the Lewis classification is found in the catalytic ability of Lewis acids. It has been shown that in certain cases the apparent catalytic character of the Lewis acid can be attributed to impurities that lead to the formation of a hydrogen ion.[8] And, in general, it is found that the reactions catalyzed by Lewis acids are not catalyzed by the common protonic acids. This is of some concern inasmuch as Lewis considered catalytic behavior one of the four criteria of acid character. However, more recently, reactions have been found where Lewis acids were shown to serve as better catalysts than the protonic acids. Bell and Skinner[9] studied the catalytic depolymerization of paraldehyde in ether using both protonic acids and Lewis acids, and the Lewis acids were found, in general, to be better catalysts for this reaction than the protonic acids. Nevertheless, Bell points out that this particular reaction is rather unique in that it requires only a rearrangement of electrons rather than the migration of atoms. In any case, there can be no question that the Lewis acids do behave as catalysts in many reactions.

## THE USANOVICH DEFINITION

Although the definitions of Lewis and Bronsted are the most commonly used, there have been several additional classifications proposed for the interpretation of acid-base behavior. Of these, perhaps the most significant is the *positive-negative* definition of Usanovich,[10] in which an acid is defined as *any species capable of giving up cations, combining with anions or electrons, or neutralizing a base to give a salt. A base is defined as any species capable of giving up anions or electrons, combining with cations, or neutralizing an acid to give a salt.* It is obvious that the Usanovich definition includes all previous acid-base definitions, as well as including oxidation-reduction reactions as a special class of acid-base reactions.

In accounting for the acid-base character of a substance, Usanovich places considerable emphasis on the degree of coordination unsaturation of the central atom in the compound as well as emphasizing the general trends of the periodic table. If we consider the coordination unsaturation of an electropositive ion, we note that it can exhibit its acid function by adding an anion, and if an electronegative ion is coordinatively unsaturated, it can exhibit its basic function by adding a cation. For instance, in sulfur dioxide the central atom can be considered to be sulfur, and it is coordinatively unsaturated. Thus, it is capable of accepting an anion such as $O^{2-}$, thereby acting as an acid. It is in terms of such qualitative arguments as these, along with the general periodic trends, that the Usanovich definition is capable of correlating a vast number of reactions under the classification of acid-base.

In support of the Usanovich definition, it must be admitted that it is the most general of all the acid-base definitions thus far proposed. In addition, it appears to satisfy the phenomenological criteria of Lewis about as well as does the Lewis definition. From this latter standpoint, it would appear that the Usanovich definition should certainly be accepted as a legitimate acid-base definition. On the other hand, one wonders how far this can go. If we accept the Usanovich definition, we find that virtually all chemical reactions fall into the acid-base category, and one begins to wonder at the purpose of using any name other than chemical reaction.

There have been several other acid-base definitions proposed, but they add little to the fundamental principles embodied in the ideas already discussed. The Ebert-Konopik *donor-acceptor* acids and bases[11] amount to little more than a new nomenclature for the Usanovich system, and the Lux-Flood definition[12] for oxide reactions can be handled by either the Lewis or the Usanovich definition. Bjerrum[13] has made an attempt at reconciling the protonic and Lewis approaches, in which he proposes that an acid be defined as a *proton donor* while a conventional Lewis acid be referred to as an *antibase*. This contributes nothing but a change in terminology. Finally, it is of interest to note that a quantum mechanical treatment of acids and bases has been developed by Mulliken.[14] Although much more sophisticated, it is essentially as inclusive as the Usanovich definition.

All of the various definitions of acids and bases have a certain utility, and where it may be convenient to use one particular definition it may not be convenient to use another. From the basic nature of a definition, it is obvious that the question of validity does not enter. The problem is one of choosing which definition will be of the greatest value for the particular problems at hand.

## HARD AND SOFT ACIDS AND BASES

Although there may be weaknesses in the Lewis definition of acids and bases, its generality has made it particularly useful in discussing and predicting the course of many chemical reactions. The importance of the Lewis definition in this respect can be appreciated when it is realized that all chemical reactions can be classified as either acid-base or oxidation-reduction. Luder and Zuffanti[15] were the first to use the Lewis definition to classify reaction types, and on this basis they proposed that all reactions fall into three categories: acid-base, oxidation-reduction, and free radical. More recently, Pearson[16] has reduced these to simply generalized acid-base and oxidation-reduction by treating free radical reactions as a form of oxidation-reduction. In this respect, he considers oxidation-reduction reactions to be of two types: (1) electron transfer, and (2) atom or group transfer. A free radical reaction is considered to be of the second type and might be represented as

$$A\cdot + B\cdot \rightleftharpoons A^+{:}B^-$$

If we assume a classification such as that proposed by Pearson to be valid, it is rather obvious that we can think of a large number of chemical reactions in terms of the relative acidity and basicity of the reacting species.

### Lewis Acid-Base Strengths

It was pointed out earlier that a consistent ordering of acid and base strengths has not yet been achieved in terms of the Lewis definition. This, of course, offers a serious limitation in the correlation of Lewis acid-base reactions. In approaching the problem, we might consider the relative strengths in terms of the reactions

and

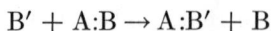

If the reactions proceed to the right as indicated, we can say that for these systems A′ is a stronger acid than A, and B′ is a stronger base than B. The relative strengths could then be expressed in terms of the relative stabilities of the acid-base complexes.

Carrying this idea further, the relationship between the acid and base strengths and the stability of the complex can be expressed by the equilibrium equation

$$(9\text{-}4) \qquad \log K = S_A S_B$$

Here, $S_A$ and $S_B$ are quantities related to the relative acid and base strengths, respectively, and are referred to as strength factors by Pearson. Now if $S_A$ and $S_B$ can be evaluated it would seem we would have a means of expressing a scale of acidity and basicity.

Of course, we have already seen that this procedure has not as yet proved successful. Apparently, the problem is more complex than implied by Eq. (9-4). Other factors must contribute to the formation of the acid-base complex, and their respective contributions will depend on the particular species that combine. This implies that $S_A$ and $S_B$ are either variables themselves or, if they are constant, other terms must occur in the equilibrium expression.

## Lewis Acid-Base Classifications

In spite of the inability to define an acid-base scale, certain qualitative correlations have been observed. It was pointed out by Ahrland, Chatt, and Davies[17] that metal ions can be separated into two general categories: class (a) acceptors and class (b) acceptors. The class (a) metal ions form their most stable complexes with the first member of each of the nonmetal groups in the periodic table, whereas the class (b) metal ions form their most stable complexes with the heavier members of the given group. Thus, we find the distinction

$$\text{Class (a)} \quad N \gg P > As > Sb > Bi$$
$$\text{Class (b)} \quad N \ll P > As > Sb > Bi$$
$$\text{Class (a)} \quad O \gg S > Se > Te$$
$$\text{Class (b)} \quad O \ll S \sim Se \sim Te$$
$$\text{Class (a)} \quad F > Cl > Br > I$$
$$\text{Class (b)} \quad F < Cl < Br < I$$

In this classification, the class (a) metal ions turn out to be small and not very easily polarized, whereas the class (b) metal ions turn out to be essentially the opposite in character. The small, nonpolarizable class (a) ions tend to combine with a nonmetal having quite similar properties, and the large, polarizable class (b) ions tend to combine with nonmetals having the same general characteristics as they.

These observations have been expanded and generalized by R. G. Pearson along with a change in terminology to *hard* and *soft* acids and bases.[16] Pearson defines a soft base as "one in which the donor atom is of high polarizability and of low electronegativity and is easily oxidized or is associated with empty, low-lying orbitals." A hard base is defined as one with the opposite properties. "The donor atom is of low polarizability and high electronegativity, is hard to reduce, and is associated with empty orbitals of high energy." The acceptor atom of a soft acid is defined by Pearson to be one that has "one or more of the following properties: low or zero positive charge, large size, and several easily excited outer electrons." A hard acid, on the other hand, is distinguished by "small size, high positive oxidation state, and the absence of any outer electrons which are easily excited to higher states."

The correlating principle on which acid-base reactions can here be discussed is simply that *hard acids prefer to coordinate with hard bases, and soft acids prefer to coordinate with soft bases.* Thus, in general it can be stated that for the reaction

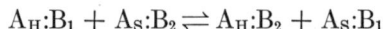

$$A_H:B_1 + A_S:B_2 \rightleftharpoons A_H:B_2 + A_S:B_1$$

$B_1$ is softer than $B_2$ if $K > 1$. On this basis, a listing of hard and soft acids and bases are given in Tables 9–3 and 9–4, respectively.

---

### TABLE 9–3.  Classification of Lewis Acids.[a]

| Hard | Soft |
|---|---|
| $H^+$, $Li^+$, $Na^+$, $K^+$ | $Cu^+$, $Ag^+$, $Au^+$, $Tl^+$, $Hg^+$ |
| $Be^{2+}$, $Mg^{2+}$, $Ca^{2+}$, $Sr^{2+}$, $Mn^{2+}$ | $Pd^{2+}$, $Cd^{2+}$, $Pt^{2+}$, $Hg^{2+}$, |
| | $\quad CH_3Hg^+$, $Co(CN)_5^{2-}$, $Pt^{4+}$, |
| | $\quad Te^{4+}$ |
| $Al^{3+}$, $Sc^{3+}$, $Ga^{3+}$, $In^{3+}$, $La^{3+}$ | $Tl^{3+}$, $Tl(CH_3)_3$, $BH_3$, $Ga(CH_3)_3$, |
| $\quad N^{3+}$, $Gd^{3+}$, $Lu^{3+}$ | $GaCl_3$, $GaI_3$, $InCl_3$ |
| $Cr^{3+}$, $Co^{3+}$, $Fe^{3+}$, $As^{3+}$ | $RS^+$, $RSe^+$, $RTe^+$ |
| $Si^{4+}$, $Ti^{4+}$, $Zr^{4+}$, $Th^{4+}$, $U^{4+}$ | $I^+$, $Br^+$, $HO^+$, $RO^+$ |
| $\quad Pu^{4+}$, $Ce^{3+}$, $Hf^{4+}$ | |
| $UO_2^{2+}$, $(CH_3)_2Sn^{2+}$, $VO^{2+}$, | $I_2$, $Br_2$, $ICN$, etc. |
| $\quad MoO^{3+}$ | |
| $BeMe_2$, $BF_3$, $B(OR)_3$ | Trinitrobenzene, etc. |
| $Al(CH_3)_3$, $AlCl_3$, $AlH_3$ | Chloranil, quinones, etc. |
| $RPO_2^+$, $ROPO_2^+$ | Tetracyanoethylene, etc. |
| $RSO_2^+$, $ROSO_2^+$, $SO_3$ | O, Cl, Br, I, N |
| $I^{7+}$, $I^{5+}$, $Cl^{7+}$, $Cr^{6+}$ | $M^0$ (metal atoms) |
| $RCO^+$, $CO_2$, $NC^+$ | Bulk metals |
| HX (hydrogen bonding molecules) | $CH_2$, carbenes |

Borderline
$\quad$ $Fe^{2+}$, $Co^{2+}$, $Ni^{2+}$, $Cu^{2+}$, $Zn^{2+}$, $Pb^{2+}$, $Sn^{2+}$, $Sb^{3+}$, $Bi^{3+}$, $Rh^{3+}$,
$\quad$ $Ir^{3+}$, $B(CH_3)_3$, $SO_2$, $NO^+$, $Ru^{2+}$, $Os^{2+}$, $R_3C^+$, $C_6H_5^+$, $GaH_3$

[a]From R.G. Pearson, *Chem. in Britain*, **3**, 103 (1967).

**TABLE 9–4.   Classification of Bases.**[a]
The symbol R stands for an alkyl group such as $CH_3$ or $C_2H_5$.

| Hard | Soft |
|------|------|
| $H_2O$, $OH^-$, $F^-$ | $R_2S$, $RSH$, $RS^-$ |
| $CH_3CO_2^-$, $PO_4^{3-}$, $SO_4^{2-}$ | $I^-$, $SCN^-$, $S_2O_3^{2-}$ |
| $Cl^-$, $CO_3^{2-}$, $ClO_4^-$, $NO_3^-$ | $R_3P$, $R_3As$, $(RO)_3P$ |
| $ROH$, $RO^-$, $R_2O$ | $CN^-$, $RNC$, $CO$ |
| $NH_3$, $RNH_2$, $N_2H_4$ | $C_2H_4$, $C_6H_6$ |
|  | $H^-$, $R^-$ |

Borderline
$C_6H_5NH_2$, $C_5H_5N$, $N_3^-$, $Br^-$, $NO_2^-$, $SO_3^{2-}$

[a] From R. G. Pearson, *Chem. in Britain*, **3**, 103 (1967).

In order to rationalize the correlating principle, Eq. (9–4) might be reconsidered. There the equilibrium constant was expressed in terms of two strength factors. We might consider $S_A$ and $S_B$ to represent some intrinsic acid and base strength characteristic of the particular species. Then we can say that the stability of the complex will actually depend on the strength factors along with additional terms that might then be lumped together and referred to as acid and base softness factors, $\sigma_A$ and $\sigma_B$, respectively. This then gives

$$(9\text{–}5) \qquad \log K = S_A S_B + \sigma_A \sigma_B$$

Now if all four of the quantities on the right side of the equation can be evaluated, it might be possible to finally express a relative scale of acid and base strengths in the Lewis system.

## Basis for Softness Factors

It is next necessary to consider what experimental or calculable quantities contribute to the softness factors. At this stage of development it must be recognized that the concept of hard and soft character is only qualitative. Nevertheless, there are several reasonable causes for the effect. Considering the nature of the combining species, it would seem that the bonding between a hard acid and a hard base is primarily ionic, whereas the bonding between a soft acid and a soft base is primarily covalent. We can further note that hard acids have vacant orbitals and can, therefore, accept $\pi$ electrons, whereas hard bases have filled outer orbitals and can donate $\pi$ electrons. The reverse situation exists for soft acids and bases. The importance of $\pi$ bonding as an explanation of the unique classification of the metal ions into types (a) and (b) has been emphasized by Chatt,[18] but the correlation seems to hold as well for acids and bases that are not metal ions. Thus it may be that the ability to form $\pi$ bonds is a contributing factor to the validity of the general principle. It is noteworthy that both of these structural characteristics oppose the combination of either a hard acid with a soft base or a soft acid with a hard base.

In addition to ionic and $\pi$ bonding, the operation of van der Waals forces can also be considered to contribute to the stabilization of the acid-base complex. The dispersion forces will be related to the polarizability of the combining groups and in soft-soft interactions are expected to be quite large.

## Applications of the Concept

In spite of the fact that a quantitative or definitive basis for the behavior of hard and soft acids and bases cannot be given, it is possible to correlate a great amount of data. The most straightforward use of the principle is illustrated by the fact that $AgI_2^-$ is stable, but $AgF_2^-$ is not. Or, we might consider the reaction

$$LiI + CsI \rightleftharpoons LiF + CsI \qquad \Delta H = -33 \text{ kcal/mole}$$

Here we see that the soft $I^-$ ion prefers to combine with the soft $Cs^+$ ion, whereas the hard $F^-$ ion prefers to combine with the hard $Li^+$ ion.

It was stated that soft acids prefer to combine with soft bases and hard acids prefer to combine with hard bases, but this does not mean that hard-soft combinations do not tend to occur. Rather, we must refer to the relative stabilities of the complexes. This can be illustrated by the following example given by Schwarzenbach. Both $CH_3Hg^+$ and $H^+$ form stable complexes with $OH^-$ and $S^{2-}$. However, $CH_3Hg^+$ is a soft acid whereas $H^+$ is a hard acid. Of the two anions, $S^{2-}$ is a soft base and $OH^-$ is a hard base. The agreement with the combination principle is seen from the equilibrium constants for the competitive reactions,

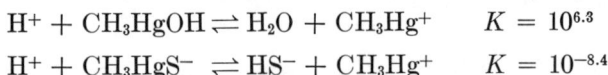

$$H^+ + CH_3HgOH \rightleftharpoons H_2O + CH_3Hg^+ \qquad K = 10^{6.3}$$

$$H^+ + CH_3HgS^- \rightleftharpoons HS^- + CH_3Hg^+ \qquad K = 10^{-8.4}$$

where the preference of $CH_3Hg^+$ for the soft base $S^{2-}$ is apparent.

It is not uncommon for a number of Lewis bases to coordinate to a single Lewis acid. As was pointed out by Jørgensen,[19] hard bases tend to group together and soft bases tend to group together on a given acid. This is particularly evident among the transition metal complexes where we find combinations such as $Co(CN)_5I^{3-}$ and $Co(NH_3)_5F^{2+}$. In the first instance, the soft $CN^-$ and $I^-$ tend to group together, and in the second instance, both the $NH_3$ and the $F^-$ are hard ligands.

Here we have seen only a few examples of the use of the hard and soft concept of acids and bases. Pearson has applied the principle to a multitude of chemical problems, including catalysis, where he points out that metal atoms are soft and therefore adsorb soft bases; and solubility, where he points out that hard solvents tend to dissolve hard solutes and soft solvents tend to dissolve soft solutes. He has also shown that it is possible to use the principle to correlate the rates of certain chemical reactions. In both electrophilic and nucleophilic substitution reactions, the speed of the reaction can be related to the hardness or softness of the acid and base centers.

It would seem, then, that there certainly are many aspects of chemical reactions that can be discussed in terms of acid-base ideas. Regardless of the terminology one wishes to use, any correlations that can be determined are worthwhile and should be adopted. It would certainly seem that the concept of soft and hard is useful in this respect, but, at the same time, one cannot help but note possible weaknesses in the approach. The extremely qualitative nature of the principle makes one wonder if it will ever amount to more than a rule of thumb, and one that has possibly been used in a less formalized manner for specific problems for years. The arbitrariness permitted in breaking up a molecule into acid and base fragments to explain an observed reaction is convenient once the reaction is already known. However, one might wonder which fragments to consider if the reaction is not as yet determined. A case in point is that of acetic acid and ethyl alcohol. The reactants can be considered

$$CH_3CO^+OH^- + C_2H_5O^-H^+$$

or

$$CH_3COO^-H^+ + C_2H_5{}^+OH^-$$

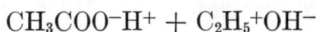

Although the first of these seems to represent the observed reaction, there seems to be no reason to have expected this.

We are also faced with the problem of interpretation of acid and base strengths. Supposedly the hard-soft factors are independent of acid and base strength, and if we are to use the hard-soft principle, it should work independently of the strengths. Yet it would seem that electron-pair donation and acceptance are intimately dependent on those factors that contribute to softness. If we consider the reaction

$$CH_3{}^+{}_{(g)} + H_{2(g)} \rightarrow CH_{4(g)} + H^+{}_{(g)} \qquad \Delta H = +86 \text{ kcal/mole}$$

the reaction would be expected to be favored as written because of the soft-soft combination of $CH_3{}^+$ and $H^-$. However, the unfavorable heat of reaction is attributed to the greater acidity of $H^+$ relative to $CH_3{}^+$. Thus, here we are using acid strength to rationalize an apparent anomalous result. It would seem, then, that the principle can be valid only if the strength factors $S_A$ and $S_B$ do not favor a particular product. We must also question if $S_A$ and $S_B$ really represent acid and base strength or do they represent some quite different but intrinsic property of a given acid and base.

Although possible questions of the significance of the hard-soft concept do arise, the correlations that have been presented in recent years are, nevertheless, extremely impressive, and the topic certainly cannot be taken lightly.

## ACID-BASE STRENGTHS

Although it is not possible to make a general determination of the relative strength of a Lewis acid, the strength of a protonic acid can be determined by its tendency to give up a proton. Quantitatively, this tendency can be expressed for the equilibrium

$$HA = H^+ + A^-$$

by means of the equilibrium constant

(9–1)     $$K = \frac{a_{H^+} a_{A^-}}{a_{HA}}$$

The actual direct measurement of such an equilibrium constant is not possible, since the hydrogen ion cannot be donated unless there is a base present to accept it. Thus, in practice, it is necessary to consider an equilibrium of the type

$$HA + B = BH + A$$

where B is some reference base. If we now choose the conjugate pair BH—B as a reference acid-base, we can measure the relative acidities of a series of acids with respect to this pair.

Most commonly, the base, B, is taken to be the solvent. If the solvent happens to be water, the reference acid-base pair is $H_3O^+$—$H_2O$, and this leads to the equilibrium relation

(9–6)     $$K' = \frac{a_{H_3O^+} a_{A^-}}{a_{HA} \; a_{H_2O}}$$

Since the activity of the water can be treated as a constant, its value can be included in the equilibrium constant, thereby giving

(9–7)     $$K = \frac{a_{H_3O^+} a_{A^-}}{a_{HA}}$$

Several methods can be used to determine the ionization constant of an acid. The simplest of these would be to assume that concentrations can be substituted for the activities of the various species in the equation. Such an approach may give a good first-order approximation of the ionization constant if the measurements are made in a sufficiently dilute solution using a solvent of relatively high dielectric constant. However, the most valid method for the determination of a thermodynamic ionization constant is based on emf measurements,[20] using a cell of the type

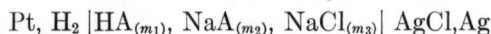

$$Pt, H_2 \,|HA_{(m_1)}, NaA_{(m_2)}, NaCl_{(m_3)}| \, AgCl, Ag$$

where HA represents the weak acid and NaA represents its sodium salt. The use of such a cell can be understood from a thermodynamic standpoint by recognizing that the potential of this cell is given by

(9–8)     $$\text{E} = \text{E}^\circ - \frac{RT}{\text{F}} \ln m_{H^+} \gamma_{H^+} m_{Cl^-} \gamma_{Cl^-}$$

and that the ionization constant for the acid can be expressed as

(9–9)     $$K = \frac{m_{H^+} \gamma_{H^+} m_{A^-} \gamma_{A^-}}{m_{HA} \gamma_{HA}}$$

If Eq. (9–9) is now solved for $m_{H^+}$ and this term is substituted into Eq. (9–8), we obtain on rearrangement,

$$(9\text{--}10) \qquad \mathbf{E} - \mathbf{E}^\circ + \frac{RT}{\mathbf{F}} \ln \frac{m_{HA}\, m_{Cl^-}}{m_{A^-}} = -\frac{RT}{\mathbf{F}} \ln \frac{\gamma_{HA}\, \gamma_{Cl^-}}{\gamma_{A^-}} - \frac{RT}{\mathbf{F}} \ln K$$

If measurements are made as a function of the ionic strength of the system, the right-hand side of the equation will approach $\left(-\dfrac{RT}{\mathbf{F}} \ln K\right)$ as the ionic strength approaches zero. Thus, an extrapolation to infinite dilution will lead to the value of the ionization constant of the weak acid.

## The Leveling Effect

In determining relative strengths of the stronger acids such as HCl or $H_2SO_4$ in an aqueous system, the equilibrium constants are all found to be indeterminately large. This is due to the basicity of the water molecule. If we consider the reaction

$$HA + H_2O \rightarrow H_3O^+ + A^-$$

it is found to go essentially 100% to the right for all of these so-called strong acids. This property of the solvent is known as the *leveling effect*, and it would, of course, be expected to become more pronounced if a more basic solvent were used. (However, for another point of view, see page 546). This point can be illustrated by considering the extent of ionization of acids in liquid ammonia. For instance, benzoic acid, which is only slightly ionized in water turns out to be quite highly ionized in ammonia due to the fact that the basicity of the ammonia molecule is sufficiently great to cause the reaction

$$HA + NH_3 = NH_4^+ + A^-$$

to go essentially to completion. Thus, if we define acid strength in terms of the magnitude of a pK, benzoic acid will appear to be a strong acid in ammonia. However, it is not really possible to compare the strength of an acid in one solvent to its strength in another solvent, and from a phenomenological basis it is not possible to consider an acid such as benzoic acid to be strong in any solvent.

Nevertheless, in order to determine the relative strengths of the stronger acids, it is obviously necessary to overcome the leveling effect of the solvent. This can be done either by going to a more acidic solvent such as acetic or sulfuric acid, or by going to an aprotic solvent such as benzene. In principle, both of these approaches should lead to a differentiation of acid strengths. However, in practice, serious difficulties are likely to be encountered in either type of solvent.

One might think that an aprotic solvent would be quite ideally suited as a medium for studying relative acid-base strengths, since the solvent itself is completely inert and will not enter in the determination by accepting or donating a proton. This, of course, would require the addition of some reference acid or base. Measurements along these lines have been made, but considerable question exists as to their interpretation. Primarily, the difficulty with the use of such solvents arises from their low dielectric constant, which is usually around 2 or 3. Consequently, if charged species are present in solution, ion pair formation and higher-order association effects will be large. Under these circumstances, the usual methods for studying acid strengths are of questionable validity, and the equilibrium relationships are not of the simple form that is found in solvents of higher dielectric constant. For these reasons, very little worthwhile information has been obtained on relative acid-base strengths in these solvents.

In spite of the fact that acidic solvents are in some ways as difficult to use as aprotic solvents, they do offer some advantages and have been much more intensively studied. Although the problem of low dielectric constant exists with acetic acid, most of the other acid solvents have quite high values for their dielectric constant. This can be considered a distinct advantage over the aprotic solvents. Unfortunately, many of these solvents react with the solute to give more complicated products than would be obtained from a simple proton transfer. Nevertheless, relative strengths of numerous acids have been determined in acidic solvents, and the order of decreasing strengths of the stronger mineral acids has been found to be

$$HClO_4 > HBr > H_2SO_4 > HCl > HNO_3$$

This order of decreasing acid strengths was determined by Kolthoff and Willman[21] using conductivity measurements in acetic acid. As can be seen from the conductivity curves in Figure 9–2, $HClO_4$ is considerably stronger than the other acids. However, it is interesting to note that all of the acids have conductivity curves typical of weak electrolytes. This behavior can be

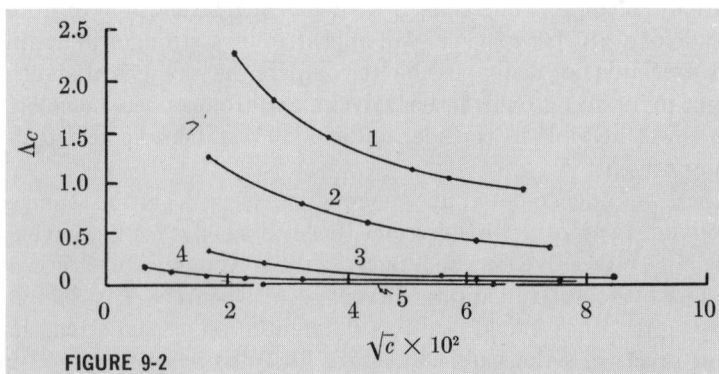

FIGURE 9-2

Relative Strengths of Mineral Acids, (1) $HClO_4$, (2) HBr, (3) $H_2SO_4$, (4) HCl, and (5) $HNO_3$ from Conductivity Measurements in Anhydrous Acetic Acid. (After I. Kolthoff and A. Willman, *J. Am. Chem. Soc.*, **56**, 1007 (1934).)

attributed to both the low dielectric constant of acetic acid and the decrease in coordinating ability of the solvent relative to water.

## GENERAL TRENDS IN ACID STRENGTH

As might be expected, various trends in acid strengths are readily apparent, and it would be desirable to understand these trends in terms of some fundamental concept or a simple theoretical picture of the system. In answer to this problem, several approaches have been offered. However, due to the complexity of the systems involved, these can give only a qualitative correlation with experiment.

### TABLE 9–5.  Approximate pK Values of Some Hydrides.

| | | | | | |
|---|---|---|---|---|---|
| $H_3N$ | 35 | $H_2O$ | 16 | HF | 3 |
| $H_3P$ | 27 | $H_2S$ | 7 | HCl | −7 |
| | | $H_2Se$ | 4 | HBr | −9 |
| | | $H_2Te$ | 3 | HI | −10 |

### Strengths of Hydro Acids

As can be seen from Table 9–5, very definite trends in acidity exist in the hydro acids. These are obvious in going across the periodic table in a given period or in going down in a given family. In the first instance, a large increase in acidity can be noted in going from left to right across the periodic table. The apparent explanation of this trend would be in terms of the electronegativities of the nonmetals concerned. Thus, the fluorine atom is more electronegative than the nitrogen atom. Although this argument is consistent with the observed horizontal trends, we find that it does not hold for the vertical trends. Here the acidity in a given family is found to increase with a decrease in electronegativity of the nonmetal. For instance, in the series HF, HCl, HBr, and HI, HI is the most acidic member of the series and HF is the least acidic, just the opposite of the horizontal trend. Thus we are forced to conclude that, although acidity might be related in some way to electronegativity, it certainly is not the only factor to be considered.

The resolution of the apparent anomaly in the relative strengths of the hydro acids lies in the recognition that the acidity is primarily associated with the ease with which the acid can give up its hydrogen atom in the form of the hydrogen ion. This is facilitated by the increase in the dipole moment of the molecule which arises from an increase in electronegativity of the nonmetal atom, but it is also facilitated by a weakening of the hydrogen-nonmetal bond. The bond strengths of the hydrogen halides at 0° Kelvin and in the gaseous state vary from 135 kcal/mole for HF to 71.4 kcal/mole for HI. This decrease in bond strength parallels the increase in acidic character. From this we might then conclude that the correlation observed between pK and electronegativity in the series $H_3N$, $H_2O$, and HF is fortuitous, and some other contributing factors to the trend should be sought.

General trends such as these are valuable rules of thumb, but a somewhat greater insight might be obtained by devising a path equivalent to the ionization process and considering each step individually. If the entropy and energy terms could be evaluated for each step, it would then be possible to calculate the pK value of an acid from thermodynamic considerations. Such a path can be represented by a cyclic process of the form

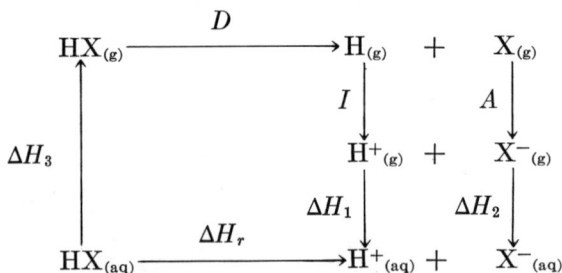

$$
\begin{array}{ccccc}
& D & & & \\
HX_{(g)} & \longrightarrow & H_{(g)} & + & X_{(g)} \\
& & \downarrow I & & \downarrow A \\
\Delta H_3 \uparrow & & H^+_{(g)} & + & X^-_{(g)} \\
& & \downarrow \Delta H_1 & & \downarrow \Delta H_2 \\
& \Delta H_r & & & \\
HX_{(aq)} & \longrightarrow & H^+_{(aq)} & + & X^-_{(aq)}
\end{array}
$$

The dissociation energy of the gaseous HX molecule is represented by $D$, the ionization energy of the hydrogen atom by $I$, and the affinity energy of the nonmetal by $A$. The enthalpy of the reaction is given by $\Delta H_r$, and the remaining $\Delta H$ terms represent enthalpies of hydration.

Although it is necessary to use a number of approximations, this treatment has been applied quite successfully by McCoubrey[22] to the halogen hydro acids. The usual difficulties arise in the distribution of solvation energies between $H_{(g)}^+$ and $X_{(g)}^-$, and $\Delta H_3$ must be approximated by comparison to a similar molecule that does not ionize to any significant extent. $\Delta H_r$ can be expressed in terms of the other steps in the cycle giving

$$\Delta H_r = \Delta H_1 + \Delta H_2 - \Delta H_3 + \Delta H_g$$

where $\Delta H_g = D + I - A$. All of the necessary quantities for this calculation have been determined and their accepted values are listed in Table 9–6 along with the calculated values for the entropies.

**TABLE 9–6.   Thermodynamic Data of the Hydrogen Halides.**

|                      | HF      | HCl     | HBr     | HI      |
| -------------------- | ------- | ------- | ------- | ------- |
| $D$                  | 134.6   | 103.2   | 87.5    | 71.4    |
| $A$                  | 82.2    | 87.3    | 82.0    | 75.7    |
| $I$                  | 315     | 315     | 315     | 315     |
| $\Delta H_{(g)}$     | 367     | 331     | 321     | 311     |
| $\Delta H_1 + \Delta H_2$ | $-381.9$ | $-348.8$ | $-340.7$ | $-330.3$ |
| $\Delta H_3$         | $-11.5$ | $-4.2$  | $-5.0$  | $-5.5$  |
| $\Delta S°$          | $-20.8$ | $-13.4$ | $-9.1$  | $-3.2$  |

The entropy term, $-\Delta S°$, is determined from the expression

(9–11)        $-\Delta S° = S°_{HXaq} - (S°_{H^+aq} + S°_{X^-aq})$

Several approximations are necessary in the evaluation of the entropy terms, but they are probably valid to within 2 or 3 eu.

Now that the enthalpy and entropy terms are known for the ionization of the acid, HX, the ionization constant of the acid can be calculated by recalling that

(9–12)        $\Delta G° = \Delta H° - T\Delta S°$

and

(9–13)        $\Delta G° = -RT \ln K$

The results of these calculations are recorded in Table 9–7 for 25°C. The

**TABLE 9–7.  Thermodynamic Calculation of the pK Values for the Hydrogen Halides.**

|      | $\Delta H_r°$ | $T\Delta S_r°$ | $\Delta G_r°$ | $pK_{calc}$ | $pK_{exp}$ |
|------|------|------|------|------|------|
| HF   | $-3.0$  | $-6$ | $3$   | $2$   | $3$   |
| HCl  | $-13.7$ | $-4$ | $-10$ | $-7$  | $-7$  |
| HBr  | $-15.2$ | $-3$ | $-12$ | $-9$  | $-9$  |
| HI   | $-14.1$ | $-1$ | $-13$ | $-10$ | $-10$ |

agreement between the calculated and experimental pK values is seen to be quite striking. This would tend to give some credence to the proposed cyclic process, and in terms of the various steps it is possible to rationalize the general trends of the hydro acids.

## Strengths of Inorganic Oxy Acids

Somewhat greater emphasis has been placed on the relative strengths of the inorganic oxy acids than those of the hydro acids, and just as with the hydro acids, various trends can be detected. From an experimental standpoint, it has been found that the pK values of these acids can be placed in four distinct groupings. These are given in Table 9–8, along with the general structures of the acids. Immediately, we can see that a correlation exists between the structure of the acid and its grouping. In this respect, it is to be noted that the oxygen atoms of acids in the first grouping are all bonded to a hydrogen atom. That is, they are all hydroxyl oxygens. However, in the second grouping there is one nonhydroxyl oxygen atom, and each successive grouping has an additional one. Corresponding to this increase in the number of nonhydroxyl oxygen atoms, we find a consistent increase in acidity. At the same time, we note that the number of hydroxyl groups in the molecule apparently has little or no effect on the relative acidity. Thus, we

### TABLE 9–8.  pK Groupings of Some Inorganic Oxy Acids.

| First Grouping | | | Second Grouping | | |
|---|---|---|---|---|---|
| Class | Example[23] | pK | Class | Example | pK |
| $X(OH)_n$ | $Ge(OH)_4$ | 8.6 | $XO(OH)_n$ | $PO(OH)_3$ | 2.1 |
| | $As(OH)_3$ | 9.2 | | $HPO(OH)_2$ | 1.8 |
| | $Te(OH)_6$ | (8.8) | | $H_2PO(OH)$ | 2.0 |
| | $Cl(OH)$ | 7.2 | | $AsO(OH)_3$ | 2.3 |
| | $Br(OH)$ | 8.7 | | $SO(OH)_2$ | 1.9 |
| | $I(OH)$ | 11.0 | | $SeO(OH)_2$ | 2.6 |
| | | | | $TeO(OH)_2$ | 2.7 |
| | | | | $ClO(OH)$ | 2.0 |
| | | | | $IO(OH)_5$ | 1.6 |

| Third Grouping | | | Fourth Grouping | | |
|---|---|---|---|---|---|
| Class | Example | pK | Class | Example | pK |
| $XO_2(OH)_n$ | $SO_2(OH)_2$ | (−3) | $XO_3(OH)_n$ | $ClO_3(OH)$ | (−10) |
| | $SeO_2(OH)_2$ | (−3) | | | |
| | $IO_2(OH)$ | 0.8 | | | |
| | $ClO_2(OH)$ | −1 | | | |
| | $NO_2(OH)$ | −1.4 | | | |

conclude that a nonhydroxyl oxygen atom is in some way significant in the determination of the strength of an oxy acid.

It can also be seen from Table 9–8 that, for a given structure, the acid strength decreases with an increase in size of the central atom. This trend can be illustrated by the example of ClOH, BrOH, and IOH, where the pK values are 7.2, 8.7, and 11.0 respectively. This principle was utilized by Cartledge in an early attempt to find a correlation between structure and acid strength. By considering acids and bases to be hydroxides of nonmetals and metals respectively, Cartledge pointed out that the *ionic potential* $\phi$, defined as the charge-to-radius ratio of the central atom, can be used to determine whether the hydroxide is acidic, basic, or amphoteric. According to Cartledge, for a value of $\sqrt{\phi} < 2.2$, the hydroxide is basic, $2.2 < \sqrt{\phi} < 3.2$, the hydroxide is amphoteric, and for $\sqrt{\phi} > 3.2$, the hydroxide is acidic.

An attempt to go beyond the simple correlation of the strengths of the oxy acids with some particular structural property of the acid was made by Kossiakoff and Harker.[23] They proposed a mechanism for the ionization process involving a series of steps and, in essence, calculated the free energy change for each step. The final expression for the free energy change obtained by Kossiakoff and Harker is

$$(9\text{–}14) \qquad \Delta G = \sum_i W_i - C + RT \ln n_O/n_H$$

in which $C$ is a constant characteristic of the solvent, $n_O$ is the number of nonhydroxyl oxygen atoms in the resultant ion, $n_H$ is the number of transferable hydrogen atoms in the acid, and $W_i$ is the electrostatic energy involved in the transfer of a single ion. This latter quantity can be determined from the classical expression for the electrostatic energy

$$(9\text{-}15) \qquad W_i = \sum_i \frac{m_j e}{\epsilon} \left( \frac{1}{r_j} - \frac{1}{r_j'} \right)$$

The distance $(r_j' - r_j)$ is the displacement of the proton with respect to the $j'$th atom in the molecule, $e$ is the electron charge, and $\epsilon$ is the solvent dielectric constant. The *formal charge* of the $j$th atom is represented by $m_j$ and is defined as the group number of the atom in the periodic table minus the number of electrons in its valence shell. In determining the formal charge of an atom, a shared pair of electrons is counted as one.

In terms of the treatment of Kossiakoff and Harker, the determination of the ionization constant of an acid involves an evaluation of $\Sigma W_i$ and $C$, which should be a constant for all of the oxy acids in a given solvent. $\Sigma W_i$ can be determined from a detailed knowledge of bond angles and bond dis-

**TABLE 9–9. Classification of Inorganic Oxy Acids.[a]**

| Class | Example | pK$_1$ | pK$_2$ | pK$_3$ | pK$_4$ |
|-------|---------|--------|--------|--------|--------|
| $m = 0$ | | $(n = 0)$ | $(n = 1)$ | | |
| | $H_4GeO_4$ | 8.6 | 12.7 | | |
| | $H_3AsO_3$ | 9.2 | — | | |
| | $H_6TeO_6$ | (6.2; 8.8) | 10.4 | | |
| | HClO | 7.2 | — | | |
| | HBrO | 8.7 | — | | |
| | (HIO) | (11.0) | — | | |
| $m = 1$ | | $(n = 1)$ | $(n = 2)$ | $(n = 3)$ | |
| | $H_3PO_4$ | 2.1 | 7.2 | 12.0 | |
| | $H_3PO_3$ | 1.8 | 6.2 | — | |
| | $H_3PO_2$ | 2.0 | — | — | |
| | $H_3AsO_4$ | 2.3 | 7.0 | 13.0 | |
| | $H_2SO_3$ | 1.9 | 7.0 | — | |
| | $H_2SeO_3$ | 2.6 | 8.3 | — | |
| | $H_2TeO_3$ | 2.7 | 8.0 | — | |
| | HClO$_2$ | 2.0 | — | — | |
| | $H_5IO_6$ | 1.6 | 6.0 | — | |
| $m = 2$ | | $(n = 2)$ | $(n = 3)$ | $(n = 4)$ | $(n = 5)$ |
| | $H_2SO_4$ | — | 1.9 | — | — |
| | $H_2SeO_4$ | — | 2.0 | — | — |
| | HIO$_3$ | 0.8 | — | — | — |
| | $H_4P_2O_7$ | 0.9 | 2.0 | 6.7 | 9.4 |
| | $H_4P_2O_6$ | (2.2) | (2.8) | 7.3 | 10.0 |
| | $H_2S_2O_4$ | 0.3 | 2.5 | — | — |

[a] From J. E. Ricci, *J. Am. Chem. Soc.*, **70**, 109 (1948).

tances of the molecules involved; however, the calculations are somewhat time-consuming. Once $\Sigma W_i$ is known for a particular acid, $C$ can be determined, using the free energy change based on the experimental evaluation of the pK of that acid. For this purpose, Kossiakoff and Harker used the first ionization constant of *ortho*phosphoric acid as a reference. Once the value of $C$ has been determined, it is possible to calculate the ionization constants of the other oxy acids. Kossiakoff and Harker have carried out these calculations for 26 acids in addition to *ortho*phosphoric acid, and they have found the overall average deviation between the calculated and the observed log $K$ values to be only 0.89.

Although the results obtained by Kossiakoff and Harker appear to be quite encouraging, a critical review of this work by Ricci[24] has shown that it is far from conclusive. As shown by Ricci, the contributions of the various structural terms are less than the average deviation between the observed and their calculated pK values. This is not to say that the factors considered by Kossiakoff and Harker are not valid, but rather that the treatment does not give a legitimate test of their validity. Ricci points out, however, that a remarkable grouping of pK values is obtained if the acids are classified in terms of the formal charge of the central atom and the number of non-hydroxyl oxygen atoms in the acid. This grouping is illustrated in Table 9–9, where the acids in each group have the same value of $m$, the formal charge of the central atom, and $n$, the number of nonhydroxyl oxygen atoms in the acid. On this basis, Ricci proposes the empirical expression

$$(9\text{–}16) \qquad \mathrm{pK} = 8.0 - m(9.0) + n(4.0)$$

for the determination of the pK of an oxy acid. Accordingly, if the structure of $H_3PO_4$ can be represented as

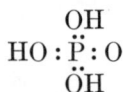

$$\mathrm{HO} : \overset{\textstyle OH}{\underset{\textstyle \ddot{O}H}{\ddot{P}}} : O$$

the pK will be $8 - 9 + 4 = 3$. The experimental value of the pK of $H_3PO_4$ is approximately 2.1. Using this equation, Ricci has calculated the pK values for 36 oxy acids with an average deviation of 0.91 between the observed and the calculated values. This agreement is comparable to that obtained by Kossiakoff and Harker, and the amount of calculation involved is considerably less.

There can be no question that an empirical relation such as that proposed by Ricci is valuable. Certainly, quite acceptable results can be obtained with a minimum of effort. However, such an approach contributes very little to our understanding of the mechanisms involved, and it is this that we are ultimately after. Consequently, if a calculation of a pK value could be made from a mechanistic picture, it would be of much greater significance, regardless of the amount of effort necessary for its determination.

# References

1. J. Bronsted, *Rec. Trav. Chim.*, **42**, 718 (1923).
2. T. Lowry, *Chem & Ind.*, **42**, 43 (1923).
3. E. Franklin, *J. Am. Chem. Soc.*, **27**, 820 (1905).
4. E. Franklin, *J. Am. Chem. Soc.*, **46**, 2137 (1924).
5. H. Cady and H. Elsey, *J. Chem. Ed.*, **5**, 1425 (1928).
6. G. Lewis, "Valence and the Structure of Atoms and Molecules," The Chemical Catalog Co., New York, 1923.
7. R. Bell, "Acids and Bases," Methuen's Monographs on Chemical Subjects, New York, 1952.
8. R. Norrish and K. Russell, *Trans. Faraday Soc.*, **48**, 91 (1952).
9. R. Bell and B. Skinner, *J. Chem. Soc.*, 2955 (1952).
10. M. Usanovich, *J. Gen. Chem.*, (USSR), **9**, 182 (1939).
11. L. Ebert and N. Konopik, *Osterr. Chem. Ztg.*, **50**, 184 (1949).
12. H. Lux, *Z. Electrochem.*, **45**, 303 (1939).
13. J. Bjerrum, *Naturwissenschaften*, **38**, 46 (1951).
14. R. Mulliken, *J. Phys. Chem.*, **56**, 801 (1952).
15. W. Luder and S. Zuffanti, "The Electronic Theory of Acids and Bases," John Wiley & Sons, Inc., New York, 1946.
16. R. G. Pearson, *Chem. in Britain*, **3**, 103 (1967); *Science*, **151**, 172 (1966); *J. Am. Chem. Soc.*, **85**, 3533 (1963).
17. S. Ahrland, J. Chatt, and N. Davies, *Quart. Rev. London*, **12**, 265 (1958).
18. J. Chatt, *J. Inorg. Nucl. Chem.* **8**, 515 (1958).
19. C. Jørgensen, *Inorg. Chem.*, **3**, 1201 (1964).
20. D. MacInnes, "The Principles of Electrochemistry," Reinhold Publishing Corp., New York, 1939.
21. I. Kolthoff and A. Willman, *J. Am. Chem. Soc.*, **56**, 1007 (1934).
22. J. McCoubrey, *Trans. Faraday Soc.*, **51**, 743 (1955).
23. A. Kossiakoff and D. Harker, *J. Am. Chem. Soc.*, **60**, 2047 (1938).
24. J. Ricci, *J. Am. Chem. Soc.*, **70**, 109 (1948).

## Suggested Supplementary Reading

R. Bell, "The Proton in Chemistry," Cornell University Press, Ithaca, N. Y., 1959.
W. Luder and S. Zuffanti, "The Electronic Theory of Acids and Bases," John Wiley & Sons, Inc., New York, 1946.
B. Douglas and D. McDaniel, "Concepts and Models of Inorganic Chemistry," Blaisdell, New York, 1965.
N. Bjerrum, "Acids, Salts, and Bases," *Chem. Revs.*, **16**, 287 (1935).
T. Moeller, "Inorganic Chemistry," John Wiley & Sons, Inc., New York, 1952.
N. Hall, "Modern Conceptions of Acids and Bases," *J. Chem. Ed.*, **7**, 782 (1930).

## Problems

1. According to which of the various definitions of acids and bases is
   (a) $NaNH_2$ in liquid ammonia a base?
   (b) HCl in water an acid?

2. Give an example of neutralization according to each of the various definitions of acids and bases.

3. Noting the discussion in Chapter 8 concerning the nonthermodynamic significance of pH, what, if any, thermodynamic meaning can be given to the concept of acidity or basicity?

4. Give an example of a solvent system acid and a solvent system base in each of the following solvents.
   (a) $NH_2OH$
   (b) $BrF_3$
   (c) $N_2O_4$

5. Why is $BCl_3$ a better Lewis acid than $BF_3$?

6. Is the reaction

$$Cu^{2+} + 4NH_3 \rightarrow [Cu(NH_3)_4]^{2+}$$

a Lewis acid-base reaction if the bonding is explained by means of the crystal field theory? (See Chapter 10.)
   (a) Consider the question in terms of the original Lewis definition.
   (b) Consider the question in terms of the ideas presented by Pearson on hard and soft acids and bases.

7. Use Eq. (9–16) to determine the pK values of the series $H_2TeO_4$, $H_4TeO_5$, and $H_6TeO_6$ and note the effect of hydration on the acid strength.

8. What must be the structures of $H_3PO_3$ and $H_3AsO_3$ in order to justify the observed pK values of 1.8 and 9.2 respectively?

9. By use of the oxychlorine acids, show the effect of increasing formal charge on acid strength.

10. Show that the use of formal charge in describing acid behavior in oxyacids excludes the presence of pi bonding.

11. Rationalize the $\Delta H$ values of the following reactions in terms of the hard-soft concept.
   (a) $CH_3CH_{3(g)} + H_2O_{(g)} \rightarrow CH_3OH_{(g)} + CH_{4(g)}$      $\Delta H = 12$ kcal
   (b) $CH_3COCH_{3(g)} + H_2O_{(g)} \rightarrow CH_3COOH_{(g)} + CH_{4(g)}$      $\Delta H = -13$ kcal
   (c) $CH_{4(g)} + CH_3OH_{(g)} \rightarrow CH_3CH_{3(g)} + H_2O_{(g)}$      $\Delta H = -12$ kcal

# Coordination
# Chemistry I

**10**

The field of coordination chemistry has grown in a half-century from a readily defined and limited area into what is now the most active research field of inorganic chemistry. In recent years it has received not only a large amount of experimental study, but also a very extensive theoretical treatment. The scope of this field has now become so broad and the compounds and kinds of compounds with which it is concerned are so numerous that we can do no more than devote this chapter to simply introducing the subject and indicating the current theoretical approaches.

For over one hundred years the study of inorganic, metal-containing compounds was largely descriptive, as indeed was much of the whole field of chemistry. Real theoretical progress toward understanding the structure and behavior of this large class of inorganic compounds could not be made until the discovery of the electron in 1897 made possible the later beginnings (1916) and development (1920's) of the electronic theory of valence. Since that time theoretical inorganic chemistry has made very rapid progress. Primarily, this has been due to the pioneering work of Lewis, Kossel, Langmuir, Sidgwick, Fajans, Pauling, Van Vleck, and many others who have extended and amplified their ideas.

As in all fields, there is a characteristic vocabulary in coordination chemistry. Generally, new terms will be defined as they arise, but there are several definitions which should be given at the outset. A *complex ion* will be understood to be a more or less stable, charged aggregate formed when an atom or ion, most commonly from a metal, becomes directly attached to a group of neutral molecules and/or ions. The latter are called *ligands* or *donor groups* and they are said to be coordinated or complexed to the *central ion* or *acceptor* in a *first coordination sphere*. The first coordination sphere of the complex ion is indicated by enclosing the formula for the ion in square brackets, for example: $[Cu(H_2O)_4]^{2+}$, $[Co(NH_3)_4F_2]^+$, etc. The number of ligand atoms arranged in a definite geometry and directly bonded to a central ion is called the *coordination number* of that ion. Whereas 6 and 4 are the most commonly encountered coordination numbers, 2, 3, 5, 7, 8, 9, and 10 have all been well established. A group that can attach to the same metal ion through more than one of its atoms is termed a *chelate* or *multidentate* ligand. Chelating agents are known which can attach to metal ions through two, three, four, five, six, or even more donor atoms, with two being the most common. The terms applied to such multidentate ligands are *bidentate*, *terdentate*, *quadridentate*, *quinquidentate*, and *sexidentate*, respectively.

A coordination compound may be one of two types. Firstly, it may be a neutral complex, which, by our definition, would make it identical to the above complex ion except that the aggregate has no net charge. This may be due to either a central metal atom in the zero oxidation state surrounded by neutral ligands, such as $[Fe(CO)_5]$ and $[Pt(RNC)_4]$, or a central metal ion surrounded by enough oppositely charged ligands to produce a neutral aggregate. The latter is much more common. Typical examples of this type of complex are $[Co(NH_3)_3(NO_2)_3]$, $[Fe(aca)_3]$ and $[Cr(gly)_3]$. Here the *aca* represents the acetylacetonate ion, and *gly* represents the glycinate ion. Secondly, the compound may consist of ions, in which case at least one of these ions must be a complex ion. This is the most common type of complex compound encountered.

One of the characteristics of the complex compound is that the complex ion or neutral complex that composes it will most often retain its identity in solution, although partial dissociation may occur. In fact, both the extent of dissociation as well as the time required for dissociation may vary from very slight to very extensive. For example, the compound originally written as $2KBr \cdot HgBr_2$ actually contains the tetrahedral ion $[HgBr_4]^{2-}$ in the crystalline solid, and this same ion persists in solution with only very slight dissociation. On the other hand, the compound originally written as $2KCl \cdot CoCl_2$ has been shown to contain the tetrahedral ion $[CoCl_4]^{2-}$ in the crystalline solid but it is found to dissociate extensively in aqueous solution into chloride ions, potassium ions, and hydrated cobalt(II) ions. These two examples illustrate the differences found in the relative thermodynamic stability of various bonds. Thus, from the preceding we would infer that the mercury(II)-bromide ion bond is more stable than the mercury(II)-water bond, whereas the cobalt(II)-water bond is more stable than the cobalt(II)-chloride ion bond.

Complexes which exchange ligands rapidly, that is, within mixing times, are generally referred to as *labile* complexes, while those which exchange ligands at a slower rate are called *nonlabile* or *inert* complexes. One must guard against confusing thermodynamic stability with kinetic stability when employing these terms. For example, the average energy per bond in $Cr(H_2O)_6^{3+}$ (122 kcal/mole) is nearly the same as that in $Fe(H_2O)_6^{3+}$ (116 kcal/mole), yet the former exchanges water molecules with the solvent water at a slow and measureable rate, whereas the latter exchanges them extremely rapidly.

Much of the chemistry of coordination compounds is determined by the electronic configuration of the central ion, by the donor and acceptor properties of the ligands, and by the nature of the linkage between the ligand and the central ion. For this reason, correspondingly more space will be devoted here to these aspects of the subject than to such items as stereo-chemistry, types of isomerism, substitution reactions, and redox or electron transfer reactions. Also we shall not discuss the great importance of co-ordination compounds in the fields of analytical chemistry, biochemistry, and electrochemistry. Closely related to, and in a sense simply an extension of, coordination chemistry is the now vast field of *organometallic* chemistry. Since this area is still largely descriptive we shall do no more with it than briefly consider some of the interesting bond types. Several excellent texts are available which may be consulted for detailed treatments of these and the many other qualitative aspects of coordination chemistry.[1-8]

## WERNER'S COORDINATION THEORY

The early study of inorganic complex compounds consisted largely of a series of attempts to explain the existence and structure of hydrates, double salts, and metal ammonia compounds. These compounds were termed molecular or addition compounds because they are formed by the union of two or more already stable and apparently saturated molecules. Early theories and explanations offered by such men as Graham (1837), Claus (1854), Blomstrand (1869), and Jorgensen (1878), are of little more than historical interest now since the coordination theory proposed by Alfred Werner in 1893 proved to be so all-encompassing in its scope. This theory, which was extended and substantiated experimentally in the quarter-century following its conception, was largely responsible for the renewed interest in, and rapid growth of inorganic chemistry around the turn of the century.

Like many great theories, Werner's coordination theory is fundamentally quite simple. The basic postulate in his own words is as follows: "Even when, to judge by the valence number, the combining power of certain atoms is exhausted, they still possess in most cases the power of participating further in the construction of complex molecules with the formation of very definite atomic linkages. The possibility of this action is to be traced back to the fact that, besides the affinity bonds designated as principal valencies, still other bonds on the atoms, called auxiliary valences, may be called into action."

The remainder of the theory is an elucidation of the *number, nature,* and *spatial arrangement* of these secondary or nonionizable valences. Thus, every metal ion possesses a fixed number of secondary valences which must be satisfied in compound formation. For example, Werner recognized that platinum(IV), cobalt(III), iridium(III), and chromium(III) all have six such valences, that is, a coordination number of 6, and platinum(II), palladium(II), copper(II), and zinc all have four such valences, that is, a coordination number of 4. Whereas the primary valences must be satisfied by negative ions, the secondary valences may be satisfied by either negative ions, neutral molecules, or occasionally even positive ions. Commonly we find a negative ion satisfying both a primary and a secondary valence. For example, in the complex $[Co(NH_3)_5Cl]Cl_2$, one chloride ion is different from the other two in that it has lost its ionic behavior and is actually spatially closer to the cobalt than the other two.

Werner's method for assigning groups to the first coordination sphere was based on their response in solution to various chemical and physical tests. To illustrate this we can consider the complex compounds of platinum(IV) and cobalt(III) shown in Table 10–1. From the corresponding conductance

**TABLE 10–1.    Molar Conductances of Some Platinum(IV) and Cobalt(III) Complex Compounds.**

| Werner Formula | Molar Conductance 0.001$N$ solution, 25°C | Number of Ions Indicated |
|---|---|---|
| $[Pt(NH_3)_6] Cl_4$ | 523 | 5 |
| $[Pt(NH_3)_5Cl] Cl_3$ | 404 | 4 |
| $[Pt(NH_3)_4Cl_2] Cl_2$ | 228 | 3 |
| $[Pt(NH_3)_3Cl_3] Cl$ | 97 | 2 |
| $[Pt(NH_3)_2Cl_4]$ | 0 | 0 |
| K $[Pt(NH_3)Cl_5]$ | 108 | 2 |
| $K_2 [PtCl_6]$ | 256 | 3 |
| $[Co(NH_3)_6] Cl_3$ | 432 | 4 |
| $[Co(NH_3)_5NO_2] Cl_2$ | 246 | 3 |
| $[Co(NH_3)_4(NO_2)_2] Cl$ | 98 | 2 |
| $[Co(NH_3)_3(NO_2)_3]$ | 0 | 0 |
| K $[Co(NH_3)_2(NO_2)_4]$ | 99 | 2 |

data it can be seen that the indicated number of ions produced in solution substantiates the Werner formula shown. Further support is obtained from chemical evidence showing how many chlorine atoms are tightly bound to the platinum and how many are ionizable. Thus, the last three platinum complexes in the table give no test for ionic chloride whereas the first four precipitate $\frac{4}{4}$, $\frac{3}{4}$, $\frac{2}{4}$, and $\frac{1}{4}$, respectively, of their chlorine with silver nitrate. Furthermore, the fact that the ammonia molecules are tightly bound to both the platinum and the cobalt is indicated by their inability to neutralize strong acids and the failure of strong alkali to drive off the ammonia from aqueous solutions of the complexes. The primary valence of 4 for the Pt(IV) and 3 for the Co(III), and the secondary valence of 6 for both of the metals are maintained throughout the series. Many other complex ions were studied in this manner, and a great mass of experimental data was accumulated by Werner and his students to substantiate the postulation of two kinds of valence. It should be pointed out, however, that Werner had no theoretical justification for his two types of valence, and widespread acceptance of the theory did not come until the electronic theory of valence was able to present a self-consistent explanation of valence types.

## STEREOCHEMISTRY OF WERNER COMPLEXES

As important as the postulation of two types of valence by Werner was his fundamental postulate that the secondary valences are directed in space about the central ion, not only in the solid state, but also when the complex ion is dissolved in solution. Thus he postulated, and later obtained considerable experimental evidence, mainly chemical, to prove that a grouping of six such valences is directed toward the apices of a regular octahedron that is imagined to be circumscribed about the metal ion. On the other hand, he was able to show that a grouping of four such valences may be arranged in either a square planar configuration or, less often, in a tetrahedral configuration. The stereochemical postulation has made possible the correlation of a large body of facts concerning known complexes and the prediction of the structures of many new and as yet unprepared inorganic compounds. Perhaps the most striking support for the directed valence idea came from its ready explanation of both the various isomeric compounds which were known at that time as well as the huge number of new isomers prepared in the succeeding years.

### Coordination Number Four

Werner demonstrated that in the case of tetracoordinated platinum(II) complexes, for compounds with the general formula $[Ma_2b_2]^{n+}$, *cis-trans* isomerism exists.* This is made possible if the four ligand atoms lie in or nearly in the same plane, as illustrated for $[Pt(NH_3)_2Cl_2]$ in Figure 10–1. Cleverly devised chemical reactions were carried out to determine which of a pair of isomeric compounds was *cis* and which was *trans* for a large variety of compounds of this basic structure. Werner pointed out that the isolation of two isomers of any compounds having this general formula is a good indication of a planar structure, since only one compound is possible if the configuration is that of a regular tetrahedron. Further proof of the planar structure of platinum(II) and palladium(II) complexes was obtained by the synthesis of two geometrical isomers of complexes of the general type

* $M$ will often be used to represent a central metal ion, lower-case letters will stand for unidentate ligands, and capital letters will be used for multidentate ligands, for example, $AA$ will represent a symmetrical bidentate group; $AB$, an unsymmetrical one, etc.

**FIGURE 10-1**

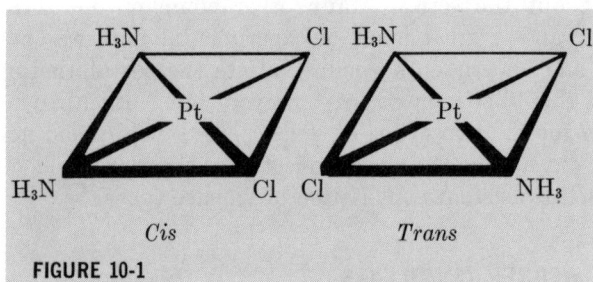

*Cis* and *Trans* Isomers of the Square Planar Complex Dichlorodiammine Platinum(II), [Pt(NH₃)₂Cl₂].

$[M(AB)_2]$, neither of which could be resolved into optical isomers (see (a) below). A tetrahedral structure allows only one geometrical structure that is asymmetric and must give rise to optical isomers as shown below. Many different methods, both chemical and physical, have now firmly established the planar structure of four-coordinated platinum(II) complexes, as well as

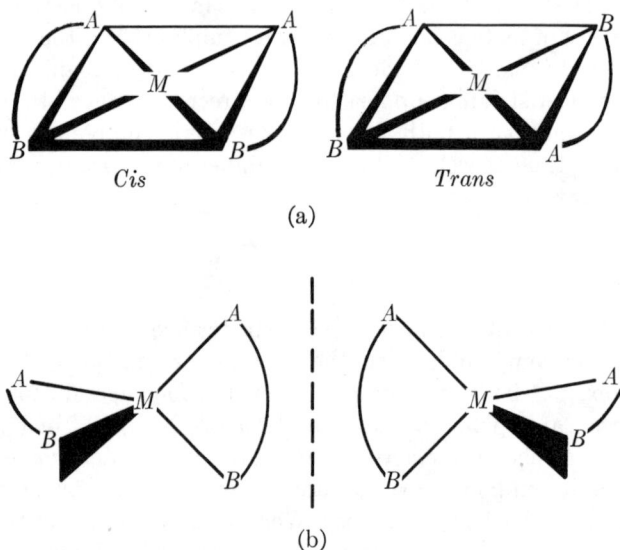

(a)

(b)

those of Pd(II), Ni(II), Ag(II), Cu(II), Au(III), Rh(I), and Ir(I), although Ni(II) also forms tetrahedral complexes. Furthermore, a considerable amount of recent evidence has been accumulating that many, if not most, square planar complexes should really be considered as having tetragonal structures. That is, they should be considered to have a fifth and more often a sixth group also coordinated in a strongly tetragonally distorted octahedral array. Perhaps it is more correct to say that these groups are *located* at a longer distance from the central ion than are the four planar ligands and they make a much smaller contribution to the total bond energy than a planar group. For example, it is likely that in solution, or in the solid obtained from a solution, square planar ions may have solvent molecules or even other anions in a fifth and possibly sixth position completing a strongly distorted octahedron about the central ion. The ions $[PdCl_5]^{3-}$ and $[Ni(CN)_5]^{3-}$ are examples of this for which spectroscopic evidence has been found. In addition, solid complexes of the type $[M(AA)_2X]ClO_4$, where $M$ is Pd(II) or Ni(II) and $X$ is Cl, Br, or I, have been isolated. These complexes exhibit the conductivity in nitrobenzene of univalent electrolytes, which supports their formulation as five-coordinated species. Even Au(III), which is isoelectronic with Pt(II), has been observed to form five- and six-coordinated complexes of the types $[Au(AA)_2X]^{2+}$ and $[Au(AA)_2X_2]^+$.

There is now a considerable body of evidence available to show that the tetrahedral configuration is also important in four-coordinate complexes. The existence of optical activity in compounds of the type $[M(AB)_2]$, where $M$ is Be(II) or B(III), is direct chemical proof of this configuration, or more precisely, it is proof that the structure cannot be planar. Other metals for which there is strong evidence for the tetrahedral structure of many of their complexes are Cu(I), Au(I), Zn(II), Cd(II), Hg(II), Al(III), Ga(III), In(III), Fe(III), Co(II), and Ni(0). Most of the evidence for the structure of these complexes has come from X-ray crystallography, but in recent years it has been found that structural evidence may also be obtained from the results of certain magnetic and spectral studies. This has been particularly well-illustrated, for example, in the case of Co(II) complexes. Also in recent years there have been increasing numbers of reports of the preparation of tetrahedral complexes of all of the first-row transition elements in the 2+ as well as certain other less familiar oxidation states. The conditions under which the tetrahedral complexes form in preference to other geometries are rather restrictive and will be considered in more detail later. Indeed, this configuration must still be regarded as relatively rare for the transition elements where octahedral, tetragonally distorted octahedral, and square planar configurations are far more common. The reasons for this will be made clear later in terms of the modern theories used to treat the coordinate bond. However, we should point out here that most of the metals in their very high oxidation states, in which they behave like nonmetals and in which oxide ion is the most common ligand, form tetrahedral oxyanions, such as $VO_4^{3-}$, $CrO_4^{2-}$, $MnO_4^-$, $MnO_4^{2-}$, $FeO_4^{2-}$, etc.

## Coordination Number Six

Werner also demonstrated the existence of *cis-trans* and optical isomerism in hexa-coordinated complexes, thereby substantiating his prediction of an octahedral configuration for these compounds. Figure 10–2 illustrates the

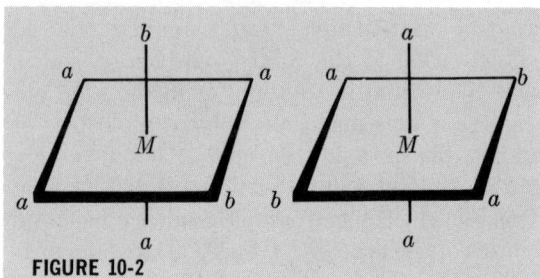

FIGURE 10-2

*Cis* and *Trans* Isomers of the Octahedral Complexes Having the General Formula $[Ma_4b_2]^{n\pm}$.

geometrical isomers for a typical complex having the general formula $[Ma_4b_2]^{n\pm}$. In Figure 10–3 an example of the complex type $[M(AA)a_2b_2]^{n\pm}$ is given and several of the terms introduced thus far are illustrated. Werner's proof of the octahedral configuration also rested upon the demonstration of the existence of optical isomers for complexes with the general formulas *cis*-$[M(AA)_2ab]^{n\pm}$ (*a* may be identical with *b*) and $[M(AA)_3]^{n\pm}$. The optical isomers for these complex ion types are illustrated in Figure 10–4. The *trans* form of the first complex that is shown has a plane of symmetry, and hence

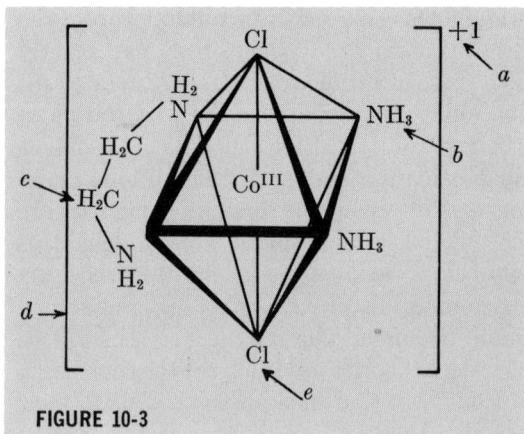

FIGURE 10-3

Example of the Complex Type $[M(AA)a_2b_2]^{n\pm}$, Indicating: (a) the charge on the entire complex ion, (b) a unidentate neutral ligand in a *cis* position to the other $NH_3$ and *cis* to both Cl ligands, (c) a bidentate neutral ligand (spans *cis* positions only), (d) square brackets, enclosing the metal ion and its first coordination sphere, and (e) a unidentate negative ligand in a *trans* position to the other Cl and *cis* to the remaining ligand atoms.

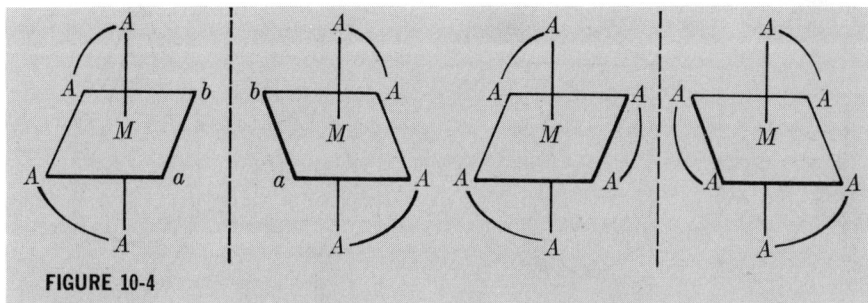

**FIGURE 10-4**

The Optical Isomers of the Complex Types $[M(AA)_2ab]^{n\pm}$ and $M(AA)_3{}^{n\pm}$.

nonsuperimposable mirror images cannot exist for this geometrical isomer. The *trans* form of the second complex is structurally impossible. The two other symmetrical six-cornered geometries, the planar hexagon and the trigonal prism, are clearly eliminated by the optical resolution of such complexes, since no asymmetric structures are possible with these configurations for the case of the above complex when $a$ and $b$ are identical. Recently, however, the first complexes having a trigonal prismatic geometry were prepared. These are mainly complexes of bidentate sulfur ligands such as

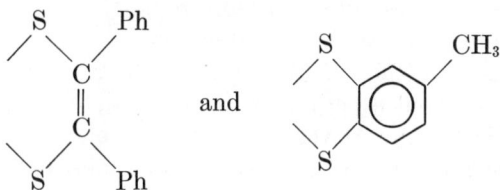

with transition metals such as V, Mo, W, Re and Cr. Additional examples are sure to be forthcoming now that the conditions under which trigonal prismatic coordination is favored over octahedral are being systematically delineated.

Six-coordinated complexes which we term loosely as octahedral are actually often distorted strongly from this very high symmetry, class $O_h$. For example we find the *trigonal distortion*, in which the octahedron is compressed or extended along one of its 3-fold axes, to yield the trigonal antiprism geometry of symmetry class $D_{3d}$ as shown in Figure 10–5(a). But more often we find the *tetragonal distortion*, in which the octahedron is elongated, as shown in Figure 10–5(b), or flattened (not shown) along a 4-fold axis to yield the tetragonally distorted octahedral geometry of symmetry class $D_{4h}$.

Six-coordinated complexes of at least 16 different metals, some in more than one oxidation state, have been successfully resolved into optical isomers. The rates and possible mechanisms of the *racemization* of many of these optically active complexes have for many years been the subject of a great deal of the total research effort in the general area of rates and mecha-

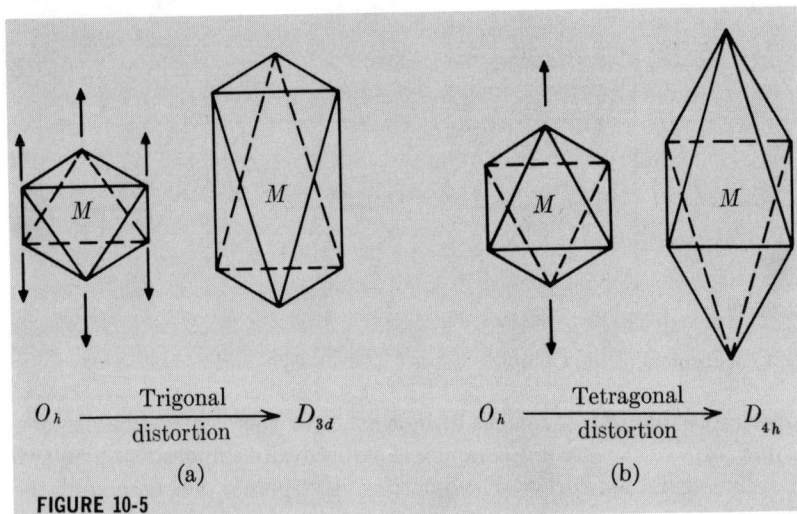

FIGURE 10-5

The (a) Trigonal and (b) one possible Tetragonal Distortion of the Octahedron.

nisms of complex ion reactions. The latter area has received a great amount of attention in the past two decades, and most of the studies have been made with six-coordinate species, although more recently four-coordinate complexes have come in for increasing investigations.

Much of the research effort of Werner and his many students went into devising and carrying out experiments to substantiate the stereochemical implications of his theory. At that time such proofs involved primarily chemical methods. Today there is an overwhelming amount of X-ray diffraction data as well as other physical data providing unequivocal support for the structural aspects of the theory. In addition, both the stereochemical concepts and the two kinds of valence envisioned by Werner are now explicable in terms of modern valence theories, as we shall see later. As a tribute to Werner's great insight and his untiring research and teaching efforts he was awarded the Nobel Prize in 1913.

## CLASSIFICATION OF COORDINATION COMPOUNDS

Because of the great variety of compounds that are properly considered coordination compounds, a classification is difficult, even for study purposes. In fact, no one method of classification has been found to be completely satisfactory. However, it is at least very instructive to examine some possible methods of classification as these will lead to a deeper understanding of the nature and scope of the coordination compounds.

(1) If we choose to classify complexes according to coordination number of the central ion, then perhaps 98% of all complexes will be found under the numbers 6 and 4, since all metal ions exhibit one or both of these coordina-

tion numbers in some of their complexes. In this connection it would be valuable to study Table 4–15, in which we have listed all oxidation states for the 3$d$, 4$d$, and 5$d$ metals along with all of the known or suspected coordination numbers for each oxidation state. In Table 10–2 we have listed

**TABLE 10–2.  Complexes with Coordination Numbers Other Than Four or Six.**

| Coordination Number | Examples[a] |
|---|---|
| 2 (rare) | $M(CN)_2^-$, $[M = Cu(I), Ag(I), Au(I)]$, $Hg(NH_3)_2^{2+}$, $AuCl_2^-$, $Ag(NH_3)_2^+$, $Cu(tu)_2^+$ |
| 3 $\left(\begin{array}{c}\text{very}\\\text{rare}\end{array}\right)$ | $M(R_3Z)_2I$, $[M = Cu(I), Ag(I), Au(I); Z = P, As]$ $Ag(R_3P)_3^+$, $Ag(R_2S)_3^+$, $Ag(tu)_3^+$, $HgI_3^-$ |
| 5 | $Fe(CO)_5$, $M(terpy)Cl_2$ $[M = Cu(II), Zn, Cd]$ $Ni(DMG)_2Br$, $Ni(tas)X_2$ $[X = Cl,Br]$, $Ni(Et_3P)_2Br_3$, $Zn(aca)_2(H_2O)$ $M(das)_2I^+$ $[M = Ni(II), Pd(II), Au(III)]$ $Co(CN)_5^{3-}$, $Co(CNR)_5$ |
| 7 (rare) | $ZrF_7^{3-}$, $MF_7^{2-}$ $[M = Nb, Ta]$, $MoF_7^-$ $UO_2F_5^{3-}$, $UF_7^{3-}$, $NbOF_6^{3-}$ |
| 8 | $Mo(CN)_8^{4-}$, $TaF_8^{3-}$ $M(aca)_4$ $[M(IV) = Ce, Zr, Hf, Th, U, Pu]$ $M(ox)_4^{4-}$ $[M(IV) = Zr, Hf, Th, U, Sn]$ $M(oxine)_4$ $[M(IV) = Th, Pu]$ $TiCl_4(das)_2$ |
| 9 | $Nd(H_2O)_9^{3+}$, $La(OH)_3$, $UCl_3$, $ReH_9^{2-}$ |
| 10 | $M_2[M'(CN)_8X_2]\cdot4H_2O$  $[M = Cd, Mn; M' = Mo, W; X = H_2O, NH_3, N_2H_4]$ |

[a] Symbols used are: $tu$ = thiourea; $terpy$ = 2,2′,2″-terpyridine; $DMG$ = dimethylglyoximate ion; $tas$ = $(CH_3)_2As(CH_2)_3AsCH_3(CH_2)_2As(CH_3)_2$; $Et$ = ethyl; $das$ = $o$-phenylenebisdimethylarsine; $aca$ = acetylacetonate ion; $ox$ = oxalate ion; $oxine$ = 8-hydroxyquinolinate ion. See E. L. Muetterties and C. M. Wright, *Quart. Rev.*, **21**, 109 (1967) for many more examples of molecular polyhedra of high coordination number.

some examples of complexes with coordination numbers from 2 to 10, excluding 6 and 4. The number of metallic complexes with a true *odd* coordination number is relatively small. However, more examples have been turning up in recent years than was ever expected just a few years ago, particularly with coordination number 5. For example, it has recently been reported that the number of complexes of the dioxocation, $UO_2^{2+}$, having a coordination number of 5 (excluding the two oxo-oxygens) may be as great as the number having coordination number 4 or number 6. Valuable reviews of coordination number 5, the most common of the odd coordination numbers, are to be found in reference 9 of Chapter 7 and in reference 9 of this chapter. However, it is still true that odd coordination numbers generally prevail only in unusual circumstances. These special conditions include unusual ligand stereochemistry, unusual coordinate bond character, and the possession by the central atom of a particular $d^n$ configuration. Since most complexes have either six or four ligands attached to the central atom, a further breakdown for classification purposes is desirable.

(2) It might be suggested that we classify complexes in terms of the kind of central metal atom. However, there are about 84 elements whose atoms may be considered to act in the capacity of central atom in a complex, and most of these display several oxidation states each of which must be considered an entirely different central species. In fact, several transition metals exhibit different central species within the *same* oxidation state. This is a consequence of the ability of many of these metals to form oxocation species[10,11], such as $MO_x^{n+}$, where $x$ may be 1, 2, or 3 and $n$ may vary integrally from 0 to 5. Thus, for example, molybdenum in the 5+ oxidation state may be found as a central species of a complex ion in any of the following three mononuclear forms: Mo(V), $MoO^{3+}$, or $MoO_2^+$, and even more dinuclear forms, such as $Mo_2O_2^{6+}$, $Mo_2O_3^{4+}$, $Mo_2O_4^{2+}$, etc. The multiple-bonded oxygens greatly alter the properties of the central metal. Again, reference to Table 4–15 will be helpful in understanding why this second criterion for the classification of complexes does not represent much of an improvement over the one discussed above.

(3) A classification of complexes according to the kind of ligand atom is a much more realistic approach and is the one taken, for example, by Bailar[1] in an excellent, if somewhat out-of-date, survey of complexes and by Jørgensen[7] in a more recent and voluminous treatment. The elements whose atoms, in appropriate molecules or ions, may be attached directly to metal atoms are shown below:

|   |   |   | H  |
|---|---|---|----|
| C | N | O | F  |
| — | P | S | Cl |
| — | As| Se| Br |
| Sn| Sb| Te| I  |

The list does not include the many atoms which in recent years have been found to be capable of forming metal-metal bonds in complexes. These metal atoms occupy a coordination site in each other's first coordination sphere even though the nature of the metal-metal bond is basically different in character from the usual type of coordinate bond. A brief review containing several references to other reviews and original research in the rapidly expanding area of metal-metal bond chemistry is presented in the excellent advanced inorganic chemistry textbook by Cotton and Wilkinson[10] and a longer review with more references has been prepared by Lewis and Nyholm.[11] The subject will not be pursued further in this book.

Note that the above list of ligand atoms includes all of the true nonmetals, with the exception of boron and the inert gas elements. Those to the right of the vertical line are simple ligands that only coordinate as the simple uninegative ion. The remainder usually coordinate when they are part of a molecule or polyatomic ion. However, we should really not exclude oxide  sulfide, selenide, and nitride simple ions from being termed ligands. In Table 10–3 we have listed some examples of commonly encountered ligands grouped according to the donor atom.

**TABLE 10–3.  Some Typical Ligands Grouped According to the Donor Atom.**

| Donor Atom | Examples[a] |
|---|---|
| C | CO, CN⁻, $R$NC |
| N | $NH_3$, $C_5H_5N$, $NO_2^-$, $R$NH$_2$, $NH_2CH_2CH_2NH_2$ ,  $N_3^-$, $R$CN, $NCS^-$, , $NO$, $N_2H_4$, $\phi$—N=N—$\phi$ |
| P | $R_3P$, $PX_3$  ($X$ = F, Cl, Br, NCS),  $\phi_2PCH_2CH_2P\phi_2$ |
| As | $R_3As$,  $AsCl_3$ |
| O | $H_2O$, $OH^-$, $CO_3^{2-}$, $SO_4^{2-}$, $S_2O_3^{2-}$, $R$COO⁻, $R_3PO$, $R_3AsO$, $R_2SO$, $ONO^-$, $C_2O_4^{2-}$, $CH_3CCH_2CCH_3$ (with two =O groups) |
| S | $R_2S$, $SCN^-$, $S_2O_3^{2-}$, $R$S—$CH_2CH_2$—S$R$ |
| O and N | ,  $H_2NCH_2COO^-$, ,  $EDTA$ |

[a] $R$ represents an alkyl or aryl radical and $\phi$ is used specifically for the phenyl radical.

In general it is found that there is a very great difference between the coordination preferences of the elements in the first short period and their respective congeners[13], for example, specifically between N and P, O and S, and F and Cl. In fact it is possible to recognize two extreme classes of acceptor metals: *Class a*, those which form their most stable complexes with the first ligand atom of each family (N, O, and F), and *Class b*, those which form their most stable complexes with the second or subsequent ligand atom of each family. However, it should be recalled in this connection that each oxidation state of a metal must be regarded as a different acceptor and for this reason many metals are better placed in a third class intermediate between the extremes of *a* and *b*. Most of the metals with empty or completely filled $(n - 1)d$ orbitals fall into *Class a*, as do the transition elements with one, two, or three *d* electrons, providing these latter elements are not in oxidation states below $2+$. Only a small number of elements fall clearly and solely into *Class b*. This is shown in Table 10–4. It is interesting to point out

**TABLE 10–4.  Classification of Metals According to Their Coordinating Affinities for the Elements of the First Short Period or Their Respective Congeners.**

| Class *a* | Intermediate | Class *b* |
|---|---|---|
| R1, R2, R3<br>T4, T5 families<br>Al<br>Ga  Ge<br>    Cr, U<br>In  Sn | Mn, Fe, Co, Ni, Cu, Zn<br>Mo, Tc, Ru, Rh      Cd<br>W, Re, Os, Ir<br><br>Tl, Pb, Bi, Te, Po | Pd, Ag<br>Pt, Au, Hg |

that this classification of metals is relevant to the distinction between *soft* and *hard* acids and bases suggested by Pearson and discussed in detail in Chapter 9. Thus, the *Class a* metals are hard acids, whereas the *Class b* metals are soft acids. Of course, more significant than the classification *per se* are the reasons for the existence of the classes and their particular membership. As we shall see later, perhaps the most important single factor is the electronic configuration of the metal atom.

(4) A fourth method by which we may classify complexes is according to the type or nature of the coordinate bond which is formed. Since a given metal atom may form several different types of bonds, even within the same complex, it will be more convenient to consider the possible bond types in terms of the electronic makeup of the various ligands. In Table 10–5 we have attempted to categorize unidentate ligands by their electronic structure, which is relevant to their behavior as ligands. A polydentate coordinating agent may have all ligand atoms alike, as regards bond type, for example, $NH_2CH_2CH_2NH_2$, $C_2O_4^{2-}$, $CH_3COCHCOCH_3^-$,

$$As(CH_3)_2, \quad NH_2CH_2CH_2NHCH_2CH_2NH_2, \text{ etc.,}$$

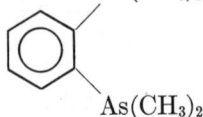

$As(CH_3)_2$

or one or more of the ligand atoms may be a different type, for example,

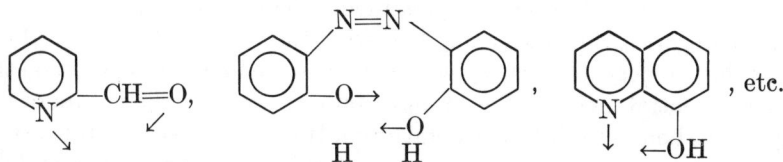

$$\text{—CH=O,} \qquad \text{N=N} \\ \text{—O→} \qquad \qquad \text{, etc.}$$

## TABLE 10–5.  Types of Unidentate Ligands and Some Examples.

1. Those in which the ligand atom contains only a $\sigma$-bonding lone pair, no available $\pi$ electrons, and no vacant orbitals; e.g.,

    $H^-$, $NH_3$, neutral aliphatic amines, $SO_3^{2-}$, aliphatic $\sigma$-bonding ligands such as $CH_3^-$ (generally good Lewis base molecules having a donor atom from the first short period).

2. Those in which the ligand atom contains

    a. at least three lone pairs which split under the influence of bonding to a metal atom into two orbitals of higher energy, both $\pi$ orbitals, and one orbital of lower energy, which is the $\sigma$ orbital; e.g.,

    $N^{3-}$, $O^{2-}$, $F^-$, $OH^-$, $NH^{2-}$, $S^{2-}$, $Cl^-$, $Se^{2-}$, $Br^-$, $I^-$ (the higher energy filled $\pi$ orbitals *may* become involved in $(L{\rightarrow}M)\pi$ bonding, but this tendency drops off rapidly in the listed series of ions after $F^-$).

    b. two lone pairs, one of which becomes $\sigma$-bonding, the other of which *may* become $\pi$-bonding, e.g.,

    $H_2O$, $NH_2^-$, $R_2S$, $R_2O$

3. Those in which the ligand atom contains a $\sigma$-bonding pair plus low-lying, *empty* $\pi$-antibonding orbitals which can accommodate back-donation from metal to ligand; e.g.,

    $CO$, $R_3P$, $R_3As$, $Br^-$, $I^-$, $CN^-$, py, dipy, ophen, picolinic acid N-oxide, $aca^-$, etc.

4. Those without unshared electron pairs, but with electrons already involved in intramolecular $\pi$ bonding; e.g.,

    alkenes, alkynes, benzene, cyclopentadienyl ion, etc. (The term unidentate has no meaning for these ligands.)

5. Those that may act as *bridging* ligand atoms between two metal atoms, and which are therefore capable of forming two $\sigma$ bonds to separate metal atoms; e.g.,

    $OH^-$, $Cl^-$, $NH_2^-$, $O_2^{2-}$, $CO$, $F^-$, $SO_4^{2-}$, $O^{2-}$, etc.

There are perhaps more examples of the first kind, but both kinds of chelating agents are quite numerous. It should be noted that in this classification all coordinate bonds are assumed to be covalent and that in every case donation of electrons, both to and from metal atoms, is mentioned or implied. As we shall see later, it will not always be necessary or convenient to treat the coordinate bond as a covalent one, but rather to consider that in certain cases it arises from almost purely electrostatic interactions. Notice that in the type of classification presented in Table 10–5 the ligands may overlap categories, thus weakening somewhat the value of the classification. Indeed the category in which a given ligand may be found depends upon the particular central metal and even, in certain cases, upon the nature of the *other ligands* bound to the same metal atom. It is easy to see how the former of these situations may arise. Thus, a ligand which may partake in metal-to-ligand pi bonding, $(M{\rightarrow}L)\pi$, with, for example, Ni(0) or Rh(I) or Pt(II), cannot be so involved with V(V) or Ti(III) or Mn(II) or any representative metal, since the latter species do not bond by this mechanism. Furthermore, some central metal species are capable of forming ligand-to-metal pi bonds, $(L{\rightarrow}M)\pi$. The early transition elements, particularly in their high oxidation states, fall into this category. On the other hand, the later transition metals in low oxidation states, as well as the representative metals, do not form such bonds.

The situation in which a given metal-ligand bond depends upon or is influenced by another metal-ligand bond to the same metal atom is less easily documented. However, it is not at all unreasonable or difficult to see that in the case of the coordination to a given metal atom of one ligand capable of $(L{\rightarrow}M)\pi$ bonding and another ligand capable of $(M{\rightarrow}L)\pi$ bonding the two should enhance the respective bonding capacities of each other. Thus, while the first ligand adds charge to the metal atom, decreasing its electronegativity, the second ligand removes charge from the metal atom, increasing its electronegativity, leading to mutual enhancement.

(5) Finally we might try to place all metallic complexes into one of four categories which have as their sole criterion the *electronic configuration* of the metal atom or ion in question. The members of each category are shown in Figure 10–6 with the oxidation states given for the first three categories.

## Category I

In this category we place all metal ions which in their complexes possess a valence shell inert gas configuration, that is, $1s^2$ or $ns^2np^6$, where $n$ has values from 2 to 6. These ions are all spherically symmetrical with the element being in its highest possible oxidation state. The latter situation will not be true for the $4f$ and $5f$ series of elements yet we can include them in their $3+$ state. The incomplete $4f$ energy levels are buried sufficiently far below the valence shell to result in relatively minor effects on the nature of metal-ligand bonds. However, in the actinide series, the $5f$ orbitals are closer to the surface of the ions and should be included in any rigorous covalent bonding as well as symmetry considerations. Thus, these elements fit least well into this category and perhaps should be considered in Category IV, to be described below. The elements included in this first category are shown in Figure 10–6(a).

In general the central atoms in this category are *Class a* acceptors. As we proceed from left to right through these elements the bond type goes from highly ionic to highly covalent in a nearly regular way. Sigma bonds are expected to increase in strength in a regular way from left to right. As we move into the transition elements an ever-increasing contribution to the metal-ligand bond from $(L{\rightarrow}M)\pi$ bonding is expected whenever the ligand is capable of this type of bonding. It is for this reason that oxygen becomes an increasingly important and increasingly strong ligand for these central atoms as we go from left to right. Thus, oxygen is capable of forming double and perhaps even triple bonds with the transition elements in this category by virtue of its capacity to delocalize its $\pi$-bonding electrons into the empty $d\pi$ orbitals of the central atom, resulting in $p\pi{\rightarrow}d\pi$ bonding. The fluoride ion, an excellent ligand for the metals in this category, may also form stronger bonds via this bonding mechanism (see Chapter 7, page 329).

The stereochemistry of the complexes formed by the metal atoms in this category is in general that predicted by VSEPR theory (Chapter 7), and all complexes are diamagnetic.

## Category II

In this category we place the metal atoms shown in Figure 10–6(b), which in their complexes have a valence shell *pseudo-inert gas* configuration, that is, $(n - 1)d^{10}$, where $n$ is 4, 5, or 6. These central atoms are also spherically symmetrical species, and included are some metals in negative oxidation states. All complexes formed by these species are highly covalent. As we proceed from *right to left* in this category, particularly from the coinage metals (Cu, Ag, Au), the contribution to the metal-ligand bond from $(M{\rightarrow}L)\pi$ bonding is believed to be of increasing importance. Indeed this is a logical mechanism by which the central metal atoms are able to delocalize some of the increasing negative charge which would otherwise accumulate on them.

From the coinage metals rightward, an increasing contribution to the metal-ligand bonds from $(L{\rightarrow}M)\pi$ bonding is expected and what was said above about this kind of bonding for Category I metals applies here as well. However, the big difference between the two categories is that for species in the first category, the $d\pi$ acceptor orbitals are of lower principal quantum number than the $\sigma$-bonding $s$ and $p$ orbitals, whereas in this second category the $\pi$ acceptor orbitals are outer $d$ orbitals.

The stereochemistry of the complexes formed by these metal atoms is also explained by VSEPR theory (Chapter 7) and all complexes are diamagnetic.

## Category III

In this category we place the metal atoms which in their complexes have a *pseudo-inert gas plus two* configuration, $(n - 1)d^{10}ns^2$, where $n$ is 4, 5, or 6. These are shown in Figure 10–6(c). These metal atoms are quite interesting

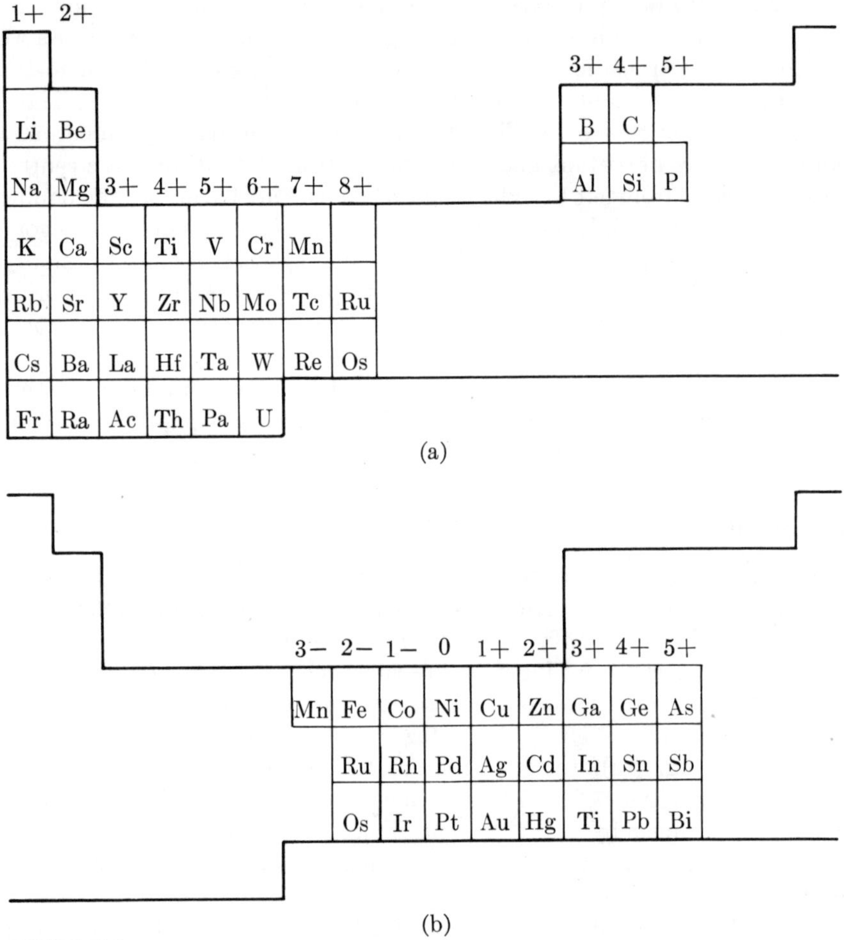

1+ 2+

3+ 4+ 5+

| Li | Be | | | | | | | | B | C | |
|----|----|----|----|----|----|----|----|----|----|----|----|
| Na | Mg | 3+ | 4+ | 5+ | 6+ | 7+ | 8+ | | Al | Si | P |
| K | Ca | Sc | Ti | V | Cr | Mn | | | | | |
| Rb | Sr | Y | Zr | Nb | Mo | Tc | Ru | | | | |
| Cs | Ba | La | Hf | Ta | W | Re | Os | | | | |
| Fr | Ra | Ac | Th | Pa | U | | | | | | |

(a)

3− 2− 1− 0 1+ 2+ 3+ 4+ 5+

| Mn | Fe | Co | Ni | Cu | Zn | Ga | Ge | As |
|----|----|----|----|----|----|----|----|----|
| | Ru | Rh | Pd | Ag | Cd | In | Sn | Sb |
| | Os | Ir | Pt | Au | Hg | Ti | Pb | Bi |

(b)

**FIGURE 10-6**

(a) Category I Metals; (b) Category II Metals; (c) Category III Metals; and (d) Category IV Metals. The oxidation states of the elements in the first three categories are given; but in the fourth category the elements have all oxidation states allowed them except those which would place them in prior category.

in that they are spherically symmetrical in terms of their $d$ electronic distribution, but the outer pair of $s$ electrons, while they may remain nonbonding, nevertheless should require a spatial position in any molecule. Relatively little data is available to support this expectation but, for example: $SnCl_3^-$ is pyramidal; $GeF_2$, shown in Figure 7–5, is distorted to make room for the lone pair; $S(Se,Te)X_4$ compounds have geometries based upon the lone pair occupying a stereochemical site; and the same is true of compounds of Br(V), I(V), Xe(VI), etc., even though these are not generally considered to be central *metal* atoms.

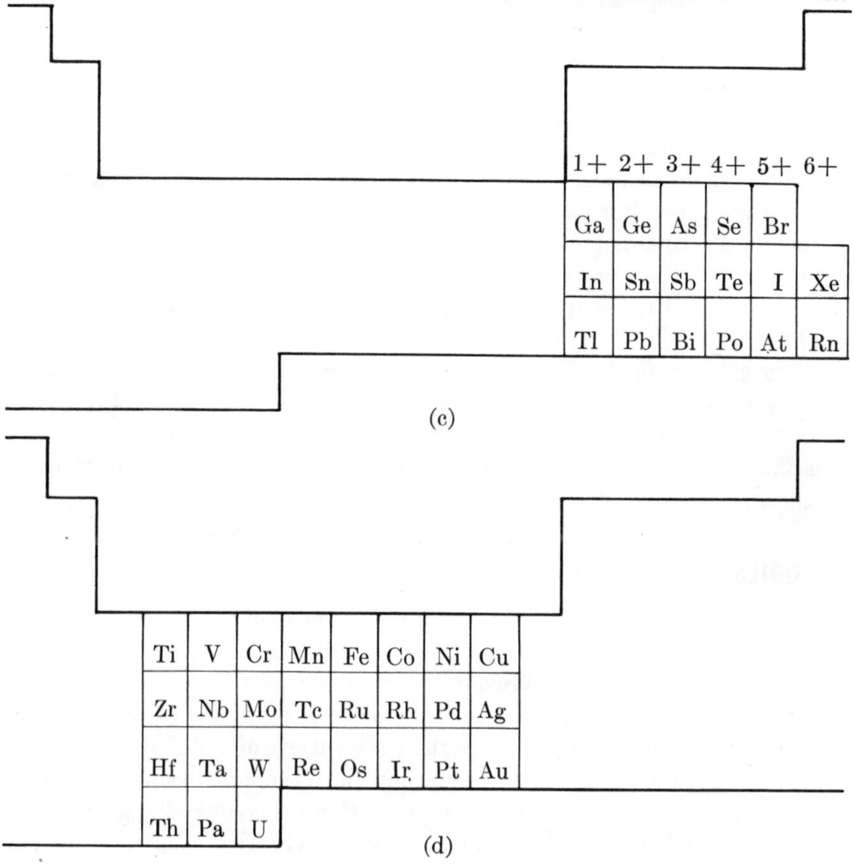

| 1+ | 2+ | 3+ | 4+ | 5+ | 6+ |
|----|----|----|----|----|----|
| Ga | Ge | As | Se | Br |    |
| In | Sn | Sb | Te | I  | Xe |
| Tl | Pb | Bi | Po | At | Rn |

(c)

| Ti | V  | Cr | Mn | Fe | Co | Ni | Cu |
|----|----|----|----|----|----|----|----|
| Zr | Nb | Mo | Tc | Ru | Rh | Pd | Ag |
| Hf | Ta | W  | Re | Os | Ir | Pt | Au |
| Th | Pa | U  |    |    |    |    |    |

(d)

**FIGURE 10-6 (Continued)**

## Category IV

In this category we place the metal atoms which in their complexes have occupied but incompletely filled $d$ orbitals, $(n - 1)d^{1-9}$, where $n$ is 4, 5, or 6. This group of central atoms, shown in Figure 10–6(d), is by far the largest and most diverse since it includes all of the transition metals in all of their many oxidation states except those which would place them in a previous category. Because of the size of this group, as well as for more fundamental reasons, it would be desirable to divide this category still further according to the detailed electron configuration in the $d$ orbitals. In order to more fully understand and explain such a subdivision, however, we will have to introduce some additional ideas, for example those of the crystal field theory. Suffice it to say at this point that the further subdivision of this category will depend primarily upon two things, the number of $d$ electrons and the nature of the bond formed with a particular ligand. Thus, in this category we have coordinate bonds which range from ionic to highly covalent, the latter including both $(M{\rightarrow}L)\pi$ and $(L{\rightarrow}M)\pi$ bonding. From the point of view of the structural symmetry of the complex ion, the subdivisions could be set up to separate perfectly regular structures, such as octahedral, tetrahedral, and dodecahedral, slightly distorted structures, and grossly distorted structures. For example, in this connection we might point out that high-spin $d^5$ systems (that is, all five $d$ electrons unpaired) have spherical symmetry and the complexes formed have perfectly regular structures predicted by VSEPR theory. These matters will become clearer in the next chapter.

After examining the foregoing five methods by which we might classify complexes, we see that no one method stands out clearly as best and none of them is totally satisfactory. However, the mere attempt to find a suitable classification system has hopefully led the reader to a greater appreciation of the broad scope of the field and the many facets of it to be explored.

## THEORIES OF THE COORDINATE BOND

We require of a theory that it be capable of providing at least two things. It should be able to explain experimental facts and it should be capable of predicting new experimental results. For complexes, we wish to explain and predict their thermodynamic, kinetic, spectral, stereochemical, and magnetic properties. When we consider the wide variety of metal atoms, ligands, and resultant complex types we can appreciate the magnitude of this task. We shall find then, not unexpectedly, that there is no one all-encompassing theory but rather several partially successful theories each with its particular merits for dealing with selected complex systems. It has already been pointed out that most complexes are either six-coordinated and approximately octahedral or tetragonal, or four-coordinated and approximately planar or tetrahedral in configuration. Therefore, in our introductory type of treatment, we shall confine most of our further considerations to those complexes which have coordination numbers of 6 and 4.

Before discussing the several theories of the coordinate bond, it might be well to point out that a theory is no more than an approximation to reality. The more examples which can be found to conform to the theory the better the approximation. However, when some exceptions to the theory arise, they do not necessarily invalidate the whole concept. More likely they simply point up our failure to have given a rigorously satisfactory treatment. And this usually requires only that the theory be modified or extended so as to include these exceptions. Close cooperation and communication between theoreticians and experimentalists is obviously necessary in this connection. The *valence bond* approach to the coordinate bond is an example of a theory which served us long and well until the exceptions to it and, more importantly, the inadequacies of it, caused its decline in favor of newer approaches. However, it is not basically less sound a theory than any of the others, but it does need modifications and extensions to be of service to the modern coordination chemist.

Also, it often happens that two or more theories can explain the same set of phenomena. Here we should search for some more fundamental concept or approach common to both theories, for this will likely be a still better approximation to reality. This has been the situation existing with the *crystal field* and *molecular orbital theories* of bonding as they are applied to complexes. Growing out of these has been a more useful and more widely applicable approach known as *ligand field theory*.

The electronic theory of valence, clearly enunciated by Lewis in 1916 and interpreted and extended for many systems by Langmuir in 1919 and others in the succeeding decade, enabled chemists to express Werner's valence concepts in terms of electrons. Much credit for this particular application of the new theory of valence should go to Sidgwick[14] and Lowry.[15] The primary valences of Werner were interpreted as arising from electrovalence, having resulted from complete electron transfer, and the secondary valences were regarded as arising from covalence or electron-pair sharing. A primary valence may or may not be ionic. Thus, if a negative ion is present in the first coordination sphere, as, for example, the chlorine in chloropentaam-minechromium(III) nitrate, $[Cr(NH_3)_5Cl](NO_3)_2$, it satisfies simultaneously both a primary and a secondary valence. In this case the chlorine has lost its effective ionic character. The nitrate ions satisfy only the primary valence and thereby retain their ionicity. The quantum mechanics of Schroedinger and Heisenberg, developed in the late 1920's, was to provide the basis for the modern treatments of the coordinate bond in electronic terms. But before we get to that story we shall first briefly consider the approach suggested by Sidgwick prior to the advent of quantum mechanics.

## Sidgwick Model

Sidgwick extended the Lewis concept of the two-electron covalent bond between two atoms in a molecule by introducing the *coordinate bond* for the case in which both electrons in the shared pair originate on the same atom. Noting that all molecules and ions that become attached to metal atoms have at least one unshared pair of electrons, he suggested that this free electron pair is partially donated to the metal ion in the formation of the bond. The bond thus formed is also called a *dative* or *semipolar* bond, and it is sometimes represented by an arrow, $M{\leftarrow}L$, to indicate that the donor group, $L$, has supplied both electrons in the bond to the acceptor, $M$.

Sidgwick further suggested that metal atoms will tend to accept electron pairs from donors until they have obtained a sufficient number of electrons such that the metal atom in the resulting complex ion has an *effective atomic number* (EAN) of the succeeding inert gas. This may be illustrated with hexaammineplatinum(IV) chloride, $[Pt(NH_3)_6]Cl_4$:

| | |
|---|---|
| Pt(IV) contains | 74 electrons |
| 6 $NH_3$ groups donate | 12 electrons |
| The EAN of Pt(IV) in complex | 86 electrons, the number in Rn |

Although this rule is followed by many of the very stable known complex ions, there are now known a much larger number of exceptions. For example, all of the metal ions that have more than one coordination number must violate this rule. This often depends upon the nature of the ligand and it includes almost all central atoms. Some metal atoms such as Fe(III), which is 4-coordinate in $[FeCl_4]^-$ and 6-coordinate in $[Fe(CN)_6]^{3-}$, never obey the rule. Whereas Sidgwick's effective atomic number concept has little more than historical significance now, it is at least noteworthy that almost all known metal carbonyls, most of their derivatives, such as carbonyl halides, carbonyl hydrides, carbonyl nitrosyls, and related compounds involving $\pi$-acceptor ligands in general, do obey this simple rule. In fact, the few compounds of these types which violate the rule, for example $V(CO)_6$ and $Rh_6(CO)_{16}$, are much less stable than those which conform to the rule.

The concept of a coordinate bond between two atoms seems entirely satisfactory for explaining the formation and stability of the ammonium ion (from $H^+$ and $:NH_3$), *addition compounds* such as $(CH_3)_3B:NH_3$, and many other related simple *Lewis acid-base adducts*. However, in the case of a hexa-coordinated metal atom, the implied donation of six pairs of electrons to the central species poses the serious problem of the accumulation of excess negative charge on the metal atom. Although it would be nice to be able to retain the simplicity of the Sidgwick model, we must recognize that modifications in the theory are required to remove not only the objection just cited but also other shortcomings which will become apparent later.

### The Four Current Models

At the present time, four more or less distinct approaches to the theoretical treatment of the bonding and properties of coordination compounds are recognized. Chronologically, but not according to increasing complexity, they are (1) the *electrostatic* theory with its more recently employed *crystal field* (CFT) modifications, (2) the *valence bond* theory (VBT), (3) the *molecular orbital* theory (MOT), and (4) the *ligand field* theory (LFT). It should be realized, of course, that none of these theories, except LFT, was specifically designed to treat only complex compounds, yet they have all achieved considerable success in this respect.

Before we proceed to consider each of these theories in some detail, it will be useful to review some of the pertinent knowledge relating to the atomic orbitals belonging to the central atoms as they would exist in the gaseous ion free of perturbing or bonding ligands, and to the orbitals of the ligands which become involved in metal-ligand bonds.

Polar diagrams for the angular portion of the wave functions of the $s$, $p$, and $d$ orbitals are presented in Figure 2–9. It should be made clear again that in such diagrams the distance from the origin is proportional to the absolute value of the angular portion of the wave function, $\psi_{(\theta,\phi)}$. Thus, although this gives an approximate picture of the angular distribution of electron density it is not, as is often mistakenly assumed, a boundary surface enclosing a certain portion of the charge density, and neither is the corresponding polar plot of the square of the angular portion of the wave function $|\psi_{(\theta,\phi)}|^2$.

The valence electrons in transition metal central atoms move primarily in the $d$ orbitals, and for this reason they will be of particular concern to us. The $d_{xy}$, $d_{xz}$, and $d_{yz}$ orbitals are mutually perpendicular (orthogonal), and each has four alternating plus and minus lobes determined by the sign of $\psi_{(\theta,\phi)}$, in the respective quadrants. Three other orbitals, designated as $d_{x^2-y^2}$, $d_{z^2-y^2}$, and $d_{z^2-x^2}$ are likewise mutually perpendicular, each having four lobes lying along the respective axes in one of the three perpendicular planes. However, of these three equivalent orbitals, only two are independent and the usual procedure is to take a hybrid, for example, a normalized linear combination of the latter two listed above and label it simply $d_{z^2}$. This orbital will have large positive lobes along the $z$ axis and a doughnut-shaped negative belt around the $z$ axis which is symmetric in the $xy$ plane (see Figure 2–9). For reasons which will be made clear later, the first three $d$ orbitals described above are grouped together and designated as $t_{2g}$ or $d_\epsilon$, whereas the latter two, $d_{z^2}$ and $d_{x^2-y^2}$, are designated $e_g$ or $d_\gamma$ orbitals. In the absence of a magnetic or electric field, the five $d$ orbitals of a given quantum level are of equal energy, as are the three $p$ orbitals.

In the pre- and post-transition elements the orbitals that may be involved in bonding are the $ns$, $np$, and $nd$. As we shall see, for the pre-transition metals the ionic radius and the effective nuclear charge are of greater importance than covalent bonding orbitals in the formation of metal-ligand bonds. For post-transition metals, ionic radius and effective nuclear charge are important but so also are such factors as polarizability and covalent bonding involving the outermost $s$, $p$, and $d$ orbitals. For the transition elements all of the aforementioned factors come into play to greater or lesser degrees, but additionally for these elements there are as well sigma and pi bonding involving the $(n-1)d$ orbitals and electrons. With the lanthanide and particularly with the actinide elements the $4f$ and $5f$ orbitals and electrons, respectively, may also become involved in metal-ligand bonding, but this is never as important as $d$-orbital involvement.

With regard to the various ligands, the bonding interactions with the central metal ions may range from ionic to covalent. In the first instance, the ligand atom radius, charge, polarizability, and permanent dipole moment play the major roles. In the covalent case, the availability and energy of the sigma-bonding electron pair is of most importance followed by the availability of either suitably situated empty $\pi$-acceptor orbitals or filled $\pi$-donor orbitals.

We shall now proceed to examine the four theories separately. The valence bond theory will be considered first, since it is conceptually the simplest. And although it is now seldom used, it served coordination chemistry admirably for the interpretation of certain properties of complex molecules for over a quarter of a century. Next, we shall discuss the electrostatic theory with emphasis on the crystal field approach. Although the crystal field theory had its origin in 1929, it generated great interest among chemists in the ten years between 1952 and 1962. With a beautifully simple model it correlated many physical-chemical observations of coordination compounds. But the realization of its many limitations and necessary approximations spurred the extension of the purely electrostatic CFT to include the effects of covalent bonding. This produced a very useful current approach, the ligand field theory, which we shall consider after we have briefly discussed the application of the more general molecular orbital theory to complexes. The molecular orbital theory really encompasses all the others, and they become special cases of it. It is thus the most general and sophisticated approach, but because of the complexity involved in dealing with many-atom systems by this method, it is difficult to obtain exact treatments. Thus, MOT as applied to complexes must be considered still in the formative stages, but it is surely the most promising one of all for the future.

## THE VALENCE BOND THEORY

The application of the valence bond theory to complexes is due originally and mainly to Pauling.[16] It deals with the electronic structure of the ground state of the central metal atom and, as we shall see, is concerned primarily with the kind of bonding, stereochemistry, and gross magnetic properties present in complexes. The orbitals of the complex are designated only in terms of the central atom orbitals and the hybridization of these to produce bonding orbitals. The very simple means devised by Pauling for presenting the bonding picture involves the following assumptions:

(1) The central metal atom must make available a number of orbitals equal to its coordination number for the formation of covalent bonds with suitable ligand orbitals. The latter orbitals are not specified precisely by this model but they are presumed to be filled $\sigma$-bonding orbitals.

(2) A covalent $\sigma$ bond arises then from the overlap of a vacant metal orbital and a filled $\sigma$ orbital of the donor group. The metal orbital will be a hybrid orbital formed from the available $s$, $p$, and $d$ orbitals. The donor group must, therefore, be a chemical species which contains at least one lone pair of electrons. The resulting coordinate bond is seen to be simply a covalent bond involving the characteristic overlap of two directed orbitals.

Of course, it possesses a considerable amount of polarity because of the mode of its formation.

(3) In addition to the $\sigma$ bond, there is also allowed in later VB treatments the possibility that a $\pi$ bond may form providing that suitable $d$ orbitals and electrons are present on the metal atom and that overlap with ligand $\pi$ orbitals can occur. This kind of bond, if it is $(M{\rightarrow}L)\pi$, will alter the charge distribution on both the metal atom and the ligand in such a way as to strengthen the $\sigma$ bond. If it should be $(L{\rightarrow}M)\pi$-bonding, usually not considered in the early VB treatments, the $\sigma$ bond might be weakened, but the overall bond strength would increase.

The strongest covalent bond will be formed when the orbitals overlap one another as much as possible. In order to satisfy this criterion, it has been shown that the original atomic orbitals should be hybridized to form a new set of equivalent bonding orbitals possessing definite directional properties. Since hybridization of orbitals has already been treated in Chapter 5, we have given in Table 10–6 only the results of the calculations which deter-

**TABLE 10–6. The Shapes and Relative Strengths of the Important Hybrid Orbitals.**

| Coordination Number | Orbital Configuration | Spatial Configuration | Relative VBT Bond Strengths |
|---|---|---|---|
| — | $s$ | — | 1.000 |
| — | $p$ | — | 1.732 |
| 2 | $sp$ | linear | 1.932 |
| 3 | $sp^2$ | trigonal | 1.991 |
| 4 | $sp^3$, $d_\epsilon^3 s$ | tetrahedral | 2.000 |
| 4 | $(d_{x^2-y^2})sp^2$ | square planar | 2.694 |
| 5 | $d_{z^2}sp^3$ | trigonal bipyramid | — |
| 5 | $(d_{x^2-y^2})sp^3, d^4s$ | square pyramid | — |
| 6 | $d_\gamma^2 sp^3$ | octahedral | 2.923 |

mine the shape and the relative strength of several hybrid bond orbitals important for many complexes. The convenient representation of bonding devised by Pauling is illustrated with some examples in Table 10–7 for the three most commonly encountered hybridizations, $d^2sp^3$, $sp^3$, and $dsp^2$.

The $d_{x^2-y^2}$ and $d_{z^2}$ orbitals are directed toward the ligands in an octahedral complex and they are therefore the ones used in the formation of the $d^2sp^3$ hybrid orbitals. The square planar hybrid orbitals, $dsp^2$, utilize only the $d_{x^2-y^2}$ orbitals. Five-coordinate metal atoms employing $dsp^3$ or $sp^3d$ hybrid orbitals employ the $d_{z^2}$ orbital for a trigonal bipyramidal geometry, but they can use either the $d_{x^2-y^2}$ or the $d_{xy}$ orbital for a square pyramidal geometry. Both of these stereochemistries are found.

The magnetic moments, in Bohr magneton units, listed in Table 10–7 were calculated using the *spin-only* formula (Chapter 11). It is seen that

**TABLE 10–7.  Examples of Commonly Encountered Hybridizations Showing Electron Distribution and Calculated (Spin Only) and Experimental Magnetic Moments.**

| Complex and Its Gross Geometry | Electron Distribution | | | $\mu_{calc}$ | $\mu_{exp}$ |
|---|---|---|---|---|---|
| | **3d** | **4s** | **4p** | | |
| [Co $en_3$]$^{3+}$ octahedral | ⇅ ⇅ ⇅ ⇅ ⇅ | ⇅ | ⇅ ⇅ ⇅  ($d^2sp^3$) | 0 | ~0 |
| [Mn(CN)$_6$]$^{4-}$ octahedral | ⇅ ⇅ ↑ ⇅ ⇅ | ⇅ | ⇅ ⇅ ⇅  ($d^2sp^3$) | 1.73 | 1.80 |
| [Ni(CN)$_4$]$^{2-}$ square planar | ⇅ ⇅ ⇅ ⇅ ⇅ | ⇅ | ⇅ ⇅ □  ($dsp^2$) | 0 | 0 |
| [Cu(NH$_3$)$_4$]$^{2+}$ square planar | ⇅ ⇅ ⇅ ⇅ ⇅ | ⇅ | ⇅ ⇅ ↑  ($dsp^2$) | 1.73 | 1.89 |
| [Ni($Et_3$P)$_2$(NO$_3$)$_2$] tetrahedral | ⇅ ⇅ ⇅ ↑ ↑ | ⇅ | ⇅ ⇅ ⇅  ($sp^3$) | 2.83 | 3.05 |
| [MnCl$_4$]$^{2-}$ tetrahedral | ↑ ↑ ↑ ↑ ↑ | ⇅ | ⇅ ⇅ ⇅  ($sp^3$) | 5.92 | 5.95 |
| Ni($Et_3$P)$_2$Br$_3$ square pyramidal | ⇅ ⇅ ⇅ ↑ ⇅ | ⇅ | ⇅ ⇅ ⇅  ($dsp^3$) | 1.73 | 1.8 |

agreement between calculated and observed values for the particular complexes listed is quite satisfactory. The deviations, although small, can be accounted for in other theories, but not by VBT. This will be explained in detail in the next chapter. It can also be seen that the known stereochemistries of the complexes are consistent with the spatial arrangements required by the hybrid orbitals which are employed.

## Defects in the Early Model

Difficulties are encountered in the original Pauling approach when the number of central atom orbitals needed to accommodate all of the ligand $\sigma$-bonding electrons is too low. This situation can result from the occupancy of the needed orbitals by either paired or single electrons originating on the central metal atom. As an example, we can consider the complex ion [FeF$_6$]$^{3-}$, which has a magnetic moment that corresponds to five unpaired electrons, and the complex ion $[Ni(NH_3)_6]^{2+}$, which has two unpaired electrons. In neither of these complex ions is it possible for $d^2sp^3$ hybridization to occur. As shown below, there are but four orbitals of approximately the same energy beyond the occupied $d$ orbitals.

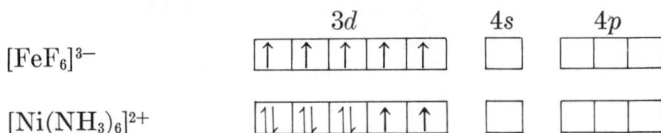

|  | $3d$ | $4s$ | $4p$ |
|---|---|---|---|
| $[FeF_6]^{3-}$ | ↑ ↑ ↑ ↑ ↑ | ☐ | ☐ ☐ ☐ |
| $[Ni(NH_3)_6]^{2+}$ | ↑↓ ↑↓ ↑↓ ↑ ↑ | ☐ | ☐ ☐ ☐ |

Pauling originally overcame this problem by assuming that such complexes are primarily *ionic*. That is, he assumed that the metal orbitals are left free to accommodate only metal atom electrons, and the bonding electrons are contained in separated orbitals located primarily on the ligands. Alternatively he suggested that the outer empty orbitals might be used for forming covalent bonds which would resonate with purely ionic bonds. We shall see that the CFT, LFT, and MOT all provide a much more satisfactory treatment for complexes such as these.

The introduction of the term "ionic" to describe these complexes was unfortunate since many of them, such as the five-unpaired-electron complex $[Fe(aca)_3]$ ($aca$ = acetylacetonate anion), behave like typical covalent compounds. Another alternative was proposed in 1937 by Huggins.[17] He suggested that *outer d* orbitals might be utilized in covalent bond formation. However, at the time, it was considered by many chemists that the $nd$ orbitals could not hybridize satisfactorily with the $ns$ and $np$ orbitals to give strong covalent octahedral bonds. Later calculations by Craig, et al.,[18] showed that with highly electronegative ligands the use of $4d$ orbitals probably leads to stronger bonds than those formed with $4sp^3$ orbitals alone. Thus, the complexes originally described as ionic by Pauling came to be pictured later as employing *upper level* covalent bonds as opposed to *lower level* covalent bonds for those termed "covalent" by Pauling. Other terms that were introduced to distinguish the apparent two types of complexes are: *outer* and *inner orbital* (Taube), *spin-free* and *spin-paired* (Nyholm), and *high-spin* and *low-spin* (Orgel). In the third edition of his classic book[16] "The Nature of the Chemical Bond," Pauling introduced the terms *hypoligated* and *hyperligated*. Most of these terms have lost their original value now and the only ones generally still retained are the truly descriptive ones, namely, high-spin–low-spin and spin-free–spin-paired. We shall use only these in our further discussions.

## Multiple Bonding

So far in our discussion of the VBT it has been assumed that all ligands will possess an accessible and readily donated pair of $\sigma$-bonding electrons, that is, that all ligands are good Lewis bases. This is far from the true situation. There are many common ligands, such as CO, $R$NC, $PX_3$

($X$ = halogen), $PR_3$, $AsR_3$, $SR_2$, $C_2H_4$, benzene, etc., which are very poor electron pair donors, and yet they form many stable complexes. Furthermore, we have not yet discussed the problem of accumulation of unusually high negative charge which would be placed upon a central atom by six negative groups if each donated a pair of electrons.

It was partially to explain the latter situation that Pauling suggested that the atoms of the transition elements are not restricted to the formation of single covalent bonds. Rather, he proposed that they are capable of forming multiple bonds with electron-acceptor ligands by utilizing electrons in $\pi$ orbitals below the valence shell. To illustrate, we can consider the hexacyanoferrate(II) ion, $[Fe(CN)_6]^{4-}$, in which it appears that a 4— charge has been placed upon the iron atom. If partial double bonding between iron and the carbon atoms is allowed, this unlikely concentration of charge may be more suitably distributed. The structure pictured below represents one Lewis-valence bond configuration which shows how the charge on the iron atom might be reduced to a more reasonable value of 1—. Other similar *resonance* structures can be drawn.

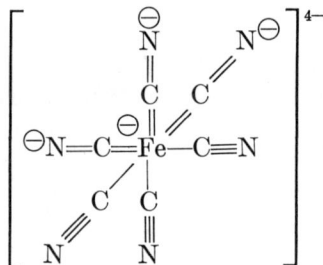

Further reduction of the negative charge on the iron atom will occur if ionic resonance structures are also allowed. Alternatively, if we do not worry about being able to write a Lewis structure, we can simply distribute the 4— charge throughout the molecule with the greater share of it residing on the most electronegative atoms. This could allow the iron atom to become neutral or even slightly positive as we might expect it to be.

Recently there have been a great many efforts made to apply semi-empirical molecular orbital theory to calculate a variety of molecular properties of complexes, such as stereochemistries, electric-dipole moments, transition moments, electronic spectral transitions, ionization potentials and charge distributions.[19,20,21] One such self-consistent charge and configuration MO calculation, reported for 32 octahedral and tetrahedral 3$d$ metal complexes containing halide and calcogenide ligands,[19] reveals that a small positive charge always resides on the central metal atom. In the particular complexes investigated the charges ranged from +0.13 for tetrahedral $CoS_4{}^{6-}$ to +1.12 for octahedral $TiF_6{}^{3-}$. Pauling's *electroneutrality principle* would then be satisfied for these cases. This principle states that electrons will distribute themselves in a molecule in such a way as to leave the residual charge on each atom zero or very nearly zero, except that the most electro-positive atoms may acquire a partial positive charge and the most electro-negative atoms may acquire a partial negative charge. For example, in the cationic complex $[Fe(H_2O)_6]^{3+}$ or $[Co(NH_3)_6]^{3+}$, the formal charge on the iron or cobalt atom would be $3-$, since the $3+$ ion receives a one-half share in 12 electrons. But if we assume that the water or ammonia molecules are attached simply by ion-dipole forces, the charge on the metal atoms will be $3+$. Neither of these extreme situations is at all likely. It is more reasonable to expect that the negative charge will distribute itself to a minimum energy state within the system of static positive centers in such a way as to leave the $3+$ charge on the entire ion distributed over the hydrogen atoms. The latter may be pictured as being on the surface of a sphere, which is a good approximation to the volume of the complex ion, and it is well-known that a charge on a sphere distributes itself uniformly over the surface.

Multiple or $\pi$ bonding in metallic complexes is now generally recognized as an important factor in the formation of many stable metal-ligand bonds. If we ignore for now the metal-metal bonds in certain complexes and the highly delocalized bonding arising in the large class of complexes having ligands such as olefins and aromatic systems, it is easy to classify the remaining types of $\pi$ bonds found in metal complexes:

(1) $M(d\pi) \rightarrow L(p\pi)$, or donation of metal atom $d\pi$ electrons to empty ligand atom $p\pi$ orbitals.

(2) $M(d\pi) \rightarrow L(d\pi)$, or donation of metal atom $d\pi$ electrons to empty ligand atom $d\pi$ orbitals.

(3) $L(p\pi) \rightarrow M(p\pi)$, or donation of ligand atom $p\pi$ electrons to empty metal atom $p\pi$ orbitals.

(4) $L(p\pi) \rightarrow M(d\pi)$, or donation of ligand atom $p\pi$ electrons to empty metal atom $d\pi$ orbitals.

These four types of pi bonds may be pictured as shown in Figure 10-7, which also gives some examples of ligands and metal atoms that are believed to engage most often in the various types of $\pi$ bonding. The orbitals shown are in the pre-bonding state as pure atomic orbitals and the shaded and non-shaded areas differentiate filled from empty orbitals, respectively. Bonding of type (3) is relatively quite rare. There are some transition metal atoms that can bond by types (1) and (2) and by type (4) as well, depending upon the ligands. Likewise there are a few ligands, for example $Cl^-$, $Br^-$, $I^-$, $RS^-$, pyridine, dipyridyl, which may partake in $\pi$ bonding via either a type (2) or type (4) bonding, depending upon the central atom. These are ligands that obviously possess both stable filled $\pi$ orbitals and less stable empty $\pi$ orbitals. The molecular orbital theory will be seen to be a superior approach to such delocalized bonding and will be considered in this connection later in more detail.

Thus, we have seen that $\pi$ bonding is able to remove the excess charge that might otherwise accumulate on the central metal atom or for that matter on a ligand atom. However, there still remain several important

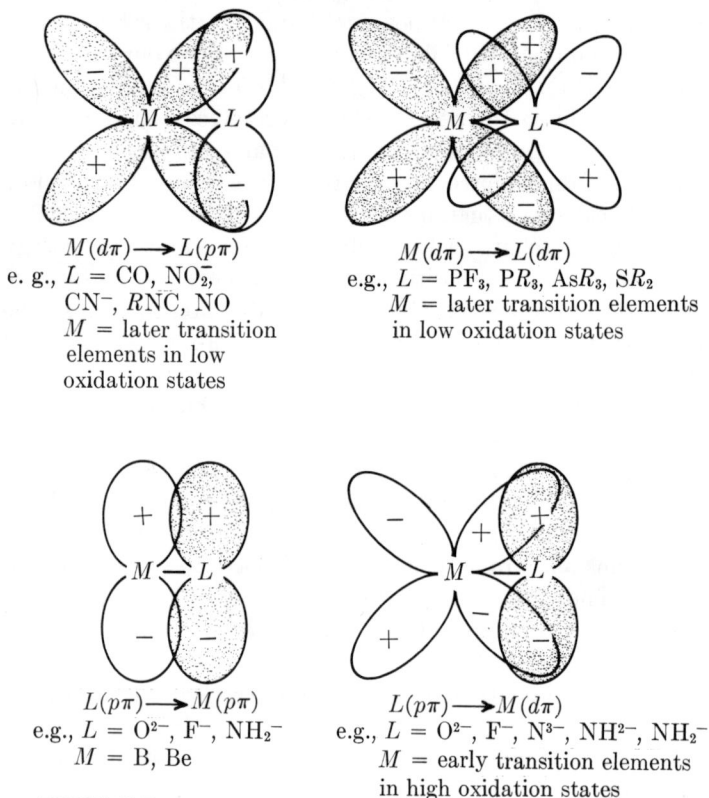

$M(d\pi) \longrightarrow L(p\pi)$
e. g., $L$ = CO, $NO_2^-$,
CN$^-$, $R$NC, NO
$M$ = later transition
elements in low
oxidation states

$M(d\pi) \longrightarrow L(d\pi)$
e.g., $L$ = PF$_3$, P$R_3$, As$R_3$, S$R_2$
$M$ = later transition elements
in low oxidation states

$L(p\pi) \longrightarrow M(p\pi)$
e.g., $L$ = O$^{2-}$, F$^-$, NH$_2^-$
$M$ = B, Be

$L(p\pi) \longrightarrow M(d\pi)$
e.g., $L$ = O$^{2-}$, F$^-$, N$^{3-}$, NH$^{2-}$, NH$_2^-$
$M$ = early transition elements
in high oxidation states

**FIGURE 10-7**

Diagrammatic Representation of Four Types of $\pi$ bonds Which by the VBT May Form between Metal and Ligand Atoms. The respective orbitals are pictured in their pre-bonding state as pure atomic orbitals.

shortcomings in the valence bond theory which are not easily overcome. Although the VBT served long and well during the early development of the theory of bonding in complexes, we no longer rely upon it because of the following major defects: (a) it offers only qualitative explanations; (b) it permits no accounting of detailed magnetic properties; (c) it cannot explain or predict the relative energies of different stereochemistries, that is, no thermodynamic properties are accessible; (d) it cannot begin to interpret or predict spectral properties; and (e) indeed none of the properties resulting from or dependent upon the energy splittings of the $d$ orbitals can be explained or predicted. Here we see that the only differentiation made by VBT in the $d$ orbitals is between those which are employed in the formation of hybrid orbitals and those which are not.

## ELECTROSTATIC THEORY — CRYSTAL FIELD THEORY

### Simple Electrostatic Model

The application of a simple electrostatic theory to the bonding in metallic complexes was carried out primarily by VanArkel and DeBoer[22] and by Garrick[23] around 1930. They simply applied the well-known potential energy equations of classical electrostatics to their bonding model. This required a knowledge of such variables as the charge and size of the central atom, and the charge, permanent dipole moment, polarizability, and size of the ligand. It is relatively easy to show that if a purely electrostatic model is adopted, then, with identical ligands, regular configurations are expected for all coordination numbers. For example, for the common coordination numbers of 2 ,4, 6, and 8 the configurations would be linear, tetrahedral, octahedral, and square antiprismatic, respectively, since these will reduce the electrostatic repulsion between ligands to a minimum. Using this simple model, it is possible to calculate average bond energies for a few complexes which compare quite favorably with the experimentally determined values.[2]

Nevertheless, one does not have to search far to find many phenomena that are inexplicable in terms of the simple electrostatic model. There are also many inconsistencies which arise from its application. For instance, the existence of the numerous square coplanar complexes cannot be justified by this model since only tetrahedral configurations are predicted for four-coordinated complexes. The stability of complexes involving nearly non-polar ligands, such as CO, $PF_3$, etc., as well as those involving unsaturated organic molecules or positive ion ligands, is untenable. In addition, the ions of the $4d$ and $5d$ series form many complexes of greater stability than those of the $3d$ series, yet because of the greater size of the heavier ions their complexes would be predicted to be less stable. Finally, along with the other shortcomings, the simple theory is of no help in predicting and explaining magnetic, spectral, and kinetic properties of complexes.

### Crystal Field Theory

The crystal field theory, like the simple electrostatic approach, proposes to treat metal complexes as if the only interaction between the central atom and the surrounding ligands is a purely electrostatic one. Unlike the simple electrostatic theory, however, it deals with orbitals and electrons, if only those of the central atom. The orbitals of the central atom are considered separated from the ligand orbitals and the latter are ignored. Indeed the ligands are considered to be merely *point charges* or *point dipoles*. Thus, we see from the start that the CFT employs a formalism that does not correspond to reality, since ligand atoms are not points and they do possess orbitals and electrons and sizes comparable to the metal atoms. Nevertheless, it is extremely valuable to discuss and understand the use of the crystal field theory, not only because it is a necessary prelude to the more general ligand field theory, but also because by itself it can provide useful results.

Crystal field theory is not new. Ionic bonding in complexes was suggested by Langmuir in 1919 and the quantum mechanical theory of ionic bonding, crystal field theory, was developed by Bethe a decade later.[24] The first application to transition metal complexes was made in 1932 by Schlapp and Penney[25] and by VanVleck[26], who used the theory to calculate magnetic susceptibilities. In 1935, VanVleck summarized and compared the three theories, VBT, CFT and MOT.[27] But in the succeeding years, up until the early 1950's, only a few physicists, mainly VanVleck and his students, used CFT, and this was primarily for the study of some of the fine details of magnetochemistry and electronic absorption spectra. The more recent interest in CFT stems from the 1951 work of the chemists Ilse and Hartmann[28], who applied it to the weak, visible absorption band shown by hexaaquotitanium(III) ions. After that, Orgel[29] probably had more to do with pointing out to chemists the usefulness of CFT in the study of transition metal complexes. Significant contributions to the theory and particularly to its applications were made by Ballhausen, Jørgensen, Griffith, Nyholm, Owen, Cotton, and many others. These same chemists have been responsible in large measure for the subsequent development and application of the ligand field theory, to be considered later.

In crystal field theory the complex is looked upon as a single isolated molecule in which the electrons of the central metal atom, and in particular those moving in incompleted $d$ orbitals, are subjected to an electrostatic field generated by the surrounding ligands. Although applications of CFT have been made to complexes in which the degeneracy of the $f$ orbitals is assumed to be affected by the ligand field, the greatest success and widest application has been achieved with complexes in which the outer electrons move in $d$ orbitals. We shall be concerned here with only the latter complexes. The theory asks the question: What effect does this field, which is characterized by a particular *strength* and a particular *symmetry*, have upon the five $d$ orbitals of the central ion? The answer may be formulated quantum mechanically by adding a new term, V, actually an operator, to the free-ion Hamiltonian. This term represents the perturbing potential caused by the presence of the ligands in the neighborhood of the ion. Approximate solutions to the Schroedinger equation are then obtained in order to obtain a specific answer to the question. That is, we solve

$$H\psi = E\psi$$

where now

$$H = H_F + V$$

and

$$H_F = -\frac{\hbar^2}{2m} \sum_i \nabla_i^2 - \sum_i \frac{ze^2}{r_i} + \frac{1}{2} \sum_{i \neq j} \frac{e^2}{r_{ij}} + \sum_i \xi_i(r) l_i \cdot s_i$$

in which the first two terms are the usual ones and the latter two result from perturbations due, respectively, to interelectronic repulsion and spin-orbit coupling interaction. It is important to know how the new perturbing potential compares in magnitude with the other two perturbing terms in the Hamiltonian. At least three cases arise:

(1) $$\text{V} < \xi_i(r)l_i \cdot s_i < \frac{e^2}{r_{ij}}$$

This is the situation closely approximated by complexes of the lanthanides in which the incomplete level, the $4f$ subshell, is well-shielded from the perturbing field of the ligands.

(2) $$\xi_i(r)l_i \cdot s_i < \text{V} < \frac{e^2}{r_{ij}}$$

This is the situation which prevails in many complexes of the $3d$ series, and it is often referred to as the *weak field* case.

(3) $\dfrac{e^2}{r_{ij}} < \text{V}$   (The spin-orbit coupling term varies widely, but is less than V in this case.)

This is the situation prevailing in many $3d$ complexes and almost all $4d$ and $5d$ complexes, and it is often referred to as the *strong field* case.

At this point the mathematics of quantum mechanics must be used in order for us to obtain the $d$ electron eigenvalues from approximate solutions to the wave equation. This is outside the scope of our book and so we shall refer the reader to the fine books by Ballhausen[30] and by Figgis[31] where the mathematical details of CFT and LFT may be found. The accomplished mathematician will find yet another reference, the book by Griffith,[32] to be the most comprehensive treatise on the subject.

We are more interested here in the qualitative effects on the metal atom $d$

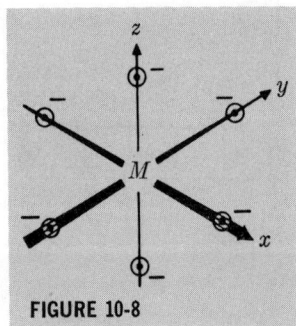

**FIGURE 10-8**

Central Metal Atom, $M$, Surrounded by an Octahedral Array of *Point Charges* Representing Six Identical Ligands.

orbitals caused by the crystal field perturbing potential. Consider a central atom, $M$, surrounded by an array of six point charges or dipoles representing six identical ligands. These charged points will locate themselves, because of electrostatic interactions, at the corners of an octahedron and may therefore be placed for convenience along the Cartesian coordinates as shown in Figure 10–8. Now it can easily be seen that the two $d$ orbitals which lie with the maxima in their lobes along these coordinates, namely, $d_{z^2}$ and $d_{x^2-y^2}$, and thus point directly at the charges, must assume a different and higher energy than the three $d$ orbitals whose lobes point between the coordinates, namely, $d_{xy}$, $d_{xz}$, and $d_{yz}$. In other words, an electron occupying a $d_\gamma$ orbital will suffer greater repulsion than an electron occupying a $d_\epsilon$ orbital, since the former points towards the negative charges and the latter points between the charges. In terms of energy levels, the original degeneracy of the $d$ orbitals in the field-free ion or atom $in\ vacuo$ is removed and split into two parts for the octahedral case as is illustrated in Figure 10–9. Above (a) we have the level of energy of the five degenerate $d$ orbitals in a free gaseous ion. Above (b) we have the energy level of the $d$ orbitals in the presence of the

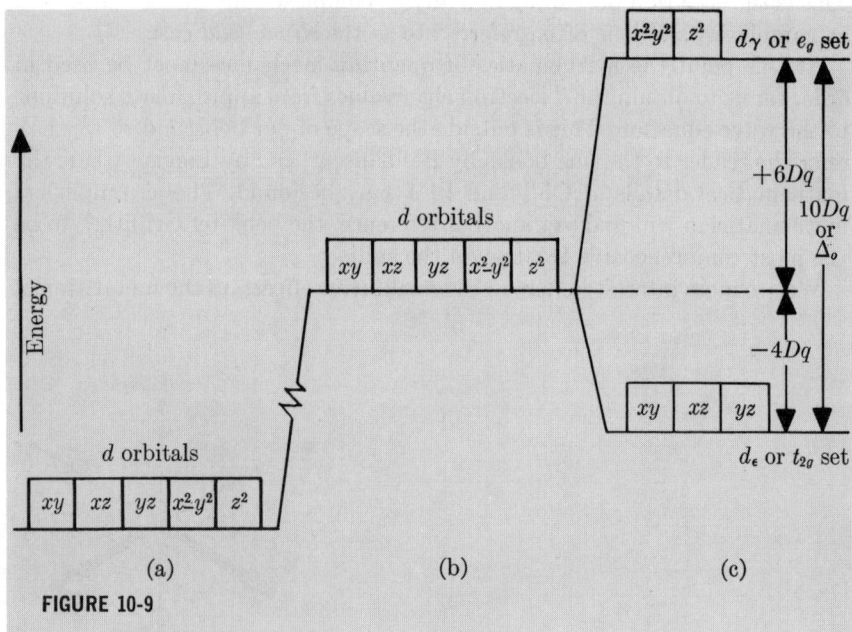

**FIGURE 10-9**

Energy Level Diagram Illustrating the Splitting of the 5-fold $d$-orbital Degeneracy in an Octahedral Field. (a) Field-free ion $d$ orbitals; (b) $d$ orbitals in spherical field, raised above the field-free level primarily by repulsion of the central ion electrons by the ligand charges; and (c) the $d$-orbital splitting in a field of $O_h$ symmetry. The energy gap in (c) between the $d_\epsilon$ and $d_\gamma$ levels is actually much smaller than that between (a) and (b). It has been exaggerated for clarity and because the gap between (a) and (b) is of an unknown energy.

six point-charge ligands but with the field assumed to be spherically symmetric. The rise in energy of all the $d$ levels, by an unknown but relatively large amount, is a result of the repulsion of the central ion electrons by the negative ligand charges. Allowing the point charges to localize at octahedral sites as shown in Figure 10–8, producing a field of $O_h$ symmetry, then causes the $d$-orbital splitting into the two sublevels shown above (c) in Figure 10–9. The higher of these is doubly degenerate and is designated by $d_\gamma$ or $e_g$, and the lower is triply degenerate and is designated as $d_\epsilon$ or $t_{2g}$. It should be pointed out, although it will not be shown here, that *from symmetry considerations alone*, using group theory, the set of $d$ orbitals is split in an $O_h$ environment into a triply degenerate set, labeled $T_{2g}$ and a doubly degenerate set, labeled $E_g$. (Lower-case group theory symbols are often used to label *orbitals* and upper case symbols for *states*.) The group theoretical treatment, however, does not give us either absolute or relative energy information. This must be obtained by other methods such as we have outlined previously. Of course group theory will also permit determination of the splitting of various sets of orbitals in environments of other symmetries often found in complexes, such as $T_d$ (tetrahedral), $D_{4h}$ (tetragonal), $D_{2d}$ (dodecahedral), $D_{3h}$ (trigonal bipyramidal), $C_{4v}$ (square pyramidal), etc. We shall briefly consider these fields of lower symmetry later on, but the excellent book by F. A. Cotton,[33] "Chemical Applications of Group Theory," may be consulted for details.

The energy difference between the two $O_h$ split levels is measured in terms of a parameter $\Delta_o$ or $10\ Dq$. A theorem of quantum mechanics requires that so long as the forces are purely electrostatic and the degenerate set of energy levels being split is well-removed from other levels with which it might interact, the average energy of the perturbed levels must remain unchanged. Therefore, the upper level must be $6Dq$ or $3/5\Delta_o$ above the baricenter and the lower level $4Dq$ or $2/5\Delta_o$ below it. The gain in energy achieved by the preferential filling of the lower-lying $d$ levels, over the energy of a completely random occupancy of all five $d$ levels, is called the *crystal field stabilization energy* (CFSE). We shall have more to say later about the magnitude of $\Delta_o$, the factors which affect its magnitude, and the methods by which it is obtained.

### Distribution of Electrons in the $d$ Orbitals in Octahedral Complexes

We have stated that the lower-lying $d$ orbitals will be preferentially filled since this should lead to a lower energy for the system than if all $d$ orbitals were equivalently occupied. But the picture is not really so simple and we must look more carefully at the factors that determine the distribution of the $d$ electrons among the $e_g$ and $t_{2g}$ orbitals. Considering the ground state we find that there are at least two important, and opposing, factors which determine the population distribution of $d$ electrons under the influence of a purely electrostatic crystal field. One is the tendency for electrons to occupy,

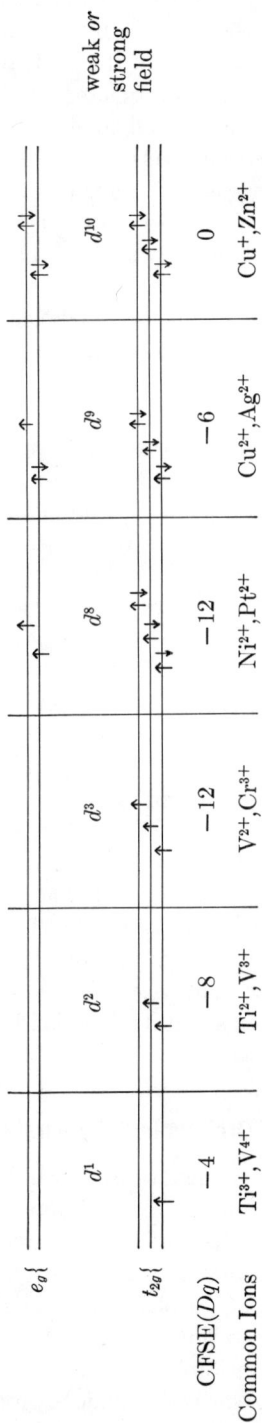

| | $d^1$ | $d^2$ | $d^3$ | $d^8$ | $d^9$ | $d^{10}$ | |
|---|---|---|---|---|---|---|---|
| $e_g$ | | | | | | | weak or strong field |
| $t_{2g}$ | | | | | | | |
| CFSE($Dq$) | $-4$ | $-8$ | $-12$ | $-12$ | $-6$ | $0$ | |
| Common Ions | $Ti^{3+}, V^{4+}$ | $Ti^{2+}, V^{3+}$ | $V^{2+}, Cr^{3+}$ | $Ni^{2+}, Pt^{2+}$ | $Cu^{2+}, Ag^{2+}$ | $Cu^+, Zn^{2+}$ | |

as far as possible, the lower-energy orbitals. The other is the tendency for electrons to enter different orbitals with their spins parallel in accord with Hund's first rule. The latter occurrence lowers the Coulombic repulsive energy among the electrons and at the same time allows a more favorable quantum mechanical exchange energy. If there are 1, 2, 3, 8, 9, or 10 $d$ electrons, there are no uncertainties as to where the electrons will go, regardless of the magnitude of $\Delta_o$ or of the total interelectronic repulsion energy. This may be seen on the preceding page.

On the other hand, for $d^{4-7}$ systems we have two extreme possibilities for each system depending now upon the relative magnitudes of the crystal field splitting energy, $\Delta_o$, and the mean pairing energy of the electrons. This leads to the necessary consideration of two limiting situations known as the *strong field* or *low-spin* case and the *weak field* or *high-spin* case, as shown below:

| | $d^4$ | $d^5$ | $d^6$ | $d^7$ | weak field |
|---|---|---|---|---|---|
| CFSE($Dq$) | $-6$ | $0$ | $-4$ | $-8$ | |
| Common ions | $Cr^{2+}, Mn^{3+}$ | $Mn^{2+}, Fe^{3+}$ | $Fe^{2+}, Co^{3+}$ | $Co^{2+}, Ni^{3+}$ | |

| | $d^4$ | $d^5$ | $d^6$ | $d^7$ | strong field |
|---|---|---|---|---|---|
| CFSE($Dq$) | $-16 + P$ | $-20 + 2P$ | $-24 + 2P$ | $-18 + P$ | |

Here $P$ is used to represent a *mean energy of pairing* of $d$ electrons per unit of $\Delta_o$ or 10 $Dq$. $P$ may be considered as being composed of two parts. $P_c$, the coulombic part, is not very different among the several $d^n$ systems of given principal quantum number, but it should decrease somewhat in the order $3d > 4d > 5d$, since the larger $d$ orbitals may accommodate an electron pair with less interelectron repulsion existing between the members of the pair. The other major contribution to $P$, $P_e$, is due to the loss of quantum mechanical exchange energy. Now notice that the differences in CFSE values between the strong field and the weak field, which is what must be considered in determining whether the high-spin or low-spin state will prevail, are $(-10 + P)$, $(-20 + 2P)$, $(-20 + 2P)$, and $(-10 + P)$ for $d^4$, $d^5$, $d^6$, and $d^7$, respectively. Thus, we see that in order for spin-pairing to occur, that is, the low-spin state to prevail, the value of $(10\,Dq - P)$ or $(\Delta_o - P)$ should be greater than zero.

Values of $P$ may be calculated from interelectron repulsion parameters determined for gaseous free ions. As we shall see later these values are probably 10% to 30% too high for use in connection with complexes of the ions. Nevertheless, we have listed in Table 10–8 some ions whose $P$ values

**TABLE 10–8.**    Octahedral Splittings, $\Delta_o$, for Some Aqueous Ions, and Mean Pairing Energies, $P$, for the Free Metal Ions.[a]

| $d^n$ | Ion | $\Delta_o$ | $P$ | Ion | $\Delta_o$ | $P$ |
|-------|-----|-----------|------|-----|-----------|------|
| $d^4$ | $Cr^{2+}$ | 13,900 | 23,500 | $Mn^{3+}$ | 21,000 | 28,000 |
| $d^5$ | $Mn^{2+}$ | 7,800 | 25,500 | $Fe^{3+}$ | 13,700 | 30,000 |
| $d^6$ | $Fe^{2+}$ | 10,400 | 17,600 | $Co^{3+}$ | 13,000[c] | 21,000 |
|       |     | 33,000[b] |      |     | 23,000[d] |      |
| $d^7$ | $Co^{2+}$ | 9,300 | 22,500 |     |           |      |

[a] The values of $P$ may be 10 to 30% too high since they have been calculated from interelectronic repulsion parameters for the free gaseous ions.
[b] Value for $Fe(CN)_6^{4-}$.
[c] Value for $CoF_6^{3-}$.
[d] Value for $Co(NH_3)_6^{3+}$.

are known and whose $\Delta_o$ values have also been determined so that direct comparison of these values is possible. In every case in which $P > \Delta_o$ spin-free or high-spin complexes are found and where $\Delta_o > P$ spin-paired or low-spin complexes are found, as predicted.

### The Crystal Field Splitting Parameter, $\Delta$

As we have already indicated, the magnitude of $\Delta_o$ for a given complex will depend largely upon the magnitude of the electrostatic field presented by the set of ligands. The ligand properties which influence this strength will include the classical factors such as size, charge, permanent dipole moment, $\mu_o$, and polarizability, $\alpha$. The latter controls the induced dipole moment, $\mu_i$, since $\mu_i = E\alpha$, where $E$ is the polarizing field, which for the ligands is that which is set up by the central atom. The total dipole moment is given as $\mu = \mu_o + \mu_i$. Of course, the $\sigma$-bonding strength, as well as the possible supplemental $\pi$-bonding strength, of the ligands will affect $\Delta_o$, but these factors cannot be taken account of in CFT. When we discuss the MOT we shall see the effects of these factors upon $\Delta_o$.

$\Delta_o$ has been obtained by tedious and very difficult calculations using the complete Hamiltonian and solving the requisite secular equations. However,

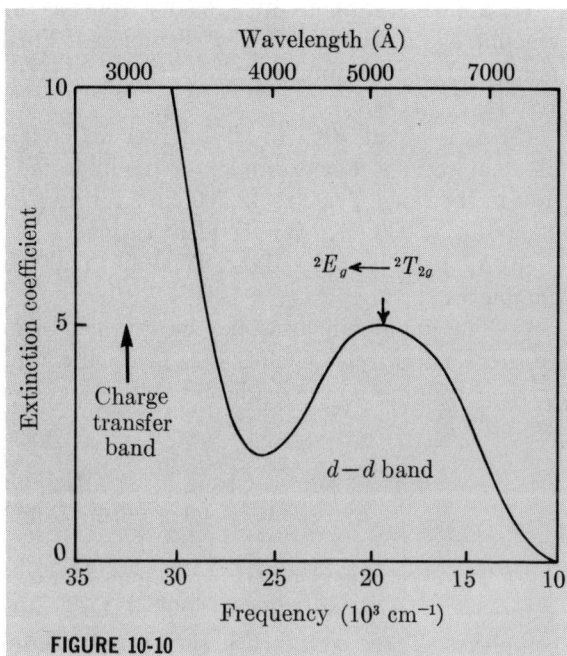

**FIGURE 10-10**

A Portion of the Optical Spectrum of the Ion $Ti(H_2O)_6^{3+}$. Maximum of the broad visible band shown is at 20,400 cm$^{-1}$. The asymmetry of the band is due to Jahn-Teller splitting as discussed later in the text.

$\Delta_o$ is usually treated as a semiempirical parameter and hence obtained from experimental data, most often from spectra, but also from magnetic and thermodynamic data. Although we shall have much more to say about electronic spectra later, we shall illustrate here, using the simplest example, namely, the optical spectrum of $Ti(H_2O)_6^{3+}$, how $\Delta_o$ is obtained from spectral data. The visible spectrum of this species is shown in Figure 10–10. The energy of the maximum of this broad visible absorption band, 20,400 cm$^{-1}$, is interpreted as being just the energy required to raise the lone $3d$ electron from the $t_{2g}$ to the $e_g$ level. If this is the case, then $\Delta_o = 20,400$ cm$^{-1}$ and $Dq = 2040$ cm$^{-1}$. Since $-4\ Dq$ is the CFSE for a $d^1$ system, then CFSE $= -4 \times 2040 = -8160$ cm$^{-1}$, which is roughly $-23$ kcal/mole. The asymmetry which may be seen in the absorption band is believed to arise from an important effect known as the *Jahn-Teller splitting*, which will be considered in some detail later.

On the basis of a large mass of experimental data, we may make a few generalizations concerning the magnitude of $\Delta_o$. These are most valid for high-spin complexes and especially those of metal ions in their normal oxidation states.

(1) $\Delta_o$ increases about 30% to 50% from $3d^n$ to $4d^n$ and by about the same amount again from $4d^n$ to $5d^n$ complexes. For example, consider the T9 family trivalent ammines: $Co(NH_3)_6^{3+}$, 23,000; $Rh(NH_3)_6^{3+}$, 34,000; and $Ir(NH_3)_6^{3+}$, 41,000.

(2) $\Delta_o$ is about 40% to 80% larger for complexes of trivalent than for divalent cations. For example, for the hydrated cations in the $3d^n$ series it falls in the range 7,500 to 12,500 $cm^{-1}$ for divalent cations and in the range 13,500 to 21,000 $cm^{-1}$ for trivalent cations.

(3) $\Delta_o$ varies between about 8,000 and 14,000 $cm^{-1}$ for most divalent $3d^n$ complexes.

(4) The common ligands may be arranged in a regular order known as the *spectrochemical series*, such that $\Delta_o$ for their complexes with most metal ions, in their common lower oxidation states, increases along the sequence. For example:

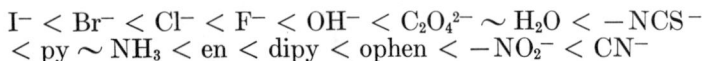

$$I^- < Br^- < Cl^- < F^- < OH^- < C_2O_4^{2-} \sim H_2O < -NCS^-$$
$$< py \sim NH_3 < en < dipy < ophen < -NO_2^- < CN^-$$

(5) The crystal field splitting parameter for tetrahedral complexes, $\Delta_t$, has values which are about 40% to 50% of the $\Delta_o$ value for similarly constituted complexes. These values are surprisingly close to the theoretical value, which from pure electrostatic CFT is $\Delta_t = -4/9\ \Delta_o$.

It might be appropriate at this point to emphasize that while CFSE's may be important in explaining and predicting *differences* in energies existing between various ions in a series, they actually comprise only about 2% to 10% of the *total* binding energy of a given complex. But CFSE's, while not the main source of binding energy in complexes, do seem to be critical for certain thermodynamic and kinetic changes. We shall support this statement later. And certainly it is already possible for the reader to see the improvement of CFT over VBT in the treatment of certain limited spectral data, namely that arising from essentially $d$-$d$ transitions. Although we shall discuss magnetic properties also in detail later, it is useful at this point to show at least that CFT predicts a gradual change in magnetism rather than the abrupt change predicted by VBT. This may be seen in Figure 10–11, in which we picture between $a$ and $b$ the effect upon the $d$ orbitals of gradually increasing the electrostatic field in an octahedral case. With such a model it is understandable that for certain complexes in which the $\Delta_o$ value is very close to the actual value of $P$ for the metal atom in the complex, simple temperature changes may effect magnetic changes. This has indeed been observed experimentally. Also shown in Figure 10–11, between $b$ and $c$, is the further splitting up of the $d$ orbitals as an octahedral disposition of ligands becomes increasingly tetragonally distorted, that is, as two *trans* ligands are moved progressively further away and the remaining four equatorial ligands are moved closer to the central atom. This brings us to the subject of ligand fields having other symmetries than $O_h$.

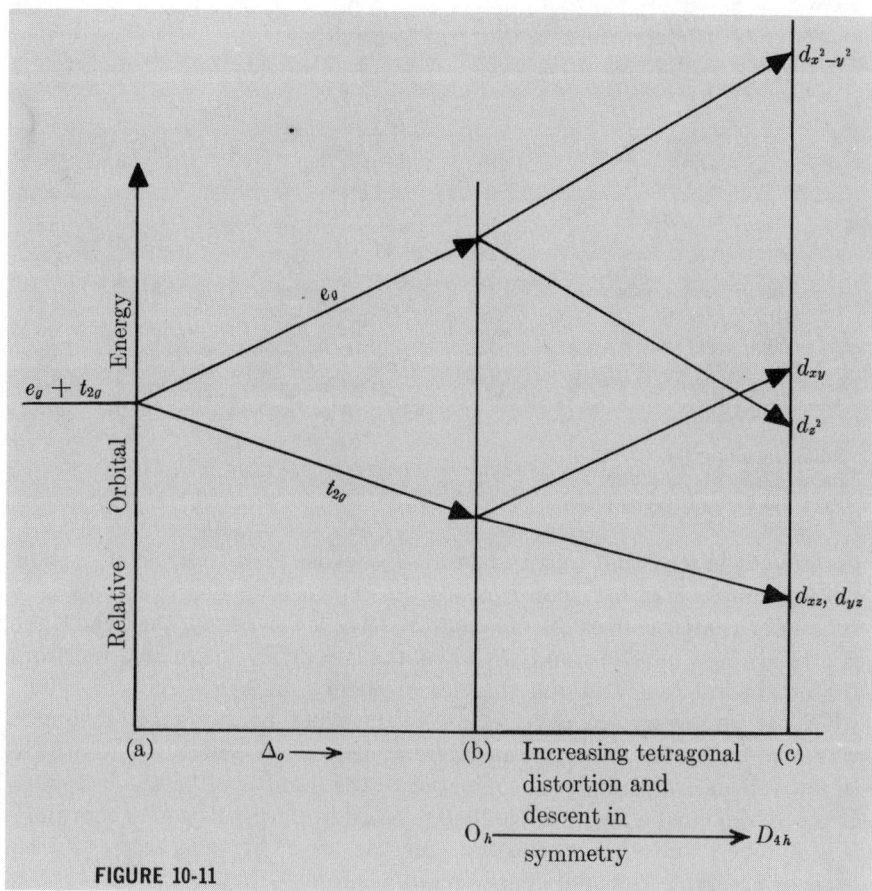

**FIGURE 10-11**

The Effect of Increasing the Electrostatic Field Strength, $\Delta_o$, upon the $t_{2g}$-$e_g$ Splitting in $O_h$ Symmetry, Shown between (a) and (b). The effect of an increasing tetragonal distortion of the octahedral ligand array, that is, moving two *trans* ligands further out and the four equatorial ligands further in, shown between (b) and (c).

## Fields Other Than Octahedral

Until now we have considered only octahedral complexes. When other geometries for the electrostatic field are considered we obtain the relative crystal field $d$ orbital splittings shown in Figure 10-12. In Table 10-9 we have listed the calculated *one-electron* $d$-orbital energies for various crystal fields of importance. The very important effects of interelectron repulsion have been ignored in the calculation of these numbers, so they cannot be assumed to be valid for real multielectron systems, and they must be used with care for detailed comparison purposes. They are useful as rough guidelines to the relative effects of stereochemistry on $d$-orbital splittings. For

## TABLE 10–9. The Calculated One-electron $d$-Orbital Energy Levels in Crystal Fields of Different Symmetries.

| CN | Structure | $d_\gamma$ | | $d_\epsilon$ | | |
|---|---|---|---|---|---|---|
| | | $d_{x^2-y^2}$ | $d_{z^2}$ | $d_{xy}$ | $d_{xz}$ | $d_{yz}$ |
| 2 | linear[a] | −6.28 | 10.28 | −6.28 | 1.14 | 1.14 |
| 3 | trigonal[b] | 5.46 | −3.21 | 5.46 | −3.86 | −3.86 |
| 4 | tetrahedral | −2.67 | −2.67 | 1.78 | 1.78 | 1.78 |
| 4 | square planar[b] | 12.28 | −4.28 | 2.28 | −5.14 | −5.14 |
| 5 | trigonal bipyramid[c] | −0.82 | 7.07 | −0.82 | −2.72 | −2.72 |
| 5 | square pyramid[c] | 9.14 | 0.86 | −0.86 | −4.57 | −4.57 |
| 6 | octahedron | 6.00 | 6.00 | −4.00 | −4.00 | −4.00 |

[a] Bonds lie along the $z$ axis.
[b] Bonds lie in the $xy$ plane.
[c] Pyramid base in the $xy$ plane.

example, it is seen that in tetrahedral, $T_d$, fields, the two $d_\gamma$ orbitals, now labeled simply $e$ rather than $e_g$ since the tetrahedron does not possess a symmetry center as does the octahedron, have lower energy than the three $d_\epsilon$ orbitals, now labeled simply $t_2$. Thus, the energy levels are inverted from the octahedral case and the theoretical energy difference $\Delta_t$ is $-4/9\Delta_o$, which, as we saw earlier, is very close to the experimental facts. Of course, we can also understand from this difference between $\Delta_t$ and $\Delta_o$ why there are so few tetrahedral complexes compared to the number of octahedral ones. Later we shall make use of Table 10–9 to assist in the prediction of coordination numbers and stereochemistries for various $d^n$ systems solely on the basis of favorable $d$-orbital energy relationships.

## Uses and Limitations of CFT

Perhaps the most notable aspect of the CFT is that despite its obvious artificiality, it is still useful and relatively simple for supplying a theoretical basis for quantitative correlations between empirical and semiempirical numbers. In this respect it stands in marked contrast to the VBT. It can supply a theoretical basis for understanding and predicting the variation of magnetic moments with temperature as well as the detailed magnetic properties of most complexes. This also stands in contrast to VBT, which cannot predict or explain magnetic behavior beyond the level of specifying the number of unpaired electrons. CFT also makes possible clearer understanding of stereochemical properties and, as we shall see later, allows a more detailed approach to this subject than does VBT. We shall see later that certain thermodynamic as well as certain kinetics problems can be handled by CFT, if only semiquantitatively. Most importantly, perhaps, CFT can treat certain limited spectral problems; for example, intra-$d$-level

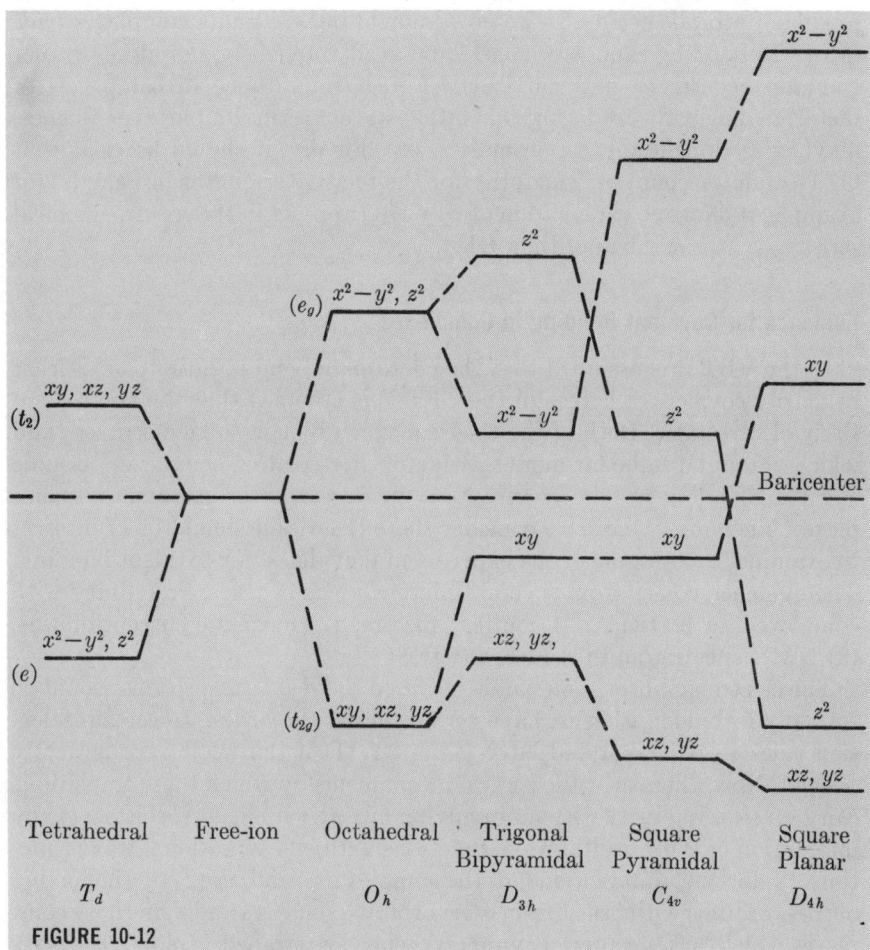

**FIGURE 10-12**

Relative One-electron $d$-orbital Splittings for Crystal Fields of Several Different Symmetries.

transitions, which necessarily involve excited energy states, hence depend upon the $d$-orbital splittings. Again, this is beyond the scope of VBT as it is usually applied to complexes. When we come to the discussions of the applications of the several bonding theories we shall demonstrate and illustrate some of the foregoing uses of CFT.

What are the shortcomings and limitations of the CFT? Perhaps most important is the total emphasis it places upon the metal orbitals without giving any consideration to the ligand orbitals. Therefore, all properties dependent upon the ligand orbitals and their alteration or interaction with the metal orbitals will necessarily be outside the scope of CFT. In particular, charge transfer bands cannot be treated at all. And treatment of $d$-$d$ transitions, particularly their intensities, is not possible when even moderate mixing of orbitals occurs, as we now know it does in many complexes (*vide infra*). Pi bonding cannot be considered at all, obviously, despite its rather common occurrence and particularly great importance in complexes of metals in unusually low or high oxidation states and in complexes of alkenes, alkynes, cyclopentadienes, aromatics, etc. Finally, it should be clear that CFT cannot account satisfactorily for the relative strengths of ligands; for example, it offers no explanation of why $H_2O$ appears in the spectrochemical series as a stronger ligand than $OH^-$.

## Evidence for Covalent Bonding in Complexes

In the VBT it is assumed that the coordinate bond is entirely or at least primarily covalent, whereas in the CFT it is assumed that the bond is entirely electrostatic. Both of the theories which remain to be discussed, and which are found to be far more satisfactory for treating complexes, assume covalent bonding to be occurring to a greater or lesser extent in all complexes. Therefore, before we consider these theories it should be of interest to examine briefly some of the experimental evidence for covalent bonding. The evidence comes from several different areas of research activity with complexes; in particular (1) optical spectra, (2) magnetic susceptibilities, (3) NMR spectra, and (4) ESR spectra.

The electronic absorption bands ascribed to "$d{\rightarrow}d$" transitions would be formally forbidden, that is, have zero intensity, according to certain selection rules to be elaborated later (Chapter 11) if the CFT were perfectly obeyed. Now, there are at least two mechanisms by which these transitions can gain *some* intensity without involving mixing with ligand orbitals; (1) by interaction of the $d$ orbital wave functions with odd vibrational wave functions ("vibronic" interactions) of the complex ion, and/or (2) by the mixing of the $d$ orbitals with other *metal atom* orbitals, such as $s$ and $p$, in those complexes which lack a center of symmetry, such as tetrahedral ones. However, there are many examples of complex ions for which it is reasonably certain that neither of these two mechanisms can account for the observed spectral intensities. By the assumption of covalent bonding, that is, overlap and hence mixing of metal atom $d$ orbitals with various *ligand atom* orbitals, it becomes quite easy to account for observed intensities.

Spectroscopic data for complexes are often compared with calculated energy level diagrams so that electronic transitions may be appropriately assigned. Calculations of this type, assuming a CFT model, use the same interelectron repulsion parameters for the complexed metal ion as are known for the free gaseous ion. The repulsion parameters, as we shall see later, measure separations between the various Russell-Saunders states, so that in effect these separations are assumed unchanged in the CFT from free ion to complexed ion. However, fittings of experimental and calculated data are usually not good and in fact are often very poor. The fit can *always* be improved by assuming that the separations of the Russell-Saunders states are smaller in the complexed ion than in the free ion, that is, that the inter-electronic repulsions operative between the $d$ electrons are decreased. If the latter has occurred, it suggests that the mean distance between $d$ electrons has increased, and this can logically be attributed to an increase in the size of the several $d$ orbitals. The presently accepted explanation for this swelling of the $d$ electron clouds, or the so-called *nephelauxetic* (from the Greek and meaning "cloud expanding") *effect*, is that at least part of it is due to the overlap of the metal atom $d$ orbitals with ligand atom orbitals. This thereby provides a mechanism for reducing the effective nuclear charge on the metal atom and for partial removal of $d$ electrons from the central atom. Indeed a ligand series, more or less independent of the metal atom, as is the case with the spectrochemical series, may be set up and it has been called the *nephel-auxetic series*. For example, such a series of the order of increasing ability to produce $d$ cloud expansion is, in part, $F^- < H_2O < NH_3 < C_2O_4^{2-} \sim CN^- < Br^- < I^-$. In one sense, then, this series measures a tendency towards increasing covalency in the metal-ligand bond.

Further evidence of the existence of covalent bonding may be deduced from the *antiferromagnetic* behavior of certain "ionic" compounds, most notably the oxides, VO, MnO, FeO, CoO, and NiO. This form of magnetic behavior is ascribed to a substance which follows the Curie or Curie-Weiss law at high temperatures but which below a certain temperature, the *Neel point*, exhibits a *decreasing* rather than increasing magnetic susceptibility as the temperature is lowered further. Neutron diffraction studies have demonstrated that this effect is due *not* to the pairing of electrons within the individual ions, but rather to a tendency of half of the ions to have their magnetic moments aligned antiparallel to those of the other half of the ions. This phenomenon is pictured as occurring in the following manner. Consider the system $M^{2+}$—$O^{2-}$—$M^{2+}$. If each metal ion possesses an unpaired elec-tron in a $d$ orbital that can overlap a filled $\pi$ orbital on the oxide ion, then it is possible for an electron from the oxygen to move so as to partially occupy that $d$ orbital. Since the spin of the transferred electron must be antiparallel to that of the original $d$ electron, the other oxide $\pi$ electron must have its spin parallel to that of the $d$ electron. Now if that other oxide $\pi$ electron moves partially into the $d$ orbital of the second metal ion it will force the lone $d$ electron therein to have its spin antiparallel to that of the $d$ electron on the first metal ion. Even if the $\pi$ orbital overlaps are not large, they are almost certainly occurring. If they lead to a lower energy state of the system, it is then understandable why the metal oxide lattice will go over into this state as the temperature is decreased.

Nuclear magnetic resonance studies of complexes have also contributed to the substantial amount of evidence that now exists in support of covalent metal-ligand bonds. For example, it is found that the resonance frequency of the ring protons, $H_\alpha$, in *tris*(acetylacetonato)vanadium(III) is sub-stantially shifted from its position in an analogous diamagnetic complex, such as the Al(III) compound (Figure 10–13(a)). To account for the large

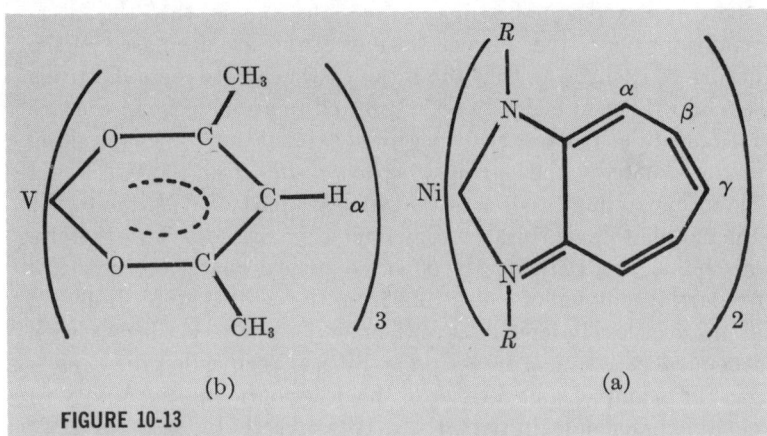

**FIGURE 10-13**

(a) *Tris*(acetylacetonato)vanadium(III), Showing the "Ring Proton," $H_\alpha$, Which Has Its Nuclear Resonance Line Strongly Shifted Presumably by Electron Spin-nuclear Spin Coupling with the $d$ Electrons of V(III). (b) *Bis*(N,N'-dialkylaminotroponeiminato)nickel(II), Showing the Three Different Proton Positions, $\alpha,\beta,\gamma$, for Which Large Nuclear Resonance Shifts Caused by Ni(II) $d$ Electrons Have Allowed the Estimation of Nonzero Spin Densities at These Positions.

magnitude of the shift, it must be deduced that some unpaired electron spin density, restricted by the CFT to the $t_{2g}$ orbitals of V(III), must actually enter into the $\pi$ system of the ligand so as to find its way into the $1s$ orbitals of the hydrogen atoms. Analogous, but much more extensive recent studies of aminotroponeiminate complexes of Ni(II) (Figure 10–13(b)) have revealed the transfer of unpaired electron spin density from nickel atoms into the ligand $\pi$ system. Indeed the various large shifts in the positions of the different ($\alpha$, $\beta$, $\gamma$) proton nuclear magnetic resonances has permitted determination of the spin density residing on each different carbon atom. For example, in the analogue in which the $R$ groups are each $C_2H_5$ (that is, the N,N'-diethylaminotroponeiminate system), a calculation which assumes the transfer of 0.10 electrons from Ni(II) to the ligands yields spin densities at the $\alpha$, $\beta$, and $\gamma$ carbon atoms of $+0.038$, $-0.023$, and $+0.057$, respectively, whereas the experimental spin densities are $+0.041$, $-0.021$, and $+0.057$, respectively. Finally we should mention that even in $MF_6^{2-}$ complexes (where, for example, $M$ = Mn, Fe, Co), in which we might surely expect as high a degree of ionic bonding as can be found, fluorine NMR spectra have been interpreted to show that delocalization of the $d$-electron spin density takes place to at least an extent of 2% to 5%.

Perhaps the most direct evidence we have of covalent bonding via metal-ligand orbital overlap comes from electron spin resonance studies. In 1953, Owen and Stevens found that the ESR spectrum of $IrCl_6^{2-}$ (actually 0.5% in solid $Na_2PtCl_6 \cdot 6H_2O$) shows a complex hyperfine structure which could only be interpreted as originating from the overlap of Ir atom $d$ orbitals with certain orbitals of the chloride ions to such an extent that the single unpaired $d$ electron is not localized solely on the Ir atom. The hyperfine splitting arises from the interaction of the unpaired electron spin with the *nuclear spin* of the chlorine nucleus, and the magnitude of the splittings is proportional to the fraction of electronic charge located on the chloride ions. This turns out to be about 5% on each chloride ion, leaving about 70% of the electronic charge remaining on the iridium atom.

Numerous other ESR results have been obtained since 1953 supporting the importance of overlapping of orbitals in metal-ligand bonding. These include examples from the $d^1$ systems, such as $VO(aca)_2$ ($aca$ = acetylacetonate ion) and $Mo(CN)_6{}^{3-}$, to the $d^9$ systems of the $Cu^{2+}$-salicylaldimine complex and similar Cu(II) and Ag(II) complexes.

Finally, in this section we should mention that other modern experimental techniques such as nuclear quadrupole resonance and Mössbauer spectroscopy have also been of value in firmly establishing the presence of orbital overlap. All of the techniques mentioned in this section are very adequately introduced in the excellent book by Drago,[34] and there is a particularly valuable recent article by Greenwood[35] covering the many chemical aspects of Mössbauer spectroscopy.

## MOLECULAR ORBITAL THEORY

We shall now consider the qualitative aspects of the application of molecular orbital theory to complexes. The discussion will really include also the ligand field theory since this is perhaps best pictured as a theory, or even a series of theories, which combines the convenience and basic simplicity of CFT with the rigorousness and generality of MOT in whatever relative amounts seem to be necessary for the particular complex molecule or problem at hand. What is sometimes referred to as LFT is a modification of the CFT that allows certain parameters to be empirically altered to take account of covalence effects without explicitly introducing orbital overlap. This approach has been aptly termed *adjusted crystal field theory*, ACFT, by Cotton and Wilkinson[10] and we shall examine it briefly later.

As we have seen in Chapter 5, the molecular orbital theory can accommodate at one extreme the completely electrostatic situation which involves no orbital overlap to the maximum overlap at the other extreme, as well as all intermediate degrees of overlap. Of all the theoretical approaches to bonding this, then, is the most sophisticated and general and correspondingly the most difficult. Many approximations must be made in its quantitative application to the many-atom multielectron complex ion systems. Its usefulness in this area is very recent and is certain to increase in the future as better wave functions and computational methods become available.

The molecular orbital method employs the same central atom orbitals as does the VBT, but additionally it considers the available orbitals of the coordinated ligand atoms. Thus, excluding initially the $\pi$-bonding ligand orbitals, for the case of six ligands around a central transition metal atom there will be a total of fifteen orbitals available for molecular orbital construction. These arise from nine orbitals on the metal atom and six orbitals from the ligands. For a particular complex, it is first necessary to establish which orbital overlaps are possible. Simply because of the inherent symmetry properties of the orbitals in the problem at hand some cannot overlap. It is only meaningful and proper to mathematically combine, for example by the LCAO method, orbitals that possess the same symmetries. In Table 10–10, we have listed according to symmetry class the metal atom orbitals for a $3d$ transition metal atom and the composite "symmetry" orbitals of the ligands for the regular octahedral case. The individual ligand orbitals are identified by the appropriate cartesian coordinate subscript designation from the following figure:

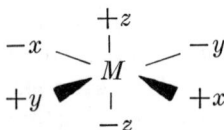

$$-x \underset{+y}{\overset{+z}{\underset{\big|}{\overset{\big|}{\diagdown}}}} M \overset{-y}{\underset{+x}{\diagdown}}$$
$$-z$$

Thus, the nine significant metal atomic orbitals are designated as $\Phi_{3d_{xy}}, \Phi_{3d_{xz}}, \ldots \Phi_{4s}, \ldots \Phi_{4p_z}$. These nine orbitals fall into one of four symmetry classes, labeled according to their group theoretical origin: nondegenerate totally symmetric $A_{1g}$, a single orbital having the full symmetry of the molecule; doubly degenerate $E_g$, two orbitals equivalent except for spatial orientation; triply degenerate $T_{1u}$, three orbitals equivalent except for spatial orientation; and triply degenerate $T_{2g}$, three orbitals equivalent except for spatial orientation. The three $T_{2g}$ orbitals are spatially oriented so as to be suitable only for $\pi$ bonding in the octahedral system, that is, there are no ligand $\sigma$ orbitals formed from orbitals having the $T_{2g}$ symmetry, but the remaining six metal atom orbitals are all suitable for $\sigma$-bonding purposes. The subscripts $g$ (from the German *gerade* meaning *even*) and $u$ (from the German *ungerade* meaning *uneven*) are used to indicate whether the orbital is centrosymmetric or noncentrosymmetric, respectively.

The six ligand $\sigma$ orbitals must first be combined so as to form a set of six composite "symmetry" orbitals, each constructed to effectively overlap with a particular one of the six $\sigma$-bonding metal atom orbitals. These are sometimes designated as *ligand group orbitals, LGO's*, and methods for obtaining them may be found in two recent articles by Kettle.[36] The ligand composite $\sigma$ orbitals are designated $\Sigma_a, \Sigma_{z^2} \ldots \Sigma_z$, and these fall into the symmetry classes $A_{1g}$, $E_g$, and $T_{1u}$. (See Figure 10–14.) Then each metal atom orbital is combined with its matching symmetry ligand orbital by the LCAO method to yield a bonding and an antibonding MO pair of orbitals. (See Table 10–11 and Figure 10–15.) It is seen that the $T_{2g}$ central atom orbitals remain nonbonding since they are not matched in symmetry by any composite ligand $\sigma$ orbitals. However, we shall see shortly that it is just

TABLE 10-10. Symmetry Classification of Orbitals for Regular Octahedral Complexes.

| Symmetry Class | Metal Atomic Orbitals | Composite of Ligand $\sigma$ Orbitals | Composite of Ligand $\pi$ Orbitals |
|---|---|---|---|
| $A_{1g}$ | $\Phi_{4s}$ | $\Sigma_a = \dfrac{1}{\sqrt{6}}\left(\sigma_x + \sigma_{-x} + \sigma_y + \sigma_{-y} + \sigma_z + \sigma_{-z}\right)$ | — |
| $E_g$ | $\Phi_{3d_{z^2}}$ | $\Sigma_{z^2} = \dfrac{1}{2\sqrt{3}}\left(2\sigma_z + 2\sigma_{-z} - \sigma_x - \sigma_{-x} - \sigma_y - \sigma_{-y}\right)$ | — |
| | $\Phi_{3d_{x^2-y^2}}$ | $\Sigma_{x^2-y^2} = \dfrac{1}{\sqrt{2}}\left(\sigma_x + \sigma_{-x} - \sigma_y - \sigma_{-y}\right)$ | — |
| $T_{1u}$ | $\Phi_{4p_x}$ | $\Sigma_x = \dfrac{1}{\sqrt{2}}\left(\sigma_x - \sigma_{-x}\right)$ | $\pi_x = \dfrac{1}{\sqrt{2}}\left(\pi_{x,y} + \pi_{x,-y} + \pi_{x,z} + \pi_{x,-z}\right)$ |
| | $\Phi_{4p_y}$ | $\Sigma_y = \dfrac{1}{\sqrt{2}}\left(\sigma_y - \sigma_{-y}\right)$ | $\pi_y = \dfrac{1}{\sqrt{2}}\left(\pi_{y,x} + \pi_{y,-x} + \pi_{y,z} + \pi_{y,-z}\right)$ |
| | $\Phi_{4p_z}$ | $\Sigma_z = \dfrac{1}{\sqrt{2}}\left(\sigma_z - \sigma_{-z}\right)$ | $\pi_z = \dfrac{1}{\sqrt{2}}\left(\pi_{z,x} + \pi_{z,-x} + \pi_{z,y} + \pi_{z,-y}\right)$ |
| $T_{2g}$ | $\Phi_{3d_{xy}}$ | — | $\pi_{xy} = \dfrac{1}{\sqrt{2}}\left[\pi_{y,x} - \pi_{y,-x} + \pi_{x,y} - \pi_{x,-y}\right]$ |
| | $\Phi_{3d_{xz}}$ | — | $\pi_{xz} = \dfrac{1}{\sqrt{2}}\left(\pi_{z,x} - \pi_{z,-x} + \pi_{x,z} - \pi_{x,-z}\right)$ |
| | $\Phi_{3d_{yz}}$ | — | $\pi_{yz} = \dfrac{1}{\sqrt{2}}\left(\pi_{z,y} - \pi_{z,-y} + \pi_{y,z} - \pi_{y,-z}\right)$ |

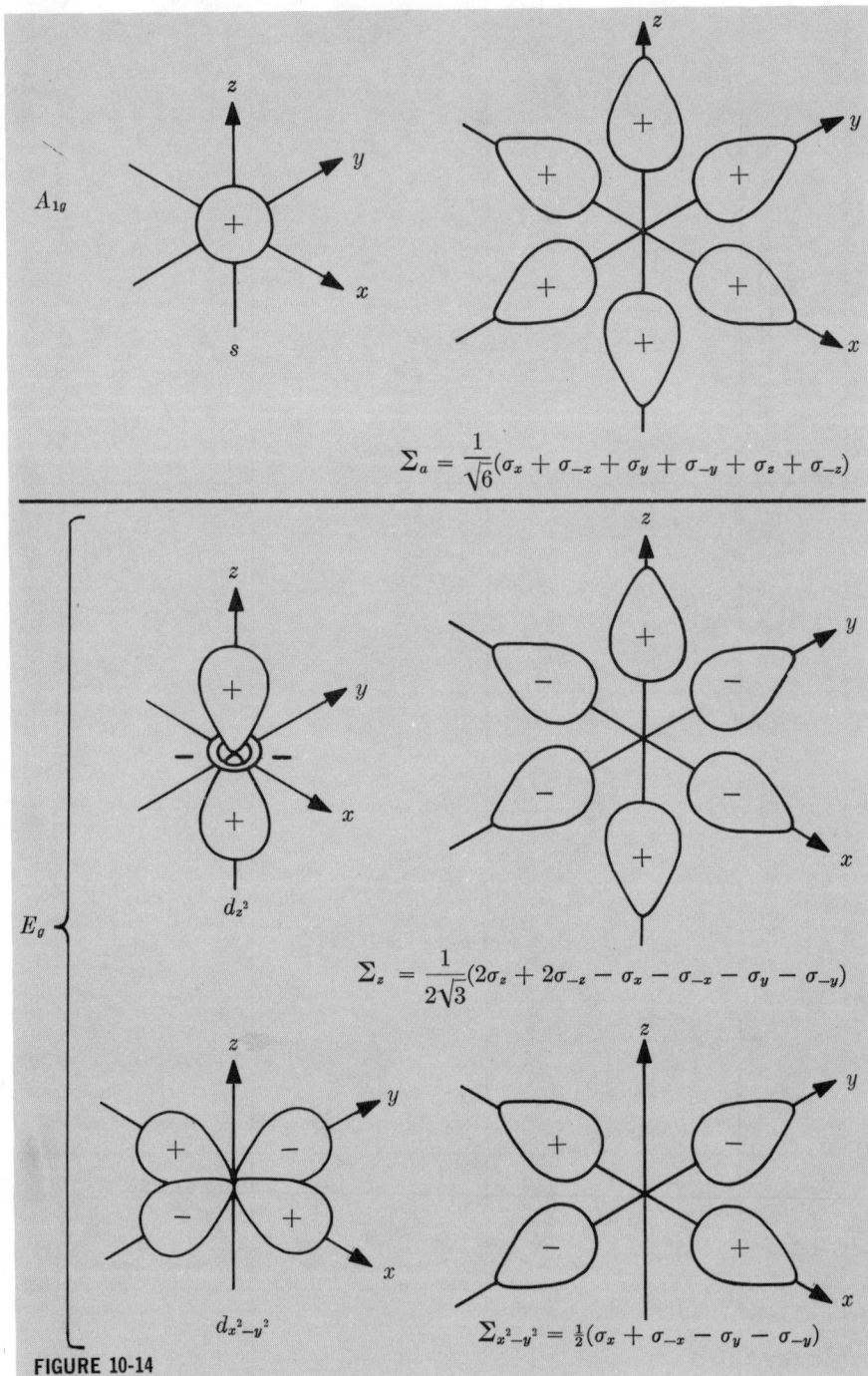

$$\Sigma_a = \frac{1}{\sqrt{6}}(\sigma_x + \sigma_{-x} + \sigma_y + \sigma_{-y} + \sigma_z + \sigma_{-z})$$

$$\Sigma_z = \frac{1}{2\sqrt{3}}(2\sigma_z + 2\sigma_{-z} - \sigma_x - \sigma_{-x} - \sigma_y - \sigma_{-y})$$

$$\Sigma_{x^2-y^2} = \tfrac{1}{2}(\sigma_x + \sigma_{-x} - \sigma_y - \sigma_{-y})$$

**FIGURE 10-14**

The Six Metal Atom $\sigma$ Orbitals, on the Left in Each Case, and Their Matching Ligand Composite Symmetry Orbitals, on the Right in Each Case.

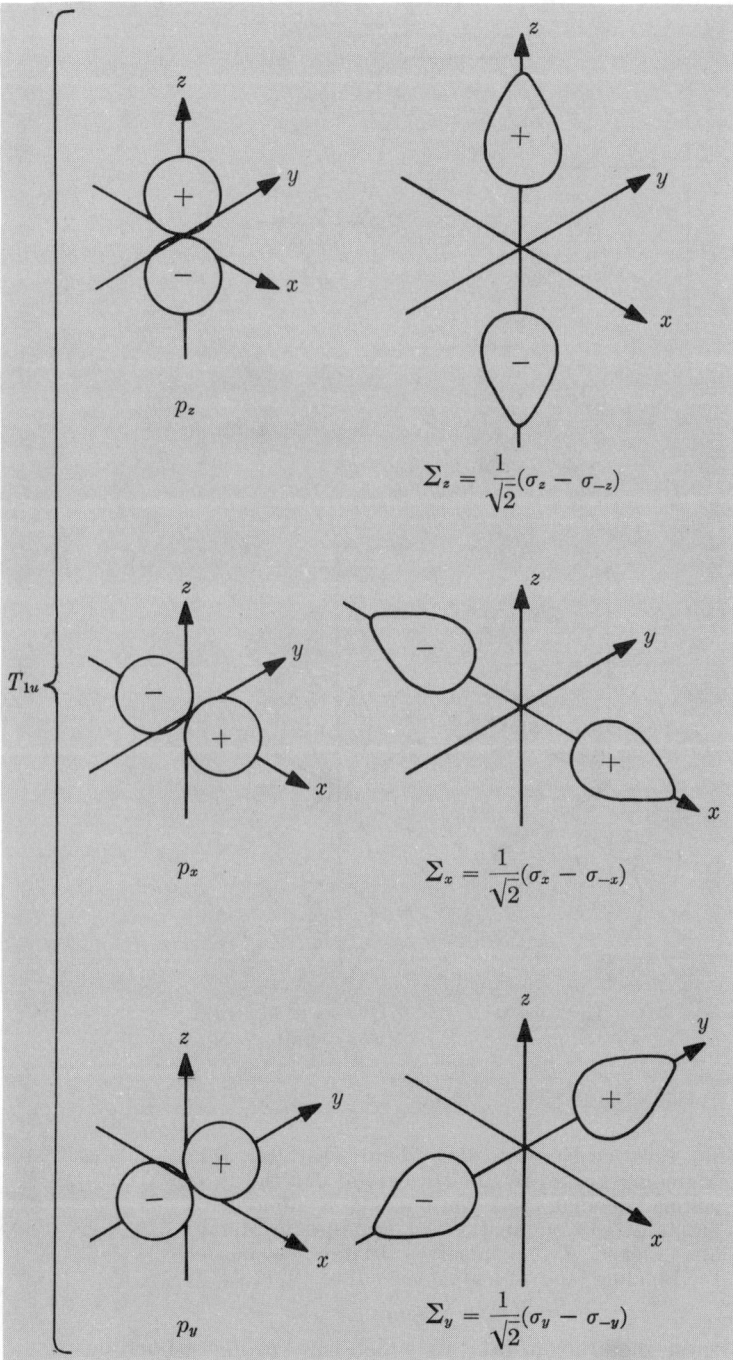

$$\Sigma_z = \frac{1}{\sqrt{2}}(\sigma_z - \sigma_{-z})$$

$$\Sigma_x = \frac{1}{\sqrt{2}}(\sigma_x - \sigma_{-x})$$

$$\Sigma_y = \frac{1}{\sqrt{2}}(\sigma_y - \sigma_{-y})$$

FIGURE 10-14 (Continued)

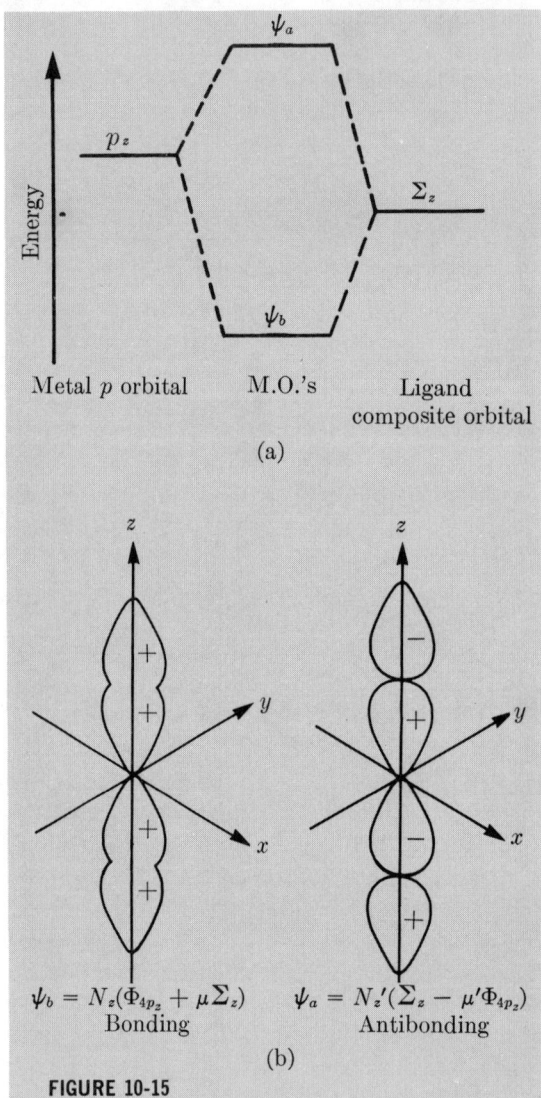

FIGURE 10-15

(a) Relative Energies of a Metal $p$ Orbital, Its Matching Ligand Symmetry Orbital and the Resulting Bonding and Antibonding $\sigma$ Molecular Orbitals. (b) Combinations of Appropriate Metal and Ligand $T_{1u}$ Symmetry Orbitals to Form the Bonding (above) and Antibonding $MO$'s.

these metal atom orbitals which can become $\pi$-bonding provided that the ligands possess matching symmetry $\pi$-bonding orbitals.

A molecular orbital (MO) energy level diagram, which results from a

**TABLE 10–11. Molecular Orbitals for Regular Octahedral Complexes Neglecting $\pi$ Bonding.**

| Symmetry Class | Bonding Orbitals | Nonbonding Orbitals | Antibonding Orbitals |
|---|---|---|---|
| $A_{1g}$ | $N_a(\Phi_{4s} + \lambda\Sigma_a)$ | — | $N_a'(\Sigma_a - \lambda'\Phi_{4s})$ |
| $T_{1u}$ | $N_x(\Phi_{4p_x} + \mu\Sigma_x)$ $N_y(\Phi_{4p_y} + \mu\Sigma_y)$ $N_z(\Phi_{4p_z} + \mu\Sigma_z)$ | — — — | $N_x'(\Sigma_x - \mu'\Phi_{4p_x})$ $N_y'(\Sigma_y - \mu'\Phi_{4p_y})$ $N_z'(\Sigma_z - \mu'\Phi_{4p_z})$ |
| $E_g$ | $N_{x^2-y^2}(\Phi_{3d_{x^2-y^2}} + \nu\Sigma_{x^2-y^2})$ $N_{z^2}(\Phi_{3d_{z^2}} + \nu\Sigma_{z^2})$ | — — | $N'_{x^2-y^2}(\Sigma_{x^2-y^2} - \nu'\Phi_{3d_{x^2-y^2}})$ $N'_{z^2}(\Sigma_{z^2} - \nu'\Phi_{3d_{z^2}})$ |
| $T_{2g}$ | — — — | $\Phi_{3d_{xy}}$ $\Phi_{3d_{xz}}$ $\Phi_{3d_{yz}}$ | — — — |

mathematical treatment of the type outlined qualitatively above, for a hypothetical regular octahedral complex ignoring $\pi$ bonding, is shown in Figure 10–16. The exact ordering of the strongest, that is, lowest energy, bonding levels is uncertain due to the uncertainties in obtaining the necessary exchange integrals. In general it may be assumed that, to a first approximation, the energies of the bonding and antibonding $MO$'s lie equal energy distances below and above, respectively, the mean value of the energies of the combining orbitals. Furthermore, it may be assumed that if an $MO$ is much nearer to one of the $AO$'s or $LGO$'s used in its construction than to the other one, it will have much more the character of the nearer one. Thus, the six $\sigma$-bonding $MO$'s are considered to have more the character of ligand atom orbitals than metal atom orbitals, and we therefore consider electrons in these orbitals to be mainly "ligand electrons." Likewise, any electrons occupying any of the antibonding $MO$'s are considered to be predominately "metal electrons," and any electrons in the non-$\sigma$-bonding $T_{2g}$ orbitals will be purely metal electrons, providing, of course, that no ligand $\pi$ orbitals exist to overlap these $T_{2g}$ orbitals.

We see then from the MO diagram that the $T_{2g}$ and $E_g{}^*$ levels, both containing mainly metal atom orbitals, are split apart (qualitatively) in the same manner they were by the purely electrostatic arguments from the CFT. All that has changed in this limited portion of the energy diagram is that in the MOT the $E_g{}^*$ orbitals are not *pure* metal atom $d$ orbitals. Furthermore, in the CFT the splitting arises from only *electrostatic* and symmetry considerations whereas in the MOT the splitting arises from *covalent bonding* and symmetry considerations.

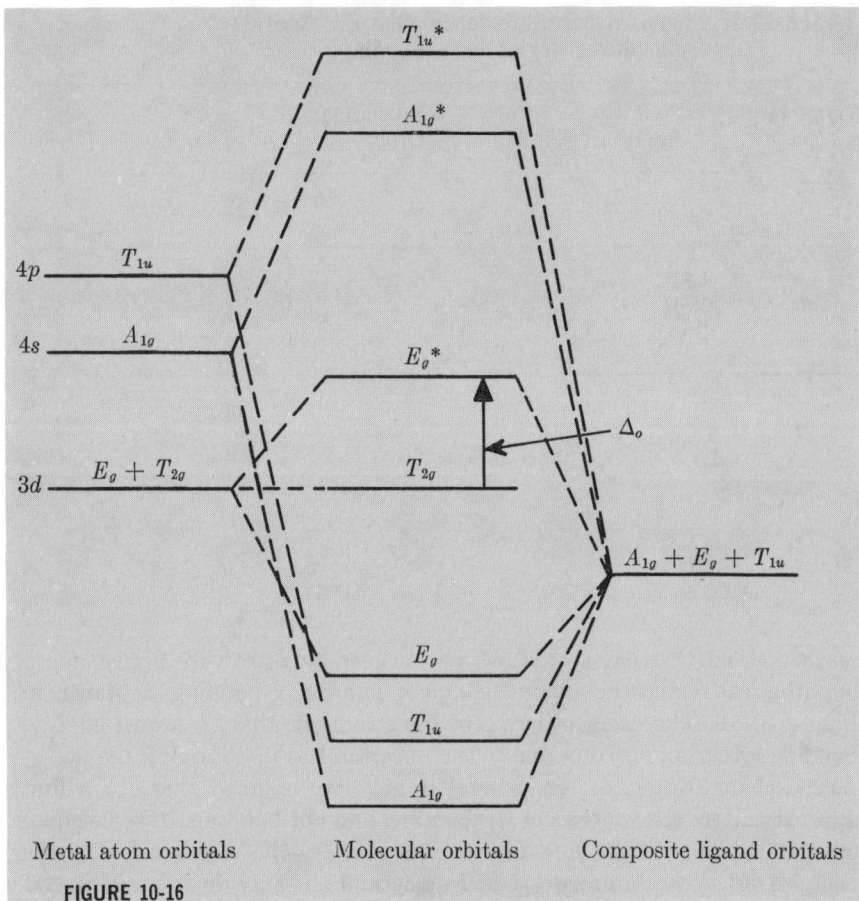

**FIGURE 10-16**

Energy Level Diagram for a Hypothetical Regular Octahedral Complex Formed between a $3d$ Metal Atom and Six Identical Ligands Which Do Not Possess $\pi$ Orbitals.

What is more important, however, is that the MOT energy level diagram contains much more information than the CFT one. For example, the anti-bonding levels above the $E_g^*$ set also represent terminal levels for electronic transitions originating in $T_{2g}$ (or below), and electrons in the bonding levels may be excited into higher levels yielding so-called *charge-transfer* bands. Neither of these phenomena can be dealt with by the CFT. And of course, the CFT also cannot consider $\pi$ bonding and all of its consequences.

### Inclusion of $\pi$ Bonding

Most ligand atoms possess $\pi$ orbitals which may be either filled or unfilled and which therefore may interact with the metal $T_{2g}$ $d$ orbitals, that is, the set $d_{xy}$, $d_{xz}$, $d_{yz}$. Still considering the octahedral case, we may consider each ligand to possess a pair of mutually perpendicular $\pi$ orbitals, giving rise to a total of 12 $\pi$ orbitals. It is found from group theory that these 12 orbitals may be combined into four triply degenerate sets belonging to the four symmetry classes $T_{1g}$, $T_{2g}$, $T_{1u}$, $T_{2u}$. Those orbitals in the $T_{1g}$ and $T_{2u}$ classes must remain rigorously nonbonding, at least in the metal-ligand bond area, for the simple reason that the metal atoms do not possess orbitals corresponding to these symmetries with which interaction might occur. The $T_{1u}$ ligand orbitals might interact with the $T_{1u}$ metal atom orbitals, that is, its $p$ AO's, but this interaction cannot be very important since these same metal atom orbitals are presumably already engaged in $\sigma$ bonding to the ligand atoms. The remaining $T_{2g}$ LGO's may, however, combine with the $T_{2g}$ AO's of the central atom, and if this happens $\pi$ bonding results.

We may conveniently classify the types of $\pi$ bonds according to the nature of the ligand $\pi$ orbitals: (1) simple $p\pi$ orbitals, always filled, as in $O^{2-}$, $RO^-$, $F^-$, $Cl^-$, $Br^-$, $I^-$, $RS^-$, etc., (2) simple $d\pi$ orbitals, always empty, as in phosphines, arsines, sulfides, etc., and (3) molecular $\pi$ orbitals of certain polyatomic ligands such as $NO_2^-$, $CN^-$, $CO$, py*, ophen*, dipy*, aca$^-$*, and unsaturated organic molecules.

It is relatively simple to picture the overlapping of metal atom $\pi$ orbitals

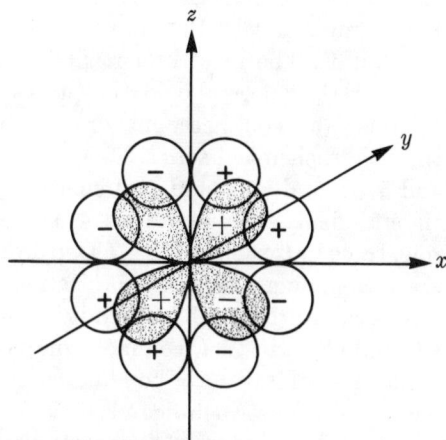

**FIGURE 10-17**

Overlap of a Metal $T_{2g}$ Symmetry Orbital, $d_{xz}$, with a Ligand Composite $T_{2g}$ Symmetry Orbital, $\pi_{xz} = \dfrac{1}{\sqrt{2}}(\pi_{z,x} - \pi_{z,-x} + \pi_{x,z} - \pi_{x,-z})$. The $d_{xy}$ and $d_{yz}$ metal orbitals will overlap the ligand composite orbitals $\pi_{xy}$ and $\pi_{yz}$, respectively, in an entirely analogous manner.

* Abbreviations for pyridine, o-phenanthroline, dipyridyl and acetylacetonate, respectively.

with ligand $p\pi$ or $d\pi$ orbitals. See, for example, Figure 10–7. However, it is not so easy to visualize the combination of molecular $\pi$ orbitals with the metal atom orbitals. But we can recognize two important cases that arise depending upon the energy of the ligand $\pi$ orbitals relative to the energy of the central atom $T_{2g}$ orbitals and whether the ligand orbitals are filled or empty. These cases will also apply to the simple $p\pi$ and $d\pi$ orbital ligands. (1) The ligand $\pi$ orbitals are unoccupied and less stable than the metal atom $\pi$ orbitals. In this case, the metal atom $T_{2g}$ orbitals are stabilized by the interaction, relative to the $E_g{}^*$ orbitals, and the value of $\Delta_o$ is increased (see Figure 10–18(a)). Phosphine and arsine ligands surely fall into this category, but so do many other ligands that have both empty higher-energy $\pi$ orbitals *and* filled lower-energy $\pi$ orbitals, such as $Cl^-$, $Br^-$, $I^-$, $CO$, $CN^-$, py, etc. (2) The ligand $\pi$ orbitals are filled and more stable, that is, of lower energy, than the metal atom $T_{2g}$ orbitals. In this case the latter orbitals are destabilized relative to the $E_g{}^*$ orbitals and hence the value of $\Delta_o$ is diminished (see Figure 10–18(b)). Oxide and fluoride ions are almost certainly in this category as well as many other ligands, such as the $\beta$-diketones, for which evidence is much more tenuous.

As we have already noted, there are in fact many ligands that possess both empty and filled $\pi$ orbitals, and it is not always easy to predict which will contribute more to the bonding picture and by what relative amount. For example, in $Cl^-$, $Br^-$, and $I^-$ there are both filled $p\pi$ orbitals and empty $d\pi$ orbitals. The most important interaction, at least with the $3d^n$ metal atoms in their normal oxidation states, seems to be with the stable filled $p\pi$ orbitals. This could account for their appearance at the weak ligand end of the spectrochemical series. However, there is evidence that with later $4d^n$ and $5d^n$ metal ions, and particularly Pt(II), Pd(II), Hg(II), and Au(III), the stabilizing effect of the use of the unfilled $d\pi$ orbitals plays the dominant role. In ligands such as $CN^-$, $CO$, py, ophen, aca$^-$, etc., the empty $\pi$ orbitals are antibonding $\pi$ molecular orbitals and the filled $\pi$ orbitals are the bonding $\pi$ molecular orbitals. Although it is normally assumed that ligands such as $CO$ and $CN^-$, in their common complexes, utilize mainly the unfilled antibonding $\pi$ orbitals for $\pi$ bonding, in general it is very difficult to make predictions for most such ligands.

We shall not attempt to present here any of the quantitative aspects of the application of the MOT to complexes. Suffice it to say that although there are currently many serious efforts in this area, it must still be considered to be in the early stages of development. We shall only give reference to a few of the more recent and somewhat successful efforts in this area as well as to their critical evaluation.[19-21,38-40]

## Adjusted Crystal Field Theory or Ligand Field Theory

For many purposes of correlating and understanding experimental data pertaining to a large variety of complexes of metal atoms in their common

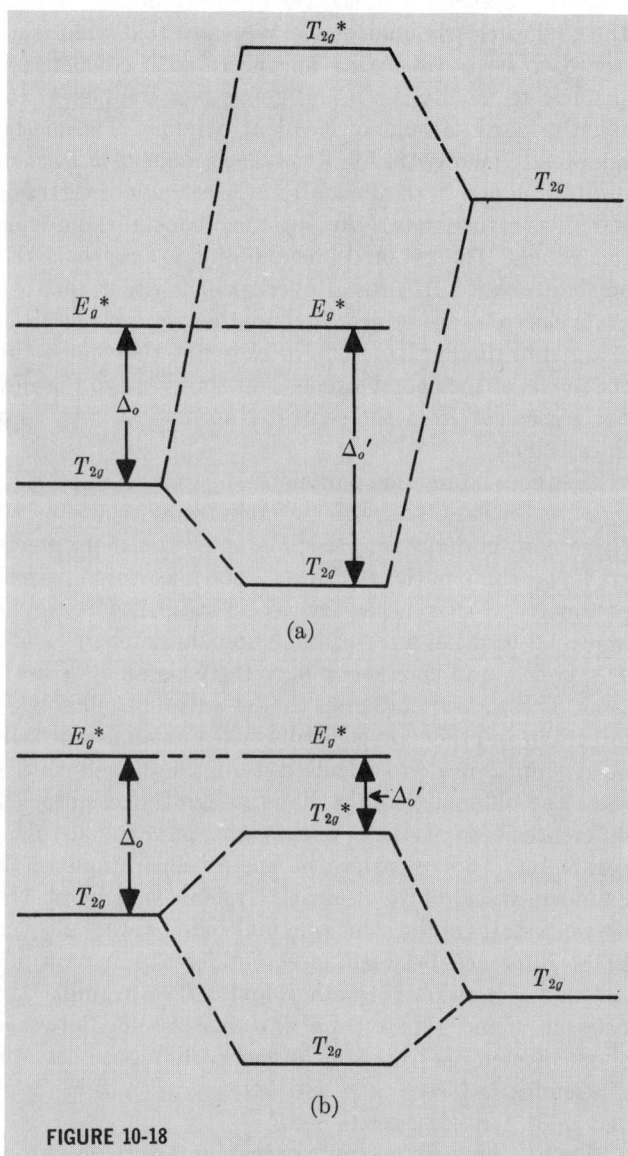

**FIGURE 10-18**

Energy Level Diagrams Illustrating the Effects of $\pi$ bonding on the $\Delta_o$ Value. (a) In this case the ligand $\pi$ orbitals (at right) are empty and of higher energy than the filled or partially filled metal atom $\pi$ orbitals, and $\Delta_o$ *increases* to $\Delta_o'$. (b) Here the ligand $\pi$ orbitals are filled and of lower energy than those of the metal atom, and $\Delta_o$ *decreases* to $\Delta_o'$. In the literature, and specifically for particular complex ion systems, the energy levels pictured here are incorporated into the more general energy level scheme which includes all of the $\sigma$-bonding and antibonding orbitals.

oxidation states, it is not necessary to resort to the rigors of the MOT, yet the CFT is clearly inadequate. We know that what is wrong with the latter approach is in essence its unconcern with covalence effects, and so it is possible to modify it without going over completely to a model which explicitly takes account of covalent bonding. The method most commonly adopted to modify the CFT to take account of at least some of the effects of orbital overlap is to allow all interelectronic interaction parameters to be variables rather than assuming them constant and equal to the free metal ion values. The rationale behind this approach is relatively simple and straightforward. If orbital overlap does occur then certainly the central atom electrons are subject to more than simply the electrostatic field of the charged or dipolar ligands. For example, they come under the influence of the nuclei of the ligand atoms. This should have the effect of drawing them out somewhat from the central atom and thereby lowering their mutual interactions.

The three interaction parameters of greatest importance are the *spin-orbit coupling constant*, $\xi$, and the interelectronic repulsion parameters which are either certain *Slater integrals, $F_n$*, or certain more precise quantities which are linear combinations of these, known as *Racah parameters, B and C*. One parameter, $F_2$, is necessary to describe the energy difference between multiplet terms of a $p^n$ system, two parameters, $F_2$ and $F_4$, are required for $d^n$ systems, and correspondingly three parameters are required for $f^n$ systems. These $F$'s are integrals that arise from Coulomb $(J)$ and exchange $(K)$ integrals, and they are generally taken as semiempirical parameters. Thus, no attempt is made to calculate them. The Racah parameters are particular sums and differences of the Slater integrals chosen so as to make the energy differences between states of the same spin explicitly dependent on only one parameter. They measure the energy separations of the various Russell-Saunders states of an electronic system. In general, the energy difference between states of the same spin multiplicity are multiples of $B$ alone, whereas the differences between states of different spin multiplicity are formulated as sums of multiples of both $B$ and $C$. For example, the energy difference between $^3F$ and $^3P$ in either $d^2$ or $d^8$ systems and between $^4F$ and $^4P$ in either $d^3$ or $d^7$ systems is $15B$. But the energy difference between $^3F$ and $^1D$ in $d^2$ or $d^8$ systems is $(5B + 2C)$, and between $^4F$ and $^2G$ in $d^3$ or $d^7$ systems, is $(4B + 3C)$. In all cases $C \sim 4B$.

Now the spin-orbit coupling constant is very important in the determination of detailed magnetic properties of many transition metal complexes. For example, it may be used to account for the deviations of real magnetic moments from the calculated spin-only values, and to account for the temperature dependence of some magnetic moments. However, it is most significant that in order to bring theoretical (CFT) and experimental values into closer agreement, the $\xi$ value for the complexed metal atom must usually be taken as only 70% to 85% of the free-ion value. In an entirely analogous way, but usually from electronic spectral data, it is found that

CFT and experiment can be brought into excellent agreement when the Racah parameters for the complexed ion are reduced by about the same factor from their free-ion values. Thus, in general, we find that:

$$B'/B \sim C'/C \sim 0.7 - 0.8$$

and the values of $B'/B = \beta$ for a series of complexes of different ligands with a given metal atom will fall into the ligand order which is called the *nephelauxetic* ligand series (see p. 435). Alternatively, we might compare the $\beta$'s for a series of complexes of different metal atoms with a given ligand and arrive at a nephelauxetic metal atom series.

With this brief and qualitative survey of the modern approaches to coordinate bonding, we have at least laid the groundwork for the meaningful consideration and study of some of the current research areas in coordination chemistry. These are discussed in the light of the foregoing material in the next chapter. Ample references to more extensive and more intensive coverage of the theories and their applications have been given at the end of this and the succeeding chapter. The very broad field of organometallic chemistry is still mainly experimental. But the application of the several theories presented here, in particular the semiempirical molecular orbital theory, holds great promise for establishing a firm theoretical basis for explaining and predicting the properties of this rapidly growing class of compounds.

## References

1. J. C. Bailar, Jr., Ed., "The Chemistry of the Coordination Compounds," Reinhold Publishing Corporation, New York, 1956.
2. F. Basolo and R. G. Pearson, "Mechanisms of Inorganic Reactions," 2nd ed., John Wiley & Sons, Inc., New York, 1967.
3. J. Lewis and R. G. Wilkins, "Modern Coordination Chemistry," Interscience Publishers, Inc., New York, 1960.
4. F. P. Dwyer and D. P. Mellor, Eds., "Chelating Agents and Metal Chelates," Academic Press, Inc., New York, 1964.
5. A. A. Grinberg, "The Chemistry of Complex Compounds," Pergamon Press, Inc., New York, 1962.
6. M. M. Jones, "Elementary Coordination Chemistry," Prentice-Hall, Inc., Englewood Cliffs, N.J., 1964.
7. C. K. Jørgensen, "Inorganic Complexes," Academic Press, Inc., New York, 1963.
8. S. Chaberek and A. E. Martel, "Organic Sequestering Agents," John Wiley & Sons, Inc., New York, 1959.
9. R. S. Nyholm and M. L. Tobe, in "Essays in Coordination Chemistry," Birkhauser Verlag, Basel, 1964, pp. 112–127.
10. F. A. Cotton and G. Wilkinson, "Advanced Inorganic Chemistry," 2nd ed., Interscience Publishers, Inc., New York, 1966.
11. J. Selbin, *J. Chem. Ed.*, **41**, 86 (1964); *Angew. Chem., Internat. Ed. Engl.*, **5**, 712 (1966).

12. J. Lewis and R. S. Nyholm, *Sci. Prog.*, **52** (208), 557 (1964).
13. S. Ahrland, J. Chatt, and N. Davies, *Quart. Revs.*, **12**, 265 (1958).
14. N. V. Sidgwick, *J. Chem. Soc.*, **123**, 275 (1923); *Trans. Faraday Soc.*, **19**, 469 (1923); *Chem. Ind.*, **42**, 901, 1203 (1923); "The Electronic Theory of Valence," Clarendon Press, Oxford, 1927.
15. T. Lowry, *Chem. Ind.*, **42**, 316 (1923).
16. L. Pauling, "The Nature of the Chemical Bond," 3rd ed., Cornell University Press, Ithaca, N.Y., 1960.
17. M. Huggins, *J. Chem. Phys.*, **5**, 527 (1937).
18. D. Craig, A. Maccoll, R. S. Nyholm, L. E. Orgel, and L. Sutton, *J. Chem. Soc.*, **1954**, 332.
19. H. Basch, A. Viste, and H. B. Gray, *J. Chem. Phys.*, **44**, 10 (1966), and many references contained therein.
20. R. F. Fenske, K. G. Caulton, D. D. Radtke, and C. C. Sweeney, *Inorg. Chem.*, **5**, 951, 960 (1966).
21. F. A. Cotton and C. B. Harris, *Inorg. Chem.*, **6**, 369, 376 (1967).
22. A. VanArkel and J. DeBoer, *Rec. trav. chim.*, **47**, 593 (1928).
23. F. Garrick, *Phil. Mag.*, **9**, 131 (1930); **10**, 71, 76 (1930); **11**, 741 (1931); **14**, 914 (1932).
24. H. Bethe, *Ann. Physik.*, **3**, No. 5, 133 (1929).
25. W. Penney and R. Schlapp, *Phys. Rev.*, **41**, 194 (1932); **42**, 666 (1932).
26. J. H. VanVleck, *Phys. Rev.*, **41**, 208 (1932).
27. J. H. VanVleck, *J. Chem. Phys.*, **3**, 803, 807 (1935).
28. F. Ilse and H. Hartmann, *Z. Physik. Chem.*, **197**, 239 (1957).
29. L. Orgel, *J. Chem. Soc.*, **1952**, 4756; *J. Chem. Phys.*, **23**, 1819 (1955); "An Introduction to Transition Metal Chemistry, Ligand Field Theory," 2nd ed., John Wiley & Sons, Inc., New York, 1966.
30. C. J. Ballhausen, "Introduction to Ligand Field Theory," McGraw-Hill, Inc., New York, 1962.
31. B. N. Figgis, "Introduction to Ligand Fields," John Wiley & Sons, Inc., New York, 1966.
32. J. S. Griffith, "The Theory of Transition Metal Ions," Cambridge University Press, New York, 1961.
33. F. A. Cotton, "Chemical Applications of Group Theory," John Wiley & Sons, Inc., New York, 1963.
34. R. S. Drago, "Physical Methods in Inorganic Chemistry," Reinhold Publishing Corporation, New York, 1965.
35. N. N. Greenwood, *Chem. in Britain*, **3**, 56 (1967), 81 refs.
36. S. F. A. Kettle, *J. Chem. Ed.*, **43**, 21, 652 (1966).
37. C. J. Ballhausen and H. B. Gray, "Molecular Orbital Theory," W. A. Benjamin, Inc., New York, 1964.
38. H. Basch and H. B. Gray, *Inorg. Chem.*, **6**, 365 (1967).
39. R. F. Fenske and C. C. Sweeney, *Inorg. Chem.*, **3**, 1105 (1964).
40. R. F. Fenske, *Inorg. Chem.*, **4**, 33 (1965).

## Problems

1. Discuss the formation and existence of carbonyl complexes as evidence for covalent bonding in coordination compounds. List several other common ligands

whose very existence as ligands constitutes evidence for covalence in their complexes.

2. Compare and critically contrast the VBT, CFT, and MOT in their treatment of magnetic and stereochemical properties of the complexes of $Cu^{2+}$. Do the same for $Ti^{3+}$, $Fe^{2+}$, and $Ni^{2+}$.

3. Give the names of and illustrate all of the types of isomerism that are possible in an octahedral complex composed of one cobalt(III) ion, two ethylenediamine molecules, two chloride ions and one $NO_2^-$ ion.

4. For the paramagnetic complex anion $[CoF_6]^{3-}$, what is the order of increasing magnitude of the three energy factors: crystal field splitting, spin-orbit coupling, and electron-electron repulsion?

5. Compare and/or contrast, using the parlance of MOT, the effect on $\Delta$ of the two ligands, acetate anion and dithioacetate anion, $CH_3CS_2^-$.

6. From the viewpoint of MOT what is responsible for $\Delta$ in the complex $Mo(CO)_6$?

7. List as many factors as you can which you feel are of importance in determining the very widespread occurrence and importance of coordination number 6 among the transition elements. Do the same for coordination number 4, but here also point out why this is not as common as is coordination number 6.

8. Name the following compounds (See W. C. Fernelius, *J. Chem. Doc.*, **5,** 200 (1965)):

   $[Fe(CN)_2(CH_3NC)_4]$
   $Rb[AgF_4]$
   $[Fe(\pi-C_5H_5)_2]Cl$
   $[Co(NCSe)NH_3)_5]Cl_2$
   $[Ir(NCO)Co(Ph_3P)_2]$
   $[CrBr(ONO)en_2]_2[Co(NO_2)_4(H_2O)_2]$

9. The oxide $Fe_3O_4$ exists as an *inverse spinel* structure, whereas $Mn_3O_4$ exists as a *normal spinel* structure. Explain. *Hint:* Consider CFSE's.

10. Picture and discuss the various important contributions to the coordinate bonding in each of the following *six* cases:

    (a) $Ti^{3+}$ with $F^-$, $H_2NCH_2CH_2NH_2$, and $P(CH_3)_3$; separately.

    (b) $Cu^{2+}$ with $H_2O$, a $\beta$-diketone such as acetylacetonate anion, $CH_3CCH=C-CH_3$, and $P(CH_3)_3$; separately.

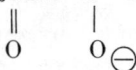

# 11

# Coordination
# Chemistry II

Now that we have introduced, in the previous chapter, the several bonding theories used to deal with coordinate bonds, we are in a position to examine the current areas of application of these theories. The properties of complexes which coordination chemists are most interested in today, and which therefore they wish to be able to explain and make predictions for, are the following: (1) stereochemical, (2) thermodynamic, (3) kinetic, (4) spectral, and (5) magnetic properties. It will be the purpose of this chapter to present an introduction to these subjects with emphasis on the appropriate theories for dealing with them.

## STEREOCHEMISTRY

One of the more successful features of the valence bond approach was its ability to predict and at least describe, if not explain, the stereochemistry of most complex ions known up until about 1950. Thus, if the set of hybridized orbitals utilized by a central atom could be determined by some means, usually by magnetic susceptibility measurements, the stereochemistry generally followed. However, using the VBT, we are not always capable of accurately predicting under what circumstances a particular geometry is expected. And in many cases we are forced by the VBT to conclude that ionic rather than covalent bonding occurs in certain complexes where, for example, the central ion does not possess the necessary number of equivalent energy orbitals for hybridization to accommodate all of the ligand electrons. The CFT, which ignores ligand orbitals and electrons, and the MOT, which places the ligand electrons into orbitals still primarily associated with the ligands, both obviate this difficulty. The CFT alone is more reliable and detailed than VBT for stereochemical predictions and it allows us a greater insight into the important factors which govern the detailed stereochemistry of transition metal complexes.

## Regular Symmetries

In Chapters 6 and 7 the factors which govern the shapes of inorganic molecules derived from non-transition elements were discussed. In particular, these are the size and charge of the central atom, the presence of non-bonding lone pair electrons, the possibility of expansion of the bonding shell beyond the octet limit of the first short period, the capacity for $\pi$ bonding, the steric requirements of the ligands, and of no lesser importance than all these, the operation of the Pauli exclusion principle. When we are dealing with a spherically symmetrical central atom, as is the case with complexes of all the main group metallic elements without lone pair electrons, we predict and we find regular molecular shapes. Thus, coordination numbers 2, 3, 4, 5, 6, 7, 8, and 9 give molecular species that are linear, triangular planar, regular tetrahedral, trigonal bipyramidal, regular octahedral, pentagonal bipyramidal, square (or Archimedes) antiprismatic, and tripyramidal, respectively. So we may expect that whenever we have a spherically symmetrical nonbonding electron shell with a central transition element atom, the stereochemistry will be regular and determined only by the coordination number. Therefore, we can write out the electronic configurations which are expected to lead to perfectly symmetrical complexes. For the most common coordination numbers, 6 and 4, we have the following:

$$\text{Spin-free (Weak Field)} \quad \text{Spin-paired (Strong Field)}$$

| | Spin-free (Weak Field) | Spin-paired (Strong Field) |
|---|---|---|
| Perfect Octahedron | $d^0;\ d_\epsilon^3;\ d_\epsilon^3 d_\gamma^2;$ $d_\epsilon^6 d_\gamma^2;\ d^{10}$ | $d^0;\ d_\epsilon^3;\ d_\epsilon^6;\ d^{10}$ |
| Perfect Tetrahedron | $d^0;\ d_\gamma^2;\ d_\gamma^2 d_\epsilon^3;$ $d_\gamma^4 d_\epsilon^3;\ d^{10}$ | $d^0;\ d_\gamma^4;\ d^{10}$ |

All other configurations necessarily lead to distorted geometries or, in certain cases, to the square planar configuration. However, even the latter may be pictured as arising from a strongly tetragonally distorted octahedron in which axial groups, say on the $z$ axis, have moved away from the central atom to a point where they no longer contribute to the effective ligand field or to the bonding MO's, and are therefore not considered to be bonded to the central atom. Thus, all stereo arrangements that are not regular as outlined above may be considered distortions of the regular arrangements ranging, as we shall see, from small to very large. Our main interest then is to determine when and how these distortions arise. And it is obvious that we need consider only those electronic configurations that are not listed above.

## Distortions From Regular Symmetries

Whenever the $d_\epsilon$ orbitals, which point *between* the ligands in the octahedral case, contain 1, 2, 4, or 5 electrons we shall expect to find only slight distortions from the regular octahedron. Thus, depending upon the ligand

field strength we shall expect to find slightly distorted octahedra for the
following configurations:

| Weak Field | Strong Field |
|:---:|:---:|
| $d_\epsilon^1$; $d_\epsilon^2$; $d_\epsilon^4 d_\gamma^2$; $d_\epsilon^5 d_\gamma^2$ | $d_\epsilon^1$; $d_\epsilon^2$; $d_\epsilon^4$; $d_\epsilon^5$ |

Whenever the $d_\gamma$ orbitals, which point *directly at* the ligands, are unsym-
metrically occupied in a weak field or occupied at all (but not filled) in a
strong field, we shall expect strong distortions, leading to tetragonal and
even to square planar complexes. These configurations are:

| Weak Field | Strong Field |
|:---:|:---:|
| $d_\epsilon^3 d_\gamma^1$; $d_\epsilon^6 d_\gamma^3$ | $d_\epsilon^6 d_\gamma^1$; $d_\epsilon^6 d_\gamma^2$; $d_\epsilon^6 d_\gamma^3$ |

In order to understand these foregoing distortions more clearly it is necessary to consider the Jahn-Teller effect.

## The Jahn-Teller Effect

An interesting theorem proved by Jahn and Teller in 1937, although only
qualitative, has important stereochemical significance. It states, in effect,
that *when the orbital state of an ion is degenerate for symmetry reasons, the
ligands will experience forces distorting the nuclear framework until the ion
assumes a configuration of both lower symmetry and lower energy, thereby removing the degeneracy.* More generally the theorem, as extended by Jahn,
can be formulated as follows: A nonlinear molecule possessing a degenerate
state in either orbit or spin must distort in order to remove as much degeneracy as possible.

The general theory is best illustrated with a specific example. The $Cu^{2+}$
ion is particularly desirable for it is here that the effect is most notable. The
$Cu^{2+}$ ion is a $d^9$ system and has the configuration $d_\epsilon^6 d_\gamma^3$. In an octahedral environment this gives rise to a doubly degenerate ($E_g$) ground state since the
following two assignments of the $d_\gamma$ electrons are possible: $d^2_{z^2} d^1_{x^2-y^2}$ and
$d^1_{z^2} d^2_{x^2-y^2}$. Note that the problem could be approached by considering $Cu^{2+}$
to be a $d^1$ *positron* system, according to the so-called *hole formalism,* for
which case the $d_\gamma$ degenerate levels comprise the lower-energy set, and the
equivalent problem then becomes one of finding in which of the two $d_\gamma$
orbitals the positron or hole resides.

Let us now suppose that the $d_\gamma$ electrons are distributed according to the
first of the above possibilities, namely, $d^2_{z^2} d^1_{x^2-y^2}$. The $Cu^{2+}$ ion is now placed
in an already set-up regular octahedral environment of ligands. The $d_{z^2}$ orbital, which is *filled* and points at the ligands on the $z$ axis, now offers greater
shielding of the $Cu^{2+}$ nucleus than the *half-filled* $d_{x^2-y^2}$ orbital, which points
towards the ligands in the $xy$ plane. The latter, which *see* a higher effective
nuclear charge, are drawn in closer to the $Cu^{2+}$ nucleus, whereas the ligands

on the $z$ axis move further out since they experience a lower effective nuclear charge. We should therefore observe four short and two long bonds. In other words, if a $d_{x^2-y^2}$ electron is removed from a $d^{10}$ system, the spherical symmetry of the latter is lost and there is greater nuclear screening along the $z$ axis than along the $x$ and $y$ directions.

If, on the other hand, the configuration of the $Cu^{2+}$ ion were $d^1{}_{z^2}d^2{}_{x^2-y^2}$, we should expect an exactly opposite distortion. That is, the ligands in the $xy$ plane would move out and the $z$ axis ligands would move in from their equilibrium positions in the hypothetical regular octahedron, and we would expect to find two short and four long bonds.

Now the theory requires that if the undistorted nuclear framework structure has a center of symmetry, the distorted structure must also have such a symmetry center. However, the elementary theory *does not* indicate which of the possible degeneracy-removing distortions should lead to the greatest stabilization of the complex.

To further elaborate the energy relationships arising from a specific Jahn-Teller distortion, we can consider the manner in which the $d$-orbital energy levels change in the $Cu^{2+}$ $d^9$ system when the regular octahedron distorts. Taking the commonly observed case in which the $d_{z^2}$ orbital is apparently doubly occupied and $d_{x^2-y^2}$ is singly occupied, the schematic diagram of the splittings caused by the resultant distortion of the octahedron can be drawn as shown in Figure 11–1. In the figure we see not only the splitting of the $d_\gamma$ levels by an amount $\delta_1$, but also the splitting that should occur in the $d_\epsilon$ levels, $\delta_2$. Both splittings obey the baricenter rule. Further we note that $\delta_1$ is much larger than $\delta_2$, although both are very small compared to $\Delta_o$. None of these three splittings is shown in the figure to proper scale, but that has been sacrificed for the sake of clarity. In any case, it may be seen that in the $Cu^{2+}$ system there will be no net gain in energy from the split $d_\epsilon$ levels since the

$$4\left(-\frac{1}{3}\delta_2\right) = -\frac{4}{3}\delta_2$$

*energy gain* is counterbalanced by

$$2\left(\frac{2}{3}\delta_2\right) = +\frac{4}{3}\delta_2$$

*energy loss.* However, the

$$2\left(-\frac{1}{2}\delta_1\right) = -\delta_1$$

energy gain minus the

$$1\left(\frac{1}{2}\delta_1\right) = +\frac{1}{2}\delta_1$$

energy loss leaves a *net energy gain* of $-\frac{1}{2}\delta_1$, which might be called the Jahn-Teller stabilization energy. We can picture this gain in orbital energy as

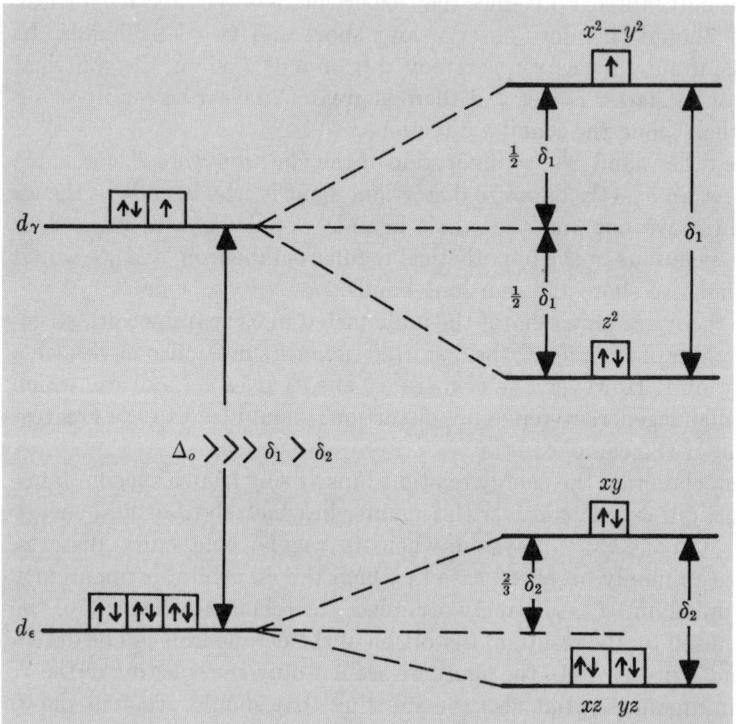

**FIGURE 11-1**

A Schematic Diagram of the Splittings Caused by the Operation of the Jahn-Teller Effect in a $d^9$ System. Elongation has occurred along one $C_4$ axis of an octahedron. The several splittings are not drawn to scale, $\Delta_0$ being much larger than $\delta_1$, which in turn is larger than $\delta_2$.

providing the driving force which causes the observed distortion. In Table 11–1, a number of $Cu^{2+}$ compounds are listed with their interatomic distances, which substantiate the Jahn-Teller effect.

**TABLE 11–1     Interatomic Distances in Some Jahn-Teller Distorted Compounds.**

| Compound | Interatomic Distances, Å |
|---|---|
| $CuO$ | 4O at 1.95 |
| $CuF_2$ | 4F at 1.93, 2F at 2.27 |
| $CuCl_2$ | 4Cl at 2.30, 2Cl at 2.95 |
| $CuBr_2$ | 4Br at 2.40, 2Br at 3.18 |
| $CsCuCl_3$ | 4Cl at 2.30, 2Cl at 2.64 |
| $CuCl_2(H_2O)_2$ | 2O at 2.01, 2Cl at 2.31, 2Cl at 2.98 |
| $CuCl_2(py)_2$ | 2N at 2.02, 2Cl at 2.28, 2Cl at 3.05 |
| $Cu(NH_3)_2Cl_2$ | 2N at 1.95, 4Cl at 2.76 |
| $Cu(NH_3)_2Br_2$ | 2N at 2.03, 4Br at 2.88 |
| $CrF_2$ | 4F at 1.98–2.01, 2F at 2.43 |

Also listed in Table 11–1 is a $Cr^{2+}$ compound which is seen to be similarly distorted. Indeed, from the foregoing discussion it should be clear that whenever the $d_\gamma$ set of levels is occupied by an odd number of electrons we should get large Jahn-Teller distortions. We expect the distortions to occur then for the following real systems:

$$d^4: \quad d_\epsilon^3 d_\gamma^1 \quad \text{high-spin } Cr^{2+}, Mn^{3+}$$

$$d^7: \quad d_\epsilon^6 d_\gamma^1 \quad \text{low-spin } Co^{2+}, Ni^{3+}$$

$$d^9: \quad d_\epsilon^6 d_\gamma^3 \quad Cu^{2+}, Ag^{2+}$$

Structural evidence for all of these examples, with the exception of the case of low-spin $Co^{2+}$, now exists, and almost invariably there are four short and two long bonds. For many $Cu^{2+}$ compounds (see Table 11–1) the distortion of the octahedron is so large that for practical purposes the complexes are best considered to be square planar.

Jahn-Teller splittings are also predicted for six-coordinated complexes in which the $d_\epsilon$ levels are not either empty, half-filled, or filled. Thus, distortions should occur for the cases of 1, 2, 4, or 5 $d_\epsilon$ electrons. However, the splittings are very small, since these electrons are concentrated in regions between the ligands and are not involved with bonding. Accordingly there is very little experimental confirmation of them.

The Jahn-Teller theorem also applies to excited electronic states. In fact, there is hardly a configuration which does not have the conditions for the effect in at least one state, ground or excited. But the effect in the latter case is complicated by the short lifetime of the excited state. This does not allow the attainment of a stable equilibrium configuration of the nuclei of the complex. Nevertheless, the phenomenon has apparently been observed in the spectral studies of such species as $Ti(H_2O)_6^{3+}$, as noted in the previous chapter, as well as $Fe(H_2O)_6^{2+}$ and $CoF_6^{3-}$. The latter two ions both have the ground state configuration $d_\epsilon^4 d_\gamma^2$, and the excited state $d_\epsilon^3 d_\gamma^3$. As we shall observe later, when there is present some other mechanism for removing the degeneracy of the ground state, the Jahn-Teller effect is not needed!

## Octahedral versus Square Planar and Tetrahedral Coordination

From the crystal field theory we can gain some insight into a few of the additional pertinent factors that control the stereochemistry in various $d^n$ complex ion systems. For example, consider the figures in Table 11–2, which are the calculated one-electron CFSE differences, $\delta Dq$, in both weak and strong fields, between octahedral and both the square planar and the tetrahedral configurations. We note that the $\delta Dq$ values for the case square planar minus octahedral, are large for weak field $d^4$ and $d^9$ and for strong field $d^7$, $d^8$, and $d^9$ complexes. However, before we use this information to make stereochemical predictions, we must take into account the following additional factors: (1) The parameter $Dq$ is in itself geometry-dependent and is generally somewhat larger for the square planar case. (2) The mutual ligand-ligand repulsion of four groups is most likely less than that of six

**TABLE 11–2. Differences in One-electron CFSE's in Weak and Strong Fields for the Structure Differences, Square Planar and Octahedral, Octahedral and Tetrahedral.**

| $d^n$ | Weak Field | | Strong Field | |
|---|---|---|---|---|
| | $\delta Dq$ (square planar-octahedral) | $\delta Dq$ (octahedral-tetrahedral) | $\delta Dq$ (square planar-octahedral) | $\delta Dq$ (octahedral-tetrahedral) |
| $d^0$ | 0 | 0 | 0 | 0 |
| $d^1$ | 1.14 | 1.33 | 1.14 | 1.33 |
| $d^2$ | 2.28 | 2.66 | 2.28 | 2.66 |
| $d^3$ | 2.56 | 8.44 | 2.56 | 3.99 |
| $d^4$ | 6.28 | 4.22 | 3.70 | 5.32[b] |
| $d^5$ | 0 | 0 | 4.48 | 11.10 |
| $d^6$ | 1.14 | 1.33 | 5.12 | 16.88 |
| $d^7$ | 2.28 | 2.66 | 8.84 | 12.66 |
| $d^8$ | 2.56 | 8.44 | 12.56[a] | 8.44 |
| $d^9$ | 6.28 | 4.22 | 6.28 | 4.22 |
| $d^{10}$ | 0 | 0 | 0 | 0 |

[a] One electron must be paired in the square planar case but not in the octahedral case.
[b] Two electrons must be paired in the tetrahedral case, but only one must be paired in the octahedral case.

groups, even if the four groups can approach the central atom more closely. (3) The total bond energy of six ligands is much greater than that for four ligands. Factors (1) and (2) favor the square planar configuration, but the much larger factor (3) very strongly favors the octahedral arrangement. Therefore, it is observed that the $\delta Dq$ value must heavily favor the square planar structure in order for it to be realized.

It is also seen from Table 11–2 that as far as simple CFSE values are concerned, the tetrahedral configuration will never be favored over the octahedral configuration, and this is roughly in accord with our knowledge of transition metal complexes. The ever-increasing number of tetrahedral, and distorted tetrahedral complexes being discovered among the transition metals still does not alter the great preponderance and greater ease of preparation of octahedral complexes. It is only under very special circumstances, not all fully understood yet, that tetrahedral complexes occur. It is clear from the table that except for $d^0$, $d^{10}$, and weak field $d^5$ systems the octahedral structure should always be favored over the tetrahedral structure, especially when the ligand field is strong. Thus, the weak field ligand $Cl^-$ yields the tetrahedral ions $FeCl_4^-$ and $MnCl_4^{2-}$, but with the strong field ligand $CN^-$ we obtain the octahedral ions $Fe(CN)_6^{3-}$ and $Mn(CN)_6^{4-}$. Examples of tetrahedral coordination among transition metals include, in addition to the foregoing, $CoX_4^{2-}$ ($X = $ Cl, Br, I, NCS), which are stable even in aqueous solutions; $VX_4^-$, $MnX_4^{2-}$, $NiX_4^{2-}$, and $CuX_4^{2-}$ ($X = $ Cl, Br, I), which are stable in the solid state with very large cations such as

$R_4N^+$, $R_4P^+$, or $R_4As^+$ or in noncoordinating solvents, but unstable in co-ordinating solvents; $MX_3L^-$ or $MX_2L_2$ ($M =$ Mn(II), Co(II), Ni(II); $X =$ halide; $L =$ neutral ligand such as $H_2O$, $R_3PO$, or $R_3AsO$); and a few cationic species $ML_3X^+$ and $ML_4^{2+}$, mainly with $M =$ Co(II).

Tetragonal distortions of tetrahedral complexes, although apparently more widespread and common than previously suspected, have been studied in only a few cases. Consequently, the subject will not be pursued here. Suffice it to say that these distortions have been found where predicted by a simple CFT model. For example, in certain $d^8$ $Ni^{2+}$ complexes the distortions are in the direction of *elongation* of the tetrahedron. In certain $d^9$ $Cu^{2+}$ complexes the distortions are in the direction of *flattening* of the tetrahedron toward a square plane. This distortion is required to maintain tetragonal symmetry when both the $d_{xz}$ and $d_{yz}$ orbitals are filled and there is either a hole or only one electron in the $d_{xy}$ orbital which therefore attracts the ligands towards it.

### Linear Coordination of $d^{10}$ Ions

Ions such as $Cu^+$, $Ag^+$, $Au^+$, and $Hg^{2+}$ would be expected by any simple electrostatic theory to possess high coordination numbers. Yet many of their compounds, particularly those of $Au^+$ and $Hg^{2+}$, have these ions linear-ly coordinated. If this were to be explained, as has often been done, by assuming $sp$ covalent bonding rather than ionic bonding, it is difficult to understand why $Hg^{2+}$ prefers $sp$ bonding but $Zn^{2+}$, $Cd^{2+}$, or $Tl^{3+}$ do not. The $s$-$p$ separation energies are not different enough between the $d^{10}$ ions which do and those which do not form linear bonds to explain the foregoing obser-vation. Orgel has offered an interesting explanation that is based upon the use of $d$-$s$ hybridized orbitals. That is, he assumes that the two electrons, which in a $d^{10}$ ion would normally occupy the $d_{z^2}$ orbital, are placed in the $\sqrt{\frac{1}{2}}(d_{z^2} - s)$ hybrid orbital. This transfers charge from the $z$ axis into the $xy$ plane and permits strong bonds to form along the $z$ axis. Of course, to produce this $d$-$s$ mixing, electrons must be promoted from the $(n - 1)d$ to the $ns$ level, and so the entire process will be energetically favorable only if the $d$-$s$ *separation energy* is small. Examination of these energy values, that is, the energy of the lowest $d^9s$ excited state above the $d^{10}$ ground state for the ions in question supports the Orgel explanation for the observed stereo-chemical preference. In eV units the energies are:

| | | | | | |
|---|---|---|---|---|---|
| $Cu^+$ | 2.7 | $Zn^{2+}$ | 9.7 | $Tl^{3+}$ | 9.3 |
| $Ag^+$ | 4.8 | $Cd^{2+}$ | 10.0 | | |
| $Au^+$ | 1.9 | $Hg^{2+}$ | 5.3 | | |

### Ionic Radii of Transition Metal Ions

Although stereochemistry generally implies a concern with bond *angles*, we should also be interested in bond *lengths* since most of the factors which

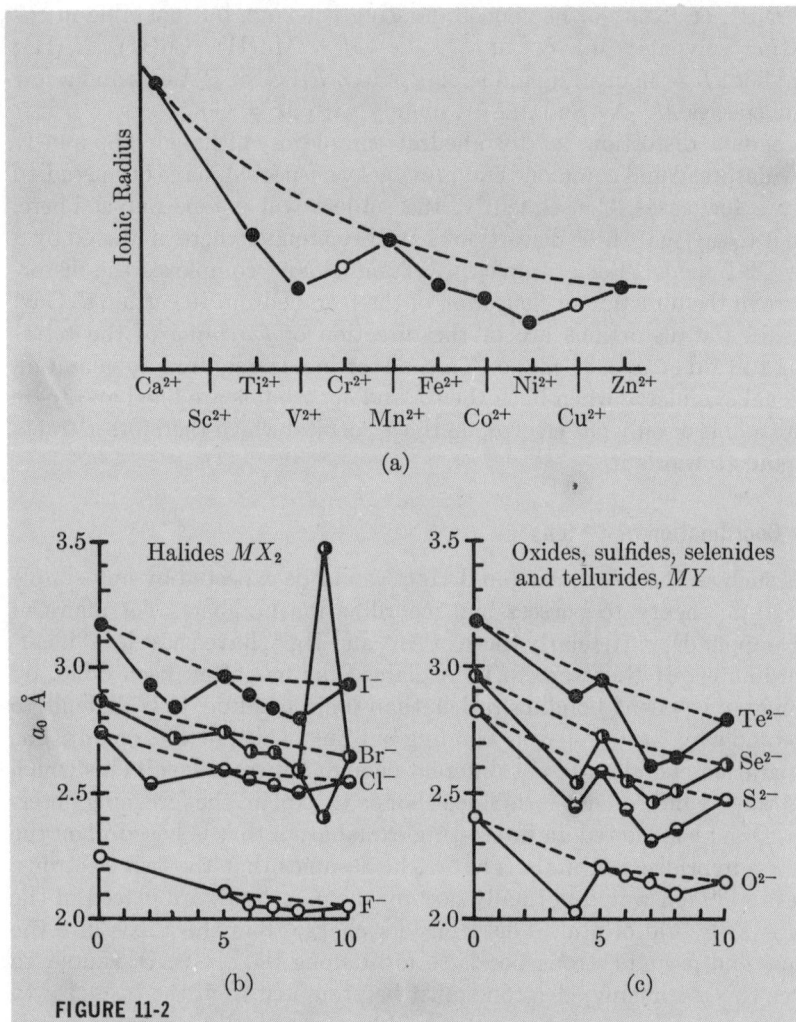

**FIGURE 11-2**

(a) A Plot of the Relative Octahedral Ionic Radii of the $3d$ Series Divalent Ions. (b) and (c) Plots of the Interatomic Distances Versus the Number, $n$, of $d$ Electrons in the $3d^n$ Divalent Halides and Chalcogenides, Respectively. The dotted curves in each figure pass through the points for the spherically symmetrical $d^0$, $d^5$ and $d^{10}$ ions.

affect the one affect the other. In this section we shall examine the effect upon bond lengths, or more specifically upon the central ion radii, of the nonspherical distribution of $d$ electrons caused by the splitting of the $d$ orbitals in a ligand field. For simplicity, and because of its far greater importance, we shall deal only with ions in an octahedral field.

Consider the octahedral radii of the divalent ions of the $3d$ series of elements as shown in Figure 11–2. If there were no crystal field effects, that is, no $d$-orbital splittings, the radii of the ions would be expected to show a uniform decrease similar to the decrease in trivalent radii found between La and Lu in the $4f$ series (see Chapter 4). Thus, we might expect the radii to fall along the dotted line shown in Figure 11–2(a), which connects the spherically symmetrical ions $Ca^{2+}$, $d^0$; $Mn^{2+}$, $d^5$; and $Zn^{2+}$, $d^{10}$. The reason for expecting the uniform contraction of ionic radius across the series is simply that the nuclear shielding of one $d$ electron for another is inadequate to fully counterbalance the increasing nuclear charge that comes with increasing atomic number. As is seen, the actual intervening radii values always fall below the dotted line and an explanation may be given in terms of the simple crystal field model.

The $t_{2g}$ electrons concentrate in spatial regions *between* the ligands, whereas $e_g$ electrons concentrate in spatial regions which point *at* the ligands. Therefore, addition of a $t_{2g}$ electron in going from left to right across the series provides less nuclear shielding of the increased nuclear charge than would be provided by either a spherically symmetrical $d$ electron charge or an $e_g$ electron. The ligand negative charge is therefore pulled toward the metal ion more strongly, thereby reducing the effective ion radius. The large radius decreases observed for the addition of the first three $t_{2g}$ electrons are thus accounted for. The fourth $d$ electron added in a weak field goes into an $e_g$ orbital and effects a relative increase in the radius since it now shields the increased nuclear charge *more* effectively than a spherically symmetrical $d$ electron would. Actually the radii for $d^4$, $Cr^{2+}$, and for $d^9$, $Cu^{2+}$, shown as clear circles in Figure 11–2(a), are not directly comparable with the other radii. As we have seen earlier, these ions cannot exist in an octahedral environment, but only in a strongly tetragonally distorted environment forced by the operation of the Jahn-Teller effect. The addition of the second $e_g$ electron, at $Mn^{2+}$, again produces a spherically symmetrical $d$ shell, and the radius point lies on the smooth curve. A similar pattern is observed in the second half of the series with the addition of the next five electrons, as seen in the figure. Although not as well documented, similar effects are expected with comparable ions in the $4d$ and $5d$ series and with trivalent ions in octahedral environments, as well as with ions in tetrahedral environments. Also, it should be clear that as we go from high-spin to low-spin complexes of a given ion the radius should contract since ligand-repelling $e_g$ electrons are being dropped into non-ligand-repelling $t_{2g}$ orbitals.

The molecular orbital theory can interpret the radius changes equally clearly. The placing of electrons in a non-$\sigma$-bonding $t_{2g}$ orbital should not affect the metal-ligand bond strength very much, but the increased nuclear charge should, thereby causing a radius decrease. On the other hand, the placing of electrons into *antibonding* $e_g$ orbitals will lower the bond strength and lead to a lengthening of the metal-ligand bond which may or may not be as large as the average decrease in ionic radius that occurs between consecutive $3d$ members. Likewise spin-pairing, the change from spin-free to spin-paired complexes, occurs by transfer of antibonding electrons into nonbonding orbitals and thereby leads to a decrease in the apparent ionic radius.

There is another area of stereochemistry which we have not touched upon at all, but which is nevertheless of current interest. That area concerns the stereochemical changes that complexes undergo during *substitution, isomerization*, or *racemization* reactions. We shall make no attempt to discuss this large field. We only mention here that the CFT has been used in this area with moderate success, particularly in the interpretation of kinetic data and the formulation and understanding of inorganic reaction mechanisms.[1,2]

## THERMODYNAMIC PROPERTIES

The subject of the stability of metallic complexes is indeed a large and varied one.[3-8] The many variables associated with the central metal atom, $M$, and the ligand, $L$, in addition to the variables that arise from different solvent and solid lattice conditions and temperature, serve to greatly complicate the study of this subject. The only reasonable approach to the study of stability is to maintain as many variables constant as possible and then to examine a small area of the whole subject. Furthermore, we should recognize from the start that there are two quite different kinds of stability, *thermodynamic* stability and *kinetic* stability. When we are concerned with the former, we deal with metal-ligand bond energies, stability constants, and the several thermodynamic variables which are derivable from them, or with oxidation-reduction potentials which measure valence state stabilization (see Chapter 8). When we are interested in kinetic stability, and this is primarily for complex ions in solutions, we deal with the rates and mechanisms of chemical reactions (*substitution, isomerization, racemization*, and *electron or group transfer* reactions), as well as with the thermodynamic variables involved in the formation of intermediate species or activated complexes. In the kinetic sense it will be more proper to speak of complexes as being *inert* or *labile* rather than stable or unstable. Too often these terms are confused or used incorrectly. Thus, a thermodynamically stable complex may be labile or it may be inert. Correspondingly, unstable complexes, although usually labile, may be inert. For example, we find that $[Fe(H_2O)_6]^{3+}$ and $[Cr(H_2O)_6]^{3+}$ have roughly the same energy per bond, 116 and 122 kcal/mole, respectively. But the former is labile and exchanges its ligands very rapidly while the latter is inert and exchanges its ligands only slowly $(t_{1/2} \sim 3.5 \times 10^5 \text{ sec})$. Perhaps even more dramatic an illustration of the difference between kinetic and thermodynamic stability is afforded by the following two complexes whose dissociation *equilibrium constants* are given:

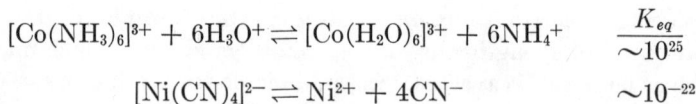

$$[Co(NH_3)_6]^{3+} + 6H_3O^+ \rightleftharpoons [Co(H_2O)_6]^{3+} + 6NH_4^+ \qquad \begin{matrix} K_{eq} \\ \sim 10^{25} \end{matrix}$$

$$[Ni(CN)_4]^{2-} \rightleftharpoons Ni^{2+} + 4CN^- \qquad \sim 10^{-22}$$

Yet the *thermodynamically unstable* hexaammine complex will persist in acid solution for days due to its *kinetic inertness*, whereas the tetracyano complex, despite its great *thermodynamic stability*, displays a large *kinetic lability* by exchanging its $CN^-$ ions immeasureably fast with added isotopically labeled $CN^-$.

## Thermodynamics of Coordinate Bond Formation

Ideally, if we wish to know the absolute coordinate bond energy we must have thermodynamic data on reactions in the *gas phase*, such as

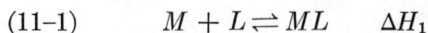

(11–1)        $M + L \rightleftharpoons ML$      $\Delta H_1$

Recognizing that reactions occur in a stepwise fashion, we may write further

$$ML + L \rightleftharpoons ML_2 \qquad \Delta H_2, \text{ etc.},$$

or in general

(11–2)        $ML_{(n-1)} + L \rightleftharpoons ML_n$      $\Delta H_n$

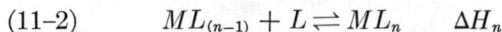

where $L$ is a monodentate ligand, $n$ is the coordination number of the metal atom, $M$, and the $\Delta H_i$ are reaction enthalpies and $\Delta H_1 \neq \Delta H_2 \neq \ldots \Delta H_n$. Actually very little direct data of this kind are available and for most cases can never be obtained. However, as is so often done in thermodynamic studies, we may make a satisfactory approximation to the desired information by an alternative but measurable path. Thus, we can study formation constants for complex ions in solution where, for example, the reactions in aqueous media may be represented as follows (ignoring the solvation of the ions and molecules):

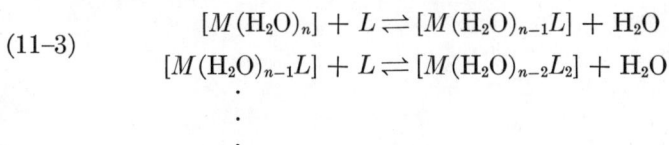

(11–3)
$$[M(H_2O)_n] + L \rightleftharpoons [M(H_2O)_{n-1}L] + H_2O$$
$$[M(H_2O)_{n-1}L] + L \rightleftharpoons [M(H_2O)_{n-2}L_2] + H_2O$$
$$\vdots$$

or overall:

$$[M(H_2O)_n] + nL \rightleftharpoons [ML_n] + nH_2O$$

In addition to the heat change involved in the gas phase reaction, any one of these reactions also includes the heats of hydration of both complexes, the ligand and the water.

In measuring the heats of these reactions, we actually determine the difference in bond energies between those of the coordinated water molecules and those of the coordinated ligands, $L$. Therefore, in order to evaluate metal-ligand bond energies, it is necessary that we know the heats of hydration of the gaseous metal ions as well as the heats of aquation of the various species involved in the reaction. These quantities are generally available or in most other instances can be estimated quite well. Thus, it should be possible to obtain a reasonably good approximation of the desired metal-ligand bond energies.

For each of the foregoing equilibria (11–3), we can write an equilibrium constant expression in accordance with the law of mass action. Omitting the solvent water, whose activity does not change if low concentrations are assumed, we obtain the following expressions:

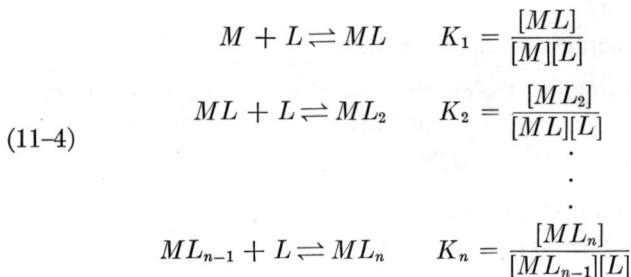

$$M + L \rightleftharpoons ML \qquad K_1 = \frac{[ML]}{[M][L]}$$

$$ML + L \rightleftharpoons ML_2 \qquad K_2 = \frac{[ML_2]}{[ML][L]}$$

(11–4)

$$\vdots$$

$$ML_{n-1} + L \rightleftharpoons ML_n \qquad K_n = \frac{[ML_n]}{[ML_{n-1}][L]}$$

where the constants $K_1, K_2, \ldots K_n$ are called the *stepwise stability* or *formation constants* and the bracketed quantities represent the activities of the enclosed species. The overall formation constant, $\beta$, is the product of the successive formation constants,

(11–5) $$M + nL \rightleftharpoons ML_n \qquad \beta_n = \frac{[ML_n]}{[M][L]^n}$$

$$\beta_n = K_1 K_2 \ldots K_n = \prod_{i=1}^{i=n} K_i$$

and it is this value that is used to determine the thermodynamic functions.

The standard free energy change, $\Delta G°$, is related to the equilibrium constant, $\beta$, by the relation

(11–6) $$\Delta G° = -2.303 \, RT \log \beta$$

Recalling that

(11–7) $$\Delta G° = \Delta H° - T\Delta S°$$

we see that by measuring $\beta$ at several temperatures we may obtain $\Delta H°$ by a graphical solution of the equation

(11–8) $$2.303 \, R \, \log \beta = \left( \Delta S° - \frac{\Delta H°}{T} \right)$$

Other refinements are necessary when $\Delta H°$ varies appreciably with temperature, but this complication will be ignored here. $\Delta S°$ is then obtained from Eq. (11–7). Additional information concerning the experimental methods and the detailed significance of the thermodynamic variables may be found in reference 4.

From Eq. (11–8) it is obvious that complex formation (like all reactions) is favored by negative enthalpy changes and positive entropy changes. In many instances both changes are found to favor complex formation, but there are also many examples where only one of these quantities is favorable. The relative importance of these effects is found to be dependent on variations in the ligand as well as the central metal ion. From the stepwise formation constants we may obtain stepwise enthalpy changes. In aqueous solution, these generally have values in the range of $+5$ to $-5$ kcal/mole for complexation involving ionic ligands, and 0 to $-5$ kcal/mole for complexation involving neutral monodentate ligands. For multidentate ligands the values may run over $-20$ kcal/mole. Various properties of both the ligand and the metal ion have a marked influence on the overall heat of formation of a complex, and we shall consider these properties shortly.

Entropy changes involving monodentate ligands are complicated by several factors in addition to those that affect heat changes. One might think that the ordering process associated with the formation in solution of one ionic species at the expense of two or more others would lead to an entropy decrease. Actually, this causes an overall entropy increase due to the whole or partial neutralization of charge and the freeing of several bound solvent molecules from the solvation spheres of each of the reactants. By the same reasoning, the entropy changes occurring when complexation involves neutral ligands will not be as favorable. Additional factors are involved when multidentate ligands coordinate. Loss of vibrational and rotational as well as translational entropy will generally be larger, but this will be compensated for by the displacement per ligand of more solvent molecules from the solvation spheres. Overall entropy changes generally range from small negative values of the order of $-2$ e.u. to large positive values of the order of 20 to 60 e.u. for multidentate ligands. Generally, the higher the charge and the greater the number of donor atoms of the chelating ligand, the larger will be the entropy increase.

## Factors Which Determine Stability

In considering the effect of the metal atom on the stability of a complex it is convenient to separate the central atoms into those categories which were outlined in the previous chapter (pages 406–410), and it would be useful at this point to read the discussion there of each category. The reason for this is simply that the number of major factors which affect the stability of complexes formed by the metal ions falling into categories I through III is less than the number for metal ions of category IV. Thus, in the first three categories the central ions are generally spherically symmetrical, they have completely filled nonbonding levels and sublevels, and thus the stability of their complexes will be primarily dependent upon their effective ionic radius and effective nuclear charge. Covalent bonding, and in particular $\pi$-bonding, will be of varying importance in these first three categories as previously discussed. If we define the *ionic potential*, $\phi$, as the ratio of effective cationic charge to effective cationic radius, we can then make the generalization that for small and for highly charged *ionic* ligands, and with most multidentate ligands, the complex stability will increase with increasing $\phi$. The following observed orders of complex ion stabilities bear out this statement:

Li > Na > K > Cs;   Be ≫ Mg > Ca > Sr > Ba > Ra;   B > Al > Sc > Y > La; Lu > ... Gd > ... La, within families where formal charge remains constant, and: $Th^{4+}$ > $Y^{3+}$ > $Ca^{2+}$ > $Na^+$; $La^{3+}$ > $Sr^{2+}$ > $K^+$ within series where charge changes but size is relatively constant.

Apparent anomalies arise when we compare complex stabilities of similarly sized and charged members which are in category I and categories II and III, respectively, for example $Na^+$ and $Cu^+$; $Ca^{2+}$ and $Cd^{2+}$ or $Sn^{2+}$; or $Sc^{3+}$ and $Ga^{3+}$ or $Sb^{3+}$. The elements in categories II and III invariably form the more stable complexes with a given ligand, and the reason must be traced to the fundamental electronic difference between category I and categories II and III. The members of the former have the inert gas configuration, $s^2p^6$, whereas the members of the latter two categories have the pseudo-inert gas, $s^2p^6d^{10}$, or pseudo-inert gas plus two, $s^2p^6d^{10}s^2$, configurations. We have already considered earlier (Chapter 4) that the latter two outer configurations are much poorer shielders of the excess positive charge located in the nucleus of a positive ion. So, for example, the effective nuclear charge of $Cu^+$ is much larger than that of $Na^+$ or that of $K^+$, or even of the category III ion $Ga^+$. The latter has two additional $s$ electrons to shield the nuclear charge. Thus, in general we would predict that for complexes of similarly sized and charged ions of category II and category III, the former would be the more stable.

One measure of effective nuclear charge may be found in the ionization energy. Whereas $Na^+$ and $Cu^+$ are nearly identical in size (0.95 vs. 0.93 Å), the $I_1$ value for Na is 118.5 kcal/mole while that for Cu is 178 kcal/mole. This difference may be interpreted as representing the much larger electron affinity or affinity energy of the $Cu^+$ over the $Na^+$ ion. This affinity is surely translated into greater attraction for the electrons or negative dipoles offered by ligands. Also, greater *penetration* and *polarization* of the more diffuse 18-electron cloud undoubtedly is partly responsible for the higher degree of covalent character found with the category II species.

Thus far we have considered only the central metal ions, which are in the minority when it comes to the formation of most of the known complexes. By far the most complexes are found for the central species of category IV. However, the major contributing factors to the stability of category IV complexes are still charge, size, and polarizability. And, as we shall see, the effects of the incompleted $d$ sublevel are generally much smaller and far more subtle. Before we discuss these effects we shall turn to the larger effects of the properties of the ligands.

### Properties of the Ligand Which Affect Stability

We have already touched briefly in several places upon the general properties of the donor ligand atoms which affect the stability of complexes. Thus, we are quite certain that complex ion stability increases with the Lewis base strength and increasing $\pi$-bonding capabilities of the ligands.

For monodentate negative ions their size and charge as well as the availability of a lone $\sigma$-bond electron pair are all important. For neutral molecule ligands their size, dipole moment and polarizability are of significance as well as the nucleophilicity of the $\sigma$ pair of electrons. Steric factors are generally not very important for monodentate ligands, but when they are, the protonic base affinities of the ligands will not reflect their coordinating abilities toward metal ions.

The picture becomes somewhat more complicated in the case of multidentate ligands. Here we must consider such additional factors as chelate ring size and strain, the number of rings, and substituents present in or conjugated to the ring system. In general, if the donor atom remains constant, and hence its properties nearly so, the formation of chelate rings enhances the stability of complexes. This is sometimes referred to as the *chelate effect*, and it is illustrated in Table 11–3, where the overall stability con-

**TABLE 11–3.  Stability Constants for Some Ammonia and Polyamine Complexes.**

| Metal Complex | No. of Chelate Rings | Log $\beta$ | | | | | | |
|---|---|---|---|---|---|---|---|---|
| | | $Mn^{2+}$ | $Fe^{2+}$ | $Co^{2+}$ | $Ni^{2+}$ | $Cu^{2+}$ | $Zn^{2+}$ | $Cd^{2+}$ |
| $M(NH_3)_4$ | 0 | — | ~3.7 | 5.31 | 7.79 | 12.59 | 9.06 | 6.92 |
| $M(en)_2$ | 2 | 4.9 | 7.7 | 10.9 | 14.5 | 20.2 | 11.2 | 10.3 |
| $M(trien)$ | 3 | 4.9 | 7.8 | 11.0 | 14.1 | 20.5 | 12.1 | 10.0 |
| $M(tren)$ | 3 | 5.8 | 8.8 | 12.8 | 14.0 | 18.8 | 14.6 | 12.3 |
| $M(dien)_2$ | 4 | 7.0 | 10.4 | 14.1 | 18.9 | 21.3 | 14.4 | 13.8 |
| $M(penten)$ | 5 | 9.4 | 11.2 | 15.8 | 19.3 | 22.4 | 16.2 | 16.8 |

en     = ethylenediamine, $NH_2CH_2CH_2NH_2$
trien  = triethylenetetramine, $NH_2CH_2CH_2NHCH_2CH_2NHCH_2CH_2NH_2$
tren   = triaminotriethylamine, $(NH_2CH_2CH_2)_3N$
dien   = diethylenetriamine, $NH_2CH_2CH_2NHCH_2CH_2NH_2$
penten = tetrakis(aminoethyl)ethylenediamine,
         $(NH_2CH_2CH_2)_2NCH_2CH_2N(CH_2CH_2NH_2)_2$

stants for several ammonia and amine complexes are listed. As is seen, the greater the number of chelate rings in the complex, the greater will be the complex stability. A voluminous amount of analogous stability constant data has confirmed the generality of the chelate effect. Examination of the published data[3,9] will further reveal that four-membered rings, including the central atom, are extremely rare and therefore presumably quite unstable. Five-membered rings are by far the most common and stable except when conjugation or delocalized ($\pi$) bonding is possible in the ring system, and then six-membered rings appear to be the most stable. Larger rings are decreasingly stable and therefore very uncommon. This is probably to be attributed primarily to strain set up in the heterocyclic rings and also to the greater ease or probability of a long-chain multidentate ligand bonding to more than one positive center. Here it would be acting in a bridging rather than in a chelating capacity.

Finally it is clear that certain ligands, in particular the very poor Lewis bases, owe much of their ligational strength to $\pi$ bonding. Thus, ligands such as CO, NO, $R_3P$, $R_3As$, $R_2S$, alkenes, alkynes, aromatics, etc., are not expected to form stable complexes with metal ions having filled, tightly bound electron levels that are incapable of taking part in $\pi$ bonding. The experimental facts bear out this prediction.

### Category IV Metal Atoms and the Crystal Field Theory

The stability of complexes formed by this, the largest and most varied category of metal atoms, depends not only upon ionic potential, but upon such variables as crystal field stabilization energy, CFSE, electron pairing and exchange energy, availability, that is, relative energy of empty $d\pi$ orbitals for *acceptance* $(L \rightarrow M)\pi$ bonding, and the availability of filled $d\pi$ orbitals for *back-donation* $(M \rightarrow L)\pi$ bonding. We could summarize all of these additional factors by saying simply that over and above ionic potential the complex ion stability depends significantly upon the particular number of $d$ electrons in the central atom in question. Now the CFT totally ignores covalent bonding, but as we have said in the previous chapter it can deal very nicely with the energy relationships which arise within the $d$ orbitals, and the CFSE makes a small but calculable contribution to the total bonding energy of a $d^n$ complex.

In the study of thermodynamic properties of complexes and their theoretical interpretation we are mainly concerned with the ground state of lowest thermally populated energy levels of the central atoms. Therefore, we shall briefly examine the lowest energy states for various $d^n$ systems and their further splittings in a crystal field. We shall have to examine in more detail what the crystal field stabilization energy is, how it is obtained, and just what significance and limitations it has in the overall picture of thermodynamic stability.

**TABLE 11–4.** The Terms Which Arise from the $d^n$ Configurations, with the Ground Term Indicated Separately.[a]

| Configuration | Ground Term | Higher Energy Terms |
|---|---|---|
| $d^1, d^9$ | $^2D$ | |
| $d^2, d^8$ | $^3F$ | $^3P, {}^1G, {}^1D, {}^1S$ |
| $d^3, d^7$ | $^4F$ | $^4P, {}^2H, {}^2G, {}^2F, 2 \times {}^2D, {}^2P$ |
| $d^4, d^6$ | $^5D$ | $^3H, {}^3G, 2 \times {}^3F, {}^3D, 2 \times {}^3P, {}^1I, 2 \times {}^1G, {}^1F, 2 \times {}^1D, 2 \times {}^1S$ |
| $d^5$ | $^6S$ | $^4G, {}^4F, {}^4D, {}^4P, {}^2I, {}^2H, 2 \times {}^2G, 2 \times {}^2F, 3 \times {}^2D, {}^2P, {}^2S$ |

[a] $2 \times {}^{2S+1}L$ means that the $^{2S+1}L$ term occurs twice.

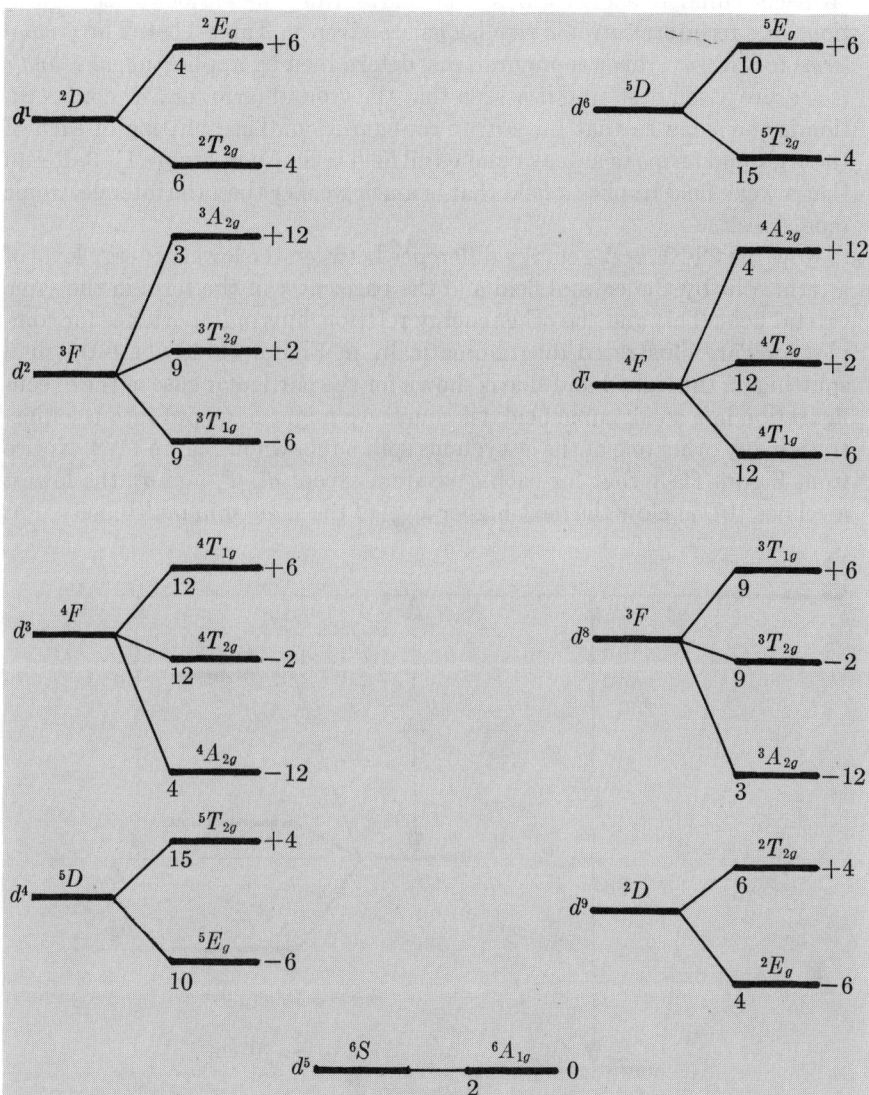

**FIGURE 11-3**

Ground Term Splittings and Relative Energies in $Dq$ for the $d^n$ Configurations in a Weak Octahedral Field. The numbers beneath each level indicate the total degeneracy, that is, spin times orbital, of that level. The orbital degeneracy of each level is determined from the group theoretical symbols $A$, $E$, and $T$, which define 1-fold, 2-fold, and 3-fold degeneracies, respectively. The splittings for a tetrahedral field may be obtained simply by inverting each split set, dropping the $g$ subscript, and recalling that the theoretical $\Delta_t$ magnitude is $(4/9)\Delta_0$.

First let us see how the CFSE arises when we take a field-free $d^n$ ion and place it in a weak field octahedral environment. All of the atomic *terms* (Russell-Saunders energy levels) which arise from the various $d^n$ configurations due to interelectronic repulsions are given in Table 11–4. The *ground terms* for the several $d^n$ configurations, determined by application of *Hund's Rules*, are listed first, and it is seen that the ground term for a $d^n$ configuration is the same as that for a $d^{10-n}$ configuration. The splitting of each of these ground terms in a weak octahedral field is shown in Figure 11–3. Recall that a weak field implies a field that is much weaker than the interelectronic repulsions.

Now the CFSE is the energy difference between the lowest energy level of a term split by the crystal field and the baricenter of the term in the same crystal field. This and the other energy relationships of importance for complex ions are illustrated diagrammatically in Figure 11–4, where the final splitting on the right-hand side is shown for the particular case of the $d^2$ configuration in a weak cubic field. Of course we might have put into the illustration any other one of the $d^n$ systems split as shown in Figure 11–3. Notice from Figure 11–3 that for each $d^n$ system except $d^2$, $d^7$, and $d^5$, the lowest level lies $10Dq$ below the next-higher level of the same spin multiplicity. For

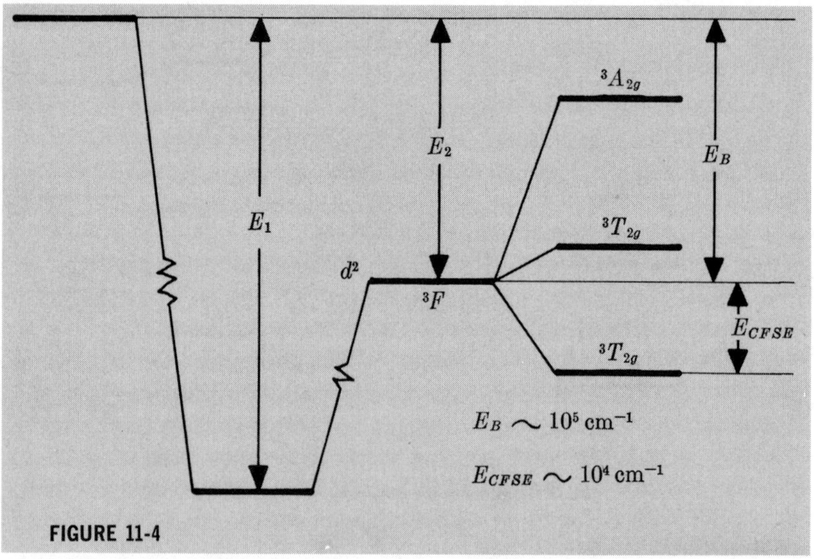

**FIGURE 11-4**

Diagrammatic Illustration of Some Energy Relationships in the Formation of a Complex, Using the $d^2$ Weak-Field Octahedral System as the Example. $E_1$ represents the energy of attraction between the central ion charge and the ligand charges or dipoles; $E_2$ represents the repulsion of the central ion electrons by themselves (interelectronic repulsion) and by the ligand electrons. $E_B$ is the overall bond energy and $E_{CFSE}$ is the crystal field stabilization energy. Since the drawing is not to scale the relative magnitudes of the latter two energies are given.

the $d^2$, $d^7$ systems the value is $8Dq$ and for $d^5$ it is 0, there being no higher levels of the same spin multiplicity in the latter case.

In a strong octahedral field, that is, when the splitting parameter $Dq$ is very large, the energies of the $d$ electrons, literally the torques exerted on their orbital motions, are determined primarily by the strength and symmetry of the field rather than by the interelectronic interactions. In this limit the energy of a term is obtained simply by counting the number of electrons in the $t_{2g}$ and $e_g$ orbitals with the respective octahedral splitting energy values of $-4Dq$ and $+6Dq$. However, we must remember that for certain $d^n$ configurations stabilization is achieved at the expense of some promotional energy (electron pairing). The CFSE's then are given in Table 11–5. The correction term for the $d^2$ configuration as well as part of

**TABLE 11–5. Ground Terms, and Relative Energies for the $d^n$ Configurations in Strong Octahedral Fields.**

| Atomic Configuration | Complex Configuration | | Ground Term | Octahedral Field Stabilization Energy |
|:---:|:---:|:---:|:---:|:---:|
| | $t_{2g}$ | $e_g$ | | |
| $d^1$ | 1 | — | $^2T_{2g}$ | $4\,Dq$ |
| $d^2$ | 2 | — | $^3T_{1g}$ | $8\,Dq - (3F_2 - 15F_4)$ |
| $d^3$ | 3 | — | $^4A_{2g}$ | $12\,Dq$ |
| $d^4$ | 4 | — | $^3T_{1g}$ | $16\,Dq - (6F_2 + 145F_4)$ |
| $d^5$ | 5 | — | $^2T_{2g}$ | $20\,Dq - (15F_2 + 275F_4)$ |
| $d^6$ | 6 | — | $^1A_{2g}$ | $24\,Dq - (5F_2 + 255F_4)$ |
| $d^7$ | 6 | 1 | $^2E_g$ | $18\,Dq - (7F_2 + 105F_4)$ |
| $d^8$ | 6 | 2 | $^3A_{2g}$ | $12\,Dq$ |
| $d^9$ | 6 | 3 | $^2E_g$ | $6\,Dq$ |

the correction for $d^7$ arises from *configuration interaction*, which we shall not go into here, but the corrections for the $d^4$ to $d^7$ cases are necessary to take account of the additional interelectronic interactions present in these systems. The mathematical details involved in obtaining the weak and strong crystal field energies, as well as the more involved problem of *intermediate fields*, are very lucidly presented in the book by Ballhausen[10] and are outside the intended scope of the presentation here.

One of the most satisfying applications of CFT has been its ability to correlate certain thermodynamic properties, particularly in terms of CFSE's. Consider, for example, the heats of hydration of the divalent and trivalent transition metal cations. These are the enthalpies of the following processes:

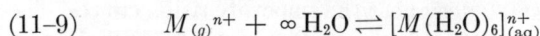

$$(11\text{-}9) \qquad M_{(g)}{}^{n+} + \infty H_2O \rightleftharpoons [M(H_2O)_6]_{(aq)}^{n+}$$

If all of the ions of a given charge were spherically symmetrical as are $Ca^{2+}$, $Mn^{2+}$, and $Zn^{2+}$, we would expect a rather smoothly rising curve for a plot of

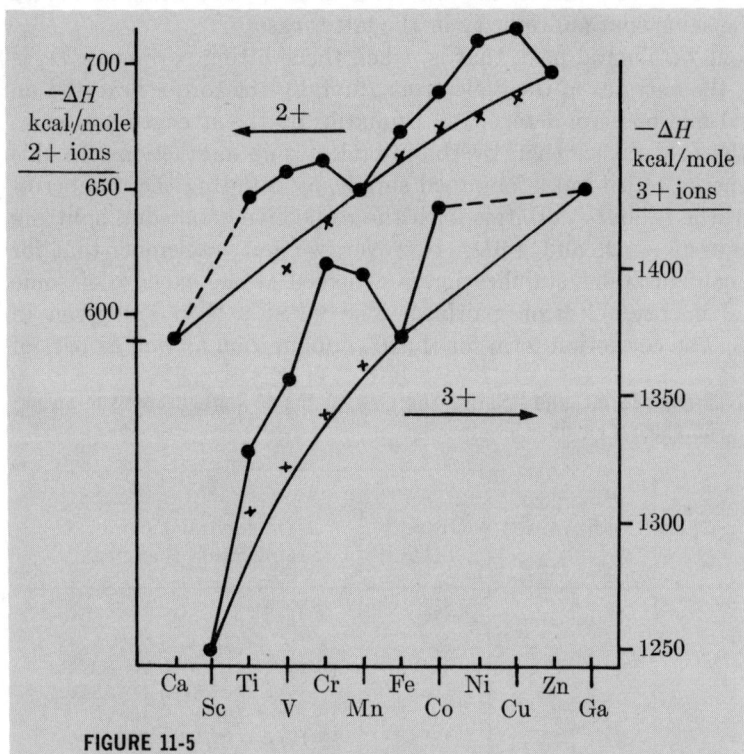

**FIGURE 11-5**

Heats of Hydration of Bivalent (Upper Curves) and Tervalent (Lower Curves) Transition Metal Cations. The upper curve in each set joins experimental points; the lower curve connects the spherically symmetrical $d^0$, $d^5$, and $d^{10}$ ions. The values designated by $x$ result from subtracting the CFSE from each $\Delta H$ value.

hydration enthalpies as a function of atomic number. The reason we would expect this is because of the steady decrease in cation size with increase in atomic number. However, we must recall that the size contraction is *not* smooth for the $d^n$ series, as discussed earlier in this chapter, but that it is smooth for both the $4f$ and $5f$ series of $3+$ and $4+$ ions as discussed in Chapter 4. If we plot the experimental data, as is done in Figure 11–5, we see that, in fact, a smoothly rising curve goes through only the experimental points for the spherically symmetrical ions, that is, $d^0$, $d^5$, and $d^{10}$, which have no CFSE. Points for all other ions lie above this curve. However, if the CFSE energies, determined from electronic spectral studies, are subtracted from the actual hydration energies, the values shown by the $x$'s in the figure are obtained, and these fall very near the smoothly rising curves. The characteristic double-humped curves shown in Figure 11–5 for hydration energies are obtained also when the values plotted are *ligation* energies arising from ligands other than $H_2O$, or if the values (see Figure 11–6) are crystal lattice energies of, for example, the divalent halides. Obviously CFSE $Dq$ values might be estimated from the thermochemical data as well as spectral data. These correlations represent most of the usefulness of the CFT, and although we are aware of its total artificiality, nevertheless the basic importance and correctness of the $d$-orbital splittings is verified. The MOT leads to the same predictions, but not nearly so simply.

Of particular interest is the fact that, although CFSE's account for no more than 5% to 10% of the *total* energies of the metal-ligand combinations, they are calculable and critical in explaining the subtle *differences* found in various thermodynamic energies. This happens mainly because other thermodynamic controlling factors either remain fairly constant, as for example the entropy changes, or vary uniformly, as for example the radii of spherically symmetrical ions.

One final point which we have so far neglected is that it would appear from the foregoing consideration of CFSE values that the peaks of the two humps in the upper (2+ ion) curve in Figure 11–5 should occur at $d^3$ and $d^8$, not at $d^4$ and $d^9$ as is observed. The explanation undoubtedly lies in the fact that the regular octahedral structure is not possible for $d^4$ and $d^9$ configurations such as are found in $Cr^{2+}$ and $Cu^{2+}$, but rather these ions adopt a strongly tetragonally distorted octahedral structure as a result of the operation of the Jahn-Teller effect, discussed earlier. The latter leads actually to a greater stabilization than predicted by simple electrostatic theory, and in fact one estimate of the "extra" Jahn-Teller stabilization energy for the hydrated $Cu^{2+}$ is 8 kcal/mole.

We might now summarize some of the predictions of the effect of the number of $d$ electrons upon the stability of complexes in terms of the stability constant, $\beta_n$, and the negative enthalpy, $\Delta H_n$, for the general reaction

$$(11\text{–}10) \qquad M(H_2O)_n + nL \rightleftharpoons ML_n + nH_2O$$

$\beta_n$ will increase and $-\Delta H_n$ will increase according to the sequence

$$d^0 < d^1 < d^2 < d^3 \gtrsim d^4 > d^5 < d^6 < d^7 < d^8 \gtrsim d^9 > d^{10}$$

for the formation of bi- and tervalent metal complexes. The sequence $d^3 > d^4$ and $d^8 > d^9$ will be expected when the Jahn-Teller stabilization is low and the coordination number is 6. The sequences $d^3 < d^4$ and $d^8 < d^9$ will be expected when the Jahn-Teller stabilization is large and the coordination number is 4.

## KINETIC PROPERTIES

### Substitution Reactions

Substitution reactions of inorganic complexes have been receiving intensive experimental investigation ever since the excellent 1952 review by

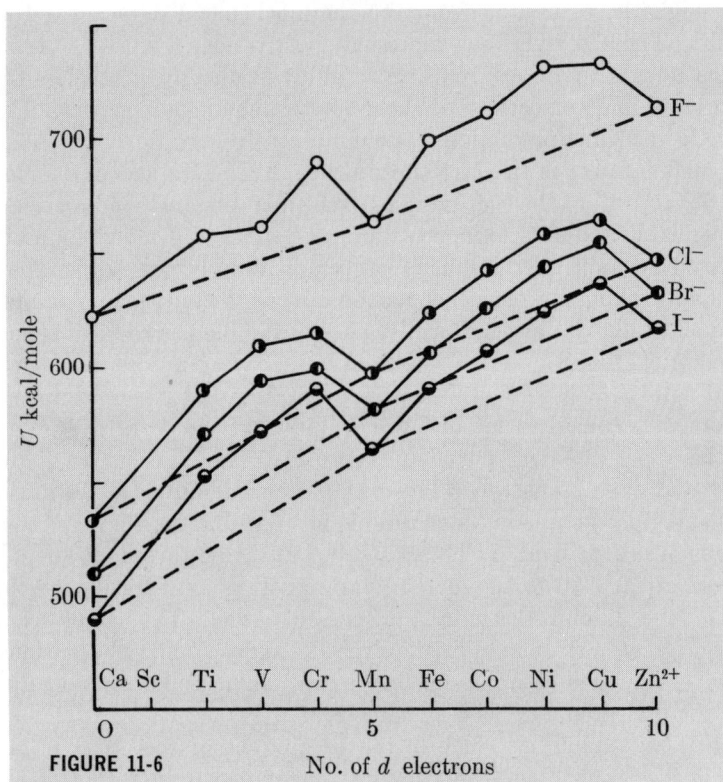

**FIGURE 11-6**    No. of $d$ electrons

Lattice Energies, $U$, of the Divalent $3d$ Series Halides and Chalcogenides, as a Function of the Number of $d$ Electrons.

Taube,[11] in which he rationalized the relative reactivities of different metal ion complexes in terms of the valence bond theory. Since that time, many hundreds of research papers as well as several thorough review articles[12-16] have appeared reporting and interpreting kinetics data. Along with these, at least two very important books have appeared which are devoted primarily to the subject. The first, which came out in 1958 and was revised in 1967, by Basolo and Pearson,[1] emphasizes the use of CFT for interpretation and explanation of kinetics results, and the second, which appeared in 1966, by Langford and Gray,[2] attempts to apply MOT to the same task.

Closely related to and generally considered along with substitution reactions of complexes is the class of reactions involving oxidation-reduction and generally referred to as electron-transfer reactions. This is an area in which the experimental science[17,18] is still well in advance of the theoretical, so we shall have very little to say about it. In fact, we shall only have space to present but a very cursory consideration of the broad subject of substitu-

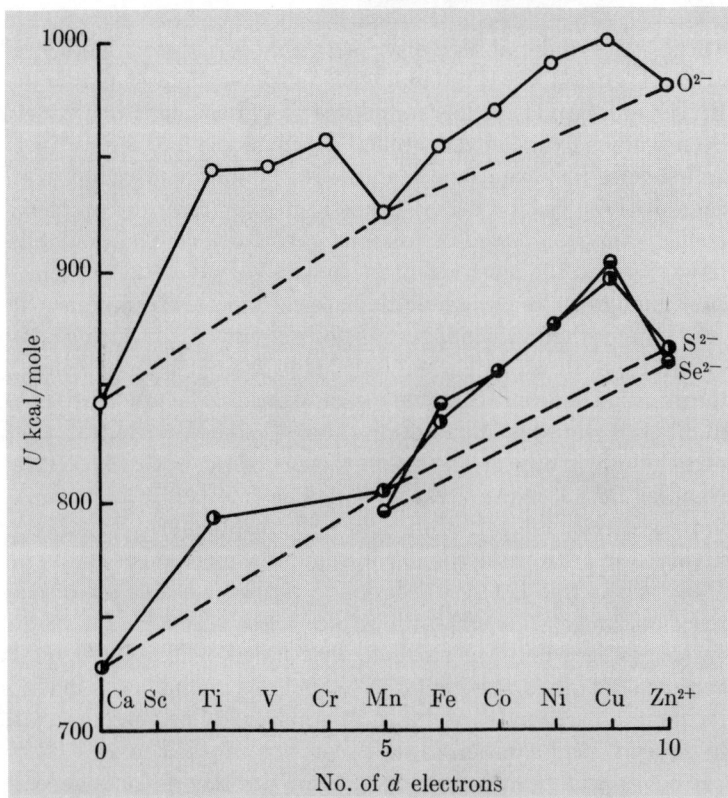

**FIGURE 11-6 (Continued)**

tion reactions, sufficiently simple to point out in what ways theory has recently been applied and in what respects the theory lags behind the experimental knowledge of the field.

A generalized equation for ligand substitution reactions may be written quite simply as

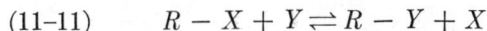

$$(11\text{--}11) \qquad R - X + Y \rightleftharpoons R - Y + X$$

in which $R$ stands for the central metal atom *plus* all of the attached ligands not undergoing substitution, and $X$ and $Y$ are any two ligands. For example, consider $[Co(NH_3)_5Br]^{2+} + H_2O \rightleftharpoons [Co(NH_3)_5(H_2O)]^{3+} + Br^-$, in which $R = [Co(NH_3)_5]^{3+}$, $X = Br^-$, and $Y = H_2O$. In practice we may make the following generalizations concerning Eq. (11–11).

(1) The central atom may be a transition or a non-transition metal atom, but most studies have involved the former and indeed only a few of these have been investigated in detail.

(2) The forward reaction rate may range from very fast ($t_{1/2}$ of $10^{-1}$ to $10^{-7}$ sec at R.T.) to very slow ($t_{1/2}$ of the order of days or weeks at R.T.).

(3) Most of the substitution reactions that have been studied system-

atically and in detail have been with six-coordinated octahedral complexes of Co(III) and Cr(III), and with four-coordinated planar complexes of Pt(II) and Au(III). A few complexes of a great many other metal ions have been studied over the years but the bulk of the data exists for the four central species just mentioned, and this is undoubtedly due to the fact that each of these central atoms forms a great variety of complexes and is extremely *inert* or *nonlabile* towards substitution. The nonlabile complexes have received the most attention to date simply because their reactions are slow enough to be accessible to classical kinetics techniques.

(4) Stereochemical changes may or may not occur during a given reaction of the type shown in Eq. (11–11). However, it is important to know this information before reasonable mechanisms can be proposed, not only for the number of steps but for the steric course of the reaction. In the special case of a kinetic study of the racemization of an optically active octahedral complex $RX$, $Y$ is usually a solvent molecule and it may or may not be involved in the racemization. The establishment of whether or not it is involved is critical for the elucidation of a mechanism.

(5) When $Y$ is $H_2O$ the reaction is termed an *aquation* or *acid hydrolysis* reaction, and if $Y$ is $OH^-$, the reaction is a *base hydrolysis* reaction. These two types have been the most widely studied of all of the several types of reactions of octahedral complexes. When $Y$ is an anion the reaction is termed an *anation* reaction. However, if $X$ is also an anion then the anation reaction in aqueous acid solution always seems to proceed in at least two steps. It has never been observed that one anion directly replaces another anion in a single step. The original anion is replaced by $H_2O$ in the first step, then the $H_2O$ is replaced by the entering anion in the second step.

(6) When $X = Y$ we are dealing with an *exchange reaction*, and in order to study the rate in this case either $X$ or $Y$ must be "labeled," for example $X = CN^-$ and $Y = {}^{14}CN^-$, or $X = Cl^-$ and $Y = {}^{36}Cl^-$, or $X = SC(NH_2)_2$ and $Y = {}^{35}SC(NH_2)_2$. In practice, of course, it may be the $X$ ligand which is labeled, with the $Y$ ligand being normal.

(7) The variables or factors that may influence reaction rates of complexes, aside from the usual factors such as temperature, catalysts, etc., are:

(a) the nature of the central metal ion
(b) the nature of the leaving ligand
(c) the nature of the entering ligand
(d) the nature of the other ligands attached to the central atom
(e) the charge on the complex
(f) the nature of the solvent
(g) the presence of other ligands or metal ions in the solvent

A thorough kinetic study will include the investigation of the effects of most of these variables, and of course the more of these that can be studied the more firmly based will be a proposed mechanism of the reaction.

(8) For octahedral substitution reactions we may visualize two limiting

reaction sequences or pathways which may be written and labeled as follows:

$$(A) \quad R - X + Y \rightarrow Y - R - X \rightarrow R - Y + X$$

This type can be referred to as either *displacement* or *association* or $S_N2$ (*lim*) or $A$ mechanism, where $S_N2$(lim) stands for *substitution* (S), *nucleophilic* (N), *bimolecular* (2) and *lim* implies for this case the existence, even though transient, of a true *seven*-coordinated intermediate.

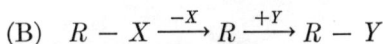

$$(B) \quad R - X \xrightarrow{-X} R \xrightarrow{+Y} R - Y$$

This type of reaction is referred to as either a *dissociation* or $S_N1$ (*lim*) or $D$ mechanism, where $S_N1$(lim) stands for *substitution, nucleophilic, mono-molecular*, and the *lim* in this case implies the existence of a true, even though transient, *five*-coordinated intermediate.

Of course, one could visualize one or more mechanisms that lie somewhere between (A) and (B), and indeed this is one of the complicating features of kinetics studies. The approach taken by Langford and Gray[2] to this problem appears at present to be a reasonable and workable one, but it will not be discussed here. What will be considered is the kind of approach permitted by CFT to the problem of choosing one extreme mechanism or the other and of explaining observed inertness or lability based upon the preferential occupancy of lower-lying $d$ orbitals in the split $d$-orbital energy level scheme and upon the resulting loss or gain in CFSE during the formation of certain transition state configurations.

Consider first the (A) mechanism in which $Y$ attacks an octahedral complex $MX_6 (= RX)$. It is reasonable to suppose that $Y$ will approach the central metal atom along one of the four $C_3$ axes of the octahedron, as illustrated in Figure 11–7. This might be expected since these project through the center of the "empty" triangular faces of a regular octahedron. Now it is presumed that the three $t_{2g}$ orbitals have their maxima near the four $C_3$ axes and that they point between the ligands, whereas the two $e_g$ orbitals lie along the three $C_4$ axes on which the six ligands sit. If the $t_{2g}$ orbitals contain either three or six electrons, as in Cr(III) and Co(III), respectively, we should expect a higher activation energy to be required to clear or make available one of these orbitals, for example by promoting an electron from $t_{2g}$ to $e_g$. This then provides an empty orbital for attachment by the attacking seventh ligand. Although this is an oversimplified approach, it is significant that of all of the common $3d$ metal ions, the complexes of Cr(III) and especially those of Co(III) are by far the most nonlabile. In fact, all the others are labile in varying degrees.

Of course, we have considered only one factor among a complicated array of factors which may affect substitution reactions of complex ions. And even that was considered qualitatively and involved several implicit approximations. Ideally we should like to be able to calculate the activation energies for different complex ion systems undergoing reactions via different mecha-

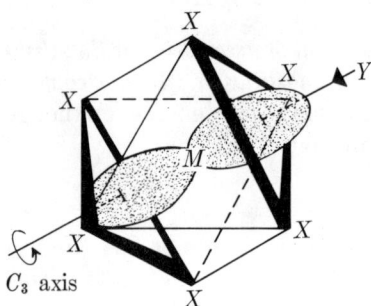

**FIGURE 11-7**

Octahedral Complex, $MX_6$, Being
Attacked by the Ligand $Y$ on One of
the Triangular Faces of the Octahedron, that is, along One of the Four
$C_3$ Axes. The orbital lobes pictured
belong to a $t_{2g}$ orbital which, if occupied, will serve to repel $Y$, but if
unoccupied will permit greater attraction of $Y$ by the $M$ nucleus.

nisms. Although this is not yet possible, the (one-electron) crystal field
energy contributions to the activation energy have been computed by
Basolo and Pearson,[1] and are recorded in Tables 11–6 and 11–7. It may be
seen from the $\Delta E_a$ values that for the $S_N2$ reactions (Table 11–6) the $d^n$ systems showing the most resistance to change are strong field $d^3$ and $d^6$ (for
which configuration there is a maximum energy change), $d^7$ and $d^8$, and
weak field $d^3$ and $d^8$. For the alternative reaction mechanism, $S_N1$ (Table
11–7), the most energetic changes are required for strong field $d^3$, $d^6$, (again
the maximum change configuration) and $d^8$, and weak field $d^3$ and $d^8$. Assuming that either mechanistic extreme should be accessible to most octahedral complexes, we deduce that the strong field $d^6$ and the $d^3$ and $d^8$
systems in general should be the most nonlabile systems and decreasingly
so in that order. This is roughly borne out by the facts, although $d^8$ octahedral systems, $Ni^{2+}$ in particular, are fairly labile. However, this is probably
due to the greater ease of expulsion of a ligand attached to a metal ion that
possesses $e_g$ electrons. Whether one considers these electrons as being purely
on the metal, as in CFT, and pointing at and repelling the ligands, or as
being antibonding electrons only partly on the metal as in MOT, the predicted result is the same, namely, a weakening of the metal-ligand bond
which should allow for a lower-energy $S_N1$ path.

**TABLE 11-6. Crystal Field Activation Energies for $S_N2$ Reactions.[1]**

| $d^n$ System | Octahedral → Pentagonal Bipyramid | | | | | |
| --- | --- | --- | --- | --- | --- | --- |
| | Strong Fields | | | Weak Fields | | |
| | Octa-hedral | Pentagonal Bipyramid | $\Delta E_a$ | Octa-hedral | Pentagonal Bipyramid | $\Delta E_a$ |
| $d^0$ | $0Dq$ | $0Dq$ | $0Dq$ | $0Dq$ | $0Dq$ | $0Dq$ |
| $d^1$ | 4 | 5.28 | −1.28 | 4 | 5.28 | −1.28 |
| $d^2$ | 8 | 10.56 | −2.56 | 8 | 10.56 | −2.56 |
| $d^3$ | 12 | 7.74 | 4.26 | 12 | 7.74 | 4.26 |
| $d^4$ | 16 | 13.02 | 2.98 | 6 | 4.93 | 1.07 |
| $d^5$ | 20 | 18.30 | 1.70 | 0 | 0 | 0 |
| $d^6$ | 24 | 15.48 | 8.52 | 4 | 5.28 | −1.28 |
| $d^7$ | 18 | 12.66 | 5.34 | 8 | 10.56 | −2.56 |
| $d^8$ | 12 | 7.74 | 4.26 | 12 | 7.74 | 4.26 |
| $d^9$ | 6 | 4.93 | 1.07 | 6 | 4.93 | 1.07 |
| $d^{10}$ | 0 | 0 | 0 | 0 | 0 | 0 |

**TABLE 11-7. Crystal Field Activation Energies for $S_N1$ Reactions.[1]**

| $d^n$ System | Octahedral → Square Pyramid | | | | | |
| --- | --- | --- | --- | --- | --- | --- |
| | Strong Fields | | | Weak Fields | | |
| | Octa-hedral | Square Pyramid | $\Delta E_a$ | Octa-hedral | Square Pyramid | $\Delta E_a$ |
| $d^0$ | $0Dq$ | $0Dq$ | $0Dq$ | $0Dq$ | $0Dq$ | $0Dq$ |
| $d^1$ | 4 | 4.57 | −0.57 | 4 | 4.57 | −0.57 |
| $d^2$ | 8 | 9.14 | −1.14 | 8 | 9.14 | −1.14 |
| $d^3$ | 12 | 10.00 | 2.00 | 12 | 10.00 | 2.00 |
| $d^4$ | 16 | 14.57 | 1.43 | 6 | 9.14 | −3.14 |
| $d^5$ | 20 | 19.14 | 0.86 | 0 | 0 | 0 |
| $d^6$ | 24 | 20.00 | 4.00 | 4 | 4.57 | −0.57 |
| $d^7$ | 18 | 19.14 | −1.14 | 8 | 9.14 | −1.14 |
| $d^8$ | 12 | 10.00 | 2.00 | 12 | 10.00 | 2.00 |
| $d^9$ | 6 | 9.14 | −3.14 | 6 | 9.14 | −3.14 |
| $d^{10}$ | 0 | 0 | 0 | 0 | 0 | 0 |

Langford and Gray[2] have summarized our knowledge of Co(III) substitution reactions in the following way. The major substitution reaction mechanism in acid solution involves essentially an $S_N1$ process in which the Co(III) complex is assumed to accumulate most of the necessary activation energy without significant bonding to the entering group. Three main lines of experimental evidence support this hypothesis: (a) the accessibility to kinetic study of only two reactions, acid hydrolysis and its reverse, anation; (b) the systematic failure of all attempts to identify selective reagents for attack of the Co(III) center, even in nonaqueous solutions; and (c) the accumulated evidence on *leaving-group* effects, such as charge effects, steric effects, chelation effects, and nonlabile ligand effects. On the other hand, the major mechanism believed operative in rapid base hydrolysis involves the *conjugate base* (CB) of the complex in a reaction, termed $S_N1CB$, which is analogous to acid hydrolysis. The main evidence here is (a) the requirement of an acid proton for *rapid* base hydrolysis and (b) the separability of the hydroxide-dependent step from the product-determining step.

Other than for octahedral Co(III), the greatest mass of experimental data exists for the nonlabile square planar Pt(II), a $d^8$ system. Qualitatively it is easy to see why this system might be inert. The preferred mechanism for substitution reactions of four-coordinate complexes will be $S_N2$. The $d_{x^2-y^2}$ orbital will be empty, whereas along with the three $d_\epsilon$ orbitals, the $d_{z^2}$ orbital lying along the $C_4$ axis of the square plane will be filled and will serve to partially repel incoming attacking ligands. Before we make some generalizations about Pt(II) substitution reactions it should be noted that most square planar substitution reactions in solution follow a two-term rate law:

$$(11\text{-}12) \qquad \text{Rate} = \frac{-d[RX]}{dt} = (k_1 + k_2[Y])[RX]$$

where $k_1$ is a first-order rate constant, $k_2$ is a second-order rate constant, $[RX]$ is the concentration of the complex and $[Y]$ is the concentration of the entering ligand.

Now the major activation process is assumed to be $S_N2$, via a trigonal bipyramidal intermediate. The main evidence, all accounted for by this mechanism, has been summarized by Langford and Gray.[2] For the $k_2$ path the evidence is (a) a unified interpretation of substituent effects based on the $S_N2$ model; (b) the large decrease in rate observed on blocking the attack positions (*cis* blocking being more effective than *trans* blocking); (c) the large entering-group effect which parallels closely the *trans*-effect (*vide infra*) order of ligands; and (d) the observation that no [Pt(dien)OH]$^+$ is formed during the reaction of [Pt(dien)Br]$^+$ with $Y^-$ in the presence of OH$^-$. This shows that there is no intervention of the solvent along the $k_2$ pathway. The main evidence for the $S_N2$ trigonal bipyramidal mechanism *involving* solvent as the reagent in the $k_1$ term is (a) the large decrease in the $k_1$ rate on blocking the attack positions in some Pt(II) complexes; (b) the observation that the $k_1$ rate is relatively insensitive to changes in the net charge on the complex; and (c) the solvent-effect experiments which show that good coordinating solvents enhance the $k_1$ term, for example, $k_1$ (DMSO) $> k_1(H_2O)$, in reactions of [Pt$L_2$Cl$_2$] complexes.

## Trans Effect

We mentioned above the so-called *trans effect* and this phenomenon, first recognized by Werner (1893), elaborated on by Chernuyaev (1926), and studied by many others since then (see reference 2), is extremely important for a full understanding of the kinetic behavior of square planar complexes. The *trans* effect refers to the very substantial effect of nonlabile ligands upon the lability of ligands *trans* to themselves, and the effect spans several orders of magnitude. For example, in the substitution reaction

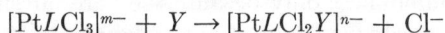

$$[PtLCl_3]^{m-} + Y \rightarrow [PtLCl_2Y]^{n-} + Cl^-$$

certain ligands, $L$, cause the *trans* product to be formed more rapidly than the *cis* product. Thus, we may arrange such ligands in order of their relative rates. A recent series, proposed by Langford and Gray,[2] is as follows:

relative increasing rates

| $\sim 10^{-1}$ | 1 | $10^2$ | $\sim 10^5$ |
|---|---|---|---|
| small | moderate | large | very large |
| py, $NH_3$, | $Cl^-$, $Br^-$ | $I^-$, $NCS^-$, | CO, $CN^-$, |
| $OH^-$, $H_2O$ | | $NO_2^-$, $SO_3H^-$, | $C_2H_4$, NO, |
| | | $SC(NH_2)_2$ | $PMe_3$, $PEt_3$, $H^-$ |
| | | $CH_3^-$, $C_6H_5^-$ | |

The most important theories of the *trans* effect are the electrostatic-polarization theory of Grinberg (1927) and the $\pi$-bonding theory of Chatt (1955) and Orgel (1956). The latter theory proposes that the increasing order of *trans* effect is roughly the order of increasing $(M{\rightarrow}L)\pi$-bonding capacity of the ligands. The basis for this idea is that charge should be removed from the central metal atom in the transition state of the complex in order to make the approach easier for the incoming ligand. And double bonding of this type removes more charge or removes it more effectively from the region near the *trans* ligand than from close to the *cis* ligands. Furthermore, there is experimental evidence which indicates that the strength of the bond *trans* to a strongly *trans*-directing ligand is in fact weakened. The $\pi$-bonding theory, however, cannot account for the recently discovered very large *trans* effect of $H^-$. Therefore, Langford and Gray[2] have proposed that both $\sigma$- and $\pi$-electronic effects are important, and they present a quite reasonable MOT approach to what they separate as a $\sigma$-*trans* effect and a $\pi$-*trans* effect.

## Electron Transfer Reactions

Oxidation-reduction reactions involving complexes of different metal ions proceed at vastly different rates and apparently by several different mechanisms.[1,17,18] Although a great deal of experimental data exists on the rates, very little is actually known about the mechanisms. Nevertheless, we can divide the redox reactions into two main classes: (a) those in which the electron transfer results in no net chemical change, or the so-called *electron exchange* processes such as, for example, $[Fe(CN)_6]^{3-}/[Fe(CN)_6]^{4-}$, $[Co(NH_3)_6]^{3+}/[Co(NH_3)_6]^{2+}$, $MnO_4^-/MnO_4^{2-}$, $[Mo(CN)_6]^{2-}/[Mo(CN)_6]^{3-}$, $[Fe(ophen)_3]^{3+}/[Fe(ophen)]^{2+}$, etc.; and the majority, which are (b) those in which there is a net chemical change, for example, $[Cr(NH_3)_5X]^{2+}/Cr^{2+}(aq)$. Those in the first class have been subjected to an amount of study far out of proportion to their commonness only because they are far simpler to deal with theoretically and mechanistically, if not experimentally.

Redox may occur by *atom transfer* in which a free radical moves from one coordination sphere to the other, but most redox reactions are believed to take place by one of two rather well-established general mechanisms for electron transfer. In the *outer-sphere activated complex* or *tunnelling* mechanism each reacting complex molecule retains its own entire coordination shell in the activated complex so that there is no ligand common to each central atom, and the electron, or more precisely the equivalent of one electronic charge, is assumed to "tunnel" through both of the coordination shells. A few generalizations concerning this mechanism have emerged from experimental studies on electron exchange reactions:

(1) The range of second-order rate constants is very large, extending perhaps from $\sim 10^{-4}$ to as high as $\sim 10^9$ liter mole$^{-1}$ sec$^{-1}$. For example, consider

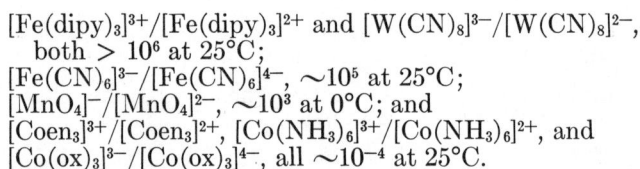

$[Fe(dipy)_3]^{3+}/[Fe(dipy)_3]^{2+}$ and $[W(CN)_8]^{3-}/[W(CN)_8]^{2-}$,
  both $> 10^6$ at 25°C;
$[Fe(CN)_6]^{3-}/[Fe(CN)_6]^{4-}$, $\sim 10^5$ at 25°C;
$[MnO_4]^-/[MnO_4]^{2-}$, $\sim 10^3$ at 0°C; and
$[Coen_3]^{3+}/[Coen_3]^{2+}$, $[Co(NH_3)_6]^{3+}/[Co(NH_3)_6]^{2+}$, and
$[Co(ox)_3]^{3-}/[Co(ox)_3]^{4-}$, all $\sim 10^{-4}$ at 25°C.

(2) Fast electron transfer is expected, and in general found, if there is little change in the molecular dimensions accompanying the reaction.

(3) Fast electron transfer is facilitated if electrons can reach the surface of a reactant molecule from the central atom via a conjugated (delocalized) electronic system or through a single atom.

(4) The outer-sphere mechanism is sure to be the correct one when both of the reactant molecules exchange their ligands more slowly than they undergo electron transfer.

The other general electron transfer mechanism is called the *inner-sphere* or *ligand-bridged activated complex* mechanism. The bridging may occur via a single monatomic ligand as in

$$[(NH_3)_5Co\text{—}Br\text{—}Cr(H_2O)_5]^{4+}$$

or through a more extended polyatomic bridging ligand as in

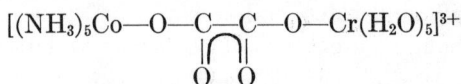

$$[(NH_3)_5Co\text{—}O\text{—}C\underset{\underset{O}{\|}}{}\text{—}C\underset{\underset{O}{\|}}{}\text{—}O\text{—}Cr(H_2O)_5]^{3+}$$

where the delocalization in the oxalato ligand is shown. Here the conjugated system provides a pathway for electron exchange between Co(III) and Cr(II) which is of the order of 100 times more rapid than a nonconjugated bridging ligand. Taube and his co-workers[19,20] have studied a great many redox reactions, for example,

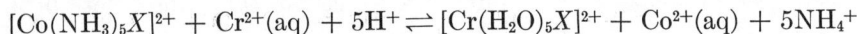

$$[Co(NH_3)_5X]^{2+} + Cr^{2+}(aq) + 5H^+ \rightleftharpoons [Cr(H_2O)_5X]^{2+} + Co^{2+}(aq) + 5NH_4^+$$

where $X$ = F⁻, Cl⁻, Br⁻, I⁻, NCS⁻, N₃⁻, CH₃COO⁻, C₃H₇COO⁻, SO₄²⁻, PO₄³⁻, P₂O₇⁴⁻, crotonate, succinate, oxalate, maleate, fumarate, o-phthalate, p-phthalate, etc. Now the Co(III) complex is nonlabile, but the Cr(II) ion is labile, whereas the product Cr(III) is nonlabile and the Co(II) is labile. Therefore, since the $X$ ligand is transferred quantitatively from Co(III) to Cr(II) during the redox reaction, which transforms these into Co(II) and Cr(III), respectively, the bridged activated complex mechanism is very elegantly demonstrated. All of the experimental data on reactions presumed to proceed via this *inner-sphere* mechanism supports the generalizations that (a) electron transfer occurs at a reasonably fast rate through polyatomic molecules only if they are conjugated systems and (b) reactions which involve large changes in molecular dimensions proceed slowly.

Thus, for both types of electron transfer mechanisms outlined above we may conclude that reactions will be fast if the equilibrium configurations of the reactant molecules do not have to change much in the course of the reaction. Otherwise they will be slow. Now since $e_g$ electrons have a much greater effect upon internuclear distances than $t_{2g}$ electrons, the slowest redox reactions between octahedral complexes should be the ones that involve the transfer of $e_g$ electrons. In general this is certainly what is observed experimentally.

Oxidation and reduction of ligands due to their interaction with their central atoms has also received some attention, but we shall simply refer the interested reader to a recent review[21] of this subject.

## MAGNETIC PROPERTIES

Considerable emphasis and importance has been placed upon the determination of magnetic properties of transition metal complexes.[6,22,23] Such studies have contributed much to our understanding of the chemistry as well as to the characterization of these compounds, particularly in regard to oxidation states, stereochemistry of the central atoms, and their bond types. Before we consider just how this has been possible it will be of value to examine the origins and types of magnetic behavior exhibited by transition metal complexes.

### Origins of Magnetic Behavior

Magnetic phenomena in chemical substances may arise from both electrons and nucleons. However, the magnetic effects due to electrons are of the order of $10^3$ times greater than those due to nucleons and nuclei. There-fore, except for the chemical information obtainable from nuclear magnetic and quadrupole resonance spectroscopy and from the hyperfine nuclear spin-electron spin interaction which may be observed in electron spin reso-nance spectroscopy, the magnetic phenomena associated with or arising from electrons alone will be of interest to the chemist. Only the latter will be discussed here.

An electron may be considered, in effect, an elementary magnet. The origin of its magnetism is most easily described in classical, that is, pre-wave mechanical terms, where we may picture the electron as a hard negatively charged sphere that is both spinning on its axis and traveling in a closed path about a nucleus. The former motion gives rise to the *spin moment* and the latter to the *orbital moment* of the electron, and some combination of these two moments results in the paramagnetic moments found for certain atoms, ions, or molecules. These are expressed in *Bohr Magneton* (BM) units, where

$$(11\text{–}13) \qquad 1 \text{ BM} = \beta = \frac{eh}{4\pi mc} = 9.27 \times 10^{-21} \text{ erg/gauss}$$

Here $e$ is the electronic charge, $h$ is Planck's constant, $m$ is the electron rest mass, and $c$ is the speed of light. The magnetic moment of a free electron, $\mu_s$, that is its spin-only moment, is given by wave mechanics as

$$(11\text{–}14) \qquad \mu_s \text{ (in BM units)} = g\sqrt{s(s+1)}$$

where $s$ is the absolute value of the spin quantum number and $g$ is called the *gyromagnetic ratio* or *Lande splitting factor*. Since the factor $\sqrt{s(s+1)}$ is the value of the angular momentum for the electron, $g$ is seen to be the ratio of the magnetic moment to the angular momentum. For the free electron $g$ has the value 2.0023, which is generally taken simply as 2.00. The value of $g$ for an unpaired electron in a gaseous atom or ion for which Russell-Saunders coupling is applicable, is given by the following expression

$$(11\text{–}15) \qquad g = 1 + \frac{S(S+1) - L(L+1) + J(J+1)}{2J(J+1)}$$

where $S$, $L$, and $J$ have their usual meaning (see Chapter 2). Note that for the free electron with no orbital angular momentum (that is, $L = 0$), $J = S$, and then $g = 2.00$. The added 0.0023 for the actual value for a free electron is due to a relativistic correction and is of no importance to us here. Using the value of 2.00 for $g$ in Eq. (11–14) we can calculate a "spin-only" moment of 1.73 BM for a single electron ($S = \frac{1}{2}$). However, if the electron is part of a chemical unit there may or may not be a contribution to its mag-netic moment due to its orbital motion, and there may or may not be a contribution from so-called *temperature-independent magnetism* (TIP). We shall consider these matters later after we have examined the types of mag-netic behavior found in bulk compounds.

## Magnetic Susceptibility and Types of Magnetic Behavior

There are several types of bulk magnetic behavior, but not all of these are of importance in metallic complex molecules. *Ferromagnetism, antiferromagnetism*, and *ferrimagnetism* are relatively rare phenomena in complexes and are of importance only in special cases. For this reason, they will not be considered here. Of much greater significance for our purposes are *normal paramagnetism* and *diamagnetism*, and it is these that we shall discuss in some detail. But first let us briefly define *magnetic susceptibility*.

When a substance is placed in a magnetic field of strength $H$, the flux or magnetic induction, $B$, within the substance is given as

$$(11\text{–}16) \qquad B = H + 4\pi I$$

where $I$ is termed the *intensity of magnetization*. Dividing both sides of Eq. (11–16) by $H$ yields

$$(11\text{–}17) \qquad \frac{B}{H} = 1 + 4\pi \frac{I}{H} = 1 + 4\pi\kappa$$

where the ratio $B/H$ is called the *magnetic permeability* of the substance and $\kappa$ is the *magnetic susceptibility per unit volume*. The latter is a measure of the susceptibility of the substance to magnetic polarization.

Now when a substance is placed in an inhomogeneous magnetic field, it experiences a force, $F$, which is proportional to the field strength, $H$, to the gradient of the field, $\partial H/\partial y$, and to the sample volume, $V$. Mathematically we may express this as follows:

$$(11\text{–}18) \qquad F = \kappa V H \frac{\partial H}{\partial y}$$

where the constant of proportionality is the same volume magnetic susceptibility as defined above. Since we more often deal with weights of solids rather than their volumes, we find it useful to define the *specific* or *gram susceptibility*, $\chi$, and the *molar susceptibility*, $\chi_M$, by the following relations:

$$(11\text{–}19) \qquad \chi = \frac{\kappa}{\rho} \qquad \text{and} \qquad \chi_M = \frac{\kappa M}{\rho}$$

where $\rho$ is the density and $M$ the formula weight of the compound. $\chi_A$ and $\chi_{A\pm}$ may be analogously defined as the atomic and ionic susceptibilities, respectively.

## Diamagnetism

Diamagnetism is a property possessed by all atoms regardless of what other type of magnetic behavior they may exhibit. It arises from the interaction of the applied magnetic field with the field induced in the completed shells of electrons. This field must necessarily oppose the applied field. The effect of this interaction is to cause the diamagnetic sample to move away from the applied field in order to diminish the interaction. Hence, the *diamagnetic susceptibility* is a negative quantity, and from both a classical and a quantum mechanical treatment the *diamagnetic susceptibility* of any poly-electron atom is found to be a function of the average squared radius of its electrons:

$$(11\text{--}20) \qquad \chi_A = -\frac{Ne^2}{6mc^2} \sum_i \overline{r_i^2} = -2.83 \times 10^{10} \sum_i \overline{r_i^2}$$

From Eq. (11--20) it is apparent that the diamagnetism will be very sensitive to changes in the value of $\overline{r^2}$. Thus, both an increase in the size of an atom or ion as well as an increase in the number of electrons, $i$, will increase the magnitude of the diamagnetic susceptibility. Temperature has no effect upon diamagnetism, nor does the magnitude of the applied field, but note that the susceptibility is *negative*. When we deal with molecules, it is assumed that atomic susceptibilities are simply additive. Any deviations from additivity are assumed to arise from susceptibilities due to the electrons in the bonds between the atoms, and these too are assumed to be additive. Lists of such atom and bond diamagnetic susceptibilities are available for organic molecules (Pascal's constants) and for inorganic ions and radicals.

The magnitude of the diamagnetic effect is normally 10 to 1000 times less than that of the paramagnetic effect which results from unpaired electrons. Therefore, its importance in most inorganic systems is primarily as a correction required for precise work. Diamagnetism is observed, then, only when all electrons in a system are paired, and it is operative only when the external field is applied. All substances that do not have the necessary electronic and structural features to give rise to para-, ferro-, antiferro-, or ferrimagnetism will necessarily still exhibit diamagnetism.

### Paramagnetism

Atomic, molecular, free radical, or ionic systems which contain one or more unpaired electrons will possess a permanent magnetic moment that arises from the residual spins and orbital angular momenta of the unpaired electrons. These may be $s$ electrons, as in R1 and T11 metal atom vapors; $p$ electrons, as in $O_2$, NO, $ClO_2$, $I^+$, organic free radicals, etc.; $d$ electrons, as in the $3d$, $4d$ and $5d$ transition series of interest here; and $f$ electrons, as in the lanthanides and actinides. All substances with permanent magnetic moments display normal paramagnetism. Now when a paramagnetic substance is placed in an external magnetic field, the individual atomic or molecular permanent magnets will align themselves in the same direction as the field and thus be attracted to it. This produces a *positive* magnetic susceptibility which is independent of the magnitude of the applied field, but which must be dependent upon temperature, since thermal agitation will oppose orientation of the magnetic dipoles. Hence the effectiveness of the magnetic field will diminish with increasing temperature. Mathematically this dependence has been expressed by either the *Curie law:*

$$(11\text{--}21) \qquad \chi = \frac{C}{T}$$

or more often more accurately by the *Curie-Weiss law:*

$$(11\text{--}22) \qquad \chi = \frac{C}{T - \theta}$$

where $C$ is the *Curie* constant and $\theta$ is the *Weiss* constant. These quantities are properties of each individual substance and must be determined by temperature-dependent studies of the particular paramagnetic material.

When $\chi_M$ is determined for a paramagnetic substance, it is necessary for precise work to correct the value for the diamagnetic effects of the constituent atoms. Correction should be made also for a weak *temperature-independent paramagnetism*, TIP, sometimes called *Van Vleck high-frequency paramagnetism*. This occurs because the applied magnetic field can effect a change in the ground state of the molecule or ion. That is, it is due to a mixing of the ground and higher energy excited states caused by the field, over and above the mixing which may already be present before the field is applied. Even systems without unpaired electrons may exhibit this effect. In any case, after the appropriate corrections for diamagnetism and for TIP have been made, we then write the susceptibility as $\chi_M{}^{corr}$.

A quantum mechanical treatment of the interaction between an applied magnetic field and the elementary permanent moments yields the following expression for the total molar magnetic susceptibility

$$(11\text{--}23) \qquad \chi_M{}^{corr} = \frac{N\mu^2}{3kT}$$

where $N$ is Avogadro's number, $k$ is Boltzmann's constant and $\mu$ is the permanent magnetic moment in BM units. On rearranging Eq. (11–23) and evaluating the constants, the magnetic moment can be expressed as

$$(11\text{--}24) \qquad \mu_{eff} = 2.84\sqrt{\chi_M T} \text{ BM}$$

### Types of Paramagnetic Behavior

Our first interest in the magnetic properties of transition metal complexes is usually concerned with the determination of the number of unpaired electrons present in the complex molecule system. Thus, it is necessary to derive an expression which relates the experimentally determinable magnetic moments (Eq. (11–24)) to the number of unpaired electrons. As previously mentioned, paramagnetism originates in the spin and orbital angular momenta of the unpaired electrons. According to the usual symbolism, $S$ is the total spin angular momentum quantum number $\left( S = \sum_i M_{s_i} \text{ where } M_s \text{ may} \right.$

be $+\frac{1}{2}$ or $-\frac{1}{2}$, and $L$ is the total orbital angular momentum quantum number $\left(L = \sum_i l_i,\right.$ where the $l_i$ are the individual electron angular momentum quantum numbers $\left.\right)$, and these must be added according to the quantum rules for addition of vectors. Now, normal paramagnetism of a complex ion is dependent upon at least three factors: (a) the number of unpaired electrons; (b) the spectroscopic ground state,* and the upper states if the separation is of the order $kT$; and (c) the strength and symmetry of the electrostatic field arising from the ligands present in the first coordination sphere. In order to see how the paramagnetism of transition metal complexes depends upon these factors, we shall find it convenient to subdivide paramagnetic behavior into the following four general types.[24]

**Large multiplet separation.** This case arises if the unpaired electrons are well-shielded from external ligand field or covalent bonding forces, and if the ground state of the atom is separated from the next-higher excited state by an energy difference, $h\nu$, which is large compared to $kT$ ($\sim 200$ cm$^{-1}$ at ordinary temperatures). Under these circumstances spin-orbit coupling is significant, and for a given state of $L$ and $S$, $J$ will take on all values from $L + S$ to $L - S$, so there will be $2L + 1$ or $2S + 1$ values of $J$, whichever is the smaller. (See Figure 11–8.) The one that gives the ground state is either $L + S$ or $L - S$ depending upon whether the electronic subshell is more or less than half full. The magnetic moment in this case is given by

$$(11\text{--}25) \qquad \mu_{eff} = g\sqrt{J(J + 1)} \text{ BM}$$

In a magnetic field of strength $H$, each $J$ level is split into $2J + 1$ components, each separated from its immediate neighbor by an energy $g\beta H$ (see Figure 11–8), where $\beta$ is the Bohr magneton.

For this type of magnetic behavior, $\mu$ is independent of the stereochemical environment and the amount of magnetic dilution, a magnetically dilute substance being one in which the paramagnetic atoms or ions are effectively separated from each other by large numbers of diamagnetic atoms, molecules, or ions. This first general type of magnetic behavior is found primarily with the lanthanide ions in which the incompleted energy level lies well-shielded from the surface of the ion and spin-orbit coupling energies are large ($\sim 1000$ cm$^{-1}$).

Most lanthanide ions have a ground state with a single well-defined $J$ value, and the next-lowest $J$ state is several times $kT$ above the ground state and therefore not populated at ordinary temperatures. Eq. (11–25) gives values in close agreement with experimental values for all lanthanide ions except $Sm^{3+}$ and $Eu^{3+}$. (See Figure 11–9.) For the latter ions, the first

---

* The ground state is generally understood to be those energy levels which are appreciably populated at normal temperatures and which therefore lie within approximately $kT$ of the lowest possible energy level.

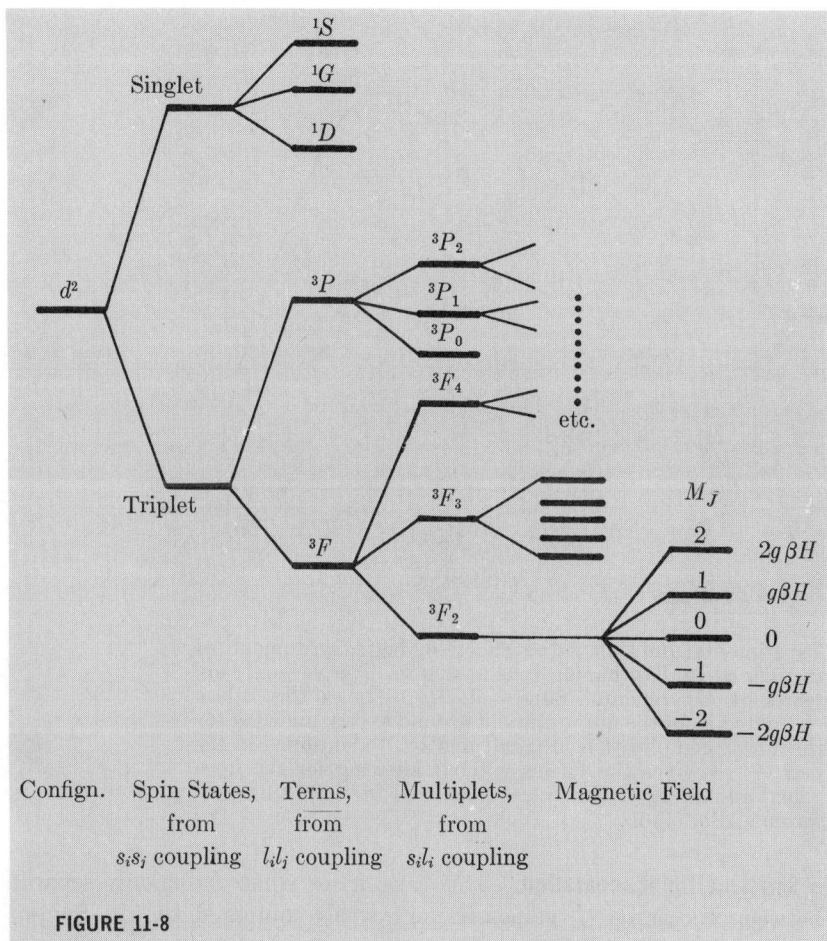

**FIGURE 11-8**

The Lifting of the Various Degeneracies of a Gaseous $d^2$ Ion, Showing on the Far Right the Ground Term Splitting (Greatly Magnified Relative to Other Splittings Shown) in a Magnetic Field. The unperturbed state, $M_J = 0$, is assumed to have zero energy. $g$ is the Landé splitting factor, $\beta$ is the Bohr magneton, and $H$ is the field strength.

excited $J$ states (actually the first three excited $J$ states for $Eu^{3+}$) are within roughly $kT$ of the ground state and are therefore somewhat populated. Thus, they yield magnetic moments in excess of those calculated assuming only ground state population. In the $5f$ series, $Pu^{3+}$ and $Am^{3+}$ show analogous behavior, but, in general, among the actinides the magnetic properties are much more difficult to interpret. This is perhaps due to the inadequacy of the Russell-Saunders coupling scheme for $5f^n$ ions and also to more subtle ligand field effects which involve $5f$ orbitals to a greater extent than the $4f$ orbitals are involved in bonding in the lanthanide complexes.

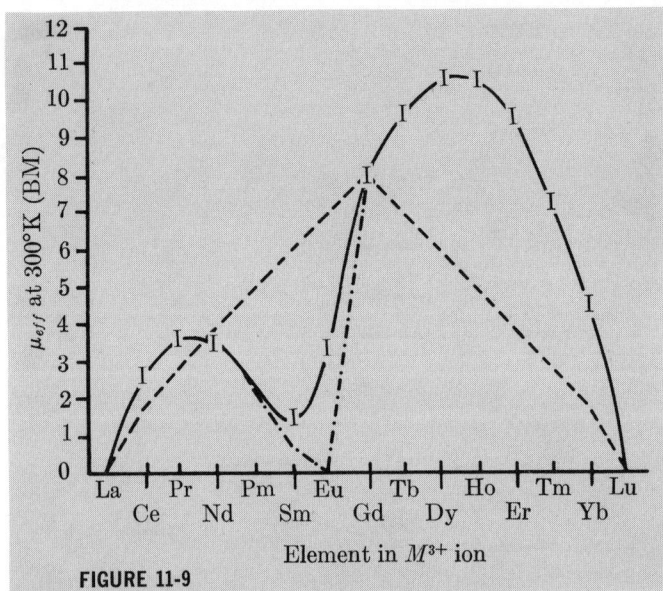

**FIGURE 11-9**

Experimental and Calculated Effective Magnetic Moments of the Tervalent Lanthanide Ions at 300°K. The vertical bars represent experimental ranges of $\mu_{eff}$; — gives the values calculated from the appropriate $J$ ground states, allowing for the added complication for $Sm^{3+}$ and $Eu^{3+}$ mentioned in the text; —·— gives the values without allowing for the foregoing; and ---- gives the values calculated from the spin-only formula, Eq. (11-27).

**Small multiplet separation.** This case arises when the energy separation between successive $J$ values is very small compared to $kT$. Spin-orbit coupling is now negligible and may be ignored. Under these circumstances $L$ and $S$ each interacts independently with an external magnetic field and wave mechanics shows that

$$(11\text{–}26) \qquad \mu_{eff} = \mu_{S+L} = \sqrt{4S(S + 1) + L(L + 1)} \text{ BM}$$

At least this is the limit approached by $\mu$ as the energy difference $h\nu_{J_i-J_j}$ approaches zero. This type of magnetic behavior is important as the limiting case in the first transition series when orbital contributions to the magnetic moment are not quenched. Such quenching, we shall see, occurs completely for ions such as $Mn^{2+}$, $Fe^{3+}$, and $Gd^{3+}$, whose ground states are $S$ states ($L = 0$), and these then obey the *spin-only* formula (see below). But ions having $D$ or $F$ ground states do possess some orbital angular momentum, the origin, limitations, and quenching of which we shall consider later. Cases involving intermediate multiplet separation are actually rare and therefore of relatively little importance.

**Spin-only.** For many of the first transition series ions, particularly those in the first half of the series, experimental results have indicated that the orbital contribution to the magnetic moment may be completely ignored. Thus, only the spin angular momentum determines the magnetic moment. This leads to the simple expression

$$(11\text{–}27) \qquad \mu_{eff} = \mu_S = \sqrt{4S(S+1)} \text{ BM}$$

which is called the *spin-only formula*. Recalling that $S = n/2$, where $n$ equals the number of unpaired electrons, we may then write

$$(11\text{–}28) \qquad \mu_S = \sqrt{n(n+2)} \text{ BM}$$

This, then, is seen to relate the magnetic moment directly to the number of unpaired electrons.

How well the spin-only moment agrees with actual experimental moments, $\mu_{eff}$, for several $3d$ element ions in both weak and strong ligand fields may be seen from the data in Table 11–8. It should be pointed out that the spread of experimental values listed in the table for certain ions represents the range observed for many different complexes of the given metal ion, and this includes some having differing stereochemistries. This is particularly true for $Fe^{3+}$, $Co^{2+}$, and $Ni^{2+}$, which, along with $Mn^{2+}$ and $Cu^{2+}$, have perhaps received the most attention to date. It is also seen from Table 11–8 that even when the experimental values exceed the $\mu_S$ value, they seldom come close to the $\mu_{S+L}$ value. As we have already noted, this is due to whole or partial "quenching" of orbital moments.

The quenching of orbital moments occurs by a mechanism which in effect is due to the ligands having restricted the orbital motion of the metal ion electrons in the complex. More explicitly, we may picture orbital momentum quenching in the following descriptive manner. An electron will possess a moment around a given axis, say the $z$ axis, if it is possible to transform the orbital that it occupies into an entirely *equivalent* and *degenerate* orbital by a simple rotation about the axis. In a free atom or ion having $d$ electrons, therefore, an electron in an $x^2-y^2$ orbital will contribute $\pm 2(h/2\pi)$ units of orbital angular momentum about the $z$ axis, since a 45° rotation will carry it into the equivalent $xy$ orbital. In the same manner a 90° rotation about the $z$ axis will carry an electron in an $xz$ orbital into an equivalent $yz$ orbital, contributing thereby $\pm 1(h/2\pi)$ units of angular momentum. An electron in a $z^2$ orbital contributes zero angular momentum about the $z$ axis and it cannot be rotated into an $x^2-y^2$ orbital. From this simple analysis, it is easy to see how a ligand field may quench part or all of the orbital moment, since the field effects a removal of the required degeneracy of rotationally related orbitals. Therefore, in the presence of a ligand field of any symmetry (except spherical) the degeneracy of the $d$ orbitals is lifted. The $x^2-y^2$ and $xy$ orbitals are no longer equivalent in energy, so their orbital contribution is completely destroyed. Only the $t_{2g}$ orbitals may be degenerate in an octahedral field, but they cannot contribute orbital momentum when either

**TABLE 11–8. Theoretical and Observed Magnetic Moments of Some First-row Transition Element Ions.**

| Ion | Config. | Ground State Free Ion Term | $\mu_S$ | $\mu_{S+L}$ | Obsd. $\mu_{eff}$ | Spin-Paired | |
|---|---|---|---|---|---|---|---|
| | | | | | | $\mu_S$ | $\mu_{eff}$ Obsd. |
| $Ti^{3+}$ $V^{4+}$ | $d^1$ | $^2D$ | 1.73 | 3.00 | 1.7–1.85 1.7–1.8 | — | — |
| $V^{3+}$ | $d^2$ | $^3F$ | 2.83 | 4.47 | 2.6–2.9 | — | — |
| $V^{2+}$ $Cr^{3+}$ $Mn^{4+}$ | $d^3$ | $^4F$ | 3.88 | 5.20 | 3.8–3.9 3.7–3.9 3.8–4.0 | — | — |
| $Cr^{2+}$ $Mn^{3+}$ | $d^4$ | $^5D$ | 4.90 | 5.48 | 4.7–4.9 4.9–5.0 | 2.83 | 3.2–3.3 3.2 |
| $Mn^{2+}$ $Fe^{3+}$ | $d^5$ | $^6S$ | 5.92 | 5.92 | 5.6–6.1 5.7–6.0 | 1.73 | 1.8–2.1 2.0–2.5 |
| $Fe^{2+}$ $Co^{3+}$ | $d^6$ | $^5D$ | 4.90 | 5.48 | 5.1–5.7 ~5.4 | 0 | ~0 |
| $Co^{2+}$ $Ni^{3+}$ | $d^7$ | $^4F$ | 3.88 | 5.20 | 4.3–5.2 — | 1.73 | 1.7–2.0 1.8–2.0 |
| $Ni^{2+}$ | $d^8$ | $^3F$ | 2.83 | 4.47 | 2.8–4.0 | — | — |
| $Cu^{2+}$ | $d^9$ | $^2D$ | 1.73 | 3.00 | 1.7–2.2 | — | — |

filled or half-filled. Thus, for octahedral complexes we will expect all orbital contributions to be quenched and hence to find *spin-only* moments for the following ground configurations with the ground state term of the complex in parentheses:

spin-free:  $t_{2g}^3(^4A_{2g})$, $t_{2g}^3e_g^1(^5E_g)$, $t_{2g}^3e_g^2(^6A_{1g})$,
$t_{2g}^6e_g^2(^3A_{2g})$, $t_{2g}^6e_g^3(^2E_g)$

spin-paired: $t_{2g}^6(^1A_{1g})$, $t_{2g}^6e_g^1(^2E_g)$

All remaining ground state configurations, that is, those with 1, 2, 4, or 5 $t_{2g}$ electrons, or significantly *all* those which possess a $^{2S+1}T_{ng}$ ground term, should have some residual orbital contribution. To a first approximation this will account for the differences between experimental and spin-only predicted moments. Of course, other ligand field symmetries may be treated analogously. Note that we may generalize that *orbital angular momentum is completely quenched for ions with* A *and* E *ground terms but only partially quenched for those having* T *ground terms*. However, although the magnetic moments of ions with $A$ and $E$ ground terms come close to the spin-only values and are essentially temperature independent, they often do depart from these $\mu_S$ values by small amounts. The amount of departure may be shown to depend upon the relationship between the spin-orbit coupling and the magnitude of the ligand field splitting.

Spin-orbit coupling can mix (hybridize) some of the higher energy levels having the same $S$ value as the ground state. To take this into account we may write the following equation for the effective magnetic moment:

$$(11\text{--}29) \qquad \mu_{eff} = \mu_S\left(1 - \alpha\frac{\lambda}{\Delta}\right)$$

where $\alpha$ is a constant which depends upon the spectroscopic ground state and the number of $d$ electrons, being 2 for $^2D$ or $^5D$, 4 for $^3F$ or $^4F$, and 0 for $^6S$; $\Delta$ is the separation between the ground level and the level being mixed in; and $\lambda$ is the spin-orbit coupling constant. The last is a positive quantity for the $3d$ ions with less than a half-filled $d$ shell, and it is negative for those with more than a half-filled shell. Note that for the case of a half-filled shell the ground term is $^6A_{1g}$, and all higher states have a lower multiplicity so that the moments are very close to the spin-only value. Since $\alpha$ and $\Delta$ are positive quantities it is seen that the observed moment should be greater or lesser than the spin-only moment as $\lambda$ is negative or positive. In this way we can explain the generally lower than spin-only moments observed for $Cr^{3+}(t_{2g}^3)$, spin-free $Cr^{2+}(t_{2g}^3e_g^1)$, and $V^{3+}(t_{2g}^2)$ and $V^{4+}(t_{2g}^1)$. Also we can explain the generally higher than spin-only moments observed for spin-free $Fe^{2+}(t_{2g}^4e_g^2)$, $Co^{2+}(t_{2g}^5e_g^2)$, $Ni^{2+}(t_{2g}^6e_g^2)$, and $Cu^{2+}(t_{2g}^6e_g^3)$. It is pertinent to note further that the greatest deviations from spin-only values are found with $Co^{2+}$ and $Fe^{2+}$, both of which also possess some unquenched orbital contribution in addition to the contribution from the spin-orbit coupling mechanism.

The magnetic moments of complexes which possess $T$ ground terms generally vary with temperature and depart significantly from the spin-only value. The treatments required to bring theory and experiment closer together here are somewhat involved and have to do with the effects of $t_{2g}$ electron delocalization and lower symmetry ligand field components.[6,23] We shall not attempt to pursue this complication further here.

**Heavy atoms.**  For second and particularly third transition series elements we find mainly spin-paired complexes. In fact, when there is an odd number of electrons a magnetic moment corresponding roughly to only one unpaired electron is generally found, and when there is an even number of electrons diamagnetism is generally observed. This greater tendency on the part of $4d$ and $5d$ elements to form low-spin complexes probably arises for the following reasons. Interelectron repulsion is not as large in the spatially larger $4d$ and $5d$ orbitals as it is in the $3d$ orbitals, and therefore double occupancy is less energetically unfavorable in the former cases. Furthermore, the ligand field- or MO-induced splitting of the $d$ orbitals is larger in the order $5d > 4d > 3d$, which also favors the low-spin case.

Even allowing for the foregoing factors, magnetic moments are still difficult to interpret for $4d$ and $5d$ element complexes. Values obtained at room temperature often fall well below spin-only moments, and with but few exceptions susceptibility data on these elements cannot be used to determine numbers of unpaired electrons, oxidation states, and relative $d$-orbital energy levels. Mainly this difficulty arises because of the high spin-orbit coupling constants that exist for these ions with heavier nuclei. The latter serve to align the $L$ and $S$ vectors in opposite directions, resulting in destruction of much of the paramagnetism which might otherwise be expected from the unpaired electrons.[23,25]

## Application to Cobalt(II) Complexes

One of the most fruitful applications of magnetic measurements has been realized in the study of $Co^{2+}$ complexes. In terms of this $d^7$ ion system it is possible to illustrate several ideas already presented. In Table 11–9, a

### TABLE 11–9. Magnetic Moments of $Co^{2+}$ Complexes.

| Stereochemistry | Spin-Free | | | Spin-Paired | | |
|---|---|---|---|---|---|---|
| | Confign. | Ground Term | $\mu_{eff}$ Obsd.[b] | Confign. | Ground Term | $\mu_{eff}$ Obsd.[c] |
| Octahedral, $O_h$ | $t_{2g}^5 e_g^2$ | $^4T_{1g}$ | 4.7–5.2 | $t_{2g}^6 e_g^1$ [a] | $^2E_g$ | 1.8–1.85 |
| Tetrahedral, $T_d$ | $e^4 t_2^3$ | $^4A_2$ | 4.4–4.8 | — | — | — |
| Square Planar, $D_{4h}$ | $t_{2g}^5 e_g^2$ | $^4A_{2g}$ | 4.8–5.2 | $t_{2g}^6 e_g^1$ and $t_{2g}^5 (z^2)^2$ | $^2A_{2g}$ | 2.2–2.9 |
| | (very likely tetragonal) | | | | | |
| Five-Coordinate[d]? | — | — | — | — | | 1.9–2.4 |

[a] Strong Jahn-Teller distortion most probably present, but structure not investigated in, e.g., $[Co(NO_2)_6]^{4-}$, $[Co(diars)_3]^{2+}$, and $[Co(triars)_2]^{2+}$ (*diars* and *triars* represent a bidentate and a terdentate arsenic donor chelate, respectively).
[b] Calculated spin-only value is 3.89.
[c] Calculated spin-only value is 1.73.
[d] For example a $CN^-$ and a $CH_3NC$ complex.

summary is presented of most of the magnetic data on $Co^{2+}$ complexes along with the predicted or proven stereochemistry, the ground state configuration, and the ground state term. The observed moments for the spin-free octahedral and tetragonally distorted octahedral complexes, which are in excess of the spin-only value by 0.8 to 1.3 BM, undoubtedly arise from the unquenched orbital contribution of both the ground state, $t_{2g}{}^5e_g{}^2$, and the first excited state, $t_{2g}{}^4e_g{}^3$. The latter is certainly expected to be mixed-in to some extent, the exact magnitude being dependent upon the ligand field strength. The observed moments for the spin-paired octahedral complexes are much closer to the calculated spin-only value since the ground state configuration, $t_{2g}{}^6e_g{}^1$ allows no orbital contribution. That the values *are* in fact often in excess of 1.73 BM must be due to a contribution from some unquenched orbital momentum in the first excited level, $t_{2g}{}^5e_g{}^2$. Note that strong Jahn-Teller distortion of the ground state will assure that these complexes are not in fact octahedral.

The observed high moments for the tetrahedral complexes must also be caused by mixing in of higher levels, for example, $e^3t_2{}^3(4p)^1$. This postulation is necessary, since the symmetrical ground state, $e^4t_2{}^3$, does not permit any orbital contribution. However, it is interesting to note that because the more heavily populated ground state does not make an orbital contribution to the total moment, the observed values are somewhat lower than those for the octahedral or tetragonal cases where the ground state does contribute to the orbital moment. The argument that some of the spin-free $Co^{2+}$ complexes with moments in the range 4.8 to 5.2 BM are square planar is not conclusive, and in all probability these complexes are tetragonal. Finally the spin-paired $Co^{2+}$ complexes having moments in the range 2.2 to 2.9 BM are more certainly square planar in structure and the mixing in of the first excited level, $t_{2g}{}^5(z^2)^2$, with its partly unquenched orbital contribution, probably accounts for the high moments found here.

## SPECTRAL PROPERTIES

Most of the complexes of the transition elements are highly colored compounds, which means they are capable of absorbing radiant energy in the visible region of the spectrum. Indeed, studies of the absorption spectra of these compounds in the pure solid state, in some solid matrices, in solutions, or even in the vapor phase, reveal that they absorb energy in the infrared (IR) and ultraviolet (UV) regions as well as the visible. The IR absorptions, which occur and have been extensively studied[26,27] in the 3500 to 200 $cm^{-1}$ region, are due mainly to the various possible fundamental vibrations occurring in the coordinated ligands. Valuable information concerning structure and bonding has come from the interpretation of IR spectra. And metal-ligand vibrations, which occur generally in the 700 to 200 $cm^{-1}$ region, are being increasingly studied. But we shall be interested here only in *electronic* spectra, inasmuch as the modern bonding theories outlined in the previous chapter may be used to explain and predict such spectra, whereas they tell us nothing about *vibrational* spectra. This is not to say that information obtained from the IR spectral data is of no value in increasing our knowledge of the structure and bonding in complexes. On the contrary, we may often gain very valuable information about the bonding present in complexes by studying the frequency shifts of ligand vibrations resulting from complexation of the ligand, and the appearance or absence of certain IR bands has been used to deduce structural information. But these topics will not be pursued here.[26,27] In fact, we shall soon see that it is necessary to call upon vibrations of complex molecules to enable us to more fully understand and explain certain electronic transitions which occur in these systems.

The study of electron absorption spectra of transition element complexes has constituted one of the major efforts in recent years by both the experimentalist and the theoretician to better characterize and understand the nature of the electronic structure and bonding in these compounds. Thousands of research papers have reported and many of these have attempted to interpret spectral data on a great variety of complexes of virtually all metallic elements in virtually all of their oxidation states. All we shall attempt here is a brief introduction to the field, limiting ourselves primarily to $3d$ transition elements and referring the student to more comprehensive treatments to be found elsewhere.[6,8,10,28-31]

Electronic transitions occur when electrons within the molecule or ion move from one energy level to another. Therefore, in absorption spectral studies, two questions immediately arise: (a) which are the energy levels that are populated (the ground state) and which are the nearby empty energy levels into which electrons may be excited (the excited states); and (b) what are the probabilities for the various possible absorptions to occur. We can answer both of these questions in a very general way for the $d^n$ complexes under consideration here.

First, we can list four general types of electronic transitions that may be observed with $d$-element complexes.

(1) Transitions may occur between the split $d$ levels of the central atom, giving rise to the so-called $d$-$d$ or *ligand field spectra*. The spectral region where these bands occur spans the near IR, visible, and UV, and thus, these transitions are the ones primarily responsible for the great variety of color found in transition element complexes. In practice the region generally scanned is 10,000 to 30,000 cm$^{-1}$ (1000 to 333 m$\mu$), although some $d$-$d$ transitions do occur beyond both ends of this range. However, the lower frequencies are often not accessible experimentally, and the higher frequencies, which *are* accessible, generally to 50,000 cm$^{-1}$ (200 m$\mu$), are almost invariably covered over with much more intense charge transfer and/or intraligand transitions (*vide infra*).

In the CFT model the $d$-$d$ transitions are considered to be totally localized within the central ion and the number and energies of the transitions are established simply by the number of $d$ electrons and the strength and symmetry of the electrostatic ligand field. The LFT produces much closer agreement between experimental and calculated data by allowing certain parameters, namely, the interelectron repulsion parameters and the spin-orbit coupling constant, to vary with the properties of the field. In the MOT treatment the transitions are between molecular orbitals, the excited levels now being antibonding MO's, but all levels still possess primarily metal orbital character. We shall concentrate most of our attention on these $d$-$d$ type transitions. They have been by far the most widely and the most often studied. In addition, they supply the greatest amount of valuable information for theory formulation and testing.

(2) Transitions may occur from molecular orbitals located primarily on the ligands, the metal-ligand bonding $\sigma$ or $\pi$ molecular orbitals, to nonbonding or antibonding molecular orbitals located primarily on the metal atom. Such transitions are termed *ligand-to-metal charge transfer* transitions. In a sense the energies of such electronic transitions should reflect the thermodynamic tendency for redox to occur between a ligand and its central metal ion, specifically the reduction of the central ion by the ligand. This is indeed found to be the case, but we shall not pursue the point here. These transitions cannot be treated at all by CFT, but semiempirical MOT is quite capable of supplying reasonable models. A great deal of work must and probably will be done in this area in the near future.

(3) The transitions which involve electrons being excited from nonbonding or antibonding orbitals located primarily on the metal atom to antibonding orbitals located primarily on the ligands are termed *metal-to-ligand charge transfer* transitions. Again, the tendency for the central metal atom to reduce the ligands should be reflected in the energies of these electronic transitions. The bands generally occur in the UV region, but occasionally are found tailing off into or even peaking in the visible region.

(4) Finally, we should mention the electronic transitions which involve electrons being excited from one ligand orbital into another ligand orbital. These *intraligand* transitions generally occur with energies found in the UV region and, as they are often little affected by the coordination, their bands can usually be disentangled from the equally intense charge transfer bands in their neighborhood. Obviously, the more affected by coordination they are, the more difficult it is to assign them with any certainty. This also is a relatively unexplored research area with a bright future.

## Selection Rules

Next we must consider the second question which we asked earlier, that concerning the expected probabilities or intensities of the various types of transitions. This requires an examination of certain quantum mechanical *selection rules*.[6,10,28] These rules arise from the particular quantum mechanical properties possessed by the wave functions which represent the various eigenstates or energy levels. First, transitions for which $\Delta S \neq 0$ are termed *spin* or *multiplicity-forbidden*. However, this spin selection rule is not completely valid in the presence of spin-orbit coupling. Therefore spin-forbidden bands do appear in the spectra of many transition metal complexes, but they are generally one to two orders of magnitude weaker than the spin-allowed bands. And their intensities increase with increasing spin-orbit coupling constants. Recall that the latter increase in the orders: $d^1 < \cdots < d^9$ and $3d < 4d < 5d$.

Now the only mechanism of importance for the absorption of light by complex ions is the *electric dipole* mechanism. Accordingly a transition between two energy states $a$ and $b$ can occur as electric dipole radiation only if the *transition moment integral*, $\int_{-\infty}^{+\infty} \psi_a r \psi_b \, d\tau$, sometimes written more simply as $\langle \psi_a | r | \psi_b \rangle$, is different from zero. For the case of interest here in which both $\psi_a$ and $\psi_b$ are $d$ orbital wave functions and hence are both "even" to inversion (that is, are $g$ states rather than $u$ states) the integral turns out to be zero. According to *Laporte's rule*, only transitions between an even state and an odd state are allowed as electric dipole transitions, that is, $g \leftrightarrow u$, but $g \not\leftrightarrow g$ and $u \not\leftrightarrow u$. A more applicable statement of the *Laporte-forbidden* transitions is possible with the use of Group Theory. For the students familiar with this valuable mathematical tool,[32] we can state that if $\psi_a$ and $\psi_b$ transform as the irreducible representations $\Gamma_a$ and $\Gamma_b$ of the group to which the molecule belongs, and $r$ transforms as $\Gamma_r$ of that group, then the integral will be zero unless the reducible representation, $\Gamma_a \times \psi_r \times \Gamma_b$, contains the totally symmetric representation of the Group, $A_{1(g)}$.[6] Thus, it would appear that $d$-$d$ transitions must be Laporte-forbidden, and if they occur at all their intensities should be very low.

The intensity of a band is most accurately measured in terms of the quantity known as the oscillator strength, $f$. The latter is the area under the band in a plot of extinction coefficient, $\epsilon$, versus frequency, $\nu$. For a band measured in cm$^{-1}$ from $\nu_1$ to $\nu_2$

$$f = 4.32 \times 10^{-9} \int_{\nu_2}^{\nu_1} \epsilon d\nu$$

It can further be shown that $f$ is related to the transition moment integral, $P_E$, by the relation

$$f = 1.096 \times 10^{11} \, \nu P_E^2$$

Now actual bands in transition metal complexes do have finite intensities for reasons which we shall now briefly explore. Approximate intensities for various types of bands ranked according to the decreasing restrictions imposed by the selection rules and according to the mechanisms for gaining intensity, in spite of the formal selection rules, are listed in Table 11–10. We should at least recognize that *magnetic dipole* and *electric quadrupole* transitions are allowed for $d$-$d$ systems, but the intensities in these cases are an order of magnitude or more lower than the normally observed intensities. The former mechanism has been observed to operate in a few cases, however.

## Mechanisms for the Breakdown of Selection Rules

Now that we have briefly considered the important selection rules under which the $d$-$d$ transitions must operate, it is necessary to examine the

**TABLE 11–10.**  Approximate Values of the Intensities of Various Types of Bands in Transition Metal Complexes.

| Type of Electronic Transition | Approximate Oscillator Strength, $f$ | Approximate Molar Extinction Coefficient, $\epsilon$ |
|---|---|---|
| Spin-forbidden, Laporte-forbidden | $10^{-7}$ | $10^{-1}$ |
| Spin-allowed, Laporte-forbidden | $10^{-5}$ | 10 |
| Spin-allowed, Laporte-forbidden, but with $d$-$p$ mixing (e.g., $T_d$ symmetry) | $10^{-3}$ | $10^2$ |
| Spin-allowed, Laporte-forbidden, but with "intensity stealing" | $10^{-2}$ | $10^3$ |
| Spin-allowed, Laporte-allowed, (e.g., charge-transfer) | $10^{-1}$ | $10^4$ |

mechanisms by which band intensity may be enhanced, as we know that it must, in order to produce the spectra which we do observe. As we have already stated, in the absence of spin-orbit coupling the total spin quantum number cannot change during the absorption of radiation. However, since the spin and orbital motions are coupled, even though weakly, the transition moment integral should be determined by the spin-orbit wave functions for the ground and upper states, which mix a little of each spin state into the other in an amount dependent upon the energy difference in the orbital states and the magnitude of the spin-orbit coupling constant. Therefore, electronic transitions which do occur between states having different spin multiplicity may be thought of as transitions between *components* of each orbital state that have the same multiplicity. For example, if the ground state were 99% *singlet* and 1% *triplet* (due to spin-orbit coupling) and the excited state were 1% *singlet* and 99% *triplet*, then the intensity would derive from the *triplet-triplet* and *singlet-singlet* components of the transition. The very weak intensity bands in octahedral spin-free $Mn^{2+}$ and $Fe^{3+}$ (both $d^5$) complexes, in which the ground term is $^6S$ in the free ion or $^6A_{1g}$ in the complexed ion and all excited states are of lower spin multiplicity, must gain what little intensity they have ($\epsilon \sim 0.01$) via this mechanism. (See Table 11–11 for the observed bands and their intensities for $Mn(H_2O)_6^{2+}$.) The origin of the calculated frequencies and their assignments, which are also given, will be considered later.

Next we must ask how the Laporte selection rule, which would hold rigorously for the free metal ion, has been at least partially broken down in the metallic complex so as to allow the electric dipole mechanism to be operative. The answer is that if we could change the character of $\psi_a$ and $\psi_b$ so that they were no longer pure $d$ orbital wave functions, but rather were mixed with some "odd" $p$ character, then the transition moment integral, $P_E$, would no longer be zero. Thus, mathematically, we look for the possibility of having

$$\psi_a' = \psi_a(3d) + \alpha\psi(4p)$$

**TABLE 11-11. Spectral Bands for $Mn(H_2O)_6{}^{2+}$ in Aqueous Solution.**

| Calcd.<br>Frequency, $kK^a$ | Exptl.<br>Frequency, $kK^a$ | $\sim\epsilon$ | Assignment<br>$^6A_{1g} \rightarrow$ |
|---|---|---|---|
| 19.40 | 18.87 | 0.013 | $^4T_{1g}(G)$ |
| 22.80 | 23.12 | .009 | $^4T_{2g}(G)$ |
| 25.20 | 24.96 | .03 | $^4E_g(G)$ |
| 25.20 | 25.27 | .014 | $^4A_{1g}(G)$ |
| 28.20 | 27.98 | .018 | $^4T_{2g}(D)$ |
| 29.90 | 29.75 | .013 | $^4E_g(D)$ |
| 35.00 | 32.96 | .02 | $^4T_{1g}(P)$ |
| 40.70 | 40.82 | — | $^4A_{2g}(F)$ |
| 41.90 | — | — | $^4T_{1g}(F)$ |

$^a$ $1 \ kK = 1000 \ cm^{-1}$.

where $\alpha$ is the mixing coefficient. Physically, the $3d$ and $4p$ wave functions may be mixed if we can remove the inversion center (symmetry element $i$) about the central atom. There are two important mechanisms by which this removal of $i$ occurs: (1) when the central metal ion is placed in a statically distorted field; or (2) when odd vibrations of the surrounding ligands create the distorted field for a time that is long compared to the time necessary for the electronic transition to occur (this being the *Franck-Condon principle*).

Perhaps the most important case of the statically distorted or asymmetric field is the case of a tetrahedral complex. This symmetry, $T_d$, has no inversion center, so the Laporte rule does not hold. Thus, we expect and we find that because of $d$-$p$ mixing, which is possible in tetrahedral complexes, the electronic transitions are much more intense, often by a factor of 100, than in analogous octahedral complexes. For example, in the tetrahedral complexes $MX_4{}^{2-}$ ($M$ = Ni, Co, Cu; $X$ = Cl, Br, I) the $\epsilon$ values are in the 200 to 1200 range, or roughly 100 times greater than those for the octahedral aquo complexes of the same metal ions.

One can also observe the effect of the non-centrosymmetric field by comparing the intensities of, for example, the visible bands in *cis* and *trans* isomers of $[Coen_2X_2]^+$ ($X$ = F or Cl) in aqueous solution. The bands of the *trans* isomer, which is centrosymmetric, are some three to four times less intense than those of the *cis* isomer, which does not possess an inversion center. Other examples of this sort abound.

The other mechanism by which an electronic $d$-$d$ transition may gain some intensity, and apparently a very important one for centrosymmetric complexes, is known as *vibronic coupling*. Mathematically this implies the coupling of *vib*rational and elect*ronic* wave functions. Physically it is easy to see that certain vibrations (see Figure 11–10) will remove the center of symmetry. The vibronic wave function will have a symmetry representation determined by the direct product of electronic (*el*) and vibrational (*vib*)

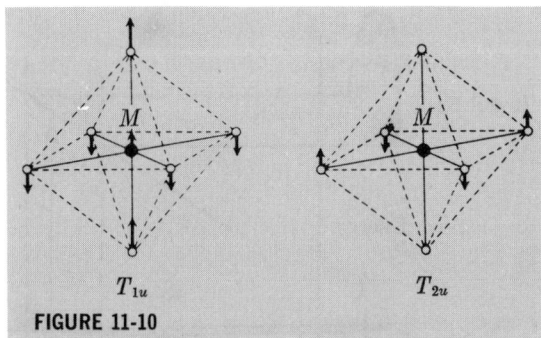

$T_{1u}$                          $T_{2u}$

**FIGURE 11-10**

Two of the Normal Vibrational Modes of an Octahedral Complex, $MA_6$, Each Triply Degenerate, in Which the Atom Displacements Shown Destroy the Centrosymmetry of the Molecule.

wave functions, that is, $\psi_a = \psi_{a,el} \times \psi_{a,vib}$. Since there will always be some "odd" vibrational modes accessible to both the ground ($a$) and excited ($b$) states of a complex, for which $\langle \psi_{a,vib} | r | \psi_{b,vib} \rangle \neq 0$, the transition will be allowed at least to some extent. From the group theoretical point of view again, there will be some vibration for which the product $\Gamma_{a,vib} \times \Gamma_r \times \Gamma_{b,vib}$ contains the totally symmetric representation, $A_{1(g)}$.

Related to the foregoing mechanism for breaking down of the Laporte selection rule by vibronic coupling is a mechanism that has been termed *intensity stealing*. Thus, if the forbidden excited term lies energetically nearby a fully allowed transition, which would produce a very intense band, there is, generally, a vibrational mode of such symmetry that it can mix the electronic wave functions of the excited-forbidden and the excited-allowed levels. Intensity stealing by this mechanism decreases in magnitude with increasing energy separation between the excited term and the allowed level. It is often observed that $d$-$d$ bands which lie near the allowed charge transfer bands possess unusually high intensities. Reference should now be made again to Table 11–10, where the relative intensities of various transitions are compared.

## Band Widths and Shapes

Room-temperature solid and solution bands of $d$ transition metal complexes are, in general, quite wide, $\sim 1000$ cm$^{-1}$; and only occasionally are narrow bands with widths $< 100$ cm$^{-1}$ found. This implies that the energy terms in the complexes do not correspond to a single energy level, but that they are spread over a range of energies comparable with the band widths which are found. There are several factors that may serve to remove the

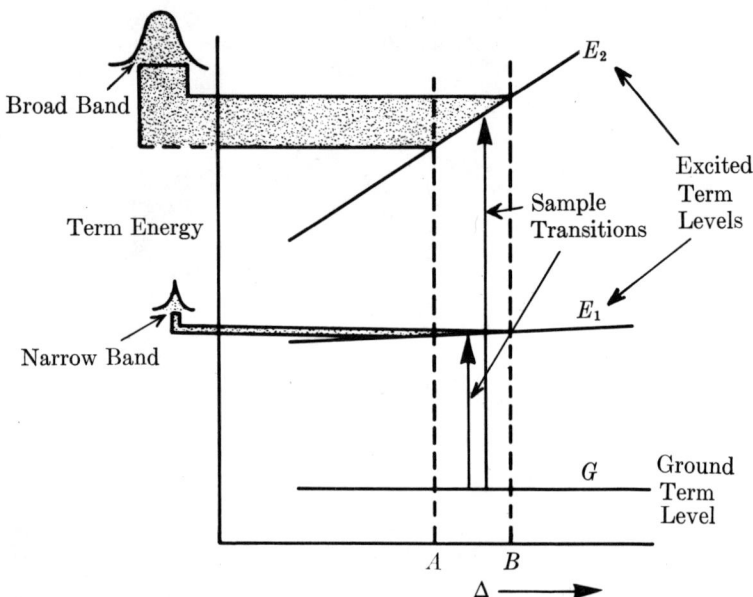

**FIGURE 11-11**

Diagrammatic Illustration of Band Broadening by the Vibrational Mechanism. As the ligands vibrate about an equilibrium position to the extremes ($A$ and $B$) the ligand field fluctuates in response to that vibration from $\Delta_A$ to $\Delta_B$. Electronic transitions may occur at any particular $\Delta_i$ value giving rise to bands which may be narrow, as when $G$ and $E_1$ differ little in their slopes, or bands which may be broad, as when $G$ and $E_2$ differ greatly in their slopes.

degeneracies of the terms and thereby produce broad bands. These include symmetry lowering by (1) molecular vibrations, (2) spin-orbit coupling, and (3) the Jahn-Teller effect.[6,10] All of these factors lead to asymmetrically shaped bands, as discussed elsewhere.[6] Band broadening by the vibrational mechanism is the most easily seen physically and its origin may be understood by reference to Figure 11–11. Both spin-orbit coupling and the Jahn-Teller effect may serve to broaden bands by lifting the degeneracy of otherwise degenerate, for example $E$ and $T$, ground terms. It might be noted here that if the degeneracy has already been lifted by a low-symmetry static ligand field (for example, a complex with nonequivalent ligands) or by spin-orbit coupling, the Jahn-Teller effect is not necessary or operative.

With this admittedly brief and totally qualitative introduction to the intensities and widths of $d$-$d$ bands, their selection rules, and the breakdown of the selection rules, we now proceed to consider the pertinent questions of how many $d$-$d$ bands are to be expected and what are their energies for the various $d^n$ configurations. That is, how may we interpret the spectra of $d$ transition element complexes?

### Energy Level Diagrams and $d$ Complex Spectra

As in the case of so many of the broad topics presented in this book, we shall have to limit ourselves to a rather superficial and generalized treatment of what is really a detailed and complex subject. We hope that this brief introduction to the topic will entice students to read some of the more comprehensive accounts, which are numerous, current, and lucid.[6,10,28,29]

The information necessary for the most general interpretations (which ignore spin-orbit coupling and a few other details) of the spectra of transition metal complexes may be found in certain energy level diagrams. In the extreme of weak ligand fields we employ so-called *Orgel diagrams*[33] and in the extreme of strong ligand fields we employ *Tanabe-Sugano diagrams*.[34] But for the more commonly found fields of intermediate strength, we may use either of these.

**Orgel diagrams — weak fields.** What is done first in the weak field approximation is to write down the free-ion Russell-Saunders states for the various $d^n$ configurations, as was done in Table 11–4, and to arrange them in order of increasing energy as determined from atomic spectroscopy. This is illustrated in Figure 11–12(a) for the $d^2$ system in an octahedral field. The remainder of this figure will become clear shortly. Next we consider, either by using quantum mechanics[10] or more simply by using Group Theory,[32] what will happen to all of the free-ion term levels when we place the free ion at the center of a weak electrostatic field of $O_h$ symmetry. This is shown above (b) in Figure 11–12 for the $d^2$ case. Figure 11–3, which was seen earlier, shows only what happens to the *ground* state term of the various $d^n$ configurations under these conditions. And although the lowest energy transitions may arise from just these split ground levels, often levels from higher atomic terms may fall low enough in energy to be involved also in the low-energy transitions. Now the weak field case implies a ligand field splitting energy (that is, perturbation of the free-ion levels) which is small in comparison with the interelectron repulsion energies. And it is these that split the $d^n$ configuration into the term levels to begin with. Therefore, we must also consider the terms arising from the higher free-ion terms, since transitions to these states, at least to the lower-lying ones, may be observable. In any case, we wish to be able to predict the number, kind, and relative energies of *all* transitions, even if there may be no hope of finding the higher-energy ones experimentally due to their being covered by much more intense bands of other origin.

What we plot, therefore, in an Orgel diagram is the quantum mechanically calculated energy of the term level (as *ordinate*) against an increasing value of $\Delta$, the ligand field splitting parameter (as *abscissa*). Such diagrams are illustrated in Figure 11–13 for the simple $d^1$ and $d^9$ cases. In these specific configurations there are no interelectronic repulsions; hence only a single free-ion term exists, and its splitting in the ligand field is pictured. Note that the diagram for the $^2D$ $d^1$ system in $O_h$ also holds for the $^2D$ $d^9$ system in

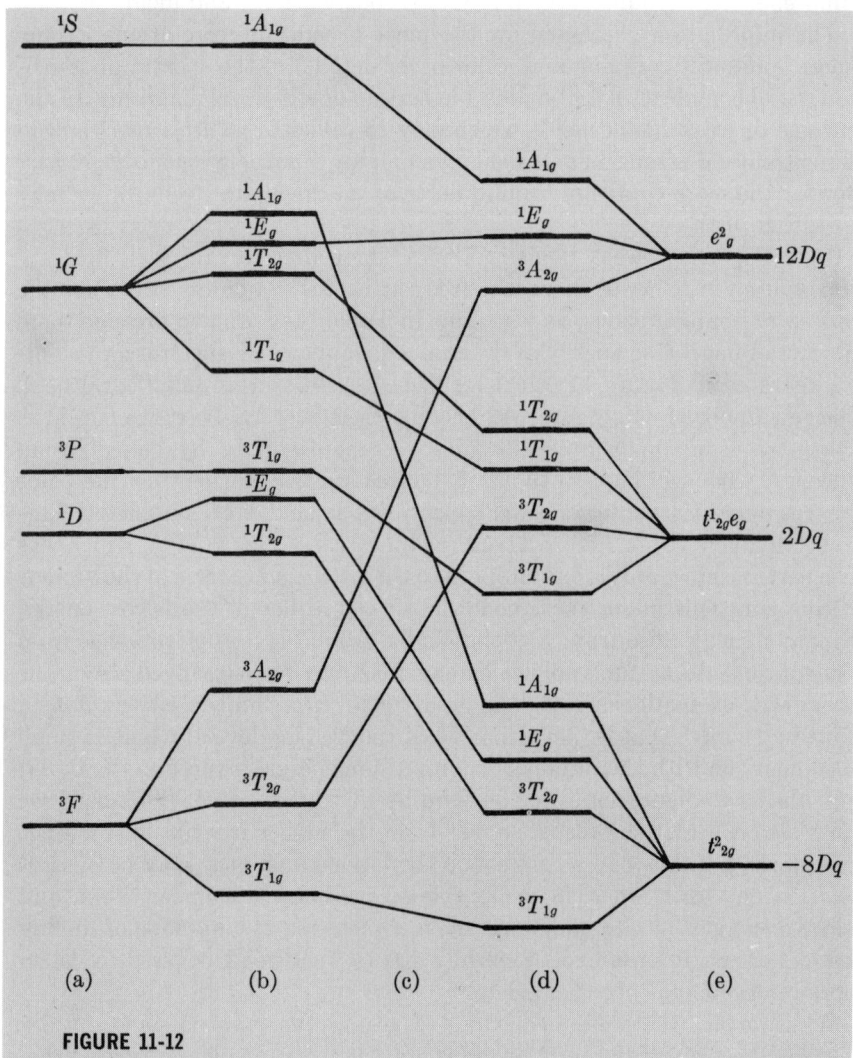

**FIGURE 11-12**

Correlation Diagram for the Free-ion (a) Strong-field Configuration for the System $d^2$ in an Octahedral Field (or for a $d^8$ System in a Tetrahedral Field). Not to scale. (a) The free-ion terms. (b) The terms of an ion in a weak crystal field. (c) Fields of intermediate strength. (d) The terms of an ion in a strong field. (e) The strong-field configurations.

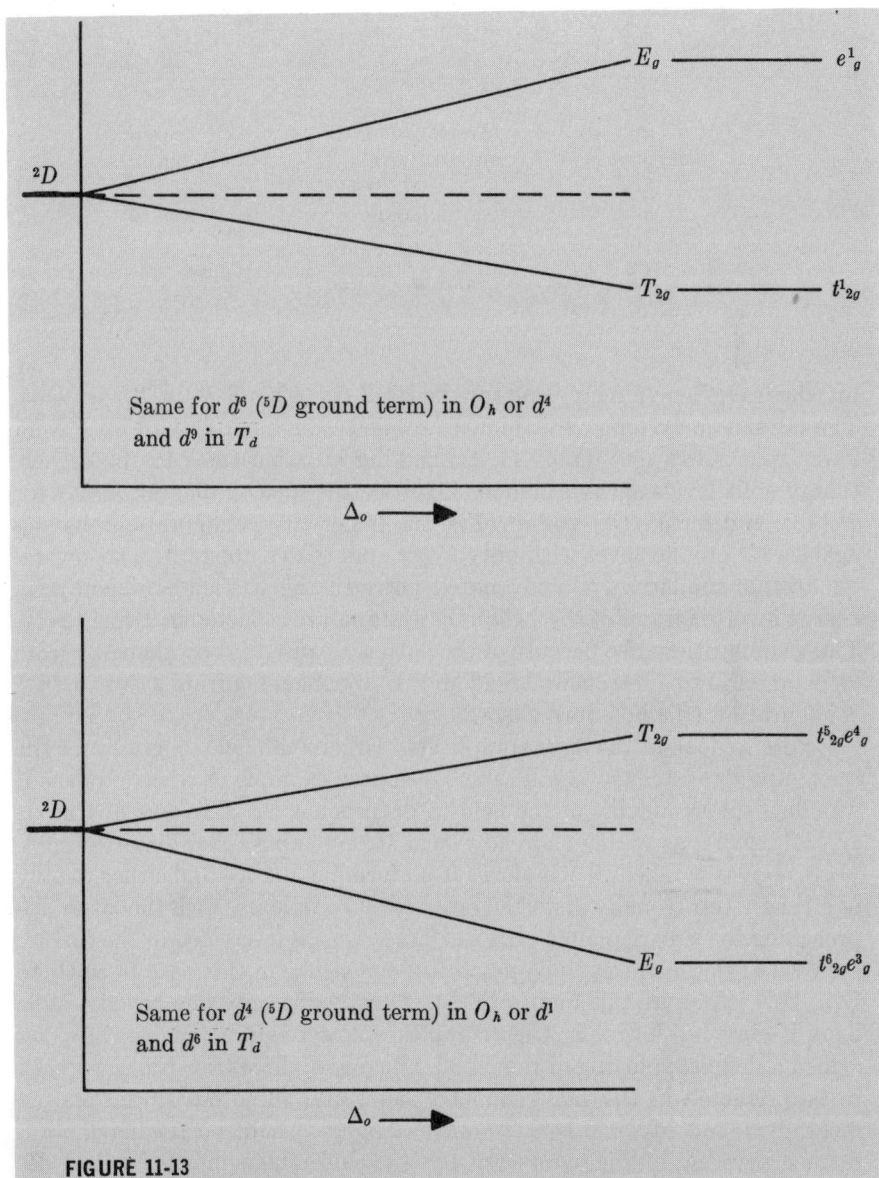

**FIGURE 11-13**

Orgel Diagrams for $d^1$ (Upper) and $d^9$ (Lower) Electronic Systems in $O_h$ Field. Note the inversion relationship between $d^1$ and $d^9$ (the hole formalism). To the far right are given the orbital configurations associated with the specific term levels.

**TABLE 11-12. The Relations Between Term Energy Level Diagrams for the $d^n$ Configurations in $O_h$ and $T_d$ Electrostatic Fields.**

| $O_h$ | $T_d$ | | $O_h$ | $T_d$ |
|---|---|---|---|---|
| $d^1$ and $d^9$ | reverse (inverted) | | $d^9$ and $d^1$ | |
| $d^2$ and $d^8$ | order of levels | | $d^8$ and $d^2$ | |
| $d^3$ and $d^7$ | from each free | | $d^7$ and $d^3$ | |
| $d^4$ and $d^6$ | ion state | | $d^6$ and $d^4$ | |

$d^5$, same in $O_h$ and $T_d$

tetrahedral ($T_d$) symmetry and for the $^5D$ $d^6$ (in $O_h$) and $d^4$ (in $T_d$) systems. The latter two systems have higher atomic terms but those all arise from lower spin states (see Table 11-4), making all transitions to the higher-energy split levels spin-forbidden. Likewise the $d^9$ ($O_h$) diagram holds for $d^1$ ($T_d$), and for $d^4$ ($O_h$) and $d^6$ ($T_d$), the latter two configurations having again a $^5D$ ground level with only lower spin states above it. The several qualitative similarities already noted between the Russell-Saunders term energy level diagrams of the various $d^n$ systems are collected in Table 11-12. These similarities arise because of the patterns caused (a) by changing from an octahedral to a tetrahedral field and (b) by changing from a $d^n$ to a $d^{10-n}$ configuration (the hole formalism).

Before we look at the more complicated Orgel weak field diagrams for the remaining $d^n$ systems, we will take one simple example of what happens if we allow the symmetry of the field to drop below $O_h$. The case of $Cu^{2+}$ is typical, where, as we have already seen, strong Jahn-Teller distortion prevents regular octahedral complexes from forming and we find either weakly or strongly tetragonally distorted octahedral complexes, with the latter approaching the square planar extreme. Thus, we are interested in the further removal of degeneracies upon *descent in symmetry*, in this case from $O_h$ to $D_{4h}$. This is pictured in Figure 11-14. Thus, while only one broad visible band is observed for many $Cu^{2+}$ complexes, lower-symmetry ones have revealed a definite splitting of this band. Of course, three transitions are seen to be predicted in the $D_{4h}$ symmetry field, and these have been experimentally found in some cases. Four $d$-$d$ bands would be predicted by a further lowering of the symmetry to $D_{2h}$, since this splits the 2-fold degeneracy of the $E_g$ state. Indeed four $d$-$d$ bands have been observed for certain $Cu^{2+}$ complexes, confirming the basic correctness of this model.

The general Orgel diagram for the maximum multiplicity states of $d^2$ and $d^7$ systems in octahedral fields and for $d^3$ and $d^8$ systems in tetrahedral fields

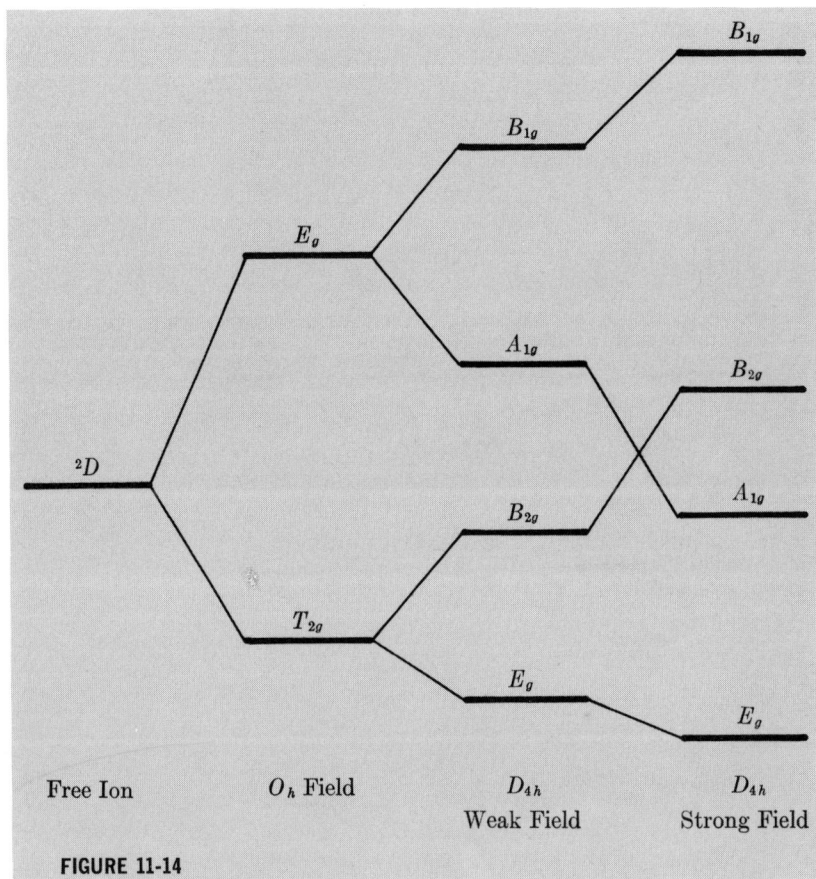

**FIGURE 11-14**

Term Diagram Illustrating Descent in Symmetry, from $O_h$ to $D_{4h}$, for the $d^9$ Cu²⁺ System. Not to scale.

is shown in Figure 11–15, to the right-hand side; and the same for $d^3$ and $d^8$ in $O_h$ and for $d^2$ and $d^7$ in $T_d$ is shown on the left-hand side of that figure. The specific Orgel diagram for the $d^7$ Co²⁺ system, showing the actual calculated energy values, is given in Figure 11–16, with the same for the $d^8$ Ni²⁺ system in Figure 11–17. Figure 11–18 shows the Orgel diagram for the Mn²⁺ $d^5$ ion. In the latter three figures the extreme right-hand side of the figure (the left side also in Figure 11–16) has identified an electron *orbital* configuration with each level. This is the orbital occupancy configuration which may be safely written for the ion in question in the limit of extreme strong field. That is, in the strong field, which is by definition energetically more important in determining the electron distribution than interelectron repulsion energies, the electrons may be clearly assigned by giving maximum

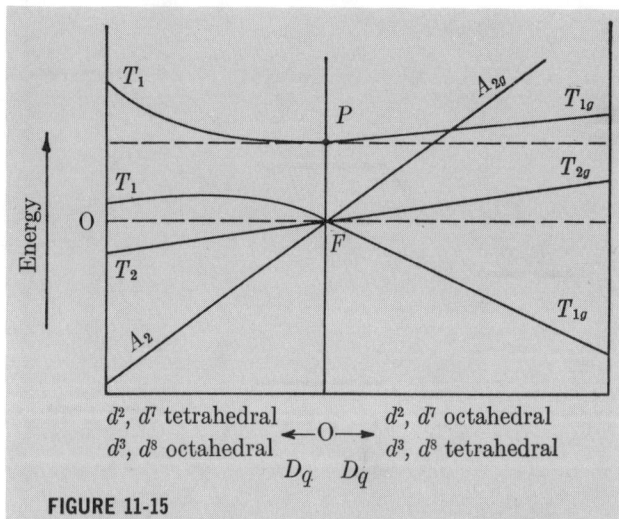

**FIGURE 11-15**

Generalized Orgel Diagram Illustrating the Splitting of the Field-free ion $F$ and $P$ Terms (the Maximum Multiplicity States) Arising from $d^2$, $d^7$, $d^3$ and $d^8$ in Both $O_h$ and $T_d$ Fields.

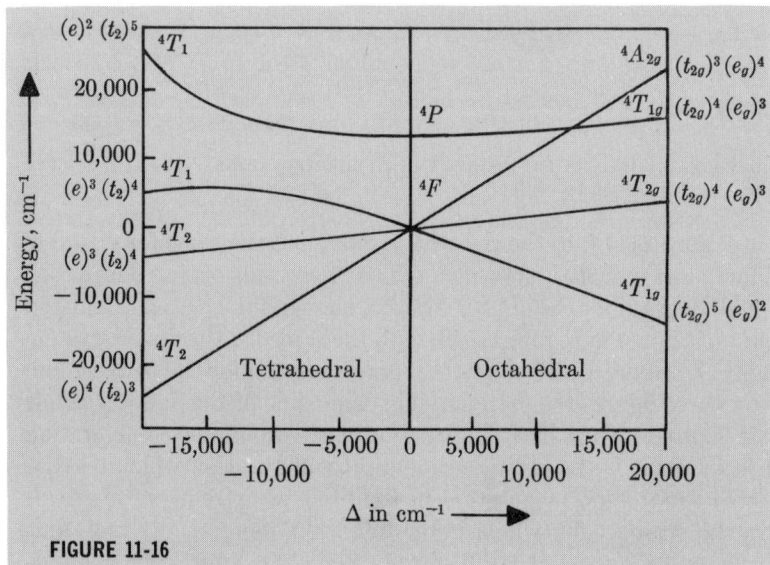

**FIGURE 11-16**

Orgel Diagram Showing the Energy Levels for the $Co^{2+}$ $d^7$ System in Both $O_h$ and $T_d$ Fields.

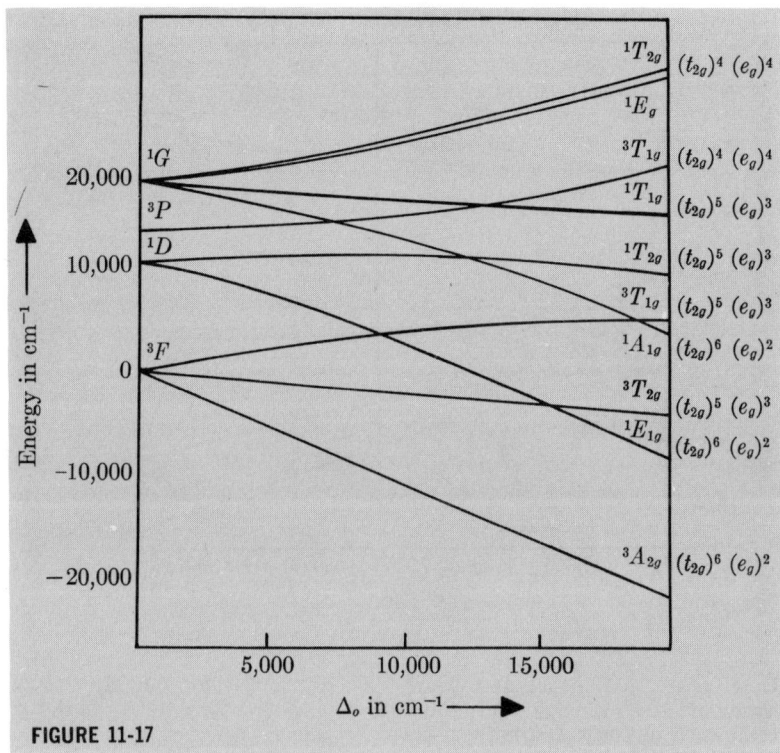

FIGURE 11-17

Orgel Diagram Showing the Energy Levels for the $Ni^{2+}$ $d^8$ System in an $O_h$ Field. The $^1S$ state is at higher energy and is not shown.

occupancy to the $t_{2g}$ orbitals in the ground state, with the first excited state having one less $t_{2g}$ electron and one more $e_g$ electron, etc.

The bands of $Mn(H_2O)_6^{2+}$ listed in Table 11-11 may now be seen to arise from transitions from the $^6A_{1g}$ ground level to the respective quartet levels shown in Figure 11-18 along a vertical line at the $\Delta_0$ value of about 8000 $cm^{-1}$. The transitions for $Mn(en)_3^{2+}$ would be predicted from a vertical line at about $\Delta_0 = 10,000$ $cm^{-1}$. In Figure 11-19 we show just enough of the $Ni^{2+}$ energy diagram to indicate the most likely assignment of the observed bands in the three complex ions: $Ni(H_2O)_6^{2+}$, $Ni(NH_3)_6^{2+}$ and $Ni(en)_3^{2+}$. The respective $\Delta_0$ values are 9000, 11,000 and 12,000 $cm^{-1}$, which illustrates the manner in which the spectrochemical series of ligands (see Chapter 10) is arrived at. In an analogous manner, of course, one may arrive at a spectrochemical series of metal ions. This would be obtained by ranking metal ions according to the order of increasing $\Delta_0$ developed with any fixed ligand. Such a series has been established, and some of the metal ions fall in the following order: $Mn^{2+} < Ni^{2+} < Co^{2+} < Fe^{2+} < V^{2+} < Fe^{3+} < Cr^{3+} < V^{3+} < Co^{3+} < Mn^{4+} < Mo^{3+} < Rh^{3+} < Ru^{3+} < Pd^{4+} < Ir^{3+} < Re^{4+} < Pt^{4+}$.

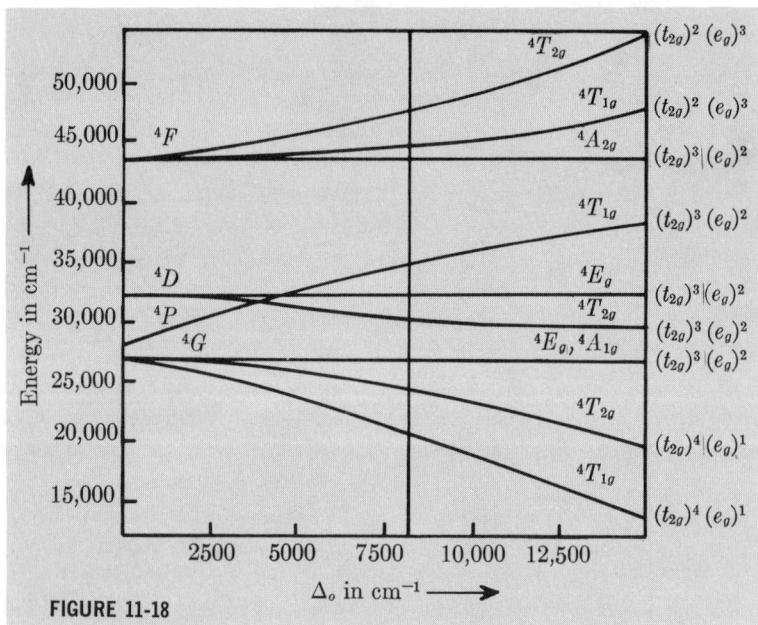

**FIGURE 11-18**

Orgel Diagram Showing the Quartet Energy Levels for the $Mn^{2+}$ $d^5$ System in an $O_h$ Field. The ground state $^6S$ in the free ion and $^6A_{1g}$ in the complexed ion, is taken here to have the zero energy and no doublet states are included. The vertical line at $\Delta_0 = 8000$ $cm^{-1}$ gives a close approximation to the electronic transitions found for $Mn(H_2O)_6^{2+}$.

**Tanabe-Sugano diagrams — strong and intermediate fields.** In the strong-field limit no electron distribution information is available from the free-ion term values, but rather one places the electrons into one-electron $t_{2g}$ and $e_g$ orbitals first and then tries to evaluate the interelectronic repulsion. That is, the Racah $B$ and $C$ parameters are estimated.[34,35] They can be evaluated for free ions, but their evaluation for the complexed ions, other than by empirical means, is extremely difficult.

The Tanabe and Sugano diagrams are plots of the energies of the levels in a $d^n$ system in units of $B$, that is, $E/B$, (as *ordinate*) against the ligand field strength, in units of $Dq/B$ (as *abscissa*). The ground state of the metal ion is always plotted as the abscissa base line in these diagrams. And since two parameters, $B$ and $C$, are necessary to fully describe the interelectronic repulsions for $d$ electrons, the diagrams must be drawn to a specified $C/B$ ratio. For systems having more than three electrons and less than eight, a change in ground state can occur as we progress from weak to strong fields. The critical $Dq/B$ value, where this changeover takes place, is indicated in the diagram by a vertical line. Sample diagrams for the $d^2$, $d^3$, $d^5$, and $d^6$

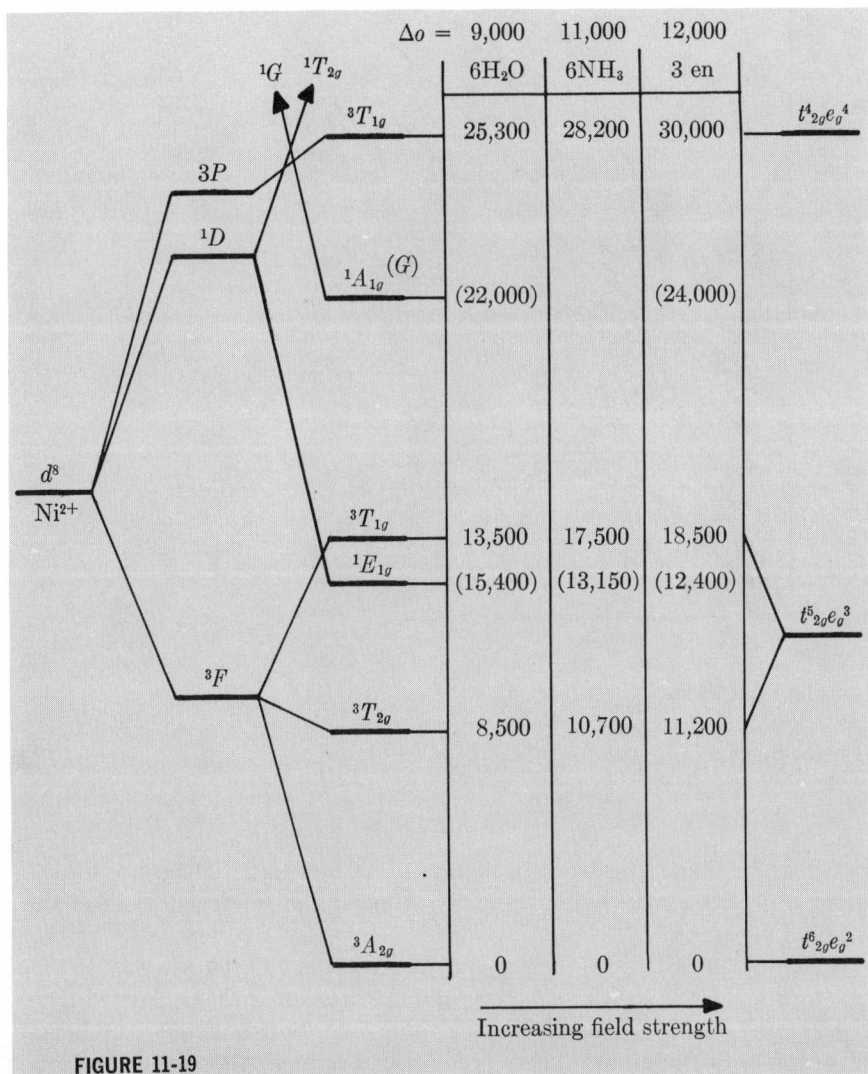

**FIGURE 11-19**

Observed Transitions in Three Complexes of $Ni^{2+}$, with Probable Assignments, Including Probable Spin-forbidden Transitions in Parentheses, Illustrating the Effect of Increasing the Octahedral Ligand Field Strength in the Order $H_2O <$ $NH_3 < en$.

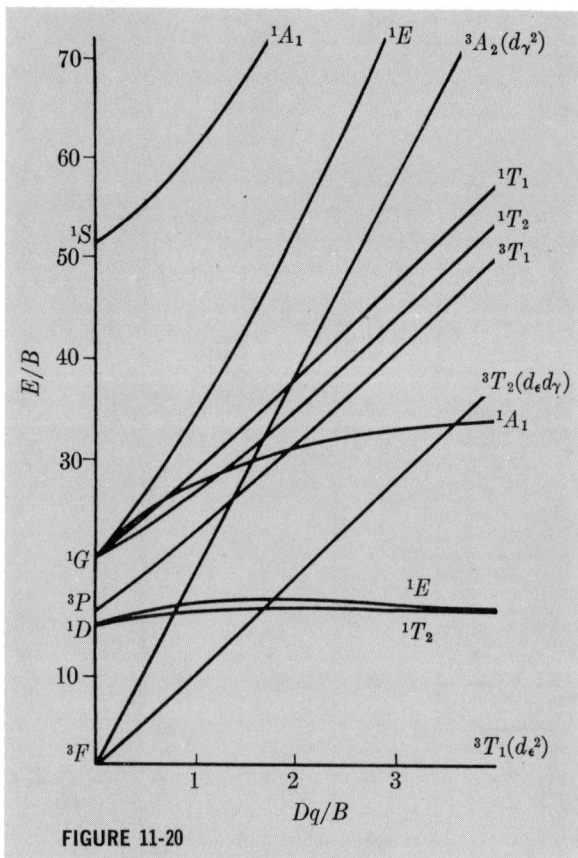

**FIGURE 11-20**

Tanabe-Sugano Diagrams for the $d^2$ (Above) and $d^3$ (Opposite) Configurations in an $O_h$ Field.

octahedral systems are shown in Figures 11–20 and 11–21. Diagrams for the other $d^n$ systems may be found in the Appendix of reference 36 or in the original paper.[34]

The problem with such diagrams is that there is no accurate way to calculate $B$ and $C$ for a given complex, and the diagram must be plotted for a fixed $C/B$ value. It is known that the value of these parameters, and hence all of the term values, are generally lower in a complex than in the free ion, yet neither the Orgel nor the Tanabe-Sugano diagram accounts for this.

The *nephelauxetic series* of ligands, based upon the ability of a ligand to reduce the value of $B$ in the complexed ion from its free-ion value, has already been referred to in the previous chapter. The same criterion may be

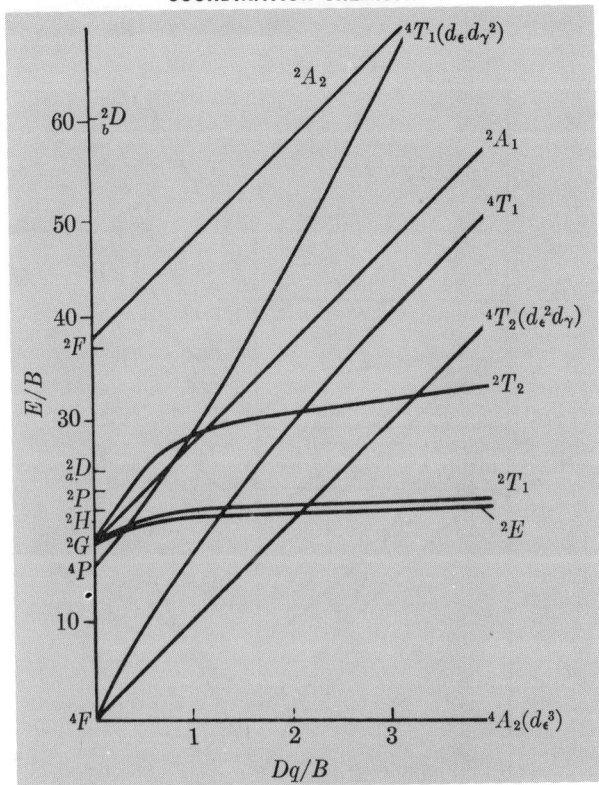

**FIGURE 11-20 (Continued)**

used to establish a metal ion nephelauxetic series. Such a series, for increasing $B$ value in the complex compared to the free-ion value, is as follows: $Mn^{2+} \sim V^{2+} > Ni^{2+} \sim Co^{2+} > Mo^{3+} > Re^{4+} \sim Cr^{3+} > Fe^{3+} \sim Os^{4+} > Ir^{3+} \sim Rh^{3+} > Co^{3+} > Pt^{4+} \sim Mn^{4+} > Ir^{6+} > Pt^{6+}$.

**Charge transfer spectra.** There have been relatively few attempts made to deal systematically with charge transfer spectra, except for some hexahalo complexes of some $4d$ and $5d$ elements and some cyano complexes. However, the subject is being increasingly investigated experimentally, and the recent applications of MOT to complexes offers great possibilities for developing a sound theoretical approach to the problem, which will undoubtedly unfold within several years.

Some useful correlations are already emerging, and the one we choose to mention is the consideration of the CT spectra as reflecting redox tendencies of both the central ions and the ligands. Thus, it is possible to classify and rank metal ions according to their spectrally determined "oxidizing power," and to rank ligands according to their "reducing power." For the hexahalo series, $MX_6^{n-}$, we have,[29] for the metal ions: $Rh^{4+} > Ru^{4+} > Cu^{2+} > Os^{4+} > Fe^{3+} > Ru^{3+} > Pd^{4+} > Re^{4+} \sim Os^{3+} \sim Pd^{2+} \sim Pt^{4+} \sim Rh^{3+} > Pt^{2+}$

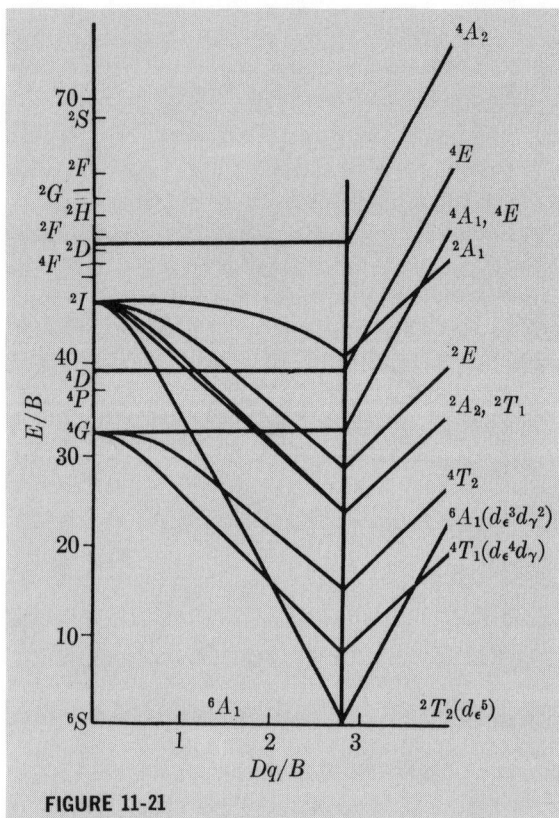

**FIGURE 11-21**

Tanabe-Sugano Diagrams for the $d^5$ (Above) and $d^6$ (Opposite) Configurations in an $O_h$ Field.

$> \mathrm{Ti}^{4+} \sim \mathrm{Ir}^{3+}$, and for the halogens, as certainly expected: $\mathrm{I}^- > \mathrm{Br}^- > \mathrm{Cl}^- > \mathrm{F}^-$.

Thus, we may formulate an approximate rule that *the greater the oxidizing power of the metal ion and the greater the reducing power of the ligand, the lower the energy at which the charge transfer bands ($L{\rightarrow}M$ type) appear.*

## References

1. F. Basolo and R. G. Pearson, "Mechanisms of Inorganic Reactions," 2nd ed., John Wiley & Sons, Inc., New York, 1967.
2. C. H. Langford and H. B. Gray, "Ligand Substitution Processes," W. A. Benjamin, Inc., New York, 1966.
3. S. Chaberek and A. E. Martell, "Organic Sequestering Agents," John Wiley & Sons, Inc., New York, 1959.
4. F. J. C. Rossotti, in "Modern Coordination Chemistry," J. Lewis and R. G. Wilkins, Eds., Interscience Publishers, Inc., New York, 1960, Chapter 1.

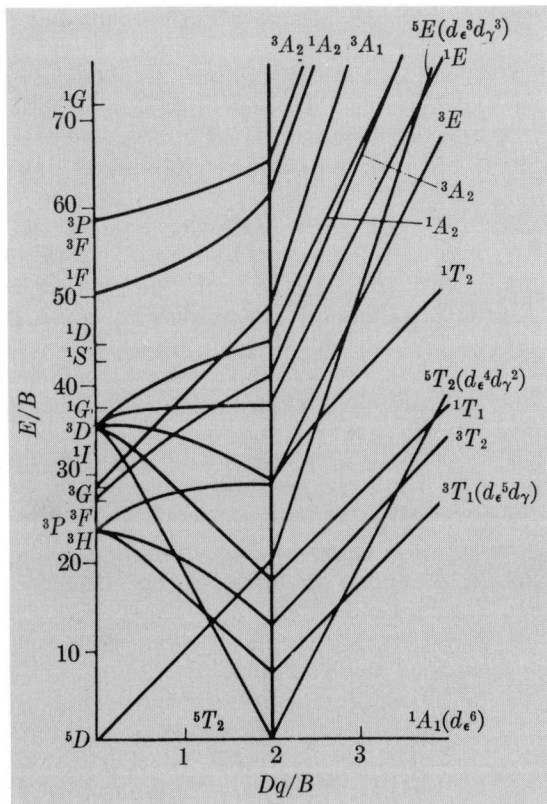

**FIGURE 11-21 (Continued)**

5. M. M. Jones, "Elementary Coordination Chemistry," Prentice-Hall, Inc., Englewood Cliffs, N.J., 1964, Chapters 8 and 12.

6. B. N. Figgis, "Introduction to Ligand Fields," Interscience Publishers, Inc., New York, 1966.

7. F. J. C. Rossotti and H. Rossotti, "The Determination of Stability Constants," McGraw-Hill, Inc., New York, 1961.

8. T. M. Dunn, D. S. McClure, and R. G. Pearson, "Some Aspects of Crystal Field Theory," Harper and Row, Publishers, New York, 1965.

9. L. G. Sillen and A. E. Martell, "Stability Constants of Metal-Ion Complexes," The Chemical Society, London, 1964.

10. C. J. Ballhausen, "Introduction to Ligand Field Theory," McGraw-Hill, Inc., New York, 1962.

11. H. Taube, *Chem. Rev.*, **50**, 69 (1952).

12. R. G. Pearson, *J. Chem. Ed.*, **38**, 164 (1961).

13. R. G. Wilkins, *Quart. Rev.*, **16**, 316 (1962).

14. F. Basolo and R. G. Wilkins, *Adv. Inorg. Chem. Radiochem.*, **3**, 1 (1961).

15. F. Basolo, *Survey Progr. Chem.*, **2**, 1 (1964).

16. H. B. Gray, *Progr. Transition Metal Chem.*, **1**, 239 (1965).

17. H. Taube, *Adv. Inorg. Radiochem.*, **1**, 1 (1959).

18. J. Halpern, *Quart. Rev.*, **15**, 207 (1961).

19. H. Taube, in "Mechanisms of Inorganic Reactions," ACS Advances in Chemistry Series, **49,** American Chemical Society, 1965, Chapter 5.
20. E. S. Gould and H. Taube, *J. Am. Chem. Soc.,* **86,** 1318 (1964).
21. M. Anbar, in "Mechanisms of Inorganic Reactions," ACS Advances in Chemistry Series, **49,** American Chemical Society, 1965, Chapter 6.
22. P. W. Selwood, "Magnetochemistry," Interscience Publishers, Inc., New York, 1956.
23. B. N. Figgis and J. Lewis, *Prog. Inorg. Chem.,* **6,** 37 (1964).
24. R. S. Nyholm, *Record Chem. Progr.,* **19,** 45 (1958).
25. B. N. Figgis and J. Lewis, in "Modern Coordination Chemistry," J. Lewis and R. G. Wilkins, Eds., Interscience Publishers, Inc., New York, 1960, Chapter 6.
26. F. A. Cotton, in "Modern Coordination Chemistry," J. Lewis and R. G. Wilkins, Eds., Interscience Publishers, Inc., New York, 1960, Chapter 5.
27. K. Nakamoto, "Infrared Spectra of Inorganic and Coordination Compounds," John Wiley & Sons, Inc., New York, 1963.
28. T. M. Dunn, in "Modern Coordination Chemistry," J. Lewis and R. G. Wilkins, Eds., Interscience Publishers, Inc., New York, 1960, Chapter 4.
29. C. K. Jørgensen, "Absorption Spectra and Chemical Bonding in Complexes," Pergamon Press, Inc., New York, 1962.
30. R. L. Carlin, *J. Chem. Ed.,* **40,** 135 (1963).
31. J. Ferguson, *Rev. Pure Appl. Chem.,* **14,** 1 (1964).
32. F. A. Cotton, "Chemical Applications of Group Theory," Interscience Publishers, Inc., New York, 1963.
33. L. E. Orgel, *J. Chem. Phys.,* **23,** 1004, 1824 (1955).
34. Y. Tanabe and S. Sugano, *J. Phys. Soc. Japan,* **9,** 753, 766 (1954).
35. D. S. McClure, in "Solid State Physics," F. Seitz and D. Turnbull, Eds., Academic Press, Inc., New York, 1959, Vol. 9.
36. R. S. Drago, "Physical Methods in Inorganic Chemistry," Reinhold Publishing Corporation, New York, 1965.

## Problems

1. What are the $d^n$ configurations which would be predicted to lead to (a) weak and (b) strong Jahn-Teller distortions in otherwise tetrahedral complexes? Have any of these been observed? What will be the nature of the molecular distortion?

2. Make a plot of tetrahedral radii of $3d^n$ ions versus $n$, pointing out and explaining its significant features.

3. The triplet Russell-Saunders states that arise from the $d^2$ configuration are $^3F$ and $^3P$, which in turn give rise in a weak octahedral field to the terms $^3T_{1g}(F)$, $^3T_{2g}$, $^3T_{1g}(P)$ and $^3A_{2g}$, in order of increasing energy. If we write out the $d$-orbital occupancies for the triply orbitally degenerate triplet ground level, $^3T_{1g}(F)$ as follows: $(xy)(xz)$, $(xy)(yx)$, $(xz)(yz)$, what will be the corresponding $d$-orbital occupancies for the three remaining triplet levels?

4. Discuss the spectral features which would be anticipated for complexes of Fe(III) in (a) octahedral and (b) tetrahedral environments of ligands. Would analogous complexes of Os(III) display the same spectral features? Explain.

5. The lone absorption band in the visible spectrum of $[Ti(H_2O)_6]^{3+}$ is of low intensity, it is somewhat broad, and it is unsymmetrical. Explain these three observations.

6. The following electronic bands (in $cm^{-1}$) have been observed for the indicated V(III) complexes. Arrange the ligands in their appropriate spectrochemical series order and assign the observed transitions assuming octahedral ligand environments in all complexes.

| | | | |
|---|---|---|---|
| $V(H_2O)_6^{3+}$ | 17,800 | 25,700 | 38,000 |
| $VF_6^{3-}$ | 14,800 | 23,000 | — |
| $V(urea)_6^{3+}$ | 16,200 | 24,200 | — |
| $V(en)_3^{3+}$ | 21,300 | 29,500 | 39,000 |
| $V(NCS)_6^{3-}$ | 16,400 | 25,400 | 35,150 |

7. How might you account for the fact that with most metal ions $H_2O$ lies to the "stronger ligand" side of $OH^-$ in the spectrochemical series?

8. Suggest an explanation for the fact that Pd(II) and Pt(II) form almost exclusively square planar complexes whereas very few Ni(II) complexes have this stereochemistry. (See reference 10, pp. 712–713, at end of Chapter 10.)

9. Starting with the point-charge CFT model, show that if it is assumed that the ligands possess orbitals which overlap metal orbitals, one might deduce the reverse stability order of, for example, the $e_g$ and $t_{2g}$ orbitals in an octahedral complex. (See reference 10, pp. 698–699, at end of Chapter 10.)

10. The energy difference between the ground state triplet term, $^3F$, and the first excited triplet level above it, $^3P$, in the $Ni^{2+}$ free ion is $(E_P - E_F) = 16,000$ $cm^{-1}$. Certain tetrahedral $NiX_4^{2-}$ complexes display three optical bands which are generally assigned as follows:

$$\nu_1 \quad ^3T_2 \quad \leftarrow ^3T_1(F)$$
$$\nu_2 \quad ^3A_2 \quad \leftarrow ^3T_1(F)$$
$$\nu_3 \quad ^3T_1(P) \leftarrow ^3T_1(F)$$

$\nu_3$ for $NiCl_4^{2-}$ is 14,000 $cm^{-1}$ and the crystal field calculated energy of $\nu_3$ is $[(E_P - E_F) + 6/5\Delta_t]$. Show that theory and experiment are incompatible here, and with the information that $\Delta_t$ is approximately 2500 $cm^{-1}$ show how the theory may be improved.

11. Explain the following data:

| | $\log \beta$ |
|---|---|
| $Ni^{2+} + 2NH_3$ | 5.00 |
| $Ni^{2+} + 4NH_3$ | 7.87 |
| $Ni^{2+} + 6NH_3$ | 8.61 |
| $Ni^{2+} + en$ | 7.51 |
| $Ni^{2+} + 2en$ | 13.86 |
| $Ni^{2+} + 3en$ | 18.28 |

12. Explain the following data:

| | $\log K_1$ for EDTA* complex |
|---|---|
| $Ca^{2+}$ | 10.6 |
| $V^{2+}$ | 12.7 |
| $Mn^{2+}$ | 13.4 |
| $Fe^{2+}$ | 14.2 |
| $Co^{2+}$ | 16.1 |
| $Ni^{2+}$ | 18.5 |
| $Cu^{2+}$ | 18.4 |
| $Zn^{2+}$ | 16.2 |

* EDTA = ethylenediaminetetraacetate ion.

13. Show with diagrams that if $Y$ replaces $X$ in the complex trans-$[Coen_2(A)(X)]$ by an $S_N1$ lim mechanism, if the intermediate is a square pyramid the product will be 100% trans isomer, but if the intermediate is a trigonal bipyramid a mixture of 33% trans and 67% cis isomers is theoretically predicted.

14. Consider the following data for the tetrahedral Co(II) complexes, $CoX_4{}^{2-}$

| $X^-$ | $\Delta_t$ | $\mu$ eff., exptl. |
|-------|------------|--------------------|
| NCS   | 4500       | 4.4                |
| Cl    | 3100       | 4.6                |
| Br    | 2800       | 4.7                |
| I     | 2600       | 4.8                |

Using $\alpha = 4$ and $\xi = -177$, calculate the effective magnetic moment for each of the above complexes and compare these values with the experimental ones. Is the theory supported?

# 12

# Nonaqueous Solvents

In the days of the alchemist, the search for the universal solvent was almost as ardent as that for the philosopher's stone. Needless to say, the search proved fruitless in both instances. Strangely enough, however, after several hundred years we find that water, which is the oldest and most convenient of all solvents, is still the nearest thing we have to a universal solvent. And with regard to convenience and versatility, it is not likely that it will ever be replaced. In fact, before 1900, it was quite generally felt that water is unique in its ability to dissolve ionic substances. We now recognize the fallacy of such a view and in retrospect may wonder why it persisted so long. However, since the turn of the century, we have made great strides in the use and understanding of nonaqueous solvents. Primarily, the present-day advances can be attributed to a realization that the differences between water and other media are, in general, only differences in degree, and they can usually be correlated with a small number of solvent parameters such as the dielectric constant and the coordinating ability of the solvent. Yet, in spite of all our efforts, we have only scratched the surface, and the study of nonaqueous solvents offers one of the remaining frontiers of chemistry.

There are numerous factors that give water its favored position, and its availability is not the least of these. A more or less allied factor is found in its ease of purification. Additionally, such factors as the dipole moment, amphoteric character, and the high dielectric constant all contribute to this unique position of water as a solvent. In spite of its many desirable properties, there are, nevertheless, instances where a nonaqueous medium can serve certain needs that an aqueous one cannot. These vary from studies of the Debye-Huckel theory to use as a medium for organic reactions. In fact, we have even come to use such an unlikely solvent as anhydrous HF on an industrial scale. In this light, it seems quite surprising to find that so little is known of the many possible nonaqueous solvents.

In choosing a solvent, various practical factors must be considered. For instance, if a particular solvent is to be worthy of development, it should be reasonably available, have physical properties which permit its relatively convenient study and it should show a sufficiently unique character to justify its use. When extremes of temperature or pressure are necessary to obtain a liquid phase, or when various hazards are encountered, one cannot help but wonder if the chemical properties of the solvent are such as to warrant further study.

In spite of serious handicaps, the study of some solvents has proved to be extremely profitable. With regard to health hazards, anhydrous HF and anhydrous HCN are obviously quite dangerous. Yet a considerable amount of work has been done with each of these solvents. One of the more common problems with many solvents is that of obtaining a liquid phase. Liquid ammonia has been the most extensively studied of all nonaqueous solvents, and its normal boiling point is $-33.35°C$. Anhydrous $SO_2$ is somewhat more favorable in this respect, inasmuch as its normal boiling point is $-10.2°C$. But in either case, it is necessary to work at lower temperatures or under higher pressures than normal. At the other extreme, we find that very high temperatures are necessary to obtain the liquid phases for fused salt solvents such as fused $KNO_3$ and fused NaCl. A very large number of the common nonaqueous solvents such as methanol, ethanol, acetic acid, and sulfuric acid, of course, have liquid ranges that include normal room temperatures.

## SOLVENT CLASSIFICATION

A variety of approaches have been used in the classification of nonaqueous solvents. However, none of these can be considered to be generally acceptable. From an historical standpoint, much of the interest in nonaqueous solvents has been centered around acid-base concepts, and for this reason classifications of these solvents are frequently based on acid-base properties. Such a classification is rather restrictive, but no more so than the other conventional classifications. It is generally found that the manner of classification one chooses will depend on the particular solvent properties of interest.

One of the simplest and also most obvious solvent classifications is in terms of *polarity*. The extreme differences in the properties of polar and nonpolar solvents certainly justify such a division. There will be many instances where such a classification can prove valuable, but it would appear that it is too broad to be of general utility.

A classification that is closely allied to the Bronsted-Lowry concept of an acid is that which distinguishes *protonic* and *nonprotonic solvents*. Frequently, it is useful to refer to a solvent that will yield a solvated proton on autoionization, such as water, hydrofluoric acid, ammonia, and low molecular-weight alcohols. However, a more valuable breakdown along these same lines is in terms of the *protophylic* character of the solvent. Here we can define four general types of solvents: (1) *acidic solvents* such as sulfuric acid, hydrofluoric acid, and acetic acid, which have a strong tendency to donate protons; (2) *basic solvents* such as ammonia, pyridine, and hydrazine, which have a strong affinity for protons; (3) *amphoteric solvents* such as water and low molecular-weight alcohols, that can act as either acids or bases; and finally (4) *aprotic solvents*, which are those that are inert to proton transfer. Solvents such as benzene and carbon tetrachloride are typical of this last class. A significant weakness of this classification is that we are again emphasizing the acid-base relationships of the solvent. A second, less serious shortcoming of the classification is inherent in the protonic concept of an acid and a base. According to the protonic concept, acidity and basicity are dependent upon the specific reaction. Thus, all of the solvents except the aprotic solvents can act as either acidic or basic solvents depending on the particular solute. Nevertheless, the distinction of the four classes is ordinarily valid, and such a classification is quite valuable.

Another very useful classification of solvents is in terms of the so-called *parent solvent* concept. A particular solvent can be considered to be a parent solvent from which a system of compounds can be derived, and the behavior of analogous compounds should be related in their respective parent solvents. Examples of a few such analogous groups are shown in Table 12–1. As an illustration, we observe that the $OH^-$ group in the parent solvent

### TABLE 12–1. Formally Analogous Groups in Different Solvents.

| Solvent | Analogous Group | |
|---|---|---|
| $H_2O$ | $H_3O^+$ | $OH^-$ |
| $NH_3$ | $NH_4^+$ | $NH_2^-$ |
| $NH_2OH$ | $NH_3OH^+$ | $NHOH^-$ |
| $(CH_3CO)_2O$ | $CH_3CO^+$ | $CH_3CO_2^-$ |
| $H_2S$ | $H_3S^+$ | $SH^-$ |

$H_2O$, is analogous to the $NH_2^-$ group in the parent solvent $NH_3$. We would thus expect these two groups to behave similarly in their respective parent solvents, and this is found to be the case. In terms of this concept, it has been possible to successfully correlate a vast number of chemical reactions in different solvents.

## TYPE REACTIONS

Although our knowledge of nonaqueous systems is such that we cannot always be confident of our understanding of many reactions, we find that some classification of reactions is possible. Needless to say, the same reaction types are defined in nonaqueous media as are defined in aqueous media. However, the specific reactions are certainly quite different from one solvent to another, and it is for this reason that nonaqueous solvents are of such interest.

## Metathetical Reactions

The interest in this type of reaction is usually associated with precipitation both from a preparative and an analytical standpoint. Just as in an aqueous medium, the course of a metathetical reaction in a given non-aqueous solvent can be predicted from the solubilities of the products in that solvent. These are normally of sufficient difference in various solvents as to lead to quite divergent reaction paths. For this reason it is often found that separations and analyses can better be effected in nonaqueous media than in water.

## Acid-Base Reactions

Probably the most thoroughly studied phase of nonaqueous chemistry is that of acid-base reactions. Actually, acid-base reactions have been rather significant in initiating and furthering studies in the field of nonaqueous chemistry, primarily through the solvent-system concept of acid-base behavior. According to the solvent-system definition, an acid can be considered to be any substance that, by direct dissociation or reaction with the solvent, gives the cation characteristic of the solvent; and a base to be any substance that, by direct dissociation or reaction with the solvent, gives the anion characteristic of the solvent. In the case of protonic solvents, the cation is nothing more than a solvated proton, and under these conditions, the protonic concept of an acid is essentially equivalent to the solvent system concept of an acid, as long as a given solvent is used. As an example, we would expect some typical neutralization reactions in ammonia to be

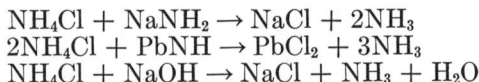

$$NH_4Cl + NaNH_2 \rightarrow NaCl + 2NH_3$$
$$2NH_4Cl + PbNH \rightarrow PbCl_2 + 3NH_3$$
$$NH_4Cl + NaOH \rightarrow NaCl + NH_3 + H_2O$$

Sight should not be lost of the fact that $NH_4Cl$ is equivalent to a solvated HCl molecule, and we can consider the reaction of $NH_4Cl$ with NaOH to be analogous to the reaction of $H_3OCl$ with NaOH.

In nonprotonic solvents, it may again be convenient to use the solvent system concept for acid-base behavior. For an acid-base reaction in anhydrous $SO_2$, we find that a typical reaction is

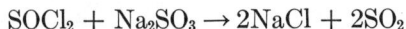

$$SOCl_2 + Na_2SO_3 \rightarrow 2NaCl + 2SO_2$$

In addition, we should recognize that a variety of other types of acid-base reactions can be considered in these solvents if we accept either the Lewis or the protonic definition. In such cases, we can consider the solvent to be nothing more than a medium in which the reaction takes place. This, of course, is a necessity in an aprotic solvent such as benzene or carbon tetrachloride. These solvents show no autoionization, and the solvent system concept has no significance here. Yet it is still possible to carry out acid-base type reactions in such media.[1]

## Oxidation-Reduction Reactions

It would be expected that the same types of oxidation-reduction reactions occur in nonaqueous media as occur in an aqueous system, and this is generally true. However, at the same time, it is found that the oxidizing and reducing ability of a given oxidant or reductant will vary from one solvent to another. Thus the magnitudes of the oxidation potentials will vary with the solvent, and it is likely that the actual order of potentials will change to some extent also. For instance, lithium has the highest standard potential in an aqueous medium. By means of the Born-Haber type of cycle shown in Chapter 8, it was seen that this is primarily due to the relatively high hydration energy of the lithium ion. In a solvent of different polarity, a quite different order might be expected. Unfortunately much of the available data along these lines is of questionable validity as can be seen from the values presented in Table 12–2 for the standard state emf of the Ag,AgCl

**TABLE 12–2.  Reported Values of the Oxidation Potential for the Ag,AgCl Half-cell in Anhydrous Ethanol at 25°C.[a]**

| Investigator | $E_m°$ (volts) |
|---|---|
| Danner | 0.0559 |
| Harned and Fleysher | 0.0442 |
| Lucasse | 0.0365 |
| Woolcock and Hartley | 0.0883 |
| Taniguchi and Janz | 0.08138 |
| LeBas and Day | 0.079 |
| Tezé and Schaal | 0.0723 |

[a] Based on H. Strehlow, Electrode Potentials in Non-aqueous Solvents, "The Chemistry of Non-Aqueous Solvents," Academic Press, N. Y., 1966.

half-cell in anhydrous ethanol. The rather pronounced disagreement among the various values can most reasonably be attributed to the presence of trace quantities of water in the ethanol. In spite of the difficulties, oxidation potentials have been determined in a number of different solvents. In Table 12–3, the values for the alkali and alkaline earth metals in anhydrous ammonia are given. Here it is seen that lithium is still at the top of the series, but at the same time, the order of the remaining metals is not the same as in aqueous media. It might, however, be pointed out that there is always sufficient question in the magnitudes of nonaqueous potentials to warrant proposal of a different order from that listed here.

Although redox reactions follow a rather normal pattern in most instances, a quite anomalous behavior is noted for solutions of the alkali, alkaline earth, and some of the rare earth metals in a variety of coordinating solvents characterized by liquid ammonia. These metals dissolve readily in liquid ammonia, and on evaporation, the alkali metals are obtained in their

**TABLE 12–3.  Oxidation Potentials for Metals in Liquid Ammonia Corrected to 25°C.**

| Couple | $E°$ (volts) |
|---|---|
| Li $= $ Li$^+ + e^-$ | 2.34 |
| Sr $= $ Sr$^{2+} + 2e^-$ | 2.3 |
| Ba $= $ Ba$^{2+} + 2e^-$ | 2.2 |
| Ca $= $ Ca$^{2+} + 2e^-$ | 2.17 |
| Cs $= $ Cs$^+ + e^-$ | 2.08 |
| Rb $= $ Rb$^+ + e^-$ | 2.06 |
| K $= $ K$^+ + e^-$ | 2.04 |
| Na $= $ Na$^+ + e^-$ | 1.85 |
| Mg $= $ Mg$^{2+} + 2e^-$ | 1.74 |

original form, whereas the alkaline earth metals form crystalline ammonates of the form $M(NH_3)_6$. All of these solutions are metastable and will eventually decompose according to the reaction

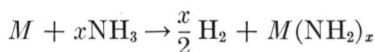

$$M + xNH_3 \rightarrow \frac{x}{2} H_2 + M(NH_2)_x$$

However, if pure reagents and clean equipment are used at a sufficiently low temperature, the decomposition rate is very low. The stability will vary, of course, among the different elements. For some time liquid ammonia was thought to be unique in this ability, but a rather large number of solvents have now been observed to show more or less similar behavior. These include the lighter aliphatic ammines, various ethers such as tetrahydrofuran, and the lighter alcohols.

There is some question about the specific structure of these solutions, but the basic interpretation of their general behavior can be attributed to Kraus.[2] It is assumed that the metal atoms are dissociated into metal ions and free electrons, and these are solvated. Although this would appear to be an oversimplification, we might represent the process as

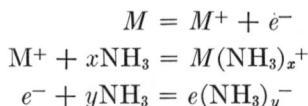

$$M = M^+ + e^-$$
$$M^+ + xNH_3 = M(NH_3)_x{}^+$$
$$e^- + yNH_3 = e(NH_3)_y{}^-$$

Actually an equilibrium exists in which the electron reacts with the ammonia according to the reaction

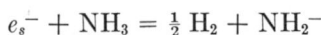

$$e_s^- + NH_3 = \tfrac{1}{2} H_2 + NH_2^-$$

Using electron spin resonance to determine the electron concentration, the equilibrium constant for this reaction has been determined by Jolly[3] to be $K = 5 \times 10^4$ at 25°C.

Depending on the metal concentration, three distinct situations can be distinguished in these solutions. In the very dilute region, the metal ions and electrons can be considered to be well separated as distinct entities. Both the metal ions and the electrons are, of course, solvated. The latter can be considered to occupy holes within the solvent structure with the hydrogen atoms of the surrounding ammonia molecules pointed towards the electrons. In these solutions, the conductance is quite high, being about five to six times greater than the expected ionic conductance of comparably sized ions. Consequently, it is felt that some alternative mechanism must be operative, and this would most reasonably involve the movement of the electrons through the system. Quantum mechanically this can be treated by imagining that there are regions in the solution that are energetically favorable to the electron. These will depend on the distribution of the ammonia molecules in the solution. To pass from one to another of these favored regions, the electron must pass through or over the potential barrier separating the regions. Because of the wave properties of the electron, it can "tunnel" through the barrier, and it is this mechanism that seems to be the most successful in treating the conductance of these relatively dilute solutions.

In the intermediate range of concentrations, the solvated metal ions appear to be bound together into clusters. But in the regions of high concentration, greater than about one molar, the behavior approaches that of a metal. Here the ammoniated metal ions are considered to be held together by a "sea of electrons" in a manner analogous to that of a metal. Although electrical conductance is abnormally high in all concentration ranges of these solutions, it is particularly high in the concentrated range where it approaches that of a metallic conductor. With regard to redox reactions, a solution containing free electrons would be expected to show unusually great reducing ability, and ammonia solutions of the alkali and alkaline earth metals show just such behavior. Much of the interest in the solvent properties of ammonia has been centered around the theoretical as well as the experimental nature of these solutions.[4]

## Solvolytic Reactions

In order to differentiate between solvation reactions and solvolytic reactions, we can define a *solvolytic* reaction as one in which the dissolved solute reacts with the solvent in a manner such as to change the normal anion and cation concentrations of the solvent. In an aqueous medium, an example of a solvolytic reaction would be

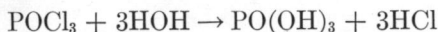

$$POCl_3 + 3HOH \rightarrow PO(OH)_3 + 3HCl$$

In the resultant solution, we have HCl and $H_3PO_4$. Both of these compounds are strong acids, and the pH of the solution will obviously show a sharp decrease. As an example of a solvolytic reaction in a nonprotonic solvent, we can consider the reaction

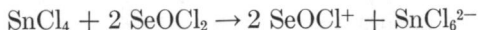

$$SnCl_4 + 2 SeOCl_2 \rightarrow 2 SeOCl^+ + SnCl_6^{2-}$$

The cation characteristic of the solvent is $SeOCl^+$ and its concentration is seen to be increased as a result of the reaction.

One great advantage in considering solvolytic-type reactions results from the similarity of reactions in terms of the parent solvent concept. If for instance, we consider reactions of protonic solvents with a compound such as $RCOCl$, we would expect to obtain products that are formally related in terms of their own parent solvents. In the case of the solvent water, we would obtain the reaction

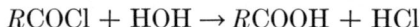

$$RCOCl + HOH \rightarrow RCOOH + HCl$$

If now we were to use other protonic solvents such as alcohols, ammonia, or a primary amine, we should observe analogous type reactions. For instance, we would expect

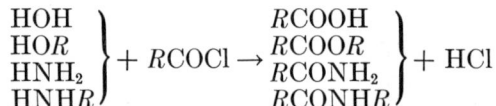

$$\left.\begin{array}{c} HOH \\ HOR \\ HNH_2 \\ HNHR \end{array}\right\} + RCOCl \rightarrow \left.\begin{array}{c} RCOOH \\ RCOOR \\ RCONH_2 \\ RCONHR \end{array}\right\} + HCl$$

Each of these resultant compounds would be expected to behave in an analogous manner in its own parent solvent. Thus, it would be expected that $RCOOH$ will behave in water in a manner similar to $RCONH_2$ in ammonia. It is for this reason that solvolytic reactions are of rather significant value as a means of reaction classification.

## THE DIELECTRIC CONSTANT

In attempting to rationalize experimental data, it is frequently found that several apparently different models of a system can be used with equal success. However, these are usually found to be manifestations of some more fundamental concept. This is particularly true of solution chemistry in nonaqueous solvents. We can attribute our observations to ion-pair formation, deviations in activity coefficients, ionic interactions of various types, and so on, but these can all be related through the dielectric constant of the solution. Consequently, it would be expected that many solvent characteristics should be understandable in terms of the dielectric constant of the solvent; and this is found to be the case.

The most familiar usage of the dielectric constant is found in the expression for the force of attraction or repulsion between two charged bodies,

$$(12\text{-}1) \qquad F = \frac{QQ'}{\epsilon r^2}$$

Here, $Q$ and $Q'$ are two charges separated by a distance, $r$, in a homogeneous dielectric. In this particular case, we can think of the dielectric constant, $\epsilon$, as a constant of proportionality dependent on the particular solvent medium. For a vacuum, we define $\epsilon$ to be unity, and for all other media, it is found to have a value greater than unity. Alternatively, we can obtain a mechanistic picture by considering the dielectric constant in terms of the charging of the plates of a parallel condenser. If we imagine a condenser having two parallel plates with surface areas that are large compared to the distance between them, and we allow them to become charged electrically to the extent of $+\sigma A$ and $-\sigma A$, we will find that there is an essentially homogeneous electric field established inside the condenser. This field will be directed perpendicular to the surface of the plates, and if the medium between the plates is a vacuum, the field will have a magnitude given by

$$(12\text{-}2) \qquad E_0 = 4\pi\sigma$$

Now, if a homogeneous dielectric material of dielectric constant $\epsilon$ is placed between the condenser plates, we will find that the field strength will be decreased to

$$(12\text{-}3) \qquad E = \frac{4\pi\sigma}{\epsilon}$$

The decrease in field strength in the presence of the dielectric is the result of the displacement of electric charges in the body of the dielectric. These displacements are of two general types. If the molecules have a permanent dipole, they will arrange themselves in a preferential position in the field. This is referred to as *orientation polarization*. If no permanent dipoles exist in the molecules, a form of polarization known as *distortion polarization* will still result from the displacement of the charges in the molecules. In either case, we can recognize that the energy of the condenser is stored as a result of the alignment of the solvent dipoles in such a manner as to neutralize the effects of the charges on the condenser plates.

The importance of the dielectric constant in ionic solutions becomes apparent when we note its relationship with respect to the interactions of adjacent ions. It can be seen from Eq. (12–1) that the force of attraction between two ions is critically dependent on the dielectric constant of the medium. As the dielectric constant decreases in magnitude, the force of attraction between adjacent ions increases. Consequently, in solvents with low dielectric constants the ionic interactions extend over much greater distances than in solvents with high dielectric constants. Since nonideality can usually be attributed to ion-ion interactions, it follows that systems with high dielectric constants should show greater conformity to ideal behavior than those of the same ionic concentration having low dielectric constants.

By now, the qualitative relationship between the dipole moment of a solvent molecule and the dielectric constant of the solvent should be obvious. If a solvent molecule has a large dipole moment, the solvent will have a correspondingly large dielectric constant. By virtue of their large dipole moment, such solvent molecules are readily capable of absorbing the energy of an electric field. On the other hand, if the solvent molecules are symmetrical or, at best, have only a small dipole moment, they will not be so successful in neutralizing an electric field, and the dielectric constant will be small. These trends can be appreciated from the list of dielectric constants given in Table 12–4.

**TABLE 12–4.  Dielectric Constants of Some Typical Solvents.**

| Solvent | Dielectric Constant | Temperature (°C) |
|---|---|---|
| Hydrogen cyanide | 118.3 | 18 |
| Sulfuric acid | 110 | 20 |
| Formamide | 109 | 25 |
| Water | 78.5 | 25 |
| Methanol | 31.5 | 25 |
| Ethanol | 24.2 | 25 |
| Ammonia | 22 | −34 |
| Acetone | 20.4 | 25 |
| Pyridine | 12.30 | 25 |
| Tetrahydrofuran | 7.39 | 25 |
| Diethyl ether | 4.26 | 25 |
| Benzene | 2.275 | 25 |
| Dioxane | 2.213 | 25 |
| Cyclohexane | 2.05 | 25 |

## THE ACTIVITY COEFFICIENT

From a strictly thermodynamic point of view, it is not necessary to relate the activity coefficient of a solute to any particular mechanistic picture of solute behavior. On the other hand, we cannot help but wonder why such deviations from ideality do occur. Thus, in the light of our present understanding of solution chemistry, it has been possible to show that the activity coefficient of an electrolyte is intimately dependent on a variety of factors, most of which can be discussed in terms of electrostatic interactions. These are factors involving, primarily, ion-ion interactions such as ionic association and ion-ion repulsions. Virtually all such factors are dependent on the dielectric constant of the solvent, and we should, therefore, expect a direct correlation between the solvent dielectric constant and the activity coefficients of the solute.

A common means of determining the activity coefficients of an ionic solute is by use of a galvanic cell. The relationship between the molar activity coefficient and the measured potential of the cell is given for a uni-univalent electrolyte by the Nernst equation in the form

$$(12\text{–}4) \qquad \mathbf{E} = \mathbf{E}^\circ - \frac{RT}{\mathbf{F}} \ln C^2 f_\pm^2$$

Thus, it can be seen that, if the standard potential of the cell $\mathbf{E}^\circ$ is known, it is possible to evaluate the activity coefficients of the electrolyte, $f_\pm$, for a given molar concentration $C$. The most convenient cell to use for this purpose is one without liquid junction, and one of the most commonly used cells of this type is the cell

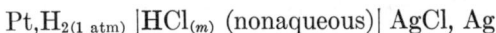

$$\text{Pt,H}_{2(1 \text{ atm})} \,|\text{HCl}_{(m)} \text{ (nonaqueous)}| \,\text{AgCl, Ag}$$

Although there is always some question of the validity of such data because, in particular, of trace impurities of water, there appears to be a definite correlation between the activity coefficients of the solute and the solvent dielectric constant.

One rather significant use of anhydrous solvents as well as mixed solvents is found in the testing of the Debye-Huckel theory. According to the Debye-Huckel theory, the molar activity coefficients of the solute will be given by the general expression

$$(12\text{-}5) \qquad -\log f_{\pm} = \frac{A\sqrt{C}}{1 + \beta a_0 \sqrt{C}}$$

where $C$ is the molar concentration of the solute, $a_0$ is the apparent ionic radius of the solute ions, and $A$ and $\beta$ are constants dependent on the temperature and the solvent. In more detailed form, the expression for the molar activity coefficients is given by

$$(12\text{-}6) \qquad -\log f_{\pm} = 1.8123 \times 10^6 \frac{\sqrt{C}}{(\epsilon T)^{3/2}} \cdot \frac{1}{1 + 50.288 \times 10^8 (\epsilon T)^{-1/2} a_0 \sqrt{C}}$$

where $\epsilon$ is the dielectric constant of the solvent and $T$ is the absolute temperature. It can readily be seen from Eq. (12-6) that the dielectric constant does enter into the theoretical calculations of the activity coefficients. The extent of this dependence can be appreciated from Figure 12-1, where the log of the molal activity coefficient is plotted against the square root of the molal concentration of hydrochloric acid in different mixtures of dioxane and water. The dielectric constant, of course, decreases with an increase in the dioxane concentration. The straight lines represent the Debye-Huckel limiting slope. Thus, in general, we can state that for a given ionic concentration, the deviation of the solute activity coefficient from unity increases with a decrease in solvent dielectric constant. This point may be better illustrated by comparing the activity coefficients of an electrolyte directly with the square root of the concentration of that electrolyte. Thus, in Figure 12-2, we see a plot of experimental activity coefficients versus the square root of the molal concentration of hydrochloric acid in three solvents having quite different values for their dielectric constants. It is apparent that the activity coefficients are affected in a consistent manner by the change in solvent dielectric constant.

If we assume that the nonideality of a solute is reflected in its activity coefficient, then it can be concluded that for a given ionic concentration, a decrease in solvent dielectric constant leads to an increase in solute nonideality. This is evident since an ideal electrolyte would ordinarily have an

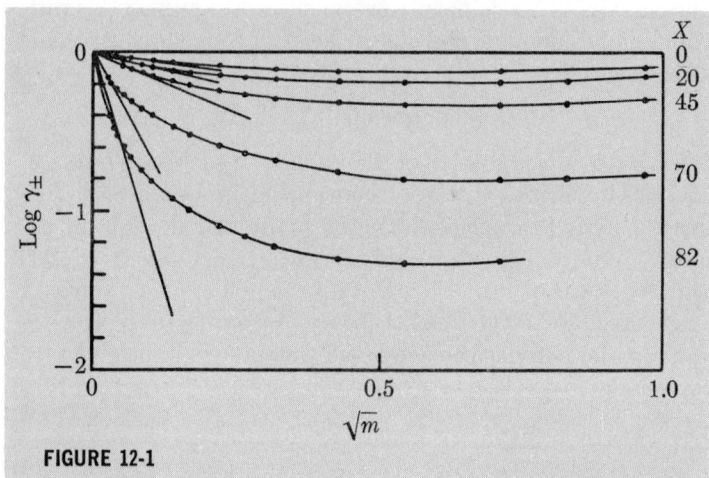

**FIGURE 12-1**

Comparison of the Molal Activity Coefficients of HCl to the Limiting Slope of the Debye-Huckel Theory in Various Mixtures of Dioxane and Water. The weight percent of dioxane is given under $X$. (After H. Harned, J. Morrison, F. Walker, J. Donelson, and C. Calmon, *J. Am. Chem. Soc.*, **61**, 49 (1939).)

activity coefficient of unity, and from Figure 12–2 it can be seen that the deviation of the activity coefficient from unity is greatest for the solvent with the lowest dielectric constant. This result is quite understandable when we recall the method used to determine the standard state electrode potential of a cell. The standard state is defined in such a manner that the activity coefficient approaches unity as the solute concentration approaches zero. This standard state is one in which the only interactions of the solute ions are those with the solvent. That is, no ion-ion interactions will be present at infinite dilution. Consequently, at finite concentrations, ion-ion interactions become significant and lead to deviations in the activity coefficients from unity. Of course, the lower the solvent dielectric constant, the more pronounced will be these ionic interactions.

## SOLUBILITY

As a general rule, we expect ionic and polar solutes to dissolve in polar solvents and nonpolar solutes to dissolve in nonpolar solvents. From a qualitative point of view, this behavior can be attributed to the types of forces that exist between a solute and a solvent molecule as compared to those that exist between two solvent molecules. If, for instance, the possibility of dissolving carbon tetrachloride in water is considered, we note that the only interaction between a carbon tetrachloride molecule and a water molecule involves van der Waals forces. Yet the interaction of a water molecule with another water molecule involves the considerably greater

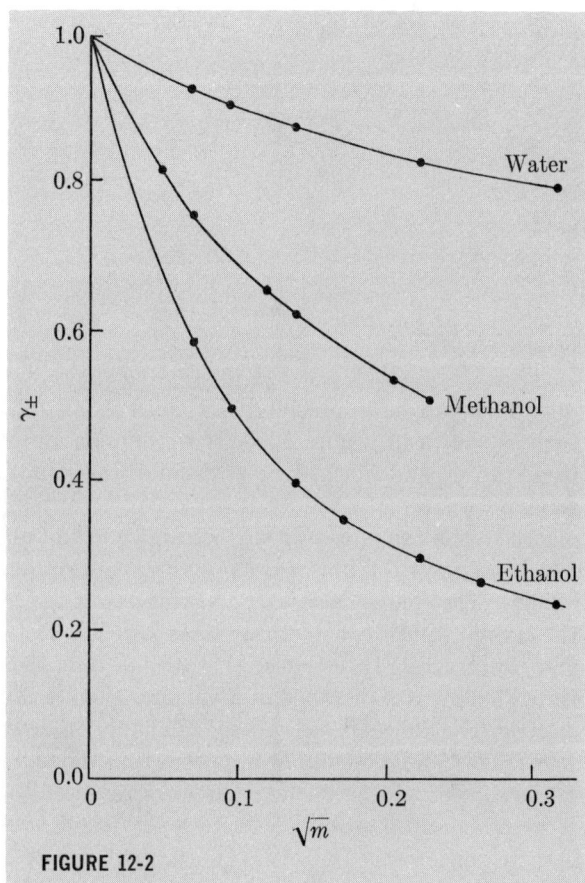

**FIGURE 12-2**

Dependence of the Molal Activity Coefficients of HCl on the Solvent Dielectric Constant.

energy of a hydrogen bond. Thus, from the standpoint of energetics, we would not expect carbon tetrachloride to be water-soluble. On the other hand, if we consider the solubility of carbon tetrachloride in a solvent such as benzene, it can be seen that both solute and solvent can offer only van der Waals-type forces. Consequently, it would seem reasonable to expect that the benzene molecule would have little preference between a carbon tetra-chloride molecule and another benzene molecule. Finally, if we raise the temperature of an immiscible system, it might be expected that thermal energies, which are of the order of $kT$, would be of sufficient magnitude to overcome hydrogen bonding effects and lead to an increase in solubility.

Attempts have been made to calculate the solubility of an ionic species in various solvents, but they have met with only limited success. In carrying out such a calculation, we must first recognize that in a solvation process involving ionic species, the dielectric constant of the solvent plays a major

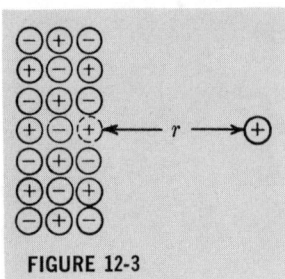

**FIGURE 12-3**

Removal of an Ion from a
Crystal Lattice.

role. This can be appreciated by considering the dissolution of a crystal such as shown in Figure 12–3. If we remove an ion to an infinite distance from the crystal surface in a vacuum, a potential energy curve of the general form seen in Figure 12–4(a) will be obtained. If the crystal is now placed in the presence of a dielectric, we would obtain a curve of the type seen in Figure 12–4(b), for the potential energy change as the ion is taken to infinity. The energy necessary to bring about the separation of the ion from the crystal will be considerably less when a dielectric exists between the ion and the crystal. In essence, this means that the ion can see the crystal through a greater distance in a vacuum than it can in the presence of a dielectric. The dielectric, of course, will be the solvent in which the dissolution takes place, and it should be apparent that the higher the dielectric constant of the solvent, the less will be the work required to separate the ions. Thus, it would be considerably easier to separate two ions in water, where the dielectric constant is 78.5 at 25°C, than it would be in ethanol, where the dielectric constant is 24.2 at this temperature. This, then, results in a general

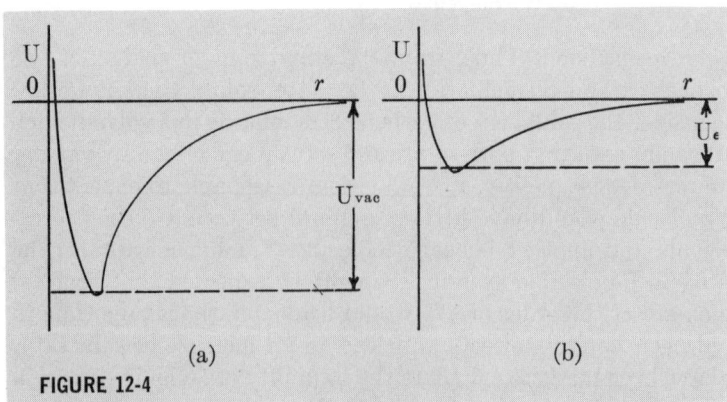

**FIGURE 12-4**

Potential Energy Diagrams for the Removal of an Ion from a Crystal Lattice (a) in a Vacuum, and (b) in the Presence of a Dielectric.

decrease in solubility of a salt with a decrease in the dielectric constant of the solvent.

## Application of the Born Equation

A detailed calculation of the solubility of a salt can be attempted strictly on the basis of electrostatic interactions. The starting point of the calculation is the Born equation[5] for the electrical work involved in charging a sphere of radius $a$, in the presence of a uniform dielectric. Basically, the equation can be developed by considering a uniform field of intensity $E$ in a vacuum. The energy associated with the field will be $E^2/8\pi$ per unit volume. If we now allow the field to be nonuniform, its total energy can be obtained by multiplying the energy per unit volume by the volume increment, $dv$, and integrating from the surface of the sphere to infinity. The field intensity will be given by $q/r^2$, and by the usual method, the integration can be made by considering the volume of a spherical shell of thickness $dr$ at a distance $r$ to $r + dr$ from the center of the sphere. This increment is given by $4\pi r^2 dr$, thereby leading to

$$(12\text{–}7) \qquad W = \int \frac{E^2}{8\pi}\, dv = \int_a^\infty \frac{q^2 4\pi r^2}{8\pi\, r^4}\, dr = \frac{q^2}{2a}$$

In the presence of a dielectric, the equation can be shown to be

$$(12\text{–}8) \qquad W = \frac{q^2}{\epsilon 2a}$$

If we make the assumption that an ion can be treated as a sphere with a fixed radius and that the dielectric constant of the solvent is uniform, then our calculations for the work involved in the charging of the sphere can be equated to the free energy of the ion with respect to the uncharged species in the same solvent.

In applying the Born equation to solubility problems, we will express the work of charging an ion in the more general form

$$(12\text{–}9) \qquad W = \frac{1}{\epsilon} \frac{z_i^2 e^2}{2r_i}$$

where $z_i$ is the number of unit charges on the ion and $e$ is the electron charge. If the same form of the equation holds for both ion types, the total work for $N^2$ ions of each type will be

$$W_+ + W_- = \frac{N\nu_+ z_+^2 e^2}{2r_+\epsilon} + \frac{N\nu_- z_-^2 e^2}{2r_-\epsilon}$$

$$(12\text{–}10) \qquad = \frac{Ne^2}{2r\epsilon} (\nu_+ z_+^2 + \nu_- z_-^2)$$

in which $r$ is the average ionic radius of the two ion types. In determining solubility relationships, we will be interested in the solubility of a salt in one solvent with respect to another solvent. Now, the work necessary to charge the ions of a given electrolyte in solvent 1 with respect to the work necessary in another solvent, 2, can be expressed as

$$(12\text{--}11) \qquad W_t = (W_+ + W_-)_1 - (W_+ + W_-)_2$$

$$= \frac{Ne^2}{2r_1\,\epsilon_1}\,(\nu_+ z_+{}^2 + \nu_- z_-{}^2) - \frac{Ne^2}{2r_2\,\epsilon_2}\,(\nu_+ z_+{}^2 + \nu_- z_-{}^2)$$

$$= \frac{Ne^2}{2r}\,(\nu_+ z_+{}^2 + \nu_- z_-{}^2)\left(\frac{1}{\epsilon_1} - \frac{1}{\epsilon_2}\right)$$

where $r$ is the average ionic radius of the electrolyte in solvents 1 and 2.

The final step in the development of the Born equation as it applies to solubility relationships involves equating the energy of transferring the ions from solvent 1 to solvent 2, to the ideal relation between the free energy and the solute concentration,

$$(12\text{--}12) \qquad \Delta G = \Delta G^\circ + RT \ln C$$

If the same standard state is used in both solvents, we can further say that

$$(12\text{--}13) \qquad \Delta G_1 - \Delta G_2 = RT \ln C_1 - RT \ln C_2$$

Now, by equating the work of transferring the electrolyte, $W_t$, to Eq. (12–13) and recognizing that there will be formed $\nu_+$ positive ions and $\nu_-$ negative ions from the ionization of the electrolyte, we can obtain

$$(12\text{--}14) \qquad \frac{Ne^2}{2r}\,(\nu_+ z_+{}^2 + \nu_- z_-{}^2)\left(\frac{1}{\epsilon_1} - \frac{1}{\epsilon_2}\right) = RT \ln \frac{(\nu_+ C_1)^{\nu_+}(\nu_- C_1)^{\nu_-}}{(\nu_+ C_2)^{\nu_+}(\nu_- C_2)^{\nu_-}}$$

We can further note that due to the symmetry of $z$ with respect to $\nu$,

$$(\nu_+ z_+{}^2 + \nu_- z_-{}^2) = z_+ z_-(\nu_+ + \nu_-)$$

This fact can be coupled with the rearrangement of the right side of Eq. (12–14) to $RT(\nu_+ + \nu_-)\ln C_1/C_2$ to give, on division by $RT(\nu_+ + \nu_-)$,

$$(12\text{--}15) \qquad \ln C_1 = \ln C_2 + \frac{Ne^2}{2RTr}\,z_+ z_-\left(\frac{1}{\epsilon_1} - \frac{1}{\epsilon_2}\right)$$

Here we have a relationship between the concentration of an electrolyte and the dielectric constant of the solvent, with the ionic radius as a parameter. If it were possible to assign reasonable values to the ionic radii, then we would be able to calculate the solubility of an electrolyte in a solvent of known dielectric constant.

Actually, the Born equation is greatly over-idealized. For one thing, we have completely ignored all ion-ion interactions, and for this reason the Born equation could not be hoped to be adequate except at extremely low concentrations. This factor would be particularly important in solvents of low dielectric constant. With regard to this point, the Born equation can be improved by correcting for ion-ion interactions with the Debye-Huckel theory. If activity is substituted for concentration in Eq. (12–15) we obtain

$$(12\text{–}16) \qquad \ln a_1 = \ln a_2 + \frac{Ne^2}{2RTr}\, z_+z_-\left(\frac{1}{\epsilon_1} - \frac{1}{\epsilon_2}\right)$$

But, $a = fC$, so we can state that

$$(12\text{–}17) \qquad \ln C_1 + \ln f_1 = \ln C_2 + \ln f_2 + \frac{Ne^2}{2RTr}\, z_+z_-\left(\frac{1}{\epsilon_1} - \frac{1}{\epsilon_2}\right)$$

Since the molar activity coefficient, $f$, can be expressed at low concentrations, by the Debye-Huckel theory, a theoretically more sound equation relating the solubilities should be obtained.

## Empirical Equation for Salt Solubility

In spite of the detail embodied in our final equation, the agreement with experiment is not favorable. A careful study of the effect of dielectric constant on the solubility of slightly soluble electrolytes has been made in dioxane-water mixtures by Davis and Ricci[6] in which the dielectric constant of the solvent was varied from that of pure dioxane to that of pure water. The extent of the agreement between the calculated and the observed solubilities of several salts is shown in Figure 12–5. As is evident, the agreement is valid only in solvent mixtures where the dielectric constant is of the order of 60 or greater.

Although the theoretical approach to electrolyte solubility based on the Born equation is not overly successful, an empirical approach proposed by

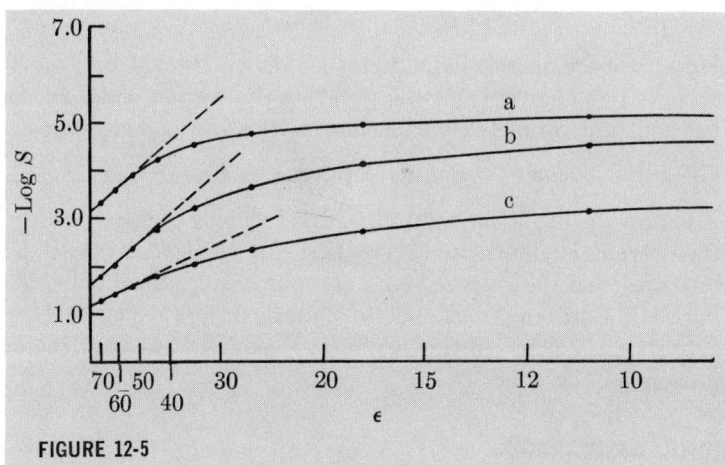

**FIGURE 12-5**

Molar Solubilities of (a) Silver Acetate, (b) Silver Sulfate, and (c) Barium Iodate-Hydrate in Water-Dioxane Mixtures. (After T. Davis and J. Ricci, *J. Am. Chem. Soc.*, **61**, 3274 (1939).)

Ricci and Davis[7] has proved to be somewhat better. They noted the interesting fact that the activity coefficient of a slightly soluble electrolyte in its saturated solution is essentially constant regardless of solvent. This can be seen from Table 12–5 to be the case for silver acetate in several different

**TABLE 12–5. Mean Ionic Activity Coefficient of Silver Acetate at Saturation.**

| Solvent | $\epsilon$ | $f_\pm$ |
|---|---|---|
| Water | 78.55 | 0.800 |
| Acetone, % | | |
| 10 | 73.0 | 0.804 |
| 20 | 67.0 | 0.805 |
| 30 | 61.0 | 0.798 |
| Dioxane, % | | |
| 10 | 69.7 | 0.786 |
| 20 | 60.8 | 0.771 |

solvents. Thus, based on the general validity of this observation, we could say that the solubilities of such an electrolyte in two different solvents can be related by the expression

$$(12\text{–}18) \qquad -\log f_1 = -\log f_2$$

Assuming that the solutions are sufficiently dilute to apply the Debye-Huckel limiting law, we can further say that for a given temperature,

$$(12\text{–}19) \qquad A\ \epsilon_1^{-3/2}\ C_1^{1/2} = A\ \epsilon_2^{-3/2}\ C_2^{1/2}$$

Here, we are merely equating the Debye-Huckel limiting relations for solvent 1 and solvent 2, but separating the solvent dielectric constant from the constant term $A$. By canceling $A$, this can be rearranged to give

$$(12\text{–}20) \qquad \log C_1 = \log C_2 + 3(\log \epsilon_1 - \log \epsilon_2)$$

A plot of the log of the solubility versus the log of the dielectric constant of the solvent is shown in Figure 12–6 for $Ba(IO_3)_2 \cdot H_2O$ in dioxane-water mixtures, and the observed data are compared to the calculated curves from both the Born equation and the empirical approach based on a constant activity coefficient. In this instance, the agreement with the empirical approach is quite good.

## IONIC ASSOCIATION

If a solution contains a sufficiently large number of ions, it is only reasonable to expect their electrostatic interactions to be appreciable. We, of course, are quite familiar with such interactions and ordinarily account for

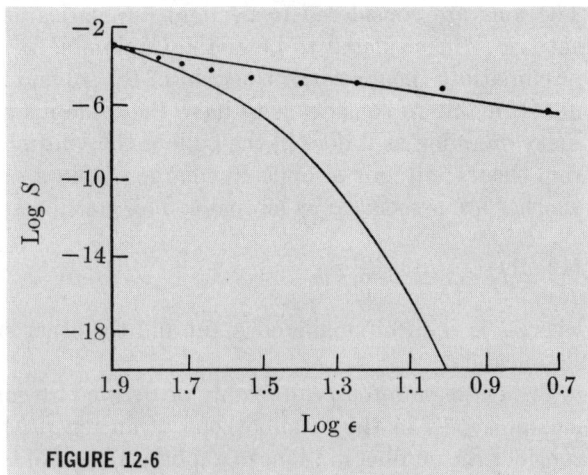

**FIGURE 12-6**

Solubility of $Ba(IO_3)_2 \cdot H_2O$ in Dioxane-Water Mixtures. The straight line represents the solubilities predicted on the basis of constant activity coefficients, and the curve represents solubilities predicted on the basis of the Born equation. The experimental data are given by the dots. (After J. Ricci and T. Davis, *J. Am. Chem. Soc.*, **62**, 407 (1940).)

them in terms of an activity coefficient. Additionally, it should be recognized that these effects are dependent on the dielectric constant of the system. As the dielectric constant decreases it is expected that the ionic interactions will increase. In fact, if the ions come sufficiently close together, we might expect a point to be reached where their electrostatic attraction is greater than the thermal energies tending to disorder them. If this should be the case, a new entity could be formed in which the two ions are combined to form a neutral species known as an *ion pair*. This would not be the same as an undissociated molecule inasmuch as the operative forces would be purely electrostatic in character. This difference may be indicated schematically as

$$A^+ + B^- = [A^+B^-]^0$$

where $[A^+B^-]^0$ represents the ion pair. Unless the ion concentration were very great, ion-pair formation would not be expected to be particularly significant in a solvent such as water. But in solvents having low dielectric constants, this would not be the case. In fact, we would expect ion-pair formation to be the rule rather than the exception in many nonaqueous solvents.

## The Bjerrum Theory

The basic theoretical treatment of ionic association was developed by Bjerrum in 1926.[8,9] This model is the simplest possible for such a system.

The ions are considered to be rigid unpolarizable spheres, and all inter-actions are considered to be of the Coulombic type. As an additional ap-proximation, the dielectric constant of the solvent is used even though it is unreasonable to consider it to have the same magnitude in the near prox-imity of an ion as it does in the bulk of the solution. According to the Bjer-rum theory, all ions of opposite charge within a particular distance of one another are associated as ion pairs. This particular distance, $q$, is given by

$$(12\text{-}21) \qquad q = \frac{z_i z_j e^2}{2 \epsilon k T}$$

where $k$ is the Boltzmann constant and the other symbols have their usual significance.

The expression for $q$ can readily be derived by considering the number of ions of type $i$ in the neighborhood of an ion $j$. More specifically, we can consider the number of $i$-ions in a spherical shell of thickness $dr$ at a distance, $r$, from a chosen $j$-ion. According to the Boltzmann distribution law, this number will be

$$(12\text{-}22) \qquad dn_i = n_i \, e^{+U/kT} \, 4\pi r^2 dr$$

Here $n_i$ is the number of ions of type $i$ per cc of solution, $U$ is the work neces-sary to separate an $i$-ion at a distance of $r$ to infinity from the particular $j$-ion, and $4\pi r^2 dr$ is the volume of a spherical shell of thickness $dr$ at a dis-tance $r$ to $r + dr$ from the center of the $j$-ion. As a good approximation, $U$ can be replaced by the Coulombic potential of two charges in the presence of a dielectric

$$(12\text{-}23) \qquad U = -\frac{z_i z_j e^2}{\epsilon r}$$

and making this substitution into the distribution equation then gives

$$dn_i = n_i \left[ \exp \left( -\frac{z_i z_j e^2}{\epsilon r k T} \right) \right] 4\pi r^2 dr$$

or

$$(12\text{-}24) \qquad \frac{dn_i}{dr} = 4\pi r^2 n_i \left[ \exp \left( -\frac{z_i z_j e^2}{\epsilon r k T} \right) \right]$$

At this point it is convenient to make a macroscopic approximation in which the differentials are allowed to become deltas. This will permit us to make an approximate calculation of the probability of finding an $i$-ion in a particular shell at a distance $r$ from the chosen $j$-ion. It is thus necessary to assign a thickness to the spherical shells, and for the sake of calculation, we might consider this to be of the order of 0.01 Å.

By carrying out such a calculation for oppositely charged ions, the curve shown in Figure 12–7 can be obtained, where the time average number of $i$-ions per shell is plotted against the distance, $r$, of the shell from the center of the chosen $j$-ion. From this it can be seen that there is a distance of mini-

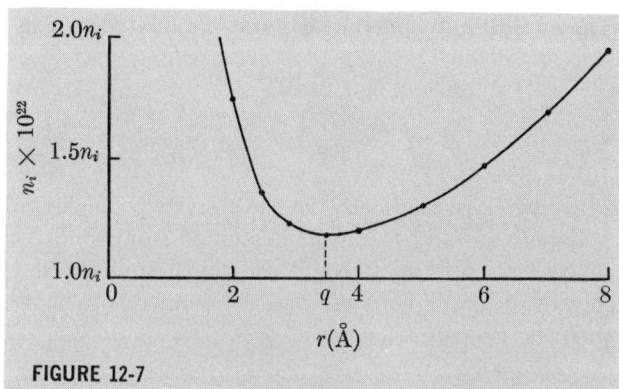

**FIGURE 12-7**

Distribution of Oppositely Charged Ions Around a Central Ion. (Based on N. Bjerrum, *K. Danske Vidensk, Selsk. Mat.-fys. Medd.*, **7**, No. 9 (1926).)

mum probability of finding an $i$-ion in a spherical shell surrounding the central $j$-ion. The position of this minimum is the distance, $q$, and it can be evaluated by differentiating Eq. (12–24) with respect to $r$

$$\frac{d}{dr}\left(\frac{dn_i}{dr}\right) = \frac{d}{dr}\left[4\pi n_i r^2 \exp\left(-\frac{z_i z_j e^2}{\epsilon k T r}\right)\right]$$

and equating it to zero. It also can be noted that with a decrease in $r$ less than $q$, the probability of an $i$-ion being in the neighborhood of a $j$-ion increases rapidly. On the other hand, it is seen that there is a less rapid increase in the probability as $r$ increases past $q$. Bjerrum arbitrarily assumed that two ions are associated if their separation is less than $q$.

It should be emphasized that the Bjerrum theory applies to systems where only electrostatic interactions need be considered. If covalent bonding is present, such as is found in the case of weak electrolytes, the potential term in the Boltzmann equation cannot be represented by the simple Coulombic potential $e^2/\epsilon r$.

Although the basic idea of ion pairing is quite generally accepted, the closeness of approach necessary for ion pairs to form cannot be directly measured. However, by indirect means, a general order of magnitude can be estimated. Such an approach depends on the evaluation of an equilibrium constant between the ion pair and the dissociated ions. The degree of association will be determined by the number of $i$-ions that are at a distance less than $q$ from a given $j$-ion. According to the Bjerrum treatment, if the degree of association of the ion pair is given by $(1 - \alpha)$, then

$$(12\text{–}25) \qquad (1 - \alpha) = 4\pi n_i \int_a^q \exp\left(-\frac{z_i z_j e^2}{\epsilon k T r}\right) r^2\, dr$$

That is, an ion pair can be considered to exist when the distance between the two ions lies between that of contact, $a$, and the Bjerrum critical distance, $q$. The equilibrium constant can now be related to $\alpha$ as

$$[A^+B^-]^0 = A^+ + B^-$$

$$K = \frac{[A^+][B^-]}{[AB]^0} = \frac{(\alpha C)(\alpha C)f_\pm^2}{(1 - \alpha)C} = \frac{\alpha^2 Cf_\pm^2}{(1 - \alpha)}$$

where we have assumed that the activity coefficient of the ion pair is unity.

Experimentally $K$ can be evaluated from conductance data at very low concentration. And it has been shown that the dependence of $K$ on the solvent dielectric constant and the temperature is, in general, in agreement with the Bjerrum theory.

## Conductance Studies on Ionic Association

A rather convincing experimental verification of ionic association is found in a series of conductance studies made by Kraus and Fuoss.[10] Their measurements were made on solutions of the salt tetraisoamylammonium nitrate in dioxane-water systems. By varying the relative amounts of dioxane and water, they were able to carry out the studies over a dielectric range of 2.2 for pure dioxane to 78.5 for pure water. The results are very significant in that they show quite anomalous behavior in solvent mixtures having low dielectric constants. As can be seen in Figure 12–8, a normal conductance curve is obtained only for the pure water solvent, curve a. As the dielectric constant is successively decreased it is noted that the minima in the conductance curves become quite pronounced with the greatest deviations occurring in pure dioxane, curve k. Also, it can be seen that as the concentrations are increased, the conductance curves again begin to rise.

In this classic series of papers, Kraus and Fuoss[10] interpret their conductance curves in terms of ionic association. Basically, they assume that the decrease in the conductance curves arises from the formation of ion pairs. Inasmuch as the resultant pair is effectively neutral, the ions will not contribute to the conductance of the solution. However, as the concentrations of the solutions become greater, there is a tendency to form triple ions rather than pairs, and these should now begin to increase the conductance of the system, and so on.

This interpretation is still commonly accepted. However, there has been a tendency in recent years to look for alternative answers. To a large extent this is due to the fact that direct evidence for ion triplets and higher aggregates has not yet been obtained. In one approach, a microscopic Wien effect has been proposed by Kenausis, Evers, and Kraus[11] in which it is considered that the probability of breakup of an ion pair is increased by the presence of an ionic field in the neighborhood of the ion pair. Thus, rather than form a triple ion, the approach of a positive ion to the negative end of the ion-pair dipole or a negative ion to the positive end of the dipole,

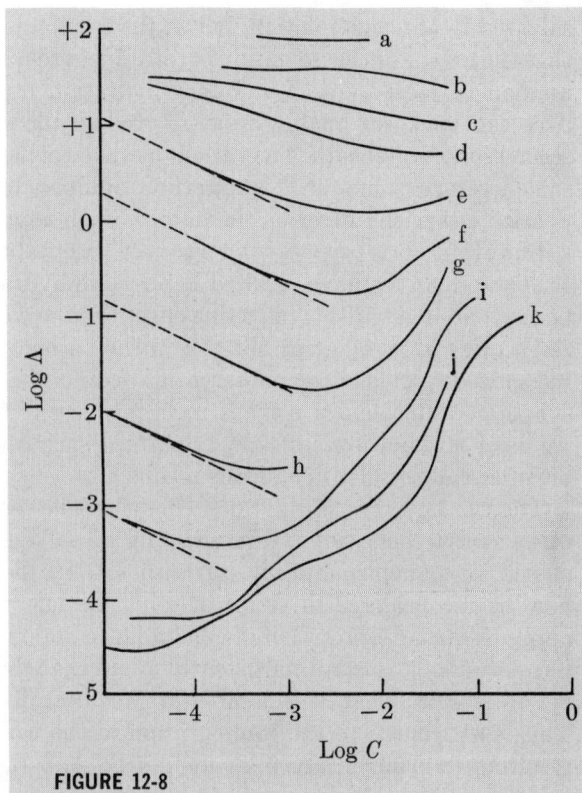

**FIGURE 12-8**

Conductance Curves of Tetraisoamylammonium Nitrate in Various Mixtures of Dioxane and Water. (After C. A. Kraus and R. M. Fuoss, *J. Am. Chem. Soc.*, **55**, 21(1933).

weakens the bond holding the ion pair together. Consequently, the probability that a colliding solvent molecule will break the ion-pair bond will be enhanced.

## ION-SOLVENT INTERACTIONS

In spite of the apparent success of the dielectric constant in correlating electrolyte behavior, we should not get the impression that solution properties can be understood in terms of this parameter alone. Specific interactions of the solvent molecules with the ions in solution are also important, and in some instances may be of greater significance than the solvent dielectric constant. A realization that the solvent molecules interact with the ions is certainly not new and, in fact, many studies have been made to determine the nature, extent, and significance of specific solvation effects on electrolyte behavior. Using a familiar example, it is well known that the $Na^+$ ion is solvated in aqueous solution, but at the same time, it is interesting to note that the extent of this solvation is yet subject to question. Depending on the method of measurement, different hydration numbers will be obtained.[9] This can most reasonably be attributed to the existence of a first and a second solvation sheath. The various methods of measurement will determine some average value for the hydration number, but each method will emphasize either the inner or the outer sheath to a lesser or greater extent.

But even this is probably not a totally acceptable explanation. Studies on $KNO_3$ and $KClO_3$ have resulted in negative hydration numbers.[12] This has been attributed to depolymerizing effects on the water structure by the ions. Regardless of whether or not this is the correct interpretation, it can be recognized that such an observation cannot be understood in terms of simple coordination of solvent molecules about the ions. And at present we have no knowledge of how important such an effect may be for NaCl or other simple salts in aqueous media.

The consequences of solvation effects can be seen in almost every aspect of electrolyte behavior and in virtually all solvents. Whereas we have discussed ion-pair formation in terms of solvent dielectric constant, it must now be reconsidered in terms of solvation effects as well. This will also apply to our interpretation of conductance and emf data where an ion size term occurs. It is quite consistently found that the apparent size of a small cation in solution is considerably greater than its crystallographic radius. This is attributed to the complexation of the cation by solvent molecules resulting in a much larger kinetic entity than the uncomplexed ion. This point is well made in a spectrophotometric study by Griffiths and Scarrow of the effect of various bromide salts on the formation of $NiBr_4^{2-}$ from $NiBr_3^-$ in acetone.[13] The ease of formation of the $NiBr_4^{2-}$ was shown to depend on the cation size and was found to be in the order

$$Pr_4N^+ > Et(Bu)_3N^+ > Bu_4N^+ > Pr(octyl)_3N^+ \gg Li^+$$

This is more clearly shown in Figure 12–9. From the trend, it would seem that the $Li^+$ ion must be larger than the other ions. Considering the lithium ion to be solvated by a monolayer of acetone molecules, Griffiths and Scarrow found that the geometrical size of the resultant species is, in fact, larger than the $Pr(octyl)_3N^+$ ion. It would thus appear that the trend in terms of ion size is, indeed, valid.

Thus far we have restricted our interest in these problems to physical chemical properties. However, solvation effects can also play an extremely important role in the rates and paths of chemical reactions, and it is here that our interests should probably be the greatest. An example of solvation effects in this respect can be seen in the Friedel-Crafts reaction

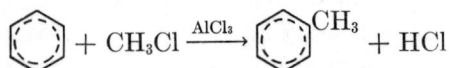

$$\bigcirc + CH_3Cl \xrightarrow{AlCl_3} \bigcirc^{CH_3} + HCl$$

which is extremely slow in the presence of ethers such as tetrahydrofuran.

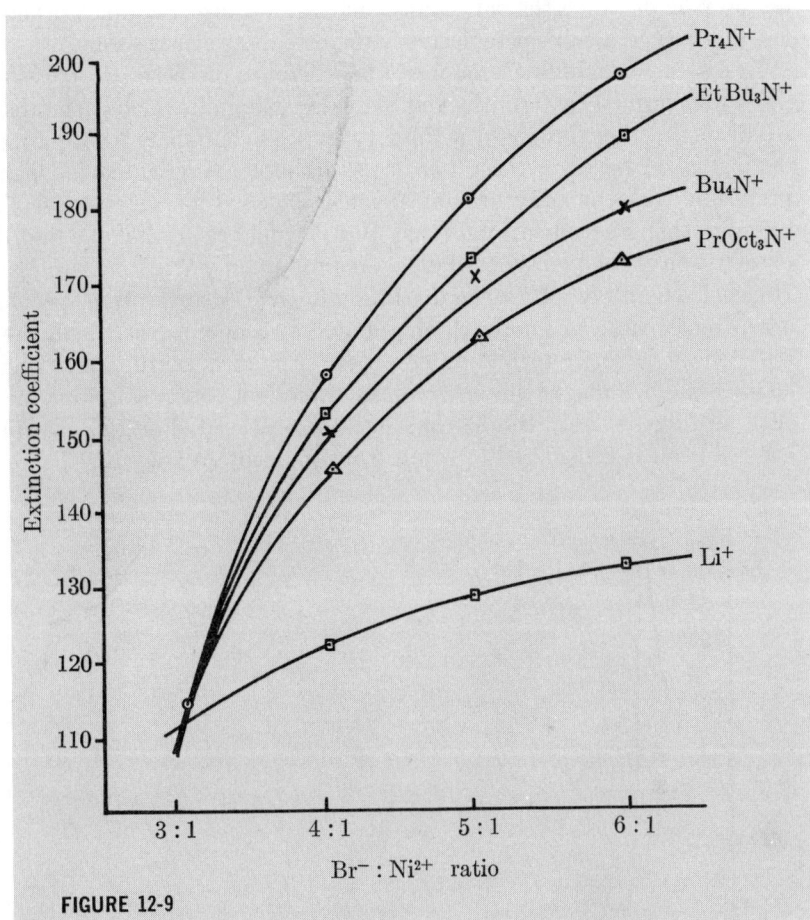

**FIGURE 12-9**

Effect of Cation Size on the Ease of Formation of $NiBr_4^{2-}$ from $NiBr_3^-$ in Acetone. (From T. Griffiths and R. Scarrow, International Conference on Nonaqueous Solvents, Hamilton, Ontario, Canada, 1967.)

Although there may be a number of explanations of this behavior, it can reasonably be attributed to the complexation of the $AlCl_3$ by the ether, thereby limiting its catalytic action. But whether this is the correct interpretation or not, it is again obvious that solvation effects are important.

## Ion Pairs

In terms of the Bjerrum concept of ion pairing, the solvent merely acts as a dielectric between the positive and negative ions. However, as early as 1955, Winstein[14] proposed the existence of two distinct types of ion pairs in order to explain the rates of various organic reactions; *contact* ion pairs and *solvent-separated* ion pairs. As the names imply, the first of these refers to an ion pair in which the cation and anion are in direct contact, whereas in the latter type a solvent molecule is imposed between the ions.

In a somewhat indirect manner, the existence of these two types of ion pairs was verified by Griffiths and Symons[15] using ultraviolet spectroscopy. Additionally they proposed a third type, a solvent-shared ion pair. They distinguished between these two types of solvent-separated ion pairs by proposing that the conventional solvent-separated ion pairs are separated by more than one solvent molecule. But it would seem that the first direct observation of the two types of ion pairs proposed by Winstein was made by Hogen-Esch and Smid,[16] when in a study of the UV absorption spectrum of alkali metal salts of fluorenyl, they noted two new peaks which they attributed to ion-pair formation. As seen in Figure 12–10, these occur at 355 m$\mu$ and 373 m$\mu$. If this spectrum is taken in a poorly solvating medium such as toluene, only the 355 m$\mu$ peak is observed. For this reason, the 355 m$\mu$ peak is considered to result from the contact ion pairs.

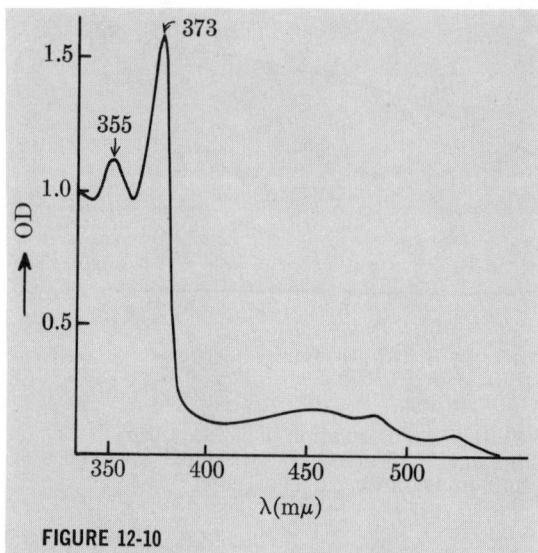

**FIGURE 12-10**

Absorption Spectrum of Fluorenyllithium in THF at 25°C. (From T. E. Hogen-Esch and J. Smid, *J. Am. Chem. Soc.*, **88**, 307 (1966).)

One possible way of looking at the two types of ion pairs is in terms of the degree of solvation of the cation. We can imagine the following equilibria to exist in the solution,

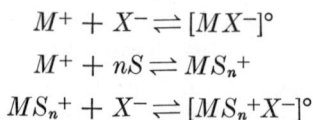

$$M^+ + X^- \rightleftharpoons [MX^-]^\circ$$

$$M^+ + nS \rightleftharpoons MS_n^+$$

$$MS_n^+ + X^- \rightleftharpoons [MS_n^+X^-]^\circ$$

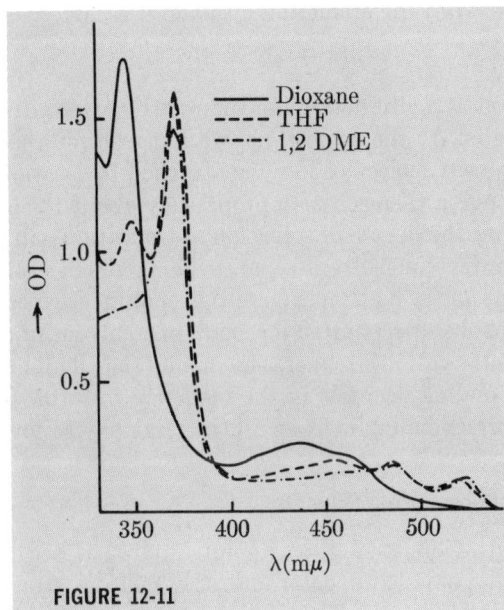

FIGURE 12-11

Fluorenyllithium Spectrum in Dioxane, THF, and DME at 25°C. (After T. E. Hogen-Esch and J. Smid, *J. Am. Chem. Soc.*, **88**, 307 (1966).)

The formation of a solvent-separated ion pair will depend strongly on two factors, the basicity of the solvent and the size of the cation. The importance of this first factor is illustrated by the data of Hogen-Esch and Smid where they observe only the absorption peak at 355 m$\mu$ if toluene is used as a solvent. But in THF, which is a quite basic solvent, both peaks are observed for the fluorenyllithium, with the solvent-separated form predominant. This point is more strongly emphasized in Figure 12–11, where the relative intensities of the two peaks for fluorenyllithium are shown in the three solvents dioxane, tetrahydrofuran (THF), and 1,2-dimethoxyethane (DME). The distribution of solvent-separated ion pairs is seen to vary here from none in dioxane to 100% in DME. This order corresponds to the basicity of the solvents, with dioxane being the least basic and DME being the most basic of the three.

With regard to the second point, the electric field around a small ion will be greater than one around a larger ion of the same charge. Correspondingly, the smaller ion will hold the solvent molecules more tightly than the larger ion. This is substantiated by the studies of Smid where he finds approximately 75% solvent-separated ion pairs in THF solutions of fluorenyllithium but only 5% solvent-separated ion pairs in the analogous solutions of the sodium salt.[17]

If we now imagine the solvent-separated ion pair to be merely an ion pair between the anion and a solvated cation, then the equilibrium

$$M^+ + nS \rightleftharpoons MS_n^+$$

will actually determine the relative concentration of the two types of ion pairs. At the same time, we must admit that one and even possibly two solvent molecules may be attached to a cation in a contact ion pair by arranging themselves appropriately about the cation. Thus, it would appear that the degree of solvation of the cation will determine the distribution of contact and solvent-separated ion pairs in a solution. The lithium ion, which has a very high ionic potential, should be expected to form a high percentage of solvent-separated ion pairs in a solvent of even rather low basicity. But as the cation size increases among the alkali metals, there should be a corresponding decrease in the tendency to form solvent-separated pairs with a corresponding increase in the tendency to form contact ion pairs.

## Differentiating Solvents

In Chapter 9, it was pointed out that because of the leveling effect, it is not possible to determine the relative strengths of the strong acids in aqueous media. The water molecule is sufficiently basic to give an indeter-

$$HX + H_2O \rightleftharpoons H_3O^+ + X^-$$

minantly large equilibrium constant. In a solvent of greater basicity, this effect should be even more pronounced, and acids that are considered weak acids in aqueous media will be quite highly ionized in the more basic media.

In order to obtain a differentiating solvent for the conventional strong aqueous acids, an acidic solvent can be used. Because of the lesser basicity of the solvent, the anion of the acid can compete more favorably for the proton. And if the solvent is sufficiently acidic, the equilibrium will be shifted to a position that will permit an evaluation of the relative degree of dissociation. It was by means of conductance studies in anhydrous acetic acid as a solvent that Kolthoff and Willman[18] determined the order of strengths

$$HClO_4 > HBr > H_2SO_4 > HCl > HNO_3$$

More recently it has been shown that quite basic solvents can also act as differentiating solvents for strong acids. In this respect a considerable amount of research has been done on ethylenediamine.[19] Since ethylenediamine is more basic than water, our previous argument cannot apply here, and it is therefore obvious that the interpretation based strictly on basicity or acidity of the solvent is, of itself, not adequate. Rather, we might more reasonably consider at least two equilibria to be involved. Proton transfer must certainly be one of these. But, in addition, the resultant solvated cation might then form an ion pair with the anion. When it is realized that the dielectric constant of ethylenediamine is only 12.5, ion pairing should be expected. Thus, we can propose

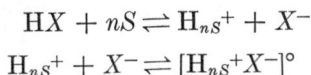

$$HX + nS \rightleftharpoons H_nS^+ + X^-$$
$$H_nS^+ + X^- \rightleftharpoons [H_nS^+X^-]^\circ$$

It is very likely that the equilibrium in the first reaction, like $HX$ in water, lies very far to the right. But the conductivity of the solution will be dependent on the second reaction. It is then the second reaction on which an interpretation of the acid strength will be based. This same type of argument might also be applied to anhydrous acetic acid which has a dielectric constant of 6.13. If an interpretation of this type is valid, one might question whether the stability of the ion pair is also a measure of the relative acidity of the acid. In aqueous media, where the dielectric constant is close to 80, ion association is negligible and the equilibrium of concern is that of proton transfer. Thus, it may be that we are talking about two altogether different things when we discuss acidity in water and acidity in a solvent of low dielectric constant. At least, this would be the case for a basic solvent of low dielectric constant. If this model is correct, then the validity of the interpretation in terms of acid strengths will depend on whether the relative affinities of the acid anions for the proton are in the same order as that of the electrostatic attractions of the anions for the solvated proton.

In the case of an acidic solvent of low dielectric constant, it is likely that measurable equilibrium constants exist for both reactions. If the ion-pair equilibrium constant can be evaluated, then the proton transfer equilibrium constant can also be evaluated. Actually these problems have been given serious study and quantitative pK values for a variety of acids and salts have been determined in anhydrous acetic acid.[20] However, the values depend on the validity of some model of ion association and this may or may not be correct.

## The Solvation Number

It has already been pointed out that an ion in solution is solvated. But the nature and extent of this solvation is not so well known. However, since the conventional solvents are basic in character, it is quite reasonable to assume that solvation occurs primarily with the cation. And since the electrical field is greater around a small ion than around a larger ion of the same charge, we should expect the extent of solvation to be greater with the smaller ion. This is borne out by a number of indirect observations such as that of Griffiths and Scarrow, which was mentioned earlier.

Some of the problems inherent in the determination of the degree of solvation of an ion can be seen from a consideration of the $Na^+$ ion in water. Assuming we have a means of looking at the first solvation sheath of the cation, we can still recognize that if an equilibrium exists, it would be desirable to vary the ratio of solvent to salt. Since water is the solvent as well as the coordinating species, this can be done only by using very high salt concentrations. In addition to the introduction of serious ion-ion effects, these solutions will still be too dilute for the study of mole ratios of the order of 1:1 or 2:1. Yet these ratios will certainly be among the most important.

If we look at spectral properties, we run into still another problem. Although the spectrum of a complexed solvent molecule would be expected to be different from that of an uncomplexed solvent molecule, the extremely large bulk of uncomplexed solvent relative to the complexed solvent will usually mask the bands of the complexed species. And although a great amount of research has been conducted along these lines, definitive answers have not been forthcoming.

To avoid the difficulties inherent in working with a complexing solvent, Kebarle and Hogg[21] have developed a means of studying complexation of ions in the gas phase. Here the complexing molecule can be added in any mole ratio relative to the cation. Using mass spectrometry as a detecting device, they have shown that the first solvation sheath of the $NH_4^+$ ion contains four ammonia molecules. However, clusters of as many as 20 complexed ammonia molecules have been observed. All but four of these, of course, are in the second or possibly higher solvation sheaths. As promising as this approach may be, the experimental limitations are considerable.

Many studies of salts in relatively noncoordinating solvents such as benzene have been made, and it would seem that this offers yet another means of attacking the problem. In principle, one could then add coordinating species in the desired ratios and the distinct properties of the coordinated molecules should be observable. The primary difficulty here is the general limitation of the salts to those having large cations. Very few salts are soluble in such solvents. The exceptions are the tetraalkylammonium salts typified by $N(i\text{-amyl})_4NO_3$ which was used by Fuoss and Kraus in their classic study on ion association. But for the study of ion-solvent interactions, the limitation to a large cation is a serious handicap.

An exception to this solubility pattern is found in the sodium aluminum tetraalkyl salts such as $NaAl(butyl)_4$. Although they are pyrophoric and therefore require the use of inert atmosphere techniques, these compounds are soluble in a large variety of nonaqueous solvents; those with alkyl groups as large as butyl being soluble in saturated hydrocarbon solvents. This latter type of solvent is of particular importance. One might question the actual inertness of an aromatic solvent, but a saturated hydrocarbon can be shown to have essentially no interaction with the ions. Thus it can be considered to be only a medium in which the ions or their associated species are distributed. Here we have an electrolyte with a small cation in an inert solvent. Complexing species can now be added and their effects on the ions can be determined. Because of reactivity many common solvents such as water and the various alcohols cannot be used, but at the same time, a large variety of basic solvents can be studied.

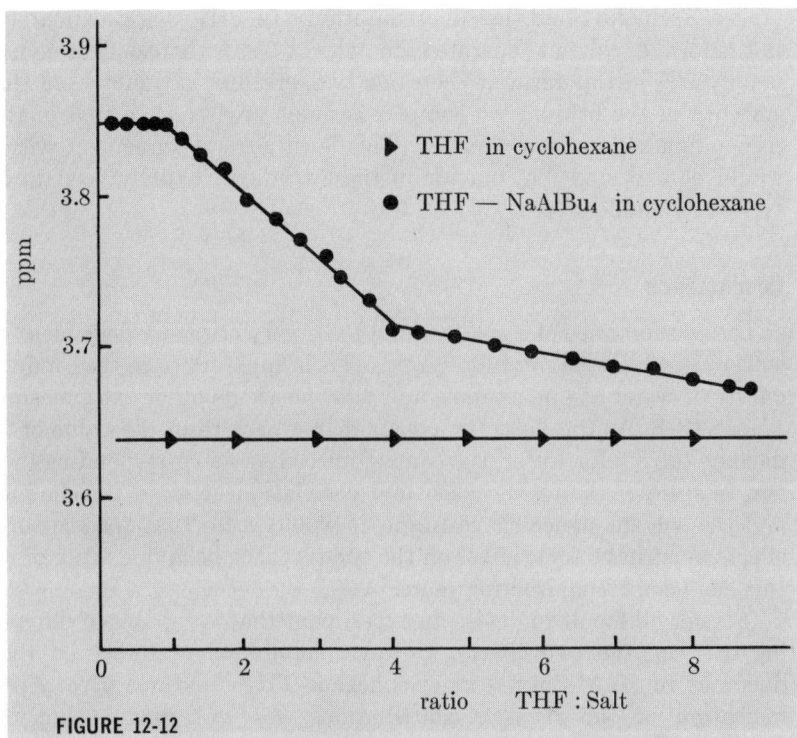

**FIGURE 12-12**

Shift in the NMR Spectrum of THF as a Function of the Ratio of THF: salt at a salt concentration in cyclohexane of 0.265 moles-liter. (Based on E. Schaschel and M. C. Day, *J. Am. Chem. Soc.*, **90**, 503 (1968).)

The complexation of the sodium ion by tetrahydrofuran is illustrated in Figure 12–12. Here the shift in the NMR spectrum of the hydrogen atoms adjacent to the oxygen atom in THF is shown as small amounts of THF are added to a solution of NaAl(butyl)$_4$ in cyclohexane. As THF is added, there is seen to be no change in the shift until a 1:1 ratio of solvent to salt is reached. This means that all of the THF molecules have the same environment up to the 1:1 ratio. This can reasonably be attributed to a strong complex with the sodium ion. It can then be seen that the next three molecules of THF are held much less tightly. This would imply that an equilibrium exists between the singly complexed sodium ion and three additional THF molecules.

Because of its smaller size, the lithium ion might be expected to show a smaller solvation number than the sodium ion, but the complexed molecules should be held more tightly. Using fluorenyllithium as the salt, it has been shown by Dixon, Gwinner, and Lini[22] that the lithium ion contains three

THF molecules in its first solvation sheath. And from the observation by Hogen-Esch and Smid that fluorenyllithium in THF is approximately 80% in the form of solvent-separated ion pairs at 25°C whereas fluorenylsodium is only 5% in this form at the same temperature, it would seem that the stability of the lithium ion complex is much greater than that of the corresponding sodium ion complex. This is in general agreement with what would be expected, but outside of these studies, virtually no data of this type is available.

## Conductance

The conductance of a salt is found to be very critically dependent on the solvent in which the salt is dissolved. This is apparent from the conductance curves of water at one extreme and dioxane at the other extreme shown in Figure 12–8. As the dielectric constant increases from the value of 2.2 for dioxane to 78.5 for water, the conductance shows a corresponding increase. But in spite of the rather consistent correlation between the conductance and the solvent dielectric constant, it would seem that specific solvation effects must have some effect on the conductance behavior. And, of course, this has been recognized for years.

It is very difficult to devise an experiment that is capable of differentiating between the two effects. However, preliminary studies on the conductance of $NaAl(butyl)_4$ in cyclohexane-THF mixtures give a possible indication of the relative contributions. As was pointed out earlier, $NaAl(butyl)_4$ is soluble in saturated hydrocarbon solvents. Additionally it shows ionic conductance in these solvents. It was also pointed out earlier that NMR studies show that a complex between the sodium ion and THF is formed in which four THF molecules can exist in the first solvation sheath about the ion. Now if THF is added to a solution of $NaAl(butyl)_4$ in cyclohexane, the conductance curve shown in Figure 12–13 is obtained. The conductance is observed to increase until the 1:1 complex is formed. There a plateau is observed, but the curve begins to rise rapidly at a ratio slightly grater than 2:1. If we assume that a tightly bound THF molecule does not contribute to the solvent dielectric constant, then the dielectric constant has not changed in going to the plateau. This increase can then be attributed to ion-solvent effects alone. If the next three molecules of THF are in equilibrium with the 1:1 complex, then after the 1:1 complex is formed, some THF molecules will be added to the solvent and thereby contribute to the dielectric constant increase. This will be slow at first because of the equilibrium, but after the 4:1 ratio is reached we can assume that all of the THF goes into the solvent and contributes to the solvent dielectric constant. At 25°C the dielectric constant of cyclohexane is 2.05 and that of THF is 7.39. Thus we see that the major change in conductance can be attributed to the dielectric constant change. Of course the 140-fold difference cannot be taken as a quantitative measure of the difference of the two

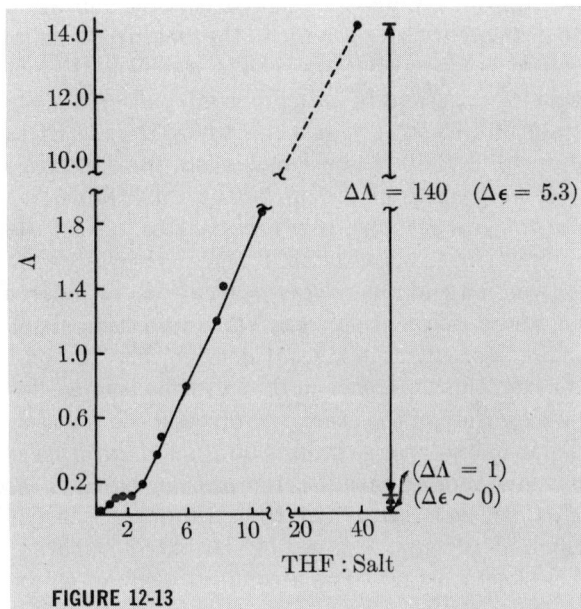

**FIGURE 12-13**

Effect of THF on the Conductance of a 0.264 Molar
Solution of NaAl(butyl)$_4$ in Cyclohexane. (From M. C.
Day, E. Schaschel, and C. Hammonds, International
Conference on Nonaqueous Solvents, Hamilton, On-
tario, Canada, 1967.)

effects because of the other factors that also contribute to conductance such
as viscosity and ion size. However, we can see that the solvent dielectric
constant is of major concern. But for reactions that depend on the presence of ions in the solution, it is apparent that ion-solvent interactions play
an important role.

Actually, it may be possible to qualitatively interpret these results in
terms of ion size. In a solvent of dielectric constant as low as that of cyclohexane, the ions are primarily in the form of ion pairs. In cyclohexane these
will be contact ion pairs. However, as THF is added, the complexation of
the sodium ion results in an increase in its effective size and a corresponding
decrease in the charge density. This should then lead to a weaker ion pair
and a corresponding increase in the number of conducting species in the
solution.

## MEDIUM EFFECTS

It has been seen that ion-solvent interactions and ion-ion interactions
will show wide variations in going from one solvent to another. And, from a
qualitative standpoint, it is possible to predict the direction that certain
solute properties will take by considering the change in the dielectric con-

stant. For example, we would expect the ionization constant of a weak acid to decrease with a decrease in the dielectric constant of the solvent. This, of course, is substantiated as can be seen from the ionization constant of acetic acid in water and in anhydrous ethanol. Acetic acid has an ionization constant of $1.75 \times 10^{-5}$ in water where the dielectric constant is 78.5, but it drops to $2 \times 10^{-11}$ in ethanol where the dielectric constant is 24.2.

If we wish to give more than a qualitative treatment of electrolytes in nonaqueous systems, it is necessary to look at the ion-ion and ion-solvent interactions in some detail. First, we recognize that the interaction between a given ion and a particular solvent will be different from that of the same ion and a different solvent. This interaction should be independent of the solute concentration since it does not involve ion-ion interactions. We can consider this difference in the solvation energies in two different solvents to be a measure of the energy involved in the transfer of an ion at infinite dilution in one solvent to infinite dilution in the other solvent. This difference of solvent property is called the *primary medium effect*. A *secondary medium effect* can be seen to arise from a difference in ion-ion interactions in two different solvents. This can be attributed primarily to a difference in solvent dielectric constant. The secondary medium effect will, of course, be concentration-dependent. We can thus break down into three steps the transfer of solute of a finite concentration in one solvent to a finite concentration in another solvent. These steps involve (1) the transfer of the solute from a finite concentration in the first solvent to infinite dilution in the same solvent, (2) the transfer of the solute to an infinitely dilute solution in the second solvent, and (3) the transfer of the solute to a finite concentration in the second solvent. This represents the *total medium effect*, which is the summation of the primary and secondary medium effects.

## Thermodynamic Representation of the Primary Medium Effect

In order to give a general thermodynamic treatment of medium effects, we can consider the transfer of a strong acid such as HCl of molal concentration $m$, from an aqueous medium to a mixed solvent such as ethanol-water. Such studies can be made by means of a galvanic cell without liquid junction of the type

$$\text{Pt, H}_{2(1 \text{ atm})} \,|\text{HCl}_{(m)} \text{ ethanol } (X), \text{ water } (Y)|\, \text{AgCl, Ag}$$

where the mole fraction of ethanol may vary anywhere from 0 to 1. If, for the moment, we consider the cell when the mole fraction of ethanol is 0, we will have the common aqueous cell

$$\text{Pt, H}_{2(1 \text{ atm})} \,|\text{HCl}_{(m)}|\, \text{AgCl, Ag}$$

and its potential can be expressed as

$$(12\text{--}26) \qquad \text{E} = \text{E}_m^\circ - \frac{2RT}{\text{F}} \ln \gamma_\pm m$$

Here the concentration of HCl is expressed in terms of molality, and the activity coefficient $\gamma_\pm$ is measured on the same scale.

To illustrate a symbolism,[9] the potential given in Eq. (12–26) can be represented as

$$(12\text{–}27) \qquad {}^w\mathrm{E} = {}^w\mathrm{E}_m^\circ - \frac{2RT}{\mathrm{F}} \ln {}^w_w\gamma_\pm m$$

The superscript $w$ indicates that the measurements were made in water as a solvent, and the subscript $w$ on the activity coefficient indicates that the activity coefficients are compared to unity at infinite dilution in water. It can now be said that the standard state potential will be given by

$$(12\text{–}28) \qquad {}^w\mathrm{E}_m^\circ = \lim_{m \to 0} \left( {}^w\mathrm{E} + \frac{2RT}{\mathrm{F}} \ln m \right)$$

The $\ln {}^w_w\gamma_\pm$ term drops out because the activity coefficient approaches unity as $m \to 0$.

If the ethanol-water mixed solvent is now considered, it is found that there exists a choice in defining the standard state. Firstly, we can consider the mixed solvent in the same manner as we did the pure solvent. That is, the cell potential can be expressed as

$$(12\text{–}29) \qquad {}^s\mathrm{E} = {}^s\mathrm{E}_m^\circ - \frac{2RT}{\mathrm{F}} \ln {}^s_s\gamma_\pm m$$

The symbolism here indicates that the measurements were made in a solvent, $s$, and the activity coefficients are relative to unity at infinite dilution in this particular solvent. Thus, the solvent is being treated just as if it were a pure solvent rather than a mixed solvent, and the expression for the standard state potential will be of the same form as that for a purely aqueous solvent, namely

$$(12\text{–}30) \qquad {}^s\mathrm{E}_m^\circ = \lim_{m \to 0} \left( {}^s\mathrm{E} + \frac{2RT}{\mathrm{F}} \ln m \right)$$

In an alternative manner, the solvent can still be considered an aqueous solvent to which a portion of ethanol has been added. We might then be prone to keep our standard reference state in pure water and thereby represent the potential as

$$(12\text{–}31) \qquad {}^s\mathrm{E} = {}^w\mathrm{E}_m^\circ - \frac{2RT}{\mathrm{F}} \ln {}^s_w\gamma_\pm m$$

Now, since the activity coefficient is measured relative to unity at infinite dilution in a purely aqueous medium, ${}^s_w\gamma_\pm$ does not approach unity as $m$ approaches zero in the mixed solvent. Consequently, the expression for the standard state potential becomes

$$(12\text{–}32) \qquad {}^w\mathrm{E}_m^\circ = \lim_{m \to 0} \left( {}^s\mathrm{E} + \frac{2RT}{\mathrm{F}} \ln m + \frac{2RT}{\mathrm{F}} \ln {}^s_w\gamma_\pm \right)$$

If the difference in the standard state potentials in water and the mixed solvent are now taken, we obtain

$$(12\text{-}33) \qquad {}^{w}E_{m}^{\circ} - {}^{s}E_{m}^{\circ} = \lim_{m \to 0} \left( \frac{2RT}{F} \ln {}^{s}_{w}\gamma_{\pm} \right)$$

Since we are actually considering the same cell with the only difference being the arbitrarily defined standard state, the experimentally determined potential, ${}^{s}E$, and the ln $m$ term will drop out. This leaves the activity coefficient of HCl at infinite dilution as measured in a mixed solvent relative to unity at infinite dilution in water. Inasmuch as ion-ion interactions are absent at infinite dilution in either solvent, we are thus measuring the effect of transferring a pair of ions from one solvent to another when only ion-solvent interactions are present. This, then, is the thermodynamic representation of the primary medium effect.

## The Secondary Medium Effect

The secondary medium effect can be determined by considering a total medium effect and then subtracting from it the primary medium effect. The total medium effect is concerned with the transfer of the electrolyte from a finite concentration in one solvent to the same concentration in another solvent. This will involve both the ion-solvent interactions and the ion-ion interactions. The thermodynamic expression for the total medium effect can be obtained by coupling together the aqueous and the mixed solvent cells that we have already considered, to give the cell

Ag, AgCl$|$HCl$_{(m)}$ water$|$Pt, H$_{2(1\text{ atm})}$$|$HCl$_{(m)}$ ethanol$(X)$, water$(Y)$$|$AgCl, Ag

The cell reaction will involve the transfer of the HCl from one solvent to the other, but if the HCl molality is the same in both solvents, the potential will not be due to an effective concentration cell. The potential of this cell will be

$$(12\text{-}34) \qquad {}^{s}E - {}^{w}E = {}^{s}E_{m}^{\circ} - {}^{w}E_{m}^{\circ} - \frac{2RT}{F} \left( \ln {}^{s}_{s}\gamma_{\pm} - \ln {}^{w}_{w}\gamma_{\pm} \right)$$

This is nothing more than the difference in the potentials of the mixed solvent cell and the aqueous cell in which the concentration of the electrolyte is the same in both solvents. Since there are no concentration changes, the energies involved are those resulting from different degrees of nonideality in the two systems along with differences in ion-solvent interactions.

It is possible to put Eq. (12–34) into a more useful form by using a different standard state. Expressing the potential of the mixed-solvent cell by Eq. (12–31) it is found that the difference in potential can be represented as

$$(12\text{-}35) \qquad {}^{s}E - {}^{w}E = -\frac{2RT}{F} \left( \ln {}^{s}_{w}\gamma_{\pm} - \ln {}^{w}_{w}\gamma_{\pm} \right)$$

If this result is now substituted along with Eq. (12–33) into Eq. (12–34), we obtain

$$(12\text{–}36) \quad -\frac{2RT}{\mathbf{F}}\left(\ln {}^s_w\gamma_\pm - \ln {}^w_w\gamma_\pm\right) = -\lim_{m\to 0}\frac{2RT}{\mathbf{F}}\ln {}^s_w\gamma_\pm$$
$$-\frac{2RT}{\mathbf{F}}\left(\ln {}^s_s\gamma_\pm - \ln {}^w_w\gamma_\pm\right)$$

On rearranging, this becomes

$$(12\text{–}37) \quad \ln\frac{{}^s_w\gamma_\pm}{{}^w_w\gamma_\pm} = \lim_{m\to 0}\ln {}^s_w\gamma_\pm + \ln\frac{{}^s_s\gamma_\pm}{{}^w_w\gamma_\pm}$$

The term on the left represents the total medium effect and it is a measure of both the ion-ion and the ion-solvent interactions in the respective solvents. More specifically, it is a measure of the energy involved in the transfer of the electrolyte from a finite concentration in one solvent to the same concentration in the other solvent. The first term on the right-hand side of the equation obviously represents the primary medium effect. The second term, however, is a measure of the relative nonideality of the solute in the different solvents at some finite concentration. That is, ${}^s_s\gamma_\pm$ is the activity coefficient of the solute measured at some finite concentration in the solvent, $s$, as compared to unity at infinite dilution in that solvent. Thus, if we can assume that ion-ion interactions are zero at infinite dilution in the solvent, $s$, then ${}^s_s\gamma_\pm$ will be unity and the nonideality expressed by ${}^s_s\gamma_\pm$ will correspondingly be zero at infinite dilution in $s$. Consequently, ${}^s_s\gamma_\pm$ at some finite concentration will then be a measure of the nonideality of the system due to the ion-ion interactions in the solvent $s$ at a given concentration. Now, the term ${}^w_w\gamma_\pm$ is of the same type as ${}^s_s\gamma_\pm$ except that it refers to water as a solvent. Therefore, the secondary medium effect is a measure of the difference in ion-ion interactions in the two solvents.

We can express the medium effects in our solutions in terms of thermodynamic concepts by empirically evaluating the activity coefficients. However, it is of considerably greater interest to attempt a calculation of medium effects and their effect on solution properties in terms of a mechanistic picture. This mechanistic picture is provided by our interpretation of the medium effects, and primarily through the Born equation and the Debye-Huckel theory, a means of calculating them is available.

## ELECTRODE POTENTIALS

If a solvent system is to be fully investigated, it is a necessity to determine a series of standard state electrode potentials. From a practical standpoint, this is of particular interest in the determination of various thermodynamic quantities such as solubility products, ionization constants, and activity coefficients. From a more fundamental standpoint, electrode potentials in nonaqueous solvents have been of considerable significance in evaluating the Debye-Huckel theory and other models of solution processes.

**FIGURE 12-14**

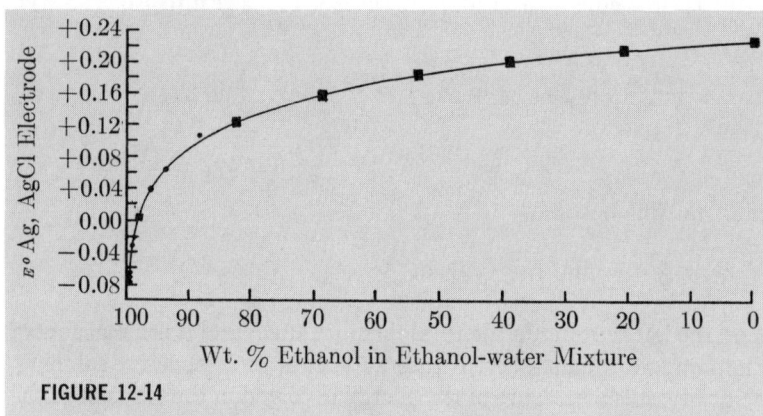

Estimated $E°$ values for the Ag,AgCl electrode in ethanol-water mixtures. (After C. L. LeBas and M. C. Day, *J. Phys. Chem.*, **64**, 465 (1960).)

As can be seen from Table 12–2, the experimental determination of a standard electrode potential in a nonaqueous solvent is extremely difficult. Basically there are two causes for the difficulty. The first of these is solvent purity. Many nonaqueous solvents are extremely hygroscopic, and one might even question whether a solvent such as hexane will not absorb enough water to affect an emf measurement. Thus, in spite of the fact that extreme caution is taken in solution preparation and subsequent measurements, solvent impurity, primarily due to water, is likely to cause a significant experimental error. This is clearly shown in Figure 12–14, where the effect of trace quantities of water on the Ag,AgCl electrode potential can be seen for the cell

$$\text{Pt, } H_{2(1 \text{ atm})} |HCl_{(m)} \text{ ethanol } (X), \text{ water } (Y)| \text{ AgCl,Ag}$$

It can be shown[23] that the effect of water on the emf of this cell is given by

$$(12\text{–}38) \qquad E_{(m_{H_2O})} - E_{(m_{H_2O}=0)} = S\sqrt{m_{H_2O}}$$

For an HCl concentration of 0.002195 molal, $S$ equals 0.077. As was pointed out by Strehlow,[24] this means that an accuracy of 0.1 mv in the $E°$ value of the cell in anhydrous ethanol requires that the molal concentration of water be less than $2 \times 10^{-6}$. In addition to contamination by water, errors can also result from reactions of the solvent with the electrodes, air, and the electrolyte. And in some instances, the solvent itself is unstable.

Along with the problem of solvent purity and instability, different values of the standard state electrode potential can result from different extrapolation procedures. In Chapter 8, a strictly empirical extrapolation method was used. Although this procedure is useful to illustrate the basic principles, it is no longer used even in aqueous systems for accurate determinations. Rather, the Debye-Huckel theory is usually assumed to be valid in the low

concentration region, and the theoretical expression for the activity co-efficient is substituted into the Nernst equation. Putting the Nernst equation in the form

$$(12\text{-}39) \qquad \frac{2RT}{F} \ln \gamma_\pm = E° - \left( E + \frac{2RT}{F} \ln m \right)$$

and substituting the Debye-Huckel expression

$$(12\text{-}40) \qquad -\ln \gamma_\pm = \frac{A'\sqrt{m}}{1 + \beta'a_0\sqrt{m}}$$

where $A'$ and $\beta'$ are constants that are dependent on the temperature and the solvent, and $a_0$ is the ion-size parameter, and then multiplying by $(1 + \beta'a_0\sqrt{m})$ we obtain

$$(12\text{-}41) \qquad E + \frac{2RT}{F} \ln m - \frac{2RT}{F} A'\sqrt{m} = E°$$
$$- \left( E + \frac{2RT}{F} \ln m - E° \right)\beta'a_0\sqrt{m}$$

If we let

$$E' = E + \frac{2RT}{F} \ln m$$

this now gives

$$(12\text{-}42) \qquad E' - \frac{2RT}{F} A'\sqrt{m} = E° - (E - E°)\beta'a_0\sqrt{m}$$

If a plot is made of $(E' - 2RTA'\sqrt{m}/F)$ against $(E - E°)\sqrt{m}$, a straight line should be obtained with a slope of $\beta'a_0$ which extrapolates to an intercept of $E°$. Obviously the application of this method requires the use of successive approximations. $E°$ occurs in the function which is to be plotted in order to obtain $E°$. Consequently, it is necessary to obtain an approximate $E°$ value, carry out the extrapolation to obtain a better $E°$ value, use this new $E°$ value to re-extrapolate, and so on until no improvement is noted. This procedure has been used by Brown and MacInnes[25] and represents the most straightforward application of the Debye-Huckel theory to the determination of a standard state electrode potential.

In solvents of low dielectric constant, the Debye-Huckel theory is not overly successful, and two general approaches have been used. The first of these is simply an extension of the method of Brown and MacInnes using the extended terms of the Debye-Huckel theory as developed by Gronwall, LaMer, and Sandved. However, it would appear that the most valid means of evaluating the standard state potential in solvents of low dielectric constant is based on the existence of ion pairs.[26] Accordingly, the effects of ionic association can be taken into consideration by assuming the validity of the Nernst equation in the form

$$(12\text{-}43) \qquad E = E° - \frac{2RT}{F} \ln \alpha m \gamma_\alpha$$

where $\alpha$ is the degree of dissociation of the acid and $\gamma_\alpha$ is the activity co-efficient of the acid as a strong electrolyte at an ionic concentration of $\alpha m$. If the equation is rearranged to the form

$$\mathbf{E} + \frac{2RT}{\mathbf{F}} \ln \alpha m \gamma_\alpha = \mathbf{E}^\circ$$

it can be seen that since $\mathbf{E}^\circ$ is a constant, a plot of the left side of the equation against some function of the molality will give a straight line of zero slope in the region where the equation is valid, and the intercept at infinite dilution will be at $\mathbf{E}^\circ$.

In utilizing this approach, the primary difficulty is seen to result from the necessity of knowing $\alpha$ and $\gamma_\alpha$. These, however, can be determined by using the extended terms of the Debye-Huckel theory in conjunction with the expression

$$(12\text{--}44) \qquad \alpha = \frac{1}{2}\left[ -\frac{K}{\gamma_\alpha^2 m} + \left( \frac{K^2}{\gamma_\alpha^4 m^2} + \frac{4K}{\gamma_\alpha^2 m} \right)^{1/2} \right]$$

which can be obtained from the simple expression for the ionization constant of the acid. Here $K$ is the thermodynamic ionization constant for the acid in the given solvent and it, of course, must be known.

It can be seen that $\alpha$ and $\gamma_\alpha$ are dependent on each other, and for this reason a method of successive approximations is again required. The usual procedure is to make a first-order approximation of $\gamma_\alpha$ from the extended terms of the Debye-Huckel theory using a particular value for the ion-size parameter. Once a $\gamma_\alpha$ has been approximated, an $\alpha$ may be determined from Eq. (12–44). This $\alpha$ is now used to recalculate $\gamma_\alpha$ by again using the extended terms, but now at a new concentration $\alpha m$. This method of successive approximations is continued until a final $\alpha$ and $\gamma_\alpha$ are obtained for a given $a_0$. These values are then used in the extrapolation to obtain an $\mathbf{E}^\circ$ value. This entire procedure is carried out for various values of the ion-size parameter until a straight line of zero slope approaching infinite dilution is obtained. This then uniquely determines the ion-size parameter, $a_0$, as well as the standard state electrode potential.

All of these methods, except the straightforward extrapolation procedure, are dependent on a particular theoretical treatment. Thus, we are assuming that the chosen theory is valid in the region in which it is applied. Of these approaches, the one based on ion pairing, but utilizing the extended term treatment of the Debye-Huckel theory appears to most adequately reflect our knowledge of ionic behavior in solvents of low dielectric constant.

## Standard Electrode Potentials from the Born Equation

We have already seen in Chapter 8 that, by means of a Born-Haber type of approach, it is possible to propose a series of steps involved in the electrode reaction and attempt to calculate the energies involved in each step.

If the entropy terms are ignored in such a calculation, only the solvation energies vary from solvent to solvent. Thus if these quantities were known, reasonable first-order approximations of the electrode potentials could be made. But as yet these have not been determined for nonaqueous systems.

As an alternative approach, a standard state electrode potential can be calculated in terms of the primary medium effect. Since the standard potential of a cell is defined in terms of ion-solvent interactions only, it is necessary to consider only the primary medium effect. Although there may be some question of its validity, a calculation of this effect can be made by means of the Born equation. Accordingly, the energy necessary to transfer an ion of radius $r$ from a solvent of dielectric constant $\epsilon_1$ to another solvent of dielectric constant $\epsilon_2$ is given by the expression

$$(12\text{--}45) \qquad W_t = \frac{e^2}{2r}\left(\frac{1}{\epsilon_1} - \frac{1}{\epsilon_2}\right)$$

If we now consider the same equation for a mole of a uni-univalent electrolyte with the cation and anion radii of $a_1$ and $a_2$ respectively, we obtain

$$(12\text{--}46) \qquad W_t = \frac{Ne^2}{2}\left(\frac{1}{\epsilon_1} - \frac{1}{\epsilon_2}\right)\left(\frac{1}{a_1} + \frac{1}{a_2}\right)$$

If this expression for the primary medium effect is now substituted into Eq. (12–33),

$$(12\text{--}33) \qquad {}^w\text{E}^\circ_m - {}^s\text{E}^\circ_m = \lim_{m\to 0} \frac{2RT}{\text{F}} \ln {}^s_w\gamma_\pm$$

by means of the relation $\quad$ AG$=-$NFE, it is seen that

$$(12\text{--}47) \qquad {}^w\text{E}^\circ_m - {}^s\text{E}^\circ_m = \frac{Ne^2}{2\text{F}}\left(\frac{1}{\epsilon_1} - \frac{1}{\epsilon_2}\right)\left(\frac{1}{a_1} + \frac{1}{a_2}\right)$$

From this expression, it is readily apparent that if the ionic radii do not vary from solvent to solvent, the E° value should be a linear function of the dielectric constant. This, however, does not appear to be the case. It is possible to propose a variety of reasons for this lack of conformity. Certainly we can question the constancy of the ionic radii in different solvents. And, of course, we must question whether the Born equation is actually a reasonable model to use. But probably the major difficulty is knowing which E° values are sufficiently accurate to include in a comparison. Again, this point is well-made in Table 12–2.

## IONIZATION CONSTANTS OF WEAK ACIDS

Qualitatively, we know that the ionic association effects that accompany a decrease in solvent dielectric constant lead to a corresponding decrease in the ionization constant of a weak acid. Quantitatively, we find that by applying the Born equation along with thermodynamic relations, it is possible to make first-order approximations of ionization constants of weak acids. Using the relation between the standard state potentials in terms of the Born equation

$$(12\text{-}47) \qquad {}^w\mathbf{E}_m^\circ - {}^s\mathbf{E}_m^\circ = \frac{Ne^2}{2\mathbf{F}} \left(\frac{1}{\epsilon_1} - \frac{1}{\epsilon_2}\right)\left(\frac{1}{a_1} + \frac{1}{a_2}\right)$$

and recalling that

$$(12\text{-}48) \qquad \Delta G^\circ = -n\mathbf{F}\mathbf{E}^\circ = -RT \ln K$$

we see that on substitution and rearrangement we obtain

$$(12\text{-}49) \qquad \ln {}^wK - \ln {}^sK = \frac{Ne^2}{2RT} \left(\frac{1}{\epsilon_1} - \frac{1}{\epsilon_2}\right)\left(\frac{1}{a_1} + \frac{1}{a_2}\right)$$

Here we have an expression that should allow us to calculate the ionization constant of a weak acid in any solvent or mixed solvent if its value in one solvent, such as water, is known. However, as usual with the Born equation, we should not expect to obtain particularly good results. It is apparent that the equation assumes that the only variable is the dielectric constant of the solvent. This would ignore such factors as have previously been mentioned. Nevertheless, useful approximations can be made.

In some instances, the calculated ionization constants are quite good, and in other instances they are not so good. We find the latter to be the case for acetic acid in ethanol. If the value of $1.75 \times 10^{-5}$ is used for the ionization constant of acetic acid in water, a value of $1.8 \times 10^{-9}$ is obtained for the ionization constant in ethanol. This value is not in very good agreement with the experimental value of $2 \times 10^{-11}$. On the other hand, it can be seen in Figure 12–15 that the calculated values for some weak acids are in quite good agreement with the experimental values. We are plotting here the term $\log {}^sK/{}^wK$ against $1/\epsilon$, and for simplicity, ${}^wK$, the acid ionization constant in water, is set equal to unity for each acid. The straight line represents the calculated curve using the Born equation with $a_1 = 3.73$ Å and $a_2 = 1.2$ Å, for the hydrogen ion and the carboxylic anions respectively. It can be seen that some grouping does occur around the calculated curve. However, it is also obvious that the results are not very consistent. The data in Figure 12–15 are for mixed solvent systems in which the major constituent is water. It is interesting to note that the agreement between theory and experiment improves as the percent of water is increased. In all such measurements, it is generally found that fair agreement is obtained between theory and experiment when the percentage of water is sufficiently large to give a dielectric constant greater than about 60.

## pH MEASUREMENTS IN NONAQUEOUS MEDIA

It is not particularly unusual to find that we have never truly appreciated the significance of a concept that we have used for years. The concept of pH falls into this category. It might at first appear that the familiar relation

$$(12\text{-}50) \qquad \mathrm{pH} = -\log C_{\mathrm{H}^+}$$

**FIGURE 12-15**

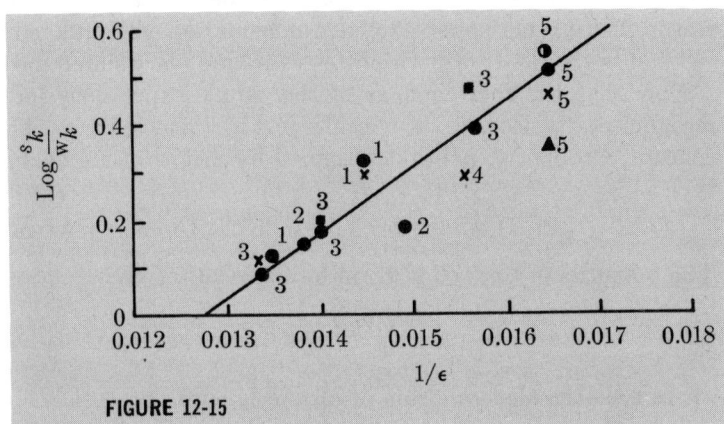

Dependence of Weak Acid Ionization Constants on Solvent Dielec‐tric Constant. The solvents are (1) methanol-water, (2) ethanol-water, (3) isopropanol-water, (4) glycerol-water, and (5) dioxane-water. The acids are: ▲ formic acid, ✗ acetic acid, ● propionic acid, ■ butyric acid, and ◑ water. (After R. A. Robinson and R. H. Stokes, "Electrolyte Solutions," Butterworth Scientific Pubs., London, 1955.)

is quite satisfactory. This was the definition given by Sorenson for the pH in 1909. Unfortunately, there are some more than trivial weaknesses in this simple expression. Basically, the difficulty lies in the fact that the conventional means of measuring pH reflect changes in activity rather than concentration. With this in mind we might now be prone to think that the way out of our difficulty is simply to define the pH as

$$(12\text{--}51) \qquad \mathrm{pH} = -\log a_{\mathrm{H}^+}$$

This looks very well, and we shall, in fact, finally use this definition of the pH. However, it is not the simple thermodynamic relationship that we might at first think. The problem that arises here is that it is not possible by any known means to measure a single ion activity. That is, $a_{\mathrm{H}^+}$ is not an experimentally observable quantity. In terms of emf measurements, this is evident if we consider the now familiar cell

$$\mathrm{Pt,H}_{2(1\ \mathrm{atm})} \mid \mathrm{HCl}_{(m)} \mid \mathrm{AgCl,Ag}$$

It should be recalled that the emf of this cell is given by

$$(12\text{--}52) \qquad \mathbf{E} = \mathbf{E}^\circ - \frac{RT}{\mathbf{F}} \ln a_{\mathrm{H}^+} a_{\mathrm{Cl}^-}$$

which can be put into the more realistic form

$$(12\text{--}53) \qquad \mathbf{E} = \mathbf{E}^\circ - \frac{RT}{\mathbf{F}} \ln a_\pm^2$$

This emphasizes the fact that emf measurements give us a mean ionic activity rather than the single ion activities. This is true for the simple reason that it is not possible to have an oxidation without a reduction, and a cell emf is the result of both processes taking place simultaneously.

Now that the shortcomings of the simple expressions for the pH are recognized, it is necessary to consider just how a meaning can be given to the concept. Ordinarily, pH is determined by means of a galvanic cell of the type

$$\text{Pt, H}_2 \mid \text{Solution } X \mid \text{Sat. KCl} \mid \text{Calomel reference electrode}$$

The potential of this cell is given by the relation

$$(12\text{--}54) \qquad \text{E} = \text{E}^\circ - \frac{2.3026\,RT}{\text{F}} \log a_{\text{H}^+}\, a_{\text{Cl}^-} + \text{E}_j$$

where $\text{E}_j$ is the algebraic sum of all the junction potentials and $a_{\text{Cl}^-}$ is the activity of the chloride ion in contact with the calomel electrode. If we now take advantage of the fact that the activity of the chloride ion in the neighborhood of the calomel electrode is a constant and further recognize that $\text{E}^\circ$ for the hydrogen electrode is zero, it is possible to rearrange Eq. (12–54) to

$$(12\text{--}55) \qquad \text{E} = \left( \text{E}^\circ - \frac{2.3026\,RT}{\text{F}} \log a_{\text{Cl}^-} \right) - \frac{2.3026\,RT}{\text{F}} \log a_{\text{H}^+} + \text{E}_j$$

and set the term in parentheses equal to the constant $\text{E}^{\circ\prime}$. This then can be rearranged to give

$$(12\text{--}56) \qquad \frac{(\text{E} - \text{E}^{\circ\prime} - \text{E}_j)\,\text{F}}{2.3026\,RT} = -\log a_{\text{H}^+} = \text{pH}$$

In retrospect, it can be seen that it is not possible to determine the pH from this cell for the simple reason that we do not know $a_{\text{Cl}^-}$ and therefore, $\text{E}^{\circ\prime}$, and we do not know $\text{E}_j$.

Various attempts have been made to circumvent these difficulties by defining pH in terms of quantities that do have thermodynamic significance. An example of such a definition is

$$(12\text{--}57) \qquad \text{pH} = -\log \gamma_\pm\, m_{\text{H}^+}$$

where $\gamma_\pm$ is the mean molal activity coefficient. Since both $\gamma_\pm$ and the molality of the hydrogen ion are determinable quantities, the pH of the system will now have thermodynamic meaning. However, the relation is not valid if there is more than one uni-univalent electrolyte or any electrolyte that is not uni-univalent present. Both of these situations are more likely to be found than not. In actuality, the definitions of pH that have thermodynamic meaning cannot be used from a practical standpoint, and we therefore find it expedient to resort to *nonthermodynamic* concepts in order to devise a useful pH scale.

Since a purely thermodynamic meaning cannot be given to the pH scale, we are faced with the problem of finding some other basis for it. We are greatly aided in this choice when we recognize that the major purpose of a pH scale is to permit a determination of the relative acidity or alkalinity of a system. Thus, we are primarily interested in a definition that will permit us to make consistent measurements of acidities even if they do not reflect the actual hydrogen ion activity. On the other hand, in order to use the concept to the fullest, we will attempt to make the pH scale conform as closely as possible to the thermodynamic definition.

If Eq. (12–56) is again considered, it might appear that most of these requirements can be met. If some means were devised for determining the pH of a standard solution and the emf of the cell was measured, the only unknown in the equation would be the constant $(E^{\circ\prime} + E_j)$. Even if it is admitted that this term might change with pH in some manner, a series of standard solutions can be prepared for different pH ranges. Finally, it might be said that even if we do lose track of our thermodynamic relations along the way, we at least have a consistent measure of something related to the acidity of the system. Basically, this is the approach that is used today for the determination of a pH scale. In the presence of a pH standard solution, Eq. (12–56) becomes

$$(12\text{–}58) \qquad pH_s = \frac{[E_s - (E^{\circ\prime} + E_j)]\ F}{2.3026\ RT}$$

If it is now assumed that the quantity $(E^{\circ\prime} + E_j)$ is the same in the presence of the unknown as it is in the presence of the standard solution, Eq. (12–58) can be subtracted from Eq. (12–56) to obtain the expression

$$(12\text{–}59) \qquad pH = pH_s + \frac{(E - E_s)\ F}{2.3026\ RT}$$

Thus, it is seen that if the term $(E^{\circ\prime} + E_j)$ truly remains constant in the region of our interests, we can feel quite confident that the activities as determined from the pH measurement are as thermodynamically meaningful as is the assigned pH of the standard solution. Unfortunately, the term $(E^{\circ\prime} + E_j)$ does not in general remain constant, and in addition, it is always possible to raise some question with regard to the determination of the pH of the reference solutions. This means that the pH scale is ultimately only a scale of relative acidities, and except for a few instances, it is void of any fundamental thermodynamic meaning.

In attempting to define a pH scale in a nonaqueous medium, we would certainly expect to encounter the same difficulties that were encountered in an aqueous medium. In fact, there is nothing in the definition itself that would restrict the pH concept to an aqueous system, and the same general approach could be used in a nonaqueous medium. Thus, pH can still be defined in terms of Eq. (12–59) and, in principle, the resultant nonaqueous pH scale should be as meaningful as the aqueous pH scale.

There are various ways in which the nonaqueous pH scale can be given some meaning. Assuming that we are considering a protonic solvent in order that the acid character can be related to the solvated proton, it would be possible to set up a cell in terms of this solvent only. The same procedure of attempting to make standard solutions of known pH that was used in the determination of an aqueous pH scale would then be followed.

It is true that this scale cannot be related to the aqueous pH scale, but this is not necessarily bad. The most serious difficulty arises from the fact that we will not be as successful in relating the pH to the thermodynamic activity of the hydrogen ion as we were in water. This will particularly be true in solvents of low dielectric constant where ionic association is important. Nevertheless, we still recognize that the aqueous pH scale is not thermodynamically valid, and there should, therefore, be no reason why a nonaqueous pH scale should be expected to be superior.

In an alternative determination of a nonaqueous pH scale, we could keep the same basic cell that was used in an aqueous medium, including the calomel reference electrode with its aqueous KCl solution. If, for a given solvent, we use aqueous standard solutions, a series of numbers can be obtained that will give a measure of something that can vaguely be associated with acidity. If the system is a mixed solvent containing water or a water-like solvent, we may even know enough about the system to relate these values to the general order of magnitude of the hydrogen ion concentration in terms of a calibration curve. However, because of the fact that the magnitude of the liquid junction potential changes from solvent to solvent, it should be rather obvious that pH in one solvent cannot be related to pH in another solvent by this approach. For instance, if the pH meter gives a reading of 5.0 for a particular solution in an ethanol-water solvent and also a reading of 5.0 for a methanol-water solvent where the same aqueous standard solution was used in both cases, it is not legitimate to conclude that the activity of the hydrogen ion is the same in the two solutions. In fact, there should be no similarity at all between the two. Primarily, this will be due to the quite different liquid junction potentials between the ethanol-water solvent boundary with the saturated aqueous KCl of the calomel electrode and the methanol-water solvent boundary with the aqueous KCl of the calomel electrode.

Although very little has yet been accomplished towards establishing pH scales that are valid in nonaqueous media, it would appear that such scales are possible, at least, for water-like solvents. Just as in aqueous media, it will be necessary for the term $(E^{\circ\prime} + E_j)$ to remain constant when the standard solution is replaced by the unknown solution. This is most likely to be the case when the solvent composition is kept constant and the concentration of solute is small. If it can be assumed that the $(E^{\circ\prime} + E_j)$ term does remain substantially constant, the problem then becomes one of developing standard solutions.

## CHELATION IN NONAQUEOUS MEDIA

Although complex ions have received little attention in nonaqueous solvents, considerable work has been conducted on neutral complexes in such solvents. This is particularly true of chelates. Because of the insolubility of a large number of chelates in a purely aqueous medium, it is necessary to use a nonaqueous or mixed solvent for their solution study. The most commonly used solvent for these studies has been dioxane-water. Chelation studies have been exceedingly rare in completely anhydrous solvents.

Of the variety of methods for the study of the stabilities of chelates, probably the most accurate is that developed by Bjerrum[27] and modified for the study of chelates in mixed solvents by Calvin and Wilson.[28] This method takes advantage of the fact that the formation of many chelates results in the displacement of a $H^+$, and this can be followed by potentiometric means. This is readily apparent if we consider the chelation of a metal ion with a $\beta$-diketone. The chelation takes place through the enol form of the molecule, and the first molecule can be considered to enter as follows:

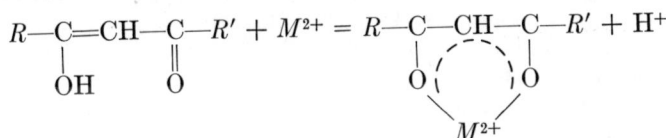

$$R-C=CH-C-R' + M^{2+} = R-C-CH-C-R' + H^+$$
$$\underset{OH}{\mid} \quad \underset{O}{\parallel} \qquad \qquad \underset{O}{\mid} \overbrace{\qquad} \underset{O}{\mid}$$
$$\underset{M^{2+}}{\diagdown}$$

More generally, if $HKe$ is the chelating agent and $M^{2+}$ the metal ion, the equilibrium for the addition of the first molecule of the chelating agent can be expressed as

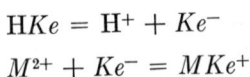

$$HKe = H^+ + Ke^-$$
$$M^{2+} + Ke^- = MKe^+$$

The first stability constant, $k_1$, is then given by

$$k_1 = \frac{[MKe^+]}{[M^{2+}][Ke^-]}$$

and the second stability constant, for the equilibrium

$$MKe^+ + Ke^- = MKe_2$$

is given by

$$k_2 = \frac{[MKe_2]}{[MKe^+][Ke^-]}$$

Fortunately the relative stability of chelates is of more concern than their thermodynamic stability constants. In solvents of low dielectric constant, the possibility of ion association even in the very dilute concentration range makes the evaluation of thermodynamic quantities questionable.

In spite of the handicaps of working in such solvents, a few potentiometric studies of chelate stabilities have been made in purely anhydrous media. In fact, a rather valid attempt to determine true chelate thermodynamic equilibrium constants has been made in both dioxane-water mixtures and in anhydrous ethanol.[29,30] In the latter instance, acetylacetonates of nickel were studied with a cell containing a Ag,AgCl electrode versus a hydrogen electrode. Of course, we should recognize that the stability constant of a chelate determined in one solvent cannot be compared to the stability constant of the chelate in another solvent. Even if we are certain that the same species are being studied in different solvents, our measurements give us a mean ionic activity, and there is no guarantee that the relative contribution by the $H^+$ ion is the same in different solvents.

## References

1. V. LaMer and H. Downes, *Chem. Revs.*, **13**, 47 (1933).
2. C. Kraus, *J. Am. Chem. Soc.*, **30**, 1323 (1908); **43**, 749 (1921).
3. W. Jolly, International Conference on Nonaqueous Solvents, Hamilton, Ontario, Canada, 1967.
4. W. Jolly, Metal-Ammonia Solutions, in "Progress in Inorganic Chemistry," Vol. I, Interscience Publishers, Inc., New York–London, 1959.
5. M. Born, *Z. Physik*, **1**, 45 (1920); H. Harned and B. Owen, "The Physical Chemistry of Electrolytic Solutions," 3rd ed., Reinhold Publishing Corporation, New York, 1958.
6. T. Davis and J. Ricci, *J. Am. Chem. Soc.*, **61**, 3274 (1939).
7. J. Ricci and T. Davis, *J. Am. Chem. Soc.*, **62**, 407 (1940).
8. N. Bjerrum, *K. Danske Vidensk, Selsk. Mat.-fys. Medd.*, **7**, No. 9 (1926).
9. R. Robinson and R. Stokes, "Electrolyte Solutions," 2nd ed., Butterworth Scientific Publs., London, 1959.
10. R. Fuoss and C. Kraus, *J. Am. Chem. Soc.*, **55**, 21, 476, 1019, 2387 (1933).
11. L. Kenausis, E. Evers, and C. Kraus, *Proc. Nat. Acad. Sci. U. S.*, **48**, 121 (1962).
12. J. Sugden, *J. Chem. Soc.*, **129**, 174 (1926).
13. T. Griffiths and R. Scarrow, International Conference on Nonaqueous Solvents, Hamilton, Ontario, Canada, 1967.
14. S. Winstein, et al., *J. Am. Chem. Soc.*, **76**, 2597 (1954); S. Winstein and G. Robinson, *J. Am. Chem. Soc.*, **80**, 169 (1958).
15. T. Griffiths and M. Symons, *Mol. Phys.*, **3**, 90 (1960).
16. T. Hogen-Esch and J. Smid, *J. Am. Chem. Soc.*, **87**, 669 (1965).
17. T. Hogen-Esch and J. Smid, *J. Am. Chem. Soc.*, **88**, 307, 318 (1966).
18. I. Kolthoff and A. Willman, *J. Am. Chem. Soc.*, **56**, 1007 (1934).
19. W. Schaap, et al., *Rec. Chem. Prog.*, **22**, 197 (1961).
20. I. Kolthoff and S. Bruckenstein, *J. Am. Chem. Soc.*, **78**, 1, 10, 2974 (1956); **79**, 1, 5915 (1957).
21. P. Kebarle and A. M. Hogg, *J. Chem. Phys.*, **42**, 668, 798 (1965); **43**, 449 (1965); *J. Am. Chem. Soc.*, **88**, 28 (1966).
22. J. Dixon, P. Gwinner, and D. Lini, *J. Am. Chem. Soc.*, **87**, 1379 (1965).
23. C. LeBas and M. Day, *J. Phys. Chem.*, **64**, 465 (1960).
24. H. Strehlow, Electrode Potentials in Nonaqueous Solvents, in "The Chemistry of Nonaqueous Solvents," Academic Press, Inc., New York, 1966.
25. A. Brown and D. MacInnes, *J. Am. Chem. Soc.*, **57**, 1356 (1935).
26. H. Harned and B. Owen, "The Physical Chemistry of Electrolytic Solutions," 3rd ed., Reinhold Publishing Corporation, New York, 1958; H. Taniguchi and G. Janz, *J. Am. Chem. Soc.*, **61**, 688 (1957).
27. J. Bjerrum, "Metal Ammine Formation in Aqueous Solution," P. Haase & Son, Copenhagen, 1941.

28. M. Calvin and K. Wilson, *J. Am. Chem. Soc.*, **67**, 2003 (1945).

29. L. Van Uitert and C. Haas, *J. Am. Chem. Soc.*, **75**, 451 (1953).

30. L. Van Uitert, W. Fernelius and B. Douglas, *J. Am. Chem. Soc.*, **75**, 3577 (1953).

## Suggested Supplementary Reading

J. Lagowski, Ed., "The Chemistry of Non-aqueous Solvents," Vol. 1, Academic Press, Inc., New York, 1966.

T. Waddington, Ed., "Non-aqueous Solvent Systems," Academic Press, Inc., New York, 1965.

R. Robinson and R. Stokes, "Electrolyte Solutions," 2nd ed., Butterworth Scientific Publs., London, 1959.

L. Audrieth and J. Kleinberg, "Non-aqueous Solvents," John Wiley & Sons, Inc., New York, 1953.

H. Sisler, "Chemistry in Non-aqueous Solvents," Reinhold Book Corporation, New York, 1961.

## Problems

1. Classify the solvents $H_2O$, THF, $AlCl_3$, HF, $(C_2H_5)_2O$, benzene, He(liq), and NaCl(liq) in terms of
   (a) polarity
   (b) protophylic character
   (c) dielectric constant
   (d) basicity.

2. Based on the values given in Table 8–1, give a rationalization of the fact that the Li/Li$^+$ couple has a higher $E^0$ value than the Cs/Cs$^+$ couple whereas the Be/Be$^{2+}$ couple has a lower $E^0$ value than that of the Ba/Ba$^{2+}$ couple.

3. What would be the expected value for the dielectric constant
   (a) of a fused salt
   (b) in the immediate neighborhood of an ion
   (c) of a 1M aqueous solution of NaCl?
   (See for (b) D. M. Ritson and J. B. Hasted, *J. Chem. Phys.*, **16**, 11 (1948), and R. M. Noyes, *J. Am. Chem. Soc.*, **84**, 513 (1962);
   (c) J. B. Hasted, D. M. Ritson, and C. H. Collie, *J. Chem. Phys.*, **16**, 1 (1948), and G. H. Haggis, J. B. Hasted, and T. J. Buchanan, *ibid.*, **20**, 1452 (1952).

4. Since the mobility of a free ion should be greater than that of an ion-triplet or higher aggregate, account for the conductance behavior illustrated by curves $g, i, j$, and $k$ in Figure 12–8.

5. Considering the Friedel-Crafts reaction on p. 542, what might be expected in a system in which hexane is used as the solvent with sufficient ether added to give an ether:AlCl$_3$ ratio of 1:1?

6. Account for the fact that the addition of dimethyl ether to a solution of AlBr$_3$ in nitrobenzene as a solvent results in a decrease in conductance until a 1:1 ratio of ether:AlBr$_3$ is attained, but as the ratio is increased pass 1:1, a rapid increase in conductance is observed. (See R. E. Van Dyke and C. A. Kraus, *J. Am. Chem. Soc.*, **71**, 2694 (1949).)

7. Propose an experiment using transference numbers to determine the existence of ion-triplets. (See A. M. Sukhotin, *Russ. J. Phys. Chem.*, **33,** 450 (1959).)

8. What part will ion-solvent interaction play in the interpretation of the data obtained in the experiment proposed in Problem 7?

9. What is the nature of the immediate environment of the $H^+$ ion in
   (a) water
   (b) diethyl ether
   (c) benzene?

10. If the effective radius of a cation is determined in water and in benzene, how would you expect the values to compare if the ion is
   (a) $Na^+$
   (b) $N(butyl)_4^+$?

# Appendix A
# The Variation Method

In discussing the variation method, we developed the expression

$$(1) \qquad \frac{\int \psi^* H \psi \, d\tau}{\int \psi^* \psi \, d\tau} = E$$

It was then pointed out that the correct Hamiltonian for a given system is usually easy to write down. However, for most systems it is necessary to guess the form of the wave function. If the correct wave function is chosen then, in principle, it should be possible to obtain the correct energy of the given system. In fact, we would define the correct wave function as the one that gives the correct energy, $E_0$. Other wave functions will then give different values of the energy. The variation theorem tells us that although many different $E_i$ may be obtained, if $E_0$ is the lowest eigenvalue of the given operator, then, using normalized wave functions

$$(2) \qquad \frac{\int \psi^* H \psi \, d\tau}{\int \psi^* \psi \, d\tau} = \int \psi^* H \psi \, d\tau \geq E_0$$

If the correct $\psi$ is chosen, the equality will hold, whereas if any other $\psi_i$ is chosen, the integral will be greater than $E_0$.

To prove this relationship, equation (2) can be rearranged to

$$(3) \qquad \int \psi^* (H - E_0) \psi \, d\tau \geq 0$$

and it will be necessary to show that, indeed, the left side of the equation is greater than or equal to zero. The wave function $\psi$ can be expanded as a linear combination of orthogonal wave functions $\varphi_i$ giving

$$(4) \qquad \psi = a_1\varphi_1 + a_2\varphi_2 + \cdots + a_n\varphi_n = \sum_i a_i\varphi_i$$

This same type of approach was used in our treatment of hybridization. We can now express Eq. (3) as

$$(5) \qquad \int \left(\sum_i a_i^*\varphi_i^*\right)(H - E_0)\left(\sum_i a_i\varphi_i\right) d\tau \geq 0$$

Now since the $\varphi_i$ are eigenfunctions of $H$, then

$$H\varphi_i = E_i\varphi_i$$

Using this fact, Eq. (5) becomes

$$(6) \qquad \int \left(\sum_i a_i^*\varphi_i^*\right)\left(\sum_i (E_i - E_0) a_i\varphi_i\right) d\tau \geq 0$$

Because of orthogonality and normalization, this simplifies to

$$(7) \qquad \sum_i a_i^*a_i(E_i - E_0) \geq 0$$

Now $a_i^*a_i$ is a positive number and the $E_i$ are the result of wave functions other than the correct wave function. They are therefore greater than $E_0$ by definition. Consequently Eq. (7) is valid and we can therefore say that

$$\int \psi^*(H - E_0)\psi \, d\tau \geq 0$$

or finally

$$\int \psi^*H\psi \, d\tau \geq E_0$$

# Appendix B
# Hermitean Operators

If two functions $\varphi$ and $\psi$ are of class $Q$, and $\alpha$ is an operator, $\alpha$ is said to be Hermitean if

$$(1) \qquad \int \varphi^*(\alpha\psi)\, d\tau = \int \psi(\alpha^*\varphi^*)\, d\tau$$

The importance of Hermitean operators in quantum mechanics is that their eigenvalues for functions of class $Q$ are always real. Thus, in the general formulation of quantum mechanics it is assumed that for every observable quantity, which, of course, must be real, there can be assigned an Hermitean operator.

In order to show that the eigenvalues of Hermitean operators are real, we can assume that $\alpha$ is an Hermitean operator and $\psi$ is a class $Q$ function and an eigenfunction of $\alpha$. Therefore,

$$(2) \qquad \alpha\psi = a\psi$$

where $a$ is the eigenvalue. Now if we take the complex conjugate of each side, we obtain

$$(3) \qquad (\alpha\psi)^* = (a\psi)^*$$

but this is the same as

$$(4) \qquad \alpha^*\psi^* = a^*\psi^*$$

From this

$$(5) \qquad \int \psi^*(\alpha\psi)\, d\tau = a \int \psi^*\psi\, d\tau$$

and

$$(6) \qquad \int \psi(\alpha^*\psi^*)\, d\tau = a^* \int \psi\psi^*\, d\tau$$

But

(7) $$\int \psi^*\psi \, d\tau = \int \psi\psi^* \, d\tau$$

and since $\alpha$ is Hermitean

(8) $$\int \psi^*(\alpha\psi) \, d\tau = \int \psi(\alpha^*\psi^*) \, d\tau$$

Therefore $a = a^*$, a condition that can exist only if $a$ is real.

# Appendix C
## The Hamiltonian Operator

In rectangular coordinates, the Hamiltonian operator has been shown to be

$$\text{(1)} \qquad H = -\frac{h^2}{8\pi^2 m}\left(\frac{\partial^2}{\partial x^2} + \frac{\partial^2}{\partial y^2} + \frac{\partial^2}{\partial z^2}\right) + V_{(xyz)}$$

And since $H$ is real, $H = H^*$. Therefore if $H$ is Hermitean,

$$\text{(2)} \qquad \int\int\int_{-\infty}^{\infty} \varphi^* H\psi\, dxdydz = \int\int\int_{-\infty}^{\infty} \psi H\varphi^*\, dxdydz$$

where $\varphi$ and $\psi$ are functions of class $Q$.

Actually it is important that $H$ is Hermitean. This is apparent from the form of the wave function

$$H\psi = E\psi$$

The energy of the system described by $\psi$ is a real quantity and the operator that gives $E$ must therefore be Hermitean. To show the Hermitean character of $H$, we can use the fact that class $Q$ functions must be finite. Thus, they must vanish at infinity. Now it should be noted that $H$ is composed of two types of operators, differential operators of the type $\dfrac{\partial^2}{\partial q^2}$ and multiplicative operators. Since the order of multiplication with the simple multiplicative operators is immaterial, these will be Hermitean. However, this is not necessarily so for the differential operators, and these are the ones that must be investigated here.

Using $\dfrac{\partial^2}{\partial x^2}$, it can be said that

$$\text{(3)} \qquad \int\int\int_{-\infty}^{\infty} \varphi^* \frac{\partial^2\psi}{\partial x^2}\, dxdydz = \int\int_{-\infty}^{\infty} \varphi^* \frac{\partial\psi}{\partial x}\, dydz \Bigg]_{-\infty}^{+\infty}$$
$$- \int\int\int_{-\infty}^{\infty} \frac{\partial\varphi^*}{\partial x}\frac{\partial\psi}{\partial x}\, dxdydz$$

Because of the restriction to class $Q$ functions, $\varphi^*$ vanishes at infinity, and additionally since $\psi$ must be of class $Q$, the slope of $\psi$, $\dfrac{\partial \psi}{\partial x}$ must be finite or zero giving

$$(4) \qquad \iiint_{-\infty}^{\infty} \varphi^* \frac{\partial^2 x}{\partial x^2} \, dx\,dy\,dz = -\iiint_{-\infty}^{\infty} \frac{\partial \varphi^*}{\partial x} \frac{\partial \psi}{\partial x} \, dx\,dy\,dz$$

Therefore, the first term on the right of equation (3) must be zero. Likewise,

$$(5) \qquad \iiint_{-\infty}^{\infty} \psi \frac{\partial^2 \varphi^*}{\partial x^2} \, dx\,dy\,dz = -\iiint_{-\infty}^{\infty} \frac{\partial \varphi^*}{\partial x} \frac{\partial \psi}{\partial x} \, dx\,dy\,dz$$

Thus, we can say that operators of the type $\dfrac{\partial^2}{\partial q^2}$ are Hermitean. And since the Hamiltonian operator is composed of sums of Hermitean operators, it is also Hermitean.

# Appendix D
# Point Symmetry Notation

In the field of crystallography, the Hermann-Mauguin or International System of expressing point symmetry is conventionally used. It is for this reason that this system of notation was presented in the chapter on crystal structure. However, the infrared spectroscopist seems to prefer the Schoenflies notation, and this influence is strongly felt in the various fields of chemistry. For this reason we will give a rather detailed comparison of the point symmetry notation in the two systems.

In the Hermann-Mauguin system, the corresponding numeral is used to indicate a rotation axis of order $n$, an $m$ is used to indicate a mirror plane, and a numeral with a bar over it is used to indicate a rotary-inversion axis. In both systems it is conventional to show the minimum number of elements necessary to define a given point group. This can lead to some confusion if the point groups are not well in mind. This is obvious from the fact that the normal class of the cubic system contains 23 symmetry elements, but it is symbolized in the Hermann-Mauguin notation by simply $m3m$.

In the Schoenflies system, the symbol $C$, for cyclic, is used to denote a rotation axis, and the order is indicated by a subscript, that is, $C_n$. The general symbol for a mirror plane is $\sigma$ and the orientation of the plane relative to the major axis is given by the subscripts $v$ (vertical), $h$ (horizontal), and $d$ (dihedral). Thus a vertical mirror plane may be designated as $\sigma_v$. However, when the mirror plane exists along with a rotation axis, the plane will be designated only by the subscript. For instance, a rotation axis of order $n$ with a vertical mirror plane will be represented as $C_{nv}$. Rather than use a rotation-inversion axis as is done in the Hermann-Mauguin notation, the Schoenflies system customarily uses a rotation-reflection axis, $S_n$. Finally, in the Schoenflies notation, specific symbols are used to indicate unique geometries. A rotation axis with $n$ diad axes normal to it is designated by $D_n$. And in the cubic system, the importance of the tetrahedral and octahedral symmetries are recognized by the symbols $T$ and $O$.

In general, the following comparisons between the two systems can be made. But it should be recognized that since our considerations are restricted here to point symmetry, we are no longer limited to 32 point groups. Consequently, all possibilities will not be considered.

(1) No rotation axes.

| Schoenflies | Hermann-Mauguin |
|---|---|
| $C_1$ | $1$ |
| $C_s\ (\sigma)$ | $\bar{2} = m$ |
| $C_i$ | $\bar{1}$ |

(2) Simple rotation axes.

| Schoenflies | Hermann-Mauguin |
|---|---|
| $C_2$ | $2$ |
| $C_3$ | $3$ |
| $C_4$ | $4$ |
| . | . |
| . | . |
| . | . |
| $C_n$ | $n$ |

(3) Rotation axes lying in a symmetry plane.

| Schoenflies | Hermann-Mauguin |
|---|---|
| $C_{1v} = C_{1h} = \sigma$ | $m = \bar{2}$ |
| $C_{2v}$ | $mm = 2m = \bar{2}m$ |
| $C_{3v}$ | $3m$ |
| $C_{4v}$ | $4m$ or $4mm$ |
| . | . |
| . | . |
| . | . |
| $C_{\infty v}$ | $\infty\, m$ |

(4) Rotation axes perpendicular to a plane of symmetry.

| Schoenflies | Hermann-Mauguin |
|---|---|
| $C_{1h}$ | $m$ |
| $C_{2h}$ | $2/m = \bar{1}m$ |
| $C_{3h}$ | $3/m = \bar{6}$ |
| $C_{4h}$ | $4/m$ |
| . | . |
| . | . |
| . | . |

(5) Rotation-reflection axes of order $n$.

| Schoenflies | Hermann-Mauguin |
|---|---|
| $S_1 = C_{1h} = \sigma$ | $m$ |
| $S_2$ | $\bar{1}$ |
| $S_3 = C_{3h}$ | $3/m$ |
| $S_4$ | $\bar{4}$ |
| $S_6$ | $\bar{3}$ |

(6) Rotation axes of order $n$ with $n$ perpendicular diad axes.

| Schoenflies | Hermann-Mauguin |
|---|---|
| $D_1 = C_2$ | 2 |
| $D_2$ | 222 |
| $D_3$ | 32 |
| $D_4$ | 422 |
| $D_6$ | 622 |

(7) Rotation axes of order $n$ with $n$ perpendicular diad axes and a horizontal mirror plane.

| Schoenflies | Hermann-Mauguin |
|---|---|
| $D_{2h}$ | $mmm = 2/mm$ |
| $D_{3h}$ | $6m = 3/mm$ |
| $D_{4h}$ | $4/mmm$ or $4/mm$ |
| . | . |
| . | . |
| . | . |
| $D_{\infty h}$ | $\infty/mmm$ or $\infty/mm$ |

(8) Rotation axes of order $n$ with $n$ perpendicular diad axes and a dihedral plane.

| Schoenflies | Hermann-Mauguin |
|---|---|
| $D_{2d}$ | $\bar{4}2m$ |
| $D_{3d}$ | $3m = \bar{6}2m$ |

(9) In the cubic system, we find tetrahedral symmetry with four triad axes and three diad axes and octahedral symmetry with four triad axes and three tetrad axes.

| | |
|---|---|
| $T$ | 23 |
| $T_h$ | $m3 = \bar{2}3 = 2/m3$ |
| $T_d$ | $\bar{4}3m$ |
| $O$ | 432 |
| $O_h$ | $m3m$ |

Some simple examples of these points groups are given below.

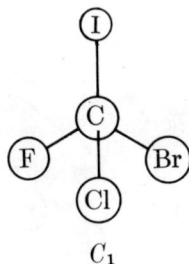

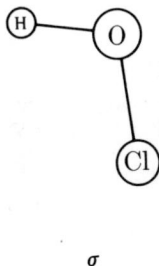

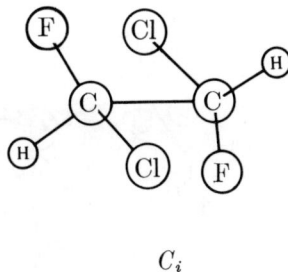

$C_1$          $\sigma$          $C_i$

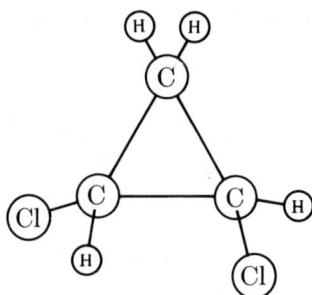

$C_2$

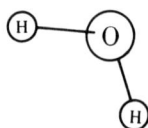

$C_{2v}$

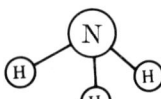

$C_{3v}$

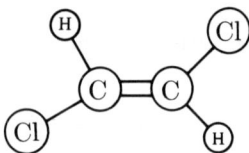

$C_{\infty v}$

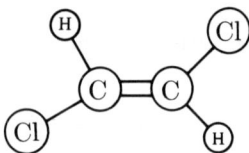

$C_{2h}$

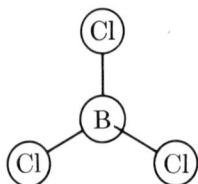

$D_{3h}$

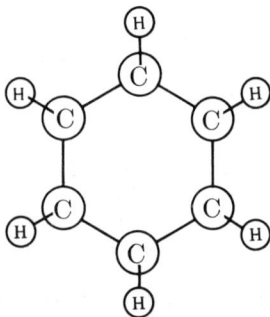

$D_{6h}$

$D_{\infty h}$

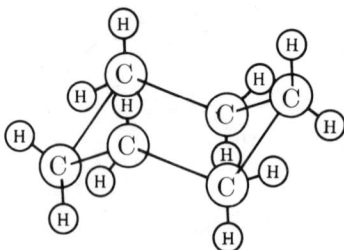

$D_{3d}$

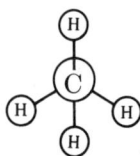

$T_d$

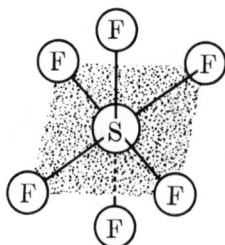

$O_h$

# INDEX

# Index